T0139860

Lecture Notes in Computer Science 13958

The series Lecture Notes in Computer Science (LNCS), including its subseries Lecture Notes in Artificial Intelligence (LNAI) and Lecture Notes in Bioinformatics (LNBI), has established itself as a medium for the publication of new developments in computer science and information technology research, teaching, and education.

LNCS enjoys close cooperation with the computer science R & D community, the series counts many renowned academics among its volume editors and paper authors, and collaborates with prestigious societies. Its mission is to serve this international community by providing an invaluable service, mainly focused on the publication of conference and workshop proceedings and postproceedings. LNCS commenced publication in 1973.

Olivier Bernard · Patrick Clarysse ·
Nicolas Duchateau · Jacques Ohayon ·
Magalie Viallon
Editors

Functional Imaging and Modeling of the Heart

12th International Conference, FIMH 2023 Lyon, France, June 19–22, 2023 Proceedings

Editors
Olivier Bernard
CREATIS, INSA-Lyon
University of Lyon
Lyon, France

Patrick Clarysse
CREATIS, CNRS
University of Lyon
Lyon, France

Nicolas Duchateau
CREATIS, IUF
University of Lyon
Lyon, Francc

Jacques Ohayon
TIMC, UGA-Grenoble
Polytech - Savoie Mont Blanc University Le
Bourget du Lac
Chambéry, France

Magalie Viallon
CREATIS
Jean Monnet University
Saint-Etienne, France

ISSN 0302-9743 ISSN 1611-3349 (electronic)
Lecture Notes in Computer Science
ISBN 978-3-031-35301-7 ISBN 978-3-031-35302-4 (eBook)
https://doi.org/10.1007/978-3-031-35302-4

This Springer imprint is published by the registered company Springer Nature Switzerland AG
The registered company address is: Gewerbestrasse 11, 6330 Cham, Switzerland

Preface

FIMH 2023 was the 12th International Conference on Functional Imaging and Modeling of the Heart, hosted by Université de Lyon from June 19 to June 22, 2023. This edition included both pre-conference and post-conference workshops and challenge. The publication of these FIMH 2023 proceedings follows the success of the past eleven conferences held in Helsinki (2001), Lyon (2003), Barcelona (2005), Salt Lake City (2007), Nice (2009), New York (2011), London (2013), Maastricht (2015), Toronto (2017), Bordeaux (2019) and Stanford (2021).

After the difficult pandemic period that prevented us from meeting in person in 2021, the 2023 edition marked a return to Lyon, 20 years after the 2nd edition and, therefore, at a privileged time to question the fields of research and technology dedicated to this fascinating organ. This demonstrates the perenniality of this biennial event and the still actual necessity to foster the discussions and interplay between research in cardiology, imaging, data processing and modeling. These 20 years have witnessed tremendous improvements and achievements in each of the topics with increased imaging resolutions, data explosion, sophistication of computational models and advent of AI frameworks, while new imaging modalities have emerged (e.g. combined PET-MRI, Spectral CT). The domain therefore remains particularly active.

FIMH 2023 attracted 80 submissions by authors from more than nineteen countries with the ambition to emphasize works by young researchers in the field. From the original submissions, and after single-blind 2 reviews per paper by international experts, 72 papers were accepted for presentation at the meeting and publication in this Lecture Notes in Computer Science Proceedings volume. Before submitting the final version of their manuscripts, authors had the opportunity to address specific concerns and issues raised by the reviewers to offer high-quality papers to the present FIMH proceedings.

The oral communications were organized in 6 oral sessions with 20 presentations addressing cardiac: 1) Multiscale Structure; 2) Electrophysiology Modeling; 3) Image and Shape Analysis; 4) Hemodynamics and CFD; 5) Biomechanics; and 6) Clinical Applications. Two poster sessions completed the program of the main conference.

Four renowned experts were invited to share their experience in keynote speeches. Dr. Marta Sitges (Hospital Clinic de Barcelona, Universitat de Barcelona, Spain) addressed the mechanisms of adaptation of the heart to exercise. Dr. Matthias Stuber (University of Lausanne, Switzerland) shared his view on facilitating the dissemination of cardiovascular magnetic resonance imaging. Dr. Franck Nicoud (Université de Montpellier, France) presented a computational framework for patient-specific blood flow simulation. As a junior keynote speaker, Dr. Alison Vander Roest (University of Michigan, USA) discussed how changes in sarcomere kinetics and protein interactions can alter cardiac mechanics. We express our extreme gratitude for their participation to this event.

This edition also included four peripheral events on specific topics. On June 19, a challenge on MYOcardial Segmentation with Automated Infarct Quantification (MYOSAIQ) and a workshop on cardiac kinematics benchmarks were organized. Two workshops, on simulation and imaging of mitral regurgitation on the one hand, and on simulation of echocardiographic sequences on the other hand took place in the afternoon of June 22.

Of note this year was the exceptional mobilization of academic and industrial partners to support this event.

It is the hope of the organizing committee that this edition contributed to motivate young students and researchers, share advancements, and maintain and even reinforce collaborative initiatives and projects in this exciting domain of functional imaging and modeling of the heart.

We would like to warmly thank all contributing authors, program committee members, additional reviewers, the organizers of the challenge and workshops, administrative support from Insavalor, our respective institutions - in particular Polytech Lyon for welcoming the conference in their building - colleagues and students at CREATIS Lab, and sponsors for their time, efforts, and financial support in making FIMH 2023 a successful event.

June 2023

Olivier Bernard
Patrick Clarysse
Nicolas Duchateau
Jacques Ohayon
Magalie Viallon

Organization

Conference Chairs

Olivier Bernard	CREATIS, INSA, Université de Lyon, France
Patrick Clarysse	CREATIS, CNRS, Université de Lyon, France
Nicolas Duchateau	CREATIS, IUF, Université Claude Bernard Lyon 1, Université de Lyon, France
Jacques Ohayon	TIMC, UGA-Grenoble Polytech - Savoie Mont Blanc University Le Bourget du Lac, France
Magalie Viallon	CREATIS, Université Jean Monnet and CHU Saint-Etienne, France

Local Organizing Committee

Damien Garcia	CREATIS, Inserm, Université de Lyon, France
Monica Sigovan	CREATIS, CNRS, Université de Lyon, France
François Varray	CREATIS, Université Claude Bernard Lyon 1, Université de Lyon, France

Program Committee

Leon Axel	New York University, USA
Peter Bovendeerd	Eindhoven University of Technology, The Netherlands
Oscar Camara	Universitat Pompeu Fabra, Barcelona, Spain
Dominique Chapelle	Inria Saclay, France
Teodora Chitiboi	Siemens Healthineers, Hamburg, Germany
Henry Chubb	Stanford University, USA
Richard Clayton	University of Sheffield, UK
Yves Coudière	Inria Bordeaux, France
Tammo Delhaas	Maastricht University, The Netherlands
Seraina Dual	KTH Royal Institute of Technology, Stockholm, Sweden
Daniel Ennis	Stanford University, USA
Miguel Fernandez	Inria Paris, France
Pierre-Marc Jodoin	University of Sherbrooke, Canada

Andrew King	King's College London, UK
Cristian Linte	Rochester Institute of Technology, USA
Axel Loewe	Karlsruhe Institute of Technology, Germany
Lasse Løvstakken	Norwegian University of Science and Technology, Trondheim, Norway
Rob MacLeod	University of Utah, Salt Lake City, UT, USA
Martyn Nash	University of Auckland, New Zealand
Steven Niederer	King's College London, UK
Mathias Peirlinck	Delft University of Technology, The Netherlands
Luigi Perotti	University of Central Florida, Orlando, FL, USA
Mihaela Pop	Inria Sophia-Antipolis, France / Sunnybrook Research Institute, Toronto, Canada
Franck Sachse	University of Utah, Salt Lake City, UT, USA
Michael Sacks	University of Texas at Austin, USA
Francisco Sahli Costabal	Pontificia Universidad Católica de Chile, Santiago, Chile
Laurent Sarry	Clermont Auvergne University, France
Andrew Scott	Royal Brompton & Harefield NHS Foundation, London, UK
Rafael Sebastian	Universitat de Valencia, Spain
Maxime Sermesant	Inria Sophia Antipolis, France
Lawrence Staib	Yale University, New Haven, CT, USA
Régis Vaillant	GE Healthcare, Paris, France
Linwei Wang	Rochester Institute of Technology, USA
Vicky Wang	University of Auckland, New Zealand
Guang Yan	Imperial College London, UK
Alistair Young	King's College London, UK
Nejib Zemzemi	Inria Bordeaux, France
Xiahai Zhuang	Fudan University, Shanghai, China
Maria Zuluaga	Eurecom, Sophia-Antipolis, France

Additional Reviewers

Gabriel Bernardino	Universitat Pompeu Fabra, Barcelona, Spain
Mathieu De Craene	Philips Research, Suresnes, France
Ewan Evain	Dassault Systèmes, Vélizy-Villacoublay, France
Damien Garcia	CREATIS, Inserm, Université de Lyon, France
Prashnna Kumar Gyawali	Stanford University, USA
Alain Lalande	University of Burgundy, CHU Dijon, France
Pablo Lamata	University of Burgundy, CHU Dijon, France
Pierre-Jean Lartaud	CHU Lyon, France

Zhiyuan Li	Rochester Institute of Technology, USA
Ryan Missel	Rochester Institute of Technology, USA
Hernán Morales	Dassault Systèmes, Vélizy-Villacoublay, France
Andreas Østvik	Norwegian University of Science and Technology, Trondheim, Norway
Simone Pezzuto	Università degli Studi di Trento, Italy
Monica Sigovan	CREATIS, CNRS, Université de Lyon, France
Erik Smistad	Norwegian University of Science and Technology, Trondheim, Norway
Yuemin Zhu	CREATIS, CNRS, Université de Lyon, France

FIMH Board Members

Leon Axel	New York University, USA
Patrick Clarysse	CREATIS, CNRS, Université de Lyon, France
Martyn Nash	University of Auckland, New Zealand
Mihaela Pop	Sunnybrook Research Institute, Toronto, Canada
Vicky Y. Wang	Stanford University, USA

Administrative Organizers

Muriel Personne	Insavalor, France
Charlotte Besnard	Insavalor, France
Pauline Jorio	Insavalor, France

Sponsors

The FIMH 2023 conference was certified by the MICCAI society and the French Society of Biomechanics (with financial support).

FIMH 2023 is particularly grateful for the support of the following academic institutions:

The following companies:

SIEMENS
Healthineers

vermon

And GRAND LYON la metropole

Contents

Cardiac Multiscale Structure

Characterization of the Septal Discontinuity in Ex-Vivo Human Hearts Using Diffusion Tensor Imaging: The Potential Structural Determinism Played by Fiber Orientation in Clinical Phenotype of Laminopathy Patients 3
Pierre Cabanis, Julie Magat, Girish Ramlugun, Nestor Pallares-Lupon, Fanny Vaillant, Emma Abell, Laura Bear, Cindy Michel, Philippe Pasdois, Pierre Dos-Santos, Marion Constantin, David Benoist, Line Pourtau, Virginie Dubes, Julien Rogier, Louis Labrousse, Mathieu Pernot, Oliver Busuttil, Michel Haissaguerre, Olivier Bernus, Bruno Quesson, Edward Vigmond, Richard Walton, Josselin Duchateau, and Valéry Ozenne

Ultrastructure Analysis of Cardiomyocytes and Their Nuclei 14
Tabish A. Syed, Yanan Wang, Drisya Dileep, Minhajuddin Sirajuddin, and Kaleem Siddiqi

Description of the Intrusion Angle of Local Cardiomyocyte Aggregates in Human Left Ventricular Free Wall Using X-ray Phase-Contrast Tomography 25
Shunli Wang, Zhisheng Wang, Zongfeng Li, Junning Cui, and François Varray

A Micro-anatomical Model of the Infarcted Left Ventricle Border Zone to Study the Influence of Collagen Undulation 34
Emilio A. Mendiola, Eric Wang, Abby Leatherman, Qian Xiang, Sunder Neelakantan, Peter Vanderslice, and Reza Avazmohammadi

Symmetric Multimodal Mapping of Ex Vivo Cardiac Microstructure of Large Mammalian Whole Hearts for Volumetric Comparison of Myofiber Orientation Estimated from Diffusion MRI and MicroCT 44
Valéry Ozenne, Girish Ramlugun, Julie Magat, Nestor Pallares Lupon, Pierre Cabanis, Pierre Dos Santos, David Benoist, Virginie Dubes, Josselin Duchateau, Louis Labrousse, Michel Haïssaguerre, Olivier Bernus, and Richard Walton

The Effect of Temporal Variations in Myocardial Perfusion on Diffusion Tensor Measurements 54
Ignasi Alemany, Pedro F. Ferreira, Sonia Nielles-Vallespin, Andrew D. Scott, and Denis J. Doorly

Ventricular Helix Angle Trends and Long-Range Connectivity 64
Alexander J. Wilson, Q. Joyce Han, Luigi E. Perotti, and Daniel B. Ennis

On the Possibility of Estimating Myocardial Fiber Architecture from Cardiac Strains 74
Muhammad Usman, Emilio A. Mendiola, Tanmay Mukherjee, Rana Raza Mehdi, Jacques Ohayon, Prasanna G. Alluri, Sakthivel Sadayappan, Gaurav Choudhary, and Reza Avazmohammadi

Cardiac Electrophysiology Modeling

The Fibrotic Kernel Signature: Simulation-Free Prediction of Atrial Fibrillation 87
Francisco Sahli Costabal, Tomás Banduc, Lia Gander, and Simone Pezzuto

Isogeometric-Mechanics-Driven Electrophysiology Simulations of Ventricular Tachycardia 97
R. Willems, E. Kruithof, K. L. P. M. Janssens, M. J. M. Cluitmans, O. van der Sluis, P. H. M. Bovendeerd, and C. V. Verhoosel

Cellular Automata for Fast Simulations of Arrhythmogenic Atrial Substrate 107
G. S. Romitti, A. Liberos, P. Romero, D. Serra, I. García, M. Lozano, R. Sebastian, and M. Rodrigo

Effect of Gap Junction Distribution, Size, and Shape on the Conduction Velocity in a Cell-by-Cell Model for Electrophysiology 117
Giacomo Rosilho de Souza, Simone Pezzuto, and Rolf Krause

Automated Generation of Purkinje Networks in the Human Heart Considering the Anatomical Variability 127
María Correas, María S. Guillem, and Jorge Sánchez

On the Accuracy of Eikonal Approximations in Cardiac Electrophysiology in the Presence of Fibrosis 137
Lia Gander, Rolf Krause, Martin Weiser, Francisco Sahli Costabal, and Simone Pezzuto

Sensitivity of Repolarization Gradients to Infarct Borderzone Properties Assessed with the Ten Tusscher and Modified Mitchell-Schaeffer Model 147
Justina Ghebryal, Evianne Kruithof, Matthijs J. M. Cluitmans, and Peter H. M. Bovendeerd

Numerical Investigation of Methods Used in Commercial Clinical Devices for Solving the ECGI Inverse Problem 157
Narimane Gassa, Vitaly Kalinin, and Nejib Zemzemi

Evaluation of Inverse Electrocardiography Solutions Based on Signal-Averaged Beats to Localize the Origins of Spontaneous Premature Ventricular Contractions in Humans 166
Yesim Serinagaoglu Dogrusoz, Nika Rasoolzadeh, Beata Ondrusova, Peter Hlivak, and Jana Svehlikova

An *in silico* Study of Cardiac hiPSC Electronic Maturation by Dynamic Clamp 175
Sofia Botti, Chiara Bartolucci, Rolf Krause, Luca F. Pavarino, and Stefano Severi

Electrocardiology Modeling After Catheter Ablations for Atrial Fibrillation 184
Simone Nati Poltri, Guido Caluori, Pierre Jaïs, Annabelle Collin, and Clair Poignard

Modeling Cardiac Stimulation by a Pacemaker, with Accurate Tissue-Electrode Interface 194
Valentin Pannetier, Michael Leguèbe, Yves Coudière, Richard Walton, Philippe Dhiver, Delphine Feuerstein, and Diego Amaro

Simulated Excitation Patterns in the Atria and Their Corresponding Electrograms 204
Joshua Steyer, Lourdes Patricia Martínez Diaz, Laura Anna Unger, and Axel Loewe

Deep Learning-Based Emulation of Human Cardiac Activation Sequences 213
Ambre Bertrand, Julia Camps, Vicente Grau, and Blanca Rodriguez

Influence of Myocardial Infarction on QRS Properties: A Simulation Study 223
Lei Li, Julia Camps, Zhinuo Wang, Abhirup Banerjee, Blanca Rodriguez, and Vicente Grau

Image and Shape Analysis

Effect of Spatial and Temporal Resolution on the Accuracy of Motion Tracking Using 2D and 3D Cine Cardiac Magnetic Resonance Imaging Data ... 235
Kateřina Škardová, Tarique Hussain, Martin Genet, and Radomír Chabiniok

Extraction of Volumetric Indices from Echocardiography: Which Deep Learning Solution for Clinical Use? ... 245
Hang Jung Ling, Nathan Painchaud, Pierre-Yves Courand, Pierre-Marc Jodoin, Damien Garcia, and Olivier Bernard

Whole Heart 3D Shape Reconstruction from Sparse Views: Leveraging Cardiac Computed Tomography for Cardiovascular Magnetic Resonance ... 255
Hao Xu, Marica Muffoletto, Steven A. Niederer, Steven E. Williams, Michelle C. Williams, and Alistair A. Young

Comparison of CNN Fusion Strategies for Left Ventricle Segmentation from Multi-modal MRI ... 265
Cylia Ouadah, Azadeh Hadadi, Alain Lalande, and Sarah Leclerc

Long Axis Cardiac MRI Segmentation Using Anatomically-Guided UNets and Transfer Learning ... 274
Andre Von Zuben, Emily Whitt, Felipe A. C. Viana, and Luigi E. Perotti

Deep Active Learning for Left Ventricle Segmentation in Echocardiography ... 283
Eman Alajrami, Preshen Naidoo, Jevgeni Jevsikov, Elisabeth Lane, Jamie Pordoy, Nasim Dadashi Serej, Neda Azarmehr, Fateme Dinmohammadi, Matthew J. Shun-shin, Darrel P. Francis, and Massoud Zolgharni

Right Ventricular Volume Prediction by Feature Tokenizer Transformer-Based Regression of 2D Echocardiography Small-Scale Tabular Data ... 292
Tuan A. Bohoran, Polydoros N. Kampaktsis, Laura McLaughlin, Jay Leb, Serafeim Moustakidis, Gerry P. McCann, and Archontis Giannakidis

Detection of Aortic Cusp Landmarks in Computed Tomography Images with Deep Learning ... 301
Luka Škrlj, Matija Jelenc, and Tomaž Vrtovec

Automatic Detection of Coil Position in the Chest X-ray Images for Assessing the Risks of Lead Extraction Procedures 310
YingLiang Ma, Vishal S. Mehta, C. Aldo Rinaldi, Pengpeng Hu, Steven Niederer, and Reza Razavi

Cardiac MRI Tagline Extraction Based on Diffeomorphic Active Contour Algorithm 320
Ruiyi Zhang, Jinchi Wei, Dnyanesh Tipre, Robert G. Weiss, Laurent Younes, and Siamak Ardekani

Weighted Tissue Thickness 329
Nicolas Cedilnik and Jean-Marc Peyrat

Strainger Things: Discrete Differential Geometry for Transporting Right Ventricular Deformation Across Meshes 338
Gabriel Bernardino, Thomas Dargent, Oscar Camara, and Nicolas Duchateau

Shape Morphing and Slice Shift Correction in Congenital Heart Defect Model Generation 347
Puck Pentenga, Ashley Stroh, Wouter van Genuchten, Wim A. Helbing, and Mathias Peirlinck

Implicit Neural Representations for Modeling of Abdominal Aortic Aneurysm Progression 356
Dieuwertje Alblas, Marieke Hofman, Christoph Brune, Kak Khee Yeung, and Jelmer M. Wolterink

Prototype of a Cardiac MRI Simulator for the Training of Supervised Neural Networks 366
Marta Varela and Anil A. Bharath

Deformable Image Registration Using Vision Transformers for Cardiac Motion Estimation from Cine Cardiac MRI Images 375
Roshan Reddy Upendra, Richard Simon, Suzanne M. Shontz, and Cristian A. Linte

Unsupervised Polyaffine Transformation Learning for Echocardiography Motion Estimation 384
Yingyu Yang and Maxime Sermesant

Automated Analysis of Mitral Inflow Doppler Using Deep Neural Networks ... 394
Jevgeni Jevsikov, Elisabeth S. Lane, Eman Alajrami, Preshen Naidoo, Nasim Dadashi Serej, Neda Azarmehr, Sama Aleshaiker, Catherine C. Stowell, Matthew J. Shun-shin, Darrel P. Francis, and Massoud Zolgharni

VisHeart: A Visualization and Analysis Tool for Multidimensional Data 403
Edson A. G. Coutinho, Bruno M. Carvalho, Selan R. dos Santos, and Leon Axel

Generating Short-Axis DENSE Images from 4D XCAT Phantoms: A Proof-of-Concept Study 412
Hugo Barbaroux, Michael Loecher, Karl P. Kunze, Radhouene Neji, Daniel B. Ennis, Sonia Nielles-Vallespin, Andrew D. Scott, and Alistair A. Young

Cardiovascular Hemodynamics and CFD

Vortex Duration Time to Infer Pulmonary Hypertension: *In-Silico* Emulation and Dependence on Quantification Technique 425
Malak Sabry, Pablo Lamata, Andreas Sigfridsson, Hamed Keramati, Alexander Fyrdahl, Martin Ugander, Magdi H. Yacoub, David Marlevi, and Adelaide De Vecchi

Modelling Blood Flow and Biochemical Reactions Underlying Thrombogenesis in Atrial Fibrillation 435
Ahmed Qureshi, Maximilian Balmus, Shaheim Ogbomo-Harmitt, Dmitry Nechipurenko, Fazoil Ataullakhanov, Gregory Y. H. Lip, Steven E. Williams, David Nordsletten, Oleg Aslanidi, and Adelaide de Vecchi

SE(3) Symmetry Lets Graph Neural Networks Learn Arterial Velocity Estimation from Small Datasets 445
Julian Suk, Christoph Brune, and Jelmer M. Wolterink

Influence of Anisotropy on Fluid-Structure Interaction Simulations of Image-Based and Generic Mitral Valves 455
Nariman Khaledian, Pierre-Frédéric Villard, Peter E. Hammer, Douglas P. Perrin, and Marie-Odile Berger

Computational Modelling of the Cardiovascular System for the Non-invasive Diagnosis of Portal Hypertension 465
M. Inmaculada Villanueva, Patricia Garcia-Cañadilla, Oscar Camara, Angeles Garcia-Criado, Genis Camprecios, Valeria Perez-Campuzano, Virgina Hernandez-Gea, Fanny Turon, Anna Baiges, Angela Lopez Sainz, Juan Carlos García-Pagan, Bart Bijnens, and Gabriel Bernardino

An Image-Based Computational Model of the Newborn Cardiovascular System with Term and Preterm Applications 475
Robyn W. May, Gonzalo D. Maso Talou, Finbar Argus, Thomas L. Gentles, Frank H. Bloomfield, and Soroush Safaei

Impact of Blood Rheological Strategies on the Optimization of Patient-Specific LAAO Configurations for Thrombus Assessment 485
Carlos Albors, Andy L. Olivares, Xavier Iriart, Hubert Cochet, Jordi Mill, and Oscar Camara

Shape-Guided In-Silico Characterization of 3D Fetal Arch Hemodynamics in Suspected Coarctation of the Aorta 495
Uxio Hermida, Milou P. M. van Poppel, Malak Sabry, Hamed Keramati, David F. A. Lloyd, Johannes K. Steinweg, Trisha V. Vigneswaran, John M. Simpson, Reza Razavi, Kuberan Pushparajah, Pablo Lamata, and Adelaide De Vecchi

Showcasing Capabilities of a Hybrid Mock Circulation Loop for Investigation of Aortic Coarctation 505
Emanuele Perra, Oliver Kreis, and Seraina A. Dual

Hemodynamics in Patients with Aortic Coarctation: A Comparison of *in vivo* 4D-Flow MRI and FSI Simulation 515
Priya J. Nair, Martin R. Pfaller, Seraina A. Dual, Michael Loecher, Doff B. McElhinney, Daniel B. Ennis, and Alison L. Marsden

Cardiac Biomechanics

Evaluating Passive Myocardial Stiffness Using *in vivo* cine, cDTI, and Tagged MRI 527
Fikunwa O. Kolawole, Vicky Y. Wang, Bianca Freytag, Michael Loecher, Tyler E. Cork, Martyn P. Nash, Ellen Kuhl, and Daniel B. Ennis

High-Speed High-Fidelity Cardiac Simulations Using a Neural Network Finite Element Approach 537
Shruti Motiwale, Wenbo Zhang, and Michael S. Sacks

Effect of Varying Pericardial Boundary Conditions on Whole Heart Function: A Computational Study 545
Justina Ghebryal, Cristobal Rodero, Rosie K. Barrows, Marina Strocchi, Caroline H. Roney, Christoph M. Augustin, Gernot Plank, and Steven A. Niederer

An Extended Generalized Hill Model for Cardiac Tissue: Comparison with Different Approaches Based on Experimental Data 555
Dennis Ogiermann, Daniel Balzani, and Luigi E. Perotti

pyheart-lib: A Python Library for LS-DYNA Multi-physics Heart Simulations 565
Martijn Hoeijmakers, Karim El Houari, Wenfeng Ye, Pierre L'Eplattenier, Attila Nagy, Dave Benson, and Michel Rochette

Evaluation of Mechanical Unloading of a Patient-Specific Left Ventricle: A Numerical Comparison Study 575
Britt P. van Kerkhof, Koen L. P. M. Janssens, Luca Barbarotta, and Peter H. M. Bovendeerd

A Modelling Study of Pulmonary Regurgitation in a Personalized Human Heart 585
Debao Guan, Yingjie Wang, Xiaoyu Luo, Mark Danton, and Hao Gao

Pump and Tissue Function in the Infarcted Heart Supported by a Regenerative Assist Device: A Computational Study 594
Koen L. P. M. Janssens, M. van der Knaap, and Peter H. M. Bovendeerd

Clinical Applications

Which Anatomical Directions to Quantify Local Right Ventricular Strain in 3D Echocardiography? 607
Maxime Di Folco, Thomas Dargent, Gabriel Bernardino, Patrick Clarysse, and Nicolas Duchateau

Biomechanical Model to Aid Surgical Planning in Complex Congenital Heart Diseases 616
Maria Gusseva, Nikhil Thatte, Daniel A. Castellanos, Peter E. Hammer, Sunil J. Ghelani, Ryan Callahan, Tarique Hussain, and Radomír Chabiniok

Automated Estimation of Left Ventricular Diastolic Chamber Stiffness: Application to Patients with Heart Failure and Aortic Regurgitation 626
Abdallah I. Hasaballa, Debbie Zhao, Vicky Y. Wang, Thiranja P. Babarenda Gamage, Stephen A. Creamer, Gina M. Quill, Peter N. Ruygrok, Satpal S. Arri, Robert N. Doughty, Malcolm E. Legget, Alistair A. Young, and Martyn P. Nash

A Computational Pipeline for Patient-Specific Prediction of the Post-operative Mitral Valve Functional State 636
Hao Liu, Natalie T. Simonian, Alison M. Pouch, Joseph H. Gorman, III, Robert C. Gorman, and Michael S. Sacks

Automatic Aortic Valve Pathology Detection from 3-Chamber Cine MRI with Spatio-Temporal Attention Maps 648
Y. On, K. Vimalesvaran, C. Galazis, S. Zaman, J. Howard, N. Linton, N. Peters, G. Cole, A. A. Bharath, and M. Varela

Uncertainty to Improve the Automatic Measurement of Left Ventricular Ejection Fraction in 2D Echocardiography Using CNN-Based Segmentation 658
Antonio Sánchez-Puente, Pablo Pérez-Sánchez, Víctor Vicente-Palacios, Alberto García-Galindo, Pedro Pablo Vara, Candelas Pérez del Villar, and Pedro L. Sánchez

Automated Estimation of Motion Patterns of the Left Ventricle Supports Cardiomyopathy Identification 668
Edmond Astolfi, Athira Jacob, Indraneel Borgohain, Akos Varga-Szemes, Puneet Sharma, and Tiziano Passerini

Assessment of the Evolution of Temporal Segmental Strain in a Longitudinal Study of Myocardial Infarction 678
Bianca Freytag, Nicolas Duchateau, Lorena Petrusca, Jacques Ohayon, Pierre Croisille, and Patrick Clarysse

Localizing Cardiac Dyssynchrony in M-mode Echocardiography with Attention Maps 688
Marta Saiz-Vivó, Isaac Capallera, Nicolas Duchateau, Gabriel Bernardino, Gemma Piella, and Oscar Camara

The Extent of LGE-Defined Fibrosis Predicts Ventricular Arrhythmia Severity: Insights from a Preclinical Model of Chronic Infarction 698
Terenz Escartin, Philippa Krahn, Cindy Yu, Matthew Ng, Jennifer Barry, Sheldon Singh, Graham Wright, and Mihaela Pop

Left Ventricular Work and Power are Constant Despite Varying Cardiac Cycle Length—Implications for Patients with Atrial Fibrillation 708
Debbie Zhao, João F. Fernandes, Stephen A. Creamer, Abdallah I. Hasaballa, Vicky Y. Wang, Thiranja P. Babarenda Gamage, Malcolm E. Legget, Robert N. Doughty, Peter N. Ruygrok, Pablo Lamata, Alistair A. Young, and Martyn P. Nash

Correction to: Simulated Excitation Patterns in the Atria and Their Corresponding Electrograms C1
Joshua Steyer, Lourdes Patricia Martínez Diaz, Laura Anna Unger, and Axel Loewe

Author Index 719

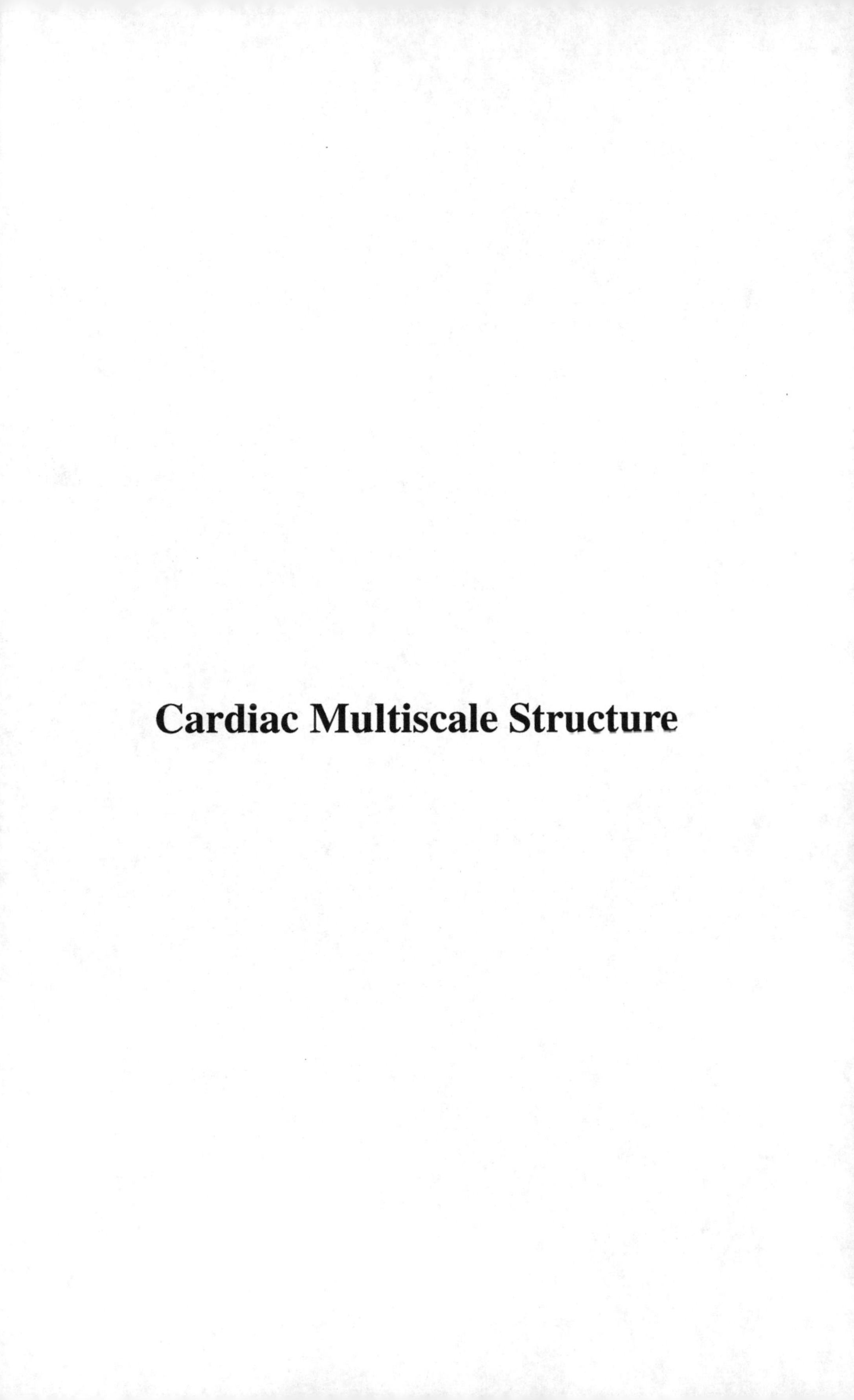

Cardiac Multiscale Structure

Characterization of the Septal Discontinuity in Ex-Vivo Human Hearts Using Diffusion Tensor Imaging: The Potential Structural Determinism Played by Fiber Orientation in Clinical Phenotype of Laminopathy Patients

Pierre Cabanis[1,2](✉), Julie Magat[1,2], Girish Ramlugun[2], Nestor Pallares-Lupon[2], Fanny Vaillant[2], Emma Abell[2], Laura Bear[2], Cindy Michel[2], Philippe Pasdois[2], Pierre Dos-Santos[2,3], Marion Constantin[2], David Benoist[2], Line Pourtau[2], Virginie Dubes[2], Julien Rogier[2,3], Louis Labrousse[2,3], Mathieu Pernot[2,3], Oliver Busuttil[2,3], Michel Haissaguerre[2,3], Olivier Bernus[2], Bruno Quesson[1,2], Edward Vigmond[2], Richard Walton[2], Josselin Duchateau[2,3], and Valéry Ozenne[1,2]

[1] Univ. Bordeaux, CNRS, CRMSB, UMR 5536, Bordeaux, France
pcabanis@rmsb.u-bordeaux.fr

[2] Liryc, Electrophysiology and Heart Modeling Institute, Fondation Bordeaux Université, Pessac-Bordeaux, France

[3] Cardiology Department, Bordeaux University Hospital (CHU), Pessac, France

1 Introduction

A key component of the cardiac conduction system is the atrioventricular (AV) node where the His bundle (HB) divides into the right (RBB) and left (LBB) bundles branches. The RBB is connected to the right ventricle (RV) via the moderator band (MB). The latter is an intra-cavitary and compartmentalized structure with various morphology [1, 2] that has been recently identified as a source for arrhythmia [3, 4]. Presently, the mechanism or the origin of the arrhythmogenesis is unknown. Although the understanding of the anatomy [5] and the cardiac conduction system [6] is well known and associated with conventional treatments of cardiac arrhythmia such as pacemaker RV lead implantation [6], the understanding of the architecture of the neighboring cardiac fibers is unclear. Indeed, the largest and closest associated « structure» is the intraventricular septum (IVS) whose fiber organization is a matter of debate in the literature. The IVS has been described either as a dual layer structure with two populations of fiber orientation by few studies [7–9] or as underlined by a recent meta review [10] as a continuous arrangement of cardiomyocytes following the helix angle rule. The latter refuting the existence of septal discontinuity or abrupt change in helical angles.

Arrhythmias and related cardiomyopathies are associated with distinct clinical phenotypes. A key hallmark of LMNA-dependent cardiomyopathy (LMNA-CMP) is the presence of fibrosis within the IVS and extracellular volume ECV with a mid-septal myocardial basal enhancement [11]. While a dense literature exists on the origin and mechanism of cardiac fibrosis [12], the location of the fibrosis has not extensively been

O. Bernard et al. (Eds.): FIMH 2023, LNCS 13958, pp. 3–13, 2023.
https://doi.org/10.1007/978-3-031-35302-4_1

investigated from a simplified anatomical viewpoint. Our hypothesis is that the alteration in fiber orientation remains asymptomatic until various factors, such as environmental influences or gene mutations, create a favorable environment for the development of ECV or fibrosis. The present study examined the morphological heterogeneity of the IVS and the MB in 9 ex-vivo human hearts, demonstrated the connectivity of RV layer of the IVS with the MB using streamlines representations, showed that the basal IVS can be divided into two layers of fibers using either usual (Helix Angle [HA]) or reference frame invariant (Myocardium Disarray Index [MDI]) metrics, and measured and compared the location of the septal discontinuity versus the location of mid-septal Late Gadolinium Enhancement (LGE) in patients with LMNA-CMP [13].

2 Methods

2.1 Sample Description

Human samples #1–6 & #8 were obtained through a human donor program (providing access to heart samples from patients under cerebral death for scientific research purposes). After written informed consent from the patient's family was obtained, the heart was collected as a human biological sample for scientific research. Hearts #7 & #9 were derived from another human donor program (providing access to heart samples from patients under cardiac transplantation) with a written informed consent of the patient. Donors' information (sex, age, size of the heart and cause of death/transplantation) are described on Table 1. All experiments were approved by the Agence Française de la Biomédecine and conducted in accordance with the declaration of Helsinki and the institutional ethics committee.

Table 1. Summary of the key characteristic of ex-vivo hearts. F, Female; M, Male.

Heart n°	Age [y]	Sex	dimensions [cm x cm x cm]	cause of death / transplantation
#1	53	F	10.9 x 8.0 x 14.1	hemorrhagic stroke
#2	56	M	8.6 x 9.4 x 10.7	cardiovascular accident, cerebral anoxia
#3	82	F	8.2 x 10.1 x 11.1	hemorrhagic stroke
#4	83	F	10.1 x 8.1 x 11.4	head trauma
#5	83	F	8.4 x 7.4 x 12.1	hemorrhagic stroke
#6	51	F	10.2 x 10.2 x 14.4	cardiac respiratory arrest
#7	47	F	8.8 x 8.3 x 10.9	cardiomiopathy (transplantation)
#8	71	F	7.3 x 10.5 x 12.6	cardiac respiratory arrest
#9	53	M	8.8 x 11.6 x 12.1	heart attack (transplantation)

2.2 Sample Preparation

Hearts were fixed with a solution containing formalin (10%) and a gadolinium-based contrast agent (Dotarem; Guerbet, Paris, France, 0.2% of total volume of perfusion), by retrograde perfusion from the aorta. The sample preparation was performed as previously described [8, 14, 15].

2.3 MRI Acquisition

- All experiments were performed at 9.4 T with an inner bore size of 30 cm (BioSpin MRI; Bruker, Ettlingen, Germany) equipped with 300 mT/m gradient insert of 200-mm inner diameter adapted for large samples. Images were acquired using a dedicated radiofrequency volume array coil with seven elements in transmit and receive. Each sample were immerged a solution of Fluorinert and placed into a plastic container during the scan. All MRI scans and experiment acquisitions were similar to those described previously in [8, 14, 15]. The acquisition workflow is summarized in Fig. 1A and detailed in the next sections. The same parameters were set for each acquisition except for the field of view and matrix size which were adjusted for each sample.
- Diffusion-Weighted Imaging (DWI). A 3D DW spin-echo sequence was used to acquire DWI (TE = 22 ms, TR = 500 ms, acquisition time ≈24 h). The DWI dataset consisted of six noncollinear diffusion encoding directions acquired with a b value of 1,000 s/mm^2 and one non-diffusion weighted (b0) image with an isotropic spatial resolution of 600 μm.
- Anatomical Imaging. The 3D anatomical image was acquired with a gradient echo sequence (with, TR = 30 ms, TE = 9 ms, acquisition time ≈18 h) at an isotropic resolution of 150 μm.

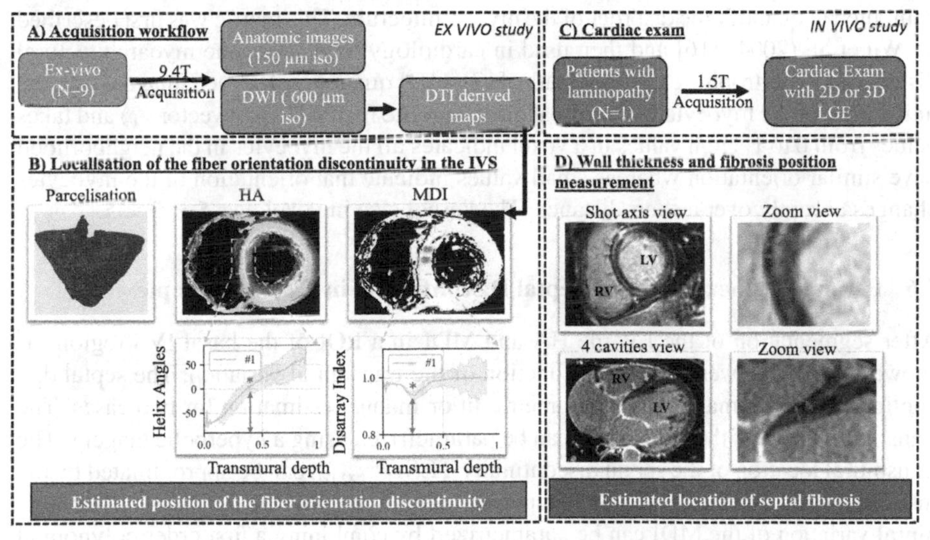

Fig. 1. Chart of the processing steps for N = 9 ex vivo human hearts and N = 20 in-vivo human cardiac exams. A) After perfusion with contrast agent and fixative, (N = 9) hearts were im-aged in a 9.4T magnet using successively gradient echo and DW spin-echo sequence. DTI derived maps (FA, eigenvectors) were calculated. B) Segmentation of the basal IVS was performed. HA and MDI which were derived from DTI maps. Average transmural variation of HA and MDI. C) Cardiac exam using LGE. D) Measurement of the wall thickness and fibrosis position in SA and 4 chambers view.

2.4 Diffusion Tensor Estimation

Each DWI was up-sampled by a factor of 2 using trilinear interpolation to reach a voxel size of 0.3 × 0.3 × 0.3 mm. DT calculations were performed prior to any registered to avoid difficult transformations of the diffusion encoding matrix. DT maps [eigenvalues: λ1, λ2, λ3, apparent diffusion coefficient, fractional anisotropy (FA), and color-coded FA (cFA), also known as red-green-blue colormap] were obtained with MRtrix3 software (https://www.mrtrix.org). The first DT eigenvector v_1 was associated with the main myofiber orientation. Tractrography processing was performed on all samples using the first DT eigenvector v1 with the FACT (deterministic) algorithm. Streamlines were generated using the following parameter: 500k seeds, a FA stopping of 0.15, maximum angle of 30° and a maximum length of 40 mm. The polar coordinate system was defined using the LA of the LV by setting two points in basal and apex location in the left cavity.

2.5 DTI-Derived Metrics Computation

HA were calculated in the polar coordinate system using the projection of the first eigenvector v_1 (from DTI metrics) on the tangential plane. To avoid non-physiological jumps in our region of interest, an unwrap filter was applied in all voxels located at mid-distance from the LV epicardium. The unwrap filter was set to subtract 180° if the angle were found superior to 45° and is used in the study unless mentioned. MDI was computed to quantify the disorder of the myoarchitecture. This metric was first described by Wu et al. (2004) [16] and then used in cardiology to evaluate the myoarchitectural disarray of hypertrophic cardiomyopathy [17]. MDI quantifies, for each image voxel, the uniformity of the myocyte longitudinal direction (the first DTI eigenvector v_1) and takes values from 0 to 1. High values in a voxel indicates all the myocytes in the neighborhood have similar orientation whereas small values indicate that orientation of the myocytes changes abruptly over a short distance. The kernel size in voxel was 5 × 5 × 5.

2.6 Estimated Location of the Septal Discontinuity in Ex-Vivo Sample

After segmentation of the LV, the HA and MDI in a ROI of the basal IVS region (as shown in Fig. 1B) were plotted as function of the transmural variation. The septal discontinuity was estimated using parametric fit or manual estimation for two cases. The transmural depth of the helix angle can be parametrized using a hyperbolic tangent. The transmural location of the septal discontinuity $TLSD^{ha}{}_{exvivo}$ can be approximated by the inflection point and calculated by computing the minimum of the derivative. The transmural variation of the MDI can be parametrized by combining a first order polynomial with a Gaussian curve. The $TLSD^{mdi}{}_{exvivo}$ was estimated by finding the minimum of this function.

2.7 Study Population and CMR Acquisition

A prospective patient study was conducted in 20 patients to undergo clinical CMR to evaluate heart failure. CMR was performed on a 1.5 T clinical scanner (MAGNETOM Aera, Siemens Healthcare, Germany) with a dedicated 32-channel spine coil and an

18-channel body coil. LGE (reconstructed spatial resolution of 1.5 × 1.5 × 4.0 mm) imaging using magnitude inversion recovery was performed 10 min after the injection of a contrast agent (gadoteric acid, 0.2 mmol/kg) using a breath-held and inversion recovery-prepared turbo FLASH sequence encompassing the ventricles in short-axis (SA), 2-chamber, 4-chamber orientations and transverse view. The imaging parameters were TR/TE 4.9/2.3 ms, flip angle 10°, 256 × 256 matrix, Bandwidth 360 Hz/pixel. Inversion time was optimized for each measurement to null the signal intensity of normal myocardium (240–360 ms).

2.8 Estimated Location of Septal Discontinuity in In-Vivo Study Population

The images were independently analyzed by two readers. In each patient, the presence and extent in wall thickness of the mid-septal myocardial LGE pattern was visually estimated. LGE was considered present only if visible in SA and LA views. The wall thickness (L1), the distance (L2) between the RV cavity and the location of the fibrosis and distance (L3) between the LV cavity and the location of the fibrosis. L1-L2-L3 being the thickness of fibrosis and L4 = L1-L2, the distance between the LV cavity and the extent of the fibrosis toward the RV. Transmural relative locations between [0, 1] were computed using R3 = L3/L1, using R4 = L4/L1. Finally, two one-sided t-tests' (TOST) procedures between the $TLSD_{exvivo}$ against R3 and R4 were done using 95% confidence limit and upper and lower equivalence bound of 0.1 and -0.1 respectively and 0.05 and −0.05 respectively.

3 Results

Figure 2 shows the anatomical images of the IVS in LA view to investigate the reproducibility across the 9 ex-vivo human samples. The top of the view is located approximately at the base of the IVS and close to the tricuspid (TV) and mitral (MV) valve while the bottom of the view shows the MB and the RV myocardium and cavity. After visual inspection, several elements are to be underlined. First (A), the RV of samples #3, 4, 6 is flattened or collapsed to the IVS. Second, the MB is clearly visible on heart #1, 2, 5, 7, 8, 9 on the anatomical view zoomed (B, orange arrows) while not well defined in the sample #4 or even identified in the sample #6. A small difference in intensity contrast between the continuity of the MB in the IVS and the IVS is visible in samples #5, 6. The FA (C) sounds homogenous in the IVS in samples #1, 2, 6, 8, 9 with the exception of samples #1, 3, 5, 7 where a black line indicating a loss of FA is visible (blue arrows). The streamline representation (D) revealed in samples #1, 2, 3, 5, 7, 8 fiber-bundles in base to apex orientation (purple or blue color) on the moderator band connecting to the epicardial (or RV) area of the IVS and circumferential fiber bundles (green color) on the endocardial or (LV) area of the IVS. An abrupt change of fiber orientation in the IVS is therefore noticeable at the vicinity of the intersection of the two streamline populations. Small fiber bundles in base to apex orientation are also visible in the endocardial part of the IVS in samples #1, 2, 3, 6, 7 but do not show fiber orientation discontinuity with the neighboring area.

Figure 3 shows in both the LA and the SA view, the MB of Fig. 2.B using the HA (C, D) and the MDI (E, F) for each heart. At first glance, the HA of each sample goes linearly

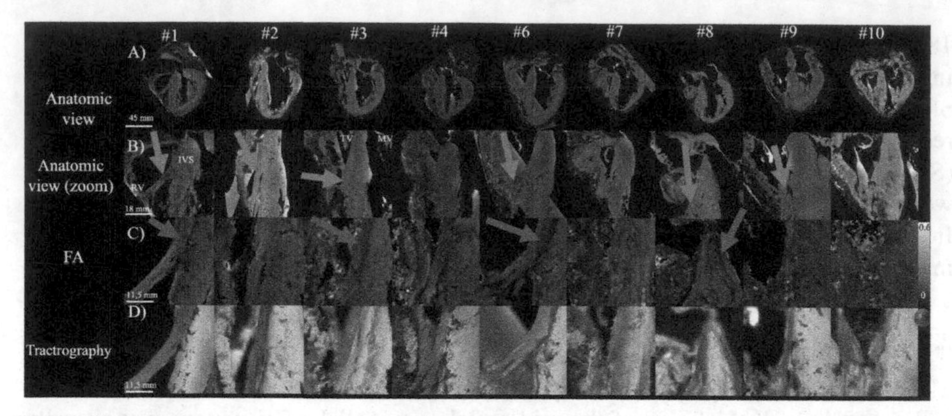

Fig. 2. Comparison across ex-vivo human hearts (N = 9) of the fiber orientation in the IVS in LA view using anatomical and diffusion tensor images. From top to bottom, A) anatomical view, B) anatomic view with a focus on the moderator band, C) Fractional anisotropy, D) tractography of the whole heart. A loss of FA (blue arrows) divided the MB (orange arrows, blue-purple streamlines) from the IVS (green streamlines). (Color figure online)

from the LV cavity to RV cavity but the HA range differs between samples with intervals such as [−60, 60] or [−90, 90] for samples #5 and #9. On the other hand, the MDI of the IVS sounds constant and equal to 1 with the exception of a decrease close to 0.5 pointed by green arrows (E, F), which means a heterogenous orientation of myocytes in the neighborhood of a voxel. 6/9 samples (66%) displayed a myofiber orientation discontinuity displaying a line in the basal and mid-ventricular IVS that vanish above the MB (e.g. #1, #5, #9). The corresponding HA map displayed an accentuated color gradient in these areas.

Figure 4 shows the variation of the mean HA unwrapped as a function of transmurality in the IVS basal zone. $TLSD^{ha}{}_{exvivo}$ (dotted line) located at the inflection point can be interpreted as the fiber orientation discontinuity. It was estimated by parametric fits or manual estimation to 0.79 ± 0,08% of the transmural depth. For sample #2–5, 7–9 the inflection point is higher pronounced than the others. Except samples #6, the HA variation is relatively monotonous and linear before and after the infection point. Similarly, Fig. 5 shows the transmural variation of the MDI. In most of the IVS, MDI is a plateau close to 1. Except for samples #7, 8, the decrease of MDI shown on Fig. 3.E, F can be recovered on these transmurally graphs. $TLSD^{ha}{}_{exvivo}$ (dotted line) was estimated using a parametric fit or manual estimation to 0.77 ± 0.09% of the transmural depth.

Figure 6 shows representative midwall LGE involving the IVS in four LMNA patients. Mid-septal myocardial LGE was found in the basal or mid-ventricular septal wall in 13/20 patients (65%). Although it depends on the readers' appreciation, the location of the LGE in the reformatted transverse views for patients #9 and #20 mimic the location of MDI in ex-vivo samples. The wall thickness distance was 10.6 ± 2.3 mm in comparison to the ex vivo groups 14.6 ± 2.4 mm ($p < 0.0001$). The location of the septal discontinuity was R3 = 0.4 ± 0.1% and R4 = 0.73 ± 0.05%. The null hypothesis of two one-sided t-tests' (TOST) procedures (the two samples have different means) between $TLSD_{exvivo}$ from MDI against R3 and R4 were found accepted and rejected respectively

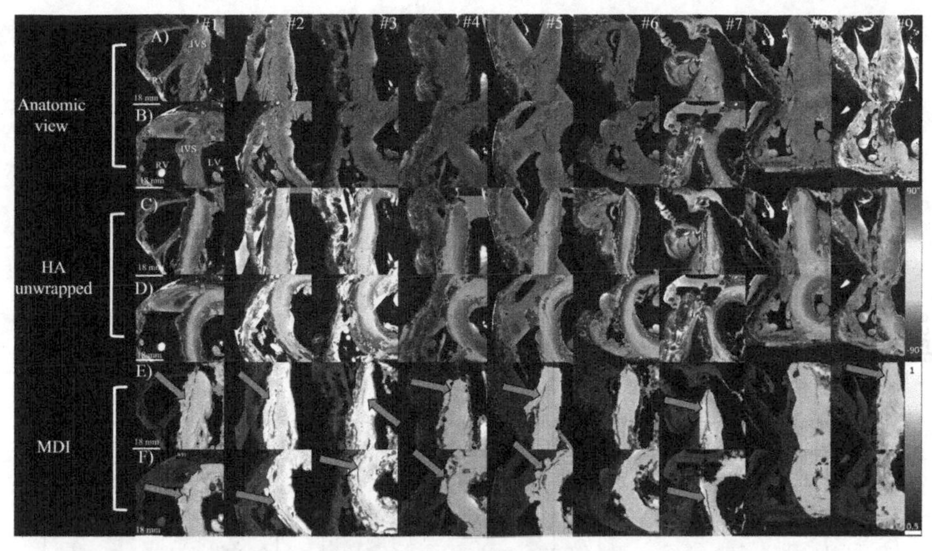

Fig. 3. Comparison across ex-vivo human hearts (N = 9) of the discontinuity in the IVS in LA and SA view using anatomical and diffusion tensor images. From top to bottom, A–B) anatomical view with a focus the moderator band, in LA and SA view respectively, C–D) unwrapped HA in LA and SA view respectively E–F) MDI in LA and SA view respectively. Green arrows show decrease of the MDI in the IVS. (Color figure online)

using a confidence interval of [−0.1, 0.1], and both accepted using a confidence interval of [−0.05, 0.05].

4 Discussions

In this study, microstructure of the IVS was investigated using ex-vivo MRI across nine whole human hearts. Heterogeneous medical conditions were included in the study due to the rarity and difficulty of such sample collections. First, after visual inspection of the anatomical image at 150 μm, the heterogeneity of the MB morphology was found in agreement with previous work of Lee and al [1] with evident visual observation (7/9, 77%) and location between the middle and the top of the septum (8/9, 88%). Lee and al found 95% and 89% respectively using N = 38 formalin-fixed human hearts. DTI-derived metrics were plotted spatially (Fig. 2, Fig. 3) or as a function of the transmural variation (Fig. 4, Fig. 5) to analyze fiber orientation and to question the existence of abrupt change in fiber orientation. Although the interpretation of the quantitative metrics remains subject to discussion, we tried to use objective criteria to define such abrupt change. Indeed, the main advantage of MDI and streamlines processing (with the exception of the color of the streamline) over the HA computation is these are their invariant properties with respect to the image referential. The second limitation of the HA is the inherent presence of wraps. The fiber connectivity of MB to the IVS and to the basal part which include the AV node was highlighted with streamline representation in agreement with the common understanding of the cardiac conduction system and previous studies

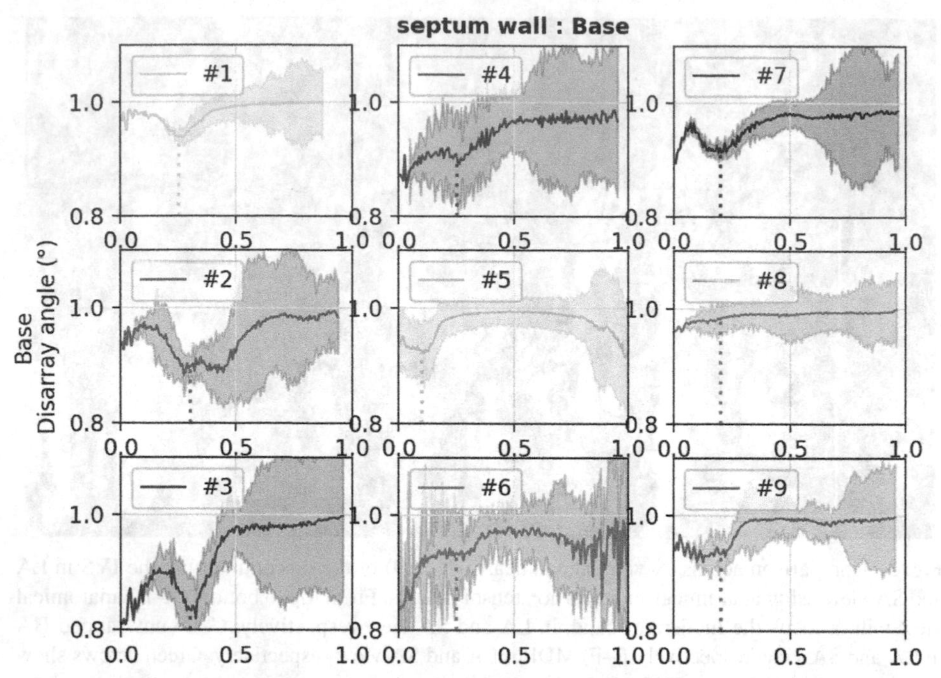

Fig. 4. Average transmural variation of helix angle on ex-vivo human hearts. For each diagram, the horizontal axis is the transmural depth, 0 means right ventricle wall and 1 is endocardium of the left ventricle; dotted line is the inflection point of the HA.

in large mammalian species [8]. Using HA and MDI maps, the basal and mid-ventricular IVS can be resoundingly described with two layers, the left layer with a circumferential orientation the myofiber and the right layer with a base to apex orientation myofiber and the moderator band (MB). The findings are therefore partially at odds with literature [11] as noticeable variations in the base-apex (Fig. 3, LA view) or anterior-posterior (Fig. 3, SA view) direction are important. In samples #1, #6, the right layer vanishes above the MB and then presents a linear HA variation. As expected, the MDI maps allow a clearer interpretation of fiber orientation change than the HA maps.

While regional heterogeneity structure or function assessment is now a clinical routine procedure, the structural or functional impact of the dual layer of the IVS is relatively sparse in the literature with few examples in EP [18], echography [19] or modeling [8, 22]. In contrast to CMR DTI, the presence of bright patterns, in LGE images, largely used for risk stratification of cardiomyopathies, do not identify myocardial disarray but an increased interstitial space and fibrosis [21]. Nevertheless, it is often admitted that myocardial fibrosis has been found in regions where myocardial disarray is expected [22]. The underlying concept is that the presence of an interface or discontinuity between distinct populations (in orientation) of myocardium cells might be a potential substrate for interstitial fibrosis. Experimentally, a decrease of FA was noticeable at the interface of the fiber bundles in samples #1, 3, 5, 7 This could indicate either cross orientation of fibers within a voxel, or also the presence of interstitial fibrosis. Histological studies of

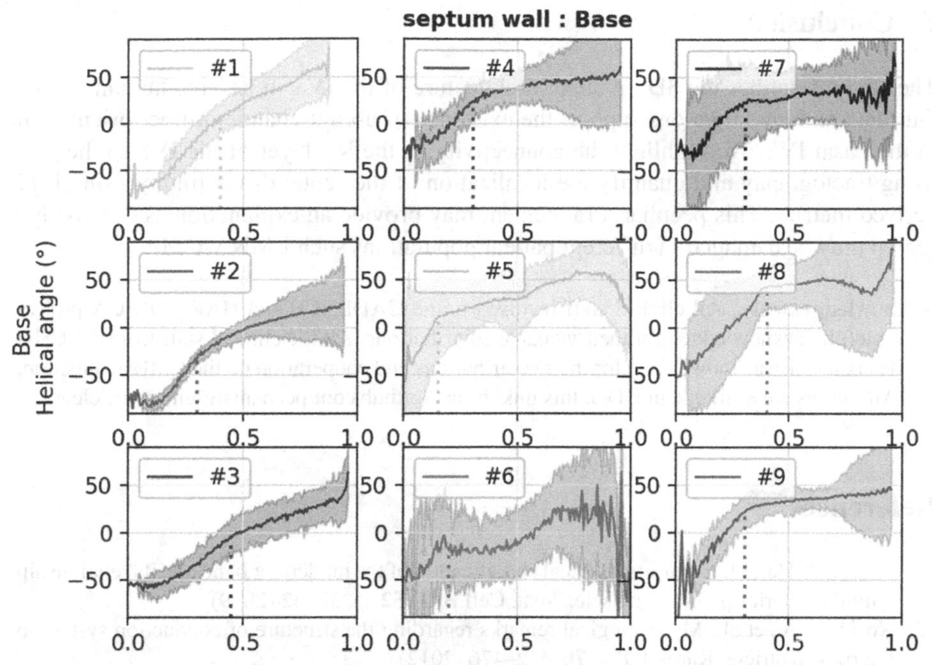

Fig. 5. Average transmural variation of disarray index on ex-vivo human hearts. For each diagram, the horizontal axis is the transmural depth, 0 means right ventricle wall and 1 is endocardium of the left ventricle; dotted line is the inflection point of MDI.

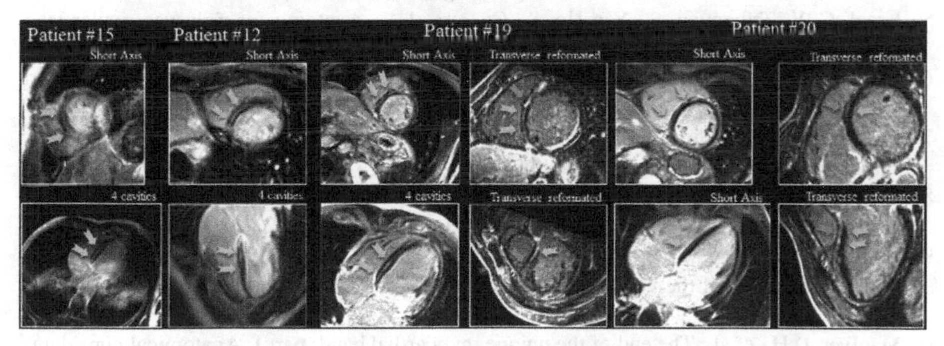

Fig. 6. CMR findings in four representative patients with LMNA-CMP with midwall LGE (yellow arrows) involving the basal IVS (top row: SA view, bottom row: four-chamber view). Additional reformatted transverse views for patients #19 and #20 show the midwall LGE and the location of the MB (blue arrows). (Color figure online)

the location and characterization of mid-septal fibrosis in humans with an overall view of the septum are scarce. One example MRI in patient with ventricular tachycardia [23] demonstrate the presence of fibrosis in the mid-wall as expected from LGE images.

5 Conclusion

The study describes the 3D myofiber architecture of the IVS in ex-vivo human hearts. Our data and investigations support the existence of abrupt change in fiber orientation in the basal IVS. We highlight the connectivity of the RV layer of the IVS to the MB using tractography and quantify the localization of the septal discontinuity using DTI derived metrics. This peculiar arrangement may provide an explanation as to why this region may be damaged in different patient populations such LMNA-CMP.

Acknowledgements. All clinical staff involved in the CADENCE and HARMONICA project are gratefully acknowledged for their valuable contributions. All the clinical staff from the CHU Bordeaux are also acknowledged for their contributions and cooperation on the CMR acquisition.

All figures are available in HD at this link: https://github.com/pcabanis/Figures_article.git.

References

1. Lee, J.-Y., Hur, M.-S.: Morphological classification of the moderator band and its relationship with the anterior papillary muscle. Anat. Cell Biol. **52**(1), 38–42 (2019)
2. Kosiński, A., et al.: Morphological remarks regarding the structure of conduction system in the right ventricle. Kardiol. Pol. **70**, 472–476 (2012)
3. Barber, M.: Arrhythmias from the right ventricular moderator band: diagnosis and management (2019). Accessed 31 Jan 2023
4. Haïssaguerre, M., et al.: Role of Purkinje conducting system in triggering of idiopathic ventricular fibrillation. Lancet Lond. Engl. **359**(9307), 677–678 (2002)
5. Israel, C.W., Tribunyan, S., Yen Ho, S., Cabrera, J.A.: Anatomy for right ventricular lead implantation. Herzschrittmachertherapie Elektrophysiol. **33**(3), 19–326 (2022)
6. Kapa, S., Bruce, C.J., Friedman, P.A., Asirvatham, S.J.: Advances in cardiac pacing: beyond the transvenous right ventricular apical lead. Cardiovasc. Ther. **28**(6), 369–379 (2010)
7. Baban, A., et al.: Cardiovascular involvement in pediatric laminopathies. Report of six patients and literature revision. Front. Pediatr. **8** (2020)
8. Rodríguez-Padilla, J., et al.: Impact of intraventricular septal fiber orientation on cardiac electromechanical function. Am. J. Physiol. Heart Circ. Physiol. **322**(6), H936–H952 (2022)
9. Kocica, M.J., et al.: The helical ventricular myocardial band: global, three-dimensional, functional architecture of the ventricular myocardium. Eur. J. Cardiothorac. Surg. **29**, S21–S40 (2006)
10. MacIver, D.H., et al.: The end of the unique myocardial band: part I. Anatomical considerations. Eur. J. Cardio-Thorac. Surg. Off. J. Eur. Assoc. Cardio-Thorac. Surg. **53**(1), 112–119 (2018)
11. Fontana, M., et al.: CMR-verified interstitial myocardial fibrosis as a marker of subclinical cardiac involvement in LMNA mutation carriers. JACC Cardiovasc. Imaging **6**(1), 124–126 (2013)
12. Krenning, G., Zeisberg, E.M., Kalluri, R.: The origin of fibroblasts and mechanism of cardiac fibrosis. J. Cell. Physiol. **225**(3), 631–637 (2010)
13. Eijgenraam, T.R., Silljé, H.H.W., de Boer, R.A.: Current understanding of fibrosis in genetic cardiomyopathies. Trends Cardiovasc. Med. **30**(6), 353–361 (2020). https://doi.org/10.1016/j.tcm.2019.09.003

14. Magat, J., et al.: 3D MRI of explanted sheep hearts with submillimeter isotropic spatial resolution: comparison between diffusion tensor and structure tensor imaging. Magn. Reson. Mater. Phys., Biol. Med. **34**(5), 741–755 (2021). https://doi.org/10.1007/s10334-021-00913-4
15. Haliot, K., et al.: A 3D high resolution MRI method for the visualization of cardiac fibro-fatty infiltrations. Sci. Rep. **11**(1), Art. no. 1 (2021). https://doi.org/10.1038/s41598-021-85774-6
16. Wu, Y.-C., Field, A.S., Chung, M.K., Badie, B., Alexander, A.L.: Quantitative analysis of diffusion tensor orientation: Theoretical framework. Magn. Reson. Med. **52**(5), 1146–1155 (2004). https://doi.org/10.1002/mrm.20254
17. Garcia-Canadilla, P., et al.: Myoarchitectural disarray of hypertrophic cardiomyopathy begins pre-birth. J. Anat. **235**(5), 962–976 (2019)
18. Vetter, F.J., Simons, S.B., Mironov, S., Hyatt, C.J., Pertsov, A.M.: Epicardial fiber organization in swine right ventricle and its impact on propagation. Circ. Res. **96**(2), 244–251 (2005)
19. Boettler, P., et al.: New aspects of the ventricular septum and its function: an echocardiographic study. Heart Br. Card. Soc. **91**(10), 1343–1348 (2005)
20. Doste, R., et al.: A rule-based method to model myocardial fiber orientation in cardiac biventricular geometries with outflow tracts. Int. J. Numer. Methods Biomed. Eng. **35**(4), e3185 (2019)
21. Moon, J.C.C., et al.: The histologic basis of late gadolinium enhancement cardiovascular magnetic resonance in hypertrophic cardiomyopathy. J. Am. Coll. Cardiol. **43**(12), 2260–2264 (2004)
22. Kuribayashi, T., Roberts, W.C.: Myocardial disarray at junction of ventricular septum and left and right ventricular free walls in hypertrophic cardiomyopathy. Am. J. Cardiol. **70**(15), 1333–1340 (1992)
23. Nishimura, T., et al.: Prognostic value of cardiac magnetic resonance septal late gadolinium enhancement patterns for periaortic ventricular tachycardia ablation: heterogeneity of the anteroseptal substrate in nonischemic cardiomyopathy. Heart Rhythm **18**(4), 579–588 (2021)

Ultrastructure Analysis of Cardiomyocytes and Their Nuclei

Tabish A. Syed[1], Yanan Wang[1], Drisya Dileep[2,3], Minhajuddin Sirajuddin[2], and Kaleem Siddiqi[1(✉)]

[1] School of Computer Science, McGill University, Montréal, Canada
siddiqi@cim.mcgill.ca

[2] Institute for Stem Cell Science and Regenerative Medicine, Bengaluru, India

[3] The University of Trans-Disciplinary Health Sciences and Technology (TDU), Bengaluru, India

Abstract. Cardiomyocytes are elongated and densely packed in the mammalian heart and connected end on end to achieve a functional syncytium. Qualitative accounts describe their nuclei as being elongated in their long axis direction, which might help to better distribute mechanical load and reduce mechanical stress during contraction. Alterations of nuclear orientation and shape have also been known to be associated with certain cardiomyopathies. Yet, to date, the alignment of cardiomyocytes and their nuclei at the cellular (micron) scale has not been assessed in a quantitative fashion. To examine this we developed 3D computer vision methods to segment myocytes and their nuclei in cleared and membrane stained thick 3D tissue sections from a wild type mouse heart, imaged using confocal microscopy. We extended a geometric flow based superpixel algorithm to 3D and then adaptively merged the resulting supervoxels to recover individual myocytes. In parallel we also applied recent popular deep learning based cell segmentation methods to the same data. Our experiments revealed a close alignment of myocyte orientation with nucleus orientation, with a median difference of approximately 10°, and also showed that most cardiomyocytes contain only one or two nuclei. These findings pave the way for future investigations of the effect of specific cardiac diseases on nuclear shape, elongation and number.

Keywords: cardiomyocytes · nuclei · tissue clearing · staining · confocal microscopy · cell segmentation · orientation analysis

1 Introduction

The morphology of nuclei within cardiomyocytes is known to play an important role in their development. Nuclei can be reshaped when these cells are under a tensile load and gene expression can be altered by changes in nuclear

Supplementary Information The online version contains supplementary material available at https://doi.org/10.1007/978-3-031-35302-4_2.

O. Bernard et al. (Eds.): FIMH 2023, LNCS 13958, pp. 14–24, 2023.
https://doi.org/10.1007/978-3-031-35302-4_2

morphology, potentially signalling aberrant patient conditions [18]. The contraction of cardiomyocytes also distributes mechanical forces within them, affecting their internal structure [4]. Nuclear rearrangement can also evolve into disorders that are manifest in elderly individuals as cardiovascular disease or hypertension [15]. Cardiomyopathies can also affect nuclei within cardiomyocytes. For example, hypertrophic cardiomyopathy is correlated with a reduction in nuclear deformation and reduction in the frequency of contractions, while dilated cardiomyopathy induced by laminopathies can alter the circularity, morphology, and orientation of nuclei [3,5]. Nuclear morphology is thus highly correlated with healthy function and the possibility of assessing cardiomyocyte and nucleus shape, structure and orientation from tissue samples could provide both diagnostic and therapeutic benefits.

Recent work has examined how tensile loads, coupled with the action of certain proteins, can result in an in situ deformation of a cell [1]. Rows of cell nuclei within the field of view are seen moving in opposite directions, changing the orientation of their long axis, and falling in and out of the focal plane. Using phase contrast/polarization microscopy the role of cytoskeletal proteins and integrins in determining cell shape and nuclear re-orientation induced strain is examined in [13]. When integrin proteins are distorted with a strain, the actin cytoskeletal filaments re-orient and the nucleus aligns with and elongates in the long-axis direction of the applied tension field. Myocyte orientation has also been studied in tissue engineering. For example, when grown on flat substrates cardiomyocytes exhibit no particular orientation preference, and the nuclei are isotropic or circular in shape [8]. However, when a complex scaffold is used as a substrate the cells self organize into oriented structures. The nuclei then become elongated along the cell long axis direction, and the cells exhibit persistent mechanical and electrical activity. In fact, the importance of orientation dependent alterations to cardiomyocyte shape, e.g., in the presence of cardiomyopathy, is recognized [5]. Incidences of polyploidy and multi nucleation have been studied across species [7,14], but a thorough quantitative analysis of the number of nuclei within cardiomyocytes has not yet been performed.

To potentially address this, workflows for segmenting individual cardiomyocytes, by isolating the interior of mitochondria and myofibrils, have been proposed [10]. Such methods rely on adaptive thresholding and on the manual delineation of myocyte boundaries. Related approaches in [2] for cell shape analysis also rely on thresholding and thus generalizations to handle large sections of heart tissue are not obvious. Thus, few methods presently exist for automated ultrastructure shape analysis of heart tissue at the micron scale of cardiomyocytes. Motivated by this, we developed methods for ultrastructure analysis of cardiomyocytes at the micron scale. We first prepared cleared heart tissue samples using a CLARITY based protocol [6] and imaged thick stacks using confocal microscopy. We then considered two different classes of algorithms. The first (bottom up) approach began with an oversegmentation using an extension of the popular geometric flow based Turbopixel [11] algorithm to 3D. This was followed by a greedy merge procedure designed to coalesce Turbovoxels that were connected at interfaces free of the WGA membrane stain and were thus likely to lie within the same cardiomyocyte. The second (top down) approach used current trained neural networks for single cell segmentation [9,17]. Both approaches provided quantitative verification of nuclear alignment with cardiomyocytes in excised wild type mouse heart tissue, and further revealed that cardiomyocytes have only a few nuclei. The methods and findings in this paper pave the way for future quantitative studies of the effects of cardiomyopathies on cardiomyocyte and nucleus shape and alignment.

2 Methods

2.1 Tissue Sample Preparation and Imaging

A mouse heart was excised after perfusion with ice-cold 1X PBS, 4% PFA, and then incubated in a hydrogel solution for several days. The hydrogel was polymerized at 37 °C, fixing the heart samples in a hydrogel matrix. The heart tissue was then allowed to passively undergo clearing passively by incubating in a clearing buffer, following the CLARITY protocol [6]. The cleared heart was sectioned by a compressotome to generate sections of width 500 µm perpendicular to its long axis. Heart sections selected from the mid-ventricular region were processed for staining with fluorescently labeled WGA (wheat germ agglutinin) and DAPI, which marked cell membranes and nuclei, respectively. We used an Olympus FV3000 confocal microscope with a 60X/1.4NA oil objective for imaging. The imaging regions were concentrated near the mid-LV wall where cardiomyocytes were largely in the plane of imaging in each short axis section, as demonstrated in Fig. 1 (top left). We selected 15 stacks of size $320 \times 320 \times 150$ voxels, with each voxel being $.66 \times .66 \times .8\ \mu m^3$ in size, for analysis.

2.2 Geometric Flow Based Segmentation

Turbovoxels. Our first method is a 3D extension of the 2D geometric flow based Turbopixel algorithm [11], which we dub "Turbovoxels". The algorithm is designed to provide a dense segmentation of the image space into compact regions while respecting boundaries, where a typical structure of interest would be comprised of a connected group of such regions. Thus it provides a useful first step in the recovery of densely packed cells whose membranes have been stained. The algorithm is implemented using a narrow-band level sets method [16], which represents the supervoxel boundaries as the zero level surfaces of an embedding function Φ. The flow based design of Turbovoxels allows for the choice of initial seed locations and their number. We first divide the 3D image volume into a uniform cuboidal grid of size $G_x \times G_y \times G_z$, based on the number of seeds to place within each grid block. Then, within each grid block, we place a seed at the location of minimum intensity of the WGA stain, to ensure that the initial seeds are distant from the brightly stained cell membranes. In our dataset containing image volumes of size $320 \times 320 \times 150$ voxels, we used a seed count of

5000. This corresponds to roughly one supervoxel per cube of length 10 µm in each of its dimensions, which is around half the width of a typical myocyte in a mouse heart. The initial level set embedding Φ is constructed from the Euclidean signed distance function of spheres of radius 3, centered at the seed locations, as illustrated in Fig. 1 (bottom, middle) for a 2D slice of a sample region. Using the standard level set framework the evolution of the embedding level set function Φ is given by

$$\frac{\partial \Phi}{\partial t} = -S||\nabla \Phi|| \tag{1}$$

where S, the speed function, determines the speed of evolution in the direction of the normal to a point on the surface of an evolving supervoxel. Following [11] we factor the speed function into a product $S = S_I S_B$ of an intensity term S_I and a boundary term S_B. The boundary term is a function of distance to the nearest supervoxel. S_B is set to zero for voxels that are equidistant from the boundaries of more than one supervoxel. This multiplicative speed function ensures that supervoxel boundaries do not cross and the resulting Turbovoxels are simply-connected. The boundary term can be evaluated efficiently by first computing the medial surface of the background and setting the term to 0 at locations coincident with the medial surface and to 1 elsewhere. In practice, for efficiency, we approximate a thick medial surface by removing background voxels iteratively while using a digital homotopy preserving method, as described in [12]. The intensity term S_I is based on a linear combination of a constant, curvature and doublet term [11]:

$$S_I = \psi(\mathbf{x}) + \alpha\kappa(\mathbf{x})\psi(\mathbf{x}) + \beta N(\mathbf{x}).\nabla\psi(\mathbf{x}).$$

In our implementation we set the weight α to 0.2 and the doublet weight β to 0.3. The curvature term ensures that the boundaries of supervoxels are smooth, while the doublet term attracts the evolving boundaries to the edges as defined by the edge strength function $\psi = e^{\frac{-|\nabla I|}{G_\sigma * |\nabla I| + \gamma}}$ [11], where I is the image intensity, G_σ is a Gaussian kernel with standard deviation σ, and γ is a scaling factor, which was empirically set to 10. The right panels in Fig. 1 depict the final boundaries of the Turbovoxels (overlayed in cyan) while the middle panels show the zero levelset of the evolving fronts at three time points of the evolution.

Merging. We then merged Turbovoxels into larger segments (Turboblobs) using a criterion that takes into account the membrane stain at the interface of two Turbovoxels. For all Turbovoxel pairs that share a common boundary we compute an affinity term that is inversely proportional to the average stain intensity at their interface. We then employ a 2-stage greedy approach. In the first stage, a Turbovoxel is merged with at most one of its neighbors, the one that it is closest to in terms of the affinity score, provided that their common surface area is above a threshold. This process is carried out in 3D over the entire volume, giving rise to larger groups of Turbovoxels, and is applied for two iterations. At the end of

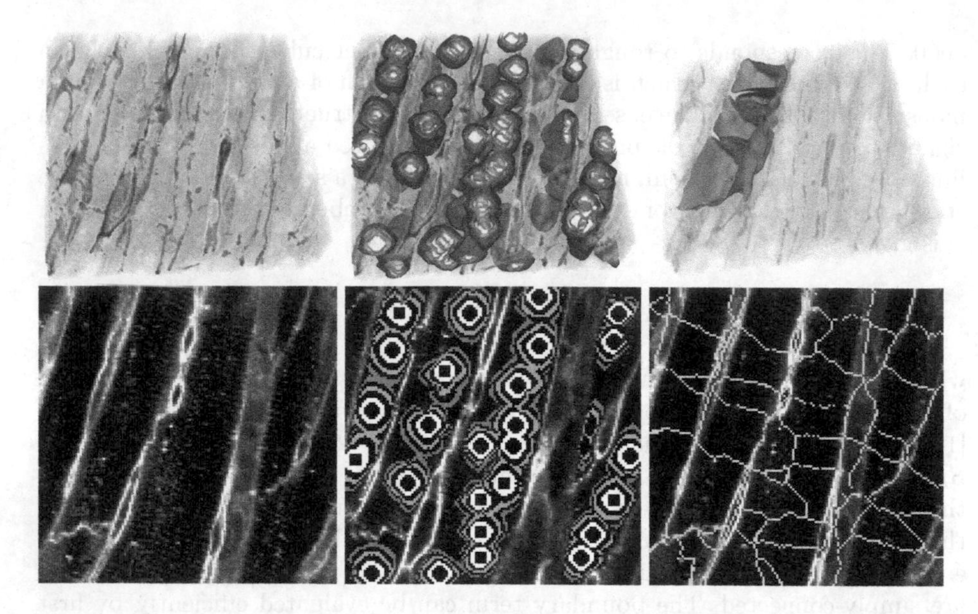

Fig. 1. TOP: A 3D stack of cleared mouse tissue, with the WGA stain highlighting membranes shown in red (left). Seeds evolve outwards, with iterations shown in green to blue (middle). The medial surface between them prevents merging, and an edge-strength based stopping term slows the fronts down in the vicinity of cell membranes. A few of the final dense Turbovoxels are overlayed on the WGA stain in cyan (right). BOTTOM: A 2D slice of the respective panels on the top. Typical results showing merged Turbovoxels are shown in Fig. 3 (bottom right).

the first stage, the segments consist of one, two or four original Turbovoxels. In the second stage, we first construct a minimum heap using all pairs of neighbouring segments with the affinity term as the sorting key. We then carry out greedy pairwise merges, in an iterative manner, resulting in larger Turboblobs. The pair with the lowest average intensity at their interface (highest affinity) is merged first. The merged Turboblob is then inserted into the heap, with the affinities of all its neighbouring blobs recomputed. The process is repeated in an iterative manner, with merges allowed only if a threshold on affinity is reached, until it converges, i.e., no additional merges can be applied.

Run Time. Our implementation of Turbovoxels has not been optimized or parallelized, and thus is not yet suited for real time processing. For a single volume it takes on the order of 4 hrs for their computation on an Intel(R) Core(TM) i9-7900X processor, with an additional 5 min for the merging step.

2.3 Neural Network Based Segmentation

Neural network based approaches for cell segmentation use pre-trained models to produce initial segmentations. Given application specific examples with labeled ground truth results, the network weights can be fine-tuned to produce final results. In the realm of myocyte segmentation from membrane stained data, few (if any) labeled data sets presently exist. We chose to experiment with two networks that represent the present state of the art: Cellpose [17] and Mesmer [9]. Cellpose uses a modified U-Net backbone with residual building blocks, predicting two pixel-wise probability maps. The first classifies a pixel as background or interior, while the second, a spatial gradient mask, learns the gradient of a distance function from the center of a cell. Cellpose also provides a 3D extension of its 2D models. The 3D extension generates 3 separate cell probability maps for three 90° rotations of the volume, and the corresponding 3 predicted 2D gradient maps.

Mesmer uses a ResNet50 backbone-based feature pyramid network and a watershed post-processing step and is pre-trained on TissueNet, which has more than 1 million (2D) labelled cells obtained through crowd-sourcing. Being a 2D method, we had to apply a stitching approach based on intersection over union between successive slices, to obtain labels in 3D.

For segmenting nuclei we found that Cellpose performed well on the DAPI channel, using its pre-trained "nucleus" model. The model was set to operate in 3D, enabling segmentation across all XY, XZ, and YZ slices. However, when applied to the original WGA stained membrane data with the "cytoplasm" option, the model gave poor results for myocyte segmentation, likely due to its bias towards interior texture in the data it is pre-trained on. Such texture is almost completely absent in the interior of elongated myocytes in a membrane channel. To improve its performance, motivated by the construction in [11], we chose to first apply the same edge-strength based speed function used for Turbovoxels, described in Sect. 2.2, to the WGA channel. This function is designed to highlight boundaries while placing texture within (Fig. 3 shows a typical case), and in practice gave far improved segmentation results.

Run Time. Running both the Cellpose and Mesmer models on an Intel(R) Xeon(R) CPU @ 2.30 GHz with a GPU Tesla T4 takes approximately 5 min per volume.

2.4 Orientation Analysis and Fractional Anisotropy

For each segmented nucleus and myocyte we estimated its orientation using an eigenvalue/eigenvector decomposition of the associated 3D covariance matrix of each binary object. The covariance matrix defined as $cov(x_i, x_j) = \mathbf{E}[(x_i - \mathbf{E}[x_i]) \cdot (x_j - \mathbf{E}[x_j])^T]$, where $\mathbf{E}()$ is the expected value of object voxel coordinates x_i. The covariance matrix quantifies the spread of object coordinates (x_i) about the centroid ($\mathbf{E}[x_i]$). Specifically, given a binary segmented object, its orientation was taken to be that corresponding to the eigenvector of the covariance matrix associated with the largest eigenvalue. Further, with $\lambda_1, \lambda_2, \lambda_3$ being the eigenvalues, a measure of its elongation from [0, 1] was obtained by computing the fractional anisotropy (FA) measure

$$FA = \sqrt{\frac{1}{2}} \frac{\sqrt{(\lambda_1 - \lambda_2)^2 + (\lambda_2 - \lambda_3)^2 + (\lambda_3 - \lambda_1)^2}}{\sqrt{\lambda_1^2 + \lambda_2^2 + \lambda_3^2}}.$$

3 Results

3.1 Elongation of Cardiomyocytes and Their Nuclei

For nuclei the mode of the FA measure over all datasets was found to be 0.84 with the median being 0.74, with a 95% confidence interval of (0.7337, 0.7419). For myocytes, the mode of the FA measure over all datasets, shown in Fig. 2, was found to be 0.95 or higher. These measurements indicate that both nuclei and myocytes are highly oriented structures.

Method	FA mode	FA median	median CI
Turbovoxels	0.9483	0.9243	(0.9201, 0.9279)
Cellpose	0.9725	0.9501	(0.9489, 0.9514)
Mesmer	0.9475	0.9085	(0.9061, 0.9105)
GT	0.9628	0.9527	(0.9453, 0.9599)

Fig. 2. Fractional Anisotropy(FA) analysis of segmented myocytes across entire dataset, using the Turbovoxels, Cellpose and Mesmer methods and also a single 3D image with hand segmented labels (GT). The Table lists the mode in column 2, the median in column 3 and the 95% confidence intervals of the respective medians in column 4.

3.2 Orientation Comparison

Our segmentations enabled us to determine the number of nuclei contained in each cardiomyocyte and to compare their orientations. Figure 4 (bottom) shows polar histograms of the magnitude of the angular difference in degrees using the 3 different methods. Turbovoxels gave a median orientation difference of 11.9° with a 95% confidence interval (CI) of (11.5, 12.4). The median orientation differences using Cellpose and Mesmer were 11.6° CI of (11.2, 12.1)) and 11.6° (CI of (11.1, 12.1)), respectively. To validate our auto-segmentations we hand segmented all the myocytes in one 3D volume. For each auto-segmented myocyte in an image volume, we picked the ground truth segment with the maximum volume overlap and computed the Dice score between them. The median Dice scores were 0.62 (Turbovoxels), 0.53 (Cellpose) and 0.05 (Mesmer). The 2D (slice by slice) median Dice score for Mesmer was much higher at 0.71, indicating that generating 3D results using stitching was a major source of error. For Mesmer the average volume of the segmented myocytes was only about half the volume of

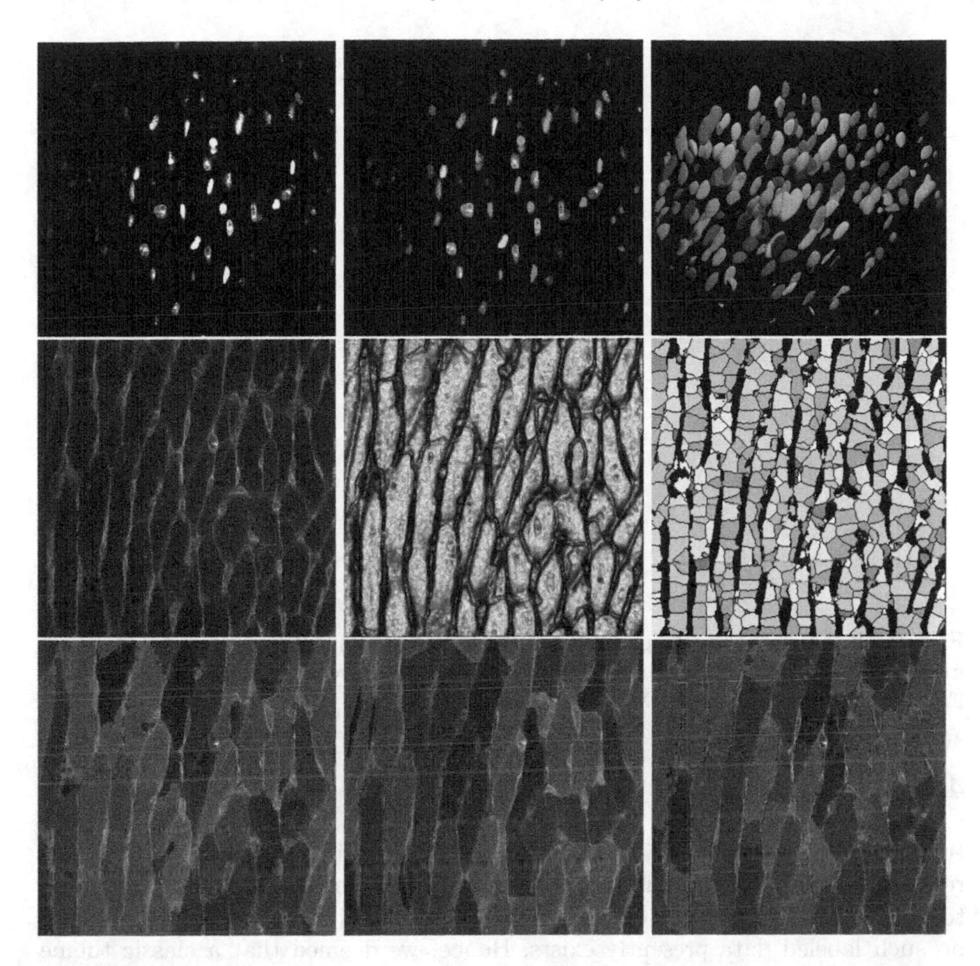

Fig. 3. TOP: A DAPI channel image (left) with a sample slice of segmented nuclei using Cellpose (middle) and rendered in 3D (right). MIDDLE: The associated WGA channel image (left), its edge-strength based speed function (middle) (see Sect. 2.2) and the initial Turbovoxels (right). BOTTOM: Myocyte boundaries segmented from the edge-strength based speed function using Cellpose (left), Mesmer (middle) and by merging Turbovoxels into Turboblobs (right).

that of the ground truth myocytes, while Turbovoxels and Cellpose gave superior results (see Supplementary material for details). The stitching errors also affects the nuclear count per myocyte. As expected, the orientation analysis is consistent across the three methods (Fig. 4). Overall, our experiments confirm both that nuclei within cardiomyocytes are elongated and that their orientations match those of the underlying myocytes quite closely, as observed qualitatively in the literature [2,10].

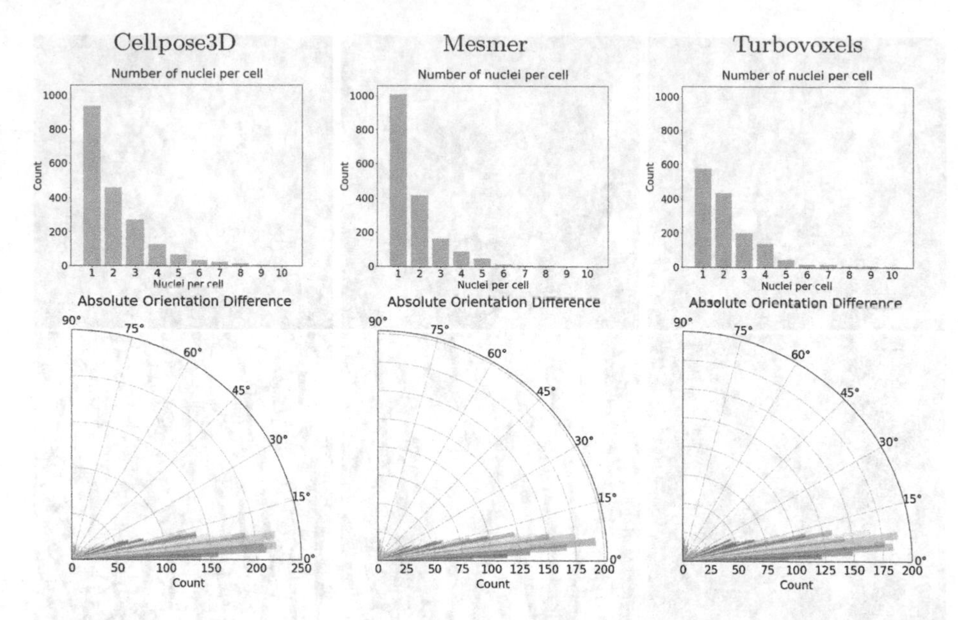

Fig. 4. The number of nuclei per segmented myocyte (top) and histograms of the magnitude of the orientation difference in degrees between myocytes and nuclei (bottom). From left to right: Cellpose3D, Mesmer, Turbovoxels.

4 Discussion

In microscopy the acquisition of labeled data to train neural network models remains a challenge, although the community is moving towards crowd-sourcing to achieve this [9,17]. For cleared excised heart wall tissue samples, virtually no such labeled data presently exists. Hence, we deemed that a classic theme in the computer vision and image analysis literature, that of following dense oversegmentation with merging, was appropriate. Our examinations using 3D confocal microscopy show that the estimated orientation differences between cardiomyocytes and their nuclei were consistent with those obtained using deep learning based general cell segmentation models and that cardiomyocytes align with their nuclei. There is some discrepancy between estimates of the number of nuclei across the methods, which could be due to the failure of 2D deep learning based models such as Mesmer to work in 3D. Qualitative inspection showed that Cellpose and Mesmer also tend to include the cell membrane as part of the segmentation result. Turbovoxels, followed by merging, is implemented as a native 3D method and it could be that it leads to larger segmented myocytes on average.

Given enough ground truth data, neural network based approaches have the advantage of having learned from many past examples and of capturing multi-scale relationships in an implicit way, due to their pyramidal architectures. Purely bottom up geometric flow based approaches such as Turbovoxels use much weaker input data. Our demonstration that they are still successful suggests that they might be able to provide pseudo-ground truth cardiomyocyte labels for training neural network models for novel datasets for heart wall ultrastructure analysis.

References

1. Arnoczky, S.P., Lavagnino, M., Whallon, J.H., Hoonjan, A.: In situ cell nucleus deformation in tendons under tensile load; a morphological analysis using confocal laser microscopy. J. Orthop. Res. **20**(1), 29–35 (2002)
2. Bensley, J.G., De Matteo, R., Harding, R., Black, M.J.: Three-dimensional direct measurement of cardiomyocyte volume, nuclearity, and ploidy in thick histological sections. Sci. Rep. **6**(1), 1–10 (2016)
3. Bera, M., Kumar, R., Sinha, B., Sengupta, K.: Nuclear deformation and anchorage defect induced by DCM mutants in lamin A. bioRxiv, p. 611665 (2019)
4. Bray, M.A.P., Adams, W.J., Geisse, N.A., Feinberg, A.W., Sheehy, S.P., Parker, K.K.: Nuclear morphology and deformation in engineered cardiac myocytes and tissues. Biomaterials **31**(19), 5143–5150 (2010)
5. Brayson, D., Ehler, E., dos Remedios, C.G., Shanahan, C.M.: Analysis of cardiomyocyte nuclei in human cardiomyopathy reveals orientation dependent defects in shape. medRxiv (2020)
6. Chung, K., et al.: Structural and molecular interrogation of intact biological systems. Nature **497**(7449), 332 (2013)
7. Derks, W., Bergmann, O.: Polyploidy in cardiomyocytes: roadblock to heart regeneration? Circ. Res. **126**(4), 552–565 (2020)
8. Entcheva, E., Bien, H.: Tension development and nuclear eccentricity in topographically controlled cardiac syncytium. Biomed. Microdevice **5**(2), 163–168 (2003)
9. Greenwald, N.F., et al.: Whole-cell segmentation of tissue images with human level performance using large-scale data annotation and deep learning. Nat. Biotechnol. **40**(4), 555–565 (2022)
10. Hussain, A., Ghosh, S., Kalkhoran, S.B., Hausenloy, D.J., Hanssen, E., Rajagopal, V.: An automated workflow for segmenting single adult cardiac cells from large-volume serial block-face scanning electron microscopy data. J. Struct. Biol. **202**(3), 275–285 (2018)
11. Levinshtein, A., Stere, A., Kutulakos, K.N., Fleet, D.J., Dickinson, S.J., Siddiqi, K.: TurboPixels: fast superpixels using geometric flows. IEEE Trans. Pattern Anal. Mach. Intell. **31**(12), 2290–2297 (2009)
12. Malandain, G., Fernández-Vidal, S.: Euclidean skeletons. Image Vis. Comput. **16**(5), 317–327 (1998)
13. Maniotis, A.J., Chen, C.S., Ingber, D.E.: Demonstration of mechanical connections between integrins, cytoskeletal filaments, and nucleoplasm that stabilize nuclear structure. Proc. Natl. Acad. Sci. **94**(3), 849–854 (1997)
14. Miko, M., Kyselovic, J., Danisovic, L., Barczi, T., Polak, S., Varga, I.: Two nuclei inside a single cardiac muscle cell. More questions than answers about the binucleation of cardiomyocytes. Biologia **72**(8), 825–830 (2017)
15. Seelbinder, B., et al.: Nuclear deformation guides chromatin reorganization in cardiac development and disease. Nat. Biomed. Eng. **5**(12), 1500–1516 (2021)
16. Sethian, J.A.: Level Set Methods and Fast Marching Methods: Evolving Interfaces in Computational Geometry, Fluid Mechanics, Computer Vision, and Materials Science, vol. 3. Cambridge University Press, Cambridge (1999)

17. Stringer, C., Wang, T., Michaelos, M., Pachitariu, M.: Cellpose: a generalist algorithm for cellular segmentation. Nat. Methods **18**(1), 100–106 (2021)
18. Versaevel, M., Grevesse, T., Gabriele, S.: Spatial coordination between cell and nuclear shape within micropatterned endothelial cells. Nat. Commun. **3**(1), 1–11 (2012)

Description of the Intrusion Angle of Local Cardiomyocyte Aggregates in Human Left Ventricular Free Wall Using X-ray Phase-Contrast Tomography

Shunli Wang[1,2], Zhisheng Wang[1,2(✉)], Zongfeng Li[1,2], Junning Cui[1,2(✉)], and François Varray[3]

[1] Center of Ultra-Precision Optoelectronic Instrument Engineering, Harbin Institute of Technology, Harbin 150080, China
zhisheng.wang@stu.hit.edu.cn, cuijunning@hit.edu.com

[2] Key Lab of Ultra-Precision Intelligent Instrumentation, (Harbin Institute of Technology), Ministry of Industry and Information Technology, Harbin 150080, China

[3] Univ Lyon, INSA-Lyon, Université Claude Bernard Lyon 1, CNRS, Inserm, CREATIS UMR 5220 U1294, Lyon, France

Abstract. A better knowledge of the radial orientation of myocytes from the left ventricular wall's endo- to epi-side may promote our understanding of the cardiac work mechanisms and diseases. The transverse angle is the most used angle, which depicts the myocyte aggregates' angles on the horizontal plane. Compared with the transverse angle, the intrusion angle (IA) can avoid the influence of projection and better describe the radial orientation of local myocyte aggregates. In this preliminary study, we used a simple empirical formula to fit the distribution of IA in three human left ventricular transparietal tissue samples. The samples were imaged using X-ray phase-contrast tomography, and reconstructed with an isotropic 3.5 μm^3 voxel size. We calculated the mean myocyte aggregates IA in each window ($112 \times 112 \times 112$ μm^3), which is regularly spaced in three-dimension and the adjacent ones have a half overlap. The results present that, in small-sized transparietal samples, the IA experiences a quadratic function along the local radial direction and linear functions along longitudinal and circumferential directions. The distribution of IA in the left ventricle is regional.

Keywords: Human left ventricular wall · myocyte arrangement · intrusion angle · X-ray phase-contrast tomography

1 Introduction

Helix angle (HA) and transverse angle (TA) are always used to quantify the orientation of local myocyte aggregates, which respectively depicts the myocytes' angles on the epicardium's tangent and horizontal plane. Along the local left ventricular (LV) wall's radial direction, HA decreases approximately linearly with wall depth spanning [11]. The

O. Bernard et al. (Eds.): FIMH 2023, LNCS 13958, pp. 25–33, 2023.
https://doi.org/10.1007/978-3-031-35302-4_3

distribution of myocyte aggregates TA is complex. When calculated with cardiac MRI, the local myocyte aggregates TA value appears small and relatively constant [6]. Mathematical modelling is assumed to experience a gradual change, varying quadratically along radial and longitudinal directions, respectively [2–5, 9, 10].

A better knowledge of the radial orientation of myocytes from the endo- to epi-side may promote our understanding of cardiac work mechanisms and diseases. In literature [1, 12], the researchers pointed out that TA is a projected angle, which can introduce a considerable bias when the local myocyte aggregates HA is large. Therefore, they suggested using the intrusion angle (IA) to describe the radial orientation of local myocyte aggregates.

We used a set of human cardiac tissue X-ray phase-contrast micro-tomography imaging data to investigate the local myocardial architecture [4, 13–15]. Compared with the standard X-ray attenuation-based imaging, the sensitivity of X-ray phase-contrast micro-tomography imaging in organic materials detection is several orders of magnitude higher [8]. The voxel size in our data is $3.5 \times 3.5 \times 3.5\ \mu m^3$. The size of cardiac myocytes is about 10 μm in diameter and 100 μm in length. So, in our data, the myocytes can be clearly observed [7].

In this paper, firstly, we present the procedures for acquiring cardiac tissue data and the strategy to measure the myocyte aggregates IA in the samples. We introduce a Diophantine function model to quantitatively analyse the myocyte aggregates IA distributions in samples. In the Results section, we present transmural distributions of the IA in each sample and compare the parameters in the proposed model. The Discussion and Conclusion end the paper.

2 Material and Method

2.1 Experiment Data

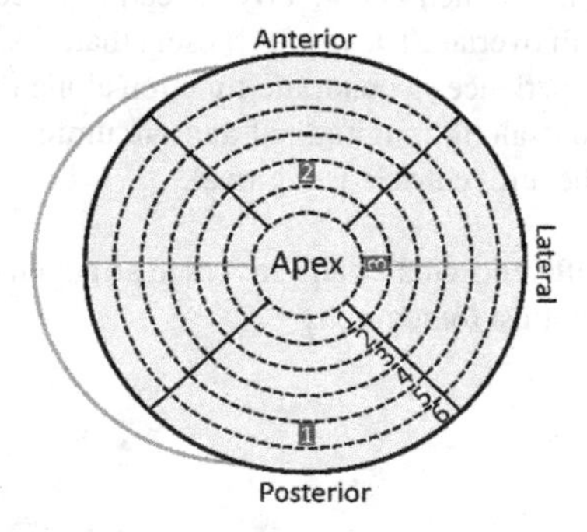

Fig. 1 Location of the samples in the left ventricular wall

The Medico-Legal Institute of Lyon IML HCL (No. DC-2012–1588) provided a series of fresh LV wall transmural tissue samples removed from two males' hearts and preserved in the 10% formalin solution less than 24 h after sudden deaths. The two males, aged 32 and 35 years, had no history of heart disease or addiction. After rehydration and ethanol gradient dehydration, the samples were imaged at the ID-19 beamline at the European Synchrotron Radiation Facility (ESRF) in Grenoble. The reconstructed volumes have an isotropic spatial resolution with each voxel's edges equaling 3.5 μm. Three samples were taken into this preliminary study, extracted from the left ventricle's posterior, anterior and lateral, respectively (Fig. 1). To avoid the influence of edge distortion, a rectangular block was extracted as the region of interest (ROI) from each sample for the study.

2.2 Measurement of the Local Cardiomyocyte Aggregates IA

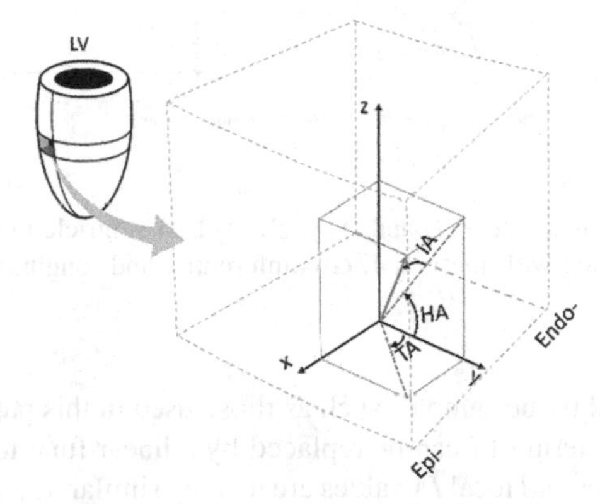

Fig. 2 Definition of the myocyte aggregates angles in local (x, y, z) coordinates

We measured the orientations of local myocyte aggregates in the samples using the algorithm proposed by Varray et al. [11] with the following five steps: (a) Cut each sample volume into a set of sub-volumes sized 32 × 32 × 32 voxels on a regular 3D grid. The adjacent sub-volumes have a half overlap; (b) Compute the spectrum in each sub-volume using 3D fast Fourier Transformation (FFT); (c) Binarize the 3D FFT result and keep the 3% highest amplitude values in each sub-volume; (d) Measure the tensor of each binarised FFT result using the principle component analyses; (e) Calculate the local myocyte aggregates HA, TA and IA in each sub-volume based the tensor's tertiary eigenvector. From Fig. 2, we can note that TA is the projection of IA in the short-axis plane. When HA is small, the values of TA and IA are similar.

2.3 Modeling of the Distribution of Cardiomyocyte Aggregates IA in Samples

In the literature [3], the distribution of the myocyte aggregates TA was expressed mathematically using simple empirical formulas, that the changes of myocyte aggregates TA can be assumed to be quadratic both along the radial and longitudinal directions in the global LV wall coordinate system (c, r, l) presented in Fig. 3 (a):

$$TA(r, l) = \begin{cases} p_1 \cdot (1 - r^2) \cdot l^2 & l > 0 \\ -p_1 \cdot (1 - r^2) \cdot l^2 & l \leq 0 \end{cases} \tag{1}$$

where $TA(r, l)$ is the value of myocyte aggregates TA calculated at points (r, l); (r, l) are normalised local coordinates in each radial-longitudinal plane; r varies from -1 at the endocardium to $+1$ at the epicardium; l varies from -1 at the apex, through 0 at the equator to $+0.5$ at the base; p_1 and p_2 are two coefficients related to circumference value. The relationship between p_1 and p_2 presented in the literature [3] is $p_1 \approx 4 \cdot p_2$.

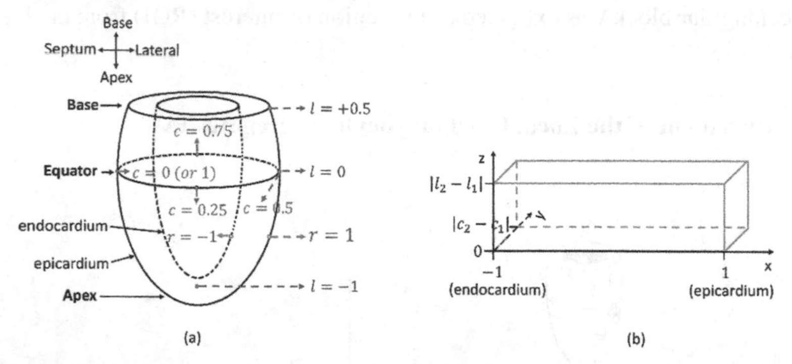

Fig. 3 Normalized coordinates (c, r, l) and (x, y, z). (a) Left ventricle model. (b) Local sample model. x, y and z coincide with the radial, circumferential and longitudinal directions in (a), respectively

In small transmural tissue samples such as those used in this paper, the range of l is small, so the quadratic term of l can be replaced by a linear function. Considering that the local mean *TA* values and local *IA* values are usually similar, Eq. (1) can be expressed as:

$$IA(x, z) = p \cdot (1 - x^2) \cdot (a + k_z \cdot z) \tag{2}$$

where $IA(x, z)$ is the value of myocyte aggregates intrusion angle calculated at points (x, z) in the local (x, y, z) coordinate system depicted in Fig. 1 (b); x is identical to r, varying from -1 at the endocardium to $+1$ at the epicardium;z is from 0 to the length of corresponding l value range (i.e. $|l_2 - l_1|$); k_z and a are longitudinal location related parameters.

Equations (1) and (2) are proposed under the assumption that, along circumferential or y direction, the distribution of *TA* and *IA* is homogeneous. However, in real LV sample data, along the circumferential direction, myocyte aggregates *TA* and *IA* may own slight changes. Considering the small y value ranges in our samples, here, a linear function of y is also added into Eq. (2):

$$IA(x, y, z) = p \cdot (1 - x^2) \cdot (a + k_y \cdot y) \cdot (b + k_z \cdot z) \tag{3}$$

where y is from 0 to the length of the corresponding c value range (i.e. $|c_2 - c_1|$); k_y and b are circumferential location-related parameters.

Based on Eq. (3), the mean IA values ($\overline{IA}$) in coordinate axis perpendicular sections can be calculated:

$$\begin{cases} \overline{IA}(x) = p \cdot \frac{\int_0^{c_2-c_1}\int_0^{l_2-l_1}(a+k_y\cdot y)\cdot(b+k_z\cdot z)dydz}{(c_2-c_1)\cdot(l_2-l_1)} \cdot (1-x^2) \; x \in [r_1, r_2] \\ \overline{IA}(y) = p \cdot \frac{\int_{r1}^{r2}\int_0^{l_2-l_1}(1-x^2)\cdot(b+k_z\cdot z)dxdz}{(r_2-r_1)\cdot(l_2-l_1)} \cdot (a + k_y \cdot y) \quad y \in [0, c_2 - c_1] \\ \overline{IA}(z) = p \cdot \frac{\int_{r1}^{r2}\int_0^{c_2-c_1}(1-x^2)\cdot(a+k_y\cdot y)dxdy}{(r_2-r_1)\cdot(c_2-c_1)} \cdot (b + k_z \cdot z) \quad z \in [0, l_2 - l_1] \end{cases} \tag{4}$$

where $\overline{IA}(x)$ is the mean IA value in each yOz section along x direction. Similarly, $\overline{IA}(y)$ and $\overline{IA}(z)$ present the mean IA values respectively along y and z directions; the $r_1, r_2, c_1, c_2, l_1$and l_2 are ROI's normalised parameter values in each sample, whose values are identified based on their locations and listed in Table 1.

Table 1. ROI's normalised parameter values in each sample

Sample number	Parameters					
	r_1	r_2	c_1	c_2	l_1	l_2
1	−0.92	0.92	0.2	0.29	0.04	0.22
2	−0.4	0.86	0.69	0.82	−0.71	−0.53
3	−0.6	0.88	0.44	0.57	−0.97	−0.8

Equation (4) can be simplified as follows:

$$\begin{cases} \overline{IA}(x) = P_x \cdot \left(1 - x^2\right) x \in [r_1, r_2] \\ \overline{IA}(y) = P_y \cdot y + A \quad y \in [0, c_2 - c_1] \\ \overline{IA}(z) = P_z \cdot z + B \quad z \in [0, l_2 - l_1] \end{cases} \tag{5}$$

Without other restrictive conditions, when the parameters (p, a, k_y, b, k_z) in Eq. (4) are calculated, this Eq. (4) is a Diophantine function. Equation (3) can be rewritten as:

$$IA(x, y, z) = p' \cdot (1 - x^2) \cdot (1 + k'_y \cdot y) \cdot (1 + k'_z \cdot z) \tag{6}$$

with

$$\begin{cases} p' = p \cdot a \cdot b \\ p \cdot k_y \cdot k_z = p' \cdot k'_y \cdot k'_z \end{cases} \tag{7}$$

3 Results

In Fig. 4, we normalised axis x, y and z in each sample's ROI to [0,1] and depicted the distribution of IA together to show their change trend along different directions. The profiles of $\overline{IA}(x)$ (i.e. the mean IA value in each yOz section) approximate quadratic functions, whose graph opens upward in Samples 1 and 2 and extends downward in Sample 3. In each sample, the $\overline{IA}(y)$ decreases, and its distribution approximates linear function. Along z direction, the distributions of IA in samples also approximate linear functions. In Sample 2, the $\overline{IA}(z)$ increases, while in Samples 1 and 3, the $\overline{IA}(z)$ slightly decreases.

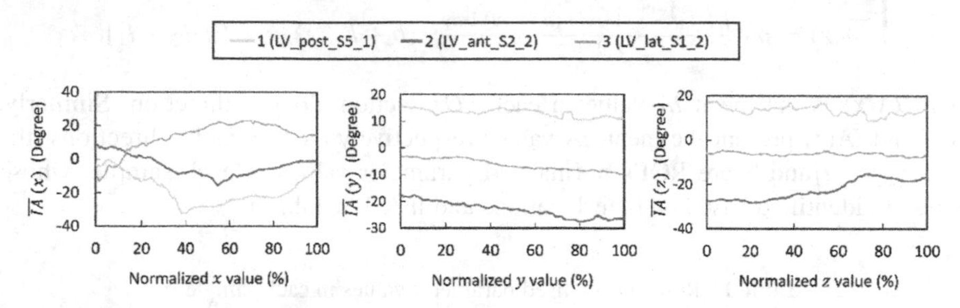

Fig. 4. Profiles of *IA* in samples' ROIs

The parameters (P_x, P_y, P_z, A, B) in Eq. (5), the parameters $\left(p', k_y', k_z'\right)$ in Eq. (6) and the term $p' \cdot k_y' \cdot k_z'$ in Eq. (7) are calculated and presented in Table 2.

Table 2. Parameters in IA distribution gradual change model

Sample number	Parameters in Eq. (5)				
	P_x	P_y	P_z	A	B
1	13.1°	−40.4°	−22.5°	14.9°	15.2°
2	27.4°	−74.0°	79.3°	−18.3°	−29.7°
3	−8.4°	−53.8°	−7.8°	−2.7°	−5.6°

Sample number	Parameters in Eqs. (6) and (7)			
	p'	k_y'	k_z'	$p' \cdot k_y' \cdot k_z'$
1	18.5°	−2.9	−1.6	85.5°
2	−31.1°	4.7	−2.9	423.9°
3	−3.3°	20.2	1.4	−92.9°

4 Discussion

This paper analysed the distributions of IA in three human LV samples. The samples' data were acquired with X-ray phase contrast tomography at ESRF ID-19 and reconstructed at an isotropic spatial resolution of 3.5 μm. The local myocyte aggregates IA was calculated in local tissue sub-volumes of size 32^3 voxels using FFT and PCA methods. In most tissue, the local myocyte aggregates IA approximates the local myocyte aggregates TA. In contrast, the IA can overcome TA's severe fluctuations around where the local myocyte aggregates HA value is about ±90° [1, 12].

As expressed in Eq. (4), the IA in small-sized transmural samples was assumed to obey the quadratic function along the x direction and the linear function along the y and z directions. The assumption that IA follows quadratic function along the x direction was based on the model proposed for TA in literature [2–5, 9, 10]. The mean IA values calculated in y–z sections along the x direction in each sample were depicted on the left of Fig. 4. Their appearances approximate quadratic function curves as assumed. Along the y and z directions, the IA was proposed to obey linear functions, which accounts for the fact that our samples' sizes in y and z directions are small. If the samples' sizes in the y and z directions were significant, we believe the functions would be unreasonable.

Along circumferential (i.e. y) direction, the values of $\overline{IA}(y)$ in all samples decrease, and all P_y are negative. Besides the myocyte architecture, the sample's location and size will introduce an IA error, angle θ depicted in Fig. 5, to enlarge this decreasing trend. When we calculated the myocyte aggregates IA, the y direction in the local coordinates system was used instead of the circumferential direction in the global coordinates system. When the sample size is large, or the local LV radius is small, the difference between these two coordinate axes is significant. Along the longitudinal (i.e. z) direction, the mean IA values increase in Sample 2 and decrease in Samples 1 and 3, which coincide with Eq. (2) and the model proposed in the literature [3].

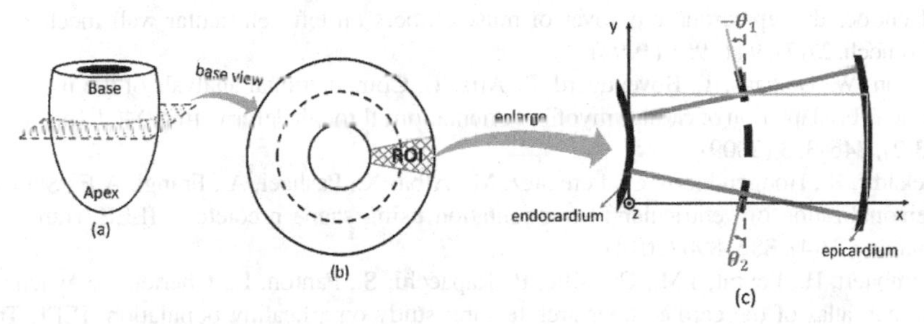

Fig. 5 A diagram to illuminate the effect of the sample's size on IA calculation along the y direction. (a) A LV model with $IA = 0°$; A base view of the section with (b); (c) the angle θ between the local y-axis and circumferential direction

Though the results in this paper are interesting, the limitation is also apparent, that the number and size of the samples are too small. In further work, more data should be assessed to investigate the distribution of the myocyte IA in LV wall. When we only analyze the gradual change trend of the orientation of local myocyte aggregates, the cardiac MRI data can be used [1].

5 Conclusion

The IA can avoid the influence of projection, and is better than TA to describe the radial orientation of local myocyte aggregates. This is the first time to use simple empirical formulas to fit the distribution of IA in LV wall. The results indicate that, in small-sized transparietal samples, the IA experiences a quadratic function along the local radial direction and linear functions along longitudinal and circumferential directions. The distribution of IA in the left ventricle is regional. This work will help us to build more accurate myocardial mathematical models, promoting the study of heart mechanical functions.

Acknowledgements. Chinese Heilongjiang Postdoctoral Grant (LBH-Z22184), National Natural Science Foundation of China (no. 52075133) and Metislab (Medical Engineering and Theory in Imaging and Signal Laboratory), CNRS LIA no.1124, INSA Lyon supported this work. This work was also performed within the framework of the LABEX PRIMES (ANR-11-LABX-0063) of Université de Lyon, within the program "Investissements d'Avenir" (ANR-11-IDEX-0007) operated by the French National Research Agency (ANR).

References

1. Agger, P., Omann, C., Laustsen, C., Stephenson, R.S., Anderson, R.H.: Anatomically correct assessment of the orientation of the cardiomyocytes using diffusion tensor imaging. NMR in Biomed. **33**(3), e4205 (2020)
2. Bayer, J.D., Blake, R.C., Plank, G., Trayanova, N.A.: A novel rule-based algorithm for assigning myocardial fiber orientation to computational heart models. Ann. Biomed. Eng. **40**, 2243–2254 (2012)
3. Bovendeerd, P.H.M., Huyghe, J.M., Arts, T., Van Campen, D.H., Reneman, R.S.: Influence of endocardial-epicardial crossover of muscle fibers on left ventricular wall mechanics. J. Biomech. **27**(7), 941–951 (1994)
4. Kroon, W., Delhaas, T., Bovendeerd, P., Arts, T.: Computational analysis of the myocardial structure: adaptation of cardiac myofiber orientations through deformation. Med. Image Anal. **13**(2), 346–353 (2009)
5. Lekadir, K., Hoogendoorn, C., Pereanez, M., Alba, X., Pashaei, A., Frangi, A.F.: Statistical personalization of ventricular fiber orientation using shape predictors. IEEE Trans. Med. Imaging **33**(4), 882–890 (2014)
6. Lombaert, H., Peyrat, J.M., Croisille, P., Rapacchi, S., Fanton, L., Cheriet, F., Ayache, N.: Human atlas of the cardiac fiber architecture: study on a healthy population. IEEE Trans. Med. Imaging **31**(7), 1436–1447 (2012)
7. Mirea, I., et al.: Very high-resolution imaging of post-mortem human cardiac tissue using X-ray phase contrast tomography. In: van Assen, H., Bovendeerd, P., Delhaas, T. (eds.) FIMH 2015. LNCS, vol. 9126, pp. 172–179. Springer, Cham (2015). https://doi.org/10.1007/978-3-319-20309-6_20
8. Momose, A., Takeda, T., Itai, Y.: Phase-contrast x-ray computed tomography for observing biological specimens and organic materials. Rev. Sci. Instrum. **66**(2), 1434–1436 (1995)
9. Rijcken, J., Bovendeerd, P.H.M., Schoofs, A.J.G., Van Campen, D.H., Arts, T.: Optimization of cardiac fiber orientation for homogeneous fiber strain during ejection. Ann. Biomed. Eng. **27**, 289–297 (1999)
10. Ubbink, S.W.J., Bovendeerd, P.H.M., Delhaas, T., Arts, T., van de Vosse, F.N.: Towards model-based analysis of cardiac MR tagging data: relation between left ventricular shear strain and myofiber orientation. Med. Image Anal. **10**(4), 632–641 (2006)
11. Varray, F., Mirea, I., Langer, M., Peyrin, F., Fanton, L., Magnin, I.E.: Extraction of the 3D local orientation of myocytes in human cardiac tissue using X-ray phase-contrast micro-tomography and multi-scale analysis. Med. Image Anal. **38**, 117–132 (2017)

12. Wang, S., Cui, J., Jing, Y., Varray, F.: Oscillation of the orientation of cardiomyocyte aggregates in human left ventricle free wall. J. Anat. **242**(3), 373–386 (2022)
13. Wang, S., Mirea, I., Varray, F., Liu, W.-Y., Magnin, I.E.: Investigating the 3D local myocytes arrangement in the human LV mid-wall with the transverse angle. In: Coudière, Y., Ozenne, V., Vigmond, E., Zemzemi, N. (eds.) FIMH 2019. LNCS, vol. 11504, pp. 208–216. Springer, Cham (2019). https://doi.org/10.1007/978-3-030-21949-9_23
14. Wang, S., Varray, F., Liu, W., Clarysse, P., Magnin, I.E.: Measurement of local orientation of cardiomyocyte aggregates in human left ventricle free wall samples using X-ray phase-contrast microtomography. Med. Image Anal. **75**, 102269 (2022)
15. Wang, S., Varray, F., Yuan, F., Magnin, I.E.: A methodology for accessing the local arrangement of the Sheetlets that make up the extracellular heart tissue. In: Ennis, D.B., Perotti, L.E., Wang, V.Y. (eds.) FIMH 2021. LNCS, vol. 12738, pp. 159–167. Springer, Cham (2021). https://doi.org/10.1007/978-3-030-78710-3_16

A Micro-anatomical Model of the Infarcted Left Ventricle Border Zone to Study the Influence of Collagen Undulation

Emilio A. Mendiola[1], Eric Wang[1], Abby Leatherman[1], Qian Xiang[2], Sunder Neelakantan[1], Peter Vanderslice[2], and Reza Avazmohammadi[1,3,4](✉)

[1] Department of Biomedical Engineering, Texas A&M University, College Station, TX, USA
rezaavaz@tamu.edu
[2] Department of Molecular Cardiology, Texas Heart Institute, Houston, TX, USA
[3] J. Mike Walker '66 Department of Mechanical Engineering, Texas A&M University, College Station, TX, USA
[4] Department of Cardiovascular Sciences, Houston Methodist Academic Institute, Houston, TX, USA

Abstract. Myocardial infarction (MI) results in cardiac myocyte death and often initiates the formation of a fibrotic scar in the myocardium surrounded by a border zone. Myocyte loss and collagen-rich scar tissue heavily influence the biomechanical behavior of the myocardium which could lead to various cardiac diseases such as systolic heart failure and arrhythmias. Knowledge of how myocyte and collagen micro-architecture changes affect the passive mechanical behavior of the border zone remains limited. Computational modeling provides us with an invaluable tool to identify and study the mechanisms driving the biomechanical remodeling of the myocardium post-MI. We utilized a rodent model of MI and an image-based approach to characterize the three-dimensional (3-D) myocyte and collagen micro-architecture at various timepoints post-MI. Left ventricular free wall (LVFW) samples were obtained from infarcted hearts at 1-week and 4-week post-MI (n = 1 each). Samples were labeled using immunoassays to identify the extracellular matrix (ECM) and myocytes. 3-D reconstructions of the infarct border zone were developed from confocal imaging and meshed to develop high-fidelity micro-anatomically accurate finite element models. We performed a parametric study using these models to investigate the influence of collagen undulation on the passive micromechanical behavior of the myocardium under a diastolic load. Our results suggest that although parametric increases in collagen undulation elevate the strain amount experienced by the ECM in both early- and late-stage MI, the sensitivity of myocytes to such increases is reduced from early to late-stage MI. Our 3-D micro-anatomical modeling holds promise in identifying mechanisms of border zone maladaptation post-MI.

Keywords: cardiac remodeling · border zone · myocardial infarction · confocal imaging · finite element modeling

O. Bernard et al. (Eds.): FIMH 2023, LNCS 13958, pp. 34–43, 2023.
https://doi.org/10.1007/978-3-031-35302-4_4

1 Introduction

Myocardial infarction (MI), caused by a lack of sufficient blood flow to the ventricular myocardium, results in cardiac myocyte death and often initiates the formation of fibrotic scar tissue in the left ventricle free wall (LVFW). In the weeks following MI, several remodeling events occur in the left ventricle (LV) myocardium inducing substantial changes in LVFW biomechanical behavior [12]. The biomechanical behavior of the LVFW, informed by structural and compositional remodeling post-MI, has been indicated as an essential determinant of the long-term functional outcome of the LV post-MI [2]. As such, there is a need to study the link between structural remodeling and the biomechanical behavior of the LV with the aim of better understanding the interaction between the two.

Given the established link between LVFW biomechanical behavior and the organ-level function [10], post-MI LV remodeling has been studied extensively. However, previous studies have focused on characterizing the remodeling of the scar region at the tissue scale [2,4] and the knowledge of the micro-environment of infarcted myocardium, in particular the infarct border zone, remains elusive. In particular, the effect of collagen fiber undulation, which governs the point at which collagen fibers are fully recruited, has been overlooked. We hypothesize that multiscale computational modeling could be a tool used to improve the understanding of the micromechanical environment in the border zone and advance the understanding of the link between microstructural remodeling in the border zone and organ-level function.

To form an understanding of the effect of collagen undulation on the micromechanical behavior of the infarcted LV myocardium, we have developed 3-D finite element (FE) models of the infarct border zone that incorporate accurate micro-anatomical structures of the myofibers (myocytes) from high-resolution confocal imaging of the microscale samples of the border zone (Fig. 1). Models were developed from 3-D imaging datasets from infarcted rat hearts at two timepoints (1-week and 4-week) post-MI. In this study, we simulated the mechanical behavior of these 3-D micro-anatomical models under biaxial stresses, representing end-diastolic loading, and quantified the myofiber and extracellular matrix (ECM) strain profiles. We then performed a parametric study to explore the effect of collagen undulation on the strains. Material properties used in the model were taken from an analytical fit to ex-vivo mechanical test data. Results from the present investigation improve the understanding of the micro-environment at early- and late-stage MI timepoints and can aid in identifying mechanisms driving border zone biomechanical remodeling.

2 Materials and Methods

2.1 Animal Model

The animals used in this work were treated in accordance with guidelines approved by the Institutional Animal Care and Use Committee (IACUC) at the Texas Heart Institute. Two male Wistar-Kyoto (WKY) rats, 8 weeks old at

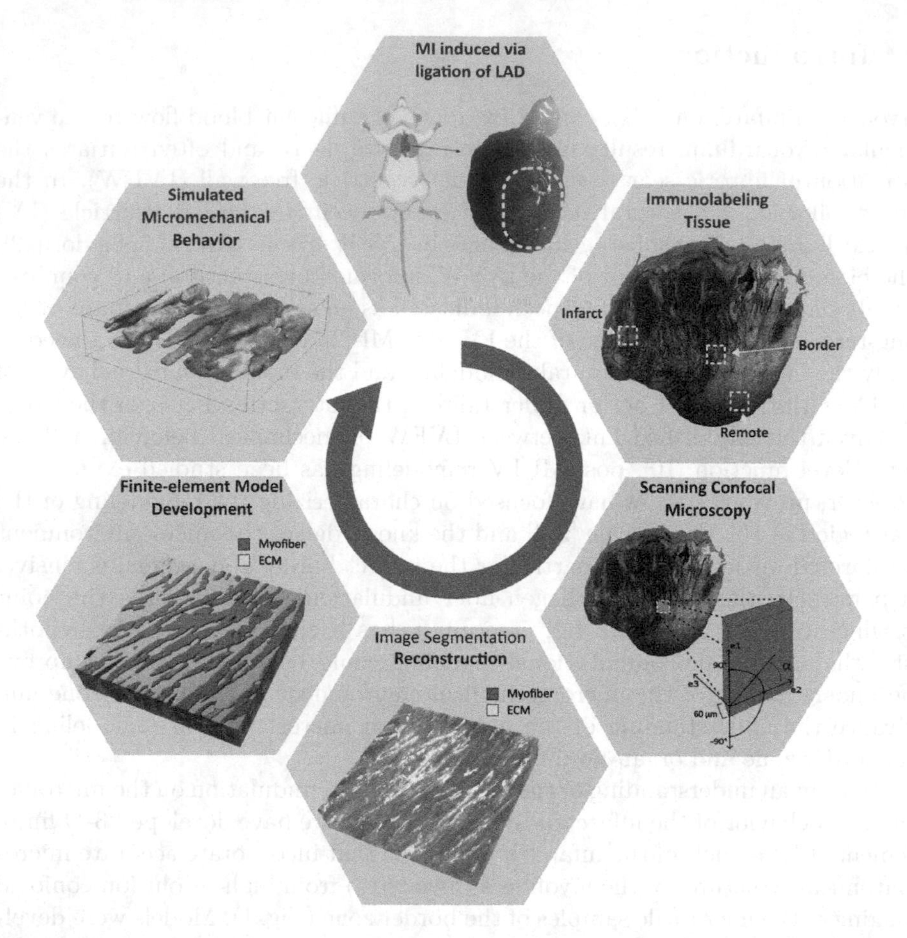

Fig. 1. Micro-structurally accurate finite-element models of the infarcted LV myocardium were developed from high-resolution imaging. The precise image segmentation process allowed for the identification of myocyte and extracellular matrix regions of the tissue, resulting in a model capable of estimating cellular-level regional kinematic behavior. LAD: left anterior descending artery; ECM: extracellular matrix; α: denotes fiber angle.

the start of the experiment, were used in this study. Anteriobasal infarct was induced in two rats (n = 2) by ligation of the left anterior descending artery near the base of the heart as previously described [11]. Animals were sacrificed at 1-week and 4-week post-MI, hereafter designated as 1-wk and 4-wk, respectively. Immediately prior to sacrifice, LV hemodynamic and anatomical measurements were obtained via left heart catheterization and echocardiography.

2.2 Preparation and Labeling of Myocardium Samples

Hearts were excised, and the ventricles were perfused and stored in 4% paraformaldehyde (PFA) solution for 3 h. The ventricles were then washed with phosphate-buffered saline (PBS). Biopsies were incubated in 30% sucrose in PBS solution and frozen to prevent tissue damage. Full-thickness rectangular specimens of the full LVFW were isolated with slab edges being aligned with the longitudinal, circumferential, and radial directions of the LV. The slabs were then frozen in optimal cutting temperature compound (Sakura Finetek Europe B.V., Alphen aan den Rijn, Netherlands) and cryosectioned into 60-micron thick slices. Mid-wall slices were washed in PBS with 0.1% Triton X-100 (PBST) and fixed in a Pelco Biowave (Ted Pella, Inc., Redding, CA) laboratory microwave processor. The tissue underwent microwave irradiation for 2 min, washed in PBST with 3% bovine serum albumin (BSA). Using direct immunofluorescence, the tissue sections were labeled with Alpha-Smooth Muscle Actin (α-SMA) Monoclonal Antibody (1A4), eFluor 570, eBioscience^{TM} for the myocytes. Alexa Fluor 647-conjugated anti-Vimentin antibodies were also applied at a concentration of 1:200. Each slice received 150 μL of antibody solution and was covered in parafilm. The tissues underwent microwave irradiation for 2 min followed by 5-minute rest, then were incubated for 8 h overnight, and then rinsed in PBS with microwave irradiation for 2 min. Next, wheat germ agglutinin (WGA)-conjugated Alexa Fluor 680 (Invitrogen, Carlsbad, California) was applied at 1:200 concentration, and each slide received 150 μL of antibody solution. DAPI (D3571, Thermo Fisher Scientific) at 1 μg/ml was distributed at 150 μL for each slide to label the nuclei. The slides were rinsed off and underwent microwave irradiation for 2 min in sodium phosphate buffer (50 mM, pH 7.0). Finally, slices were mounted in a glycerol-based medium (80% glycerol, with n-propyl gallate antifade) and allowed to dry before storing at −20 °C.

2.3 Confocal Microscopy Imaging

Prepared coverslips were imaged using a laser scanning confocal microscope Leica TCS SP8 with a 40x oil immersion objective at a resolution of 1024 × 1024 pixels and an image size of 204 × 204 μm with a voxel size of 0.196 × 0.196 × 0.333 μm. The infarct scar was identified as fibrotic tissue dense in collagen and without any myocytes. The remote region was identified far from the infarct scar, where only healthy myocytes and typical collagen structures are present. Finally, the border zone was recognized as the area between the infarct scar and the remote region where living and dead myocytes were both present [9]. A rotation of the field of view was applied before acquisition to yield a uniform myocyte orientation approximately parallel to the $\mathbf{e}_1$ axis (Fig. 2). 3-D image stacks were acquired to collect signals from DAPI, autofluorescence, WGA, α-SMA, and vimentin. Image stacks of 200–300 images covered the depth of each sample with a spacing of 200 nm between the images. To compensate for decreasing signal intensity at increased depths within the tissue, excitation compensation was applied by manually adjusting the laser intensity at points at various depths in the sample

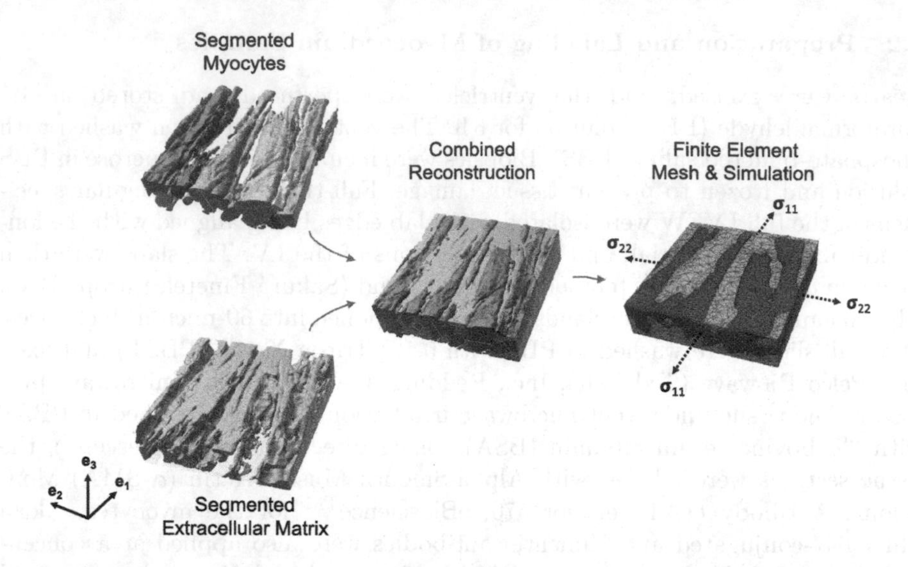

Fig. 2. Myocytes and the extracellular matrix (ECM) were segmented from high-resolution imaging data. The combined reconstruction was meshed to develop a finite-element representation of the myocardium sample consisting of separate myocyte and ECM regions. The 4-wk post-MI segmentation and model are shown here. The dashed arrows indicate the faces upon which the stress boundary condition was applied. The $\mathbf{e}_1$ and $\mathbf{e}_2$ directions are aligned with the edges of the model.

and linearly interpolating between them. The imaging software LAS X (version 1.1.0 and higher, Leica) was used for this processing.

Slides were imaged at 40x using a compound microscope to obtain images of the collagen fibers. We quantified the undulation of collagen fiber bundles as the ratio of the distance between two endpoints of the bundle to the total length of the bundle such that the undulation parameter lies between 0 and 1. The extreme undulation values of 0 and 1 correspond to an entirely coiled and a perfectly straight fiber, respectively. Fiber undulation was converted to slack stretch for use in FE constitutive modeling by taking the slack stretch λ_s as the inverse of fiber undulation.

2.4 Finite-Element Model Development

Three-Dimensional Image Reconstruction. 3-D structures of myocytes and collagen fibers were reconstructed from confocal microscopy z-stack images of the border zone and analyzed using Imaris software. We applied median filters to reduce noise in the acquired image stacks and performed threshold cutoffs and background subtraction to enhance the contrast between the background and the target tissue. The image segmentation resulted in a 3-D representation of the individual myocytes and the extracellular space within the tissue sample (Fig. 2). The individually reconstructed myocytes were represented as a consolidated *myofiber* region. The two-region reconstruction, consisting of ECM and myofiber regions, was meshed in Mimics (Materialise, Leuven, Belgium), resulting in a volumetric mesh of ~85,000 linear tetrahedral elements.

Myofiber and Extracellular Matrix Constitutive Modeling. The myofiber and ECM were modeled as hyperelastic, anisotropic, incompressible materials using constitutive models adapted from previous studies [6,7]. As individual myofiber orientations could be obtained directly from the imaging dataset, we defined the myofiber direction $\mathbf{f}_0$ as the predominant direction of the long axis of myofibers in the mesh, which was fairly constant throughout the sample with a splay of $\pm 3.5°$. The cross-fiber direction $\mathbf{s}$ was defined as the direction normal to the myofibers within the plane of the mesh.

The myofiber strain energy function (Ψ_{MYO}) was defined as a two-term invariant-based exponential constitutive form [5]. The form consisted of an isotropic matrix term stiffened by a primary fiber along $\mathbf{f}_0$ given by

$$\Psi_{MYO}(\mathbf{C}) = \frac{a}{2b}[e^{b(I_1-3)} - 1] + \frac{a_m}{2b_m}[e^{b_m(I_4-1)^2} - 1], \tag{1}$$

where $\mathbf{C}$ is the right Cauchy deformation tensor and the matrix term, involving $I_1 = tr(\mathbf{C})$ and the material properties of a and b, is related to the isotropic behavior of the myofiber regions. Also, $I_4 = \mathbf{f}_0 \cdot \mathbf{C}\mathbf{f}_0$ is equivalent to the square of the stretch along the fiber direction $\mathbf{f}_0$, and a_m and b_m characterize the transversely-isotropic behavior of the myofiber region.

Individual collagen fibers, having diameters at the scale of less than 1 μm, were bundled and consolidated together to constitute the ECM region. Similar to the case of myofiber, the ECM was modeled as an isotropic matrix reinforced by a primary fiber direction:

$$\Psi_{ECM}(\mathbf{C}) = \frac{a}{2b}[e^{b(I_1-3)} - 1] + \frac{a_c}{2b_c}[e^{b_c(\frac{I_4}{\lambda_s^2}-1)^2} - 1], \tag{2}$$

where, for simplicity, the constants related to the isotropic part were kept the same as those in the myofiber energy function, and a_c and b_c are, similarly, related to the anisotropic behavior of the ECM region. Given our interest in estimating myofiber and collagen contributions to myocardial mechanical behavior, we set $a = 0.22$ kPa and $b = 1.62$ as reference values, reflecting an inconsiderable mechanical contribution of the ground matrix in the LV myocardium [1,3], and let $\{a_m,\ b_m,\ a_c,\ b_c,\ \lambda_s\}$ to be the material parameter variables in this work. Although the chosen low values for a and b parameters are expected to lead to a minimal effect of the matrix term on the simulated LV biomechanical behavior, the inclusion of the matrix terms, involving I_1, in both Ψ_{MYO} and Ψ_{ECM}, was important to aid in the stabilization of the computational problem. Lastly, we note that the parameters $\{a, a_m, a_c\}$ have stress-like dimensions while $\{b, b_m, b_c\}$ are dimensionless. The constitutive parameters were estimated through an inverse problem approach via fitting an analytical solution for the full LVFW behavior to post-infarct biaxial data as described by Mendiola et al. [8]. The fitted parameters were used for parametric studies described below.

Finite-Element Biomechanical Simulations. To better understand the effect of collagen undulation on the passive behavior of myocardial tissue in the scar border zone, we conducted a parametric study in which the slack stretch of the collagen fibers at each timepoint was varied, and the resulting alterations in regional strain were analyzed. First, the stress in the LVFW at the end-diastolic (ED) point was estimated via the Laplace law as

$$\sigma_{ED} = \frac{P_{ED} R_{ED}}{2t} \tag{3}$$

where P_{ED} is the ED pressure in Pascals, R_{ED} is the LV radius at ED in milimeters, and t is the LVFW thickness in milimeters. The 1-wk data used for these calculations were $P_{ED} = 1033.2\,\text{Pa}$, $R_{ED} = 4.6\,\text{mm}$ $t = 1.47\,\text{mm}$ whereas the 4-wk data were $P_{ED} = 1080.7\,\text{Pa}$, $R_{ED} = 4.6\,\text{mm}$, $t = 1.22\,\text{mm}$. These data were collected via LV catheterization and echocardiography immediately before sacrifice. The estimated ED wall stress was applied to the boundary surfaces of the FE model as a stress boundary condition. Initial simulations were run with the collagen slack length measured from imaging. Subsequent *in-silico* experiments were conducted using values for λ_s that were varied by $\pm 5\%$ and $\pm 10\%$.

3 Results

The stress-strain response of the micro-anatomical models showed a stiffening behavior from 1-wk to 4-wk timepoints, similar to the biaxial test data [8]. Results are not presented here for brevity. At the maximum stress state, the strain in the $\mathbf{e}_1$ and $\mathbf{e}_2$ directions at both timepoints showed substantial heterogeneity throughout the model (Fig. 3). E_{11} tended to peak at the tips of the myocytes (noting that fiber directions were nearly aligned with $\mathbf{e}_1$) while E_{22} showed a more uniform distribution with the myocytes. Both strain components exhibited higher values near the boundaries (compared to the central region of the volume) expected due to higher values of forces at the boundaries Fig. 3).

Varying collagen undulation in the ECM resulted in distinct regional strain behavior at each timepoint. The ECM accommodated greater strain than the myocytes at both timepoints (Fig. 4). A similar trend was noted in the ECM strain response versus undulation in both 1-wk and 4-wk models: larger strain was exhibited in the ECM as the slack stretch increased. In contrast, the 1-wk myocytes accommodated a larger strain than the 4-wk myocytes, and the 1-wk myocyte strain was noted to be more sensitive to the alterations in collagen fiber undulation than the 4-wk myocytes.

4 Discussion

In this study, we have developed a micro-anatomically accurate FE model of infarcted left ventricular myocardium based on a high-resolution imaging dataset. We used a structurally motivated constitutive modeling approach to

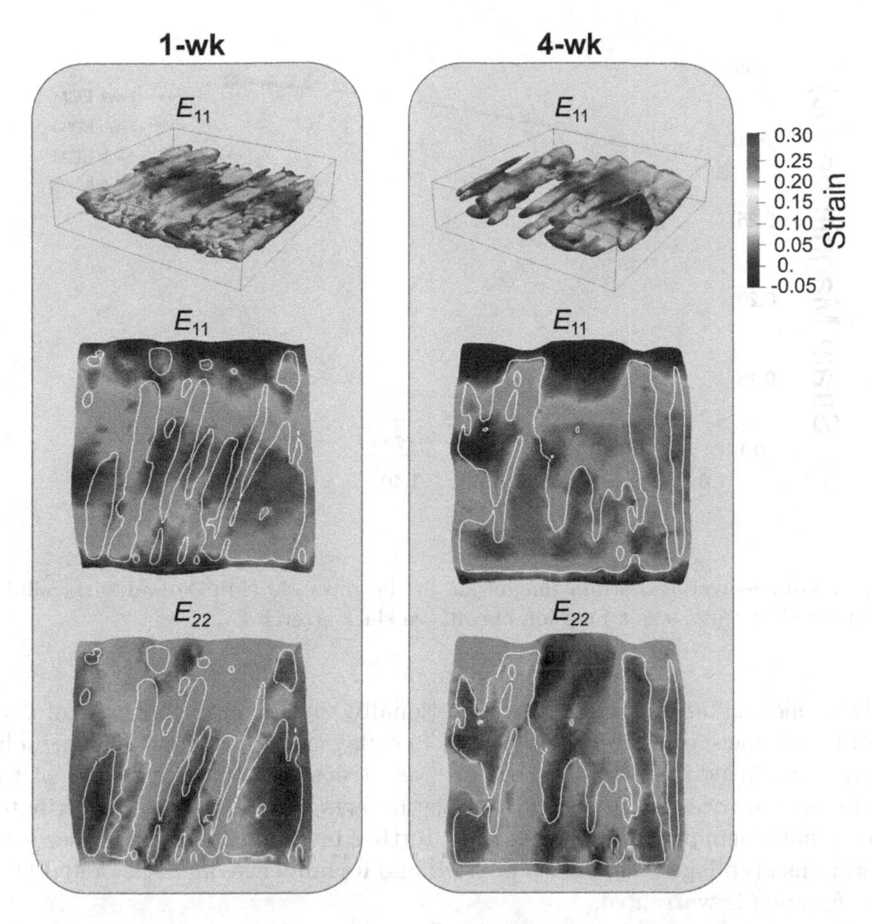

Fig. 3. 3-D visualization of E_{11} in the myofibers (top row) and 2-D cross-sections (second and third rows) of E_{11} and E_{22} at the maximum stress state in the myofiber and ECM regions at the 1-wk and 4-wk post-MI timepoints. The area within the white outlines in the 2-D cross-sections indicated the location of myofibers.

quantify the microscale strain fields in the tissue and investigate the effect of collagen undulation on the micromechanical behavior of the infarct border zone.

Interestingly, results indicate that the ECM of the border zone accommodates greater strain than the myofibers in both early- and late-stage MI. We hypothesize that this observation is a result of (i) a greater concentration of collagen within the border zone samples and (ii) the slack behavior of the collagen fibers in the ECM, allowing for the strain to develop prior to the development of large tensile forces in the fibers. The parametric study conducted regarding the effect of the slack stretch on regional strain shows the myofibers in the 1-wk model accommodate greater strain under diastolic loading than do the 4-wk myofibers. This is consistent with previous studies indicating early post-MI myocardium is more compliant than late MI myocardium, resulting in generally better diastolic

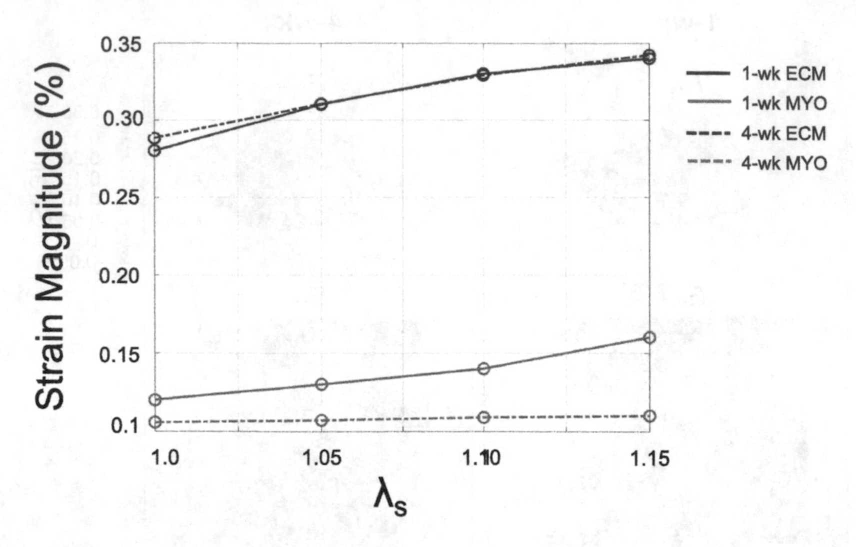

Fig. 4. Volume-averaged strain magnitude in the myocyte (MYO) and extracellular matrix (ECM) regions as a function of collagen slack stretch (λ_s).

and systolic organ-level function [8]. Additionally, the kinematic behavior of 1-wk myofibers is more sensitive to alterations in collagen slack stretch. These results imply that, in addition to the cellular-level structure and composition of the border zone myocardium, collagen undulation acts as an important contributor to regional kinematic behavior. As such, further investigation of the timecourse alterations in collagen undulation post-MI and its influence on tissue- and organ-level function is warranted.

The model presented in this work offers improved histological and imaging processes to simulate the micro-environment of the infarcted LV. Such work is necessary to better understand the elusive link between cellular-level remodeling and organ-level function post-MI. Our modeling results have indicated that the diastolic kinematic behavior of the scar border zone is highly heterogeneous and is a function of various structural and biomechanical remodeling processes, including collagen fiber undulation. Future work will focus on the development of similar micro-structurally informed models capable of connecting our cellular-level predictions with tissue- and organ-level behaviors. Ultimately, our micro-anatomical modeling approach will allow for the connection of multiscale post-MI remodeling events and potentially aid in the identification of fiber-level remodeling events driving the transition of the MI to heart failure.

Acknowledgements. This research was supported by the NIH Grant No. R00HL138288 to R.A.

References

1. Babaei, H., et al.: A machine learning model to estimate myocardial stiffness from EDPVR. Sci. Rep. **12**(5433) (2022). https://doi.org/10.1038/s41598-022-09128-6
2. Fomovsky, G., Thomopoulos, S., Holmes, J.: Contribution of extracellular matrix to the mechanical properties of the heart. J. Mol. Cell. Cardiol. **48**(3), 490–496 (2010)
3. Gao, H., Li, W.G., Cai, L., Berry, C., Luo, X.Y.: Parameter estimation in a Holzapfel–Ogden law for healthy myocardium. J. Eng. Math. **95**(1), 231–248 (2015). https://doi.org/10.1007/s10665-014-9740-3
4. Holmes, J.W., Borg, T.K., Covell, J.W.: Structure and mechanics of healing myocardial infarcts. Annu. Rev. Biomed. Eng. **7**(1), 223–253 (2005). https://doi.org/10.1146/annurev.bioeng.7.060804.100453
5. Holzapfel, G.A., Ogden, R.W.: Constitutive modelling of passive myocardium: a structurally based framework for material characterization. Philos. Trans. R. Soc. Lond. A: Math. Phys. Eng. Sci. **367**(1902), 3445–3475 (2009). https://doi.org/10.1098/rsta.2009.0091
6. Li, D.S., Mendiola, E.A., Avazmohammadi, R., Sachse, F.B., Sacks, M.S.: A multi-scale computational model for the passive mechanical behavior of right ventricular myocardium. J. Mech. Behav. Biomed. Mater. **142**, 105788 (2023). https://doi.org/10.1016/j.jmbbm.2023.105788
7. Li, D.S., Mendiola, E.A., Avazmohammadi, R., Sachse, F.B., Sacks, M.S.: A high-fidelity 3D micromechanical model of ventricular myocardium. In: Ennis, D.B., Perotti, L.E., Wang, V.Y. (eds.) FIMH 2021. LNCS, vol. 12738, pp. 168–177. Springer, Cham (2021). https://doi.org/10.1007/978-3-030-78710-3_17
8. Mendiola, E.A., et al.: Contractile adaptation of the left ventricle post-myocardial infarction: predictions by rodent-specific computational modeling. Ann. Biomed. Eng. **16**(2), 721–729 (2022)
9. Mendiola, E.A., et al.: Identification of infarct border zone using late gadolinium enhanced MRI in rats. FASEB J. **36**(S1) (2022). https://doi.org/10.1096/fasebj.2022.36.S1.R6220
10. Richardson, W.J., Clarke, S.A., Quinn, T.A., Holmes, J.W.: Physiological implications of myocardial scar structure. Compr. Physiol. **5**(4), 1877–1909 (2018). https://doi.org/10.1002/cphy.c140067
11. Samsamshariat, S.A., Samsamshariat, Z.A., Movahed, M.R.: A novel method for safe and accurate left anterior descending coronary artery ligation for research in rats. Cardiovasc. Revasc. Med. **6**(3), 121–123 (2005). https://doi.org/10.1016/j.carrev.2005.07.001
12. Sutton, M.G.S.J., Sharpe, N.: Left ventricular remodeling after myocardial infarction: pathophysiology and therapy. Circulation **101**(25), 2981–2988 (2000)

Symmetric Multimodal Mapping of Ex Vivo Cardiac Microstructure of Large Mammalian Whole Hearts for Volumetric Comparison of Myofiber Orientation Estimated from Diffusion MRI and MicroCT

Valéry Ozenne[1](✉), Girish Ramlugun[2], Julie Magat[1], Nestor Pallares Lupon[2], Pierre Cabanis[1], Pierre Dos Santos[2,3], David Benoist[2], Virginie Dubes[2], Josselin Duchateau[2,3], Louis Labrousse[3], Michel Haïssaguerre[2,3], Olivier Bernus[2], and Richard Walton[2]

[1] Univ. Bordeaux, CNRS, CRMSB, UMR 5536, IHU Liryc, 33000 Bordeaux, France
valery.ozenne@u-bordeaux.fr

[2] Univ. Bordeaux, INSERM, CRCTB, U 1045, 33000 Bordeaux, France

[3] CHU de Bordeaux, Service de Cardiologie, INSERM, U 1045, 33000 Bordeaux, France

Abstract. Description of the cardiac myofiber architecture in pathological or even physiological conditions is essential for image-based modeling in electrophysiology or mechanical studies. While diffusion tensor imaging (DTI) is one of the best modalities to capture myofiber orientation of large mammalian hearts, validations of putative myofiber's main orientation from DTI in whole hearts of large mammals is limited. First we design an experimental protocol for sheep ($N = 1$) and human ($N = 1$) whole hearts that combine a standardized sample preparation with high-resolution diffusion MRI at 600 μm^3 using low angular resolution (6 directions) followed by a tissue air-drying approach coupled with X-ray imaging at 42 μm^3. Secondly, we propose a standardized post-processing pipeline for symmetric multimodal mapping allowing the comparison of myofiber orientation computed from DTI and structure tensor imaging (STI), respectively. We then identified region-of-interest (ROI) exhibiting small or sharp spatial variations in myofiber orientation and compared the putative myofiber orientation for both methods. In conclusion, we show a good correspondence of structural features between the two imaging modalities and identify new unexpected and complex cardiomyocytes organization such as oscillating patterns or clear separation of opposing fiber-bundles.

Keywords: Cardiac myofiber architecture · Diffusion MRI · MicroCT · Large mammalian whole hearts · Registration

O. Bernard et al. (Eds.): FIMH 2023, LNCS 13958, pp. 44–53, 2023.
https://doi.org/10.1007/978-3-031-35302-4_5

1 Introduction

Cardiac remodeling is a complex alteration that involves micro and macro structural changes [1]. The understanding of cardiac myofiber alterations due to remodeling has been documented in small animal model at the cell level using histology [2], gene expression [3] or at macroscopic level using ex-vivo 3D imaging techniques like DTI [2, 4], phase contrast computed tomography (CT) [5]. However, although the overall arrangement of myocytes is well established in the left ventricle (LV) (and described by helix angle rule), 3D detailed structural descriptions are relatively sparse in whole hearts of large mammals. The latter having significant differences from small mammalian hearts [6], the impact and extent of remodeling can be difficult to assess. Typically, subtle or major changes with known myofiber orientation could be attributed to either reorganization or disorganization of the myocardial mesh but might also exist in physiological conditions. One limitation is the structural investigation techniques themselves. Indeed, available high-resolution imaging (<100 μm) techniques are invasive and applied on ex-vivo samples. Gold standard methods like histology are limited to 2D thin slices and provide both incomplete volumetric data. 3D virtual histology like confocal microscope [7] or MicroCT [8] have been applied on small samples but are not yet suited to image large mammalian whole hearts. The most common alternative is ex-vivo DTI that provides an indirect measure of myofiber orientation [9–11] at lower spatial resolution (~500 μm) with a partial volume. Alternative imaging techniques like polarized light imaging (PLI) have been compared with DTI [12] using $90 \times 90 \times 500\ \mu m^3$ spatial resolution. Recently, MicroCT imaging coupled with sample desiccation has also been proposed on large mammalian whole hearts and offers direct visualization of the cardiac myofiber architecture at an isotropic voxel resolution of 20–40 μm [13]. The study aim to: i) develop an experimental protocol for large mammalian whole hearts that combine ex-vivo MRI acquisitions followed by MicroCT acquisition. ii) perform multimodal image analysis with myofiber orientation estimation using DTI and STI followed by symmetric mapping of both modalities and associated metrics. iii) explore the heterogeneity of myofiber orientation in physiological and remodeling conditions.

2 Methods

2.1 Samples

- The sheep S1 sample and protocol used in this study were approved by the Animal Research Ethics Committee (CEEA – 050 Comité d'éthique pour l'expérimentation animale) in accordance with the European rules for animal experimentation. The sample was fixed in a pseudo diastolic phase.
- Human sample H1 was derived from the Human donor program (providing access to heart samples from patients under cardiac transplantation) approved by the "Agence Française de la Biomedecine" and with a written informed consent of the patient. The experiment was conducted in accordance with the declaration of Helsinki and the institutional ethics committee. Donor (M, 54) had a history of acute coronary syndrome. He underwent a cardiogenic shock a few days before transplantation.

2.2 Sample Preparation and Acquisition for MRI

Both hearts were fixed with a solution containing formalin (10%) and a gadolinium-based contrast agent (Dotarem; Guerbet, Paris, France, 0.2% of the total volume of perfusion), by retrograde perfusion from the aorta. Whole hearts were immersed in Fomblin and sealed in a plastic container during imaging. The sample preparation was performed as previously described [14–16]. All experiments were performed at 9.4 T with an inner bore size of 30 cm (BioSpin MRI; Bruker, Ettlingen, Germany) equipped with 300 mT/m gradient insert of 200-mm inner diameter adapted for large samples. MRI acquisitions were similar to those described previously in [14–16]. In summary, FOV was $100 \times 90 \times 110$ mm^3 and $110 \times 100 \times 140$ mm^3 for S1 and H1 respectively.

- Diffusion-Weighted Imaging (DWI). A 3D DW spin-echo sequence was used to acquire DWI (TE = 22 ms, TR = 500 ms). The DWI dataset consisted of six non-collinear diffusion encoding directions acquired with a b value of 1,000 s/mm^2 and one non-diffusion weighted (b0) image with an isotropic spatial resolution of 600 μm^3. The total acquisition time for was 18 h 59 m (S1) and 26 h 47 m (H1).
- Anatomical Imaging. The 3D anatomical image was acquired with a gradient echo sequence which was averaged 18 (S1) and 10 (H1) times (with, TR = 30 ms, TE = 9 ms) at an isotropic resolution of 150 μm^3. The total acquisition time for was 32 h 45 m (S1) and 25 h 52 m (H1).

2.3 Sample Preparation and Acquisition for MicroCT

Standard sample preparation for soft tissue MicroCT imaging relies on contrast agents. For large mammalian whole hearts, this approach results in low and homogeneous photon attenuation and thus in poor image quality (indistinguishable structural features). A novel tissue preparation technique to alleviate background photon attenuation was performed using a tissue air-drying approach [13]. The hearts were rinsed three times for 20 min in PBS and then dehydrated using ethanol and perfused with hexamethyldisilazane (HMDS) to prevent tissue deformation. Finally, hearts were hung to air-dry under the fume hood and inside small containers to reduce airflow for 7 days. Both hearts were imaged using MicroCT (SkyScan 1276, Bruker, Belgium). X-ray transmission images were acquired using X-ray source energies of 55 kV and 80 kV for sheep and human samples respectively and 200 μA. Images were reconstructed by 3D tomographic reconstruction of the raw images using the cone-beam FDK11 algorithm using the NRecon software (Bruker) with an isotropic voxel resolution of 42.409 μm.

2.4 Diffusion Tensor Estimation

Each DWI was up-sampled by a factor of 2 using trilinear interpolation to reach a voxel size of $0.3 \times 0.3 \times 0.3$ mm^3. Diffusion tensor calculations were performed prior to any registration to avoid difficult transformations of the diffusion encoding matrix. Diffusion tensor maps [eigenvalues, eigenvectors, apparent diffusion coefficient, fractional anisotropy (FA), and color-coded FA (cFA)] were obtained with MRtrix3 software [17]. The first diffusion tensor eigenvector e_1 was associated with the main myocyte orientation.

2.5 STI Estimation

3D STI was computed from MicroCT images to estimate the myofiber orientation similarly to [18]. First, the outer product of the intensity gradient vectors was computed to estimate the derivatives. The derivatives were then convolved with a Gaussian kernel of 9 voxels with standard deviation $\sigma = 2$ and assembled to generate a symmetric second-moment matrix or structure tensor per voxel. The principal directions of the structure tensor in each voxel were extracted using eigenanalysis. The third eigenvector e_3 (smallest magnitude eigenvalue) define as the minimum structural gradient was associated with the main myocyte orientation.

2.6 Spatial Registration

The MRI anatomical images, at 150 μm isotropic resolution, were first downsampled to the diffusion-weighted imaging resolution (at 300 μm). Similarly, the MicroCT data, at 42.4 μm isotropic resolution, were downsampled by a factor of 2 and 3 for sheep and human, respectively to reach an isotropic spatial resolution of 84.8 μm and 127.2 μm due to RAM limitation. First each sample was registered manually with ITK-snap using translation, rotation and scaling to obtain a similar alignment among hearts. Co-registration of MicroCT images on MRI anatomical images was performed using ANTs [19]. Anatomical images, laplacian of anatomical images and binary label of epicardial fat were used jointly using the symmetric normalization (SyN) transformation model. Then, the MicroCT images were warped to MRI space from coarse resolution (at 300 μm isotropic resolution) to the largest possible spatial resolution (84.8 μm and 127.2 μm for sheep and human sample). Finally, transforms from MicroCT to MRI were applied to the third eigenvector e_3. To determine registration quality, first, the Dice Similarity Coefficient and Jaccard index between each individual full sample label, myocardium label and fat label was computed with the different transformation models (without alignment, after manual alignment, after Affine and SyN registration).

3 Results

Figure 1 displays representative images of both whole hearts after MRI and MicroCT acquisition. The yellow arrows indicate a reduction in myocardial volume due to the desiccation. The sample H1 has been cut (purple arrow) in the LV free wall before MicroCT acquisition because of the size limitation of the system. The sample H1 has a large low intensity region (red arrows) possibly due to myocardial infarction in approximately a sixteenth of the left ventricle on the anatomical MRI images.

Table 1 summarizes the volume estimates from the MRI and MicroCT acquisition. The number of voxels is 8.1×10^6 and 13.7×10^6 at 300 μm isotropic resolution for the MRI dataset resulting in a volume of 218 cm^3 and 293 cm^3 for the sheep and human sample respectively. For the MicroCT dataset, the corresponding values are 1.0×10^9 and 1.8×10^9 voxels at 42 μm isotropic resolution resulting in 75 cm^3 and 137 cm^3. The ratio between the MicroCT and MRI volume estimates was found inferior to 0.5 for both samples. For S1, this ratio (0.47) for the segmented fat volume (mostly epicardial) was found inferior to the ratio (0.32) for the segmented myocardium.

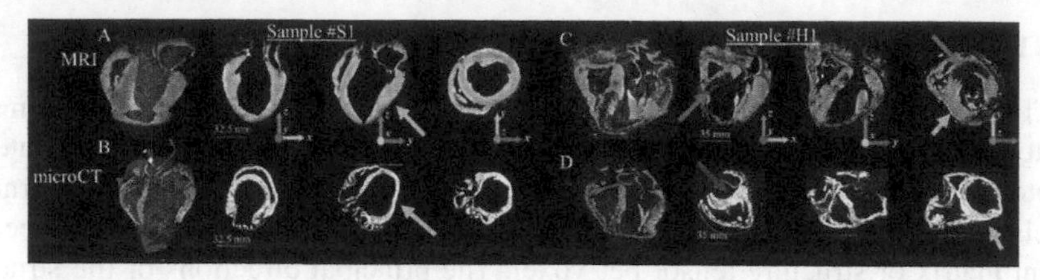

Fig. 1. Representative **images** of sheep S1 (A,B) and human H1 (C,D) ex vivo samples using MRI (A,C) and MicroCT (B,D) imaging modalities using 3D rendering and orthogonal views.

Table 1. Volume and ratio for both modalities. (Top) Volume (cm^3) on sample, tissue, and fat volume for MRI, MicroCT acquisition and MicroCT acquisition warped into MRI space. (Bottom) Ratio on sample, tissue, and fat volume between MicroCT (mCT) and MRI.

(cm3)	Vol. of Full sample			Vol. of Myocardium			Vol. of Epicardial Fat		
Sample	MRI	mCT	mCTwarped	MRI	mCT	mCTwarped	MRI	mCT	mCTwarped
S#1	218	75	211	186	59	171	34	16	40
H#1	293	137	313	213	N.A	N.A	80	N.A	N.A

ratio	Vol. mCT/Vol. IRM			(Vol. MRI)/(Full Vol. MRI)		(Vol. mCT)/(Full Vol. mCT)	
Sample	Full	Myocardium	Fat	Fat	Tissue	Fat	Tissue
S#1	0.34	0.32	0.47	0.15	0.85	0.21	0.79
H#1	0.47	N.A	N.A	0.27	0.73	N.A	N.A

In Fig. 2, through visual inspection, the SyN transformation model exhibits a better agreement than the affine transformation model. For the sample S1 (left panel, Fig. 2A), the epicardial fat (yellow arrows) is correctly coregistered in either in the LV wall (Fig. 2B) or at the posterior right ventricle insertion point (RVIP) (C). Small anatomical structures like the conduction system (Fig. 2D) with the division of the His bundle (HS) in left bundle (LB) and right bundles (RB) or the anterior leaflet of the tricuspid valve (TV) (Fig. 2E) show a good agreement between MRI and registered MicroCT images. For the sample H1 (right panel, Fig. 2F), we notice a slight fragmentation (red arrows) of the tissue sample. This effect is visible at the intersection of an RV papillary muscle with the septal myocardium (Fig. 2G), in the wall of the RV (H) or at the intersection of the roof of the RVOT with the RV (J). Nevertheless, images exhibited a good agreement for most tissue/air interfaces or for smaller structures like the mitral valve (MV) and tricuspid valve (purple arrows) close to the HS division (Fig. 2I).

Figure 3 shows the Jaccard (A) and Dice (B) overlap metrics computed between each full sample label, tissue label and fat label for different transformation models (initial, aligned, affine and SyN). Using quantitative image-based metrics, the Symmetric Normalization (SyN) transformation model showed superior performance (Dice = 0.87 $\pm$ 0.04, Jaccard = 0.77 $\pm$ 0.06) over other transformation models (Dice = 0.73 $\pm$ 0.01, Jaccard 0.58 $\pm$0.02 for the affine model).

Figure 4 compares the putative myofiber's main orientation from first diffusion tensor eigenvector e_1 (top row) using cFA maps representation to both the anatomical MicroCT images (middle rows) and to the putative myofiber's main orientation from last structure tensor eigenvector e_3 (bottom row) derived from MicroCT images. A good agreement between the putative myofiber orientation (e_1) from DTI and the MicroCT images is visible in all ROIs and a relatively good agreement was found With e_3 from STI. The

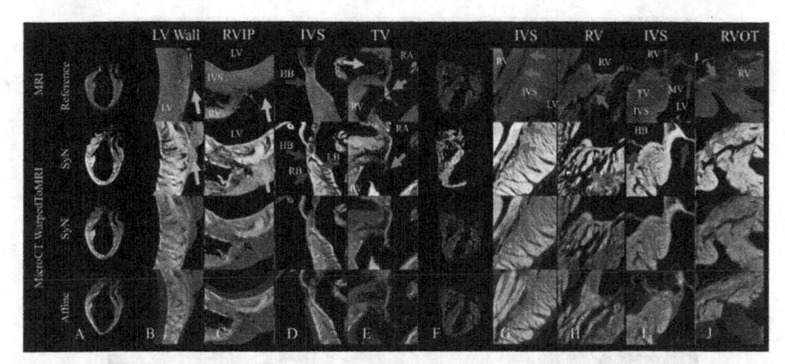

Fig. 2. Impact of the transformation model during the coregistration of MicroCT images on MRI anatomical images. Zoom-view of sheep (left panel) and human (right panel) samples in four ROIs. Glossary is in the results section. 1st row: MRI anatomical images. 2nd row: MicroCT images warped into MRI space. Overlay of MicroCT images warped into MRI space using Affine (4th row) or SyN transformation model (3rd row) using MRI anatomical images as background.

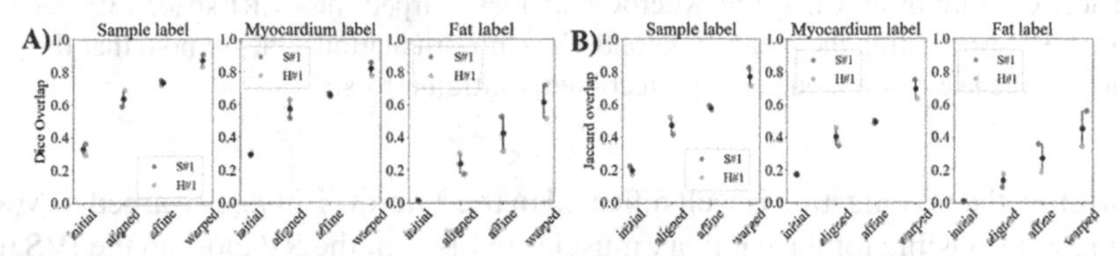

Fig. 3. Comparison of registration quality using Jaccard index and the Dice Similarity Coefficient. The box and whisker plots show the Jaccard (A) and Dice (B) metrics computed between the full sample label, the tissue label and fat label after different transformation models for S1 (red dots) and H1 (blue dots).

basal part of the IVS (Fig. 4A) is divided into two layers [16] visible in orange and pink color. The HB division with the LB (purple arrows) shows good correspondence. The posterior RVIP (Fig. 4B) includes a triangular arrangement [14] of fiber-bundles in base to apex direction starting at the inferobasal crux (IC) and ending at the middle of the inferoseptal wall of the LV. In the anterior RVIP, an anterior-septal perforator artery (ASPA) divides the ventricular myocardial extensions (ME) [14, 20] (red arrows). The posterior RV wall presents an unsual pattern with zebras or stripes indicated by the alternating of red and purple color (green arrows, Fig. 4D) and corresponds to a small oscillation of myocytes. Lastly the basal part of the LV wall includes two well-defined ribbon bundles in base to apex direction visible in blue color (indicated by yellow arrow, Fig. 4E) on the epicardium and in the myocardium originating from the LV summit and extending down to the papillary muscles.

Figure 5 compares the same information as Fig. 4 on the sample H1. Consistent with Fig. 2, the presence of abrupt variation or discontinuity in the myofiber orientation (red arrows) on the cFA maps seems associated with a fragmentation of the tissue on the MicroCT images. The circumferential myofiber (in green) of the RV wall (Fig. 5A) are separated into sub-regions by descending or ascending myofibers (in purple) in the

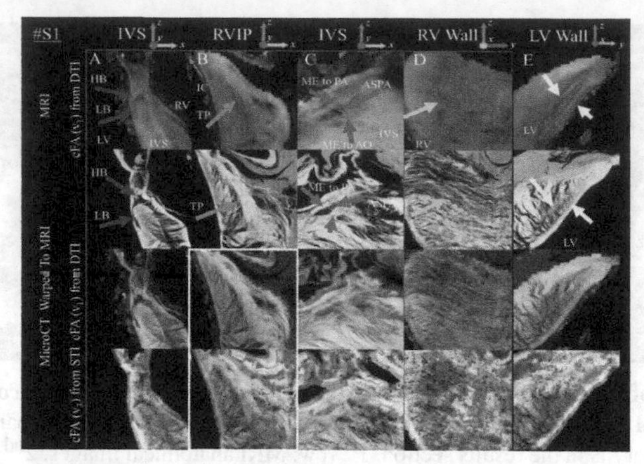

Fig. 4. Fiber orientation visualization and comparison for S1. Zoom-view of the sheep sample in five ROIs. Glossary is in the results section. 1st row: cFA maps from DTI (indicating the putative myofiber's main orientation overlaid on anatomical data. 2nd row: MicroCT images warped into MRI space. Overlay of cFA maps on MicroCT images warped into MRI space (3rd row). cFA maps from STI (indicating the putative myofiber's main orientation). Please note that the patchy black dots in the last row are due to imperfect interpolation errors.

myocardium, the organization is well reflected in the MicroCT images warped to MRI. A similar result is visible for the papillary muscle (in blue) of the RV close to the IVS in the mid-ventricular area (Fig. 5B). The basal part of the IVS (Fig. 5C) at the vicinity of the HB division is separated in two layers, the "RV" layer in purple being "connected" to the moderator band. The basal RVIP (Fig. 5D and 5E) has a complex myofiber architecture with circumferential myofibers (in green or red) interleaved with two thin planes of

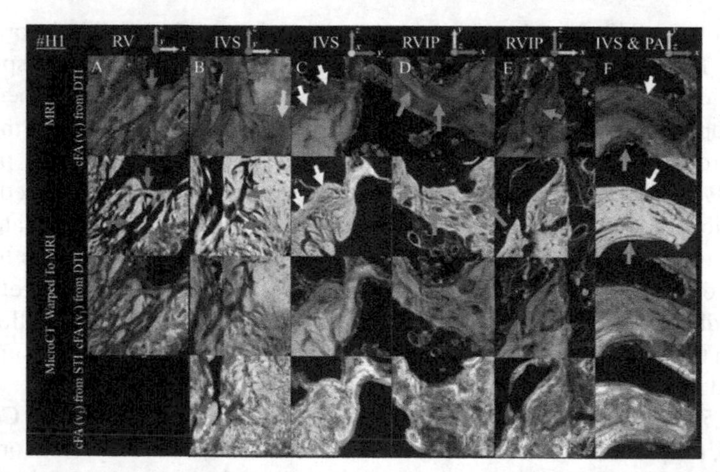

Fig. 5. Myofiber orientation visualization and comparison in 6 different ROIs for H1. Zoom-view of the human sample in six ROIs. Description and glossary are in the results section. The legend of Fig. 4 applies.

myofibers in base to apex direction (in blue). The anterior RVIP is divided into 3 layers with 2 layers with circumferential direction and a central layer with myofibers in base to apex direction. A small layer is visible in blue on the endocardium (blue arrow).

4 Discussion

In this work, we have developed an experimental protocol for large mammalian whole hearts allowing multimodal investigation of cardiac microstructure with an emphasis on myofiber orientation. A dedicated analysis pipeline including co-registration of MRI and MicroCT images and derived metrics was designed and offers a multi-scale analysis of cardiac myofiber organization. Assessment of co-registration between MRI and MicroCT images was done qualitatively by visual inspection and quantitatively using Dice and Jaccard overlap. A good agreement in total volume size between the MRI and MicroCT volume warped to MRI was found after co-registration with 3% and 7% variation for the sheep and human sample (Table 1). The reduction in sample size due to desiccation was found to be homogeneous throughout the sample. Two effects due to the desiccation process, that will need further confirmation on future studies, have been noticed: i) a tissue type influence was found. The ratio of fat/myocardium was slightly modified after desiccation from (15/85%) to (20/80%) in S1. ii) Minor structural modifications were found with a tissue fragmentation in ROI exhibiting sharp spatial variations in myofiber orientation. While agreement between image intensity from MRI and MicroCT was found good (Fig. 2 and Fig. 3), such comparison was insufficient to validate the registration. Indeed, the mapping of both modalities might create erroneous local variation in motion fields in the myocardium to match the tissue/air interface. Such a scenario would result in visual discrepancy between putative myofiber orientation from DTI and the MicroCT images. This was not the case, while being a qualitative interpretation and reader dependent, Fig. 4 and Fig. 5 show an good correspondence of structural features between putative myofiber orientation from DTI and the MicroCT images.

Additionally, the sample preparation highly preserves the overall myofiber orientation even for small variations as evidenced by the zebra pattern visible on cFA map and on MicroCT with the corresponding accordion pattern (Fig. 4D). Indeed, the symmetric mapping of both modalities at this level of resolution provides a framework for validating DTI. Similar microstructural characterization has been done in small animal studies using histology [21], tissue-clearing [22] or X-ray phase contrast [23], but to our knowledge this is the first 3D validation on large mammalian whole hearts. While a larger number of diffusion directions is recommended for a better estimation of the diffusion tensor and associated angles metrics [24], the heterogeneity of cardiac microstructure is much better depicted using a high spatial resolution without voxel anisotropy. By focusing on the spatial resolution and SNR, the proposed approach (6 directions, 1 bvalue) seems sufficient to map small spatial variations or abrupt changes in myofiber orientation. The main limitation is the absence of quantitative comparison between primary eigenvector orientation. Secondly, the absence of cellular composition information prevents us from differentiating cardiomyocytes, tendons, collagen and fat and conclude on the nature of some identified pattern (Fig. 4E).

5 Conclusion

In this work, we have developed an experimental protocol for large mammalian whole hearts allowing multimodal investigation of cardiac microstructure with an emphasis on myofiber orientation. A dedicated analysis pipeline including co-registration of MRI and MicroCT images and derived metrics was designed and offers a multi-scale analysis of cardiac myofiber organization. We show a good correspondence of structural features between the two imaging modalities and identify new unexpected and complex aggregation of cardiomyocytes.

Acknowledgements and Funding. All staff from LIRYC and CHU Bordeaux involved in the Human donor program CADENCE and HARMONICA project are gratefully acknowledged for their valuable contributions. This work received financial support from the French National Investments for the Future Programs: ANR-10-IAHU-04. HD figures are available at: https://github.com/valeryozenne/Cardiac-Structure-Database/tree/master/Article-4.

References

1. Cohn, J.N., Ferrari, R., Sharpe, N.: Cardiac remodeling—concepts and clinical implications: a consensus paper from an international forum on cardiac remodeling. JACC **35**(3), 569–582 (2000)
2. Chen, J., Song, S.K., Liu, W., McLean, M., et al.: Remodeling of cardiac fiber structure after infarction in rats quantified with diffusion tensor MRI. AJP **285**(3), H946–H954 (2003)
3. Eder, R.A., et al.: Exercise-induced CITED4 expression is necessary for regional remodeling of cardiac microstructural tissue helicity. Commun. Biol. **5**, 656 (2022)
4. Carruth, E.D., Teh, I., Schneider, J.E., et al.: Regional variations in ex-vivo diffusion tensor anisotropy are associated with cardiomyocyte remodeling in rats after left ventricular pressure overload. JCMR **22**, 21 (2020)
5. Planinc, I., et al.: Comprehensive assessment of myocardial remodeling in ischemic heart disease by synchrotron propagation based X-ray phase contrast imaging. Sci. Rep. **11**, 14020 (2021)
6. Milani-Nejad, N., Janssen, P.M.L.: Small and large animal models in cardiac contraction research: advantages and disadvantages. Pharmacol. Ther. **141**(3), 235–249 (2014)
7. Sands, G.B., et al.: Automated imaging of extended tissue volumes using confocal microscopy. Microsc. Res. Tech. **67**(5), 227–239 (2005)
8. Gonzalez-Tendero, A., et al.: Whole heart detailed and quantitative anatomy, myofibre structure and vasculature from X-ray phase-contrast synchrotron radiation-based micro computed tomography. Eur. Heart J. – Cardiovasc. Imaging **18**(7), 732–741 (2017)
9. Sosnovik, D.E., Wang, R., Dai, G., et al.: Diffusion MR tractography of the heart. J. Cardiovasc. Magn. Reson. **11**, 47 (2009)
10. Lombaert, H., Peyrat, J.M., Croisille, P., et al.: Human atlas of the cardiac fiber architecture: study on a healthy population. IEEE Trans. Med. Imaging **31**(7), 1436–1447 (2012)
11. Pashakhanloo, F., et al.: Myofiber architecture of the human atria as revealed by submillimeter diffusion tensor imaging. Circ.: Arrhythm. Electrophysiol. **9**(4), e004133 (2016)
12. Yang, F., et al.: Quantitative comparison of human myocardial fiber orientations derived from DTI and polarized light imaging. Phys. Med. Biol. **63**(21), 215003 (2018)

13. Pallares-Lupon, N., et al.: Optimizing large organ scale micro computed tomography imaging in pig and human hearts using a novel air-drying technique. Arch. Cardiovasc. Dis. Suppl. **12**, 268–269 (2021)
14. Magat, J., et al.: A groupwise registration and tractography framework for cardiac myofiber architecture description by diffusion MRI: An application to the ventricular junctions. PLoS One **17**(7), e0271279 (2022)
15. Magat, J., et al.: 3D MRI of explanted sheep hearts with submillimeter isotropic spatial resolution: comparison between diffusion tensor and structure tensor imaging. Magn. Reson. Mater. Phys., Biol. Med. **34**(5), 741–755 (2021)
16. Rodríguez-Padilla, J., et al.: Impact of intraventricular septal fiber orientation on cardiac electromechanical function. AJP **322**(6), H936–H952 (2022)
17. Tournier, J.D., Calamante, F., Connelly, A.: Robust determination of the fibre orientation distribution in diffusion MRI: non-negativity constrained super-resolved spherical deconvolution. Neuroimage **35**(4), 1459–1472 (2007)
18. Garcia-Canadilla, P., et al.: Myoarchitectural disarray of hypertrophic cardiomyopathy begins pre-birth. J. Anat. **235**(5), 962–976 (2019)
19. Avants, B.B., Tustison, N., et al.: A reproducible evaluation of ANTs similarity metric performance in brain image registration. Neuroimage **54**(3), 2033–2044 (2011)
20. Hasdemir, C.A.N., Aktas, S., et al.: Demonstration of ventricular myocardial extensions into the pulmonary artery and aorta beyond the ventriculo-arterial junction. Pacing Clin. Electrophysiol. **30**(4), 534–539 (2007)
21. Burton, R.A., Lee, P., et al.: Three-dimensional histology: tools and application to quantitative assessment of cell-type distribution in rabbit heart. Europace **16**(suppl_4), iv86–iv95 (2014)
22. Lee, S.E., et al.: Three-dimensional cardiomyocytes structure revealed by diffusion tensor imaging and its validation using a tissue-clearing technique. Sci. Rep. **8**(1), 1–11 (2018)
23. Teh, I., McClymont, D., et al.: Validation of diffusion tensor MRI measurements of cardiac microstructure with structure tensor synchrotron radiation imaging. J. Cardiovasc. Magn. Reson. **19**(1), 1–14 (2017)
24. McClymont, D., et al.: The impact of signal-to-noise ratio, diffusion-weighted directions and image resolution in cardiac diffusion tensor imaging – insights from the ex-vivo rat heart. J. Cardiovasc. Magn. Reson. **19**, 90 (2017)

The Effect of Temporal Variations in Myocardial Perfusion on Diffusion Tensor Measurements

Ignasi Alemany[1,2](✉), Pedro F. Ferreira[2,3], Sonia Nielles-Vallespin[2,3], Andrew D. Scott[2,3], and Denis J. Doorly[1]

[1] Department of Aeronautics, Imperial College London, London SW7 2AZ, UK
ia4118@ic.ac.uk

[2] Cardiovascular Magnetic Resonance Unit, Royal Brompton Hospital, London SW3 6NP, UK

[3] National Heart and Lung Institute, Imperial College of London, London SW3 6LY, UK

Abstract. The aim of this study is to investigate the impact of velocity fluctuations on the perfusion signal and tensor parameters in diffusion tensor cardiovascular magnetic resonance (DT-CMR) using numerical simulations. A sinusoidal velocity function with increasing amplitude and frequency and a physiological velocity function have been considered. Both velocity functions have been analyzed using two mean inter-capillary velocity distributions with varying levels of dispersion. The results of the perfusion simulations, along with previous diffusion results, have been utilized to analyse the impact of perfusion on the diffusion tensor. The findings indicated that MCSE effectively compensated the rapid velocity changes considered in the study, while PGSE was sensitive to temporal changes in velocity. STEAM was found to be more sensitive to variations in the mean-intercapillary dispersion rather than to temporal velocity fluctuations. These simulation results provide insights regarding the potential of dispersed perfusion velocity fluctuations to affect the DT-CMR signal.

Keywords: DT-CMR · Perfusion · Temporal velocity · Diffusion

1 Introduction

Diffusion Tensor Imaging (DTI) is a novel in-vivo MR imaging technique that provides information on tissue microstructure by measuring the self-diffusion of water molecule. DTI has been extensively used in the brain [4] and more recently translated for use in diffusion tensor cardiovascular magnetic resonance (DT-CMR) [12]. DT-CMR sequences are not only sensitive to the self-diffusion of water molecules but also to the microvascular circulation present in the myocardial capillaries.

Numerical diffusion simulations have been performed using Monte Carlo random walk solvers [9] or finite element methods in order to understand the sensitivity of the DT-CMR signal to microstructural changes underlying pathologies.

O. Bernard et al. (Eds.): FIMH 2023, LNCS 13958, pp. 54–63, 2023.
https://doi.org/10.1007/978-3-031-35302-4_6

Numerical simulations of perfusion have been carried out in the brain, where the capillaries are considered to be isotropic. However, these perfusion simulations do not accurately reflect the perfusion in the myocardium, which is characterized by anisotropy and complex capillary velocities profile due to the contraction and relaxation of the myocardium [5,7]. Recently, we [2] and others [11] have developed simulations of the effects of perfusion in DT-CMR. In our previous work [2], we extended the scope of a previous perfusion model [11] by incorporating intra-capillary velocity dispersions and studied the effect of perfusion on previous diffusion simulations [9]. It was found that variations in velocity between capillaries considerably affected the tensor outputs. Furthermore, it was shown that the commonly used STEAM sequence presented higher sensitivity to microvascular perfusion than other available DT-CMR sequences, such as PGSE and MCSE, as a result of its extended diffusion encoding time.

In this study, we improve upon our previous perfusion model [2] by considering changes in velocity over time in the capillaries. Firstly we consider the temporal velocity profile as a sinusoidal function with varying amplitude and frequency, then secondly, as an alternative, we consider capillary velocity data obtained from a canine heart [5]. The objective is to investigate the impact of temporal velocity fluctuations on the DT-CMR signal originating from the protons in the capillary network and on the subsequent diffusion tensor when the perfusion results are combined with previous diffusion simulations [2].

2 Methods

2.1 Perfusion Model

The effect of perfusion on the diffusion signal can be separately modelled as a group of particles traveling at different constant velocities v through an anisotropic capillary network [2], specifically using $N_{p,\text{perfusion}}$ independent particles that traverse k capillary segments with lengths l_k and orientations $\boldsymbol{e_k}$. Here, we extend this model by incorporating a time-dependent velocity function $v(t)$ and examine the effect of $v(t)$ on the perfusion signal. The velocity for a particle i is described by

$$\boldsymbol{v_i}(t) = v_i \frac{v(t)}{f_v} \boldsymbol{e_k}, \quad f_v = \frac{\int_0^T v(t)}{T} \tag{1}$$

where f_v is the mean velocity, $\boldsymbol{e_k}$ the capillary orientation and v_i the constant capillary specific velocity. The particle velocity $\boldsymbol{v_i}(\mathrm{t})$ explicitly affects the particle phase ϕ_i described by

$$\phi_i = \gamma \int_0^T \boldsymbol{x_i}(t) \cdot \boldsymbol{g}(t)\, dt = -\gamma \int_0^T \left(\int_0^t \boldsymbol{g}(u)\, du \right) \cdot \boldsymbol{v_i}(t)\, dt \tag{2}$$

where $\boldsymbol{x_i}(t)$ is the particle position in time, $\boldsymbol{g}(t)$ the diffusion encoding gradient waveform and T the time from the start of the first to the end of the last diffusion encoding gradient. Equation (2) can be further simplified by splitting the capillary into segments of straight lengths and summing the accumulated phase in each of the capillary segments N_{cap} as follows

$$\phi_i = -\gamma \sum_{k=1}^{N_{\text{cap}}} \int_{t_{k-1}}^{t_k} \left(\int_0^t g(u)\, du \right) v_i \frac{v(t)}{f_v} \boldsymbol{e_g} \cdot \boldsymbol{e_k}\, dt \tag{3}$$

where $\boldsymbol{e_g}$ is the gradient direction. Following this simplification, ϕ_i can be reduced by normalizing the time by T and the gradient profile $\boldsymbol{G(t)} = G_{\text{max}} h(t/T) \boldsymbol{e_g}$ by its gradient strength (G_{max}).

$$\Phi_i = \frac{\phi_i}{v_i \sqrt{bT}} = -\frac{1}{a} \sum_{k=1}^{N_{\text{cap}}} \int_{s_{k-1}}^{s_k} \left(\int_0^s h(u)\, du \right) \frac{v(s)}{f_v} \boldsymbol{e_g} \cdot \boldsymbol{e_k}\, ds \tag{4}$$

In the equation above, $s = t/T$ is the normalised time in relation to T, a is the normalised b-value $b = \gamma^2 T^3 G_{\text{max}}^2 a^2$ [11] and s_{k-1} and s_k the starting and ending normalized times in each capillary segment. At the end of the simulation, the perfusion related signal attenuation $S_{\text{perfusion}}$ is calculated as the ratio between the final and the initial magnetization vector and can be described by the following equation

$$\frac{S_{\text{perfusion}}(b, \boldsymbol{e_g})}{S_0} = \frac{1}{N_{p,\text{perfusion}}} \left| \sum_{i=1}^{N_{p,\text{perfusion}}} e^{-j\phi_i} \right| \tag{5}$$

2.2 Capillary Velocity Distribution

Experimental studies in porcine and canine hearts [5] along with several numerical models [7] have shown velocity fluctuations in capillary flow velocity with variations in amplitude and frequency during a heart beat. For this reason, as illustrated in Fig. 1, we have considered two different velocity functions $v(t)$ during a cardiac cycle of 1 s. Firstly, we have considered a simple sinusoidal modulation of the velocity with varying amplitude A and frequency ν within the cardiac cycle $\boldsymbol{v_i}(t) = v_i(1 + A\sin(2\pi\nu t))\, \boldsymbol{e_k}$ (see Fig. 1). Secondly, a more realistic velocity profile has been contemplated by digitizing a physiological velocity profile obtained in experimental data [5]. For comparison of the results, the physiological data has been normalized by its mean value f_v and scaled for each particle by the inter-capillary velocity v_i, as described in Eq. (1).

The inter-capillary velocity is modelled by using a Gaussian distribution for v_i with standard deviations of $\sigma_v = 0.001$ (effectively no velocity dispersion) and $\sigma_v = 0.15$ with a mean value of $\mu_v = 0.5\,\text{mm}\,\text{s}^{-1}$ [6]. All the perfusing spins $N_{p,\text{perfusion}}$ have been arbitrarily seeded in a plane $z = 0$ inside a voxel of $2.8 \times 2.8 \times 8\,\text{mm}^3$. The simulations have been performed with $N_p = 10^5$, which we previously found to provide a sufficiently precise perfusion signal.

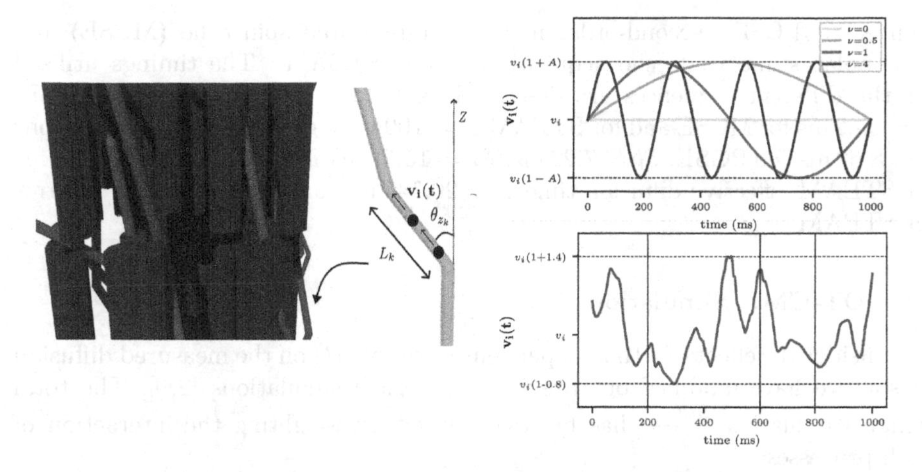

Fig. 1. Left figure: 3D render illustrating several capillaries (green) surrounding a group of cardiomyocytes (red). Each capillary is constituted of k capillary segments with specific orientation θ_{z_k} and length L_k. **Right figure**: Graphic representation of the two velocity functions $v(t)$ considered. The top figure illustrates a basic sinusoidal function during the cardiac cycle $v_i + v_i A \sin(2\pi\nu t)$ with different frequencies $\nu = 0, 0.5, 1, 4$. The bottom figure illustrates the physiological data from [5] with three different starting points (black solid lines) considered for sequences PGSE/MCSE. (Color figure online)

2.3 Capillary Network

The capillary networks in the myocardium are anisotropic with a preferred average direction aligned with the cardiomyocytes. Experimental histology studies [8] have shown that a hemispherical Dimroth-Watson axial distribution was a suitable model to represent the myocardial capillary orientation and has been considered for the angle θ_z between the capillary segment and the cardiomyocyte's long axis. The probability distribution function is outlined below

$$p(\theta_z) = \frac{1}{U_0} \sin\theta_z \exp\left(2\,K\cos^2\theta_z\right) \tag{6}$$

where K is a constant variable that controls the degree of anisotropy spanning from isotropic $K = 0$ to highly anisotropic $K >> 0$. As in [2], a default constant of $K = 3.25$ has been assumed based on experimental data from rat hearts. The distribution of capillary segment lengths was been obtained from a Weibull distribution with a mean of $\mu_L = 60\,\mu\text{m}$ and a standard deviation of $\sigma_L = 40\,\mu\text{m}$ [2].

2.4 DT-CMR Sequences

Three different diffusion encoding gradient waveforms have been considered based on typical sequences used in DT-CMR: monopolar pulsed-gradient spin echo (PGSE), second-order motion-compensated spin echo (MCSE) and monopolar stimulated echo acquisition mode (STEAM). The timings utilised for the different sequences are described as follows: $T = 31.7$ ms for PGSE, $T = 63.2$ ms for MCSE and for STEAM $T = 1004$ ms, gradient flat-top durations $\delta = 8.62$ ms for PGSE, $\delta_1 = 7.22$ ms, $\delta_2 = 15.33$ ms for MCSE and $\delta = 0.5$ ms for STEAM, effective diffusion time $\Delta = 20.547$ ms for PGSE and $\Delta = 1000$ ms for STEAM.

2.5 DT-CMR Simulation

To consider the effect of a time-dependent velocity $v(t)$ on the measured diffusion tensor, we have modified our previous numerical simulations [2,9]. The total signal attenuation $\frac{S(b,e_g)}{S_0}$ has been computed by weighting the interaction of both processes:

$$\frac{S(b,e_g)}{S_0} = f\frac{S_{\text{perfusion}}(b,e_g)}{S_0} + (1-f)\frac{S_{\text{diffusion}}(b,e_g)}{S_0} \quad (7)$$

where the perfusion fraction f refers to the fraction of the total number of detectable spins that are within the capillary network. For healthy hearts, the physiological ranges of the perfusion fraction f typically fall between 11% and 13%. As described in Eq. (8), the total signal attenuation in Eq. (7) is then employed to compute a new diffusion tensor $\boldsymbol{D}_{\text{D+P}}$.

$$\frac{S(b,e_g)}{S_0} = \exp\left(b{e_g}^T \boldsymbol{D}_{\text{D+P}}\, e_g\right) \quad (8)$$

The diffusion signal has been obtained using an in-house Monte Carlo random walk solver [2,9] for a histology-based permeable media [3]. Simulation parameters include: diffusivity of the extracellular space $D_{\text{ECS}} = 2.5\,\mu\text{m}^2\,\text{ms}^1$, intracellular diffusivity $D_{\text{ICS}} = 1\,\mu\text{m}^2\,\text{ms}^1$, cardiomyocyte membrane permeability $0.02\,\mu\text{m}\,\text{ms}^{-1}$ [10], $N_t = 10^4$ time steps and $N_{p,\text{diffusion}} = 10^5$ number of particles. The tensor is computed with a total of 7 directions including a $b = 0\,\text{ms}\,\mu\text{m}^{-2}$ and a b-value of $0.6\,\text{ms}\,\mu\text{m}^{-2}$.

3 Results

The perfusion signal attenuation is only shown in the direction of the cardiomyocytes ($\boldsymbol{E1}$) for simplicity and because the largest velocity-related changes are expected in this direction.

3.1 Amplitude and Frequency Study

By increasing the sinusoidal amplitude A of the temporal velocity function for $\sigma_v = 0.15$ and comparing to constant velocity flow, the signal intensity from the perfusing spins remains unchanged when diffusion encoding is applied along the

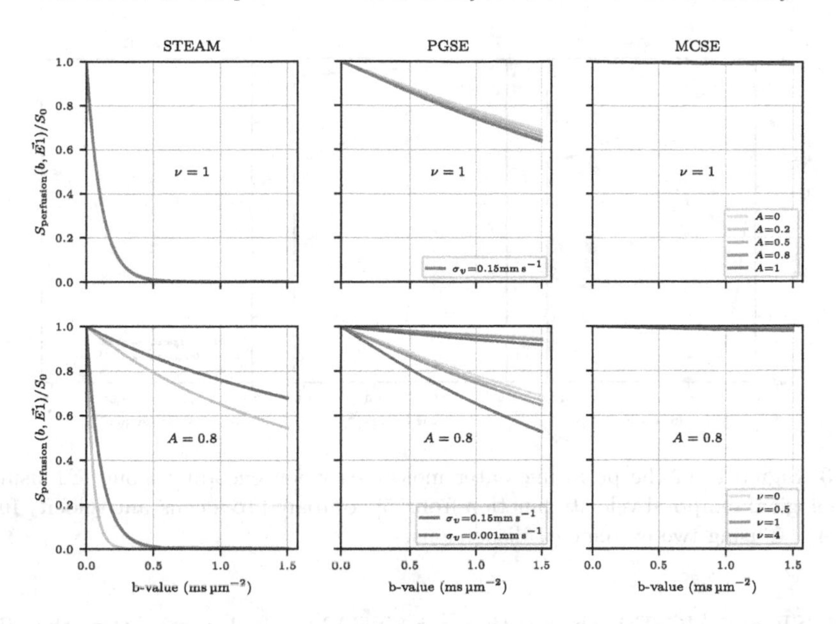

Fig. 2. Perfusion signal along $\boldsymbol{E1}$ varying the amplitude A (upper row) and frequency ν (lower row) of the sinusoidal velocity function $\boldsymbol{v_i}(t) = v_i(1 + A\sin(2\pi\nu t))\,\boldsymbol{e_k}$. Two inter-capillary velocity functions are considered $\sigma_v = 0.001\,\mathrm{mm\,s^{-1}}$ (green lines) and $\sigma_v = 0.15\,\mathrm{mm\,s^{-1}}$ (blue lines). Increasing A amplifies the magnitude of v_i while rising ν increases the number of oscillations during the cardiac cycle (see Fig. 1). STEAM shows the same signal attenuation for all ν values except for $\nu = 0.5$ while PGSE shows a gradual signal attenuation transition. (Color figure online)

cardiomyocyte's long axis for MCSE/STEAM, while a slight reduction in signal attenuation is observable for PGSE (see Fig. 2). With respect to frequency, for both σ_v it is observed that STEAM uniquely presents increased signal attenuation when the velocity function covers half an oscillation within the cardiac cycle ($\nu = 0.5$) as this results in an increase in the mean velocity over the diffusion encoding time. In contrast, PGSE shows increased signal attenuation with increased oscillation frequency while MCSE displays only small deviations from the constant velocity case.

3.2 Physiological Velocity Function

The results for both σ_v for the two different temporal velocity functions show slight variations for MCSE and STEAM, while PGSE shows the greatest sensitivity to temporal variations in perfusion, with a noticeable reduction in signal intensity due to the velocity fluctuations (see Fig. 3). Quantitatively, PGSE shows a signal attenuation reduction of 16% for $\sigma_v = 0.15\,\mathrm{mm\,s^{-1}}$ and a constant b-value of $0.6\,\mathrm{ms\,\mu m^{-2}}$ between the constant temporal velocity $A = 0$ and the physiological data while STEAM and MCSE present variations of 0.62% and 0.58%. In comparison to the cardiac cycle duration, the diffusion encoding time

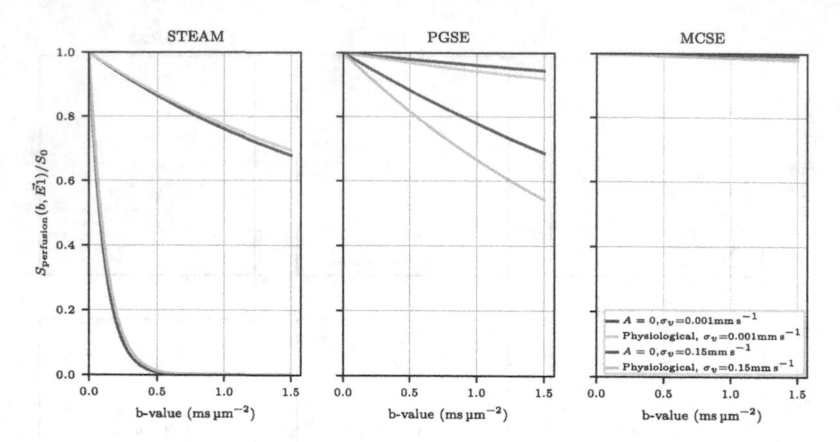

Fig. 3. Signal from the perfusing water molecules when encoding along $\boldsymbol{E1}$ using a physiological temporal velocity function from [5] compared to a constant velocity function $A = 0$ using two extreme σ_v dispersions.

for PGSE and MCSE is shorter than for STEAM, which means that the effect of the perfusing spins depends on the initial starting point T_{offset} (see Fig. 4). MCSE presents no substantial signal attenuation for the three different starting points considered ($T_{\text{offset}} = 200, 600, 800$ ms) while the largest effect for PGSE is observed at $T_{\text{offset}} = 600$ ms when the velocity is rapidly changing during the diffusion encoding.

3.3 Diffusion Tensor Results

It is shown that for STEAM, incrementing the perfusion fraction increases the calculated diffusivity along (λ_1) and the mean diffusivity (MD), while the opposite behaviour is observed for PGSE and MCSE (see Fig. 5). PGSE shows larger tensor variations between the physiological capillary flow and the constant inter-capillary velocity whereas unchanged values are observed for STEAM and MCSE. For all sequences but specifically for STEAM, reductions in λ_1 and MD are noted when increasing the inter-capillary velocity dispersion σ_v.

4 Discussion

The effects of a time-dependent capillary velocity $v(t)$ on perfusion related signal in DT-CMR have been studied using numerical simulations and compared with the results of a constant inter-capillary velocity in [2]. The results showed that sinusoidal oscillations in velocity lead to deviations in signal attenuation along $\boldsymbol{E1}$ for both STEAM and PGSE. Furthermore, STEAM was found to be sensitive only to temporal inter-capillary velocities that have a non-vanishing net area during Δ, while MCSE effectively compensated for the varying particle velocity and PGSE showed a reduction in signal with increasing frequency and amplitude.

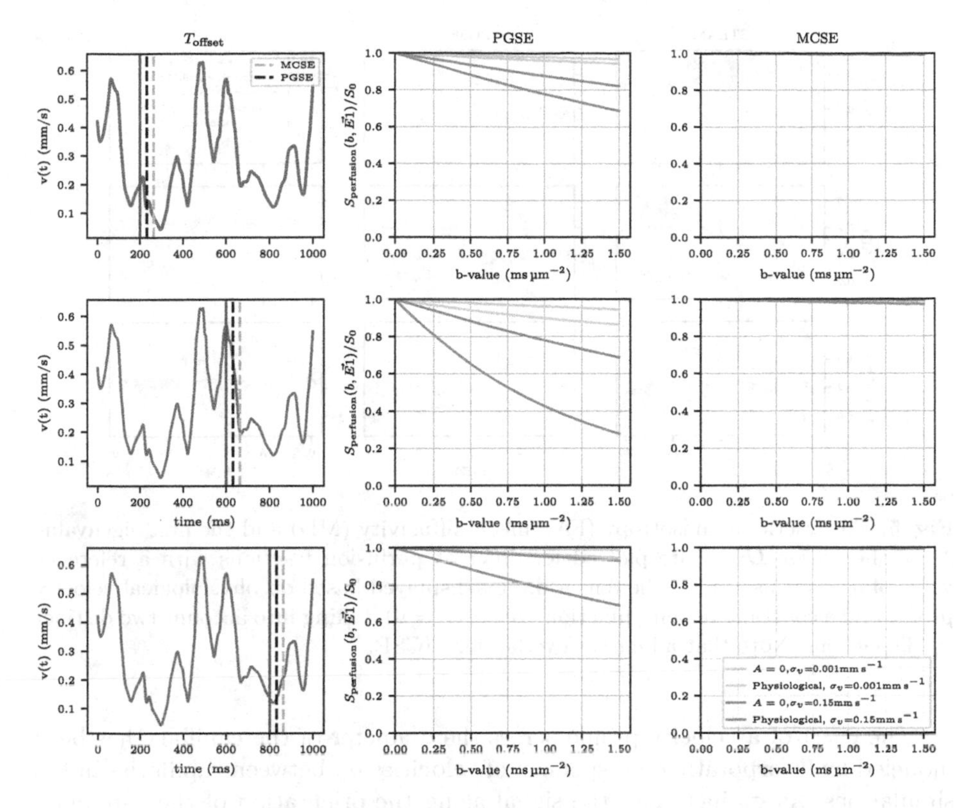

Fig. 4. Physiological velocity function from [5] plotted for $v_i = 0.5\,\mathrm{mm\,s^{-1}}$ for three different starting points T_{offset} (left column). Note that the ending points of the sequence in the velocity function are indicated with dashed lines, with MCSE in orange and PGSE in black. The corresponding perfusion related signal attenuation along $\boldsymbol{E1}$ is plotted (centre and right column) for each T_{offset} and sequence.

The temporal variation in microcirculation in the capillaries is not well characterised due to the difficulties in measuring blood flow in microscopic vessels in the contracting and expanding myocardium. The measured data on capillary blood flow in [5] provided realistic physiological blood flow data and our simulations revealed that incorporating rapid variations in the inter-capillary velocity with these data had a significant impact on short diffusion time sequences ($\Delta < 1000\,\mathrm{ms}$). However, results show that despite these rapid variations the perfusion effects on the motion-compensated sequence (MCSE) remained very small. In addition, it is important to keep in mind that the perfusion model is limited and does not consider merged/bifurcated capillaries, which may result in velocity jumps. Neither do they account for plasma skimming (the change in measured red blood cell density in the capillaries compared to larger vessels). Furthermore, we used a single velocity function $v(t)$ from [5] which does not account for potential variations between subjects or between regions of the myocardium.

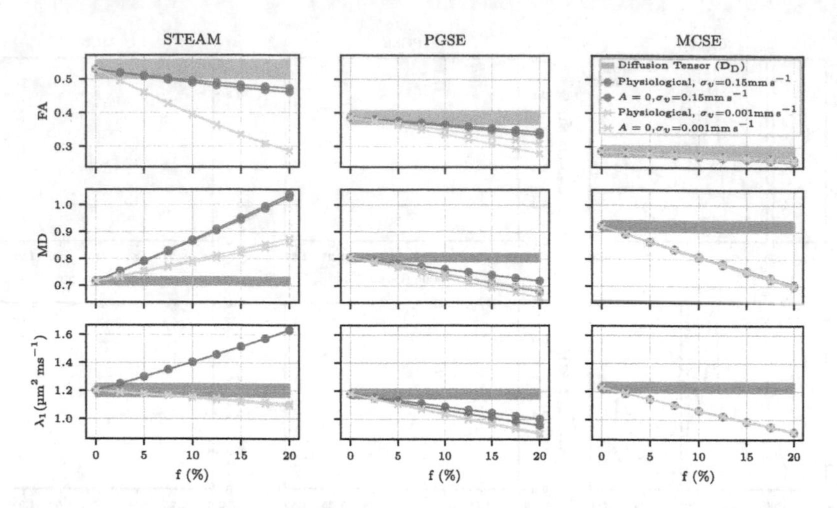

Fig. 5. The fractional anisotropy (FA), mean diffusivity (MD) and the first eigenvalue λ_1 of the tensor $\boldsymbol{D}_{\mathrm{D+P}}$ are plotted for different perfusion fractions with a reference value of $b_{\mathrm{ref}} = 0\,\mathrm{ms}\,\mu\mathrm{m}^{-2}$. The tensor data is displayed based on physiological velocity profile and a constant velocity function ($A = 0/\nu = 0$), taking into account two distinct σ_v dispersions. Note that all curves overlap for MCSE.

The effect of a velocity profile across the diameter of the capillary has been modeled by incorporating a dispersion of velocities σ_v between capillaries in the simulations. As σ_v increases, the signal along the orientation of the cardiomyocytes decreases at a faster rate as the b-value increases, which is in agreement with the results reported in [1] that showed greater signal attenuation in laminar flow compared to plug flow. Due to the relatively small effect of perfusion on the measured signal in PGSE and MCSE, adding the perfusion signal actually reduces the measured MD and λ_1 as the perfusion component decays less with increasing b value than the diffusing component. In conclusion, MCSE appears highly robust to myocardial perfusion, including inter-capillary velocity distributions and temporal fluctuations. In contrast, PGSE is highly sensitive to temporal changes in velocity due to its use of long diffusion encoding gradients and lack of motion compensation. However, PGSE is highly sensitive to bulk motion and therefore only applicable in ex-vivo cardiac imaging, where perfusion is not a concern. In contrast, STEAM is most sensitive to perfusion, but variations in the mean inter-capillary dispersion have more of an effect than temporal variations.

Acknowledgments. This work was funded by British Heart Foundation Grants RE/13/4/30184 and RG/19/1/34160.

References

1. Abdullah, O.M., Gomez, A.D., Merchant, S., Heidinger, M., Poelzing, S., Hsu, E.W.: Orientation dependence of microcirculation-induced diffusion signal in anisotropic tissues. Magn. Reson. Med. **76**(4) (2016). https://doi.org/10.1002/mrm.25980
2. Alemany, I., et al.: Realistic numerical simulations of diffusion tensor CMR: the effects of perfusion and membrane permeability (2023, under review)
3. Alemany, I., Rose, J.N., Garnier-Brun, J., Scott, A.D., Doorly, D.J.: Random walk diffusion simulations in semi-permeable layered media with varying diffusivity. Sci. Rep. **12**(1), 10759 (2022). https://doi.org/10.1038/s41598-022-14541-y
4. Bao, H., Li, R., He, M., Kang, D., Zhao, L.: DTI study on brain structure and cognitive function in patients with chronic mountain sickness. Sci. Rep. **9**(1) (2019). https://doi.org/10.1038/s41598-019-55498-9
5. Barclay, K.D., Klassen, G.A., Young, C.: A method for detecting chaos in canine myocardial microcirculatory red cell flux. Microcirculation **7**(5) (2000). https://doi.org/10.1111/j.1549-8719.2000.tb00132.x
6. Callot, V., Bennett, E., Decking, U.K., Balaban, R.S., Wen, H.: In vivo study of microcirculation in canine myocardium using the IVIM method. Magn. Reson. Med. **50**(3) (2003). https://doi.org/10.1002/mrm.10568
7. Fibich, G., Lanir, Y., Liron, N.: Mathematical model of blood flow in a coronary capillary. Am. J. Physiol. Heart Circulatory Physiol. **265**(5 34-5) (1993). https://doi.org/10.1152/ajpheart.1993.265.5.h1829
8. Poole, D.C., Mathieu-Costello, O.: Analysis of capillary geometry in rat subepicardium and subendocardium. Am. J. Physiol. Heart Circulatory Physiol. **259**(1 28-1) (1990). https://doi.org/10.1152/ajpheart.1990.259.1.h204
9. Rose, J.N., Nielles-Vallespin, S., Ferreira, P.F., Firmin, D.N., Scott, A.D., Doorly, D.J.: Novel insights into in-vivo diffusion tensor cardiovascular magnetic resonance using computational modeling and a histology-based virtual microstructure. Magn. Reson. Med. **81**(4) (2019). https://doi.org/10.1002/mrm.27561
10. Seland, J.G., Bruvold, M., Brurok, H., Jynge, P., Krane, J.: Analyzing equilibrium water exchange between myocardial tissue compartments using dynamical two-dimensional correlation experiments combined with manganese-enhanced relaxography. Magn. Reson. Med. **58**(4) (2007). https://doi.org/10.1002/mrm.21323
11. Spinner, G.R., Stoeck, C.T., Mathez, L., von Deuster, C., Federau, C., Kozerke, S.: On probing intravoxel incoherent motion in the heart-spin-echo versus stimulated-echo DWI. Magn. Reson. Med. **82**(3) (2019). https://doi.org/10.1002/mrm.27777
12. Stoeck, C.T., Von Deuster, C., GeneT, M., Atkinson, D., Kozerke, S.: Second-order motion-compensated spin echo diffusion tensor imaging of the human heart. Magn. Reson. Med. **75**(4) (2016). https://doi.org/10.1002/mrm.25784

Ventricular Helix Angle Trends and Long-Range Connectivity

Alexander J. Wilson[1,2(✉)], Q. Joyce Han[3], Luigi E. Perotti[4], and Daniel B. Ennis[1,2]

[1] Department of Radiology, Stanford University, Stanford, CA, USA
wilsonaj@stanford.edu
[2] Stanford Cardiovascular Institute, Stanford University, Stanford, CA, USA
[3] Department of Medicine, Massachusetts General Hospital, Boston, MA, USA
[4] Department of Mechanical and Aerospace Engineering, University of Central Florida, Orlando, FL, USA

Abstract. Porcine hearts (N = 14) underwent *ex vivo* diffusion tensor imaging (DTI) at 3T. DTI analysis showed regional differences in helix angle (HA) range. The HA range in the posterior free wall was significantly greater than that of the anterior free wall ($p = 0.02$), the lateral free wall ($p < 0.001$) and the septum ($p = 0.008$). The best-fit transmural HA function also varied by region, with eight regions best described by an *arctan* function, seven by an *arcsine* function, and a single region by a *linear* function. Tractography analysis was performed, and the length that the tracts spanned within the epicardial, midwall, and endocardial segments was measured. A high number of tracts span the epicardial and mid-wall thirds, with fewer tracts spanning the mid-wall and endocardial thirds. Connectivity analysis of the number of tracts connecting different ventricular regions showed a high prevalence of oblique tracts that may be critical for long-range connectivity.

Keywords: MRI · Diffusion tensor imaging · Cardiac tractography · Connectivity · Cardiomyocyte orientation · Cardiac microstructure

1 Introduction

The structure of the heart is fundamentally connected with its function. While cardiomyocytes contract only ≈14% [8], the ventricular wall achieves ≈30–40% thickening [11]. This wall thickening is facilitated by the structural organization of the myocardium: cardiomyocytes vary in orientation through the transmural span of the ventricular wall [15] and are grouped into sheetlet structures that allow meso-scale shear [5,17,19]. Additionally, cardiomyocytes are both electrochemically and mechanically coupled, and transmit electrical activation and contractile forces throughout the heart. These electrochemical and mechanical couplings are supported by the continuously branching syncytium of the myocardium comprised of cardiomyocytes that have a branching structure [18].

A. J. Wilson and Q. J. Han—The first two authors contributed equally.

O. Bernard et al. (Eds.): FIMH 2023, LNCS 13958, pp. 64–73, 2023.
https://doi.org/10.1007/978-3-031-35302-4_7

Diffusion tensor magnetic resonance imaging (DTI) measures the diffusion tensor of water within biological tissue. In the myocardium, the primary eigenvector (E1) of the diffusion tensor has been shown to correspond to the aggregate cardiomyocyte long axis orientation ("myofiber" orientation) within a voxel. The helix angle (HA) measures the aggregate cardiomyocyte orientation and can be measured using DTI even in live hearts [8]. Diffusion tractography connects adjacent E1 voxels into *tracts* to describe long-range connectivity [13]. While histology provides the ability to measure long-range connectivity along cut surfaces [6], three dimensional whole-heart DTI allows measurement of aggregate cardiomyocyte tracts across the whole heart [14].

Aggregate cardiomyocyte HA is fundamental to the structure-function of the heart, but there is still no consensus in terms of the best epicardial and endocardial HA values. Even within a single species (Wistar-Kyoto rat) HA range has been measured as $\approx 125^\circ$ (epi = -50°, endo = $+75^\circ$) using propagation-based X-ray phase contrast imaging [2] versus $\approx 180^\circ$ (epi = -90°, endo = $+90^\circ$) using DTI [3]. There is a lack of consensus within the literature as to the function that best represents transmural HA. Studies have shown different transmural HA functions, including *arctan* [7], *linear* [10], and *arcsine* [15] functions.

In this study we used *ex vivo* porcine cardiac DTI to examine HA range, median HA value, and the best-fit transmural HA function in different ventricular regions, utilizing a relatively large sample size. Additionally we performed diffusion tractography, and analyzed these tracts to measure the long-range connectivity between different regions of the heart.

2 Materials and Methods

2.1 Image Acquisition

This study utilized healthy swine (N = 14) in accordance with institutional approvals (UCLA ARC protocol # 2015-124). Subjects underwent *in vivo* cardiac MRI using a clinical 3T scanner (Siemens, Prisma), including bSSFP 2D cine image acquisitions. After *in vivo* imaging, the subjects were euthanized and the hearts explanted. Each heart was washed with saline/water and then placed in a container filled with Fomblin. Two methods were used to support the *ex vivo* hearts in a physiological configuration: six (n = 6) hearts were prepared using rapid-setting dental gel and sponges; eight (n = 8) hearts were prepared using 3D printed molds based on the cine images acquired *in vivo* and segmented at mid diastole (diastasis) [1]. Within 2–3 h from extraction, *ex vivo* cardiac DTI was performed at 3T with spatial resolution $1 \times 1 \times 1\,\text{mm}^3$, b-value $1000\,\text{s/mm}^2$, 30 diffusion directions, and 5 averages. The 3D imaging volumes encompassed both the left and right ventricles. The hearts were imaged while fresh, and were not fixed.

2.2 Image Processing

Ex vivo imaging data was manually segmented by a single observer. Myocardial segmentation was performed using the DTI images, with the papillary muscles

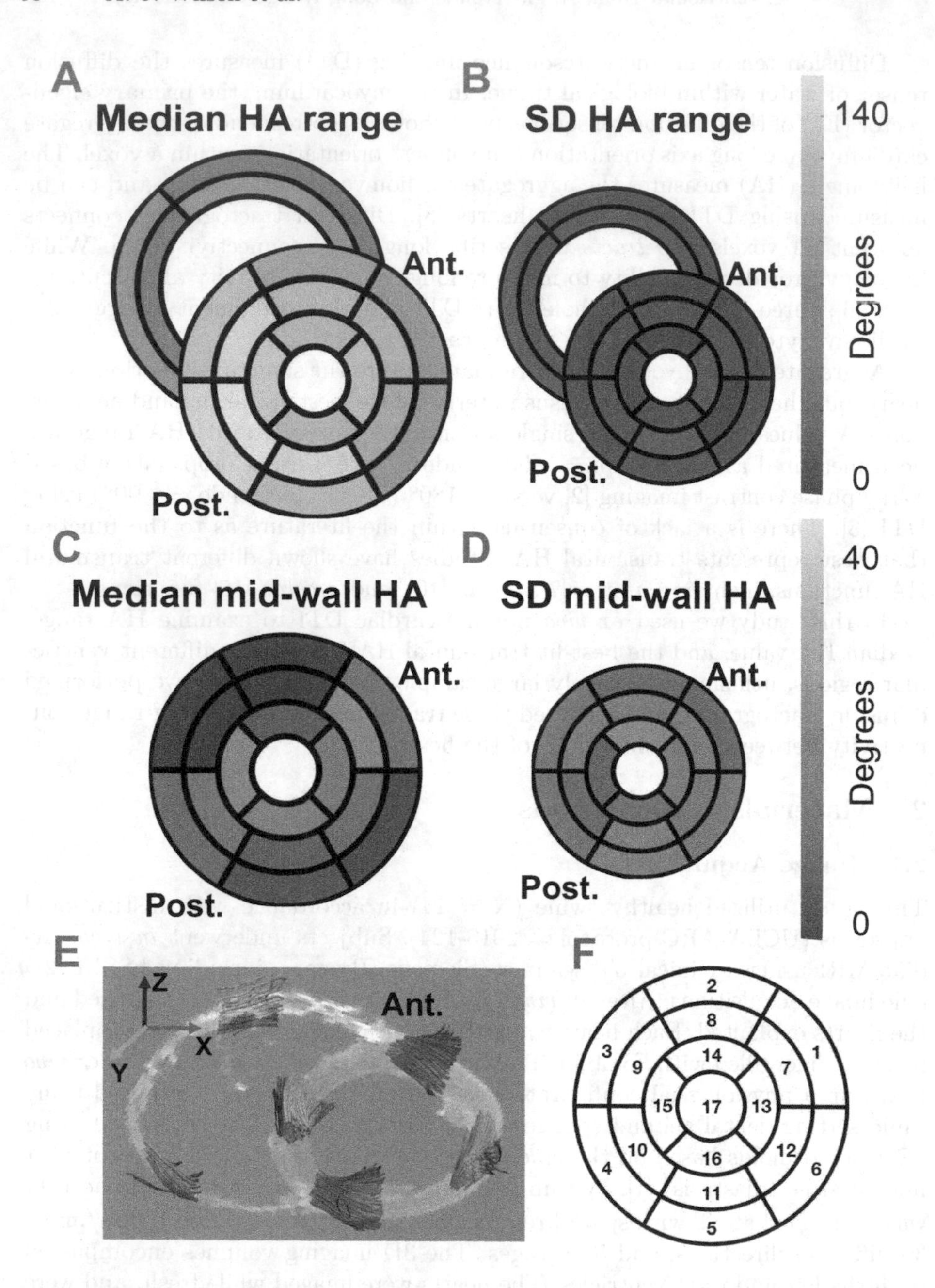

Fig. 1. Regional helix angle range and median mid-wall helix angle: Helix angle (HA) range is shown across regions of the heart (A) as well as its standard deviation (B). Also shown is the median mid-wall HA (C) and the standard deviation of mid-wall HA across all hearts (D). A 3D visualization shows the orientation of tracts within different left and right ventricular regions (E). For reference, the American Heart Association 17-segment bullseye model is also shown (F).

excluded. The cardiac long-axis was defined as the axis extending through the apex of the heart and the middle of the mitral valve. A short axis plane was defined as any plane perpendicular to this long-axis. These definitions allowed consistent prescription of short-axis planes to produce consistent HA measurements. HA was calculated as the angle between the projection of E1 onto the epicardial surface and the circumferential direction.

The LV was divided into 17 segments according to the American Heart Association (AHA) 17-segment model. The RV was divided into 4 segments: anterior base, posterior base, anterior apex, and posterior apex. For each segment of each heart, three different transmural HA functions were fit to the DTI data: (i) *linear*; (ii) *arcsine*; and (iii) *arctan*. For each region of each heart, the best-fit HA function was selected as the one with the highest coefficient of determination (R^2). Then the generalized regional best-fit HA function was determined as the mode best-fit function for that region over all of the hearts.

2.3 Tractography Analysis

Tractography was applied to E1 data using BrainSuite (Los Angeles, CA, USA) [12]. Tracts were seeded in each voxel and a step size of 1 mm was used with an angle threshold of 30°. Tracts were terminated after 500 steps. For each tract, we quantified its longitudinal (l), circumferential (c), and radial (r) spans. We also defined the transmural trajectory (endo $r = -1$ and epi $r = 1$) of a tract as the length within the endocardial ($r \in [-1, -1/3]$), mid-wall ($r \in [-1/3, 1/3]$), and epicardial ($r \in [1/3, 1]$) transmural thirds.

Connectivity analysis was performed within BrainSuite. Tracts with length greater than 100 mm were included for analysis. The connectivity (C_{AB}) between two segments (S_A, S_B) was computed as the average of two tract counts: (i) the tracts starting in S_A and ending in S_B; and (ii) the tracts starting in S_B and ending in S_A. This averaging ensured that $C_{AB} = C_{BA}$.

2.4 Statistical Analysis

The distributions of HA ranges and mid-wall HAs were non-Gaussian. We therefore used the median as the measure of central tendency. For two-group comparisons, we used a Wilcoxon signed-rank test with a least significant difference approach to correct for multiple testing. For multiple group comparisons we used a Kruskal-Wallis test.

3 Results

HA range varied across different regions of the heart (Fig. 1A). In the LV, the anterior regions tended to have a lower HA range than the posterior regions. The largest HA range was measured in the posterior free wall; this HA range was significantly larger than the HA range in the anterior free wall ($p = 0.02$), lateral free wall ($p < 0.001$), and the septum ($p = 0.008$). The RV anterior free wall had

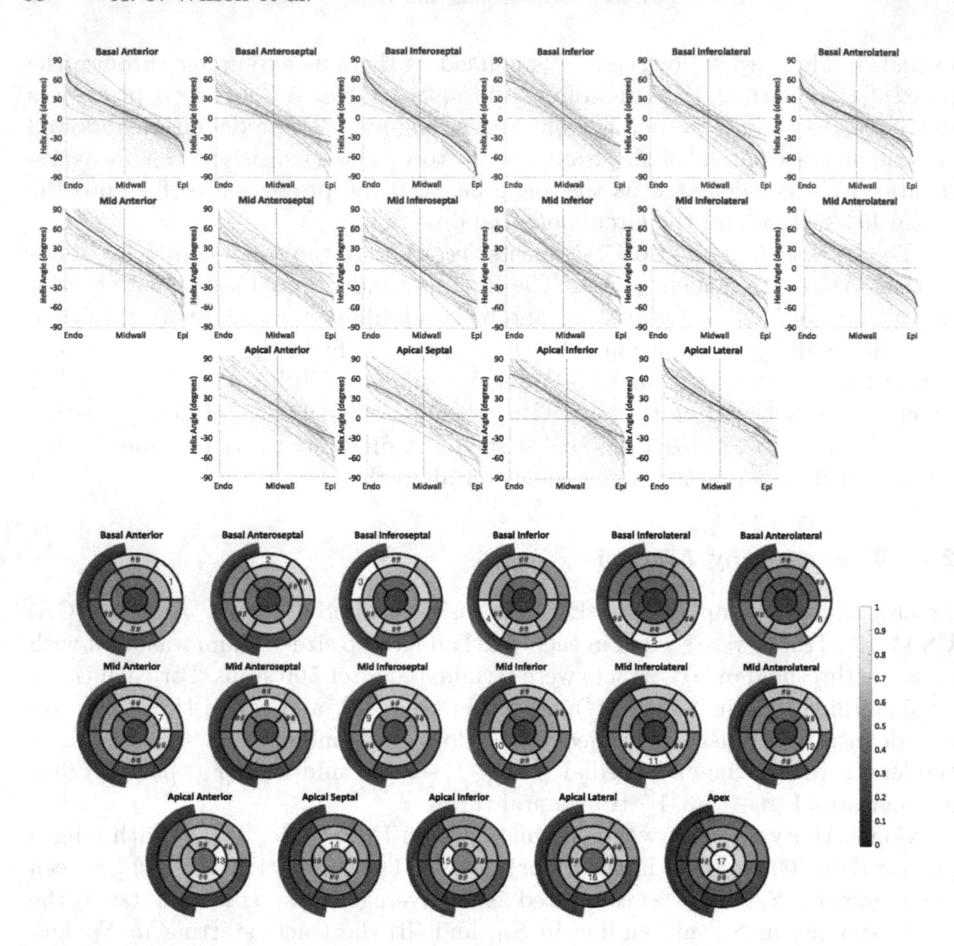

Fig. 2. Best-fit transmural helix angle function (top): For each left ventricular region, the best-fit transmural helix angle function is shown as an *arctan* function (green), an *arcsine* function (blue), or a *linear* function (red). Fits for individual hearts are shown in grey. **Connectivity analysis (bottom):** For each segment, the degree of connectivity is shown between the target region (white, numbered) and each other region. The three regions with the highest connectivity with the target region are indicated with '##'. (Color figure online)

a significantly lower HA range than the RV posterior free wall ($p = 0.03$). The standard deviation of HA range was not significantly different between segments (Fig. 1B). The median mid-wall HA was $\approx 0°$ for half of the LV regions (Fig. 1C).

The best-fit transmural HA function (Fig. 2) showed approximately an even split between regions best represented by an *arctan* function ($n = 8$) versus an *arcsine* function ($n = 7$). Only the mid anterior segment was best represented by a *linear* function. The lateral regions favored an *arcsine* function, while the septal regions tended to favor an *arctan* function.

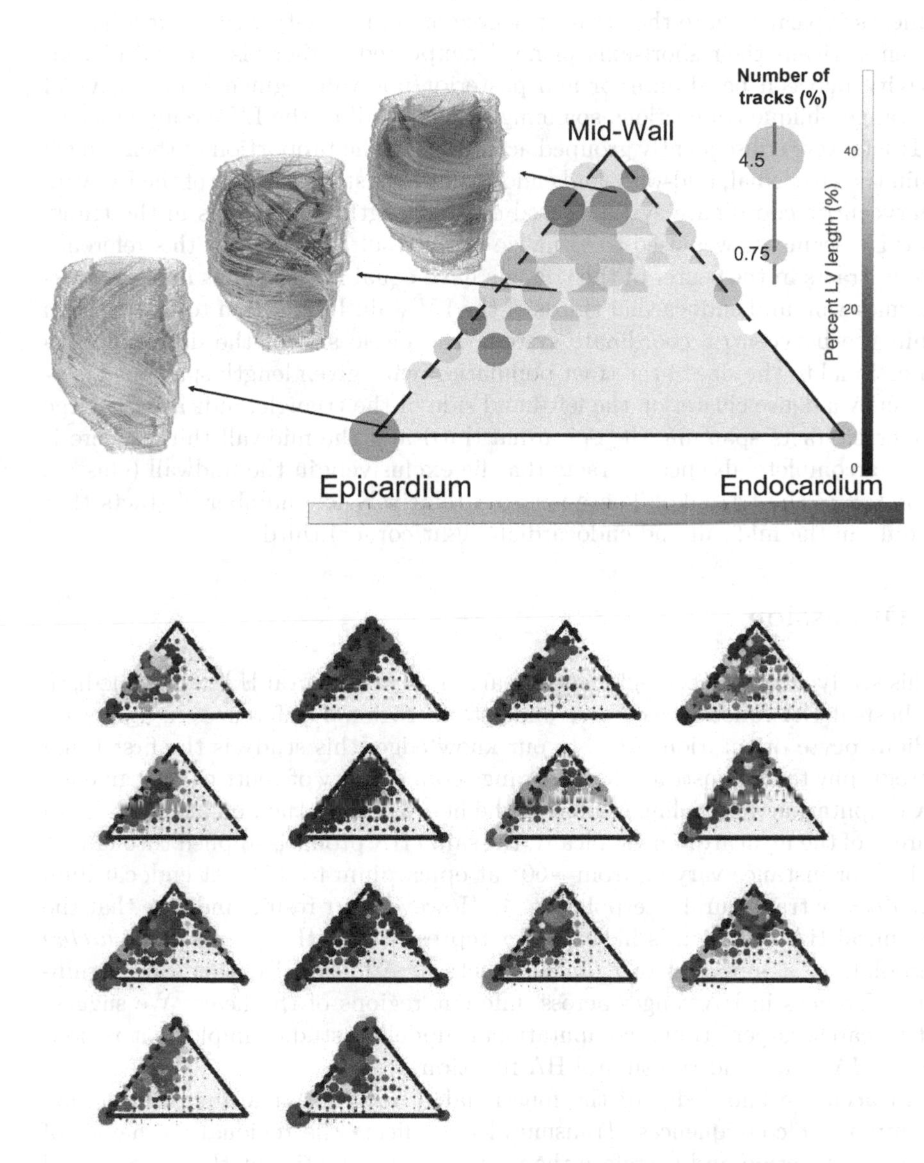

Fig. 3. Transmural analysis of the tractography data: Tracts were grouped according to their length span within each of the transmural endocardial, midwall, and epicardial thirds. Data is visualized using a barycentric coordinate system where the left, top, and right corners represent tracts spending the majority of their length in the epicardial, midwall, and endocardial thirds. An example heart is shown (top), with data on the barycentric graph corresponding with tract visualizations. The transmural spans are shown for all (N = 14) hearts (bottom).

Figure 2 shows the connectivity across the 21 segments of the LV and RV. Segments typically share the greatest degree of connectivity with their adjacent segments within their short-axis plane. Unexpectedly, there is a trend of connectivity between basal anterior and posterior midwall segments, indicative of long-range oblique connections spanning at least half of the LV circumference.

Tracts were subsequently grouped according to the proportion of their length within the epicardial, mid-wall, and endocardial transmural thirds of the LV wall. A barycentric coordinate system based on the length proportions in the transmural LV segments was used to visualize these results (Fig. 3). In this reference system, tracts in the center of the triangle have equal length spans in the epicardial, mid-wall, and endocardial thirds of the LV wall. In addition to the position within the barycentric coordinate system, the circle size of the data points is proportional to the size of the tract population with given length spans. All subjects show a dense cluster on the left-hand side of the triangle, indicating a large number of tracts spanning the epicardial third and the midwall third. There is almost a complete absence of tracts that lie exclusively in the midwall (clusters at the top of the triangle). There is also a relatively low number of tracts that run only in the midwall and endocardial (right corner) thirds.

4 Discussion

In this study we present a regional comparison of transmural HA range, median, and best-fit HA function to better understand regional differences in aggregate cardiomyocyte orientation. Also, to our knowledge, this study is the first using tractography to demonstrate the long-range connectivity of ventricular segments.

Computational modeling studies of the heart require the definition of HA in all areas of the myocardium. Typically, the same HA profile is applied throughout the LV, for instance varying from $-60°$ at epicardium to $+70°$ at endocardium with a *linear* transmural interpolation [4]. However, our results indicate that the transmural HA variation is likely better represented with an *arcsine* or *arctan* interpolation, as opposed to a *linear* function. Additionally, there are significant differences in HA ranges across different regions of the heart. We suggest that researchers performing computational modeling studies implement region-specific HA range and transmural HA function.

An accurate knowledge of the functional form of HA transmural variation has important consequences. Transmural HA affects the regional mechanics of cardiac contraction and therefore the ventricles' kinematics at the regional and global levels. Furthermore, aggregate cardiomyocyte orientation is related to the fastest direction of electrical conduction in cardiac tissue and therefore HA transmural variation directly affects cardiac electrophysiology. Different studies have shown different transmural HA functions, including *arctan* [7], *linear* [10] and *arcsine* [15] functions, and there is a lack of agreement within the literature regarding which transmural HA function is most appropriate. To our knowledge this is the first study to identify regional differences in best-fit transmural helix angle function.

The high degree of connectivity between adjacent regions within the same short-axis plane was expected. However, the high number of oblique tracts connecting anterior and posterior regions at different longitudinal locations was an unexpected finding. The transmural tract data shows a predominance of tracts within the epicardium and midwall; it is likely that the tracts connecting the anterior and posterior regions have length spans primarily in these transmural regions. The high degree of connectivity would allow for increased redundancy in both electrochemical and mechanical coupling, which indicates the importance of connections between these regions. Additionally, oblique tracts may play a key role in generating ventricular torsion, and this result is consistent with cardiac structure facilitating ventricular torsion.

We propose two reasons for the low presence of endocardial tracts relative to epicardial tracts. Firstly, there is a geometric effect – the epicardial region has a greater radius – leading to a greater three dimensional volume, and therefore a greater potential to contain tracts. This factor should increase the proportion of epicardial tracts relative to endocardial tracts (circle size in Fig. 3), but not their absence from the endocardial region. Secondly, there are organizational differences between the epicardium and endocardium. Compared with the well-defined, relatively smooth epicardial surface, the endocardium contains trabeculae and papillary muscles that complicate the mesostructural organization of endocardium. This may lead to tracts that terminate at less than 100mm, which would lead to the absence of data (lack of circles in Fig. 3).

Limitations. As the hearts were imaged *ex vivo*, their MRI properties may have changed slightly as compared to *in vivo* imaging. However, we do not expect the cardiomyocyte orientations to have changed, and so this should have minimal effect on the reconstructed diffusion tensor. In terms of segmentation of the myocardium, only a single researcher performed the segmentation leading to the possibility of observer bias. However, the single observer is experienced in this type of segmentation. One limitation of DTI is that estimates of E1 have greater uncertainty within regions that are relatively isotropic. Future work could explore this limitation further. Additionally, by imaging the hearts *ex vivo* we are only able to estimate aggregate cardiomyocyte orientation in a single cardiac phase. Work from our group has shown that HA range changes through the cardiac cycle [9]. With regards to tractography analysis of cardiac diffusion tensor data, we did not perform additional validation in this study. However, it has been previously shown that tractography of *ex vivo* cardiac diffusion tensor data aligns with tractography from high resolution synchrotron radiation imaging, and that both display the expected helix angle measurements [16].

5 Conclusion

DTI analysis of *ex vivo* porcine hearts showed regional differences of both HA range and best-fit transmural HA function. Novel connectivity analysis of diffusion tractography data revealed a high degree of long-range connectivity between oblique regions.

Acknowledgements. This material is based upon work supported, in part, by American Heart Association Grant 19IPLOI34760294 (to D.B.E.) and National Heart, Lung, and Blood Institute Grants R01-HL131823 (to D.B.E.), R01-HL152256 (to D.B.E.), and K25-HL135408 (to L.E.P.) and by the National Science Foundation under Grants 2205043 (to L.E.P.) and 2205103 (to D.B.E.).

References

1. Cork, T.E., Perotti, L.E., Verzhbinsky, I.A., Loecher, M., Ennis, D.B.: High-resolution *Ex Vivo* microstructural MRI after restoring ventricular geometry via 3D printing. In: Coudière, Y., Ozenne, V., Vigmond, E., Zemzemi, N. (eds.) FIMH 2019. LNCS, vol. 11504, pp. 177–186. Springer, Cham (2019). https://doi.org/10.1007/978-3-030-21949-9_20
2. Dejea, H., et al.: Comprehensive analysis of animal models of cardiovascular disease using multiscale X-Ray phase contrast tomography. Sci. Rep. **9**(1), 1–12 (2019)
3. Giannakidis, A., Gullberg, G.T.: Transmural remodeling of cardiac microstructure in aged spontaneously hypertensive rats by diffusion tensor MRI. Front. Physiol. **11**, 1–12 (2020)
4. Hasaballa, A.I., et al.: Microstructurally motivated constitutive modeling of heart failure mechanics. Biophys. J. **117**(12), 2273–2286 (2019)
5. LeGrice, I., Takayama, Y., Covell, J.: Transverse shear along myocardial cleavage planes provides a mechanism for normal systolic wall thickening. Circ. Res. **77**(1), 182–193 (1995)
6. Lunkenheimer, P.P., et al.: Three-dimensional architecture of the left ventricular myocardium. Anat. Rec. Part A: Discoveries Mol. Cellular Evol. Biol.: Official Publ. Am. Assoc. Anatomists **288**(6), 565–578 (2006)
7. Magat, J., et al.: 3D MRI of explanted sheep hearts with submillimeter isotropic spatial resolution: comparison between diffusion tensor and structure tensor imaging. Magn. Reson. Mater. Phys. Biol. Med. 1–15 (2021)
8. Moulin, K., Croisille, P., Viallon, M., Verzhbinsky, I.A., Perotti, L.E., Ennis, D.B.: Myofiber strain in healthy humans using DENSE and cDTI. Magn. Reson. Med. **86**(1), 277–292 (2021)
9. Moulin, K., Verzhbinsky, I.A., Maforo, N.G., Perotti, L.E., Ennis, D.B.: Probing cardiomyocyte mobility with multi-phase cardiac diffusion tensor MRI. PloS One **15**(11), e0241996 (2020)
10. Pope, A.J., Sands, G.B., Smaill, B.H., LeGrice, I.J.: Three-dimensional transmural organization of perimysial collagen in the heart. Am. J. Physiol. Heart Circulatory Physiol. **295**, 1243–1252 (2008)
11. Sallin, E.A.: Fiber orientation and ejection fraction in the human left ventricle. Biophys. J . **9**(7), 954–964 (1969)
12. Shattuck, D.W., Leahy, R.M.: Brainsuite: an automated cortical surface identification tool. Med. Image Anal. **6**, 129–142 (2002)
13. Sosnovik, D.E., Wang, R., Dai, G., Reese, T.G., Wedeen, V.J.: Diffusion MR tractography of the heart. J. Cardiovasc. Magn. Reson. **11**(1), 1–15 (2009)
14. Sosnovik, D.E., et al.: Diffusion spectrum MRI tractography reveals the presence of a complex network of residual myofibers in infarcted myocardium. Circ. Cardiovasc. Imaging **2**(3), 206–212 (2009)
15. Streeter, D.D., Spotnitz, H.M., Patel, D.P., Ross, J., Sonnenblick, E.H.: Fiber orientation in the canine left ventricle during diastole and systole. Circ. Res. **24**(8), 339–348 (1969)

16. Teh, I., et al.: Validation of diffusion tensor MRI measurements of cardiac microstructure with structure tensor synchrotron radiation imaging. J. Cardiovasc. Magn. Reson. **19**, 1–14 (2017)
17. Wilson, A., et al.: Myocardial laminar organization is retained in angiotensin-converting enzyme inhibitor treated SHRs. Exp. Mech. **61**, 31–40 (2021)
18. Wilson, A.J., Sands, G.B., Ennis, D.B.: Analysis of location-dependent cardiomyocyte branching. In: Ennis, D.B., Perotti, L.E., Wang, V.Y. (eds.) FIMH 2021. LNCS, vol. 12738, pp. 189–199. Springer, Cham (2021). https://doi.org/10.1007/978-3-030-78710-3_19
19. Wilson, A.J., Sands, G.B., LeGrice, I.J., Young, A.A., Ennis, D.B.: Myocardial mesostructure and mesofunction. Am. J. Physiol.-Heart Circulatory Physiol. **323**(2), H257–H275 (2022)

On the Possibility of Estimating Myocardial Fiber Architecture from Cardiac Strains

Muhammad Usman[1], Emilio A. Mendiola[1], Tanmay Mukherjee[1], Rana Raza Mehdi[1], Jacques Ohayon[2,3], Prasanna G. Alluri[4], Sakthivel Sadayappan[5], Gaurav Choudhary[6], and Reza Avazmohammadi[1,3,7(✉)]

[1] Department of Biomedical Engineering, Texas A&M University, College Station, TX 77843, USA
rezaavaz@tamu.edu
[2] Savoie Mont-Blanc University, Polytech Annecy-Chambéry, Le Bourget du Lac, France
[3] Department of Cardiovascular Sciences, Houston Methodist Academic Institute, Houston, TX 77030, USA
[4] Department of Radiation Oncology, UT Southwestern Medical Center, Dallas, TX 75390, USA
[5] Department of Internal Medicine, Division of Cardiovascular Health and Disease, University of Cincinnati College of Medicine, Cincinnati, OH 45267, USA
[6] Department of Medicine, Alpert Medical School of Brown University, Providence, RI 02903, USA
[7] J. Mike Walker '66 Department of Mechanical Engineering, Texas A&M University, College Station, TX, USA

Abstract. The myocardium is composed of a complex network of contractile myofibers that are organized in such a way as to produce efficient contraction and relaxation of the heart. The myofiber architecture in the myocardium is a key determinant of cardiac motion and the global or organ-level function of the heart. Reports of architectural remodeling in cardiac diseases, such as pulmonary hypertension and myocardial infarction, potentially contributing to cardiac dysfunction call for the inclusion of an architectural marker for an improved assessment of cardiac function. However, the in-vivo quantification of three-dimensional myo-architecture has proven challenging. In this work, we examine the sensitivity of cardiac strains to varying myofiber orientation using a multiscale finite-element model of the LV. Additionally, we present an inverse modeling approach to predict the myocardium fiber structure from cardiac strains. Our results indicate a strong correlation between fiber orientation and LV kinematics, corroborating that the fiber structure is a principal determinant of LV contractile behavior. Our inverse model was capable of accurately predicting the myocardial fiber range and regional fiber angles from strain measures. A concrete understanding of the link between LV myofiber structure and motion, and the development of non-invasive and feasible means of characterizing the myocardium architec-

O. Bernard et al. (Eds.): FIMH 2023, LNCS 13958, pp. 74–83, 2023.
https://doi.org/10.1007/978-3-031-35302-4_8

ture is expected to lead to advanced LV functional metrics and improved prognostic assessment of structural heart disease.

Keywords: Myocardium architecture · left ventricle · cardiac strains · magnetic resonance imaging

1 Introduction

The intricate architecture of the underlying muscle fibers of the myocardium, often characterized by the three-dimensional orientation of the myofibers, plays a crucial role in the fundamental cardiac motion, regional stress and strain development [7,11], and the pumping function of the left ventricle (LV) [16]. Although the myocardium is designed to provide optimal contractile behavior in a healthy state, the fiber architecture may undergo remodeling events in the diseased heart [5]. Myocardial architectural remodeling has been observed ex-vivo in a variety of structural heart diseases, including pulmonary hypertension [2,10], cardiomyopathy [6], and myocardial infarction [9,14]. Therefore, it is necessary to develop an understanding of the relationship between cardiac structure and function and, ultimately, include the characterization of tissue architecture in the noninvasive assessment of cardiac diseases that involve biomechanical impairment of the myocardium. Moreover, in-vivo quantification of myo-architecture enables the calculation of cardiac fiber strain in lieu of "anatomical" strains, which holds the promise of improving early diagnosis of several cardiac diseases, including cancer treatment-related cardiotoxicity by early identification of motion impairments at the myofiber level.

Recent advances in cardiac imaging modalities have allowed for the characterization of the myocardial fiber structure from whole tissues, without the need for sectioning and histology. The study of the cardiac structure using diffusion-tensor magnetic resonance imaging (DT-MRI) has accelerated the relationship between cardiac architecture and function in health and disease. However, despite recent advances in in-vivo cardiac DT-MRI [12,13], the need for additional sequencing and complex motion compensation processing to eliminate artifacts [15], a low spatial resolution, and poor accessibility have severely limited the application of cardiac DT-MRI in the clinical setting. These limitations are even more severe for small animal studies due to exasperated size, heart rate, and breath-control limits. As such, there is a need to develop methods to characterize the cardiac fiber structure from data that is easily quantified from traditional in vivo imaging techniques.

In this study, we hypothesize that regional motion anisotropy can be used to estimate fiber orientation in each region of the tissue wall and propose to estimate myofiber orientation from cardiac strains for which cine sequences have been already established and standardized in both humans and small animals [8]. We used a finite-element (FE) model of the LV, reconstructed from MRI scans of a murine heart, to investigate the relationship between myocardium structural characteristics and cardiac strain as a forward problem. Next, we present an inverse modeling approach to estimate the transmural myofiber helicity and regional fiber angle using systolic strains as inputs. As an important feature of our model, we use regional and average anatomical stains on short-axis slices from MRI rendering our mode compatible with MRI-based strain quantification. Recent studies have shown a strong correlation between strain and fiber architecture in both forward and inverse problem manners [3,4]. In the present work, we focus on developing a subject-specific structure-strain pipeline that is compatible with standard cardiac MRI acquisitions where strains can be readily calculated from short-axis MRI scans. The results from this investigation will extend the present knowledge of the influence of myocardial structure on LV contractile behavior and indicate the possibility of developing methods to accurately predict myofiber orientation measures from commonly available cardiac strain calculation modalities.

2 Materials and Methods

2.1 Left Ventricle Finite Element Model

The LV geometry used in this study was obtained from MRI scans of a healthy C57Bl/6 male mouse (n = 1). An MRI scan was acquired using a 7T Biospec system (Bruker, Billerica, MA). The imaging protocol was approved by the University of Cincinnati's Animal Care and Use Committee. A FE model of the LV was used in this study to perform forward- and inverse-model investigations of the relationship between ventricular strain and myocardial fiber architecture. Briefly, the construction of the model consisted of three main steps: (i) reconstruction (using end-diastole time point) and meshing of the 3-D ventricular geometry (using linear tetrahedral elements), (ii) assigning myofiber architecture to the FE mesh, and (iii) incorporating the passive and active constitutive laws. Detailed methods regarding model development were described previously [1]. Our development approach resulted in an animal-specific multiscale model of the LV capable of accurately reproducing organ-level hemodynamic and kinematic behavior. Several *in silico* experiments were conducted with the completed LV model to investigate the sensitivity of cardiac strains to myofiber architecture and an attempt was made to predict the fiber architecture from myocardial strains.

To investigate the myocardium fiber architecture effect on cardiac strains, synthetic fiber structures were developed for the LV model and used in forward simulations of the cardiac cycle. The resulting predicted strain data was then plotted against metrics describing the myofiber structure to determine the relationship between strain and fiber architecture. Next, an inverse model was developed to attempt to predict the myofiber architecture from cardiac strains. Finally, the effect of various fiber structures on ventricular torsion, a promising metric of LV function, was performed.

2.2 Development of Synthetic Fiber Architectures

A set of synthetic fiber architectures was created for the LV model. As part of the model development, a simple Laplace-Dirichlet boundary-value problem, using

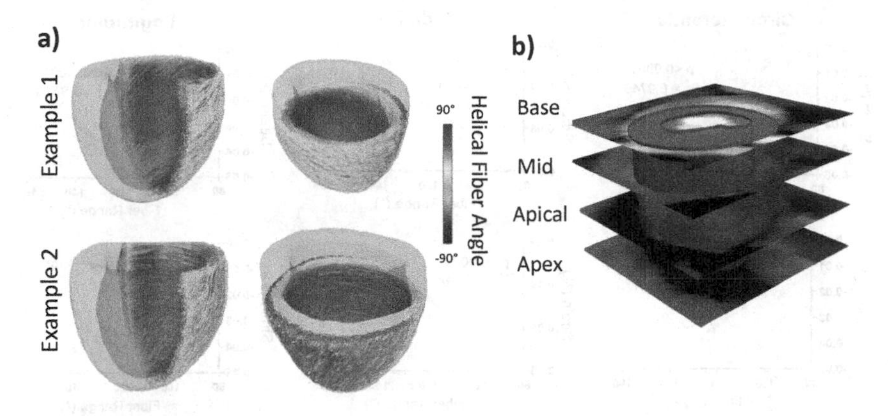

Fig. 1. (a) Examples of synthetic fiber structures with helical fiber angle ranging from $-90°$ to $90°$. (b) Short-axis slices on which strains were calculated.

fixed values at the endo- and epicardial surfaces, was used to determine three layers spanning the transmural direction of the LV wall were determined: endocardium, midwall, and epicardium. An artificial architecture was constructed by defining a linear transmural transition between the chosen fiber angles in the epicardial (θ_{epi}) and endocardial (θ_{endo}) layers (Fig. 1a). The synthetic architectures were developed, ensuring the endocardium fiber angle was between 0° and 82°, and the epicardium fiber angle was between 0° and $-62°$. Fiber range was used in this work as a metric of LV fiber helicity and defined as $(\theta_{endo} - \theta_{epi})$.

2.3 Relationship Between Fiber-Orientation and Strains: Forward Problem

To determine the relationship between cardiac strains and fiber orientation, in-silico experiments were conducted in which a synthetic fiber structure was mapped to the LV model, and a forward simulation of the cardiac cycle was completed. Average strains at the end-systolic (ES) point in the simulated cardiac cycle were calculated on four short-axis slices of the ventricle acquired in the MRI scan (Fig. 1b). The slices were labeled as base, mid, apical, and apex. The Green-Lagrange strain at ES was calculated as:

$$\mathbf{E} = \frac{1}{2}\left(\mathbf{F}^T\,\mathbf{F} - \mathbf{I}\right), \tag{1}$$

where $\mathbf{I}$ is the identity matrix, and $\mathbf{F}$ is the deformation gradient. The strains were transformed from a global coordinate system to the common {circumferential, radial, longitudinal} coordinates using:

$$\mathbf{E}_{C,R,L} = \mathbf{Q}\,\mathbf{E}\,\mathbf{Q}^T, \tag{2}$$

where C, R, L subscripts refer to circumferential, radial, and longitudinal axes, respectively, and $\mathbf{Q}$ is a transformation matrix to transform the strains from

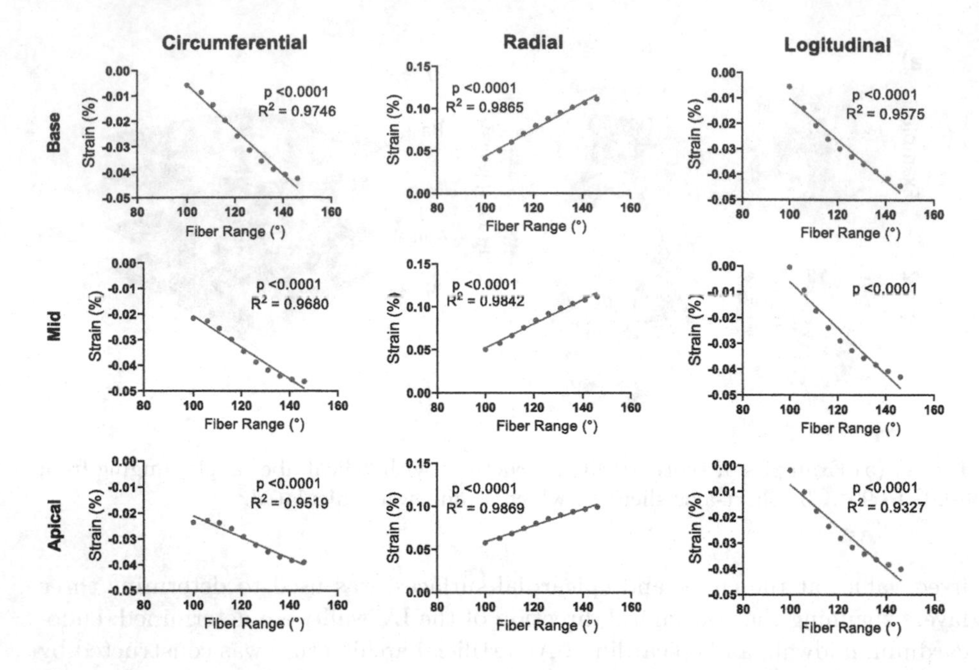

Fig. 2. Correlation analysis between full-thickness average cardiac strains and the transmural fiber range at the base, mid, and apical slices.

global coordinates to anatomical coordinates. We first studied the correlation between full-thickness averages of $\mathbf{E}_{C,R,L}$ across the base, mid, and apical short-axis slices (Fig. 1b) and the fiber range. Next, slices were divided into three transmural layers, namely, epicardial, mid, and endocardial layers, to study the correlation between regional strains and fiber angle at respective layers.

2.4 Relationship Between Fiber-Orientation and Strains: Inverse Problem

An inverse modeling approach was used to estimate the fiber range from simulated cardiac strains. An initial forward simulation with a synthetic fiber structure was conducted. Strains calculated on the top three short-axis slices were then used as target "ground-truth" for the inverse problem. A nonlinear least-squares regression, using Trust Region Reflective algorithm, was used to estimate the fiber orientations in terms of the helical angles θ_{epi} and θ_{endo} assuming a linear transition from θ_{endo} to θ_{epi}. The inverse problem iteratively ran forward simulations to minimize the collective errors between the "ground-truth" strains and predicted strains. The residual error between ground-truth and predicted strains at each iteration was defined by:

$$r_i = E_i^{true} - E_i^{pred}, \quad i = C, R, L, \tag{3}$$

where E^{true} and E^{pred} are ground-truth and predicted strains, respectively. The objective function set to be minimized was defined as the sum of all residual errors:

$$\phi = \sum_{j=1}^{N_S} r_C^2 + \sum_{j=1}^{N_S} r_R^2 + \sum_{j=1}^{N_S} r_L^2, \tag{4}$$

where j represents the number of the slices (N_S), including the base, mid, apical, and apex. Following the correlation results from our forward problem (described in Sects. 3.1 and 3.2), whole or certain terms of the objective function providing minimal errors were used in the inverse problem.

3 Results

3.1 Strains Were Correlated with Fiber Orientation

In this study, the results are presented for the fiber range spanning from 106 to 146°. This fiber range was created with the endocardial fiber angle being fixed at 84° and the epicardial fiber angle changing from −22 and −62°. Strains were calculated and averaged within the top three short-axis slices (base, mid, and apical). The correlation between the average strain and fiber range for three of the slices is presented in Fig. 2. The results indicate that circumferential and longitudinal strains are highly negatively correlated with fiber range. Radial strains exhibited a strong positive correlation with the fiber range. Interestingly, the strain ranges tended to decrease moving from the base to the apex. The radial strains showed the best correlation between fiber range and strain at all three slices (Fig. 2).

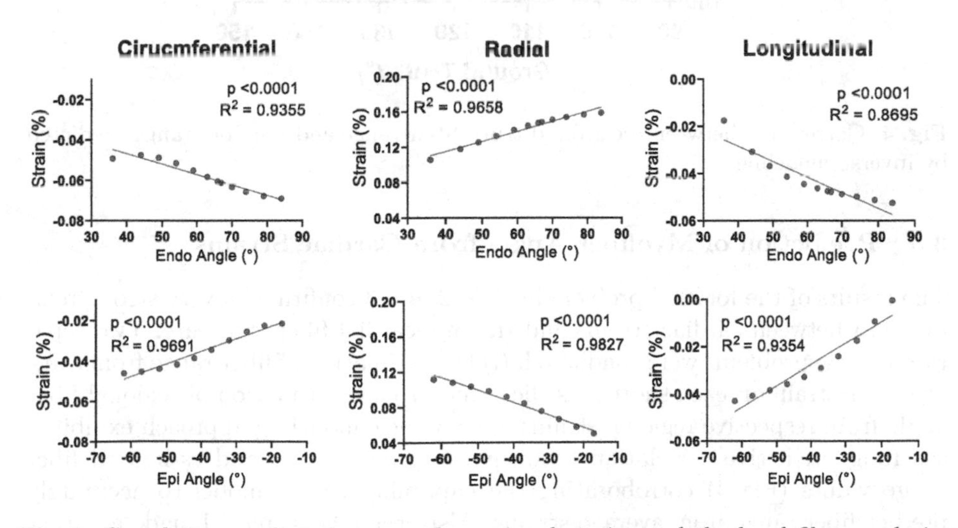

Fig. 3. Correlation analysis between the regional strains and the local fiber angle in the respective transmural layer. This analysis was performed with strains calculated at the epicardial and endocardial layers of the Mid short-axis slice.

3.2 Fiber Architecture Influences Regional Strain

A strong correlation was observed between fiber angle and regional strain at all three slices. A representative example of the correlations is displayed for the mid slice (Fig. 3). Correlation trends similar to those in Fig. 2 were observed between regional fiber angle and strains in the endo- and epicardium layers: absolute circumferential and longitudinal strains in each layer tended to increase as the fiber angle moves away from the circumferential direction (Fig. 3). Similar correlation results were also noted in the other short-axis slices. We note that our small strain magnitudes (compared to commonly reported ranges of myocardial strains) are due to a lower systolic pressure used in the LV to facilitate forward and inverse model simulations.

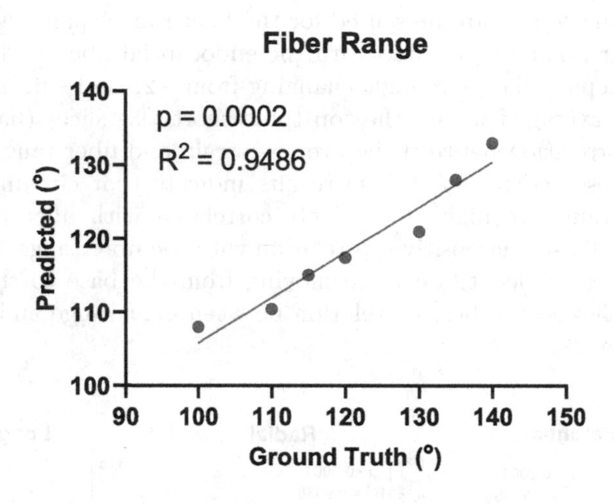

Fig. 4. Correlation between the ground-truth fiber range and the fiber range predicted by inverse modeling.

3.3 Prediction of Myofiber Angle from Cardiac Strains

The results of the forward problem in Figs. 2 and 3 confirmed a very strong relationship between cardiac strains and the myocardial fiber structure. Two separate inverse problems were conducted: (i) the estimation of fiber range from average fiber strain on each short-axis slice, and (ii) the estimation of regional fiber angle from respective regional strains. Our inverse modeling approach exhibited a strong predictive correlation between the ground-truth and estimated fiber range values (Fig. 4) corroborating the capability of the model to accurately predict fiber range from average strains. Also, the inverse model made an accurate estimation of the endo- and epicardium regional fiber angle (Fig. 5), with no significant difference between the mean ground truth and predicted fiber angles.

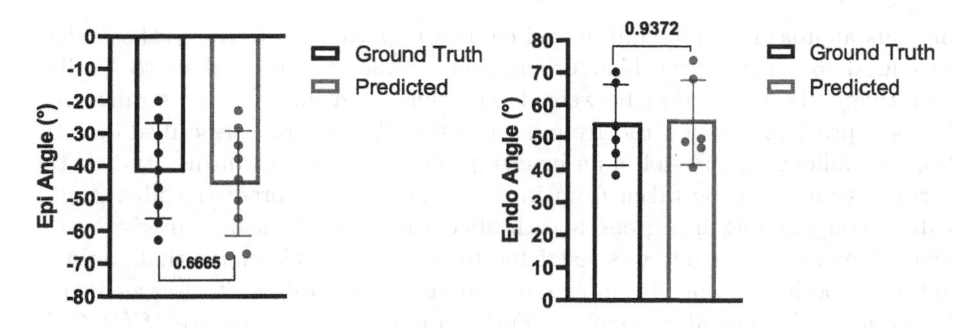

Fig. 5. Inverse modeling resulted in an accurate prediction of endocardium and epicardium fiber angle from cardiac strains.

4 Discussion and Conclusions

Myocardial fiber architecture is known to have a principal influence on LV kinematics and, thus, organ-level cardiac function. The myocardium structure is intricately designed to produce efficient contractile behavior; however, the fiber architecture often undergoes remodeling events as a result of cardiac disease. Metrics currently used in the prognosis and treatment of structural heart diseases (such as ejection fraction and relaxation time) offer global information on the state of LV function but may fail to promptly risk-stratify heart disease patients due to their limited sensitivity to tissue-level adaptations. As architectural remodeling may offer indications of the trajectory of the disease, they offer significant added value to traditional metrics such as LV ejection fraction. However, the in-vivo and high-fidelity quantification of myofiber architecture has proven challenging, prompting alternative approaches to quantify myofiber architecture.

In the present study, we investigated the relationship between cardiac strains and myofiber structure and showed architectural metrics can be accurately predicted from kinematic data using FE modeling of a murine heart. Correlation analysis between fiber orientation and cardiac strains indicated a strong association at various short-axis slices of the ventricle. Similarly, regional strains and fiber angles in distinct transmural regions of the LV were found to be highly correlated. Both these results indicate the myofiber architecture of the ventricle exerts a heavy influence on the contractile pattern of the healthy LV. Thus, it is important to develop methods to characterize the myofiber architecture of the LV non-invasively. The inverse problem approach we used to estimate architectural metrics were able to accurately predict both fiber range and regional fiber angles from cardiac strains. This promising result suggests such a method could be developed to make use of strains measured on short-axis slices of the LV in commonly used in-vivo imaging techniques, such as cardiac MRI and echocardiography.

While this study presents a promising approach to predict myocardial fiber orientation from cardiac strains, further developments are needed for translating this approach for pre-clinical and clinical applications. Here, synthetic data was used to generate the fiber orientation. While this allowed us to feasibly investigate the correlation between fiber orientation and cardiac strains in a forward problem manner, the ground truth was chosen from prescribed orientations to validate the model in an inverse problem setting. In future studies, the cardiac strains will be taken from in-vivo imaging, and predicted fiber angles will be compared against ground-truth fiber orientation measured in-vivo or ex-vivo. Moreover, our study was restricted to one healthy LV model that could be extended both in terms of sample size and the structural heart diseases known to induce architectural remodeling. Furthermore, the performance of (R, C, L) strains compared to shear strains with respect to predicting fiber orientation remains to be investigated.

Overall, we have shown here that the structure of the myocardium is a principal determinant of LV contractile motion, and a rigorous inverse model can be established between the structure and motion. We presented a version of such an inverse model that is compatible with common anatomical strain calculations from standard cine cardiac MRI acquisitions. Further investigation of this approach may result in the establishment of a feasible technique to noninvasively estimate myofiber orientation and helicity using routinely available cardiac imaging in both small animals and humans. Ultimately, such structurally-informed functional metrics could provide detailed supplemental information to the traditional gross metrics of LV function and improve prognostic and therapeutic assessments. Finally, as LV torsion has recently been introduced as a promising kinematic metric of LV functional capacity, we suggest that further studies include the association between fiber architecture and torsion. Ventricular torsion, also readily quantified from standard imaging studies, may provide another avenue to link the myocardium structure with LV kinematics.

Acknowledgements. This research was supported by the NIH Grant No. R00HL138288 to R.A..

References

1. Avazmohammadi, R., et al.: A computational cardiac model for the adaptation to pulmonary arterial hypertension in the rat. Ann. Biomed. Eng. **47**(1), 138–153 (2018). https://doi.org/10.1007/s10439-018-02130-y
2. Avazmohammadi, R., Hill, M., Simon, M., Sacks, M.: Transmural remodeling of right ventricular myocardium in response to pulmonary arterial hypertension. APL Bioengineering **1**(1), 016105 (2017)
3. Barbarotta, L., Bovendeerd, P.H.M.: A computational approach on sensitivity of left ventricular wall strains to fiber orientation. In: Ennis, D.B., Perotti, L.E., Wang, V.Y. (eds.) FIMH 2021. LNCS, vol. 12738, pp. 296–304. Springer, Cham (2021). https://doi.org/10.1007/978-3-030-78710-3_29
4. Barbarotta, L., Bovendeerd, P.H.M.: Parameter estimation in a rule-based fiber orientation model from end systolic strains using the reduced order unscented Kalman filter. In: Ennis, D.B., Perotti, L.E., Wang, V.Y. (eds.) FIMH 2021. LNCS,

vol. 12738, pp. 340–350. Springer, Cham (2021). https://doi.org/10.1007/978-3-030-78710-3_33
5. Buckberg, G., Hoffman, J.I., Mahajan, A., Saleh, S., Coghlan, C.: Cardiac mechanics revisited: the relationship of cardiac architecture to ventricular function. Circulation **118**(24), 2571–2587 (2008)
6. Ferreira, P.F., et al.: In vivo cardiovascular magnetic resonance diffusion tensor imaging shows evidence of abnormal myocardial laminar orientations and mobility in hypertrophic cardiomyopathy. J. Cardiovasc. Magn. Reson. **16**(1), 1–16 (2014)
7. Geerts, L., Kerckhoffs, R., Bovendeerd, P., Arts, T.: Towards patient specific models of cardiac mechanics: a sensitivity study. In: Magnin, I.E., Montagnat, J., Clarysse, P., Nenonen, J., Katila, T. (eds.) FIMH 2003. LNCS, vol. 2674, pp. 81–90. Springer, Heidelberg (2003). https://doi.org/10.1007/3-540-44883-7_9
8. Keshavarzian, M., et al.: An image registration framework to estimate 3D myocardial strains from cine cardiac MRI in mice. In: Ennis, D.B., Perotti, L.E., Wang, V.Y. (eds.) FIMH 2021. LNCS, vol. 12738, pp. 273–284. Springer, Cham (2021). https://doi.org/10.1007/978-3-030-78710-3_27
9. Mendiola, E., et al.: Contractile adaptation of the left ventricle post-myocardial infarction: predictions by rodent-specific computational modeling. Ann. Biomed. Eng. **16**(2), 721–729 (2022)
10. Mendiola, E.A., et al.: Right ventricular architectural remodeling and functional adaptation in pulmonary hypertension. Circulation: Heart Failure **16**(2), e009768 (2023)
11. Pluijmert, M., Delhaas, T., de la Parra, A.F., Kroon, W., Prinzen, F.W., Bovendeerd, P.H.: Determinants of biventricular cardiac function: a mathematical model study on geometry and myofiber orientation. Biomech. Model. Mechanobiol. **16**(2), 721–729 (2017)
12. Scott, A.D., et al.: An in-vivo comparison of stimulated-echo and motion compensated spin-echo sequences for 3 T diffusion tensor cardiovascular magnetic resonance at multiple cardiac phases. J. Cardiovasc. Magn. Reson. **20**(1), 1–15 (2018)
13. Stoeck, C.T., Von Deuster, C., Genet, M., Atkinson, D., Kozerke, S.: Second-order motion-compensated spin echo diffusion tensor imaging of the human heart. Magn. Reson. Med. **75**(4), 1669–1676 (2016)
14. Walker, J.C., et al.: Helical myofiber orientation after myocardial infarction and left ventricular surgical restoration in sheep. J. Thorac. Cardiovasc. Surg. **129**(2), 382–390 (2005)
15. Welsh, C.L., DiBella, E.V., Hsu, E.W.: Higher-order motion-compensation for in vivo cardiac diffusion tensor imaging in rats. IEEE Trans. Med. Imaging **34**(9), 1843–1853 (2015)
16. Zhang, X., Haynes, P., Campbell, K.S., Wenk, J.F.: Numerical evaluation of myofiber orientation and transmural contractile strength on left ventricular function. J. Biomech. Eng. **137**(4), 044502 (2015)

Cardiac Electrophysiology Modeling

The Fibrotic Kernel Signature: Simulation-Free Prediction of Atrial Fibrillation

Francisco Sahli Costabal[1(✉)], Tomás Banduc[1], Lia Gander[2], and Simone Pezzuto[3]

[1] Department of Mechanical and Metallurgical Engineering, School of Engineering and Institute for Biological and Medical Engineering, Schools of Engineering, Medicine and Biological Sciences, Pontificia Universidad Católica de Chile, Santiago, Chile
fsc@ing.puc.cl

[2] Center for Computational Medicine in Cardiology, Euler Institute, Università della Svizzera italiana, Lugano, Switzerland
lia.gander@usi.ch

[3] Laboratory of Mathematics for Biology and Medicine, Department of Mathematics, Università di Trento, Via Sommarive 14, 38123 Povo, Italy
simone.pezzuto@unitn.it

Abstract. We propose a fast classifier that is able to predict atrial fibrillation inducibility in patient-specific cardiac models. Our classifier is general and it does not require re-training for new anatomies, fibrosis patterns, and ablation lines. This is achieved by training the classifier on a variant of the Heat Kernel Signature (HKS). Here, we introduce the "fibrotic kernel signature" (FKS), which extends the HKS by incorporating fibrosis information. The FKS is fast to compute, when compared to standard cardiac models like the monodomain equation. We tested the classifier on 9 combinations of ablation lines and fibrosis patterns. We achieved maximum balanced accuracies with the classifiers ranging from 75.8% to 95.8%, when tested on single points. The classifier is also able to predict very well the overall inducibility of the model. We think that our classifier can speed up the calculation of inducibility maps in a way that is crucial to create better personalized ablation treatments within the time constraints of the clinical setting.

Keywords: Heat kernel signature · Fibrotic Kernel Signature · Atrial Fibrillation · Fibrosis · Patient-Specific Modeling

1 Introduction

Fibrosis is one of the main drivers of Atrial Fibrillation (AF), the most common cardiac arrhythmia [16]. Fibrosis significantly increases tissue heterogeneity and anisotropy in conduction, which in turn enhance AF inducibility and complexity.

O. Bernard et al. (Eds.): FIMH 2023, LNCS 13958, pp. 87–96, 2023.
https://doi.org/10.1007/978-3-031-35302-4_9

Its distribution in the atria is patient-specific and progresses with AF in a vicious loop: the more fibrosis is present in the tissue, the more AF events are likely to occur, which trigger more fibrosis deposition [13].

Ideally, therapeutic approaches to AF as catheter ablation should return the best outcome when tailored to patient fibrosis distribution, at least as claimed in recent retrospective and prospective studies [2,8,9]. These studies show that, thanks to patient-specific *in silico* models, it is possible to estimate AF inducibility for various ablation scenarios, and then select the best treatment for the patient. However, testing AF inducibility is costly, both in terms of time and required computational resources.

The assessment of AF inducibility can be understood as a classification problem. Given a patient-specific anatomy, fibrosis pattern, and ablation lines, the objective is determining whether a pacing protocol may lead to a stable AF event or not [4]. Testing a fixed number of well-distributed, pacing location is a standard protocol to estimate inducibility [2]. However, the protocol needs to be repeated when ablation lines are added or a new anatomy is to be tested. Since the total computational cost can be very high, some authors proposed an adaptive pacing protocol [1] or surrogate models of AF [14].

In this work, we propose a classifier for AF inducibility that does not require retraining when ablation lines and fibrosis change. The classifier is based on the Heat Kernel Signature (HKS) [15], a time series that effectively encodes local geometrical and topological information of a domain. Along with its variants, HKS is popular in shape analysis. Mathematically, the HKS is based on the heat (or diffusion) equation. It may be interpreted as the concentration time course of a ink drop as it diffuses throughout the domain. Here, we extend the HKS by incorporating the fibrosis pattern into the diffusion operator. In this way, once trained, the classifier only requires the HKS, which is cheap to compute compared to a standard monodomain simulation.

The manuscript is structured as follows: we review the AF modeling framework in Sect. 2.1, which has been used to generate the dataset. In Sect. 2.2 we present the fibrotic kernel signature, and apply it to the definition of the classifier, in Sect. 2.3. We conclude with results (Sect. 3) and discussion (Sect. 4).

2 Methods

2.1 Cardiac Atrial Modeling

The monodomain equation is the most common model in simulating atrial fibrillation (AF). It reads as follows:

$$\chi\Big(C_{\mathrm{m}}\frac{\partial V}{\partial t} + I_{\mathrm{ion}}(V,\boldsymbol{z}) - I_{\mathrm{stim}}(\boldsymbol{x},t)\Big) = \operatorname{div}(\boldsymbol{G}\nabla V), \tag{1}$$

where $V(\boldsymbol{x},t)$ is the transmembrane voltage as a function of the spatial position $x \in \Omega$, Ω being the active tissue of the atrial domain, and the time $t \geq 0$. The other parameters are: χ, the surface-to-volume ratio; C_{m}, the membrane capacitance; $I_{\mathrm{stim}}(\boldsymbol{x},t)$ is the stimulation current; and $\boldsymbol{G}$, the monodomain electric conductivity. The nonlinear term $f(V, \boldsymbol{z})$ encompasses all ion currents flowing through the cellular membrane, which are numerous for physiological models. Here, we consider the Courtemanche-Ramirez-Nattel model [3]. Fibrosis is modeled by reducing the intra-cellular conductivity in the cross-fiber direction, which in turn affects the tensor $\boldsymbol{G}$. Specifically, we encode the presence of fibrosis by reducing the cross-fiber conductivity in $\boldsymbol{G}$. See [5] for the parameter values and for the numerical method to solve the monodomain Eq. (1).

AF is a self-sustained, chaotic activation of the atria. There are several ways for triggering it in the atrial model. A clinically feasible approach consists in a train of stimuli delivered at some specific location, with a decreasing interval between each stimulus. We define the *inducibility function* $\mathcal{I} : \Omega \to \{0, 1\}$ as follows: $\mathcal{I}(\boldsymbol{y}) = 1$ if the stimulation protocol delivered at $\boldsymbol{y}$ successfully induced AF, and zero otherwise. We check whether AF is induced or not in the model by checking whether the integral over Ω and a window of time of the currents (diffusion, stimulus, and ionic) is non-zero.

We consider 9 different models of atrial fibrillation on a fixed geometry with: 3 different fibrotic patterns (moderate - 50%, severe case 1–70% and severe case 2–70%), and 3 ablations (no ablation, PVI, and PVI + BOX ablation). For each case, we run 100 pacing locations [4]. In total, the dataset has 900 simulations of inducibility that we will use to assess the accuracy of the proposed methodology.

2.2 Fibrotic Kernel Signature on the Atria

Simulating AF at human scale is computationally expensive. Therefore, we are interested in *learning* the classifier $\mathcal{I}$ from a sparse set of simulations. In this work, we use the heat kernel signature (HKS) [15], which is a technique to characterize points in geometries, and can be used for segmentation and shape matching among other applications. This point descriptor is based on the heat diffusion process on a given shape that captures concisely the intrinsic information from a geometry, up to isometry in an efficient, stable and multi-scaled way. The HKS is also invariant to rotations and translations.

The HKS can be computed efficiently using the heat kernel $k_t(\boldsymbol{x}, \boldsymbol{y})$, which represents evolution of the temperature over time t at point $\boldsymbol{y}$ when the initial temperature is a Dirac delta $\delta_{\boldsymbol{x}}(\boldsymbol{y})$ applied at point $\boldsymbol{x}$. Then, the HKS is a vector defined as $\text{hks}(\boldsymbol{x})_i = k_{t_i}(\boldsymbol{x}, \boldsymbol{x})$, for a finite number of time steps $t_i > 0$. These time steps are computed in a logarithmic progression and the signature is later normalized. Intuitively, the HKS represents how the temperature evolves over time in a point after applying an impulse at $t = 0$ at that location.

Instead of solving the diffusion equation, the HKS can be effectively calculated as a sum involving the eigenfunctions $\phi_i(\boldsymbol{x})$ and eigenvalues λ_i of the Laplace operator on the shape on interest:

$$k_t(\boldsymbol{x}, \boldsymbol{y}) = \sum_{i=0}^{+\infty} e^{-\lambda_i t} \phi_i(\boldsymbol{x}) \phi_i(\boldsymbol{y}) \tag{2}$$

Here, we introduce the *fibrotic kernel signature* (FKS) by incorporating information regarding the fibrotic pattern in the signature. Inspired by the diffusion operator of the monodomain Eq. (1), we first consider the elliptic operator

$\mathcal{L}u := -\operatorname{div}(\sigma_f(\boldsymbol{x})\nabla u)$ with homogeneous Neumann boundary condition, for $\sigma_f \in L^\infty(\Omega)$ and $u \in H^1(\Omega)$, that similarly to the Laplace operator has a countable spectrum of eigenvalues and eigenfunctions [12]. We remark that now the spectrum depends on the fibrosis pattern. Then, we use Eq. (2) to compute the FKS. The function $\sigma_f(\boldsymbol{x})$ is 1 where there is healthy tissue and 0.5 where there is fibrotic tissue.

To compute the FKS, we use the first 100 eigenvalues, ordered by magnitude. Since the operator $\mathcal{L}$ is symmetric, eigenvalues are real. It is also possible to show that the first eigenvalue is zero, and the others are positive. To solve the eigenproblem we use the Finite Element Method with linear Lagrange elements on a hexahedral mesh with ≈700 000 nodes. We implemented the solver in `DOLFINx` interface and `SLEPc` library on Python. We set the `SLEPc` solver for a generalized non-Hermitian case. We use the same normalization of the signature per fibrotic pattern. We compute the normalization and time steps for the base case without ablation and also apply it to the cases with the same fibrotic pattern with ablation. We represent the ablation by setting $\sigma_f(\boldsymbol{x}) = 0.001$ in the ablation lines, effectively creating a barrier for the heat.

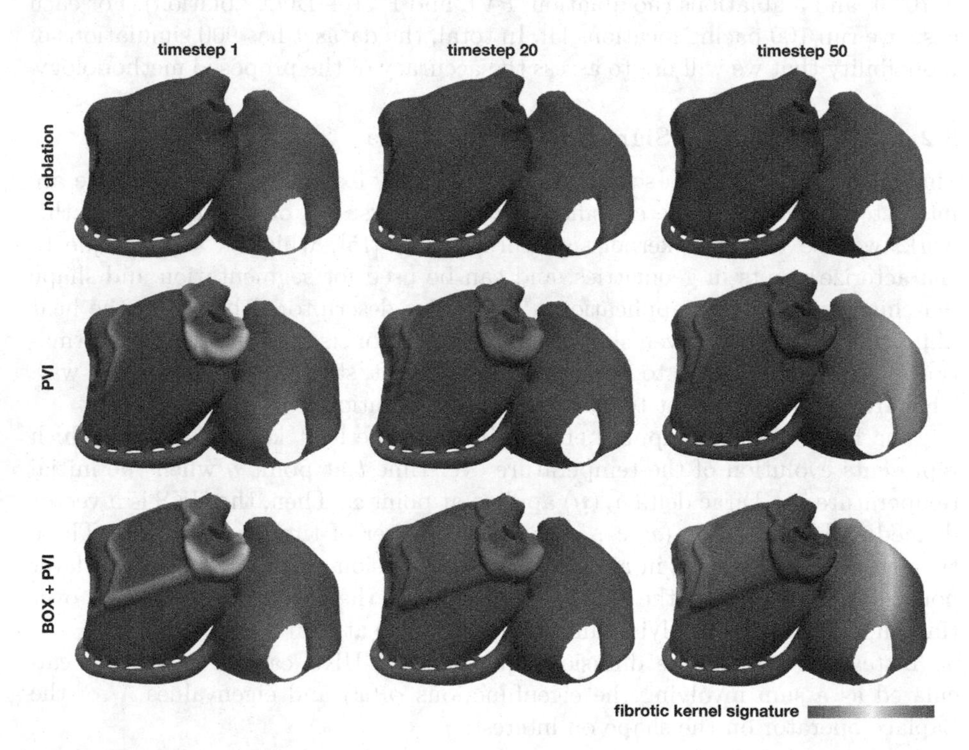

Fig. 1. Fibrotic kernel signature for moderate fibrosis case and three ablation patterns: no ablation, pulmonary vein isolation and PVI + BOX ablation. The no ablation case is taken as reference model for time-scaling and signature normalisation.

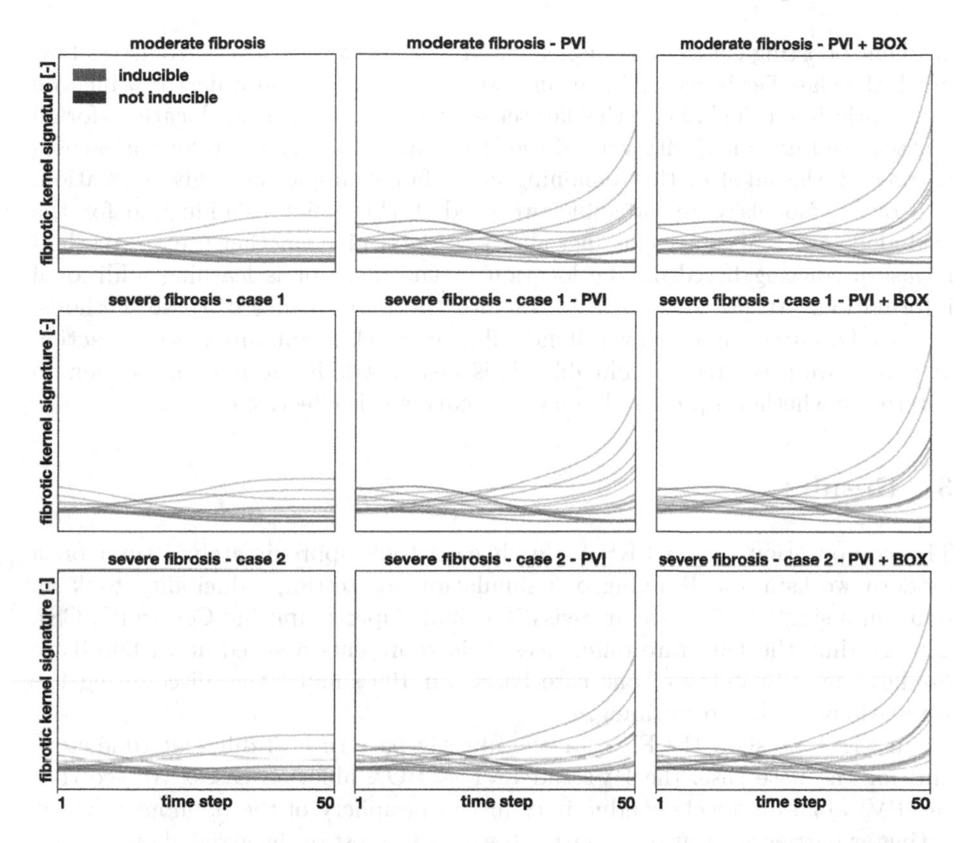

Fig. 2. Fibrotic kernel signature over time for 20 randomly selected points for 9 different cases. The red curves represent points where AF was induced and the blue curves are points where AF was not inducible. The non-ablation case is taken as reference model for time-scaling and signature normalisation. The abscissa represents the time vector, that is the input of the classifier. (Color figure online)

2.3 Prediction of Atrial Fibrillation

Once we have computed the FKS for all the cases, we can use our dataset of 900 simulations and associate the signature at those locations to its inducibility. Then, we will train machine learning classifiers to learn to distinguish between signatures associated with inducible cases and signatures where AF was not induced. Concretely, we try to approximate the inducibility function $\mathcal{I}(\boldsymbol{y}) \approx f\big(\text{fks}(\boldsymbol{y})\big)$ depending on the fibrotic kernel signature at a given location $\boldsymbol{y}$. Here $f(\cdot)$ represents some machine learning classifier. Once the classifier has been trained, we can predict the inducibility function without running simulations on a particular case. To assess the performance of this method, we perform leave-one-out cross-validation by case. We train with the simulations of 8 out of 9 cases, totalling up to 800 data points and test with 100 simulations of the unseen case. We consider three classifiers: the k-neighbours classifier, random forest, and gradient boosting classifier as implemented in `scikit-learn` [10]. To assess how much the classifier is actually learning, we propose a naïve alternative, which we call "majority voting". Since the dataset is evaluated at the same locations for all 9 cases, we take a majority vote of the labels at a given location for the 8 cases to predict the label of the remaining case. For example, at a given location, if 5 out 8 models were inducible, we predict this point as inducible for the excluded case. This naïve classifier allows to determine whether the our method is just predicting based on the location of the point or is learning additional information. We use 2 metrics to evaluate the performance of these methods: balanced accuracy and the overall inducibility, which is computed as the fraction of points are predicted as inducible. This last metric is the most important to determine whether a proposed ablation treatment is effective or not.

3 Results

The computation of the FKS for each case took approximately 10 min on a modern workstation. Running one simulation for testing inducibility took an hour on a single GPU node at Swiss National Supercomputing Centre (CSCS). Disregarding the hardware differences, this represents a speed-up of 600 if we predict the inducibility of one case based on 100 simulations, discounting the simulations needed for training.

In Fig. 1 we show the FKS for one fibrotic pattern at 3 different time steps and for the base case, the PVI and PVI + BOX ablation cases. We see that the PVI ablation affects conductivity in the periphery of the pulmonary veins, acting as barrier for heat propagation towards the rest of the atria, allowing heat accumulation on regions where base case presents lower temperatures. When adding the BOX ablation extends this accumulation of temperature in the roof of the left atrium.

Different examples of the FKS over time are shown in Fig. 2 for the 20 randomly selected locations which correspond across the 9 cases considered. We note that locations that are not inducible tend to have a higher signature than the inducible points. We also observe that the application of the ablation patterns tend to modify the signatures, especially towards the end of time.

The results regarding the prediction of AF are summarized in Figs. 3 and 4. When we analyze the balanced accuracy of our method for different levels of training data in Fig. 3, we see that in general random forest tends to perform better in all cases. We achieve maximum balanced accuracies with the classifiers ranging from 75.8% for the severe fibrosis - case 1, to 95.8% in the case moderate fibrosis + PVI + BOX. The majority voting classifier tends to show a similar performance than the classifiers based on the FKS. However, for some cases, the FKS classifiers have higher or lower accuracy than the majority voting classifier, indicating that the FKS classifiers are not simply memorizing the location of the training points and matching it the test cases.

Regarding the predictions of overall inducibility, we see bigger differences in Fig. 4. First, we note that the ablation lines applied to the models have a marked

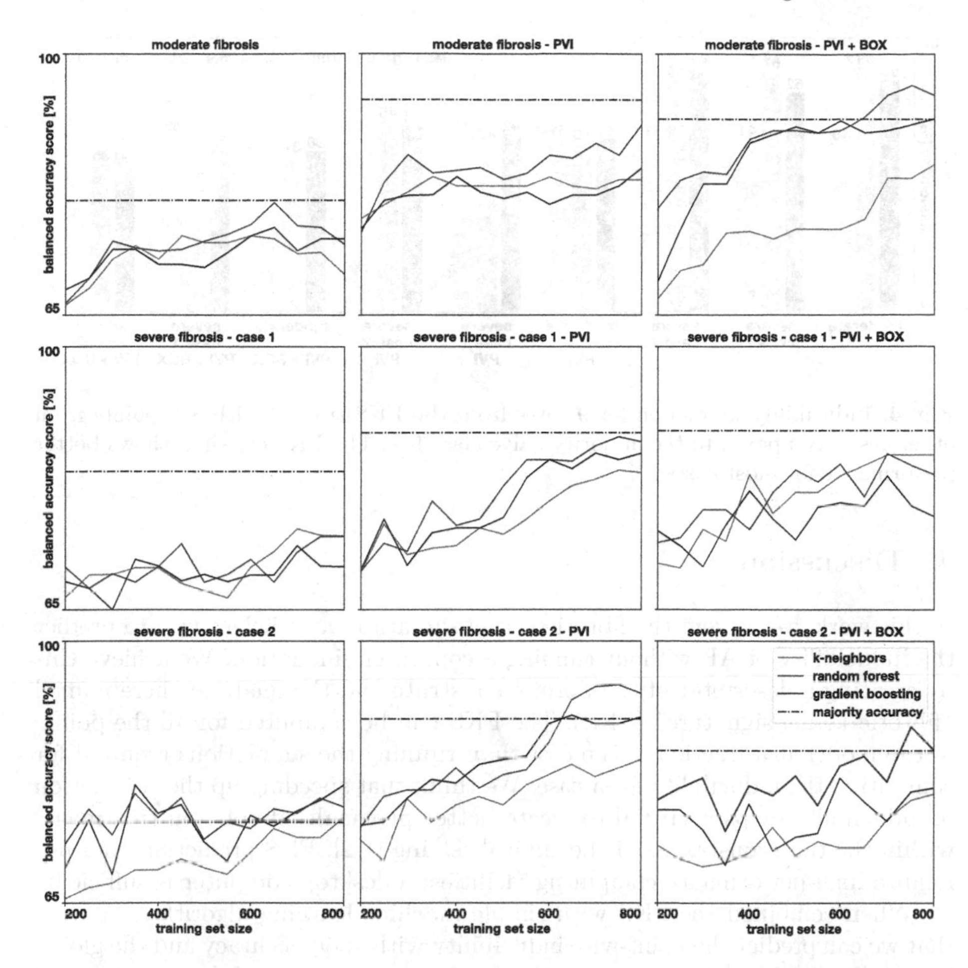

Fig. 3. Balanced accuracies for 9 different cases. We show in the solid lines the performance of 3 different machine learning classifiers as the training data increases. The dashed line represents a baseline naive classifier based on the inducibility of the other 8 cases.

effect on the inducibility. Applying PVI reduces the inducibility between 11 to 14% points and applying PVI + BOX reduces the inducibility between 16 to 19% points. The majority voting classifier, as expected, tends to predict a similar level of inducibility independent of the treatment applied to the case. For all cases, it predicts between 40 and 43% inducibility. The random forest classifier based on the FKS trained with 800 points can detect the changes in inducibility much better. The errors range between 1 to 6% points inducibility. Also, the trend that the base case is more inducible than PVI and even more inducible than PVI + BOX is also correctly predicted by the FKS-based classifier.

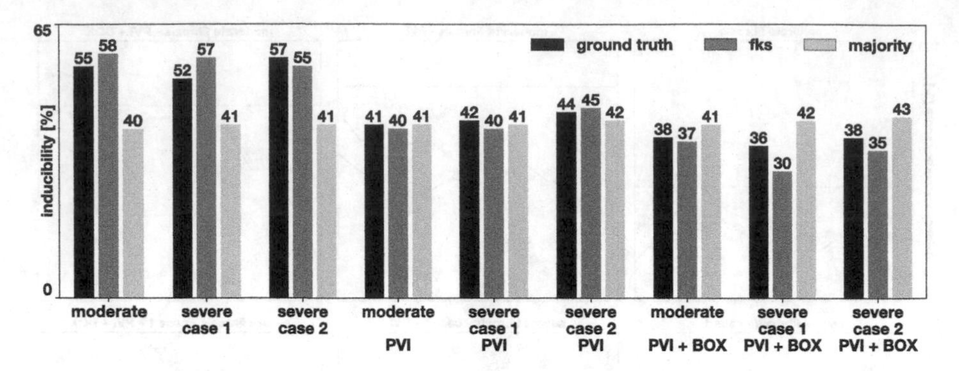

Fig. 4. Inducibility prediction for 9 cases from the FKS trained with 800 points from other cases, compared to the majority naive classifier. The FKS classifier shows better performance for most cases.

4 Discussion

In this work, we present the fibrotic kernel signature, an efficient way to predict the inducibility of AF without running a computer simulation. We achieve this by creating a descriptor of the fibrotic substrate and the anatomy, here named "fibrotic kernel signature" (FKS). The FKS can be computed for all the points in the model at a fraction of the cost than running the simulations required for computing the inducibility for a case. We think that speeding up the calculation of inducibility maps is crucial to create better personalized ablation treatments within the time constraints of the clinical setting [1,2]. FKS prediction does not require high-performance computing facilities: a desktop computer is sufficient.

When combined the FKS with simple machine learning algorithms, we see that we can predict the point-wise inducibility with good accuracy and the global inducibility with excellent accuracy. In Fig. 3 we observe that the accuracies tend to increase as more data is available. We expect that the performance would be improved as we train the classifier with more cases. Our method allows to take advantage of all the simulations that we could run for different patients in order the improve the predictions with the FKS. We remark that no mapping between different anatomies is required when evaluating the classifier since the geometrical information flows into the FKS, on which the classifier depends. Importantly, the FKS is easily extendable thus to include the local fiber direction, just by redefining the elliptic operator. For instance, the fiber direction and fibrosis, combined in the conductivity tensor, can be estimated from electroanatomical mapping system data, as recently proposed [6,7,11].

Our study has some limitations. First, we only tested our method in a single geometry. Although the different patterns of fibrosis and ablation effectively change the geometry for the monodomain equation and the FKS, we still need to verify whether our method generalizes to other patient anatomies. Another limitation is that the FKS does not consider the fiber orientation and the cell type distribution. We conducted preliminary studies including the fibers, but there was no improvement in the results. Finally, we use a mesh resolution of 0.4 mm, which we know affects the inducibility [4], when compared to finer meshes. This will affect to time to solve the eigenproblem required by the FKS, but given that we are currently using only modest hardware, we could manage the larger models. On the other hand, our current mesh resolution is in line or already finer than those from other studies [1,2]. We also did not account for uncertainty in the fibrosis pattern, which is known to be highly affected by the threshold strategy. Ideally, uncertainty could be introduced into our FKS definition to compute a mean FKS with associated covariance; this information then could be used in the classification problem.

In summary, we propose a novel method to predict atrial fibrillation without running simulations. We believe that the fibrotic kernel signature combined with machine learning techniques will enable faster and better planning of ablation treatments in atrial fibrillation patients.

Acknowledgments. FSC and TB acknowledge the support of the project FONDECYT-Iniciación 11220816. FSC also acknowledges the support of the project ERAPERMED-134 from ANID. This work was also financially supported by the Theo Rossi di Montelera Foundation, the Metis Foundation Sergio Mantegazza, the Fidinam Foundation, and the Horten Foundation to the Center for CCMC. SP also acknowledges the CSCS-Swiss National Supercomputing Centre (No. s1074).

References

1. Azzolin, L., Schuler, S., Dössel, O., Loewe, A.: A reproducible protocol to assess arrhythmia vulnerability in silico: pacing at the end of the effective refractory period. Front. Physiol. **12**, 420 (2021)
2. Boyle, P.M., et al.: Computationally guided personalized targeted ablation of persistent atrial fibrillation. Nat. Biomed. Eng. **3**(11), 870–879 (2019)
3. Courtemanche, M., Ramirez, R.J., Nattel, S.: Ionic mechanisms underlying human atrial action potential properties: insights from a mathematical model. Am. J. Physiol.-Heart Circulatory Physiol. **275**, H301–H321 (1998)
4. Gander, L., Pezzuto, S., Gharaviri, A., Krause, R., Perdikaris, P., Sahli Costabal, F.: Fast characterization of inducible regions of atrial fibrillation models with multi-fidelity Gaussian process classification. Front. Physiol. 260 (2022)
5. Gharaviri, A., et al.: Epicardial fibrosis explains increased endo-epicardial dissociation and epicardial breakthroughs in human atrial fibrillation. Front. Physiol. **11**(68) (2020)
6. Grandits, T., et al.: Learning atrial fiber orientations and conductivity tensors from intracardiac maps using physics-informed neural networks. In: Ennis, D.B., Perotti, L.E., Wang, V.Y. (eds.) FIMH 2021. LNCS, vol. 12738, pp. 650–658. Springer, Cham (2021). https://doi.org/10.1007/978-3-030-78710-3_62
7. Kotadia, I., et al.: Anisotropic cardiac conduction. Arrhythmia Electrophysiol. Rev. **9**(4), 202 (2020)
8. Loewe, A., et al.: Patient-specific identification of atrial flutter vulnerability-a computational approach to reveal latent reentry pathways. Front. Physiol. **9**, 1910 (2019)

9. McDowell, K.S., Zahid, S., Vadakkumpadan, F., Blauer, J., MacLeod, R.S., Trayanova, N.: Virtual electrophysiological study of atrial fibrillation in fibrotic remodeling. Plos One **10**(2), e0117110 (2015). https://doi.org/10.1371/journal.pone.0117110
10. Pedregosa, F., et al.: Scikit-learn: machine learning in Python. J. Mach. Learn. Res. **12**, 2825–2830 (2011)
11. Ruiz Herrera, C., Grandits, T., Plank, G., Perdikaris, P., Sahli Costabal, F., Pezzuto, S.: Physics-informed neural networks to learn cardiac fiber orientation from multiple electroanatomical maps. Eng. Comput. **38**(5), 3957–3973 (2022)
12. Salsa, S.: Partial Differential Equations in Action: From Modelling to Theory, vol. 99. Springer, Heidelberg (2016)
13. Schotten, U., Verheule, S., Kirchhof, P., Goette, A.: Pathophysiological mechanisms of atrial fibrillation: a translational appraisal. Physiol. Rev. **91**(1), 265–325 (2011). https://doi.org/10.1152/physrev.00031.2009
14. Serra, D., et al.: An automata-based cardiac electrophysiology simulator to assess arrhythmia inducibility. Mathematics **10**(8), 1293 (2022)
15. Sun, J., Ovsjanikov, M., Guibas, L.: A concise and provably informative multi-scale signature based on heat diffusion. Comput. Graph. Forum **28**(5), 1383–1392 (2009)
16. Tsao, C.W., et al.: Heart disease and stroke statistics-2022 update: a report from the American heart association. Circulation **145**(8), e153–e639 (2022)

Isogeometric-Mechanics-Driven Electrophysiology Simulations of Ventricular Tachycardia

R. Willems[1], E. Kruithof[1](✉), K. L. P. M. Janssens[1], M. J. M. Cluitmans[2], O. van der Sluis[1,2], P. H. M. Bovendeerd[1], and C. V. Verhoosel[1]

[1] Eindhoven University of Technology, 5612 AZ Eindhoven, The Netherlands
{r.willems,e.kruithof}@tue.nl

[2] Philips Research, High Tech Campus, 5656 AE Eindhoven, The Netherlands

Abstract. Computational cardiac models are progressively being used to understand, predict, and improve the treatment of cardiac diseases. These models commonly rely on the traditional finite element analysis (FEA), where the geometry description and consequent mesh generation are separate preprocessing steps that are required before conducting numerical analyses. The recent isogeometric analysis (IGA) paradigm eliminates the separate meshing step and integrates geometry construction and solution approximation using higher-order splines. In this study, we first investigate whether IGA can be efficiently used to model post-infarction left ventricular mechanics. Mechanics results from an established FEA model with a fine homogeneous mesh are used to investigate to what extent similar results could be obtained using hierarchical mesh refinement in IGA. The IGA-mechanics results show a good agreement while providing a mesh-independent geometry, but deviations are noticed close to the base, apex, and partially the endocardium. Second, both the FEA- and IGA-model results are used as input for an FEA-mechanics-driven electrophysiology model which is used for a ventricular tachycardia (VT)-inducibility study, as abnormal mechanics is believed to be a potential driving factor for the development of VT. The resulting VT propagation patterns agree visually for both mechanical inputs and differences in VT-exit points are within 1–7 mm for simulations where VT occurred. Furthermore, an agreement of 85% in the binary VT results is observed, where the 15% difference displays the electrophysiology-model sensitivity to deviations in mechanical input.

Keywords: Myocardial infarction · Numerical comparison · Myofiber strain · Arrhythmia

1 Introduction

Myocardial infarction can lead to scar-based heart rhythm disorders, like ventricular tachycardia (VT). The underlying pathophysiological mechanisms are

O. Bernard et al. (Eds.): FIMH 2023, LNCS 13958, pp. 97–106, 2023.
https://doi.org/10.1007/978-3-031-35302-4_10

not yet completely understood, but mechanical strains are considered a driving factor for electrophysiological remodeling over time [1], which may result in VTs. Mathematical models of cardiac mechanics and electrophysiology are being used to investigate these mechanisms and may ultimately be used in diagnosis and treatment planning. Typically, these models employ the finite element analysis (FEA) procedure, in which the geometry description and consequent mesh generation are separate steps. Furthermore, the meshed geometry is typically constructed from low-order polynomial elements, of which many may be required to capture curved geometrical features. As an alternative, the isogeometric analysis (IGA) paradigm is a recently developed approach that integrates geometry construction and solution approximation using higher-order splines [17]. These splines define the geometrical shape by a set of so-called control points that is referred to as the control net. By specifying the same splines to be the basis of the solution space, one effectively eliminates the need to perform a separate meshing step that is inherent to FEA. The control points that define the geometry are, therefore, practically equivalent to the nodes typically encountered in an FEA-based mesh. Splines also offer additional advantages over standard basis functions due to their high interelement continuity, smoothness, and control of continuity using the so-called knot or k-refinement. Over the past few years, IGA has been increasingly applied to a variety of biomedical problems, including the blood flow in arteries [11], the structural analysis of the aortic valves [10], and the electrophysiology of the human heart [7–9]. However, the application of IGA to cardiac mechanics is limited in literature and only recently explored in our previous work [2]. In this previous work, we demonstrated that an IGA-cardiac-mechanics model was able to capture both the hemodynamic and myofiber-strain response with good agreement when compared to an established FEA model with a fine homogeneous mesh. The IGA model employed relatively coarse high-order elements, which was possible due to the limited spatial variation in cardiac properties. However, in a diseased heart suffering from a chronic infarct, local properties of the myocardium like stiffness and contractile force vary at smaller length scales [3].

In this study, we first want to answer the question if IGA is able to capture these pathological differences using hierarchical refinement. In doing so, we consider myofiber-strain results from an established FEA model with a fine homogeneous mesh as a reference [12,13]. Second, we want to answer the question if the observed myofiber-strain differences are acceptable when simulating a mechanics-driven electrophysiological response using a well-established FEA implementation in the CARP package [6]. The electrophysiological response is solved for both the FEA- and IGA-based myofiber strain inputs which are coupled to the conduction velocity using a non-linear relation. This enables us to analyze whether the employed IGA resolution is sufficient to simulate mechanics-driven VT-inducibility and to identify regions where the electrophysiological response is different for the two numerical implementations.

2 Methods

2.1 Cardiac Mechanics Model

We employ the cardiac mechanics model of Bovendeerd *et al.* [12] with an extension of Janssens *et al.* [13] to include a chronic infarct. The model assumes an idealized left ventricular shape which is represented by a truncated ellipsoid. The local myofiber orientation is defined according to the rule-based fiber field of Bovendeerd [12]. The myocardium is modeled as a non-linear, transversely isotropic, nearly-incompressible material with an active stress component in parallel along the fiber orientation to represent myocardial contractility. A transmural infarct is included, representing the effect of occlusion of the left anterior descending artery. A border zone encompasses the infarct to represent a mixture of infarct and healthy tissue. The infarct core and border zone together occupy 11% of the myocardium. The passive stiffness in the infarct is increased 10-fold compared to the healthy tissue, while the active stress component is set to 0 in this region. Both the passive and active stress component values are linearly interpolated over the infarct border zone, which separates the infarct from the healthy area. The ventricle is coupled to a 0D closed-loop, lumped parameter model to represent the circulatory system where similar initial and boundary conditions are used as outlined in Ref. [2]. One cardiac cycle is simulated by homogeneous mechanical activation of the myocardium [12].

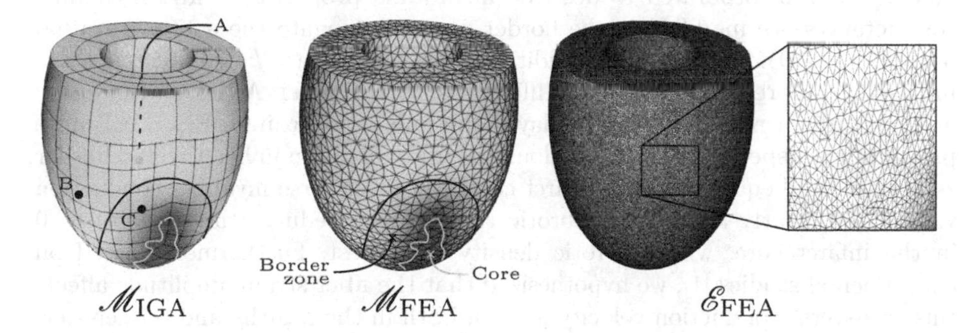

Fig. 1. Visualization of the spatial discretizations used for the cardiac mechanics, $\mathscr{M}_{\text{FEA}}$ and $\mathscr{M}_{\text{IGA}}$, and the electrophysiology computations, $\mathscr{E}_{\text{FEA}}$. Scar properties are defined in the core, indicated in red, and extend linearly through the border zone toward remote tissue, indicated in white. The locations A (far remote), B (remote), and C (border zone) are used to trace fiber-strain over time. (Color figure online)

FEA Implementation. The considered FEA implementation of left ventricular mechanics is from an established model with a fine homogeneous mesh. The mesh consists of tetrahedral elements with an average unstructured homogeneous resolution of 3 mm resulting in 23569 elements (Fig. 1). The displacement field is represented by quadratic Lagrangian basis functions which correspond to a system of 106741 degrees of freedom (dofs). The FEA simulation is performed using the FEniCS Python package [14] with time-steps of 2 ms [13].

IGA Implementation. The left ventricle is constructed as outlined in Willems *et al.* [2], and consists of 5 quadratic non-uniform rational B-spline patches. Elements are defined by knot vectors that correspond to the B-splines defined inside a patch. We employ a hierarchical refinement method, where local refinement levels are progressively introduced around the infarct and terminated once the total infarct is within 5% volumetric difference (Fig. 1). The regions more remote from the infarct are kept as coarse as possible and only refined to support a gradual transition between the different levels. This results in a coarse *base* mesh size of 15 mm and a local mesh resolution of 3.7 mm. The overall mesh is structured and consists of 1272 elements. The displacement field is approximated using cubic B-spline basis functions, resulting in a system of 7250 dofs that is solved monolithically. Simulations are performed using the Nutils Python package [15], with identical parameter settings as the reference FEA model.

2.2 Cardiac Electrophysiology Model

The cardiac electrophysiology is modeled by the monodomain equations using the methods as outlined in Arevalo *et al.* [4], where the Ten Tusscher ionic model [16] is incorporated to describe membrane properties, and ion channel conductances are modified in the border zone to elongate the action potential duration (APD). In this study, additional variations in Gk_s conductance are imposed to represent transmural differences with shorter APDs at the endo- and epicardium relative to the midlayer and sub-endocardium. Electrical pulse propagation is specified to 0.6 m/s along the fiber direction in healthy and border zone tissue and equals 0 in the infarct core. The transverse myofiber conduction velocity is linearly affected by fibrotic tissue which reduces this velocity to 0 in the infarct core, where fibrotic density is highest. Furthermore, based on experimental studies [1], we hypothesized that the fiber-strain amplitude affects this transverse conduction velocity as well, both in the healthy and border-zone tissue. More specifically, we assumed a sigmoidal relation between this velocity and the amplitude of fiber-stain during the cardiac cycle, as shown in Fig. 2. VT-risk is evaluated by virtual pacing of the ventricle applied at six endocardial locations surrounding the border zone to mimic the clinical S1-S2 protocol used to induce VT [4]. At those locations, six stimuli (S1) at a cycle length of 600 ms are applied, followed by a premature stimulus (S2). A range of S1-S2 time intervals is tested to assess the inducibility of VT.

Implementation. An unstructured tetrahedral mesh with an average resolution of 500 µm is used for the electrophysiology FEA model, resulting in 6633856 elements (Fig. 1). The membrane voltage potential, which is solved for, is approximated by linear Lagrangian basis functions which correspond to a total of

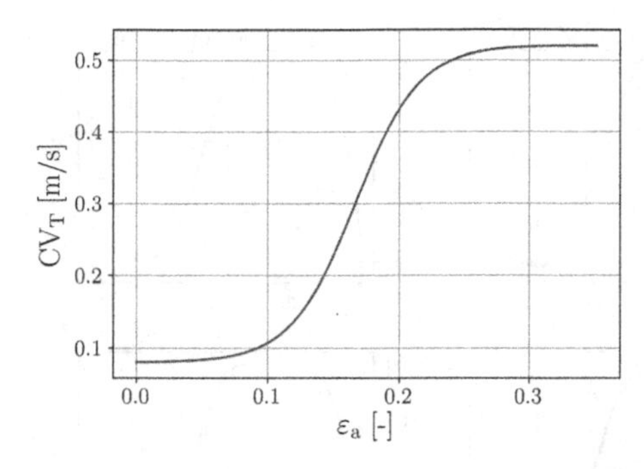

Fig. 2. Relation between the conduction velocity, CV_T, and myofiber-strain amplitude, ε_a.

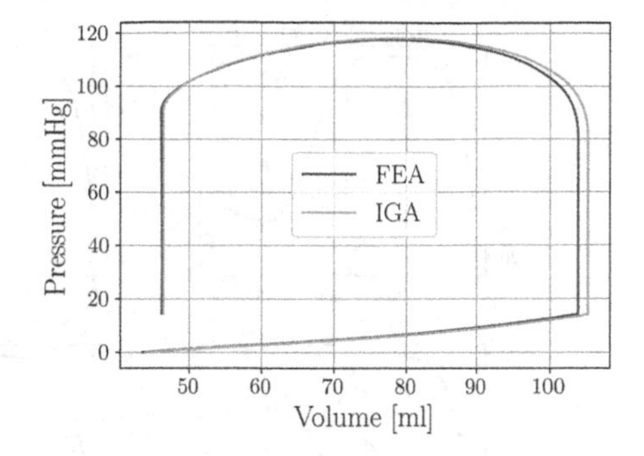

Fig. 3. Hemodynamic response of the FEA and IGA mechanics models visualized as pressure-volume loops.

1082784 dofs. The simulations are performed in the CARP package [6] with time-steps of 2 μs. The fiber-strain amplitudes, needed for the strain-dependent model of conduction velocity, were derived from the mechanics models by interpolation and nearest-point projection. The resulting desired velocities are obtained by defining the corresponding conductivities using the methods of Costa *et al.* [5].

2.3 Performed Numerical Experiments

Two mechanics simulations are performed, using the meshes as visualized in Fig. 1. The obtained numerical results are evaluated based on the hemodynamic response and local myofiber strain behavior. The comparison is continued by evaluating the strain-dependent conduction velocities of both models on the electrophysiological mesh. The subsequent electrophysiology simulations are evaluated by comparing the number of S1-S2 intervals that induce VT, the induced VT patterns, and their corresponding exit points.

3 Results

3.1 Cardiac Mechanics

The hemodynamic response is visualized in pressure-volume loops (Fig. 3) and shows good agreement. However, a difference of 1.4 ml is observed in end-diastolic volume, which might be caused by the geometric approximation error of the linear tetrahedrons that construct the FEA mesh. This behavior was also observed in the comparison of a healthy left ventricle in Ref. [2]. The local myofiber strain is monitored at three locations with identical longitudinal heights in the ventricle: remote tissue far away from the border zone, remote close to the border zone, and in the border zone itself as denoted by A, B, and C in Fig. 1. For each location, subepi-, midmyo-, and subendocardial traces are given in Fig. 4. At all three locations, the subepi- and midmyocardial myofiber strains for IGA and

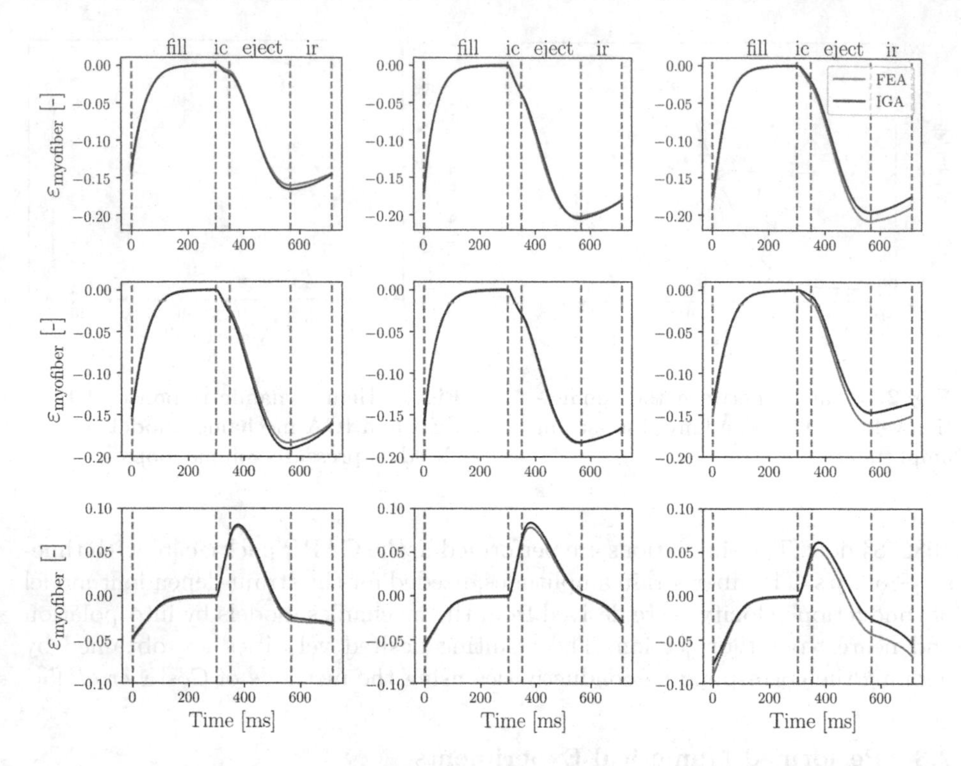

Fig. 4. Comparison between the myofiber strain results, $\varepsilon_{\text{myofiber}}$, at different locations inside the domain. The columns correspond to 3 transmural positions, from left to right: subepicardial, midmyocardial, and subendocardial. The rows correspond to the locations indicated in Fig. 1, from top to bottom: locations A, B, and C. The cardiac cycle consists of a filling (fill), isovolumetric contraction (ic), ejection (eject), and isovolumetric relaxation (ir) phase.

FEA resemble closely, but deviations are noticed near the endocardial wall. Near this subendocardial wall, the IGA results exhibit a smaller decrease in myofiber strain after contraction for all three locations with an average maximum strain difference of 0.014. While the local traces show a good agreement, larger deviations are noticed when considering the contour distribution of the difference in strain amplitude ε_{a} between the two models Fig. 5a. The largest deviations are located at the base, near the apex, and at the endocardium close to the border zone. At this latter site, IGA fiber-strain amplitudes are on average 0.06 lower than those for FEA.

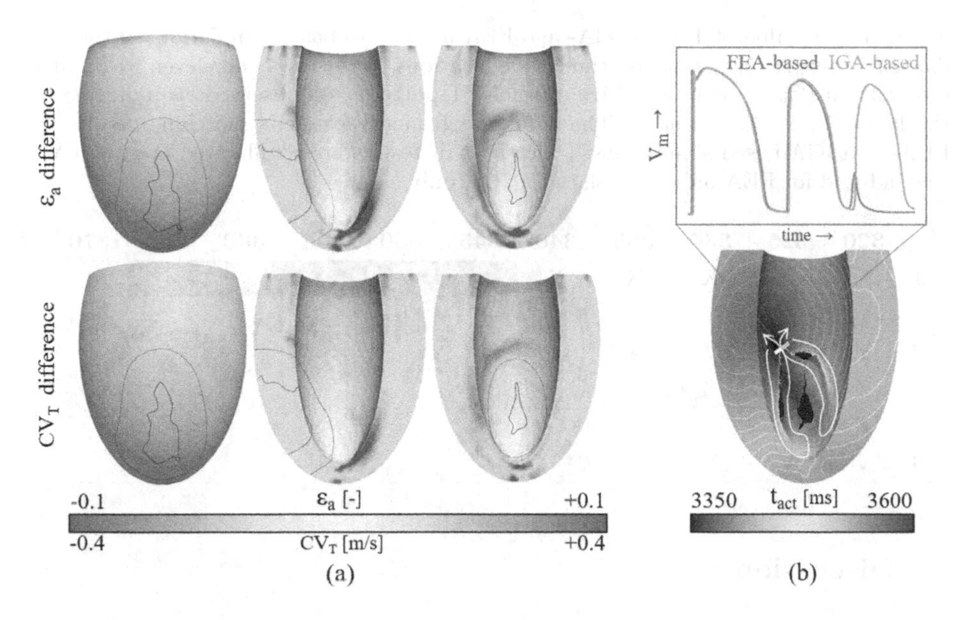

Fig. 5. (a) Difference between the FEA- and IGA-based results in terms of strain amplitudes, ε_a and conduction velocity transverse to the myofiber direction, CV_T. Black contours represent borders of the infarct tissue. (b) Contour of the activation time, t_{act}, of an X-VT simulation result, where the VT is maintained for the IGA-based simulation but blocked at the white bar for the FEA-based simulation. This block is caused by a slightly earlier re-activation for the FEA-based results, as shown in the corresponding membrane voltage, V_m, plot.

3.2 Cardiac Electrophysiology

Differences between both models in myofiber-strain amplitudes, ε_a, lead to differences in transverse conduction velocity as visualized in Fig. 5a up to a maximal local decrease in CV_T of 0.23 m/s in the endocardial border zone. The VT simulation results are given in Table 1, where each stimulation location and S1-S2 time interval combination is color-coded to indicate whether VT was induced in the simulation. The 5 observed VT patterns were identical between IGA- and FEA-based simulations, and labeled U-Z. Measured differences in exit-point locations between IGA- and FEA-based results are on average 7 mm for the U-VT, 1 mm for the V-VT, 6 mm for the X-VT, 4 mm for the Y-VT, and 1 mm for the Z-VT pattern. The VT-inducibility as based on both mechanics models resembles closely overall but deviates most noticeably for the X-VT where the number of S1-S2 time intervals resulting in VT is doubled for the IGA-based results (Table 1). The exit point of this VT is located at the endocardium close to the border zone, where differences in fiber strain and conduction velocity are observed as shown in Fig. 5a. This affected the time instance of re-activation, which results in different behavior in terms of conduction block and subsequent VT as shown in Fig. 5b.

Table 1. VT vulnerability for IGA- and FEA-mechanics-based simulations of the cardiac electrophysiology, with six stimulation locations on the rows and eleven S1-S2 time intervals [ms] on the columns. Five distinct VT patterns are observed, as indicated by the labels *U*, *V*, *X*, *Y*, and *Z*. The light background represents matching results for FEA- and IGA-based simulations. Differences in results are highlighted in gray if VT was induced for FEA only, and black for IGA only.

	320	325	330	335	340	345	350	355	360	365	370
1	X	X	X	X	X	X	X	X	X	X	
2	Y	Y	Y	Y	Y	U	U				
3	Y	Y	Y	Y	Y						
4	Z	Z	Z								
5	Z	Z	Z								
6	V	V	V								

4 Discussion

In this study, we explored if an IGA cardiac model has the ability to efficiently capture pathological differences as induced by left-ventricular infarction. Local differences in tissue properties were captured by hierarchical refinement to obtain an efficient mesh resolution for numerical analyses. The final refinement level was based on the local pathological gradient and shape of the infarct as defined in an established FEA model. Local comparison of myofiber-strain results to those of the reference FEA model showed a good agreement, but deviations are still noticed near the base, apex, and parts of the endocardial border zone. The former is also observed for the healthy case in Ref. [2] and is believed to be caused by the strong imposition of the zero-normal-displacement boundary condition at the basal plane. The apex and endocardial differences are possibly caused by a slight discrepancy in fiber orientation, combined with a relatively coarse IGA discretization. The difference is largest near the apex where all fibers converge and the deviations in mesh geometry are most evident. The remaining differences might be caused by the numerical approximation error, which should be further investigated by studying the effects of mesh refinement in future work. Furthermore, it should be investigated whether simulating multiple cardiac cycles to achieve a hemodynamic steady state would affect the strain amplitude results.

In the second part of this contribution, we focused on the question if the observed myofiber-strain differences are acceptable when simulating a mechanics-driven electrophysiological response. The myofiber-strain amplitudes of both the IGA and FEA mechanics results were coupled to an established FEA-based electrophysiology model to assess the VT risk. Here, a large difference in CV_T of 0.23 m/s is found at a small part of the endocardial border zone. This difference is located at an area with the highest strain-, and thus CV_T-gradients. The location of the high gradient is slightly shifted for IGA compared to the FEA model, which results in the observed local differences. However, these local differences are limited to a small and superficial area (Fig. 5a) and did not significantly affect the electrophysiological response, except for the X-VT for which the corresponding exit point is located in this area. This difference in X-VT inducibility is an indication of the electrophysiology model's *all-or-none* character where a small shift in activation times may alter the electrophysiological response in terms of conduction block, as shown in Fig. 5b. In general, differences in inducibility as shown in Table 1 can be used as a measure of uncertainty whether a VT would actually develop for these S1-S2 time intervals since the considered mechanics models are subject to assumptions and the corresponding solutions are numerically approximated. Overall, we conclude that both mechanics models are promising to generate input for mechanics-driven electrophysiology simulations.

The integration of geometry and solution approximation with the ability to locally refine using IGA allows for a promising addition when modeling cardiac pathologies. This might be beneficial for the workflow when exchanging information between two numerical domains which are independent of mesh size, inherent to IGA. An electrophysiology IGA implementation may be considered in future research following the outlined advantages in recent literature [7–9].

Acknowledgements. This publication is part of the COMBAT-VT project (project no. 17983) of the research program High Tech Systems and Materials which is financed by the Dutch Research Council (NWO).

References

1. Jacot, J.G., Raskin, A.J., Omens, J.H., McCulloch, A.D., Tung, L.: Mechanostransduction in cardiac and stem-cell derived cardiac cells. In: Kamkin, A., Kiseleva, I. (eds.) Mechanosensitivity of the Heart. MECT, vol. 3, pp. 99–139. Springer, Dordrecht (2010). https://doi.org/10.1007/978-90-481-2850-1_5
2. Willems, R., Janssens, K.L., Bovendeerd, P.H., Verhoosel, C.V., van der Sluis, O.: An isogeometric analysis framework for ventricular cardiac mechanics. arXiv preprint arXiv:2305.02923 (2023)
3. Holmes, J.W., Borg, T.K., Covell, J.W.: Structure and mechanics of healing myocardial infarcts. Annu. Rev. Biomed. Eng. **7**(1), 223–253 (2005)
4. Arevalo, H.J., et al.: Arrhythmia risk stratification of patients after myocardial infarction using personalized heart models. Nat. Commun. **7**(1), 11437 (2016)
5. Costa, C.M., Hoetzl, E., Rocha, B.M., Prassl, A.J., Plank, G.: Automatic parameterization strategy for cardiac electrophysiology simulations. In: Computing in Cardiology, pp. 373–376. IEEE (2013)
6. CARP. https://carpentry.medunigraz.at/carputils. Accessed Nov 2022
7. Pegolotti, L., Dedè, L., Quarteroni, A.: Isogeometric Analysis of the electrophysiology in the human heart: numerical simulation of the bidomain equations on the atria. Comput. Methods Appl. Mech. Eng. **343**, 52–73 (2019)
8. Bucelli, M., Salvador, M., Quarteroni, A.: Multipatch isogeometric analysis for electrophysiology: simulation in a human heart. Comput. Methods Appl. Mech. Eng. **376**, 113666 (2021)

9. Torre, M., Morganti, S., Nitti, A., de Tullio, M.D., Pasqualini, F.S., Reali, A.: An efficient isogeometric collocation approach to cardiac electrophysiology. Comput. Methods Appl. Mech. Eng. **393**, 114782 (2022)
10. Morganti, S., et al.: Patient-specific isogeometric structural analysis of aortic valve closure. Comput. Methods Appl. Mech. Eng. **284**, 508–520 (2015)
11. Divi, S.C., Verhoosel, C.V., Auricchio, F., Reali, A., van Brummelen, E.H.: Topology-preserving scan-based immersed isogeometric analysis. Comput. Methods Appl. Mech. Eng. **392**, 114648 (2022)
12. Bovendeerd, P.H.M., Kroon, W., Delhaas, T.: Determinants of left ventricular shear strain. Am. J. Physiol.-Heart Circulatory Physiol. **297**(3), H1058–H1068 (2009)
13. Janssens, K.L.P.M., Kraamer, M., Barbarotta, L., Bovendeerd, P.M.H.: Post-infarct evolution of ventricular and myocardial function. Biomech. Model. Mechanobiol. (2023, submitted)
14. Logg, A., Mardal, K.A., Wells, G. (eds.): Automated Solution of Differential Equations by the Finite Element Method: The FEniCS Book, vol. 84. Springer, Heidelberg (2012)
15. van Zwieten, J.S.B., van Zwieten, G.J., Hoitinga, W.: Nutils 7.0. Zenodo (2022)
16. Ten Tusscher, K.H., Panfilov, A.V.: Alternans and spiral breakup in a human ventricular tissue model. Am. J. Physiol.-Heart Circulatory Physiol. **291**(3), H1088–H1100 (2006)
17. Hughes, T.J., Cottrell, J.A., Bazilevs, Y.: Isogeometric analysis: CAD, finite elements, NURBS, exact geometry and mesh refinement. Comput. Methods Appl. Mech. Eng. **194**(39–41), 4135–4195 (2005)

Cellular Automata for Fast Simulations of Arrhythmogenic Atrial Substrate

G. S. Romitti(✉), A. Liberos, P. Romero, D. Serra, I. García, M. Lozano, R. Sebastian, and M. Rodrigo

CoMMLab, Departament d'Informàtica, Universitat de València, Valencia, Spain
giada.romitti@uv.es

Abstract. Atrial biophysical simulations can improve therapies by simulating pharmacological and ablative strategies, however their computational times are not compatible with the diagnostic ones. Discrete models such as cellular autom ata (CA) lower computational times by considering a finite number of states evaluated through restitution properties, although it is necessary to question whether this approach is sufficient to reproduce pathological simulations. The analysis of biophysical atrial simulations, under both healthy conditions and different degrees of electrical remodeling, shows an expected increase of Action Potential Duration (APD) with the previous Diastolic Interval (DI) interval. Short-term memory of atrial cardiomyocytes was observed as the dependency of the predicted APD^{+1} with the previous activation (APD^0): shorter APD^0 provoked shorter APD^{+1}, and this effect was comparable to the effect of previous DI. Independent prediction based on both APD^0 and DI allowed better estimation of APD^{+1} values, compared to using DI alone ($p << 0.01$). Finally, the comparison of re-entrant activity in CA simulations showed a closeness between biophysical and CA results, where the CA automata reproduced reentrant patterns and cycle lengths of different states of atrial remodeling. Atrial automata considering short-term memory allows to accurately reproduce the arrhythmic behavior of pathological tissue with computational times for clinical use.

Keywords: Atrial Cellular Automata · Atrial Fibrillation · Rotors

1 Introduction

Cardiac arrhythmias at atrial or ventricular level are among the main causes of disease and mortality. Atrial fibrillation (AF) is the most prevalent cardiac pathology, whose prevalence has increased by more than 30% in the last two decades [1]. Due to the lack of knowledge about the specific mechanisms initiating and perpetuating AF, choosing specific treatment for each patient is still a major clinical and economic problem, which makes pharmacological and ablative treatments sub-optimal [2]. Over the past two decades, it has been suggested that fibrillatory activity can be led by reentrant sources, which are spatially and temporally conserved over time and therefore candidates for ablation [3].

O. Bernard et al. (Eds.): FIMH 2023, LNCS 13958, pp. 107–116, 2023.
https://doi.org/10.1007/978-3-031-35302-4_11

In silico models represent a great tool for guiding personalized medicine as they provide interesting insights on the individual AF manifestation. They were confirmed to be especially useful in diagnosing pathological situations [4], and in evaluating ablative [5] and pharmacological [6] strategies results. Numerous mathematical models describe cardiac functions in detail from both an electrophysiological [7] and a hemodynamic is all together [8] point of view.

Atrial simulations can be used to quantify the response of the patient electrophysiology against different scenarios, such as changes in activation location or frequency, ionic concentration, temperature, pH, or drug effects [9]. These simulations are characterized through standard electrophysiological metrics, namely the Action Potential Duration (APD). In this regard, biophysical modelling [10] is among the most widely used and has been implemented by several solvers [11, 12], that manage to accurately describe the transmembrane potential (TP), ionic concentrations or gating variables. However, this precision brings with it a high computational cost and long simulation times, due to the huge use of parameters and to the large number of systems of ordinary differential equations to be solved, that can compromise clinical diagnostic times. For this reason, and especially in geometries with high mesh refinement, the use of electrophysiological solvers with compatible diagnostic times are demanded, such as solvers describing the cardiac electrical characteristics by discrete states, so called cellular automata (CA).

The CA concept was first introduced by von Neumann [13], and later repeated in various scenarios, including electrophysiological simulations. Some preliminary work was carried out in [14], where only five states of excitability were assigned to each cell, appreciating the very low computational cost. Gerhardt's [15] emphasized the implementation efficiency and high intuitiveness of using CA, adding the speed and curvature properties of the propagation curve. The reproduction of the basic mechanisms of an arrhythmic case such as atrial flutter, with the possibility of non-invasive detection of propagation patterns has been also confirmed [16]. More recent evidence [17] use CA to simulate complete AF episodes with stochastic patterns on the left atrium surface, while [18] use them to understand how the parasympathetic branch of the autonomic nervous system affects the heart rhythm in the right atrium. However, one of their main limitations is to adjust relevant electrophysiological parameters. More recent evidence [19] reveals that it is possible to integrate electrophysiological parameters using restitution properties, simulating healthy and pathological situations in real ventricular geometries, and succeeding in keeping computational times short thanks to the use of two excitation states.

The aim of this work is to extend a ventricular CA [19] to atrial electrophysiology, calibrating and validating it with a biophysical model [7] in different pathological scenarios.

2 Materials and Methods

2.1 Biophysical Model

Atrial biophysical simulations were performed using the Koivumäki atrial model [20] on a rectangular atrial tissue (0.3 × 2 × 0.025 cm, 2106 cells, 0.25 mm inter-node distance, 0.01 ms of temporal interval) activated from the inferior front, see Fig. 1a. Fiber direction was included in the model along the large axis of the model, the diffusion value in the short direction equals 0.12 μm^2/ms and in the long direction 0.45 μm^2/ms. For electrophysiological analysis, three midline points of the slab of tissue were considered: one for the analysis of the Transmembrane Potential curve (TP) and two for the measurement of Conduction Velocity (CV), as shown in Fig. 1a.

A total of 3000 S1-S2 pacing protocols were performed, corresponding to sequences of S1 intervals (100 ms to 1000 ms), repeated for 15/16 times before an S2 stimulus with values between 100 ms and 1000 ms. This pacing protocol was repeated under different conditions: 0%, 66%, 100% and 135% of electrical remodeling, where 100% of remodeling represents the atrial substrate of average chronic AF patients, while 0% the healthy atrial condition. Taking 100% remodeling as a reference, the less fraction of remodeled refers to a healthier case down to the normal case of 0%, while a higher fraction to a more extreme pathological case. Electrical remodeling is associated with a shortening of APD and a reduction of the diffusion in the cardiac tissue, which facilitate the initiation and maintenance of reentrant behaviors. The chronic atrial fibrillation (cAF) model (100% remodeling) is the result of the following modifications from the sinus rhythm model presented by Koivumäki: SERCA expression (−16%), PLB to SERC ratio (+18%), SLN to SERCA ratio (−40%), maximal INCX (+50%), sensitivity of RyR to [Ca^{2+}]SR (+100%), conductance of ICaL (−59%), conductance of Ito (−44%), conductance of IKur (−22%) and conductance of IK1 (+100%). Note that modifications in Ito, IKur and IK1 have been altered from the original cAF model. These modifications fit in the limits presented by Koivumäki et al. but results in a more pronounced APD shortening with the increase of remodeling ratio [21].

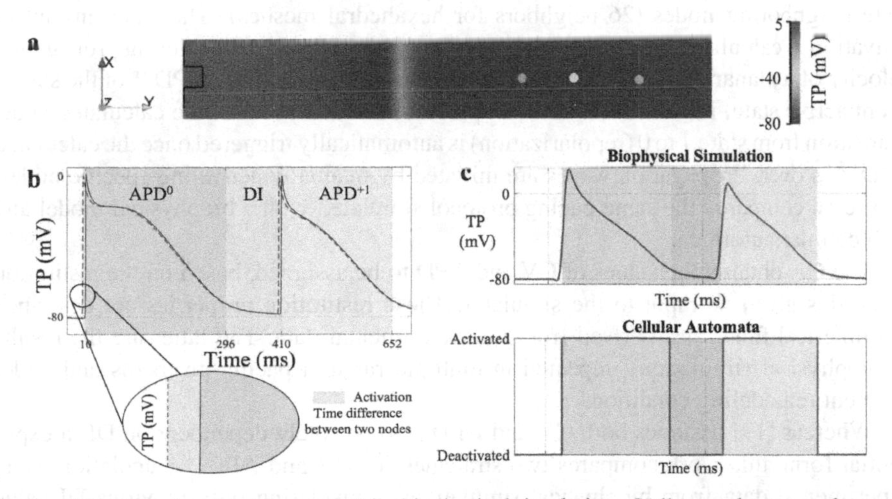

Fig. 1. a) Rectangular atrial tissue used for biophysical simulations. Yellow and green nodes were used to measure APD and CV respectively; b) Transmembrane Potential curve of two nodes (red and black lines) with highlighted sections: APD^0 (red), DI (green), APD^{+1} (yellow), and activation time difference between node (blue); c) Comparison of the same pacing protocol simulated with biophysical model (upper panel) and with cellular automata (lower panel).

For each set of simulations carried out under different scenarios, specific biomarkers were analyzed. Transmembrane Potential (TP) curves were used to measure the APD and the Diastolic Interval (DI). APD was measured as the difference between the activation time (instant of maximal positive dV/dt) and the time at which 90% of repolarization is reached, the so-called APD90. APD^0 and APD^{+1} denote the last S1 and S2 activations respectively. The DI was measured as the interval from end of APD^0 and the beginning of APD^{+1}. Figure 1b pinpoints exactly their corresponding values. Conduction velocity was measured as the space separating the measuring points (6 mm) divided by the difference in their activation times, see Fig. 1b.

2.2 Cellular Automata

The biophysical simulations just described demonstrate accuracy and ease in analyzing the results. However, the computational time they require is high and therefore they cannot be used in the clinical field. Serra's [19] presents an alternative method for modelling cardiac tissue electrophysiology, which is a spatially extended, event based, asyn-chronous cellular automaton, prepared for the analysis of ventricular tissues and geometries.

Cellular Automata (CA), unlike biophysical models, consider only two main states for each node of cardiac tissue: 0 (inactive, i.e., repolarized, and excitable) and 1 (active, i.e., able to activate the neighbors). Each event simulated (activation, repolarization) is processed at the exact time when their occur and therefore there is no need of a granular time step.

Transition from state 0 to 1 (activation) is triggered by previous activations of immediate neighboring nodes (26 neighbors for hexahedral meshes). The exact instant of activation is calculated using the Fast-Marching algorithm [22] considering propagation velocity of a planar wave. The CV for this activation, as well as the APD^{+1} of the subsequent active state, is calculated from the previous states and DI of the calculated node. Transition from state 1 to 0 (repolarization) is automatically triggered once the calculated APD^{+1} is over. Propagation waves are initiated by manually activating specific nodes. Figure 1c compares the same pacing protocol simulated with a biophysical model and with cellular automata.

The CA obtains the values of CV and APD to be assigned, based on the restitution properties given as input to the simulator. These restitution properties are described as numerical functions, derived from the experimental data. The latter are the results of biophysical simulations, repeated in multiple random pacing protocols and under different remodeling conditions.

Whereas [19] assumes both CV and APD^{+1} to be solely dependent on DI in exponential form, this work compares two strategies for CV and APD^{+1} calculation using experimental data from biophysical simulations: considering only previous DI value and summarizing the APD^{+1} values into restitution curves or considering both previous DI and APD^0 values, as descriptors of the previous state, and summarizing the APD^{+1} values into restitution surfaces.

3 Results

3.1 Biophysical Model Characterization

Characterization of the APD metrics for the biophysical simulation can be observed in Fig. 2a, where the APD^{+1} is represented as a function of the previous DI ($APD^{!1} = f(DI)$). For each of the simulated scenarios, fitting a single curve produced a greater error (2.12 ± 1.33 ms) than the one obtained by considering a surface fitting ($APD^{!1} = f(DI, APD^{\#})$, 1.41 ± 2.12 ms, p = 0.0052), as depicted in Fig. 2b. This could be explained by the model's memory or inertias of the different ionic concentrations and variables associated with the channels. A necessary observation is that in AF situations, the values of CL, APD and DI are small [23], so smaller errors are desirable.

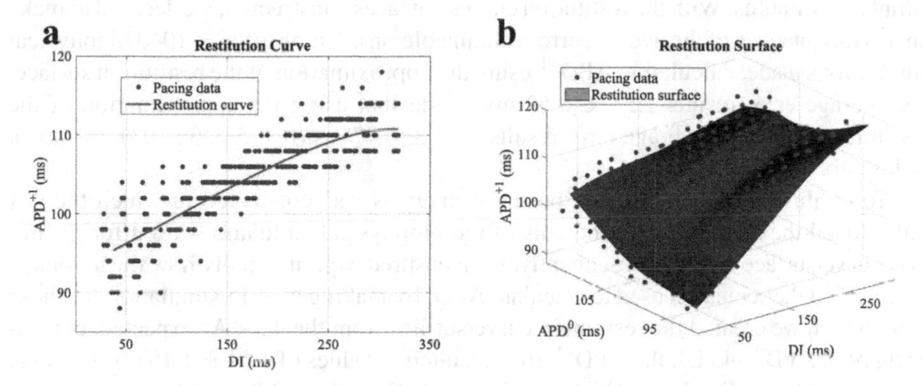

Fig. 2. a) APD Restitution Curve, b) APD Restitution Surface.

This analysis was repeated for different pathological remodeling scenarios, resulting in surfaces of different trends, as shown in Fig. 3 for three atrial remodeling scenarios. For CV characterization, no influential dependencies with respect to parameters other than DI were noted, so the use of a restitution curve was maintained.

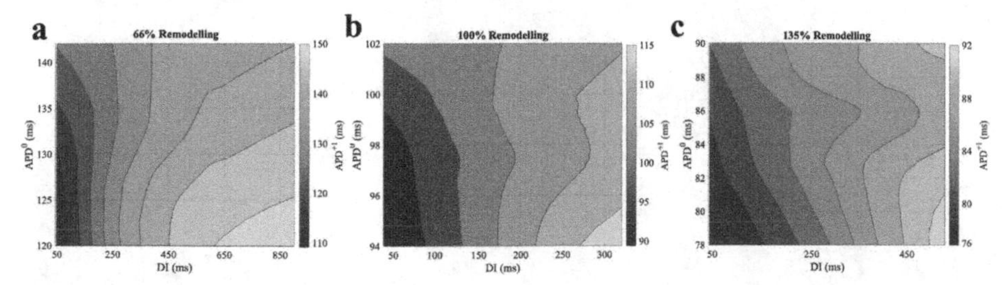

Fig. 3. Contour maps describing APD^{+1} as a function of DI and APD^{0} for simulations with a) 66% remodeled; b) 100% remodeled; c) 135% remodeled tissue.

3.2 Cellular Automata Simulations

Using the CA, we reproduced the simulations performed with the biophysical model. Computational time using the CA was in average 160 times faster than the equivalent biophysical solver: 1,9 vs. 327,3 s of computation time per second of simulated time, respectively. Analysis of the biophysical simulations took five times longer than in CA due to the need to calculate transmembrane voltage-related values: 300 s compared to 60 s.

Figure 4a shows an example of a biophysical simulation (upper panel) reproduced by the atrial CA using both strategies: calculating APD^{+1} from previous DI (restitution curve, middle panel) and calculating APD^{+1} from previous DI and APD^{0} (restitution surface, lower panel). Whereas the APD value of the CA using the APD curve change for each activation, the automata with APD surface showed better correspondence in each simulated stimulus. With the restitution curves/surfaces constructed, we decided to make an a priori analysis of the average errors obtainable on a sample of $N = 1040$ biophysical simulations made: calculating APD^{+1} using the approximation of the restitution surface, the average error results 1.25 ± 0.62 ms, while that using the approximation of the restitution curve, the average error results 1.97 ± 0.77 ms ($p = 5.233e{-}03$), as shown in Fig. 4b.

To confirm this hypothesis, a posterior analysis was conducted in which the CA reproduced the same pacing protocols of the biophysical simulations used for restitution curve/surface fitting. Subsequently, we measured experimentally how the automata reproduced the scenarios used to calculate APD, by making $N = 15$ simulations of those from which we created the restitution curve/surface from the data. As expected, considering both APD^{0} and DI, the APD^{+1} error resulted in values of 1.51 ± 1.15 ms, less than when considering DI alone, which was 2.63 ± 1.95 ms ($p = 3.34e{-}09$).

This verification confirmed the difference between the two CA strategies: considering both APD^{0} and DI the mean error was half than the one obtained while considering DI alone. This implies that considering the value of earlier APDs improves the estimation of later APDs, and therefore in simulations of long duration it is necessary to consider the short-term memory of the simulator.

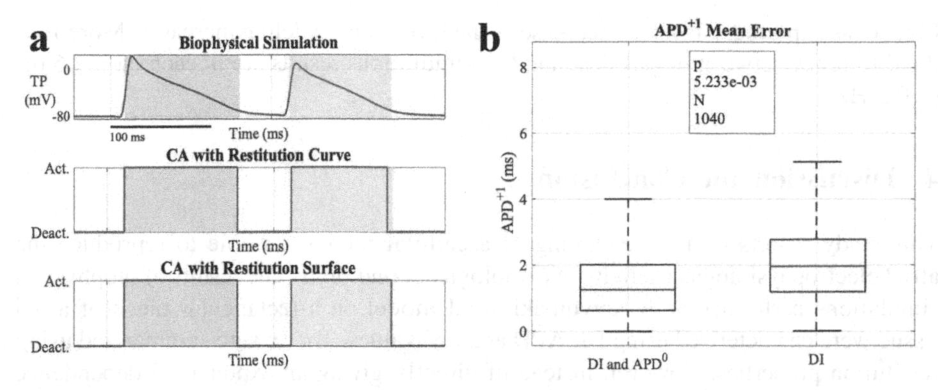

Fig. 4. a) Comparison of biophysical simulation (top panel), CA with restitution curve (middle panel), and CA with restitution surface (bottom panel). Red band: APD^0, green band: APD^{+1}; b) APD^{+1} mean error of CA obtained using restitution surface (left) and restitution curve (right).

3.3 Self-sustained Arrhythmia Simulations

The comparison between CA and biophysical model performed so far were analyzed for pacing protocols. However, the ultimate interest of the study is to decrease the computational time of simulations of pathological scenarios, to approach diagnostic times. This raises the question of whether the simulator can simulate reentrant activity and, if this is the case, how reliable are the results obtained.

A functional reentry scenario was simulated for three main remodeling scenarios in both biophysical and CA model (7.0 × 7.0 × 0.03 cm, 109512 cells, 0.30 mm inter-node distance). To do so, part of the domain was activated to simulate the interaction between a planar wave-front and an obstacle, which physiologically created a reentrant source [9].

As depicted in Fig. 5a–5b, both biophysical and CA simulations show similar reentrant patterns, highlighting how the automata can reproduce arrhythmic scenarios with non-planar wave propagation and functional reentry. A quantitative analysis was then carried out, analyzing for each node of the domain the Cycle Length (CL), namely the time interval between one activation time and the next. The results are depicted in

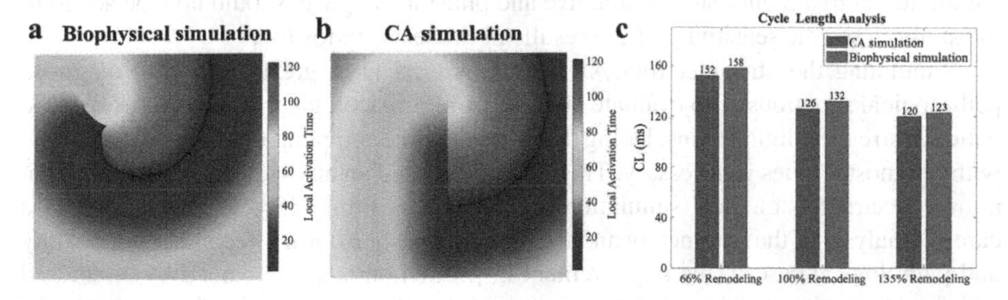

Fig. 5. a) Functional reentry simulated with biophysical model; b) Functional reentry simulated with CA; c) analysis of differences between rotor CLs obtained with CA and biophysical model at different remodeling levels.

Fig. 5c: as expected, CL values decrease as the level of remodeling increases. Moreover, the difference between biophysical and CA simulations results are in each case $\leq$6 ms (<0.3 Hz).

4 Discussion and Conclusion

This study reports on the developing of a cellular automaton able to reproduce the atrial electrophysiological activity in pathological conditions. The results of biophysical simulations performed with Koivumäki atrial model on a rectangular piece of atrial tissue were characterized using DI, APD and CV values. These were summarized using restitution properties. However, instead of directly giving an exponential dependence of the CV and APD^{+1} on the DI as in [19], we wanted to analyze the dependence of both CV and APD^{+1} on the short-term memory effect of the simulation, in this work summarized as the APD^{0}. We observed that APD^{+1} varies as a function of the APD^{0} value, while CV does not change with it. Thus, it was decided to consider APD^{+1} as a function of DI and APD^{0} and CV as a function of DI alone. This reduced the error on the estimated APD^{+1} value, making it closer to the results of biophysical simulations.

Eventually, simulations of re-entrant scenarios were carried out to approach AF situations. Of these, a qualitative and quantitative analysis was made. On one hand, it was noted how the automata faithfully reproduced the activation wave pattern also obtained with a biophysical model. On the other, it was seen how the differences concerning CL in both cases are minimal and gave results consistent with the level of remodeling.

The results obtained make it possible to have a simulator with reduced computational time ($\times$160) and with high precision, making it interesting for clinical use due to the possibility of predicting results in clinical diagnostic times. However, our investigations so far have only been applied to a piece of atrial tissue. They need to be extended to more realistic tissues including three-dimensional atrial geometries. To do this, no other requirements are necessary: the prepared CA has the information needed to reproduce simulations with Koivumäki atrial model. Computational times will increase, due to the increase in nodes to simulate on, but prior work in this regard has already been done, and an extreme reduction in computational times compared to those obtained with the biophysical model has been appreciated. Further studies, which take artificial intelligence into account instead of restitution properties, will need to be performed. Furthermore, simulation of treatments, such as ablative and pharmacological, should now be added to the study to see the sensitivity of the results in those scenarios too.

Simulating the atria electrophysiological aspects is of great interest to diagnose pathological situations or to evaluate ablative or pharmacological strategies for specific patients, so called digital twins. Doing this in reasonable computational times compatible with diagnostic times is necessary. This work presented an atria cellular automaton that allows electrophysiological simulation with accuracy fitting clinical ranges, due to a careful analysis of the parameters and the simulation short-term effects, in both healthy and pathological cases. This improvement in prediction is accompanied by a hundred and sixty times decrease in computing time, which will permit the use of patient-specific simulations in clinical timelines.

Acknowledgements. This work was funded by Generalitat Valenciana Grant AICO/2021/318 (Consolidables 2021) and Grant PID2020-114291RB-I00 funded by MCIN/https://doi.org/10.13039/501100011033 and by "ERDF A way of making Europe".

References

1. Lippi, G., Sanchis-Gomar, F., Cervellin, G.: Global epidemiology of atrial fibrillation: an increasing epidemic and public health challenge. Int. J. Stroke **16**(2), 217–221 (2021)
2. Krummen, D.E., Swarup, V., Narayan, S.M.: The role of rotors in atrial fibrillation. J. Thorac. Dis. **7**(2), 142–151 (2015)
3. Narayan, S.M., Krummen, D.E., Clopton, P., Shivkumar, K., Miller, J.M.: Direct or coincidental elimination of stable rotors or focal sources may explain successful atrial fibrillation ablation: on-treatment analysis of the CONFIRM trial (Conventional ablation for AF with or without focal impulse and rotor modulation). J. Am. Coll. Cardiol. **62**(2), 138–147 (2013)
4. Trayanova, N.A.: Mathematical approaches to understanding and imaging atrial fibrillation: significance for mechanisms and management. Circ. Res. **114**(9), 1516–1531 (2014)
5. Roney, C.H.: Patient-specific simulations predict efficacy of ablation of interatrial connections for treatment of persistent atrial fibrillation. EP Eurospace **20**, iii55–iii68 (2018)
6. Grandi, E., Maleckar, M.: Anti-arrhythmic strategies for atrial fibrillation: the role of computational modeling in discovery, development, and optimization. Pharmacol. Ther. **168**, 126–164 (2016)
7. Lopez-Perez, A.: Three-dimensional cardiac computational modelling: methods, features and applications. Biomed. Eng. Online **14**, 35 (2015)
8. Zingaro, A., Dede, L., Menghini, F., Quarteroni, A.: Hemodynamics of the heart's left atrium based on a Variational Multiscale-LES numerical method. Eur. J. Mech. – B/Fluids **89**, 380–400 (2021)
9. Jalife, J., et al.: Basic Cardiac Electrophysiology for the Clinician, 2nd edn. Wiley-Blackwell, Oxford (2009)
10. Hodgkin, A.L., Huxley, A.F.: A quantitative description of membrane current and its application to conduction and excitation in nerve. J. Physiol. **117**(4), 500–544 (1952)
11. Heidenreich, E.A., Ferrero, J.M., Doblaré, M., Rodríguez, J.F.: Adaptive macro finite elements for the numerical solution of monodomain equations in cardiac electrophysiology. Ann. Biomed. Eng. **38**(7), 2331–2345 (2010)
12. Plank, G., et al.: The openCARP simulation environment for cardiac electrophysiology. Comput. Methods Programs Biomed. **208**, 106223 (2021)
13. Von Neumann, J., Burks, A.W.: Theory of self-reproducing automata. IEEE Trans. Neural Networks **5**(1), 3–14 (1966)
14. Moe, G.K., Rheinboldt, W.C., Abildskov, J.A.: A computer model of atrial fibrillation. Am. Heart J. **67**(2), 200–220 (1964)
15. Gerhardt, M., Schuster, H., Tyson, J.J.: A cellular automaton model of excitable media including curvature and dispersion. Science **247**(4950), 1563–1566 (1990)
16. Monteiro, E.C., Miranda, L.C., Bruno, A.C., Ribeiro, P.C.: A cellular automaton computer model for the study of magnetic detection of cardiac tissue activation during atrial flutter. IEEE Trans. Magn. **34**(5), 3451–3454 (1998)
17. Lin, Y.T., Chang, E.T., Eatock, J., Galla, T., Clayton, R.H.: Mechanisms of stochastic onset and termination of atrial fibrillation studied with a cellular automaton model. J. R. Soc. Interface **14**(128), 20160968 (2017)

18. Makowiec, D., Miklaszewski, W., Wdowczyk, J., Lawniczak, A.T.: From cellular automata model of vagal control of the human right atrium to heart beats patterns. Physica D **415**, 132771 (2021)
19. Serra, D., et al.: An automata-based cardiac electrophysiology simulator to assess arrhythmia inducibility. Mathematics **10**(8), 1293 (2022)
20. Koivumäki, J.T., Korhonen, T., Tavi, P.: Impact of sarcoplasmic reticulum calcium release on calcium dynamics and action potential morphology in human atrial myocytes: a computational study. PLoS Comput. Biol. **7**(1), e1001067 (2011)
21. Koivumäki, J.T., Seemann, G., Maleckar, M.M., Tavi, P.: In silico screening of the key cellular remodeling targets in chronic atrial fibrillation. PLoS Comput. Biol. **10**(5), e1003620 (2014)
22. Ethian, J.A.: Fast marching methods. SAIM Rev. **41**, 199–235 (1999)
23. Franz, M.R., Jamal, S.M., Narayan, S.M.: The role of action potential alternans in the initiation of atrial fibrillation in humans: a review and future directions. Europace **14**(suppl_5), v58–v64 (2012)

Effect of Gap Junction Distribution, Size, and Shape on the Conduction Velocity in a Cell-by-Cell Model for Electrophysiology

Giacomo Rosilho de Souza[1], Simone Pezzuto[1,2](✉), and Rolf Krause[1]

[1] Center for Computational Medicine in Cardiology, Euler Institute, Università della Svizzera italiana, Lugano, Switzerland
{giacomo.rosilhodesouza,rolf.krause}@usi.ch

[2] Laboratory of Mathematics for Biology and Medicine, Department of Mathematics, University of Trento, Trento, Italy
simone.pezzuto@unitn.it

Abstract. Gap junction arrangement is a major determinant of cardiac conduction velocity. Importantly, structural remodeling of the myocardium may lead to pathological CV and a pro-arrhythmic substrate. In this work we aim at quantifying the side-to-side conduction velocity in a sub-micrometer model of the myocardium that accounts for gap junctions. We consider the Extracellular-Membrane-Intracellular (EMI) model, which describes the evolution of the electric potential within each cell and in the extracellular space. For the solution of the model, we propose a boundary integral formulation of the cell-to-cell model that leads to small system of ODEs. We study several configurations of lateral gap junction distribution, as well as different shapes and sizes of the cell-to-cell connection. We find that irregular positioning of gap junctions from cell to cell is of utmost importance to obtain realistic CV values, while gap junction's shape is of secondary importance.

Keywords: Electrophysiology · Cell-by-Cell Model · Gap Junctions · EMI Model · Boundary Element Method

1 Introduction

The cardiac tissue has a complex cellular and subcellular organization. Most of the myocardium is occupied by cardiomyocytes, excitable cells responsible for electric propagation and active force generation. Cardiomyocytes form the cardiac syncytium, a network of tightly connected cells that ensures a smooth propagation of the action potential [8]. Macroscopically, the propagation appears anisotropic because cell-to-cell coupling occurs mostly in the longitudinal direction, yielding a fiber bundle structure. Transverse-to-fiber conduction is generally much slower, of a ratio 1:3 to 1:6 [12]. Several factors compete for the anisotropy

O. Bernard et al. (Eds.): FIMH 2023, LNCS 13958, pp. 117–126, 2023.
https://doi.org/10.1007/978-3-031-35302-4_12

ratio. Gap junctions distribution is of utmost importance. Gap junctions are responsible for the cell-to-cell coupling, and are mostly found in the fiber direction, with only a few of them in the lateral direction. Additionally, transverse conduction is affected by geometrical factors such as cell thickness and branching.

In this work, we are interested in studying in a quantitative manner how lateral gap junction distribution and the geometry of the connection affect the macroscopic conduction velocity in the myocardium. For this, we consider an Extracellular-Intracellular-Membrane (EMI) model [18]. The EMI model is a degenerate parabolic PDE: stationary on the intra- and extra-cellular, but time-dependent on the transmembrane boundary. The EMI model is also coupled with a cellular membrane model. Within the EMI framework, the anisotropic propagation of the action potential results from the geometrical arrangement of the cells and the gap junction distribution. In contrast, the anisotropy ratio and electric conductivity in the standard bidomain model must be given *a priori*, e.g., based on experimental data. As a matter of fact, it has been shown that the bidomain model can be rigorously justified by a homogenization procedure applied the EMI model [9]. In particular, the intra- and extra-cellular conductivity tensors in the bidomain model could be estimated by solving a cell problem [11].

From a computational point of view, the EMI model is significantly more expensive than the bidomain model [10], and it requires fine meshes [18]. Since we are interested only in the transmembrane potential, we adopt here a boundary integral formulation of the problem that avoids the computation of the intra- and extra-cellular potentials. In this way, we recast the full EMI model to a set of ordinary differential equations with a structure similar to the monodomain equation [15]. The reduced membrane model, fully equivalent to the original EMI formulation, only lives on the membrane domain. Thus, we can employ an unprecedented spatial resolution (below 1 μm) for simulating gap junctions. Finally, thanks to a fast boundary element implementation, we can quickly explore multiple realizations of gap junctions distribution, thus assessing the conduction velocity in a statistical manner.

2 Methods

The EMI model considers cells represented by bounded domains $\Omega_i \subset \mathbb{R}^d$, $d \geq 2$, for $i = 1, \ldots, N$, all embedded in a extracellular space denoted by $\Omega_0 \subset \mathbb{R}^d \setminus \cup_{i=1}^N \bar{\Omega}_i$. The electric potential is in general discontinuous, with jumps across cell-to-cell boundaries and the cellular membrane. We denote by u_i, $i = 0, \ldots, N$, the electric potential in the subdomain Ω_i. We denote the interfaces with $\Gamma_{ij} = \Gamma_i \cap \Gamma_j$, $0 \leq i, j \leq N$, where $\Gamma_i = \partial\Omega_i$. The intercellular connections, named gap junctions, occur at Γ_{ij} for $1 \leq i, j \leq N$, whereas the transmembrane boundaries are Γ_{i0}. The outer boundary is $\Sigma = \partial\Omega_0 \setminus \bigcup_{i=1}^N \Gamma_{i0}$. The normals $\boldsymbol{n}_i$ point outwards to Ω_i. The EMI model reads as follows:

$$\begin{cases} -\sigma_i \Delta u_i = 0, & \text{in } \Omega_i \text{ for } i = 0, \dots, N, & \text{(1a)} \\ -\sigma_i \partial_{n_i} u_i = C_\mathrm{m} \partial_t V_i + I_\mathrm{ion}(V_i, z_i), & \text{on } \Gamma_{i0} \text{ for } 1 \leq i \leq N, & \text{(1b)} \\ -\sigma_0 \partial_{n_0} u_0 = \sigma_i \partial_{n_i} u_i, & \text{on } \Gamma_{i0} \text{ for } 1 \leq i \leq N, & \text{(1c)} \\ u_i - u_0 = V_i, & \text{on } \Gamma_{i0} \text{ for } 1 \leq i \leq N, & \text{(1d)} \\ \partial_t z_i = g(V_i, z_i), & \text{on } \Gamma_{i0} \text{ for } 1 \leq i \leq N, & \text{(1e)} \\ -\sigma_i \partial_{n_i} u_i = \kappa(u_i - u_j) & \text{on } \Gamma_{ij} \text{ for } 1 \leq i, j \leq N, & \text{(1f)} \\ -\sigma_0 \partial_{n_0} u_0 = 0, & \text{on } \Sigma. & \text{(1g)} \end{cases}$$

The electric conductivities, denoted by σ_i, $i = 0, \dots, N$, are assumed constant but possibly different from each other. The membrane electric capacitance is C_m, I_ion represents the sum of ionic currents and z_i are vectors representing the ionic model's state variable. We note that the system is stationary on the cellular and extracellular domains, as well as on the gap junctions where an algebraic condition is imposed (with permeability κ). Time dynamics occur only on the transmembrane boundary $\Gamma_m = \cup_{i=1}^N \Gamma_{i0}$.

Problem (1) has already been solved by means of the finite element method [17,18]. With the boundary element method (BEM), it has been solved only for simple geometries of isolated cells [4,6,7]. Here, we consider the BEM approach proposed in [15], with no geometrical restriction on the problem. For the numerical solution of Eq. (1) we make use of the following result (see [15] for a proof).

Theorem 1. *The BEM space discretization of* (1) *is equivalent to the ordinary differential equations system*

$$\begin{cases} C_m \dfrac{\mathrm{d}\boldsymbol{V}_m}{\mathrm{d}t} + I_{ion}(\boldsymbol{V}_m, \boldsymbol{z}) = \psi(\boldsymbol{V}_m), & \text{(2a)} \\ \dfrac{\mathrm{d}\boldsymbol{z}}{\mathrm{d}t} = g(\boldsymbol{V}_m, \boldsymbol{z}), & \text{(2b)} \end{cases}$$

where $\psi(\boldsymbol{V}_m) = \boldsymbol{\lambda}_m$ *and* $\boldsymbol{\lambda}_m \in \mathbb{R}^{M_m}$, $\boldsymbol{\lambda}_g \in \mathbb{R}^{M_g}$, $\boldsymbol{\beta} \in \mathbb{R}^N$ *are solutions to*

$$\begin{pmatrix} F_{mm} & F_{mg} & A_m G \\ F_{gm} & F_{gg} - \kappa^{-1} I & A_g G \\ G^\top A_m^\top & G^\top A_g^\top & 0 \end{pmatrix} \begin{pmatrix} \boldsymbol{\lambda}_m \\ \boldsymbol{\lambda}_g \\ \boldsymbol{\beta} \end{pmatrix} = \begin{pmatrix} \boldsymbol{V}_m \\ \boldsymbol{0} \\ \boldsymbol{0} \end{pmatrix}, \quad (3)$$

with

$$F_{mm} = A_m F A_m^\top, \quad F_{mg} = A_m F A_g^\top, \quad F_{gm} = A_g F A_m^\top, \quad F_{gg} = A_g F A_g^\top. \quad (4)$$

For the space discretization of (1) we adopt a collocation BEM with trigonometric Lagrange basis functions and, in Theorem 1, M_m and M_g represent the number of collocation nodes lying on the transmembrane boundary Γ_m and on the union of all gap junctions $\Gamma_g = \cup_{i,j=1, i \neq j}^N \Gamma_{ij}$. Variables $\boldsymbol{V}_\mathrm{m}$ and $\boldsymbol{z}$ are the spatial discretizations of V_i and z_i, $i = 1, \dots, N$, respectively. Therefore, ODE (2)

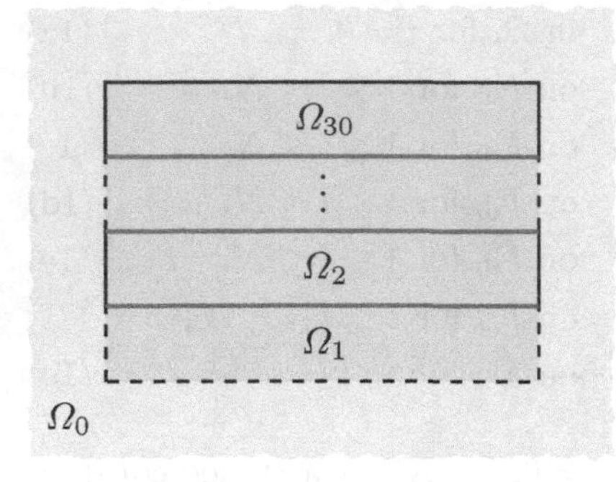

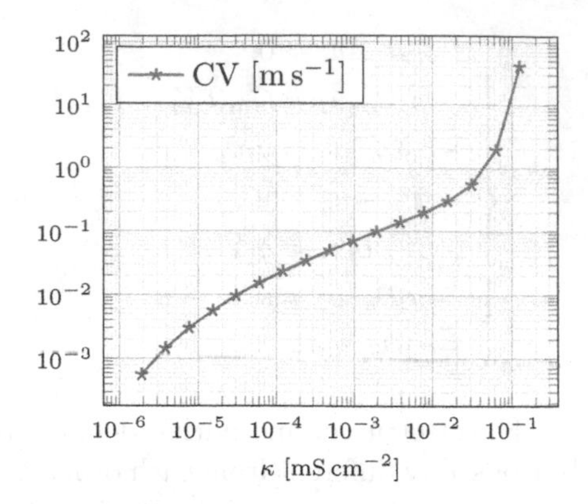

Fig. 1. Conduction velocity across a vertical array of 30 cells. Red boundaries represent gap junctions with uniform permeability κ. Each cell is 100 μm long and 10 μm thick. On the right, the CV for increasing values of κ. (Color figure online)

lives on the transmembrane boundary Γ_m and the remaining model constraints, as Laplace problems, flux continuities and gap junction algebraic conditions, are encoded into the linear map ψ defined by the linear system (3).

Every time that $\psi(\boldsymbol{V}_\mathrm{m})$ has to be evaluated, e.g. when approximating (2) with a time marching scheme, system (3) has to be solved. This system is symmetric and has size $M_m + M_g + N$ hence, when the number of degrees of freedom is not overly large, it is factorized only once. This is the approach adopted here.

Let $M = M_m + M_g$, matrices $A_m \in \mathbb{R}^{M_m \times M}$, $A_g \in \mathbb{R}^{M_g \times M}$ are projection operators, from a global system of degrees of freedom representing functions on $\Gamma_m \cup \Gamma_g$ to local ones. Matrix $F \in \mathbb{R}^{M \times M}$ encodes Laplace equations in all subdomains Ω_i, $i = 0, \ldots, N$, and flux continuities. It is based on pseudo inverses of the Dirichlet-to-Neumann maps. Matrix $G \in \mathbb{R}^{M \times N}$ enforces condition $V_i = u_i - u_0$ and as well solvability of problems involving pseudo inverses.

Finally, for the time integration of (2) we employ the mRKC method [1], which is a multirate explicit stabilized method.

3 Numerical Results

In this section we report four numerical experiments. The main goal is to investigate dependence of conduction velocities (CVs) on the distribution, size and shape of the gap junctions on the cellular boundaries parallel to the fiber direction. In all experiments, parameters are set as following [17]: $C_\mathrm{m} = 1\,\mu\mathrm{F\,cm^{-2}}$, $\sigma_e = 20\,\mathrm{mS\,cm^{-1}}$, $\sigma_i = 3\,\mathrm{mS\,cm^{-1}}$, and $\kappa = 1/R_m = 690\,\mathrm{mS\,cm^{-2}}$. For the membrane, we consider the Courtemanche-Ramirez-Nattel model [3].

In the first experiment, we evaluate the transverse CV in a vertical array of 30 cells, as shown in Fig. 1 (left panel). For the experiment, we considered the parameters given as above, except for the gap junction permeability κ. The transverse CV is defined as the difference in activation time between last and

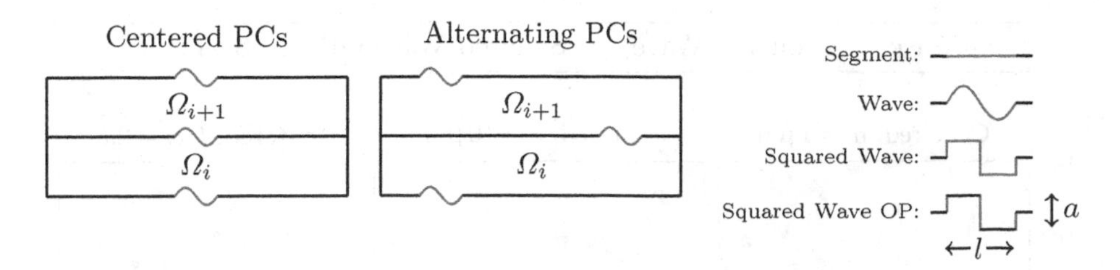

Fig. 2. Cells with non-zero permeability κ only on the Permeable Curve (PC), indicated in red. We consider aligned and centered PCs (left panel), and alternating PCs (middle panel). PC are also parametrized with different shapes (right panel), amplitude a, and length l. (Color figure online)

first cell boundary in the array, over their distance (300 µm). The activation time at x is set as the earliest time of $V_m(x,t) > -60\,\text{mV}$. As shown in the right panel of Fig. 1, when we vary the permeability κ the transverse CV increases rapidly. In particular, the value of $\kappa = 690\,\text{mS}\,\text{cm}^{-2}$ yields a very high, non-physiological value of velocity by several orders of magnitude. In fact, a physiological value of CV, e.g., $0.1\,\text{m}\,\text{s}^{1}$ to $0.2\,\text{m}\,\text{s}^{-1}$ would be achieved by a value of κ of the order of $10^{-2}\,\text{mS}\,\text{cm}^{-2}$. We conclude that only a very small portion of the boundary should be permeable.

Hence, in the second experiment we consider a similar setup of 30 vertically-stacked cells, but now the permeable portion of the boundary is small, between 1% and 15% of the whole length. We also vary the shape of the connection, from flat to wave-like, and its position along the boundary. Henceforth, we call "Permeable Curve" (PC) the permeable portion of the boundary. In Fig. 2 we report different options for the PC. The connection is parameterized with the length l and the amplitude a of the PC.

The results of the second experiment are summarized in Fig. 3. Firstly, we notice that there is a major difference in CV when moving from an aligned pattern of PCs (first row of Fig. 3) to an alternating pattern (second row). In the case of aligned PCs, the CV is always above $1\,\text{m}\,\text{s}^{-1}$, thus non-physiological, for all combinations of PC shape and size. We also tested the case of PCs are not centered but still aligned, and the results are the same (not shown). This suggests that gap junction alignment between cells can sensibly enhance the transverse CV. On the other hand, alternating PCs yields realistic values of CV for a wide range of shape parameters. A lack of alignment of gap junctions in the transverse direction is indeed possible, in contrast to the longitudinal direction. (However, in the fiber direction the cell is elongated, and we expect less end-to-end influence between gap junctions.)

It is also interesting to observe how CV varies with PC shape and size (see Fig. 3). The flat configuration always yields the fastest conduction, as expected. For large l, the flat PC is very similar to the "Wave" PC in terms of CV, but this is not the case for the other shapes. Somewhat paradoxically, the "Wave" and "Squared Wave" shapes do not always lead to similar CVs, although the

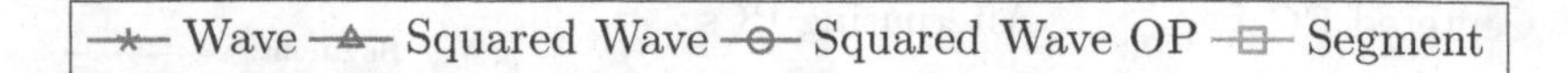

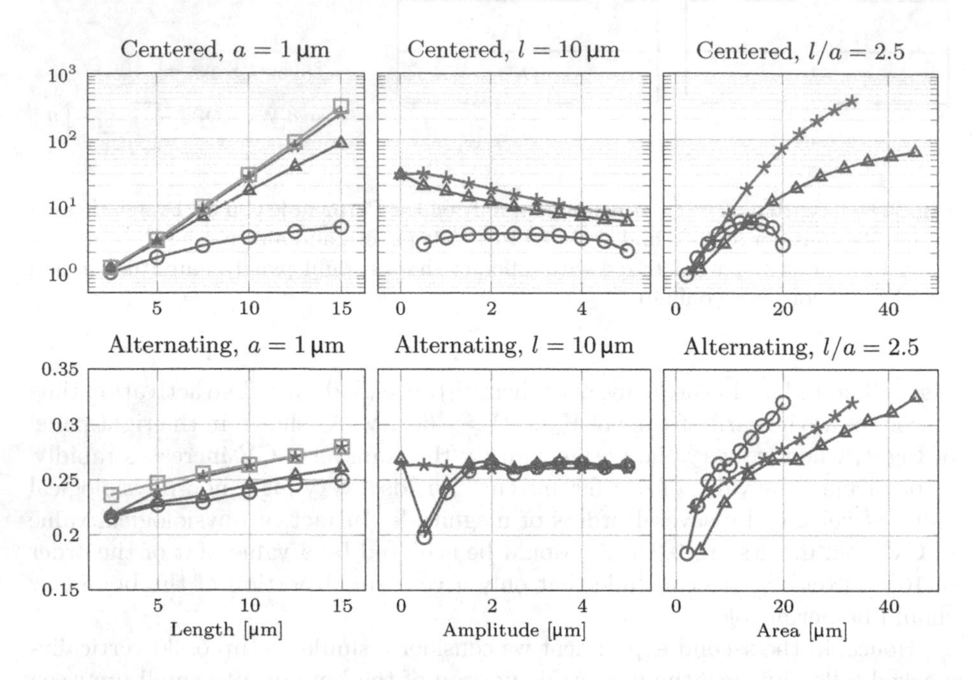

Fig. 3. Transversal CV [m s^{-1}] for different shapes and positions of the Permeable Curve (PC). For area we mean the surface area of the PC, which in $\mathbb{R}^2$ is the length. For Squared Wave, for instance, the area is $2a+l$, and for Squared Wave OP is $2a$. We vary the area by varying l and a, while keeping the ratio $l/a = 2.5$ constant.

difference is small. We conjecture that the contact area along the propagation direction, rather than the total area is what matters in determining the CV. This is quite apparent in the "Squared Wave OP" case, where the CV is generally lower than the other cases. Interestingly, here we still observe an increase in CV with l, which is due to the permeable vertical segments being better distributed along the cell's side, allowing the potential to propagate more uniformly through it. We also report that increasing the amplitude a of the PC leads to a decrease in CV for the "Wave" and "Squared Wave" shape. This is due to a decrease in smoothness of the propagation, when the amplitude of the PC is too large compared to its length. In summary, the major determinant of transverse CV in this experiment is gap junction alignment, followed by the length and smoothness of the PC.

The third experiment aims at exploring more in detail the effect of alignment of gap junctions. In fact, within a block of cells, the location of gap junctions is likely not structured. In this experiment we consider again the same array of cells, however the position of the PCs is randomized. For every shape of PC (see Fig. 2) we perform 1000 experiments. For each experiment, the PC relative

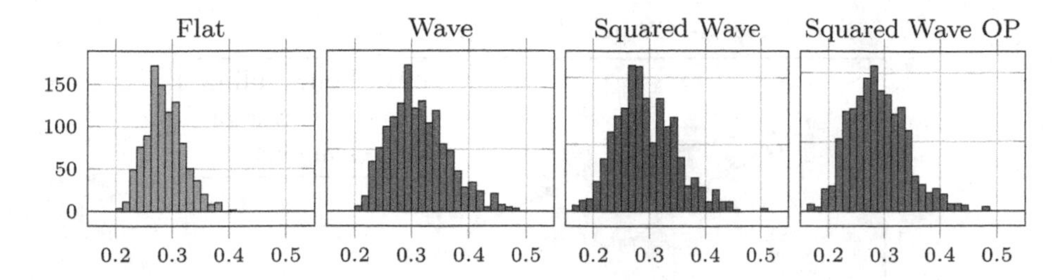

Fig. 4. Transverse CV distribution for randomly positioned PCs in a vertical array of 30 cells. For a graphical representation of PC shapes we refer to Fig. 2.

position p follows a uniform distribution $p \sim \mathcal{U}(l/c_l, 1-l/c_l)$, where $c_l = 100\,\mu\text{m}$ is the cell length. The length l and amplitude a of the PC are respectively fixed to $2.5\,\mu\text{m}$ and $1\,\mu\text{m}$, except for the Segment PC where $l = 0.5\,\mu\text{m}$. With this choices, for all PC shapes we recover a CV of $0.2\,\text{m}\,\text{s}^{-1}$ when $p = 0.2$ with alternation, similarly to Fig. 2, middle panel.

The resulting distribution of CV is displayed in Fig. 4. Interestingly, CV values are physiological in all cases, and the dispersion is low. The average CV is approximately $0.3\,\text{m}\,\text{s}^{-1}$ in all cases. The standard deviation is lower in the case of flat PC ($0.033\,\text{m}\,\text{s}^{-1}$), when compared to the other PC shapes ($0.051\,\text{m}\,\text{s}^{-1}$). Since the PC are positioned randomly, an alignment is unlikely to occur and the potential propagates following a zigzag curve, similarly to the case of alternating PC positioning of the previous experiment. The smaller standard deviation of the "Segment" PC is hard to explain. The histograms in Fig. 4 suggest that fast CVs are less likely to occur in the "Segment" case, thus reducing the dispersion.

The final experiment uses a 2-dimensional array of 20×20 cells, with a stimulus delivered at the bottom-left corner. In particular, we study how random deactivation of transversal PCs affect longitudinal, transversal and diagonal CVs. In a way, we are trying to mimic random deposition of endomysial fibrosis, as observed in the atria in patients with atrial fibrillation [13]. Here, we consider flat PCs with length $l = 0.5\,\mu\text{m}$. For fixed $P \in [0, 1]$, we run $M = 40$ experiments where transversal PCs are randomly placed and their conductivity κ is set to zero with probability P. Next, we compute the average longitudinal, transversal and diagonal CVs over the M experiments.

In Fig. 5 we display the average CVs and confidence interval of one standard deviation with respect to the probability P. As expected, a decrease in transversal PCs permeability is associated with a decrease in diagonal and transversal CV. A block in conduction occurs for a large value of $P \geq 0.8$. The CV decreases also in the longitudinal direction as P increases, and it is about a half of its original value for $P = 1$. This is because the lower fiber of cells is isolated from the rest of the tissue, since the top boundary has zero current flux. Finally, the standard deviation is very small for the transversal CV, because it is computed on the left boundary and so the signal does not cross the whole tissue. In contrast, the diagonal CV standard deviation is very high for the same reason.

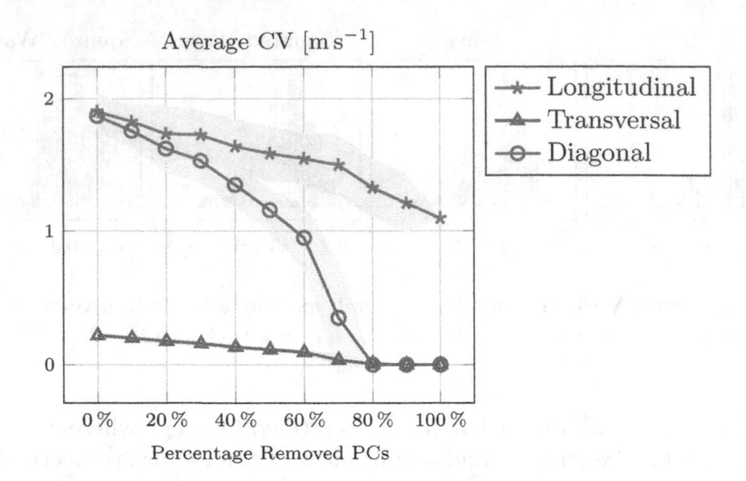

Fig. 5. Average longitudinal, transversal, and diagonal conduction velocities when transversal PCs are deactivated with probability P. The shaded area represents the confidence interval of one standard deviation.

4 Discussion

In this work we show that the transverse conduction velocity in a sub-µm microstructural model of cardiac tissue is strongly affected by lateral gap junction distribution and shape. We found that the major determinant of CV is the cell-to-cell alignment of gap junctions. The CV is non-physiologically fast when gap-junctions are vertically aligned. A lack of alignment leads to a transverse CV of $0.1\,\mathrm{m\,s^{-1}}$ to $0.3\,\mathrm{m\,s^{-1}}$, within a physiological range. The second determinant of CV is the smoothness of the cell-to-cell connection or permeable curve. These results are statistically robust, in the sense that also a random distribution of gap junctions yields similar CVs.

The EMI model can be effectively used to study the effect of gap junction remodeling in heart failure and cardiac arrhythmia [2]. It has been observed that gap junctions in a diseased myocardium may shift from a classical end-to-end (longitudinal) distribution towards a more side-to-side (transversal) arrangement, with a lower longitudinal CV [14]. In patients with history of atrial fibrillation, structural remodeling in terms of fibrosis is a major determinant of the arrhythmic substrate [13]. Here, endomysial fibrosis leads to endo-epicardial dissociation of the propagation due to a reduced transmural coupling of the myocardium [5]. This is consistent with the last numerical experiment, where transverse velocity is significantly reduced.

This work has some limitations too. We focused only on 1-D and 2-D cell arrangements, with cells of fixed shape and size. We also did not include the extracellular space between cells in all our experiments, since cell-to-cell distance is typically less than $0.2\,\mu\mathrm{m}$, see Spach *et al.* [16]. In reality, the myocardium has a complex 3-D structure with branching and extracellular space surrounding most of the cells. Reproducing this arrangement in 2-D is difficult if not impossible. We are working on extending the BEM formulation of the EMI model to the 3-D case. We believe that our implementation of the EMI model is fast and sufficiently flexible for studying complex geometries. In conclusion, our model can be used to determine CVs to be employed in the a bidomain or monodomain formulation, e.g., in cases where the tissue presents fibrosis, and the effective model parameters are hard to determine experimentally.

Acknowledgments. This work was supported by the European High-Performance Computing Joint Undertaking EuroHPC under grant agreement No 955495 (MICROCARD) co-funded by the Horizon 2020 programme of the European Union (EU) and the Swiss State Secretariat for Education, Research and Innovation.

References

1. Abdulle, A., Grote, M.J., Rosilho de Souza, G.: Explicit stabilized multirate method for stiff differential equations. Math. Comput. **91**, 2681–2714 (2022)
2. Corrado, D., Link, M.S., Calkins, H.: Arrhythmogenic right ventricular cardiomyopathy. N. Engl. J. Med. **376**(1), 61–72 (2017)
3. Courtemanche, M., Ramirez, R.J., Nattel, S.: Ionic mechanisms underlying human atrial action potential properties: insights from a mathematical model. Am. J. Physiol.-Heart Circulatory Physiol. **275**(1), H301–H321 (1998)
4. Foster, K.R., Sowers, A.E.: Dielectrophoretic forces and potentials induced on pairs of cells in an electric field. Biophys. J. **69**, 777–784 (1995)
5. Gharaviri, A., et al.: Epicardial fibrosis explains increased endo-epicardial dissociation and epicardial breakthroughs in human atrial fibrillation. Front. Physiol. **11**(68) (2020)
6. Henríquez, F., Jerez-Hanckes, C.: Multiple traces formulation and semi-implicit scheme for modelling biological cells under electrical stimulation. ESAIM: Math. Model. Numer. Anal. **52**, 659–702 (2018)
7. Leon, L.J., Roberge, F.A.: A model study of extracellular stimulation of cardiac cells. IEEE Trans. Biomed. Eng. **40**, 1307–1319 (1993)
8. Neu, J., Krassowska, W.: Homogenization of syncytial tissues. Crit. Rev. Biomed. Eng. **21**(2), 137–199 (1993)
9. Pennacchio, M., Savaré, G., Franzone, P.C.: Multiscale modeling for the bioelectric activity of the heart. SIAM J. Math. Anal. **37**(4), 1333–1370 (2005)
10. Pezzuto, S., Hake, J., Sundnes, J.: Space-discretization error analysis and stabilization schemes for conduction velocity in cardiac electrophysiology. Int. J. Numer. Methods Biomed. Eng. **32**, e02762 (2016)
11. Richardson, G., Chapman, S.J.: Derivation of the bidomain equations for a beating heart with a general microstructure. SIAM J. Appl. Math. **71**(3), 657–675 (2011)
12. Roth, B.J.: Electrical conductivity values used with the bidomain model of cardiac tissue. IEEE Trans. Biomed. Eng. **44**(4), 326–328 (1997)
13. Schotten, U., Verheule, S., Kirchhof, P., Goette, A.: Pathophysiological mechanisms of atrial fibrillation: a translational appraisal. Physiol. Rev. **91**(1), 265–325 (2011). https://doi.org/10.1152/physrev.00031.2009
14. Severs, N.J., Bruce, A.F., Dupont, E., Rothery, S.: Remodelling of gap junctions and connexin expression in diseased myocardium. Cardiovasc. Res. **80**(1), 9–19 (2008)

15. de Souza, G.R., Krause, R., Pezzuto, S.: Boundary integral formulation of the cell-by-cell model of cardiac electrophysiology, submitted. arXiv:2302.05281
16. Spach, M.S., Heidlage, J.F.: The stochastic nature of cardiac propagation at a microscopic level: electrical description of myocardial architecture and its application to conduction. Circ. Res. **76**(3), 366–380 (1995)
17. Stinstra, J.G., Hopenfeld, B., MacLeod, R.S.: On the passive cardiac conductivity. Ann. Biomed. Eng. **33**, 1743–1751 (2005)
18. Tveito, A., Mardal, K.A., Rognes, M.E.: Modeling Excitable Tissue: The EMI Framework. Springer, Heidelberg (2021)

Automated Generation of Purkinje Networks in the Human Heart Considering the Anatomical Variability

María Correas(✉), María S. Guillem, and Jorge Sánchez

ITACA Institute, Universitat Politècnica de València, Valencia, Spain
{macorga5,mguisan,jorsana4}@itaca.upv.es

Abstract. The Purkinje networks play a crucial role in the coordinated activation of the ventricular myocardium. The Purkinje networks grow in the right and left ventricles via the septum towards the apex and then upwards to the base with a different density. However, the inter-patient variability affects the growing pattern. In this study, we aimed to automatically create different Purkinje networks and evaluate the variability of their growth. We developed an algorithm that automatically generated ten different Purkinje network realizations in ten different ventricular geometries, obtaining 100 different Purkinje subendocardial networks. The growth was affected mainly by the size of the ventricles. Bigger ventricles (volume > $200\,\mathrm{cm}^3$) were highly populated with 1099 ± 43 Purkinje-Myocyte junctions compared to smaller ventricles (volume < $100\,\mathrm{cm}^3$) with 746 ± 38 Purkinje-Myocyte junctions. The Purkinje network activation sequence was also correctly verified by activating a biventricular geometry. In conclusion, we provide an algorithm that automatically produces biventricular Purkinje networks for any given ventricular geometry with a physiological activation.

Keywords: Purkinje network · Fractal growing · Cardiac modeling

1 Introduction

The Purkinje Network (PN) is a component of the ventricular conduction system which carries the electrical propagation from the atrioventricular node across the His Bundle (HB) to the ventricular myocardium. The PN's ramified network and electrophysiology enable a fast and coordinated activation of the ventricular myocardium. Significant differences in the PN's morphology between species have been outlined [23], as well as an inter-individual variability in the network branches' location, structural organization, and density [5]. Additionally, several authors have compared its morphology to fractal patterns [8,19,23].

The PN is electrically isolated from the myocardium by an insulating sheath and in humans the PN's branches can be only seen in the most internal third of the myocardial wall [16,23]. When the HB penetrates the interventricular septum, it subdivides into the left and right branches [23]. The left branch goes

O. Bernard et al. (Eds.): FIMH 2023, LNCS 13958, pp. 127–136, 2023.
https://doi.org/10.1007/978-3-031-35302-4_13

down to the apex and subdivides in the septum's superior third. Two branches grow to the base of the anterior and posterior papillary muscles, and a subendocardial network grows to the apex [16]. When the branches reach the papillary muscles and the subendocardial network reaches the apex, they begin to grow towards the base [16].

The right branch grows through the interventricular septum until reaching the base of the septal papillary muscle (PM) [12]. Afterward, it achieves the right ventricular free wall through the moderator band [1,12,23]. At this point, it subdivides into many branches giving rise to a subendocardial network [1,12,16]. The zones below the tricuspid and mitral valves are free of Purkinje fibers [1,23], and the middle basal free wall of the right ventricle (RV) is weakly covered [23].

The Purkinje-Myocyte Junctions (PMJs), located at the extremities of the network, are responsible for the electrical propagation from the PN across the PMJs to the myocardium. However, Morley et al. [13] outlined that propagation across only a subset of the existing PMJs of a PN is successful.

The study of the PN has been a demanding task since its discovery in 1845 by Jan Evangelista Purkinje [15]. The PN's anatomy is difficult to analyze because the bundles are thin and fragile. Furthermore, there is no technique to measure in detail its electrophysiological behavior, neither the number and distribution of functioning PMJs [4]. Therefore, using mathematical models opens a wide variety of tools to help understand the electrophysiology of the PN. Furthermore, computer models help understand cardiac electrophysiology by varying a wide range of parameters in a controlled environment which could be translated to the observed inter-individual variability of the PN [3,8,19,23].

In this study, we aim to investigate the anatomical variability of the Purkinje network in different ventricular anatomies and its impact on the network's activation time. Additionally, we investigate the propagation pattern of the PN and the ventricular activation sequence that arises from the PN.

2 Materials and Methods

2.1 Purkinje Tree Algorithm

The PN growth was described with an automatic fractal algorithm that adjusts to any given ventricular geometry. The algorithm's fractal growth was designed as suggested by Costabal et al. [8] and based on the current anatomical [1,12,16,23] and electrophysiological [3,9] knowledge. The algorithm was written in Python and uses the VTK (The Visualization Toolkit) library. Additionally, the proposed framework uses Cobiveco [18], a biventricular coordinate system with symmetric coordinate directions in both ventricles, to create the Purkinje network in different ventricular geometries independent of the anatomical variability.

Before the fractal algorithm starts populating the subendocardial tissue (Cobiveco tm > 0.65), the PN's main branches must be created. These branches connect two nodes located at the base of the left and right ventricles with the PM insertions. Their growth is based on the path search algorithm A* and guided by checkpoints based on anatomy. The anatomical landmarks used in the creation of these branches are summarized in Table 1. The main branches of a ventricular mesh are depicted in Fig. 1A.

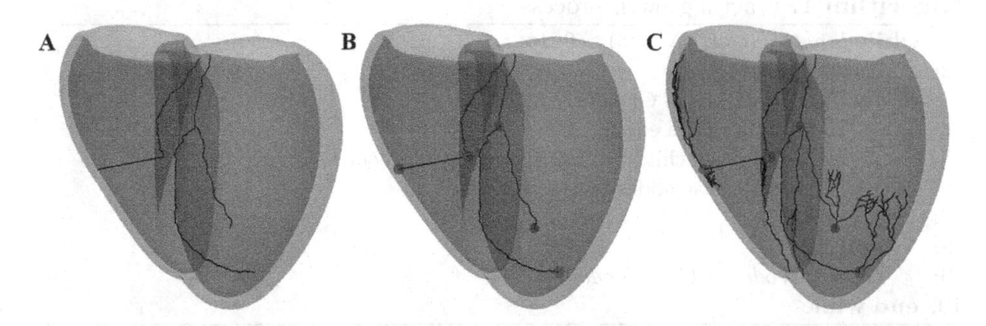

Fig. 1. Stages on the creation of the PN. A) Main skeleton of the PN. **B)** Indicated with orange circles the initial growth nodes. **C)** Beginning of the subendocardial networks which are originated in the initial growth nodes and fill the ventricular mesh. (Color figure online)

Table 1. Summary of relevant anatomical and physiological landmarks used in the PN's algorithm defined with Cobiveco.

Earliest activation sites		Anatomical structures	
Node location	ab, tm, rt, tv	Node location	ab, tm, rt, tv
Mid-posterior superior LV	0.65, 1, 0.09, 1	Posterior PM LV	0.3, 0.75, 0.11, 0
Mid-posterior inferior LV	0.55, 1, 0.18, 0	Septal PM RV	0.6, 1, 0.68, 1
Anterior basal paraseptal LV	0.9, 1, 0.57, 0	Anterior PM RV	0.5, 1, 0.47, 1
Right septal wall	0.4, 1, 0.85, 1	Initial node RV	0.99, 0.67, 0.87, 1
Free wall RV	0.45, 1, 0.15, 1	Initial node LV	0.99, 0.66, 0.87, 0
		Anterior PM LV	0.3, 0.8, 0.47, 0

Once the main branches were created, the fractal growth begins, letting as many as possible subendocardial network ramifications grow from five predefined initial growth nodes (Fig. 1B). The five initial growth nodes are located at the insertion of the papillary muscles and in the superior third of the left septal wall [23]. In Algorithm 1, the followed fractal growing iterative process is described. From each growth node, two child branches are created based on the A* algorithm. The growth nodes are labeled with a growth direction, which indicates whether the first child branch grows to the apex, to the base, or to each of the five earliest activation sites [9,10] defined in terms of the Cobiveco coordinate system (Table 1). The direction of the second branch is randomly chosen from all the points that are at a certain angle with respect to the first branch.

The extremities of the child branches will turn into new growth nodes. In Fig. 1C it can be seen the resulting PN after the first iterations of the algorithm.

Algorithm 1. Fractal growth process

```
1: while length(GrowthNodes) > 0 do
2:     shuffle GrowthNodes
3:     for NodeToGrow in GrowthNodes do
4:         if There is space enough then
5:             Create two child branches starting at NodeToGrow
6:             Add its last nodes to ChildNodes
7:         end if
8:     end for
9:     GrowthNodes = ChildNodes
10: end while
```

The algorithm follows certain growing criteria. Firstly, the nodes of the new branch can not belong to the existing network or be close to a point of the existing network. Furthermore, the middle basal free wall of the right ventricle is weakly covered, and the branches do not reach the ventricular base (Cobiveco ab < 0.95) [1,23]. If a new branch does not fulfill all the growing criteria, it can no longer grow. The network will grow until there are no more growth nodes left.

Up to this day there is still uncertainty about the inter-individual variability in the branch lengths, the angles between branches, and the distance between branches. These three parameters were adjusted to obtain the number of PMJs reported in literature [4,7,8]. They depend on their distance to the end location (e.g., apex, base, or activation sites).

2.2 Simulation

Simulations were performed to evaluate the propagation in a free-running PN and its corresponding activation sequence in a biventricular geometry. The PN mesh had ~100k nodes with an average edge length of 160 µm. The ionic model proposed by Stewart et al. [20] was used to describe the human electrophysiology of the PN. Conduction velocity across the PN was set to 2.5 m/s [3]. 10 ventricular meshes, adapted from Schuler et al. [17], were used for the evaluation of the PN algorithm. The ventricular myocadium's electrophysiology was simulated using the human myocyte model proposed by Tomek et al. [21]. Ventricular conduction velocity was set at 0.5 m/s, 0.25 m/s, 0.16 m/s [6]. The ventricular myocardium was stimulated using a free-running Purkinje network simulation. It was assumed that the PMJ excites the endocardial layer at several sites of earliest activation from which excitation spreads. Monodomain simulations were performed using the open electrophysiology simulator openCARP [14].

3 Results

A total of one hundred PNs were created to evaluate the impact of the ventricular anatomy in the developed algorithm. Ten different ventricular instances were used as a base for the networks to grow. The ratio between the ventricular volume and the length of the network was used to analize the PNs variability. The PNs variability is depicted in Fig. 2 for all 100 networks. Bigger ventricles (ventricles 2, 4 and 8, volume > 200 cm^3) were highly dense populated compared to smaller ventricles (ventricles 1 and 5, volume < 100 cm^3). The mean ratio over the 100 PN realizations is 1.19 $\pm$ 0.139. Finally, the number of PMJs was also checked. The obtained PNs presented a mean of 906.17 $\pm$ 120.83 PMJs. This value was directly proportional to the size of the ventricular mesh ($R^2 = 0.95$).

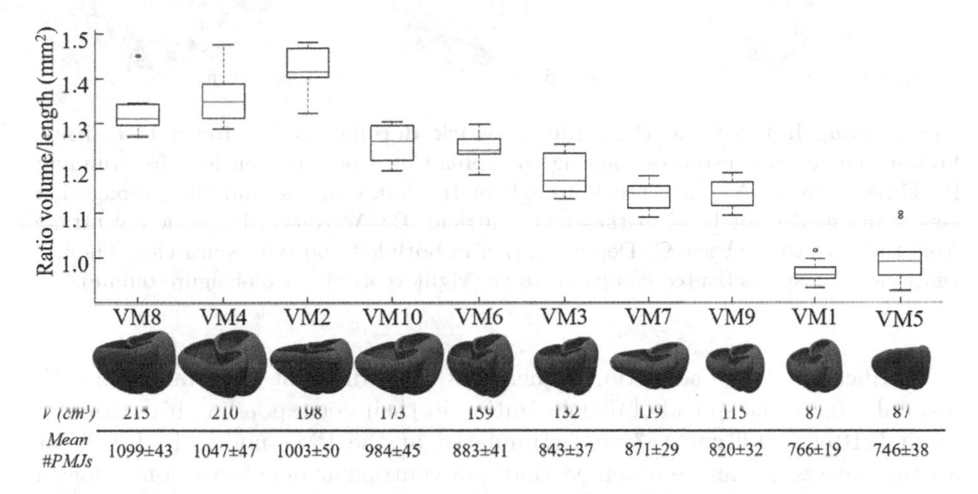

Fig. 2. Purkinje network variability density ratio. Analysis of the PNs variability as the ratio between the ventricular volume and the network length. For each ventricular mesh (VM) ten different PNs realization were obtained and the box-plot shows the dispersion of the individual ratio. Additionally, the volume (V) of each ventricular mesh, and the mean number of Purkinje-Myocyte Junctions (PMJs) were calculated.

Additionally, PN simulations were performed to analyze the electrical propagation in the network. The spread of depolarization for the PN in both ventricles is depicted in Fig. 3. Propagation started simultaneously in two nodes located at the base of the left and right ventricles (end of HB). The stimulus progressed through the left and right branches in the septum. In the RV, at 21 ms, the propagation travelled through the moderator band. At 33 ms (Fig. 3A), the propagation reached the posterior PM of the left ventricle (LV) and the apex. Then, 6 ms later, it reached the free wall of the RV and the anterior PM of the LV. At that moment, several PMJs in the LV septal wall and the apex were already activated. At 51 ms (Fig. 3B) the PN at LV posterior wall was depolarized. Later, at 69 ms (Fig. 3C), the LV anterior wall and the RV middle low external wall were reached. Finally, the whole PN was completely depolarized at 98 ms at the basal external wall of the RV.

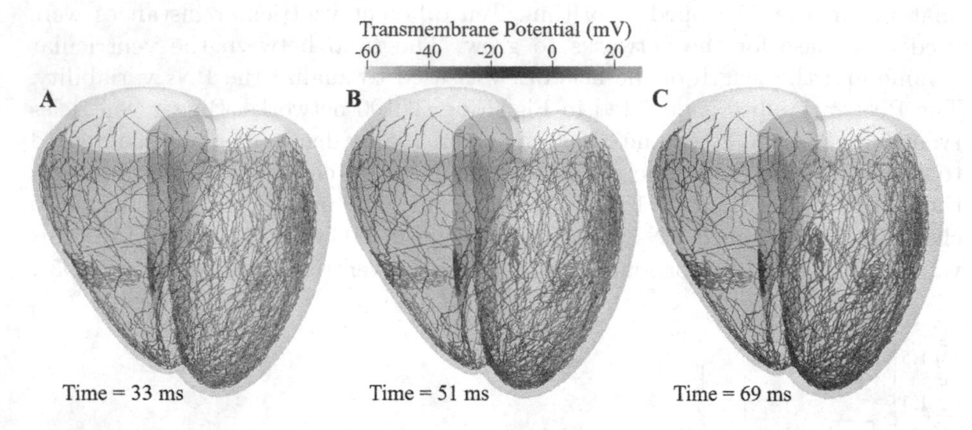

Fig. 3. Snapshots of the Purkinje network depolarization for a PN simulation. Three time instances showing the spread of depolarization in a free-running Purkinje network. **A**) First breakthrough in the left ventricle and the propagation across the moderator band in the right ventricle. **B**) Activation in the left ventricle from the apex to the base. **C**) Depolarization in both left and right ventricles. The left ventricle is mostly activated compared to the right ventricle. (Color figure online)

Verification of the activation sequence of three different PNs was done with a simulation of the spread of depolarization in their corresponding biventricular mesh (210 cm^3, 132 cm^3, 87 cm^3) stimulated at the PMJ nodes. In Fig. 4 for all three cases, it can be observed that the ventricular depolarization followed an endocardium to epicardium pattern. It started at the apex and finished at the base of the external wall of the RV. These results are in agreement with measurements in real hearts [9]. For all three biventricular simulations, the whole ventricular myocardium activated in 101 ms (Fig. 4A), 87 ms (Fig. 4B), and 85 ms (Fig. 4C).

Finally, it must be outlined that myocardial activation across 26.11% of the existing PMJs was unsuccessful. These PMJs nodes at the myocardium had already been activated by the electrical propagation traveling across the ventricular myocardium. Furthermore, 60.61% of the unsuccessfully activated PMJs were located in the basal area (Cobiveco ab > 0.75).

4 Discussion and Conclusion

Computational modeling was used to investigate the creation of PNs and the impact of ventricular anatomy variability. We present an automatic algorithm based on fractal growing to create a Purkinje network which adapts to different

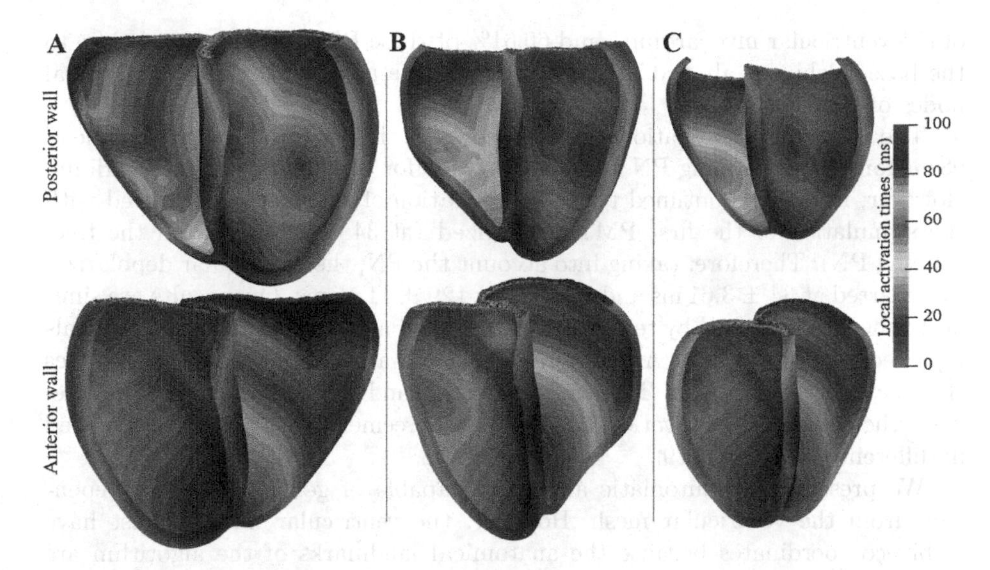

Fig. 4. Local activation maps. Ventricular activation time maps of three biventricular meshes with volume: **A**) 210 cm^3, **B**) 132 cm^3, **C**) 87 cm^3. The myocardium was stimulated at the Purkinje-myocyte junctions of the corresponding free-running Purkinje networks. (Color figure online)

ventricular geometries. Additionally, we have shown the PN's activation sequence by obtaining physiological ventricular myocardium activation times.

The 100 generated PNs had the main subdivisions of the left and right branches, as well as the subendocardial networks originated in the papillary muscles, which agrees with the 2D and 3D observations of Almeida et al. [1]. Additionally, the PNs exhibited a downward growth to the apex, followed by an upward growth to the base. It must be also outlined that, as reported in experimental studies, the PN in the LV was more populated than in the RV [2].

Even though our PN does not allow intersections compared to Costabal et al. [8] and Sebastian et al. [19], we achieved similar morphology to the previous studies. In our algorithm the number of generations of the fractal network is not predetermined [8] but our defined criterion let the networks grow freely until the submyocardium was completely populated. Additionally, the number of PMJs obtained were similar to the ones obtained by Costabal et al. [8] and Carpio et al. [7]. In rabbits, Behradfar et al. [4] found a number of PMJs proportional to the myocardium area and scaled to a human heart the number of PMJs could vary from 1000 to 5000. Moreover, we accounted for the PN's inter-individual variability by randomly changing the order of the growing nodes and the direction of the second child branch in the fractal algorithm. Additionally, our findings agree with the observations of Morley et al. [13] in which only a subset of PMJs successfully depolarized the ventricular myocardium. Our findings suggest that 26.11% of the total number of PMJs did not contribute to the depolarization of the ventricular myocardium, and 60.61% of these PMJs were located close to the base (Cobiveco ab > 0.75). These PMJ correspond to the most peripheral nodes of the network.

The verification simulations (Fig. 4) showed a similar total activation time of 98 ms for all free-running PN and 91 ± 8.72 ms for the ventricular myocardium. However, it must be outlined that the simulation of the ventricles started with the stimulation of the first PMJ depolarized (at 34 ± 3.61 ms from the free-running PN). Therefore, taking into account the PN, the ventricular depolarization started at 34 ± 3.61 ms and finished at 129 ± 11.15 ms. Our results are similar to the ones reported by ten Tusscher et al. [22] and the His-Purkinje intervals reported by Li et al. [11]. Additionally, our PN's activation sequence resembles the ones obtained by ten Tusscher et al. [22] and Behradfar et al. [4]. Likewise, the ventricular activation times are in agreement with the ones reported in different studies [8,9,19].

We presented an automatic algorithm capable of generating PN independent from the ventricular mesh. However, the ventricular meshes must have Cobiveco coordinates because the anatomical landmarks of the algorithm are defined with this coordinate system. Additionally, our proposed rules are based on data extracted from ungulates [1,16,19], rabbits [2,4], rats [5] and mice [13], which are presumably different from humans. In future works, the study can be extended to simulate all 10 ventricles and calculate the 12-lead ECG to further verify the growth conditions of the proposed algorithm. Additionally, the distribution and density of the PMJs can be further investigated to analyze the depolarization of the ventricular myocardium. Nevertheless, the created PN can be used to investigate ventricular activation in health and disease.

In conclusion, we proposed an automatic algorithm that creates Purkinje networks accounting for the inter-patient variability and provides physiological activation times.

Acknowledgements. This project is part of the grant I+D+i PLEC2021-007614, funded by MCIN/AEI/10.13039/501100011033 and by the "European Union NextGenerationEU/PRTR". We thank Inés Llorente for adapting the biventricular meshes to simulate.

References

1. De Almeida, M.C., Lopes, F., Fontes, P., Barra, F., Guimaraes, R., Vilhena, V.: Ungulates heart model: a study of the Purkinje network using India ink injection, transparent specimens and computer tomography. Anat. Sci. Int. **90**(4), 240–250 (2014). https://doi.org/10.1007/s12565-014-0255-9
2. Atkinson, A., et al.: Anatomical and molecular mapping of the left and right ventricular His-Purkinje conduction networks. J. Mol. Cellular Cardiol. **51**, 689–701 (2011). https://doi.org/10.1016/j.yjmcc.2011.05.020
3. Bayer, J.D., Sobota, V., Moreno, A., Jaïs, P., Vigmond, E.J.: The Purkinje network plays a major role in low-energy ventricular defibrillation. Comput. Biol. Med. **141** (2022). https://doi.org/10.1016/j.compbiomed.2021.105133

4. Behradfar, E., Nygren, A., Vigmond, E.J.: The role of Purkinje-myocardial coupling during ventricular arrhythmia: a modeling study. PLoS ONE **9** (2014). https://doi.org/10.1371/journal.pone.0088000
5. Bordas, R., et al.: Integrated approach for the study of anatomical variability in the cardiac Purkinje system from high resolution MRI to electrophysiology simulation. IEEE (2010). https://doi.org/10.1109/IEMBS.2010.5625979
6. Cardone-Noott, L., Bueno-Orovio, A., Mincholé, A., Zemzemi, N., Rodriguez, B.: Human ventricular activation sequence and the simulation of the electrocardiographic QRS complex and its variability in healthy and intraventricular block conditions. Europace: European Pacing, Arrhythmias, and Cardiac Electrophysiology: Journal of the Working Groups on Cardiac Pacing, Arrhythmias, and Cardiac Cellular Electrophysiology of the European Society of Cardiology **18**, iv4–iv15 (2016). https://doi.org/10.1093/europace/euw346
7. Carpio, E.F., et al.: Optimization of lead placement in the right ventricle during cardiac resynchronization therapy. A simulation study. Front. Physiol. **10**, 74–74 (2019). https://doi.org/10.3389/FPHYS.2019.00074
8. Costabal, F.S., Hurtado, D.E., Kuhl, E.: Generating Purkinje networks in the human heart. J. Biomech. **49**, 2455–2465 (2016). https://doi.org/10.1016/j.jbiomech.2015.12.025
9. Durrer, D., van Dam, R.T., Freud, G.E., Janse, M.J., Meijler, F.L., Arzbaecher, R.C.: Total excitation of the isolated human heart. Circulation **41**, 899–912 (1970)
10. Gerach, T., et al.: Electro-mechanical whole-heart digital twins: a fully coupled multi-physics approach. Mathematics **9** (2021). https://doi.org/10.3390/math9111247
11. Li, Y.G., Grönefeld, G., Israel, C., Bogun, F., Hohnloser, S.H.: Bundle branch reentrant tachycardia in patients with apparent normal His-Purkinje conduction: the role of functional conduction impairment. J. Cardiovasc. Electrophysiol. **13**, 1233–1239 (2002). https://doi.org/10.1046/j.1540-8167.2002.01233.x
12. Miralles, F.B., Garcia, I., Sebastian, R.: Inverse Estimation of the Cardiac Purkinje System from Electroanatomical Maps. Ph.D. thesis, Universitat de València (2019)
13. Morley, G.E., et al.: Reduced intercellular coupling leads to paradoxical propagation across the Purkinje-ventricular junction and aberrant myocardial activation. Proc. Natl. Acad. Sci. (2005). https://doi.org/10.1073/pnas.0500881102
14. Plank, G., et al.: The openCARP simulation environment for cardiac electrophysiology. Comput. Methods Programs Biomed. **208**, 106223 (2021). https://doi.org/10.1016/j.cmpb.2021.106223
15. Purkinje, J.: Mikroskopisch-neurologische beobachtungen. Arch. Anat. Physiol. Wiss. Med. **12**, 281–295 (1845)
16. Romero, D.: Characterization and Modelling of the Purkinje System for Biophysical Simulations. Ph.D. thesis, Universitat Pompeu Fabra (2016)
17. Schuler, S., Loewe, A.: Biventricular statistical shape model of the human heart adapted for computer simulations. Zenodo (2021)
18. Schuler, S., Pilia, N., Potyagaylo, D., Loewe, A.: Cobiveco: consistent biventricular coordinates for precise and intuitive description of position in the heart - with matlab implementation. Med. Image Anal. **74**, 102247 (2021). https://doi.org/10.1016/j.media.2021.102247
19. Sebastian, R., Zimmerman, V., Romero, D., Sanchez-Quintana, D., Frangi, A.F.: Characterization and modeling of the peripheral cardiac conduction system. IEEE Trans. Med. Imaging **32**, 45–55 (2013). https://doi.org/10.1109/TMI.2012.2221474

20. Stewart, P., Aslanidi, O.V., Noble, D., Noble, P.J., Boyett, M.R., Zhang, H.: Mathematical models of the electrical action potential of Purkinje fibre cells. Philos. Trans. R. Soc. A: Math. Phys. Eng. Sci. **367**, 2225–2255 (2009). https://doi.org/10.1098/rsta.2008.0283
21. Tomek, J., et al.: Development, calibration, and validation of a novel human ventricular myocyte model in health, disease, and drug block. eLife **8**, e48890 (2019). https://doi.org/10.7554/eLife.48890
22. Tusscher, K.H., Panfilov, A.V.: Modelling of the ventricular conduction system. Progress Biophys. Mol. Biol. **96**, 152–170 (2008). https://doi.org/10.1016/j.pbiomolbio.2007.07.026
23. Vigmond, E.J., Stuyvers, B.D.: Modeling our understanding of the His-Purkinje system. Progress Biophys. Mol. Biol. **120**, 179–188 (2016). https://doi.org/10.1016/j.pbiomolbio.2015.12.013

On the Accuracy of Eikonal Approximations in Cardiac Electrophysiology in the Presence of Fibrosis

Lia Gander[1(✉)], Rolf Krause[1,2], Martin Weiser[3], Francisco Sahli Costabal[4], and Simone Pezzuto[1,5]

[1] Center for Computational Medicine in Cardiology, Euler Institute, Università della Svizzera italiana, Lugano, Switzerland
{lia.gander,rolf.krause}@usi.ch
[2] FernUni, Brig, Switzerland
[3] Zuse Institute Berlin, Berlin, Germany
weiser@zib.de
[4] Department of Mechanical and Metallurgical Engineering, School of Engineering and Institute for Biological and Medical Engineering, Schools of Engineering, Medicine and Biological Sciences, Pontificia Universidad Católica de Chile, Santiago, Chile
fsc@ing.puc.cl
[5] Laboratory of Mathematics for Biology and Medicine, Department of Mathematics, University of Trento, Trento, Italy
simone.pezzuto@unitn.it

Abstract. Fibrotic tissue is one of the main risk factors for cardiac arrhythmias. It is therefore a key component in computational studies. In this work, we compare the monodomain equation to two eikonal models for cardiac electrophysiology in the presence of fibrosis. We show that discontinuities in the conductivity field, due to the presence of fibrosis, introduce a delay in the activation times. The monodomain equation and eikonal-diffusion model correctly capture these delays, contrarily to the classical eikonal equation. Importantly, a coarse space discretization of the monodomain equation amplifies these delays, even after accounting for numerical error in conduction velocity. The numerical discretization may also introduce artificial conduction blocks and hence increase propagation complexity. Therefore, some care is required when comparing eikonal models to the discretized monodomain equation.

Keywords: Cardiac electrophysiology · Fibrosis · Monodomain model · Eikonal model · Eikonal-diffusion model

1 Introduction

Cardiac arrhythmias, such as atrial fibrillation, are major contributors to morbidity and mortality. Arrhythmias are characterized by a chaotic electrical activ-

O. Bernard et al. (Eds.): FIMH 2023, LNCS 13958, pp. 137–146, 2023.
https://doi.org/10.1007/978-3-031-35302-4_14

ity due to re-entrant waves propagating in the cardiac tissue. The risk of occurrence of an arrhythmic event is higher in presence of structural remodeling such as fibrosis, i.e. tissue with altered conductivity and electrophysiological properties. Due to the presence of fibrosis, the conduction velocity in the cardiac tissue is non-homogeneous and the heterogeneity itself is pro-arrhythmic [1].

Computational models of the electrical activity are in use to study arrhythmias and to design treatments. The monodomain model is the most commonly used in *in silico* studies [1–3]. However, the computational demand of the monodomain solution may hinder its applicability in the clinical practice. Eikonal models are a viable alternative in this respect. Eikonal models are approximations of the monodomain model with a much lower computational footprint [4,5]. These models describe the activation time as a function of space. Here, we consider the eikonal-diffusion equation and the standard eikonal equation. The eikonal-diffusion model is more accurate than the pure eikonal model, as it accounts for curvature effects and heterogeneities [6]. On the other hand, the pure eikonal can be efficiently solved with the Fast Marching or the Fast Iterative method. Moreover, the pure eikonal model is attractive for studying arrhythmias, since its numerical solution can be obtained (with some care) by Dijkstra-like algorithms that allow extensions to describe re-entry [7,8].

The tissue conductivity plays an important role in the computational models, as it describes the tissue heterogeneity. The conductivity can be discontinuous at the boundaries between the healthy and the fibrotic tissues, for instance. An exemplary situation is shown in the numerical results of Fig. 1. In the numerical experiment, we consider a heterogeneous tissue (first panel) and we compare various models for simulating the activation times. As it can be seen, the pure

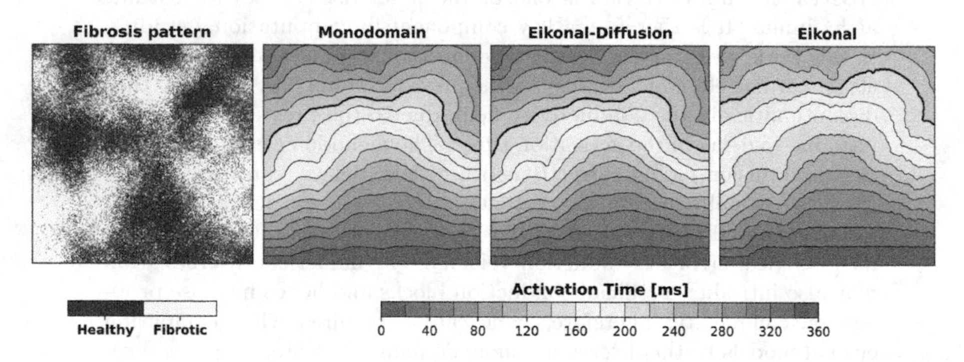

Fig. 1. Comparison between monodomain and eikonal solutions in the presence of fibrosis. The computational domain is a 2-D tissue patch of 15 cm length and height. The spatial resolution in all the models is 0.02 cm. Solutions are reported in terms of isochrones (20 ms spacing). Conduction velocities in the eikonal and eikonal-diffusion model have been adapted to compensate the numerical error in monodomain simulations. The thick contour in the last 3 panels indicates the front position after 220 ms.

eikonal approximation (fourth panel) noticeably differs from the monodomain approximation (second panel), even after adapting the conduction velocities to minimize numerical errors. Instead, the eikonal-diffusion approximation (third panel) is much more accurate. The discrepancy is due to diffusion currents, not accounted in the pure eikonal formulation. Diffusion currents are due to heterogeneities in the propagation, e.g. curvature, or in the conductivity itself, e.g. discontinuous coefficients.

In this work, we study the influence of discontinuous conductivity, e.g. due to fibrosis, in the accuracy of the monodomain and the eikonal models. Specifically, we focus on a 1-D fiber, as this enables a precise quantitative assessment of the effect of discontinuities on the overall propagation.

2 Methods

We consider a domain $\Omega \subset \mathbb{R}$ and a time interval $[0, T]$. The monodomain system for the transmembrane potential $v\colon \Omega \times (0, T) \to \mathbb{R}$ reads as follows [6]

$$\begin{cases} \frac{\partial}{\partial x}\left(\sigma(x)\frac{\partial}{\partial x}v(x,t)\right) = \beta\left(C_{\mathrm{m}}\frac{\partial}{\partial t}v(x,t) + I_{\mathrm{ion}}(v,\boldsymbol{y}) - I_{\mathrm{app}}(x,t)\right) & \text{in } \Omega \times (0,T), \\ \frac{\partial}{\partial t}\boldsymbol{y}(x,t) - \boldsymbol{F}(v,\boldsymbol{y}) = 0 & \text{in } \Omega \times (0,T), \\ \frac{\partial}{\partial x}v(x,t) = 0 & \text{in } \partial\Omega \times (0,T), \\ v(x,0) = v_0(x), \quad \boldsymbol{y}(x,0) = \boldsymbol{y}_0(x) & \text{in } \Omega, \end{cases} \tag{1}$$

where $\sigma\colon \Omega \to \mathbb{R}$ is the conductivity, $\beta > 0$ is the surface-to-volume ratio, $C_{\mathrm{m}} > 0$ is the membrane capacitance and $I_{\mathrm{app}}\colon \Omega \times (0, T) \to \mathbb{R}$ is the applied current. Moreover, $\boldsymbol{y}\colon \Omega \times (0, T) \to \mathbb{R}^m$ is a vector of gating and concentration variables of the ionic model determined by the ionic current I_{ion}, the function $\boldsymbol{F}$ and the resting states $v_0\colon \Omega \to \mathbb{R}$ and $\boldsymbol{y}_0\colon \Omega \to \mathbb{R}^m$. In this work we consider the Courtemanche-Ramirez-Nattel ionic model [9] and we set $C_m = 1$ μF/cm^2 and $\beta = 800$ cm^{-1} [2]. To numerically solve the monodomain Eq. (1), we consider a mesh with spatial resolution h and the time step Δt. Common choices for the discretization parameters are $h = 0.02$ cm and $\Delta t = 0.01$ ms [2]. For the spatial component, we use a second order finite difference method. For the time integration, we use the first order Euler method for v and the concentration variables and the Rush-Larsen method for the gating variables. We perform the monodomain simulations using the Propag-5 software [10,11].

The eikonal equations model the activation time $\phi\colon \Omega \to \mathbb{R}$ as a function of the space variable [6]. The eikonal-diffusion equation is

$$\begin{cases} \rho\sqrt{\beta\sigma(x)}\left|\frac{\mathrm{d}}{\mathrm{d}x}\phi(x)\right| - \frac{\mathrm{d}}{\mathrm{d}x}\left(\sigma(x)\frac{\mathrm{d}}{\mathrm{d}x}\phi(x)\right) = C_{\mathrm{m}}\beta & \text{in } \Omega, \\ \frac{\mathrm{d}}{\mathrm{d}x}\phi(x) = 0 & \text{in } \partial\Omega, \\ \phi(x) = 0 & \text{in } \Omega_0, \end{cases} \tag{2}$$

where $\Omega_0 \subset \Omega$ denotes the region where the action potential is initiated and $\rho \in \mathbb{R}$ is a parameter that depends on the ionic model. The parameter ρ represents the velocity of the action potential in an infinite cable with unit surface-to-volume ratio, membrane capacitance and conductivity. It is also the unique solution of the eigenvalue problem

$$\begin{cases} \rho U'(\xi) + \tilde{I}_{\text{ion}}(U(\xi)) = U''(\xi), & \xi \in \mathbb{R}, \\ \tilde{I}_{\text{ion}}(U(\xi)) \to 0, & \xi \to \pm\infty, \end{cases}$$

where $\tilde{I}_{\text{ion}}$ approximates I_{ion} during the upstroke phase, see [6]. To numerically solve the eikonal-diffusion Eq. (2), we use the finite element method on a mesh with resolution h. The non-linearity of the model is handled by a fixed-point iteration on the advection term.

By disregarding the diffusion term in Eq. (2), we obtain the standard eikonal equation, which reads as follows

$$\begin{cases} \rho\sqrt{\sigma(x)}\left|\frac{\mathrm{d}}{\mathrm{d}x}\phi(x)\right| = C_{\text{m}}\sqrt{\beta} & \text{in } \Omega, \\ \phi(x) = 0 & \text{in } \Omega_0. \end{cases} \tag{3}$$

From the eikonal equation it is possible to deduce the formula

$$\text{CV}(x) = \frac{\rho\sqrt{\sigma(x)}}{C_{\text{m}}\sqrt{\beta}} \tag{4}$$

for the conduction velocity CV: $\Omega \to \mathbb{R}$. This observation allows to compute the numerical solution of the eikonal Eq. (3) on a mesh with resolution h with a simple iterative procedure. Indeed, starting from the nodes in Ω_0, one can compute the activation times in the whole domain Ω by iteratively considering the activation times of the neighbors and the local conduction velocity.

If we assume a constant conductivity σ and we rescale Eq. (2) by the conduction velocity, i.e. if we take $\phi \mapsto \text{CV} \cdot \phi$, we obtain the equation

$$\left|\frac{\mathrm{d}}{\mathrm{d}x}\phi(x)\right| - \varepsilon\frac{\mathrm{d}^2}{\mathrm{d}x^2}\phi(x) = 1, \quad \varepsilon = \frac{\sqrt{\sigma}}{\rho\sqrt{\beta}}. \tag{5}$$

The parameter ε is an estimate of the front thickness. Thus, by Eq. (5), a lower conductivity yields a smaller front, meaning that the front is smaller in fibrotic than in healthy tissue. Moreover, a lower excitability of the membrane (lower ρ) yields a larger front. Note that Eq. (4) and Eq. (5) are still valid in 2-D or 3-D for planar waves.

To model the presence of fibrosis we follow the work of Zahid *et al.* [12]. We consider a reduction of conductivity σ only in the regions where fibrosis is present. In particular, the conductivity is a function of the spatial variable, i.e.

$$\sigma(x) = \begin{cases} \sigma_{\text{h}}, & x \in \text{healthy tissue}, \\ \sigma_{\text{f}}, & x \in \text{fibrotic tissue}, \end{cases}$$

where $\sigma_{\text{f}} < \sigma_{\text{h}}$. We do not account here for the current sink due to fibroblasts, which would be difficult to model with an eikonal approach.

3 Numerical Experiments

We perform some 1-D numerical experiments in which an action potential propagates in a tissue line from left to right. In these numerical experiments, we set $\sigma_{\mathrm{h}} = 1.5$ mS/cm [2] and we vary the reduced conductivity σ_{f}.

According to Eq. (4), a reduction in conductivity yields a reduction of conduction velocity, e.g. a 25% reduction in the conductivity corresponds to a reduction to 50% of the conduction velocity. In practice, the conduction velocity of the monodomain model is affected by a spatial discretization error and does not achieve the conduction velocity of Eq. (4). In particular, the conduction velocity decreases as the spatial resolution h increases, for a finite difference scheme as in this case [13]. This error affects the modeling of the presence of fibrosis, as the predicted reduction in conduction velocity is not achieved. Indeed, at the resolution $h = 0.02$ cm, the conduction velocity in the healthy tissue is 0.0652 cm/ms and the conduction velocity in the fibrotic tissue with $\sigma_{\mathrm{f}} = \sigma_{\mathrm{h}} \cdot 0.25 = 0.375$ mS/cm is 0.0285 cm/ms – less than the expected 0.0652 cm/ms $\cdot$ 0.5 = 0.0326 cm/ms, since the mesh width h relative to front width is larger in fibrotic than in healthy tissue. The discretization error also affects the estimate of the parameter ρ from the monodomain model. A good estimate can be obtained at the fine resolution $h = 0.005$ cm. In this case the monodomain conduction velocity in the healthy tissue is 0.0679 cm/ms, which by Eq. (4) corresponds to $\rho = 1.57$. However, at the resolution $h = 0.02$ cm, the conduction velocities in the healthy tissue and in the fibrotic tissue with $\sigma_{\mathrm{f}} = 0.375$ mS/cm respectively correspond to $\rho_{\mathrm{h}} = 1.51$ and $\rho_{\mathrm{f}} = 1.32$.

We first consider the spatial resolution $h = 0.02$ cm and we focus on the case of $\sigma_{\mathrm{f}} = 0.375$ mS/cm. We perform two numerical experiments in a tissue line of length 15 cm in which we compare the activation times of the monodomain, eikonal-diffusion and pure eikonal models. In the monodomain model the activation time is defined as the time instant when the transmembrane potential v reaches the −62 mV threshold. In the first numerical experiment, the domain consists of a random fibrotic pattern shown at the bottom of Fig. 2, panel A,

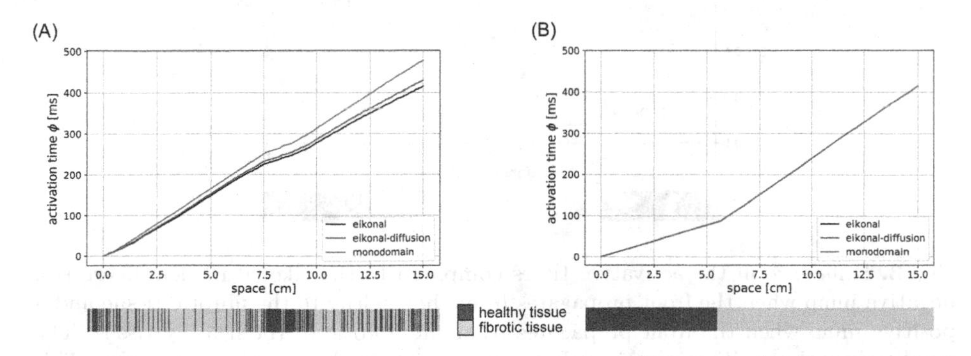

Fig. 2. Activation times in 1-D tissue with a random fibrotic pattern (panel A) and with an ordered fibrotic pattern (panel B).

with 62.4% fibrotic tissue. In the plot of Fig. 2, panel A, we can see that the activation times of the monodomain model are higher than the activation times of the eikonal-diffusion model, which in turn are slightly higher than the activation times of the pure eikonal model. This can be translated in terms of average conduction velocity, which is 0.0314 cm/ms for the monodomain model, 0.0350 cm/ms for the eikonal-diffusion model and 0.0362 cm/ms for the pure eikonal model. In the second numerical experiment, the domain consists of an ordered pattern in which the healthy tissue is on the left and the fibrotic tissue is on the right, see the bottom of Fig. 2, panel B. Note that the percentage of fibrotic tissue is the same as in the first numerical experiment. In the plot of Fig. 2, panel B, we can see that the activation times of the monodomain, eikonal-diffusion and pure eikonal models are very similar. As a consequence, the average conduction velocity is approximately 0.0362 cm/ms for the three models.

The previous observations suggest that the mismatch between the activation times of the three models is due to the discontinuities in the conductivity σ introduced at the boundaries between the healthy and fibrotic tissue. Indeed, in the first numerical experiment the domain contains 151 discontinuities and the mismatch is significant, whereas in the second numerical experiment the domain contains only one discontinuity and the mismatch is negligible. To further investigate this aspect, we perform a numerical experiment in a domain of length 6 cm consisting of three portions of length 2 cm each. The first and the last portions are healthy tissue, whereas the middle portion is fibrotic tissue. We compare the activation times of the monodomain and eikonal-diffusion models to the activation times of the pure eikonal model. Figure 3 shows the difference in the activation times compared to the pure eikonal model in the case where only two discontinuities are introduced. This plot shows two curves, one for the

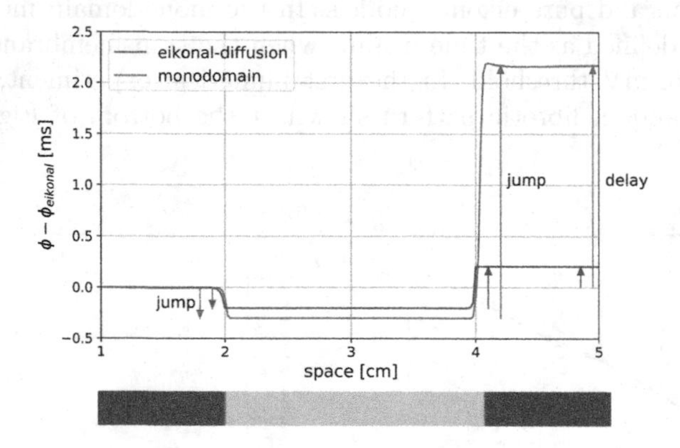

Fig. 3. Difference in the activation times compared to the eikonal model. There is a negative jump when the front propagates from the healthy to the fibrotic tissue and a positive jump when the front propagates from the fibrotic to the healthy tissue. The amplitude of the positive jump is higher than the amplitude of the negative jump. This asymmetry introduces a delay, which is the sum of the two jumps.

monodomain model and one for the eikonal-diffusion model. Note that we discard the results in the first and last portions of length 1 cm because they are affected by some boundary effects. Both curves show a negative jump as the action potential propagates from the healthy to the fibrotic tissue and a positive jump as the action potential propagates from the fibrotic to the healthy tissue. In both cases the two jumps are asymmetric, indeed the amplitude of the positive jump is higher than the amplitude of the negative jump. This asymmetry introduces a delay in the monodomain and eikonal-diffusion simulations compared to the eikonal simulations. When the conductivity σ presents many discontinuities, the total delay becomes significant, as in the numerical experiment of Fig. 2, panel A. We also observe that the amplitude of the jumps in the monodomain case is higher than the amplitude in the eikonal-diffusion case. This explains why the delay of the monodomain model is more apparent than the delay of the eikonal-diffusion model.

We now consider various spatial resolutions h and different ratios between the fibrotic conductivity σ_{f} and the healthy conductivity σ_{h}. In particular, we study 40 ratios ranging between 0.025 and 1. We consider a domain consisting of two portions with different conductivities. We denote the conductivities on the left and right portions respectively by σ_{left} and σ_{right}. When setting the healthy portion to the left and the fibrotic portion to the right, the ratio $\sigma_{\mathrm{right}}/\sigma_{\mathrm{left}}$ takes our 40 values between 0.025 and 1. Instead, when setting the fibrotic portion to the left and the healthy portion to the right, the ratio $\sigma_{\mathrm{right}}/\sigma_{\mathrm{left}}$ takes 40 values between 1 and 40. For each of these 80 cases, we simulate the propagation with the monodomain and eikonal-diffusion models and we compute the jump of the difference between the resulting activation times and the activation times of the pure eikonal model. The results are plotted in Fig. 4, panel A, against $\sqrt{\sigma_{\mathrm{right}}/\sigma_{\mathrm{left}}}$. The plot shows the monodomain curves for the coarse resolution $h = 0.04$ cm, the resolution $h = 0.02$ cm and two fine resolutions $h = 0.01$ cm and $h = 0.005$ cm. The dashed lines are vertical asymptotes that indicate the propagation failure, which occurs with the monodomain model at all resolutions.

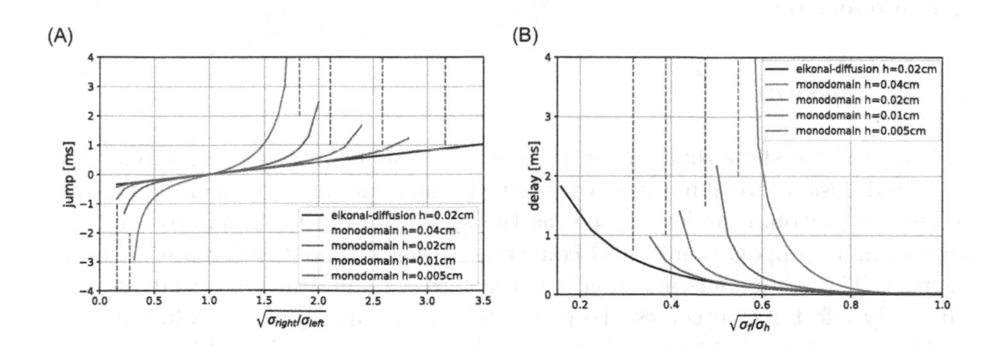

Fig. 4. Panel A: jumps of the difference in the activation times compared to the eikonal model for a front propagating from left to right. Panel B: delay introduced compared to the eikonal model. The dashed lines represent the propagation failure.

As $h \to 0$, the monodomain curve converges towards the eikonal-diffusion curve computed with $h = 0.02$ cm, which is a straight line. It is possible to show that, for a wave moving from the left to the right of the discontinuity,

$$\text{jump} \approx \frac{\varepsilon_{\text{right}} - \varepsilon_{\text{left}}}{\text{CV}_{\text{left}}} = \frac{C_{\text{m}}}{\rho^2}\left(\sqrt{\frac{\sigma_{\text{right}}}{\sigma_{\text{left}}}} - 1\right), \tag{6}$$

where the second equality follows from Eq. (4) and Eq. (5) by assuming that only the conductivity σ is discontinuous. This formula for the jump compares favourably to the numerical values obtained with the eikonal-diffusion model. For example, for the case represented in Fig. 3, the formula predicts the jumps -0.19 ms and 0.43 ms, while the numerical values are -0.20 ms and 0.41 ms. Moreover, the formula in Eq. (6) highlights the asymmetry in the jump, i.e. the amplitude of the jump at $\sqrt{\sigma_{\text{h}}/\sigma_{\text{f}}}$ is higher than the amplitude at $\sqrt{\sigma_{\text{f}}/\sigma_{\text{h}}}$. The delay introduced compared to the pure eikonal model is the sum of the jumps at $\sqrt{\sigma_{\text{f}}/\sigma_{\text{h}}}$ and at $\sqrt{\sigma_{\text{h}}/\sigma_{\text{f}}}$. These delays are plotted in Fig. 4, panel B, against $\sqrt{\sigma_{\text{f}}/\sigma_{\text{h}}}$ for the eikonal-diffusion model and the monodomain model at all resolutions h. The dashed lines again illustrate the propagation failure, which always occurs with the monodomain model. Again, there is convergence of the monodomain curve towards the eikonal-diffusion curve as $h \to 0$.

These numerical experiments show that the monodomain results converge towards the eikonal-diffusion results, therefore the eikonal-diffusion model is accurate. The difference between the monodomain and the eikonal-diffusion delays depends on both the spatial resolution h and the value of $\sqrt{\sigma_{\text{f}}/\sigma_{\text{h}}}$, see Fig. 4, panel B. The pure eikonal model does not capture the delays, therefore the error compared to the eikonal-diffusion model is given by the eikonal-diffusion delays, which depend on $\sqrt{\sigma_{\text{f}}/\sigma_{\text{h}}}$, see Fig. 4, panel B. The fibrosis model is determined by the value of $\sqrt{\sigma_{\text{f}}/\sigma_{\text{h}}}$, which is inversely proportional to the difference between the healthy and fibrotic conduction velocities. Therefore, the accuracy of the monodomain and the pure eikonal models depends on the fibrosis model. Additionally, the accuracy of the monodomain model is also affected by the spatial resolution.

4 Discussion

In this work we show that, in the presence of a high-contrast conductivity, the eikonal-diffusion model matches very well the monodomain solution, in contrast to the pure eikonal model. Indeed, as the spatial resolution tends to zero, the monodomain propagation in 1-D converges towards the eikonal-diffusion propagation. The monodomain model is subject to a numerical error that can significantly affect its outcomes. To guarantee the accuracy of the solutions, the spatial resolution of the monodomain model needs to be selected based on the fibrosis model. However, the choice determined by the fibrosis model might lead to unbearable computational costs. The pure eikonal model does not capture the delays introduced by the discontinuities in the conductivity. As a consequence, the propagation described by the pure eikonal model deviates from the eikonal-diffusion propagation.

Comparisons between eikonal and monodomain have been reported very often in the literature, see e.g. [4,5] for recent investigations. However, the effect of numerical error on the monodomain equation in the presence of highly heterogeneous conductivity has not been analyzed, to the best of our knowledge. The eikonal-diffusion model has also received little attention, in spite of its very good accuracy. The main reason is likely the computational cost, since the model is advection-dominated and this may lead to numerical issues at coarse resolutions. Indeed, the local Péclet number of Eq. (2), that is the ratio between the advective and the diffusive term, should be comparable to h^{-1} to ensure a stable solution. In other words, contrary to the pure eikonal model, the mesh requirements of the eikonal-diffusion model are as restrictive as those for the monodomain equation, unless one employs a specific numerical stabilization for the problem. The eikonal-diffusion model can capture boundary and front collision effects, but, more importantly, spatial variations due to conductivity changes. Keener [14] showed that, in the presence of conductivity variation, the conduction velocity is corrected by a term that depends on the derivative of the conductivity. This observation lead to the development of the eikonal-curvature model. A comparison between eikonal-curvature, eikonal-diffusion and pure eikonal has been provided by Pullan *et al.* [15].

This work highlights an important limitation of a pure eikonal approach when modeling cardiac arrhythmias, such as atrial fibrillation or ventricular tachycardia. The accuracy of the pure eikonal model depends on the specific choice of the fibrosis model (pattern, contrast ratio, numerical implementation), thus there is no straightforward way for adapting the conduction velocity so as to match the monodomain propagation. The eikonal-diffusion model is more suitable in this respect, because it guarantees a very good accuracy at a limited computational cost. On the other hand, the pure eikonal model is the ideal starting point in modeling re-entry phenomena [7,8], with a lot of potential thanks to its very low computational cost. In fact, in spite of the low accuracy in conduction, the pure eikonal model may still match well metrics such as inducibility of arrhythmia [3], thus ensuring its applicability in the clinical context.

Acknowledgments. This work was financially supported by the Theo Rossi di Montelera Foundation, the Metis Foundation Sergio Mantegazza, the Fidinam Foundation, and the Horten Foundation to the Center for CCMC. SP also acknowledges the CSCS-Swiss National Supercomputing Centre (No. s1074). Finally, this work was supported by the European High-Performance Computing Joint Undertaking EuroHPC under grant agreement No. 955495 (MICROCARD) co-funded by the Horizon 2020 programme of the European Union (EU) and the Swiss State Secretariat for Education, Research and Innovation.

References

1. McDowell, K.S., Zahid, S., Vadakkumpadan, F., Blauer, J., MacLeod, R.S., Trayanova, N.A.: Virtual electrophysiological study of atrial fibrillation in fibrotic remodeling. PLoS ONE **10**(2), e0117110 (2015)
2. Gharaviri, A., et al.: Epicardial fibrosis explains increased endo-epicardial dissociation and epicardial breakthroughs in human atrial fibrillation. Front. Physiol. **11**(68) (2020)
3. Gander, L., Pezzuto, S., Gharaviri, A., Krause, R., Perdikaris, P., Sahli Costabal, F.: Fast characterization of inducible regions of atrial fibrillation models with multi-fidelity Gaussian process classification. Front. Physiol. **13**, 757159 (2022)
4. Neic, A., et al.: Efficient computation of electrograms and ECGs in human whole heart simulations using a reaction-eikonal model. J. Comput. Phys. **346**, 191–211 (2017)
5. Nagel, C., et al.: Comparison of propagation models and forward calculation methods on cellular, tissue and organ scale atrial electrophysiology. IEEE Trans. Biomed. Eng. **70**(2), 511–522 (2023)
6. Colli Franzone, P., Pavarino, L.F., Scacchi, S.: Mathematical Cardiac Electrophysiology, vol. 13. Springer, Heidelberg (2014)
7. Sermesant, M., et al.: An anisotropic multi-front fast marching method for real-time simulation of cardiac electrophysiology. In: Sachse, F.B., Seemann, G. (eds.) FIMH 2007. LNCS, vol. 4466, pp. 160–169. Springer, Heidelberg (2007). https://doi.org/10.1007/978-3-540-72907-5_17
8. Gassa, N., Zemzemi, N., Corrado, C., Coudière, Y.: Spiral waves generation using an eikonal-reaction cardiac electrophysiology model. In: Ennis, D.B., Perotti, L.E., Wang, V.Y. (eds.) FIMH 2021. LNCS, vol. 12738, pp. 523–530. Springer, Cham (2021). https://doi.org/10.1007/978-3-030-78710-3_50
9. Courtemanche, M., Ramirez, R.J., Nattel, S.: Ionic mechanisms underlying human atrial action potential properties: insights from a mathematical model. Am. J. Physiol. - Heart Circulatory Physiol. **275**(1), H301–H321 (1998)
10. Potse, M., Dubé, B., Richer, J., Vinet, A., Gulrajani, R.M.: A comparison of monodomain and bidomain reaction-diffusion models for action potential propagation in the human heart. IEEE Trans. Biomed. Eng. **53**(12), 2425–2435 (2006)
11. Krause, D., Potse, M., Dickopf, T., Krause, R., Auricchio, A., Prinzen, F.: Hybrid parallelization of a large-scale heart model. In: Keller, R., Kramer, D., Weiss, J.-P. (eds.) Facing the Multicore - Challenge II. LNCS, vol. 7174, pp. 120–132. Springer, Heidelberg (2012). https://doi.org/10.1007/978-3-642-30397-5_11
12. Zahid, S., et al.: Patient-derived models link re-entrant driver localization in atrial fibrillation to fibrosis spatial pattern. Cardiovasc. Res. **110**(3), 443–454 (2016)
13. Pezzuto, S., Hake, J., Sundnes, J.: Space-discretization error analysis and stabilization schemes for conduction velocity in cardiac electrophysiology. Int. J. Numer. Methods Biomed. Eng. **32**(10), e02762 (2016)
14. Keener, J.P.: An eikonal-curvature equation for action potential propagation in myocardium. J. Math. Biol. **29**, 629–651 (1991)
15. Pullan, A.J., Tomlinson, K.A., Hunter, P.J.: A finite element method for an eikonal equation model of myocardial excitation wavefront propagation. SIAM J. Appl. Math. **63**(1), 324–350 (2002)

Sensitivity of Repolarization Gradients to Infarct Borderzone Properties Assessed with the Ten Tusscher and Modified Mitchell-Schaeffer Model

Justina Ghebryal[1](✉), Evianne Kruithof[1], Matthijs J. M. Cluitmans[2,3], and Peter H. M. Bovendeerd[1]

[1] Cardiovascular Biomechanics, Biomedical Engineering Department, Eindhoven University of Technology, Eindhoven, The Netherlands
u.ghebryal@tue.nl

[2] Philips Research Eindhoven, Eindhoven, The Netherlands

[3] Maastricht University Medical Center, Maastricht, The Netherlands

Abstract. Post-infarction ventricular tachycardia (VT) is an important clinical problem that is often caused by a re-entrant circuit located in the infarct border zone (BZ). The main changes in the BZ are in action potential duration (APD) and conduction velocity (CV), which introduce high repolarization time gradients (RTGs) and can lead to re-entry. Computational models can help in VT-risk analysis. However, the complexity of these models and the representation of the electrophysiological properties of the BZ still require investigation. In this study we conduct a sensitivity analysis in which we apply changes in APD and CV in a BZ using the detailed biophysical Ten Tusscher (TT2) model and the phenomenological modified Mitchel-Schaeffer (mMS) ionic model. First, the effect of spatial discretization on the CV is compared for both models. The TT2 model showed much larger mesh dependency for the computed CV than the mMS model. Next, we propose a tuning method to match the mMS AP shape to the TT2 AP shape. We then compare APD restitution properties. The tuned mMS showed similar APD restitution properties for large diastolic intervals (DI), but started to deviate when decreasing the DI. Finally, for both the TT2 and tuned mMS model we found that RTG is more sensitive to variation in APD than to variation in CV. When varying the APD, differences between both models were more pronounced for short than for large APDs.

Keywords: Cardiac Electrophysiology · Border Zone · Sensitivity Analysis

1 Introduction

Ventricular tachycardia (VT) is a life-threatening arrhythmia that occurs frequently in patients that have previously suffered from myocardial infarction [6].

O. Bernard et al. (Eds.): FIMH 2023, LNCS 13958, pp. 147–156, 2023.
https://doi.org/10.1007/978-3-031-35302-4_15

Clinical and experimental studies show that VT is often caused by a re-entrant wave in the infarct border zone (BZ) [10], which is a region constituting the transition between the infarct scar and healthy myocardium. Ablation is a common procedure to isolate re-entry pathways across the BZ that are responsible for VTs [9]. However, identification of critical parts of the VT re-entry circuit for ablation in the clinic is still challenging [11]. Computational models offer a powerful research tool that can provide guidance in the ablation procedure, as such models may allow to test different ablation strategies. However, the question remains how to represent the BZ in computational models [10], and how to select the appropriate ionic model, considering the trade-off between model complexity and availability of patient-specific data to personalize model parameters.

Experimental data suggest that the most prominent changes of the BZ occur in action potential duration (ADP) and conduction velocity (CV), but they also show inconsistency and variations in the changes during the chronic and post-infarction phases [10]. The changes in APD and CV both determine the spatial distribution of repolarization times (RTs) in cardiac tissue. Subsequently, regional differences in RTs yield local RT gradients (RTG). This may promote re-entry through unidirectional block [1], which can form the basis for the occurrence of VTs. In computational models, a commonly used model for the description of ionic transport across the membrane is the detailed biophysical Ten Tusscher ionic model (TT2, consisting of 19 variables and 48 parameters) [18]. As an alternative, the phenomenological modified Mitchell-Schaeffer model (mMS, consisting of 2 variables and 5 parameters) has been proposed [3]. While the TT2 model has the advantage of a more detailed description of the underlying physics, the mMS model has the advantage of a more unique translation of changes in APD and CV to model parameter modifications.

Our goal is to study whether the phenomenological mMS model can show comparable behavior as the biophysical TT2 model in studying VT-risk. First, we study the effect of spatial discretization for both ionic models. Next, a tuning method is proposed in which we aim to match the shape of the AP of mMS to the one of TT2. Finally, we conduct a comparison of the tuned mMS model with TT2 on two different levels: (i) restitution properties for APD in a 1D-cable, and (ii) RTGs in a 2D-model with an idealized infarct scar and BZ where variations in APD and CV are applied. With these model comparisons, we aim to give more insights on computational model choice and its influence on VT-risk analysis.

2 Methods

The Cardiac Arrhythmia Research Package (CARP) was used to simulate electrical activity at tissue level [20] for both ionic models. For our simulations, a monodomain representation was solved using the finite element method.

2.1 Spatial Discretization Comparison

A 1D-cable is considered with spatial resolution represented by tetrahedral element size Δx (Fig. 1**A**). Using the automatic parameterization approach [4], we set the longitudinal conductivity (σ_L) constant, leading to CV = 0.6 m/s for Δx = 100 µm. Next, a planar stimulus is applied at the left boundary of the mesh, which initiates propagation in x-direction. For our analysis, we varied the element resolution and computed the corresponding CVs for both ionic models.

2.2 AP Tuning

We propose a method to adjust the parameters of the original mMS model to match the AP shape to that of the TT2 model, allowing for comparison of RTG outcomes of both models when the AP is matched. Since the AP may differ for single-cells and tissue [17], we create matching APs in a 1D-cable (same setup as in Sect. 2.1). For the optimization process, τ_{close} in the mMS model is modified, as it has the most pronounced effect on the APD of and does not affect the CV [15]. In the original model τ_{close} is set to 150 ms. Next, we normalize the membrane voltage for both models between 0 and 1. To fit the main features of the cardiac AP, four different time instants are defined to be matched for both models:

$$T1 = t\left(\frac{dv}{dt}_{max}\right) \quad T2 = t\left(\frac{dv}{dt}_{min}\right) \quad T3 = t\left(APD_{50}\right) \quad T4 = t\left(APD_{90}\right) \quad (1)$$

$T1$ defines the time of the maximum time derivative of the first phase of the AP and expresses the moment of activation; $T2$ represents the time instant of the minimum time derivative for the last AP phase; and finally $T3$ and $T4$, depict APD at 50% and 90% of the repolarization, respectively. The time instants $T_{1,2,3,4}$ are visualized in Fig. 1**D**. We then minimize the following cost function J, which is defined as the normalized root squared mean error (NRMSE):

$$J(\tau_{close}) = \text{NRMSE} = \sqrt{\sum_{i=1}^{N} w_i \left(\frac{T_i - \tilde{T}_i}{N}\right)^2} \quad (2)$$

where T_i is the duration in time for each of the four different (N = 4) time instants for mMS, and $\tilde{T}_i$ the time instant for event i in TT2. w_i is a weight that is assigned to scale the error on the different time points. $w_{1,2,3}$ is set equal to 1 and w_4 to 2, as the main goal is to match the APD (best depicted by T4).

2.3 Restitution Properties

To compare the APD restitution properties of the tuned mMS and TT2, we use the S1-S2 protocol. The protocol consists of 20 stimuli at cycle length (CL) = 500 ms until a steady-state AP is reached; after that, CL is decreased in steps of 5 ms. The restitution curves (RCs) are obtained by plotting the steady-state APD against the diastolic interval (DI). Additionally, the slope of the APD RC is investigated, as it has been shown that a RC slope > 1 can promote conduction block and lead to re-entry [7].

2.4 Sensitivity to BZ Properties

For the sensitivity to BZ variability, we followed a similar simulation setup as proposed by Costa et al. [10]. Our simulation setup is given in Fig. 1**B**. The conductivity in the scar zone was set to approximately zero. To yield a velocity of respectively 0.6 and 0.4 m/s in the longitudinal and transversal direction, the conductivities were set using the automatic parameterization approach [4]. The tissue was stimulated 40 times with CL = 1000 ms to reach steady-state. A time step of $dt = 20$ µs is used to solve the monodomain equation.

BZ Variability. CV and APD are varied in the BZ in a realistic range as previously reported from experimental data [2,5,10]. For sensitivity to CV, the standard transversal conductivity (σ_{T_n}) that yields 0.4 m/s [4] was modified (Fig. 1**C**). For variations in APD we followed the AP tuning method described in Sect. 2.2 at CL = 1000 ms. First, the mMS model was tuned to TT2 by setting τ_{close} to 189 ms, referred to as the normal APD (indicated by the black line in Fig. 1**D**). Next, the duration of the normal APD is varied. For TT2, we modified the parameter g_{Ks}, as this has the most effect on APD for this ionic model [10]. Discrete variations were applied for the parameters g_{Ks} and τ_{close}. The resulting APs are shown in Fig. 1**D**. The maximum difference between the action potentials of both models corresponds to an NRMSE of 5.7 ms.

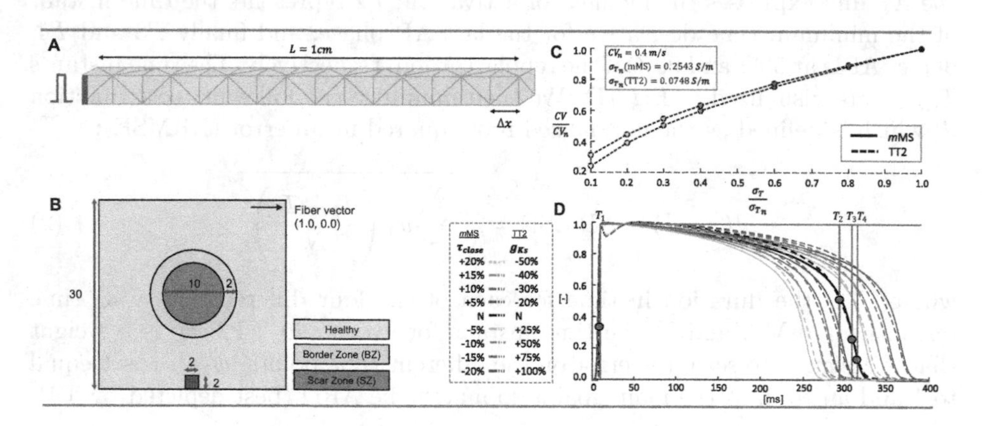

Fig. 1. A: 1D-cable with element size Δx and length (L) 1 cm. **B:** Schematic 2D-tissue setup, dimensions given in mm. **C:** variations of the transversal conductivity σ_T, with reference CV_n and σ_{T_n} given in upper left box. **D:** APD variations around the normal (N) APD (indicated in black) for mMS (dotted line) by modifying τ_{close} and TT2 (solid line) by modifying g_{Ks}. Time instants $T_{1,2,3,4}$ are marked for the normal APD.

Post-processing. We computed the repolarisation time for mMS and TT2 as the time the AP reaches a threshold of 0.1 and −70 mV, respectively. The RTG was computed as the magnitude of the spatial gradient of the repolarization time at each grid point. As a metric for RTG we compute (i) the surface area with RTG $\geq$ 3.5 ms/mm (SRTG), and (ii) the mean maximum RTG (mmRTG), defined as the mean of the 5% highest nodal RTG values (Fig. 2).

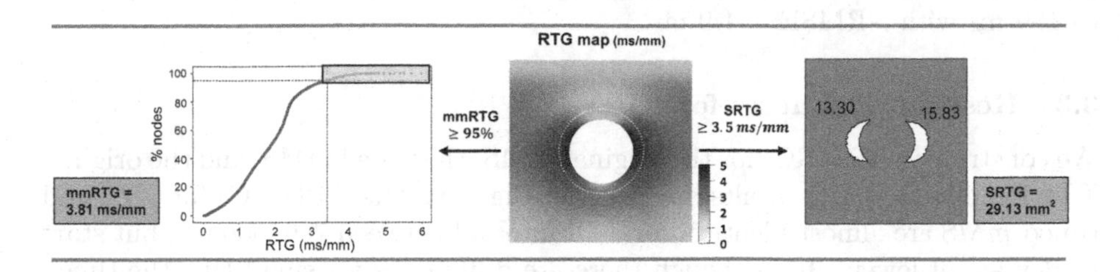

Fig. 2. Example for post-processing steps of the primary output of an RTG map. **SRTG:** surface RTG $\geq$ 3.5 ms/mm. **mmRTG:** mean maximal RTG computed as 5% highest nodal RTG values (highlighted box) from cumulative number of total nodes.

3 Results

3.1 Space-Discretization Analysis for the Conduction Velocity

Figure 3 shows the relationship between mesh resolution and CV. The TT2 ionic model exhibits a stronger dependency of CV on element size compared to *m*MS. Furthermore, it can be seen that for both the TT2 and *m*MS ionic models the solution for CV converges as the grid is refined. The deviation in CV increases about linearly with increasing element size, with a maximum deviation (for Δx = 1000µm) of 0.040 m/s (−6.7%) and 0.364 m/s (−60.7%), for the *m*MS and the TT2 model, respectively.

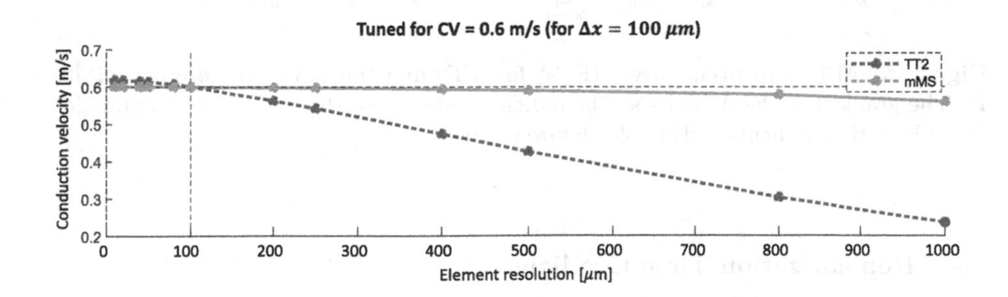

Fig. 3. Relationship of CV and element size with a constant conductivity. TT2: σ_L = 0.1312 S/m and *m*MS: σ_L = 0.5660 S S/m, resulting in CV = 0.6 m/s for Δx = 100 µm. (horizontal dotted line).

3.2 AP Tuning

For creating the RCs, the original mMS model was tuned at CL = 500 ms by a modification of τ_{close} = 192 ms (with NRMSE = 2.7 ms). For the 2D simulations, performed at CL=1000 ms, $J(\tau_{close})$ was minimized by a modification of τ_{close} to 189 ms with NRMSE = 1.9 ms.

3.3 Restitution Curves for APD

We constructed APD RCs for the original mMS, the tuned mMS, and the original TT2 ionic model. The results are given in Fig. 4**A**. The APD RCs for TT2 and tuned mMS are almost identical for a DI interval between 200–500 ms, but start to deviate at lower DIs. Although there are differences for small DIs, the tuned mMS does show better correspondence with TT2 than the original mMS. We also compared the magnitude of the APD slope (Fig. 4**B**). The time instant at which the RC slope exceeds 1, assumed to be indicative for VT risk, equals about 220 ms for the TT2 and the original mMS models. However, this is increased to about 235 ms in the mMS model.

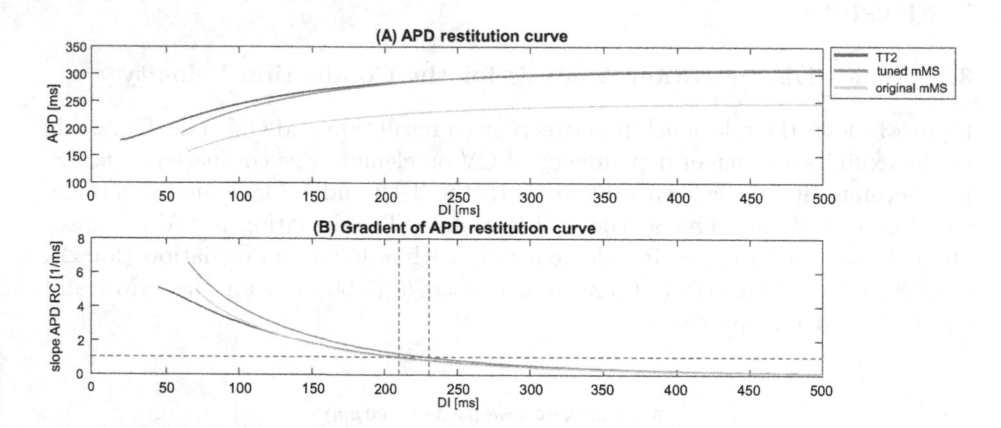

Fig. 4. **A:** APD restitution curves (RCs) for TT2 and the tuned and original mMS. **B**: The gradient of the APD RCs. The indicated green line shows the time instants for DIs where the gradients > 1. (Color figure online)

3.4 Repolarization Time Gradient

For the same variations in APD and CV, the tuned mMS and TT2 model showed qualitatively similar distribution patterns of repolarization gradients, with higher sensitivity to ΔAPD than ΔCV (Fig. 5**A**–**D**). Quantitatively prolonging the APD shows approximately the same RTG metrics, but shortening of APDs show differences up to 180 mm^2 (for SRTG) and 3.0 ms/mm (for mmRTG) between both models (Fig. 5**E**, **F**). However, mmRTG also shows differences of approximately 2.0 ms/mm when prolonging the APD with slower CVs.

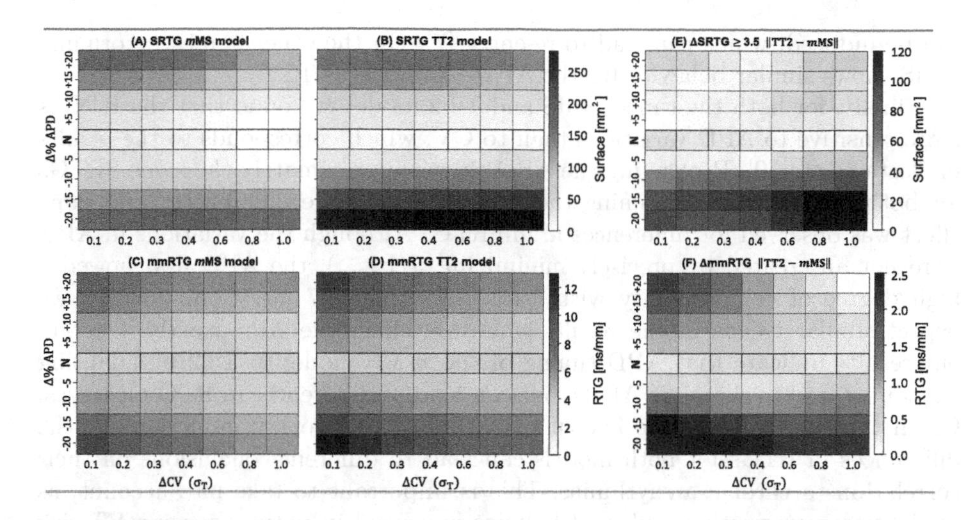

Fig. 5. Computed surface RTG $\geq$ 3.5 ms/mm (SRTG) and mean maximum RTG (mmRTG) for *m*MS and TT2 (**A–D**). Absolute difference in (**E**) SRTG and (**F**) mmRTG of the *m*MS and TT2 model

4 Discussion

First, we found that sensitivity of CV to element size is much higher in the TT2 model than in the *m*MS model. Normally, element sizes between 250 and 400 µm are used for whole heart simulations [13], with errors up to 10% considered acceptable [12]. For 400 µm, *m*MS and TT2 showed an underestimation of approximately 3% and 20%, respectively. Usually, this error in CV is compensated for by tuning the conductivity [4]. However, this conductivity tuning becomes complicated in unstructured meshes in whole heart geometries, suggesting that the use of the *m*MS model is more appropriate. We note that an important aspect in modeling EP is the requirement of high mesh resolution to capture the fast upstroke in time. This fast upstroke is related to the smallest time scale in the ionic model, which is τ_{in} for *m*MS. Additional simulations (not shown here) indeed showed when decreasing τ_{in}, the solution of CV becomes more dependent on the element resolution. This implies a relation between time and space scales, which needs to be better studied to understand the reason behind mesh sensitivity.

Next, a tuning method was proposed in which we aim to match the AP of *m*MS to TT2, by only adjusting τ_{close}. Relan et al. showed that in terms of APD personalisation, tuning τ_{close} of the MS model showed small errors of approximately 2% compared to patient data [16]. In our study, the tuned *m*MS model shows a similar APD RC as TT2 at large DIs compared to the original *m*MS model. However, when the DI is decreased, differences arise in both the computed APD and the APD RC slope. This is critical, as the APD RC slope at these short DIs can be coupled to the moment in which a conduction block can be induced, which can lead to re-entry [7]. On the other hand, the original mMS shows similar behavior for the APD RC slope as TT2.

Finally, for both the tuned mMS and TT2 model we found that the RTG is more sensitive to APD variations than to CV, which corresponds to the results in Costa et al. [10]. Prolonging the APD shows approximately the same SRTGs for both models, but shortening the APD lead to large differences. The same effect was observed for differences in mmRTG. Although the variations in APD were not all created by precisely minimizing $J(\tau_{close})$, the APD fits showed a high degree of comparability with errors less than 5.7 ms. Consequently, we expect similar trends in the results when matching the APs exactly. Overall, our results indicate that APD tuning of the mMS model to TT2 by adapting τ_{close} may lead to the same APDs, but can lead to differences in RTG measures. Given that steep RTGs are known to be linked to re-entry induction [1], the differences in RTGs for both models can lead to different conclusions on their correlation to cardiac arrhythmias. This is important to take into account, as model parameterization and model choice is crucial in patient-specific VT-risk predictions.

Overall, our results highlight the importance of carefully tuning the parameters involved in the ionic model. In future research, it can be considered to improve the tuning of the mMS model to the TT2 model by e.g. optimizing all four mMS time parameters and including restitution properties to the cost function J (Eq. 2), while taking into account the effect of time and space scales.

Limitations. When creating the RCs, we noticed for decreasing DIs that the amplitudes of the TT2 and mMS APs decreased, and for TT2 the AP shapes were changed remarkably. These changes are an expected effect of faster activation rates [8]. The ionic channels need sufficient time to recover, and too short DIs may not allow for this [14]. From literature, we could not find a commonly accepted definition for the AP to propagate or not. We defined an AP to propagate AP when the second AP reached at least the threshold of 95% of the amplitude of the normal (first) AP. For TT2 this threshold was set to approximately 10 mV and for mMS 0.95 [-].

SRTG was computed as the surface with RTG $\geq$ 3.5 ms/mm, as this value is within the previously investigated range of RTGs that can lead to conduction block. Various studies have found different values for a minimum RTG for unidirectional block to occur. These RTG values varied between 3.0 and 10.0 ms/mm, depending on subject-specific properties and the used geometry in the computational model [19]. Even though the exact threshold value is unclear, we expect that changes in this value would not affect our conclusions on the differences between the mMS and the TT2 model.

5 Conclusion

In this study, we first showed that the computed CV for the mMS model is less dependent on the mesh resolution than for the TT2 ionic model. Next, we found that the tuned mMS model approximated APD restitution curves of the TT2 model much better that the original mMS model. On the other hand, the slope of the APD RC for TT2 was approximated better by the original mMS, which is important in VT-risk analysis. Finally, both for the TT2 and the tuned mMS model we found that RTG is more sensitive to variation in APD than to variation in CV. This suggests that it is more important to accurately represent the EP properties in the BZ in terms of APD than in terms of CV. Quantitatively, differences in sensitivity of RTG to variation in APD between the two models were more pronounced for short than for large APDs in the BZ. Overall, our study provides more insights on the differences of using the biophysical TT2 and phenomenological mMS model, which can be used in deciding which model to select for VT-risk analysis.

References

1. Cluitmans, M., et al.: Noninvasive detection of spatiotemporal activation-repolarization interactions that prime idiopathic ventricular fibrillation. Sci. Transl. Med. **13**(620), eabi9317 (2021)
2. Connolly, A., Bishop, M.: Computational representations of myocardial infarct scars and implications for arrhythmogenesis. Clin. Med. Insights: Cardiol. **10**, CMC-S39708 (2016)
3. Corrado, C., Niederer, S.: A two-variable model robust to pacemaker behaviour for the dynamics of the cardiac action potential. Math. Biosci. **281**, 46–54 (2016)
4. Costa, C., Hoetzl, E., Rocha, B., Prassl, A., Plank, G.: Automatic parameterization strategy for cardiac electrophysiology simulations. In: Computing in Cardiology 2013, pp. 373–376. IEEE (2013)
5. Dangman, K., Danilo Jr., P., Hordof, A., Mary-Rabine, L., Reder, R., Rosen, M.: Electrophysiologic characteristics of human ventricular and purkinje fibers. Circulation **65**(2), 362–368 (1982)
6. Deng, D., Prakosa, A., Shade, J., Nikolov, P., Trayanova, N.: Sensitivity of ablation targets prediction to electrophysiological parameter variability in image-based computational models of ventricular tachycardia in post-infarction patients. Front. Physiol. **10**, 628 (2019)
7. Hayashi, M., et al.: Ventricular repolarization restitution properties in patients exhibiting type 1 Brugada electrocardiogram with and without inducible ventricular fibrillation. J. Am. Coll. Cardiol. **51**(12), 1162–1168 (2008)
8. Jing, L., Agarwal, A., Chourasia, S., Patwardhan, A.: Phase relationship between alternans of early and late phases of ventricular action potentials. Front. Physiol. **3**, 190 (2012)
9. Lopez-Perez, A., Sebastian, R., Izquierdo, M., Ruiz, R., Bishop, M., Ferrero, J.: Personalized cardiac computational models: from clinical data to simulation of infarct-related ventricular tachycardia. Front. Physiol. **10**, 580 (2019)
10. Mendonca Costa, C., Plank, G., Rinaldi, C., Niederer, S., Bishop, M.: Modeling the electrophysiological properties of the infarct border zone. Front. Physiol. **9**, 356 (2018)
11. Prakosa, A., et al.: Personalized virtual-heart technology for guiding the ablation of infarct-related ventricular tachycardia. Nat. Biomed. Eng. **2**(10), 732–740 (2018)

12. Prassl, A., et al.: Automatically generated, anatomically accurate meshes for cardiac electrophysiology problems. IEEE Trans. Biomed. Eng. **56**(5), 1318–1330 (2009)
13. Quarteroni, A., Lassila, T., Rossi, S., Ruiz-Baier, R.: Integrated heart-coupling multiscale and multiphysics models for the simulation of the cardiac function. Comput. Methods Appl. Mech. Eng. **314**, 345–407 (2017)
14. Ravens, U., Wettwer, E.: Electrophysiological aspects of changes in heart rate. Basic Res. Cardiol. **93**(1), s060–s065 (1998)
15. Relan, J., et al.: Coupled personalization of cardiac electrophysiology models for prediction of ischaemic ventricular tachycardia. Interface Focus **1**(3), 396–407 (2011)
16. Relan, J., Sermesant, M., Delingette, H., Pop, M., Wright, G., Ayache, N.: Quantitative comparison of two cardiac electrophysiology models using personalisation to optical and MR data. In: 2009 IEEE International Symposium on Biomedical Imaging: From Nano to Macro, pp. 1027–1030. IEEE (2009)
17. Seemann, G., Carillo, P., Weiss, D.L., Krueger, M.W., Dössel, O., Scholz, E.P.: Investigating arrhythmogenic effects of the hERG mutation N588K in virtual human atria. In: Ayache, N., Delingette, H., Sermesant, M. (eds.) FIMH 2009. LNCS, vol. 5528, pp. 144–153. Springer, Heidelberg (2009). https://doi.org/10.1007/978-3-642-01932-6_16
18. Ten Tusscher, K., Panfilov, A.: Alternans and spiral breakup in a human ventricular tissue model. Am. J. Physiol.-Heart Circulatory Physiol. **291**(3), H1088–H1100 (2006)
19. Tran, D., Yang, M., Weiss, J., Garfinkel, A., Qu, Z.: Vulnerability to re-entry in simulated two-dimensional cardiac tissue: effects of electrical restitution and stimulation sequence. Chaos Interdisc. J. Nonlinear Sci. **17**(4), 043115 (2007)
20. Vigmond, E., Hughes, M., Plank, G., Leon, L.: Computational tools for modeling electrical activity in cardiac tissue. J. Electrocardiol. **36**, 69–74 (2003)

Numerical Investigation of Methods Used in Commercial Clinical Devices for Solving the ECGI Inverse Problem

Narimane Gassa[1,3,4(✉)], Vitaly Kalinin[2], and Nejib Zemzemi[1,3,4]

[1] Institut de Mathématiques de Bordeaux, UMR 5251, Talence, France
narimane.gassa@inria.fr
[2] EP Solutions SA, Yverdon-les-Bains, Switzerland
[3] Inria centre at the university of Bordeaux, Talence, France
[4] IHU-LIRYC, Pessac, France

Abstract. Electrocardiographic Imaging (ECGI) is a promising tool to non-invasively map the electrical activity of the heart using body surface potentials (BSPs) combined with the patient specific anatomical data. In this work, we assess two ECGI algorithms used in commercial ECGI systems to solve the inverse problem; the Method of Fundamental Solutions (MFS) and the Equivalent Single Layer (ESL). We quantify the performance of these two methods in conjunction with two different activation maps to estimate the activation times and earliest activation sites. ESL provided more accurate reconstruction of the cardiac electrical activity, especially on the endocardial part of the heart. Nevertheless, both methods provided comparable results in terms of the derived activation maps and the localization of the focal origin as a clinically relevant parameter.

Keywords: Inverse problem · Electrocardiography · Equivalent Single Layer · Method of Fundamental Solutions

1 Introduction

The electrocardiographic imaging (ECGI) is a non-invasive technique that enables study of the body surface potentials for the treatment and diagnosis of cardiac arrhythmia. Heart activity is reconstructed from electrocardiograms measurements at the body surface and the patient specific heart-torso geometry. Mathematically, the problem is known as the inverse problem in electrocardiography. To date, several numerical methods have been used to solve the ECGI inverse problem. The most frequently used are the Finite Element Method (FEM; [3]), the Boundary Element Method (BEM; [2]), the Method of Fundamental Solution (MFS; [5]) and the equivalent single layer (ESL; [1]) source model. In a recent work of *Karoui et al.* fifteen algorithms for the resolution of the ECGI inverse problem were evaluated; MFS and FEM were combined with different approaches to select the optimal regularization parameter in conjunction with

O. Bernard et al. (Eds.): FIMH 2023, LNCS 13958, pp. 157–165, 2023.
https://doi.org/10.1007/978-3-031-35302-4_16

zero order and L1-Norm Tikhonov regularization. The obtained results using experimental data indicate that MFS performs as well as FEM [6]. In contrast to FEM that requires a full heart-torso discretized volume, MFS is a meshless method and does not require the torso geometry making a good compromise between accuracy and computational complexity. BEM is also a mesh-based method, where only a surface mesh is required, which reduces the computation time compared to FEM but may require more memory. On the other hand, a novel approach based on a representation of the electrical potential on the heart surface as ESL was proposed in [1] and showed promising results in the reconstruction of cardiac electrical activities.

A rigorous and theoretical justification of MFS and ESL can be found in [8]. For this study, we have implemented the MFS and ESL methods as described in the literature, and evaluated their numerical performance when used together with two different approaches for approximating activation times. Both simulated in silico data and clinical pacing data were included in our analysis.

2 Methods

2.1 Mathematical Modeling of the Inverse Problem

Let us suppose that Ω_{T} denotes the torso domain, Γ_{ext} is the external boundary, Γ is the heart-torso interface and u is the electrical potential in the torso governed by the diffusion equation. The inverse problem in electrocardiography for the geometry shown in Fig. 1 is to find the electrical potential on the heart surface Σ satisfying both Dirichlet and Neumann boundary conditions such that:

$$\nabla \cdot (\sigma \nabla u) = 0, \quad \text{in } \Omega_T, \tag{1}$$

$$\sigma \nabla u \cdot n = 0 \text{ and } u = u_{\mathrm{T}}, \quad \text{on } \Gamma_{\mathrm{ext}}, \tag{2}$$

$$u = u_{\mathrm{H}}, \quad \text{on } \Sigma, \tag{3}$$

where u_{T} is the measured body surface potential and σ denotes the torso conductivity tensor. This is known as the forward problem in electrocardiography represented by a Cauchy problem for the Laplace equation to study the electrical activity of the heart through the body surface.

Given the heart and torso geometries combined with the measurements at the body surface, we express the Problem (1)–(3) in a matrix-vector system:

$$\mathbf{A}\mathbf{x} = \mathbf{b}, \tag{4}$$

where $\mathbf{A}$ is the transfer matrix, $\mathbf{b}$ is derived from u_{T} and $\mathbf{x}$ is the unknown variable from which it is possible to reconstruct u_{H}.

2.2 Method of Fundamental Solution (MFS)

In this method, we approach the solution of the Problem (1)–(3) with a linear combination of fundamental solutions of the differential operator which is, in our

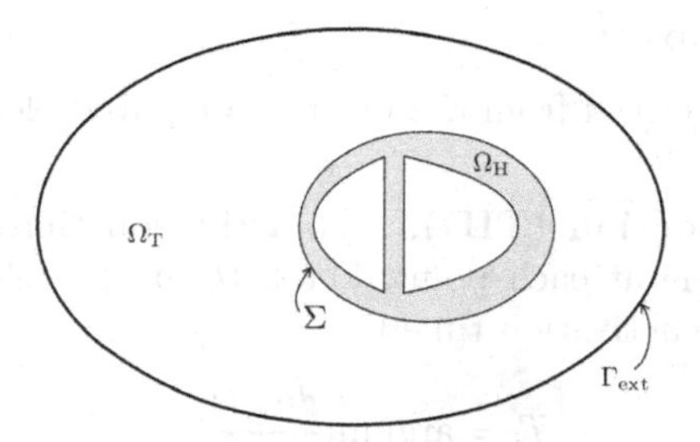

Fig. 1. The heart and torso domains: $\Omega_{\rm H}$ and $\Omega_{\rm T}$.

case, the *Laplacian*. These solutions are located on a set of virtual source points $y_j, j = 1 \ldots M$ over an auxiliary surface as shown in [5]. The electrical potential is expressed as:

$$u(x) = a_0 + \sum_{i=1}^{M} f(x, y_j)\, a_j,$$

where $x \in \Omega_T$, y_j are the virtual source points and $a_j, j = 1 \ldots M$ are their corresponding coefficients. In this context, f represents the Laplace fundamental solution, which is explicitly defined as $f(r) = \frac{1}{4\pi r}$, with r being the Euclidean distance between two points x and y. When using Dirichlet and Neumann conditions we obtain the linear system $\mathbf{Ma} = \mathbf{d}$, with:

$$\mathbf{a} = (a_0, a_1, \cdots, a_M)^T,$$
$$\mathbf{d} = (u_{x_1}, \cdots, u_{x_N}, 0, \cdots, 0)^T,$$

and $\mathbf{M}$ is the transfer matrix.

2.3 Equivalent Single Layer (ESL)

Kalinin et al. proposed a novel numerical approach to solve the inverse problem in electrocardiography. Furthermore, the paper [1] describes the following identity:

$$A = -H_{01} + G_{01} G_{11}^{-1} H_{11} \equiv 4\pi G_{01} G_{11}^{-1}, \tag{5}$$

where $q_0 = 0$, $q_1 = \frac{\partial u_1}{\partial n}$ and G_{ij}, H_{ij} are the matrices obtained from a discretization of the single- and double-layer integral operators in BEM, respectively. Given (5) we get

$$4\pi G_{01} \underbrace{G_{11}^{-1} u_1}_{w_1} = \underbrace{(H_{00} - G_{01} G_{11}^{-1} H_{10})}_{B} u_0. \tag{6}$$

Finally, the inverse problem of ECG can be formulated as follows:

$$4\pi G_{01} w_1 = B u_0. \tag{7}$$

2.4 Activation Maps

Activation times are derived from the inverse computed electrograms.

Time of Inner Deflection (TID). The activation times are defined as the intrinsic deflection time at each point. Let $u_i(t)$ be the electrical potential at point x_i at time t, the activation time is:

$$\hat{T}_i = \underset{t \in [0,T]}{\operatorname{argmin}} \frac{du_i(t)}{dt}. \tag{8}$$

Activation Direction Mapping (ADM). In this method, we first find the gradient field of the electrical potentials on the heart surface and then consider its projection on the tangential plane $g(x, t_k)$ for $x \in \Sigma$. On the heart surface, we calculate the vector function $\boldsymbol{A}(x)$ such that:

$$\boldsymbol{A}(x) = g(x, t_{k1}), \tag{9}$$

where

$$t_{k1} = \underset{t_k}{\operatorname{argmax}} \left| g(x, t_k) \right|. \tag{10}$$

From the known function $\boldsymbol{A}$, we compute the scalar function $a(x)$ that minimizes the following functional:

$$I = \int_{\Sigma} \left| \nabla a - \boldsymbol{A} \right|. \tag{11}$$

Further details of this method can be found in [4].

3 Data

In order to assess the results obtained with the two proposed ECGI methods and quantify their performance, in-silico data were provided by EP-Solutions SA. The heart-torso geometry of three patients were derived from the Computer Tomography (CT) scans. Simulations were performed using the cardiac software CHASTE [7] and three focal type electrical activation patterns were simulated: in the lateral wall of the left ventricle (LV), in the right ventricular apex (RVA), and in the right ventricle outflow tract (RVOT). In addition, we used clinical data of three patients with two different and independent induced pacings in the right and left ventricular (RV and LV). The exact locations are known from the CT scans which will allow us to evaluate the excitation origin localization.

4 Evaluation

The comparison will mainly cover: i) The relative error (RE) between the true (simulated) and inverse computed (*ic*) electrogram. ii) The Pearson's correlation coefficient (CC) between the true and *ic* activation maps (both ADM and TID). iii) The localization error (LE) of the pacing site measured as the Euclidean distance between this latter and the ECGI based earliest activation site (EAS).

5 Results

5.1 Simulated Data

For the sake of example, we present in Fig. 2 the inverse computed electrograms recorded at different locations in the heart during left ventricular pacing for geometry 1. The estimated accuracy metrics (re and cc) are indicated in Table 1.

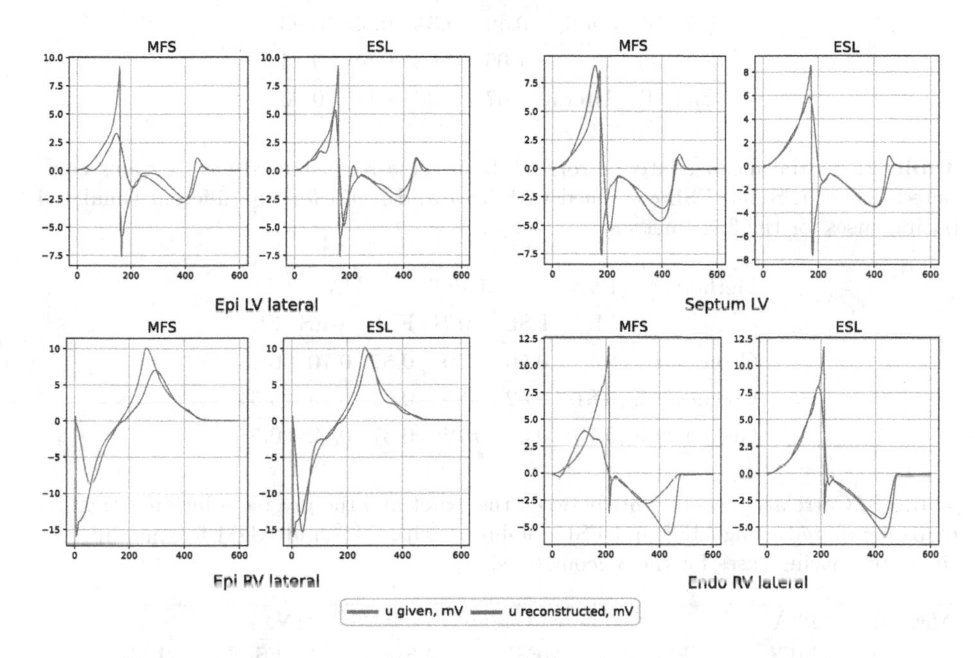

Fig. 2. Simulated (red line) and inverse reconstructed (blue line) electrograms at different sites of the endocardial and epicardial surfaces of the heart for a lateral LV pacing of geometry 1. The inverse solutions were obtained using MFS and ESL. (Color figure online)

The results for the reconstructed signals in terms of relative errors are reported in Table 2 for the various test cases. In general, ESL results in a more accurate reconstruction of electrograms. The REs in average are equal to 0.58, 0.59, 0.53 for RVA, RVOT and LV cases, respectively. While MFS is less accurate for signal reconstruction with a relative error that varies between 0.70 and 0.81.

In Table 3, we report the values of the CC between the ground-truth (as derived from simulations) and *ic* activation maps for the four methods (MFS-TID, MFS-ADM, ESL-TID, ESL-ADM), the three stimulation protocols and the three geometries. Results show that all methods provide accurate reconstruction of the activation maps with a correlation coefficient ranging from 0.86 to 0.98, except when using TID for computing the left ventricular stimulation site in geometry 1.

Table 1. Estimated relative errors and correlation coefficients of the inverse reconstructed electrograms at different sites of the endocardial and epicardial surfaces of the heart for a lateral LV pacing of geometry 1. The inverse solutions were obtained using MFS and ESL.

Method	RE		CC	
	MFS	ESL	MFS	ESL
Epi LV lateral	0.49	0.46	0.86	0.89
Epi RV lateral	0.49	0.35	0.86	0.93
Septum LV	1.06	0.68	0.68	0.74
Endo RV lateral	0.67	0.30	0.90	0.96

Table 2. Spatial mean relative errors of the inverse reconstruted electrograms computed using MFS and ESL combined with TID and ADM for the different simulated pacing cases for the 3 geometries.

Method	RVA		RVOT		LV	
	MFS	ESL	MFS	ESL	MFS	ESL
Geometry 1	0.79	0.56	0.79	0.57	0.70	0.48
Geometry 2	0.81	0.62	0.85	0.62	0.75	0.54
Geometry 3	0.70	0.55	0.72	0.57	0.72	0.57

Table 3. Correlation coefficients between the exact and the inverse solution activation maps computed using MFS and ESL combined with TID and ADM for the different simulated pacing cases for the 3 geometries.

Method	RVA				RVOT				LV			
	MFS		ESL		MFS		ESL		MFS		ESL	
Act Map	ADM	TID	ADM	TID	ADM	TID	ADM	TID	ADM	TID	ADM	TID
Geometry 1	0.90	0.96	0.92	0.97	0.89	0.96	0.96	0.95	0.86	0.37*	0.97	0.33*
Geometry 2	0.91	0.94	0.91	0.96	0.92	0.94	0.95	0.95	0.93	0.96	0.95	0.98
Geometry 3	0.93	0.96	0.94	0.97	0.92	0.93	0.96	0.95	0.92	0.95	0.97	0.98

* This CC value is significantly lower than the average values.

In Table 4, we provide the localization errors of the pacing sites with respect to the inverse solution and the activation map reconstruction approach. The Mean $\pm$ STD values of the localization errors using MFS-TID, MFS-ADM, ESL-TID and ESL-ADM methods, respectively, are 10.3 ± 4.5 mm, 10.2 ± 5.7 mm 6.8 ± 3.6 mm and 6.8 ± 3.1 mm.

Table 4. Localization errors in Millimeter between the exact and the inverse solution activation sites computed using MFS and ESL and combined with TID and ADM for the different simulated pacing cases for the 3 geometries.

Method	RVA				RVOT				LV			
	MFS		ESL		MFS		ESL		MFS		ESL	
Act Map	ADM	TID	ADM	TID	ADM	TID	ADM	TID	ADM	TID	ADM	TID
Geometry 1	6	2	8	6	9	14	11	13	13	15	3	4
Geometry 2	8	5	10	5	3	6	10	9	18	15	5	4
Geometry 3	11	5	4	5	10	18	10	10	15	12	1	6

5.2 Clinical Data

Due to the lack of electrical recordings that cover the surface of the heart in clinical cases, our only means of comparing our methods is by evaluating the localization error. In Table 5, we report the Euclidean LEs between the actual and earliest activation sites derived from the different activation maps. The Mean $\pm$ STD values of the localization errors using MFS-TID, MFS-ADM, ESL-TID and ESL-ADM methods, respectively, are 42 ± 23 mm, 21 ± 7 mm 25 ± 25 mm and 16 ± 13 mm.

Table 5. Localization errors in Millimeter between the exact and the inverse solution activation sites computed using MFS and ESL and combined with TID and ADM for the LV and RV pacing cases for the 3 patients.

Method	Lateral RV				Lateral LV			
	MFS		ESL		MFS		ESL	
Act Map	ADM	TID	ADM	TID	ADM	TID	ADM	TID
Patient 1	10	72*	3	18	23	29	26	7
Patient 2	26	22	8	17	27	33	15	16
Patient 3	16	28	7	18	29	72*	37	76*

* This LE value is significantly lower than the average values.

In Fig. 3, we show four examples of activation maps with respect to the inverse solution method and the activation map. Interestingly, ADM and TID yield visually very different maps. The use of TID based activation maps in clinical cases may induce more errors when we extract the earliest activation site, as shown in Table 5.

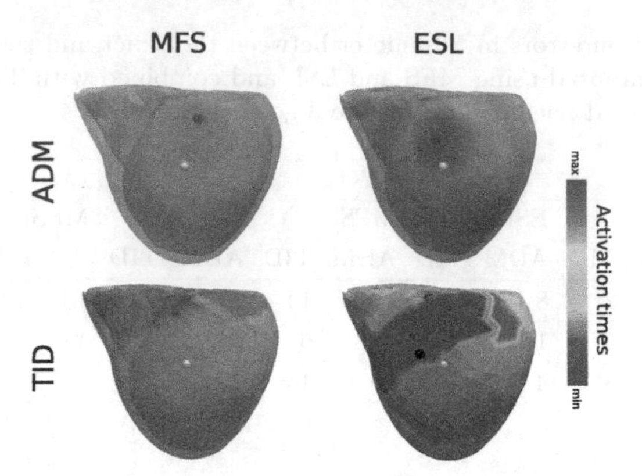

Fig. 3. Comparison of activation maps and pacing site localization with respect to the inverse solution method (MFS first column and ESL second column) and localization of early activation sites approach (ADM first row and TID second row). True activation sites are depicted in white and ECGI based activation sites are depicted in black.

6 Discussion and Conclusions

In this work, we evaluated the performance of two methods used in commercial medical devices for solving the electrocardiographic imaging inverse problem: the method of fundamental solutions and the equivalent single layer approach combined with two different approaches to compute activation times (TID and ADM). We tested the four possible combinations using simulated and clinical data. The evaluation of the different methods is based on the reconstruction of the heart surface potential and activation maps from simulated data and the localization of the pacing site from both simulated and clinical data.

For the in silico data-set, we used 3 different patient geometries and 3 cardiac paced rhythms: right-ventricular, left-ventricular and right-ventricular outflow tract pacing. In terms of signal reconstruction, ESL provided more accurate inverse solution in particular on the ventricular endocardial surface. Whereas, in terms of pacing site localization, both methods gave almost identical results with a slight difference for LV pacing.

The obtained LEs for clinical cases, indicate that ESL combined with ADM resulted in better localization accuracy and, in contrast to simulated data, TID and ADM based earliest activation sites can be very different. In fact, TID based activation maps is less robust to noisy and clinical data. This is mainly due to the fact that TID calculates activation times for each cardiac node separately (the point of maximum negative slope) without a spatial coherence. Consequently, in noisy data, it frequently provides patchy patterns in the obtained maps. However, activation direction mapping provides better results for clinical data despite the fact that it tends to over-smooth the spread of activation, which can be misleading in the excitation origin localization. The post-processing methods and in particular the activation times calculation have a significant impact on

ECGI and its usability in clinical applications. Thus, it would be interesting to work on cardiac activation maps that will better follow the temporal course of the inverse solution.

In this work, only single-paced rhythms were included. Therefore, further studies should include more complex activation patterns like ventricular tachycardia or atrial fibrillation.

Acknowledgments. This Project has received funding from the European Unions Horizon research and innovation programme under the Marie Skodowska-Curie grant agreement No. 860974 and by the French National Research Agency, grant references ANR-10-IAHU04- LIRYC and ANR-11-EQPX-0030. The study was carried out as part of the PersonalizeAF project in collaboration with EP-Solutions SA.

References

1. Kalinin, A., Potyagaylo, D., Kalinin, V.: Solving the inverse problem of electrocardiography on the endocardium using a single layer source. Front. Physiol. **10**, 58 (2019)
2. Barr, R.C., Ramsey, M., Spach, M.S.: Relating epicardial to body surface potential distributions by means of transfer coefficients based on geometry measurements. IEEE Trans. Biomed. Eng. **1**, 1–11 (1977)
3. Wang, D.: Finite element solutions to inverse electrocardiography. The University of Utah (2012)
4. Denisov, A.M., Zakharov, E.V., Kalinin, A.V.: Method for determining the projection of an arrhythmogenic focus on the heart surface, based on solving the inverse electrocardiography problem. Math. Models Comput. Simul. **4**(6), 535–540 (2012)
5. Wang, Y., Rudy, Y.: Application of the method of fundamental solutions to potential-based inverse electrocardiography. Ann. Biomed. Eng. **34**(8), 1272–1288 (2006)
6. Karoui, A., Bear, L., Migerditichan, P., Zemzemi, N.: Evaluation of fifteen algorithms for the resolution of the electrocardiography imaging inverse problem using ex-vivo and in-silico data. Front. Physiol. **9**, 1708 (2018)
7. Cooper, F.R., et al.: Chaste: cancer, heart and soft tissue environment. J. Open Source Softw. **5**(47), 1848 (2020). https://doi.org/10.21105/joss.01848
8. Kalinin, V., Shlapunov, A.: Exterior extension problems for strongly elliptic operators: solvability and approximation using fundamental solutions. arXiv preprint arXiv:2209.11009 (2022)

Evaluation of Inverse Electrocardiography Solutions Based on Signal-Averaged Beats to Localize the Origins of Spontaneous Premature Ventricular Contractions in Humans

Yesim Serinagaoglu Dogrusoz[1,2](✉), Nika Rasoolzadeh[2], Beata Ondrusova[3], Peter Hlivak[4], and Jana Svehlikova[3]

[1] Electrical-Electronics Engineering Department, Middle East Technical University, Ankara, Turkey
yserin@metu.edu.tr

[2] Institute of Applied Mathematics, Middle East Technical University, Ankara, Turkey

[3] Institute of Measurement Science, Slovak Academy of Sciences, Bratislava, Slovakia

[4] National Institute for Cardiovascular Diseases, Bratislava, Slovakia

Abstract. The signal averaging approach applied to the clinical body surface potential measurements was studied and evaluated for localizing the spontaneous premature ventricular contractions (PVC) on the data of three patients. For each patient, more than 200 beats were extracted from the ECG signal and aligned with respect to the R-peak of the QRS complex. Then, the origin of the PVC was computed from the inversely estimated heart surface potentials and activation times for each individual beat and the averaged one. It was shown that the result from the averaged beat did not differ more than 2 mm from the mean or median of the results from individual beats. Signal averaging of clinical signals leads to reliable results and saves the needed computational time.

Keywords: Inverse problem · Electrocardiographic imaging · Spontaneous PVC · Source localization · Signal averaging

1 Introduction

Localization of a premature ventricular contraction (PVC) is one of the topics of electrocardiographic imaging [1], which is a noninvasive method for reconstructing the heart's electrical activity from multiple-lead ECG measurements. Frequent occurrence of PVC beats means serious health issues and can lead to dangerous ventricular tachycardia. If pharmacological treatment fails, the location of the origin of the undesired PVC beat should be inactivated by an invasive procedure (radiofrequency ablation (RFA)) using an intracardiac catheter. The PVC origin is searched via time-demanding intraventricular mapping. Preliminary noninvasive estimation of the PVC origin position could shorten such a

O. Bernard et al. (Eds.): FIMH 2023, LNCS 13958, pp. 166–174, 2023.
https://doi.org/10.1007/978-3-031-35302-4_17

procedure and contribute to the success of ablation [2]. Several methods of solving the inverse problem of electrocardiology (ECG) were used and validated for localization of the PVC, mainly on simulated data [1] and also using animal or human experiments where the spontaneous PVC beat was replaced by a controlled pacing stimulus in the known position of the stimulating catheter [3]. But studies including spontaneous PVC in humans are scarce.

ECG signal has repeated beats morphology for its healthy condition (sinus rhythm) and for PVC beats. Theoretically, the inverse solution can be obtained from any heartbeat. On the other hand, in realistic measurements, the heartbeat signals are not precisely the same; they are slightly different because of the breathing, myopotentials, and other internal conditions of the body. Therefore it is difficult to pick a single heartbeat as a representative of ventricular activity. It is also well-known that the inverse solution is generally ill-posed and thus very sensitive to fluctuations in the input data. One approach to include beat-to-beat variability is to create a representative beat by averaging all beats with the same morphology [4,5].

A comparison of the inverse solutions from an averaged beat and from several single beats signals was studied in [4] on three patients artificially stimulated and in [5] on four patients with ventricular tachycardia. The results from the averaged signal were similar to the mean location of the results from single beats in [4] or better in [5]. Spontaneous PVC in humans may have a more complex pattern than artificially paced beats with exactly known origins. Thus, despite studies in the literature that evaluate the performance of inverse ECG solutions from the signal-averaged beats, there is still a need to look into how the signal-averaged beats from the spontaneous PVC origins would compare to the individual beat solutions.

In this study, we computed the inverse solution in the form of combined epicardial and endocardial potentials for three patients with spontaneous PVCs and compared the localization of the PVC origin from the averaged signal and individual beats which contributed to the averaging.

2 Methods

2.1 Bratislava Data

Body surface potentials (BSPs) were recorded from patients (n = 3) indicated for RFA by using the ProCardio8 measuring system [6]. 16 strips, each of which consists of 8 electrodes, were placed around the torso, resulting in 128-lead ECG measurements sampled at 1000 Hz. Baseline drift noise was removed using a high-pass filter with a 0.5 Hz cut-off frequency. A finite impulse response filter utilizing the Blackman-Harris window was used for this purpose. The R-peaks were found for all beats and these beats were clustered according to their signal morphology. Only those beats that display spontaneous premature ventricular contraction (PVC) patterns were used in this study. Chest CT scans of each patient were obtained to reconstruct patient-specific torso geometric models (Fig. 1). Information about the patients is provided in Table 1.

Table 1. Patient information summary.

Patient ID	Age	# of beats	PVC origin	RFA
P1	72	210	RV	5 days
P2	46	267	RV	2 days
P3	59	203	LV	11 months

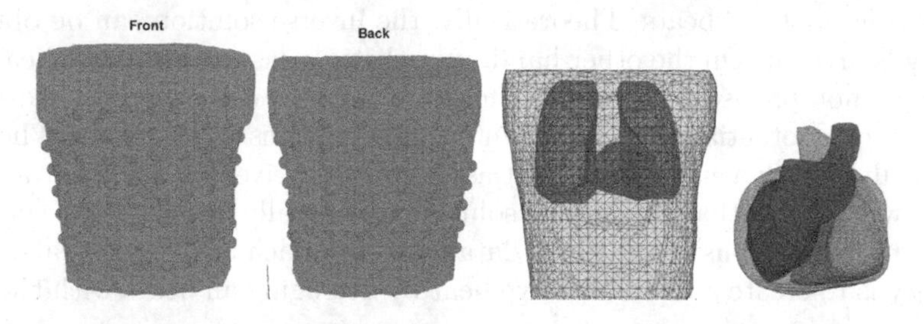

Fig. 1. An example of the front and back views of the fine torso geometry and the electrode locations; torso geometry with the lungs; EpiEndo geometry with the blood cavities.

2.2 Forward and Inverse Problem Solution

In this study, the cardiac electrical activity is represented in terms of extracellular potentials defined on a joint closed surface enclosing epicardial and endocardial surfaces (EpiEndo). These surface potentials are linearly related to the corresponding BSPs:

$$\mathbf{y}(k) = \mathbf{A}\mathbf{x}(k) + \mathbf{v}(k), \; k = 1, 2, \ldots, K, \tag{1}$$

where $\mathbf{x}(k) \in \mathbb{R}^{N\times 1}$ are the heart surface potentials and $\mathbf{y}(k) \in \mathbb{R}^{M\times 1}$ are the corresponding BSPs at time instant k. $\mathbf{A} \in \mathbb{R}^{M\times N}$ is the forward transfer matrix and $\mathbf{v}(k)$ represents noise in the measurements.

The matrix $\mathbf{A}$ was computed by solving the forward problem of ECG using the boundary element method (BEM) [7]. We calculated the patient-specific forward matrices for an inhomogeneous model including the blood cavities and the lungs. The conductivity of the blood cavities was assumed three times higher and the conductivity of the lungs four times lower than the average conductivity of the torso. The problem was initially solved using a refined torso mesh. Then, a reduced transfer matrix that relates the EpiEndo potentials to the torso ECG leads was obtained via bilinear interpolation of the values at the vertices of the triangle surrounding the position of each electrode.

The potential-based inverse problem is highly ill-posed and even small amounts of noise in the measurements yield significant errors in the solution. Here we used the well-known Tikhonov regularization method [8] to deal with this ill-posedness. This method minimizes a cost function that seeks a trade-off between a good fit to the measurements and a good fit to an *a priori* constraint on the solution:

$$\hat{\mathbf{x}}(k) = \underset{\mathbf{x}(k)}{\operatorname{argmin}} \left\{ \|\mathbf{A}\mathbf{x}(k) - \mathbf{y}(k)\|_2^2 + \lambda_k^2 \|\mathbf{R}\mathbf{x}(k)\|_2^2 \right\} \tag{2}$$

where λ_k is a regularization parameter, and $\mathbf{R}$ is a regularization matrix. Here, zero-order Tikhonov regularization was employed, hence $\mathbf{R}$ is the identity matrix. λ_k value was first determined by using the L-curve method [9]. The final solution was computed by using a single λ value equal to the median of all λ_k values over the QRS interval.

We solved the inverse problem both for a signal-averaged (SA) beat and for all individual beats. Below is a summary of how these data are processed before the inverse computation.

Signal Averaged Beat Solution: The SA PVC beat was obtained from all PVC beats that were aligned with respect to their R peaks. The QRS onset of the PVC beat was annotated manually, and the signals in all leads were corrected by a constant value such that at this onset, the BSP value became zero [10]. This signal-averaged PVC beat was then used as the input for the inverse solution.

Beat-to-beat Solution: In this approach, all beats in the PVC cluster (considered in the averaged signal) were processed individually. This processing resulted in different inverse solutions, activation times, pacing estimates, and localization error values for every beat for each patient. Individual beats had high-frequency noise; therefore, the beat-to-beat analysis was carried out for both non-filtered (No LPF) and low-pass-filtered (LPF) signals for comparison. A Butterworth filter with a cut-off frequency 45 Hz and order 5 was used for the LPF case. Additionally, a simple baseline removal was applied to all beats prior to the inverse problem solution. In this method, the average of a 20 ms window before the QRS onset (50 Hz power noise) was subtracted from each lead.

2.3 Evaluation Methods

PVC origin estimation for the inverse solutions was evaluated mainly based on the localization error (LE) metric. Ground truth PVC origins were manually marked on the EpiEndo heart geometry by the physician who applied the RFA procedure; 2–3 nodes were identified per patient as the true PVC origins. PVC origins were estimated from inverse solutions by first computing the activation times (AT) using a spatiotemporal approach [11]. The earliest activated node was then assigned as the PVC origin estimate. Finally, the average localization error (LE) for a patient was obtained by calculating the Euclidean distance of the PVC origin estimate to each ground truth PVC location annotation and then finding the mean of these LE values.

Further evaluations were also carried out to understand the relationship between the SA-beat and individual beat similarities/differences. We computed the temporal Pearson's correlation coefficient (CC) between the SA-beat and each individual beat (CC-BSP), resulting in a vector of length equal to the number of torso electrodes for each beat. Similarly, we computed the temporal

Pearson's CC between the SA-beat solutions and solutions based on each individual beat (CC-EGM), resulting in a vector of length equal to the number of EpiEndo nodes for each beat. Finally, we obtained Pearson's CC between the SA-beat-based ATs and ATs based on each individual beat (CC-AT).

3 Results

The SA-beat and the individual beats in the cluster are compared in terms of CC values introduced in Sect. 2.3. Table 2 lists the smallest CC values obtained over all beats for all metrics. Note that CC-BSP and CC-EGM values are the medians over all leads in one beat. These results reveal that the individual BSP beats, EGM reconstructions, and AT estimates are highly correlated with their corresponding SA beat counterparts.

Table 2. Comparison of the SA-beat with individual beats in the cluster in terms of the minimum CC value over all beats for all metrics. Note that CC-BSP and CC-EGM values are the medians over all leads in one beat.

Patient ID	No LPF			LPF		
	CC-BSP	CC-EGM	CC-AT	CC-BSP	CC-EGM	CC-AT
P1	0.987	0.978	0.914	0.988	0.978	0.914
P2	0.991	0.983	0.964	0.991	0.983	0.956
P3	0.994	0.993	0.968	0.994	0.993	0.968

All localization error results are summarized for all cases in Fig. 2 and in Table 3. Figure 2 displays the beat-to-beat variation of LE values for all patients. The LE value for the SA beat is marked as a black line, and the mean of LE over all beats is marked as a red line on each plot. 'NF' and 'F' labels on the x-axis stand for 'No LPF' and 'LPF', respectively. Table 3 gives a summary of all LE values. The results for the individual beat cases are presented as 'mean $\pm$ standard deviation (std)' and 'median (interquartile range (IQR))' over all beats in one patient.

Figure 3 displays the scatter plot of the 'No LPF' LE values with respect to median (over all leads in a single beat) CC-BSP variation. All individual beats in the cluster correlate with the SA beat with more than 0.98 median temporal CC value.

In Fig. 4 we plot the PVC origins for all patients for the No LPF case on the EpiEndo geometry. LPF plots are quite similar to these maps. In these plots, the ground truth PVC origins are marked as red, the SA-beat PVC origin estimate is marked as blue, and the smaller black dots indicate the individual beat PVC origin estimates. Individual beat PVC estimates for P1 are clustered around the ground truth ablation points and the SA beat-based PVC origin, also evident by the smaller LE values. P2 individual beat PVC estimates are more scattered

Table 3. Localization errors (mm) for the signal-averaged beat and individual beats (represented in terms of mean ± std and median (IQR) values)

Patient ID	SA-LE	LE-Beats, No LPF		LE-Beats, LPF	
		mean ± std	median (IQR)	mean ± std	median (IQR)
P1	19.05	18.12 ± 5.18	19.05 (6.56)	18.36 ± 4.87	19.05 (6.26)
P2	28.59	28.20 ± 2.39	28.59 (0.00)	28.26 ± 2.33	28.59 (0.00)
P3	48.99	46.36 ± 4.31	48.99 (8.18)	46.42 ± 4.32	48.99 (8.18)

compared to P1 (hence the larger LE values), but still, all are in the same anatomical region of the heart. For P3, the individual beat PVC estimates are also scattered but in different regions of the heart, and some of these individual beat PVC estimates are in irrelevant positions.

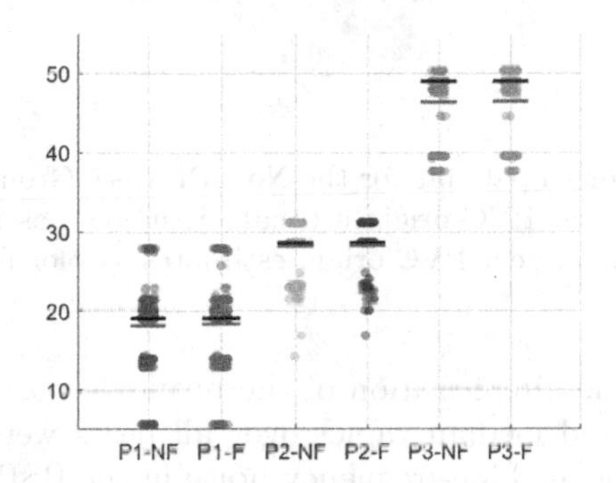

Fig. 2. Scatter plot of beat-to-beat variations of LEs for all patients for the 'No LPF' (NF) and 'LPF' (F) cases. The corresponding SA-beat LE values and mean LE values over all beats are marked as 'black' and 'red' horizontal lines, respectively, on each scatter plot. (Color figure online)

4 Discussion and Conclusions

For each patient in the study, the inverse solution from more than 200 beats and from the SA beat was computed and compared. The morphology of the SA beats for each patient was highly representative of the individual beats in the corresponding PVC cluster. The median CC-BSP values over all leads in one beat given in Table 2 were all higher than 0.98. Moreover, only 0.05%, 1.41%, and 0.49% of CC-BSP values of all beats in all leads of P1, P2, and P3, respectively, were less than 0.90.

The correlations of the inversely reconstructed EGMs and ATs were high, still, the PVC origin estimates based on the earliest AT were scattered, especially in P3. From Fig. 2, beat-to-beat LE values were clustered into groups,

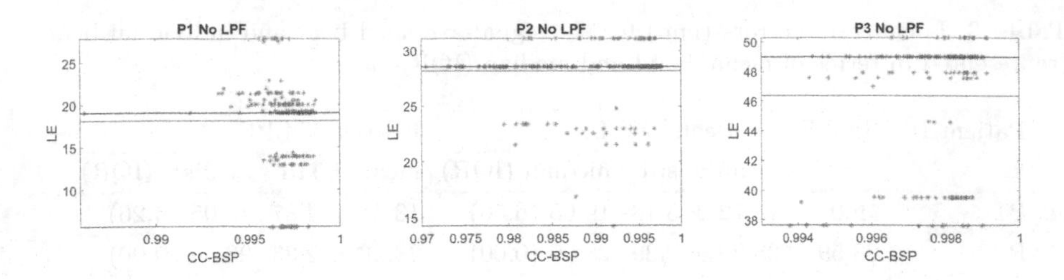

Fig. 3. Variation of beat-to-beat LE values with respect to CC-BSP. The corresponding SA-LE value and mean LE value over all beats are marked as 'black' and 'red' horizontal lines, respectively, on each scatter plot. (Color figure online)

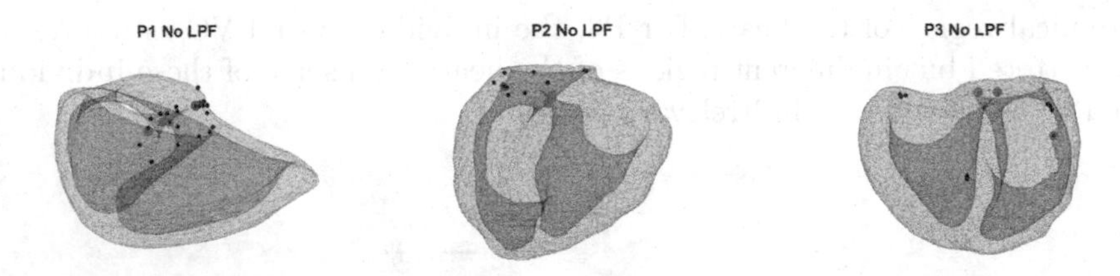

Fig. 4. PVC locations for all patients for the No LPF case. Ground truth PVC origins are marked as red, SA-beat PVC origin estimate is marked as blue, and small black dots indicate the individual beat PVC origin estimates. (Color figure online)

probably because of the discretization of the available positions for the inverse solutions. The mean and median values over all beats were not different with or without pre-filtering the high-frequency noise in the BSPs. The std and IQR values slightly decreased after applying the LPF (Table 3).

The mean LE values for patients P1 and P2 (19 and 28 mm, respectively) were smaller than the LE for P3 (49 mm), who underwent the RFA procedure significantly later than the BSP measurement was performed, after 11 months. The former two patients were ablated 2 or 5 days after BSP measurement. We can hypothesize that the long delay between the BSP measurement and the RFA procedure could negatively influence the LE of the inverse solution for P3.

For some individual BSPs, the LE is considerably smaller than the LE from the averaged signal. However, there is not a straightforward relationship between the LE values and the BSP correlation, as is visible in Fig. 3. Therefore, finding the single beat leading to the best LE would be challenging.

The obtained results from spontaneous PVCs are in agreement with previous findings in [4,5]. The inverse solution from the averaged signal was not significantly different from the average of individual solutions, evident from the high CC-EGM and CC-AT values. Moreover, LE from the averaged beat did not differ more than 2 mm from the mean or median of the LE results from individual beats. On the other hand, the computational demand of the solution from the averaged signal was significantly smaller.

PVC origin localization studies involving structurally normal hearts in the literature have reported LE values ranging from 5 to 50 mm [10,12–14]. Although the LE values obtained in this study fall within the reported LE range, they could still be improved. The Tikhonov regularization method is known to oversmooth the heart surface potential reconstructions, which in turn reduces the reconstruction accuracy. Moreover, the potential-based source model used in this study requires multiple steps in finding the origin of PVCs, which also includes finding the activation times from the reconstructed EGMs. The localization performance could be improved by choosing more robust inversion and activation time estimation methods. Other source models such as single-dipole, which estimates the PVC origin directly, could also be employed.

This study included data from a small patient group (n = 3). These preliminary findings suggest that SA beat could be used for localizing the spontaneous PVC origins. However, these findings should be further evaluated using data from a larger patient population.

Acknowledgements. This work was supported by TUBITAK (grant no: 120N200), SAS (grant no: 536057), and VEGA (grant no: 2/0109/22).

References

1. Cluitmans, M., et al.: Validation and opportunities of electrocardiographic imaging: from technical achievements to clinical applications. Front. Physiol. **9**, 1305 (2018)
2. Chrispin, J., Mazur, A., Winterfield, J., Nazeri, A., Valderrabano, M., Tandri, H.: Non-invasive localization of premature ventricular focus: a prospective multicenter study. J. Electrocardiol. **72**, 6–12 (2022)
3. Zhou, S., Sapp, J.L., Šťovíček, P., Horáček, B.M.: Localization of activation origin on patient-specific endocardial surface by the equivalent double layer (EDL) source model with sparse Bayesian learning. IEEE Trans. Biomed. Eng. **66**(8), 2287–2295 (2019)
4. Coll-Font, J., Erem, B., Stovicek, P., Brooks, D.: A statistical approach to incorporate multiple ECG or EEG recordings with artifactual variability into inverse solutions. In: Proceedings/IEEE International Symposium on Biomedical Imaging: from Nano to Macro, pp. 1053–1056 (2015)
5. Dallet, C., et al.: Combined signal averaging and electrocardiographic imaging method to non-invasively identify atrial and ventricular tachycardia mechanisms. In: 2016 Computing in Cardiology Conference (CinC), pp. 1–4 (2016)
6. Kadanec, J., Zelinka, J., Bukor, G., Tysler, M.: ProCardio 8 - system for high resolution ECG mapping. In: Proceedings of the 11th International Measurement Conference, Smolenice, Slovakia, pp. 263–266 (2017)
7. Stanley et al., P.C.: The effects of thoracic inhomogeneities on the relationship between epicardial and torso potentials. IEEE Trans. Biomed. Eng. **BME-33**(3), 273–284 (1986)
8. Tikhonov, A.N., Arsenin, V.Y.: Solutions of Ill-Posed Problems. Halsted Press, New York (1977)
9. Hansen, P.C.: The L-curve and its use in the numerical treatment of inverse problems. In: Johnston, P.R. (ed.) Computational Inverse Problems in Electrocardiography, chap. 4, pp. 119–142. WITpress, Southampton (2001)

10. Svehlikova, J., et al.: The importance of ECG offset correction for premature ventricular contraction origin localization from clinical data. Meas. Sci. Rev. **22**, 202–208 (2022)
11. Erem, B., Coll-Font, J., Orellana, R.M., St'Ovicek, P., Brooks, D.H.: Using transmural regularization and dynamic modeling for noninvasive cardiac potential imaging of endocardial pacing with imprecise thoracic geometry. IEEE Trans. Med. Imaging **33**(3), 726–738 (2014)
12. Cluitmans, M.J.M., et al.: In vivo validation of electrocardiographic imaging. JACC Clin. Electrophysiol. **3**(3), 232–242 (2017)
13. Bear, L.R., et al.: How accurate is inverse electrocardiographic mapping? Circ. Arrhythmia Electrophysiol. **11**(5), e006108 (2018)
14. Bear, L.R., et al.: The impact of torso signal processing on noninvasive electrocardiographic imaging reconstructions. IEEE Trans. Biomed. Eng. **68**(2), 436–447 (2021)

An *in silico* Study of Cardiac hiPSC Electronic Maturation by Dynamic Clamp

Sofia Botti[1,3](✉), Chiara Bartolucci[2], Rolf Krause[1,4], Luca F. Pavarino[3], and Stefano Severi[2]

[1] Euler Institute - Università della Svizzera Italiana, Lugano, Switzerland
sofia.botti02@universitadipavia.it

[2] Department of Electrical, Electronic, and Information Engineering "Guglielmo Marconi", University of Bologna, Cesena, Italy

[3] Department of Mathematics "Felice Casorati", University of Pavia, Pavia, Italy

[4] Faculty of Mathematics and Informatics, FernUni, Brig, Switzerland

Abstract. Regenerative cardiology recently advanced in patient–specific medicine by employing somatic cells to derive pluripotent stem cells and differentiate them into cardiomyocytes. Resulting populations present an immature phenotype; the Dynamic Clamp technique is a popular experimental manipulation to induce electronic maturation towards an adult phenotype.

In this work, we present a fully virtual framework to study this Dynamic Clamp technique, based on the injection of the inward-rectifier potassium current into the myocyte, taking into account six different current formulations. We investigate the effects of the current injection on the action potential morphology and on three specific biomarkers for different current percentages, and we compare resulting morphologies with the standard transmembrane potential profile of a human adult cardiomyocyte. The results of this quantitative analysis suggest that atrial–like potassium current formulations allow the cell to reach action potential features comparable with the ones of mature cells, preventing the cell to show a non physiological morphology.

Keywords: hiPSC–CMs · Virtual dynamic clamp · Cardiac action potential morphology

1 Introduction

Human–induced pluripotent stem cell–derived cardiomyocytes (hiPSC–CMs) provide a powerful tool to develop reliable human-based *in vitro* models for disease modeling and drug toxicity screening. These cells arise from differentiation protocols, that result in heterogeneous populations of immature CMs consisting predominantly of ventricular–like (VL) cells with a small percentage of atrial–like (AL) cells and nodal–like cells.

Two manipulations are widely used to push hiPSC–CMs toward more adult cardiac phenotypes. First, Retinoic Acid treatment allows to over-express atrial

O. Bernard et al. (Eds.): FIMH 2023, LNCS 13958, pp. 175–183, 2023.
https://doi.org/10.1007/978-3-031-35302-4_18

markers [4]. Then, through the Dynamic Clamp (DC) technique, the membrane diastolic potential (MDP) hyperpolarizes to values suitable for generating a mature action potential (AP) waveform, allowing the discrimination between atrial and ventricular AP phenotype. Thus, the chamber–specific AP phenotype is more pronounced and this facilitates the separation of AL and VL CMs, as described in [1,15].

Another important difference between cultured hiPSC-CMs and adult myocytes is the low, or even absence of, inward-rectifier potassium current (I_{K1}). To overcome this immature characteristic, we consider a DC technique based on the injection of a virtual I_{K1} current. This electronic maturation improves AP measurements in hiPSC–CMs and makes hiPSC–CMs a more reliable model for investigating cardiac arrhythmias (see [9]).

This study is devoted to investigating the DC technique in a fully computational setting. We carry out an *in silico* study based on Virtual Dynamic Clamp in order to analyze six different I_{K1} current formulations, considering qualitative effects on the cell and evaluating quantitatively AP features with respect to adult CMs in a perspective of cell maturation.

2 Methods

In this section, we first describe the experimental DC setup and its *in silico* rendering (virtual DC), as described in [2]. Furthermore, we will present the innovative adopted hiPSC–CMs ionic model, and the I_{K1} scaled formulations tested in the present work.

2.1 From Experimental to Virtual DC

In cultures of matured hiPSC–CMs, the I_{K1} current can be too low or even lacking, leading to unstable depolarized (MDP), if compared to the mature CMs. These immature electrophysiological conditions correspond to a spontaneous firing activity or a depolarized resting ($\simeq -20$ mV), respectively. For sake of brevity, we will take into account the worst case of lacking native I_{K1}.

DC is a valid and effective approach to overcoming immature characteristics of hiPSC–CM through the injection of a virtual I_{K1} current. In a closed–loop paradigm the transmembrane potential (V) is acquired through traditional patch clamp and used to compute the voltage-dependent I_{K1}, finally injected into the cell with the additional stimulus current, see e.g. [1,9,15]. DC allows the hyperpolarization of MDP to values suitable to generate a mature AP waveform.

The whole interface protocol and the current injection can be performed *in silico* in a fully computational setting, coupling the I_{K1} mathematical equation used in the real–time simulator with the set of ordinary differential equations (ODEs) describing the dynamics of the ionic currents and the resulting hiPSC–CM electrical activity. Then, different I_{K1} formulations can be tested, comparing the physiological responses and giving a mathematical definition of the waveforms' differences.

The physiological I_{K1} lack in experimental conditions is reached in the *in-silico* framework through the native I_{K1} current suppression. Once the depolarized membrane potential reaches the steady state, a novel I_{K1} current can be added to the total ionic current, taken from different formulations existing in the literature.

2.2 Paci 2020 Ionic Model and I_{K1} Tested Currents

hiPSC–CMs show some relevant differences concerning adult myocytes, requiring a new mathematical approach. The first ionic model based on hiPSC–CMs data was created by M. Paci, who developed a primal model in 2013 [11], creating a new line improved in 2018 [12] and 2020 [13].

Paci generation and every other single–cell model existing in literature focused on the ventricular-like phenotype, the predominant phenotype emerging during the differentiation process. Among them, it is possible to deduce a qualitative primitive atrial–specific model from the Paci2013, even if no atrial–specific current is taken into account. Since the DC technique is generally applied to unknown phenotype cells, we will base this analysis on the most recent ventricular-like model, Paci2020 [13], equipped with an improved calcium dynamic formulation with respect to previous models.

The original Paci2020 model, constrained by experimental data, simulated traces of spontaneous electrical activity. The first test we performed was about the suppression of the native I_{K1}. The lack of the potassium current leads V to a quiescent depolarized resting potential, higher than −20 mV. In DC experiments the injection of the additional current is joined with the injection of the applied current I_{app}, thus we considered an external stimulus pacing the model 1 Hz. According to Fabbri [5], the cell could be elicited because a required amount of I_{K1} could bring V to a stable and hyperpolarized MDP ($\simeq$−78 mV).

Simulations were performed in MATLAB using the ODE function ODE15S. We consider the system to be at steady state after 800 s.

Six different I_{K1} formulations available in the literature were taken into account to carry out the *in silico* DC. Four ventricular–specific formulations were tested, from Ten Tusscher (TT) [14], Fink [6], Grandi [7], O'Hara–Rudy (ORd) [10] human ventricular models. Because of the *a priori* unknown cell phenotype, also two atrial–specific formulations have been analyzed, from Koivumäki (K) [8] and Courtemanche (CRN) [3] human atrial models.

All ionic models considered are available in the CellML repository (link).

According to Fabbri et al. [5], each current formulation was scaled as in Fig. 1. The most recent model, Koivumäki, was considered as the target and every other formulation was normalized in order to obtain the same outward peak current density (0.63 pA/pF).

3 Results and Discussion

As described in [2] and [1], I_{K1} injection considering low densities gave rise to an irregular plateau. In this section, we define a novel mathematical criterion to

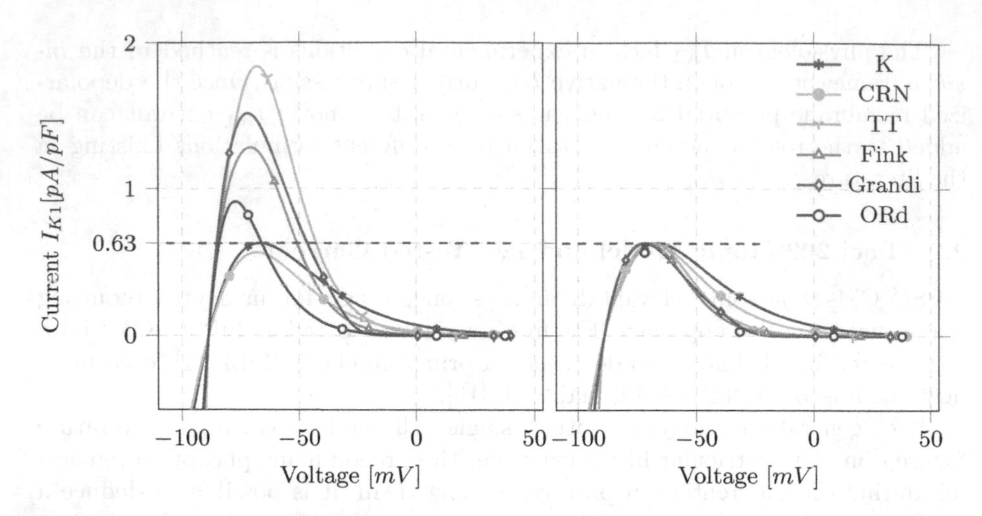

Fig. 1. I_{K1} tested formulations, normalized with respect to Koivumäki outward peak current density (dashed line). Original (left) and scaled (right) currents are shown at steady state, when considering voltages between −120 and 40 mV.

classify the AP morphology and we present resulting curves of some specific AP features (biomarkers) with respect to the injected current density. The comparison with experimental data suggests the choice of AL formulations to improve cell maturation.

3.1 A Novel AP Morphology Classifier

According to *in vitro* AP recordings, see [1], the injection of VL I_{K1} formulations with low densities highlighted a longer AP plateau. From a mathematical point of view, the abnormality corresponds to an extra inflection during phase 3 (repolarization), as depicted in Fig. 2. Thus, considering a subset of the phase 3, we derived the following definition.

Definition 1. *The AP morphology of a hiPSC-CM is physiological if*

$$\frac{d^2V(t)}{dt^2} \leq 0 \qquad \forall t \in [\text{APD}_{40}, \text{APD}_{70}], \tag{1}$$

where APD_X *is the AP duration at* $X\%$ *of repolarization, and the amplitude reference is the difference between the maximum value of* V *and the* MDP.

As described in Fig. 3, every I_{K1} formulation allows the cell to reach a physiological AP morphology, but the required amount of the current is different. Comparing the different results, we observe that atrial I_{K1} formulations, K and CRN, require a lower amount of the injected current to gain a physiological morphology, exactly equal to the normalized current (i.e. 100%). On the other hand, ventricular formulations need a much higher current density to prevent an abnormal plateau, up to 200% for TT, 300% for Fink, and 600% for Grandi and ORd. Since DC injects an external current, it is reasonable to add as little current as possible, suggesting the use of atrial formulations in the experimental real–time closed–loop.

3.2 Biomarkers Analysis

Since experimental DC is an electronic maturation method, we are interested in evaluating the rate of approximation to an adult human CM when considering the injection of different I_{K1} formulations. To this end, we take into account three different biomarkers: the APD_{30}, the $APDAPD_{90}$, and the membrane diastolic potential (MDP).

In Fig. 4 we present the dependence of these biomarkers on the injected I_{K1} percentage, discriminating between ratios inducing an abnormal AP and ratios supporting a morphological AP shape. The hiPSC–CMs' maturation is then analyzed by comparing these curves with experimental data provided by ORd, [10], referring to a human adult ventricular CM and provided as Mean ± Standard Deviation (St. Dev.).

Every presented biomarker highlights a clear partition between the behaviour of AL formulations (K and CRN) and VL ones.

First of all, MDP portrait (Fig. 4b) suggests that only AL I_{K1} formulations allow the cell to reach almost every experimental value, while the injection of VL currents leads the cell to hyperpolarized values.

Similar observations could concern the APD_{90}, in Fig. 4c: different VL formulations could not reach the experimental bound, except TT. Anyway, this model presents a gap when approaching the experimental bound, and it could perform experimental values only by injecting a huge amount of external I_{K1}.

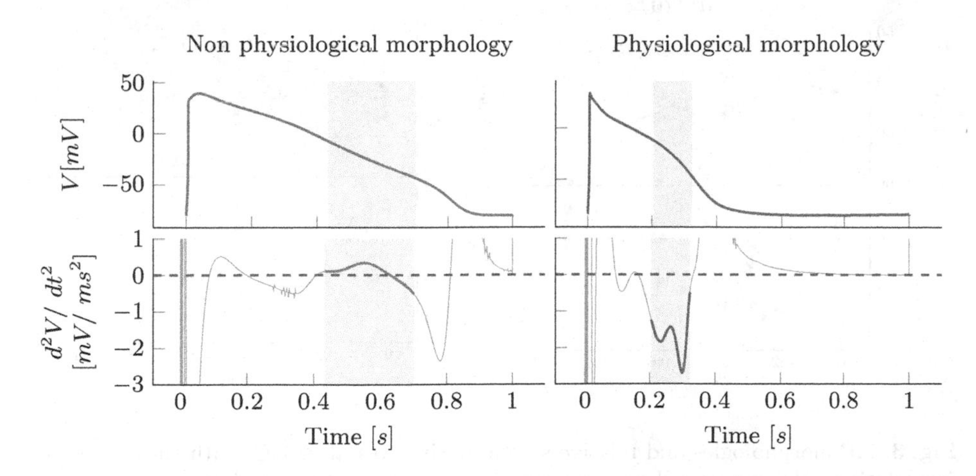

Fig. 2. AP morphology classification: the non physiological morphology (left) presents an extra inflection in the repolarization phase, the physiological morphology (right) is always convex in phase 3.

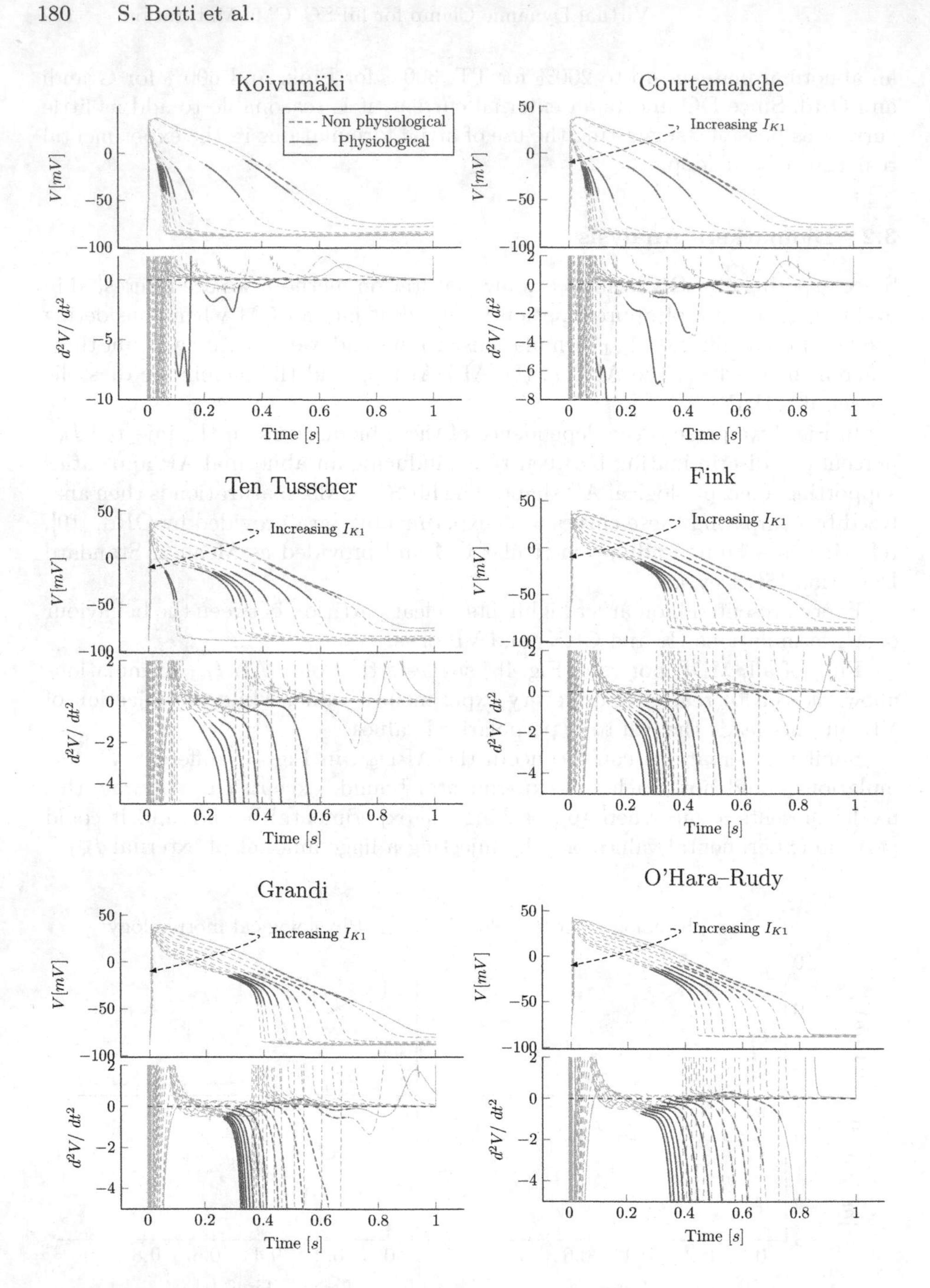

Fig. 3. AP morphologies and relative second derivatives after DC, with the six tested formulations. We considered percentages of the normalized injected current in the set {80, 100 : step 100 : 1300}, except for ORd formulation, where the set is {200 : step 100 : 1300}, since lower percentages do not trigger an action potential wave.

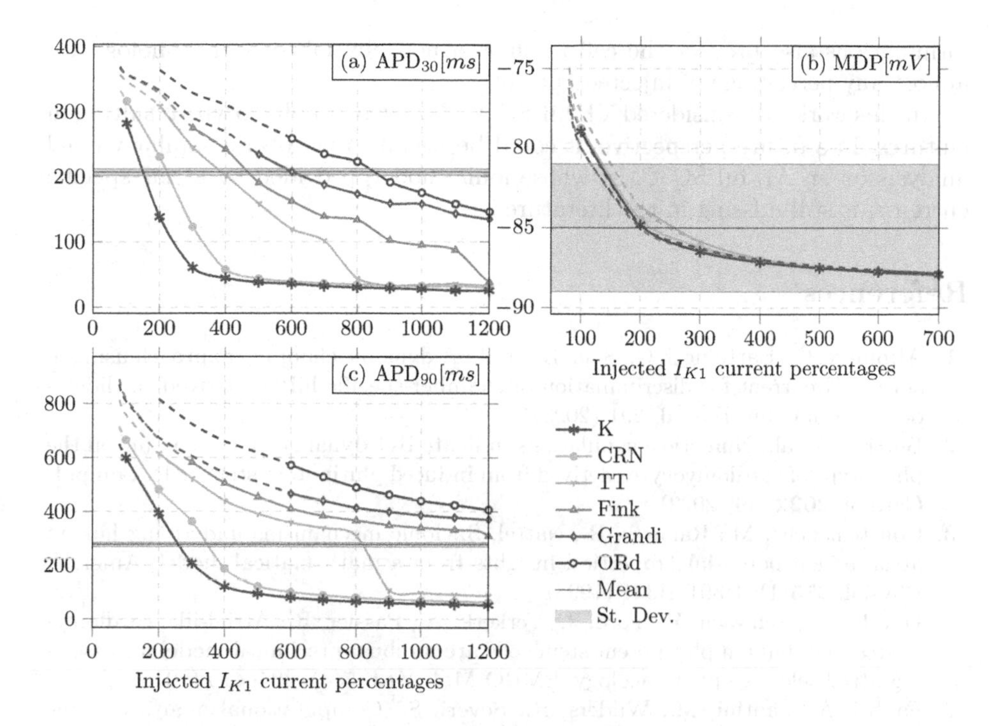

Fig. 4. Biomarkers dependence on the injected current density and comparison with experimental data. The horizontal continuous line and relative bounds stand for the mean and the interval [Mean ± St.Dev.] of ORd experimental data, respectively. Dashed lines refer to current percentages that give rise to non physiological APs.

Finally, VL formulations present the following APD$_{30}$ dynamic (Fig. 4a): they reach the experimental bound, but I_{K1} injected current percentage is too high. Otherwise, when considering both APD$_{30}$ and APD$_{90}$, K and CRN models allow to reach the experimental bound for lower values of injected current. Among them, K is much better than CRN when considering previous arguments.

4 Conclusion

Briefly, starting from the experimental DC, used for the electronic maturation of stem cells, we implemented an *in silico* tool to perform the current injection as an additional current in the Paci2020 ionic model for VL hiPSC–CMs.

Our numerical simulation shows that the AP morphology changes with respect to the injected current density and the I_{K1} formulation. Thus, we tested six different current models and we defined a mathematical classifier to discriminate a physiological and a non physiological AP repolarization phase.

In conclusion, a virtual analysis of the biomarkers suggests that K and CRN I_{K1} formulations allow the cell to reach AP features comparable with adult and mature values with a minimal amount of additional external current. These formulations also prevent the cell to show a non physiological morphology for almost any percentage of injected current.

In this work, we considered VL hiPSC–CMs, the main phenotype resulting in cultures. In a future perspective, it could be useful to perform a similar virtual analysis on an AL hiPSC–CMs, whose ionic model, provided by atrial–specific currents, is still missing in the literature.

References

1. Altomare, C., Bartolucci, C., Sala, L., et al.: A dynamic clamping approach using in silico IK1 current for discrimination of chamber-specific hiPSC-derived cardiomyocytes. Commun. Biol. **6**, 291 (2023)
2. Botti, S., et al.: Numerical simulations indicate IK1 dynamic clamp can unveil the phenotype of cardiomyocytes derived from induced pluripotent stem cells. Comput. Cardiol. **2022**, 49 (2022)
3. Courtemanche, M., Ramirez, R., Nattel, S.: Ionic mechanisms underlying human atrial action potential properties: insights from a mathematical model. Amer. J. Physiol. **275**(1), H301–H321 (1998)
4. Devalla, H., Schwach, V., Ford, J., Verkerk, A., Passier, R.: Atrial-like cardiomyocytes from human pluripotent stem cells are a robust preclinical model for assessing atrial-selective pharmacology. EMBO Mol. Med. **7**(4), 394–410 (2015)
5. Fabbri, A., Fantini, M., Wilders, R., Severi, S.: Computational analysis of the human sinus node action potential: model development and effects of mutations. J. Physiol. **595**(7), 2365–2396 (2017)
6. Fink, M., Noble, D., Virag, L., Varro,A., Giles, W.R.: Contributions of HERG K+ current to repolarization of the human ventricular action potential. Prog. Biophys. Mol. Biol. **96**(1), 357–376 (2008)
7. Grandi, E., Pasqualini, F.S., Bers, D.M.: A novel computational model of the human ventricular action potential and Ca transient. J. Mol. Cell. Cardiol. **48**(1), 112–121 (2010)
8. Koivumäki, J.T., Korhonen, T., Tavi, P.: Impact of sarcoplasmic reticulum calcium release on calcium dynamics and action potential morphology in human atrial myocytes: A computational study. PLOS Comp. Biol. **7**(1), 1–14 (2011)
9. Meijer van Putten, R.M.E., et al.: Ion channelopathies in human induced pluripotent stem cell derived cardiomyocytes: a dynamic clamp study with virtual IK1. Front. Physiol. **6**(7) (2015)
10. O'Hara, T., Virág, L., Varro, A., Rudy, Y.: Simulation of the undiseased human cardiac ventricular action potential: model formulation and experimental validation. PLoS Comput. Biol. **7**(5), 99–110 (2011)
11. Paci, M., Hyttinen, J., Aalto-Setälä, K., Severi, S.: Computational Models of Ventricular- and Atrial-Like Human Induced Pluripotent Stem Cell Derived Cardiomyocytes. Ann. Biomed. Eng. **41**(11), 2334–2348 (2013). https://doi.org/10.1007/s10439-013-0833-3
12. Paci, M., et al.: Automatic optimization of an in silico model of human iPSC derived cardiomyocytes recapitulating calcium handling abnormalities. Front. Physiol. **9**, 709 (2018)
13. Paci, M., et al.: All-optical electrophysiology refines populations of in silico human iPSC-CMs for drug evaluation. Biophys. J. **118**(10), 2596–2611 (2020)

14. Ten Tusscher, K., Noble, D., Noble, P., Panfilov, A.: A model of human ventricular tissue. Americ J. Physiol. HC. Physiol. **286**(4), H1573–H1589 (2004)
15. Verkerk, A.O., Veerman, C.C., Zegers, J.G., Mengarelli, I., Bezzina, C.R., Wilders, R.: Patch-clamp recording from human induced pluripotent stem cell-derived cardiomyocytes: improving action potential characteristics through dynamic clamp. J. Molec. Sci. **18**(9), 1873 (2017)

Electrocardiology Modeling After Catheter Ablations for Atrial Fibrillation

Simone Nati Poltri[1(✉)], Guido Caluori[2,3], Pierre Jaïs[2,3,4], Annabelle Collin[1], and Clair Poignard[1]

[1] Univ. Bordeaux, CNRS, INRIA, Bordeaux INP, IMB, UMR 5251, 33400 Talence, France
{simone.nati-poltri,annabelle.collin}@inria.fr
[2] IHU LIRYC, Electrophysiology and Heart Modeling Institute, Fondation Bordeaux Université, 33600 Pessac, France
[3] Univ. Bordeaux, INSERM UMR 1045, CRCTB, 33600 Pessac, France
[4] CHU Bordeaux, Department of Electrophysiology and Cardiac Stimulation, 33000 Bordeaux, France

Abstract. Catheter-based cardiac ablation, such as radiofrequency ablation (RFA) and pulsed electric field ablation (PFA), is the treatment of choice for atrial fibrillation (AF). However, the underlying phenomena and differences between RFA and PFA are not well understood. In this paper, we propose mathematical modeling of the cardiac electric signal of a cardiac domain containing an ablated area by RFA or PFA. Both types of ablation consist of the isolation of the pulmonary vein, but we describe them differently by using appropriate transmission conditions. More specifically, we assume that in the case of RFA, both intracellular and extracellular potentials are affected, leading to Kedem-Katchalsky type conditions at the interface. In contrast, in the case of PFA, we assume an isolation of the intracellular potential (due to the cardiomyocytes death induced by electroporation) whereas the extracellular potential is continuous. Numerical simulations in a context of AF show that PFA and RFA lead to isolation of the pulmonary vein. Our modeling also enables to propose a numerical explanation for the higher rate of fibrillation recurrence after RFA compared with PFA.

Keywords: Electrocardiology modeling · Radiofrequency ablation · Pulsed electric field ablation

1 Introduction

Cardiac arrhythmias are irregularities in the heartbeat that result in chaotic electrical waves. While most of cardiac arrhythmias are benign, some of them can directly affect the pumping function of the heart, leading then to stroke or heart failure. Isolation of the pulmonary veins by catheter ablation has become the treatment of choice for atrial fibrillation (AF). The goal is to isolate the pulmonary veins from which the fibrillation is supposed to originate by physical

O. Bernard et al. (Eds.): FIMH 2023, LNCS 13958, pp. 184–193, 2023.
https://doi.org/10.1007/978-3-031-35302-4_19

procedures such as thermal ablation (cryoablation [17] or radiofrequency ablation (RFA) [10]), and more recently pulsed electric field ablation (PFA), which is based on nonthermal irreversible electroporation.

Despite the great interest that RFA and PFA have generated in the treatment of AF, there is still a lack of understanding – and thus modelling – of the underlying biophysical phenomena of these different therapies. On one hand, it is well known that RFA ablation leads to coagulation necrosis with complete loss of cellular and vascular architecture [2] by leaving a scar composed of a fibrotic tissue. On the other hand, PFA is known to destroy mainly the cardiomyocytes, but the tissue scaffold is preserved [3,12]. Therefore, the physical properties of the cardiac tissue after RFA and PFA are very different, although they have the same goal, which is to isolate the pulmonary vein.

Recent medical studies have shown that the recurrence of atrial fibrillation with PFA [15] is on the order of 15%, compared with 30% with RFA [19]. We hypothesize that these treatment failures can somehow be explained by the long-lasting changes in the electrical properties of the tissue after ablation.

Well-designed mathematical modeling could help to better understand the effects of PFA on cardiac electrical wave and to develop numerical criteria for treatment evaluation. For example, one of the challenges is to derive electroporation models at the cellular or tissue level. This is a very interesting question, but in this work we focus on another one. More precisely, we propose a mathematical modeling of the cardiac electrical signal of a cardiac domain containing an area ablated by RFA or PFA. We study the effects of this treated area on the propagation of the electrical wave, known to correspond to the so-called bidomain model [18] in cardiac domain, through well-adapted transmission conditions across the treated region.

After a detailed presentation of the mathematical modeling in Sect. 2, we perform numerical simulations in a realistic configuration in Sect. 3. We show that our models are able to represent very well the isolation of a pulmonary vein by RFA and PFA ablation, and we propose an explanation for the higher recurrence rate of fibrillation after RFA.

2 Modeling

2.1 Geometrical Setting

For numerical purposes, it is convenient to consider the cardiac tissue as a mid-surface as suggested in [4], to avoid meshing the thin volume. This configuration has been found to be particularly well suited for the very thin wall of the atria.

Domains of Interest and Mesh. The geometric configuration is presented in Fig. 1. The left atrium, denoted by $\mathcal{D}^{\mathrm{LA}}$ is a 2D surface separated from one of the 4 pulmonary veins denoted by $\mathcal{D}^{\mathrm{PV}}$ by the interface Γ. For the sake of simplicity, we only focus on the isolation of one pulmonary vein even though in clinical cardiac ablation the 4 veins are isolated. We denote by Γ^{PV} the outer

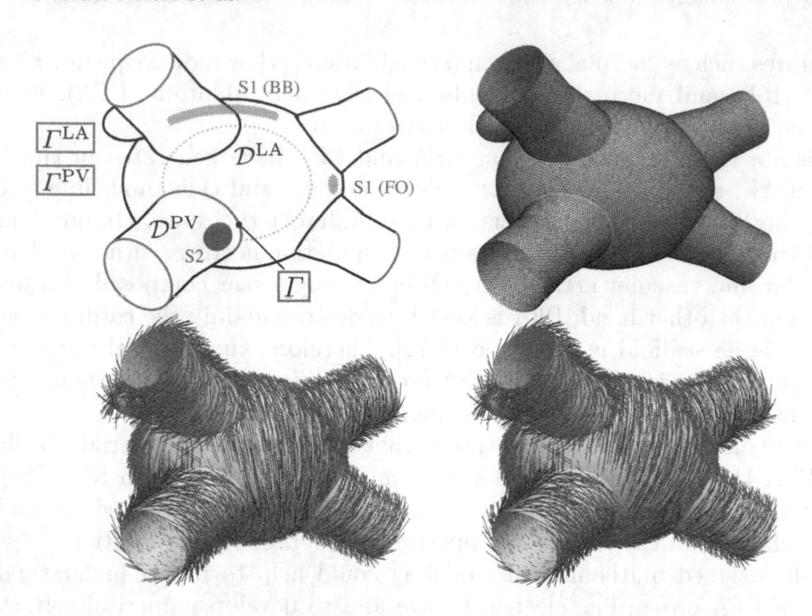

Fig. 1. Geometrical setting. Domains with interfaces (left-top), computational mesh (right-top), fibers orientation at the endocardium (left-bottom) and at the epicardium (right-bottom).

boundary of $\mathcal{D}^{\mathrm{PV}}$, while the 3 other outer boundaries of the pulmonary veins are denoted by Γ^{LA} (see Fig. 1). The whole domain of interest is denoted by $\mathcal{D}$:

$$\mathcal{D} = \mathcal{D}^{\mathrm{LA}} \cup \Gamma \cup \mathcal{D}^{\mathrm{PV}}.$$

Fibers Orientation. It is well-known that the fibers orientation impacts the propagation of the electrical wave on the heart. Chapelle et al. have proposed an efficient model of the electrical properties of the fibers [4] allowing to consider fiber variations inside the thickness of the atria. They introduce the following tensors: $\bar{\bar{I}}$ denotes the identity tensor in the tangential plane, $\bar{\tau}_0$ denotes a unit vector, linked to the fiber direction of the midsurface of the atria, and $\bar{\tau}_0^{\perp}$ is such that $(\bar{\tau}_0, \bar{\tau}_0^{\perp})$ gives an orthonormal basis to the tangent plane. Eventually, they introduce the functions $I_0(\theta) = \frac{1}{2} + \frac{1}{4\theta}\sin(2\theta)$ and $J_0(\theta) = 1 - I_0(\theta)$ to describe the effect of a variation of a 2θ angle of the direction of the fibers across the wall. Then, the intra- and extra-cellular conductivity tensors denoted by $\bar{\bar{\sigma}}_i$ and $\bar{\bar{\sigma}}_e$ respectively are defined by

$$\bar{\bar{\sigma}}_{i,e} = \sigma_{i,e}^t \bar{\bar{I}} + (\sigma_{i,e}^t - \sigma_{i,e}^l)\left[I_0(\theta)\bar{\tau}_0 \otimes \bar{\tau}_0 + J_0(\theta)\bar{\tau}_0^{\perp} \otimes \bar{\tau}_0^{\perp}\right], \qquad \text{in } \mathcal{D}. \tag{1}$$

where $\sigma_{i,e}^t$ and $\sigma_{i,e}^l$ denote the conductivity coefficients in the intracellular medium measured along and across the fiber direction, respectively.

2.2 Surface Bidomain Model in the Atrium

To model the electrical wave propagation, we consider the well-known bidomain model [18], widely studied in literature. It consists of a non linear degenerate parabolic partial differential equation (PDE), coupled with an ordinary differential equation (ODE), representing the activity of the ion channels. In particular, considering a quite simple phenomenological ionic model, equations can be rewritten in terms of the intracellular potential u_i, the extracellular potential u_e and the ionic variable w. The system of equations writes, for any $t > 0$,

$$A_m(C_m \partial_t v_m + I_{ion}(v_m, w)) - \nabla \cdot (\bar{\bar{\sigma}}_i \cdot \nabla u_i) = 0, \qquad \mathcal{D}, \tag{2a}$$

$$\nabla \cdot (\bar{\bar{\sigma}}_e \cdot \nabla u_e) + \nabla \cdot (\bar{\bar{\sigma}}_i \cdot \nabla u_i) = 0, \qquad \mathcal{D}, \tag{2b}$$

$$\partial_t w + g(v_m, w) = 0, \qquad \mathcal{D}, \tag{2c}$$

$$v_m = u_i - u_e, \qquad \mathcal{D}, \tag{2d}$$

where the functions I_{ion} and g are defined as in the model proposed by Mitchell and Schaeffer [14], A_m, the fraction of membrane area per unit volume and C_m, the membrane capacitance per unit surface. We assume that the heart is isolated, so we make the standard assumption that the extra- and intra-cellular currents do not propagate outside the heart meaning that we consider Neumann homogeneous boundary conditions on $\partial\mathcal{D}$, for any time $t > 0$,

$$(\bar{\bar{\sigma}}_i \cdot \nabla u_i) \cdot \bar{n} = 0, \quad (\bar{\bar{\sigma}}_e \cdot \nabla u_e) \cdot \bar{n} = 0, \qquad \text{on } \partial\mathcal{D}, \tag{2e}$$

$\bar{n}$ being the normal vector to $\partial\mathcal{D}$ outwardly directed from $\mathcal{D}$ towards the exterior.

2.3 Transmission Conditions Through Γ

The above system (2) has to be complemented with initial conditions for u_i, u_e and w at time $t = 0$ and a Gauge condition on u_e to fix the constant. In this paper, we impose $\int_{\mathcal{D}} u_e \, dx = 0$. More importantly, to close the system, appropriate transmission conditions through the interface Γ must be prescribed. They depend on the ablation that we consider and they are presented in the following subsections. We first introduce few notations. Denote by Γ^+ (resp. Γ^-) the interfaces

$$\Gamma^+ = \Gamma \cap \overline{\mathcal{D}^{\mathrm{PV}}}, \quad \Gamma^- = \Gamma \cap \overline{\mathcal{D}^{\mathrm{LA}}},$$

where $\overline{\mathcal{D}^{\mathrm{LA}}}$ denotes the adherence of $\mathcal{D}^{\mathrm{LA}}$, *mutatis mutandis* for $\overline{\mathcal{D}^{\mathrm{PV}}}$. For any function u defined on $\mathcal{D}$ and discontinuous through Γ, the jump of u across Γ is

$$[\![u]\!]_{|\Gamma} = u_{|\Gamma^+} - u_{|\Gamma^-}.$$

Radiofrequency Ablation. To model the effect of RFA, we consider the well-known Kedem-Katchalsky transmission conditions – initially introduced in [11] – which read

$$\alpha[\![u_e]\!]_{|\Gamma} = ((\bar{\bar{\sigma}}_e \cdot \nabla u_e) \cdot \bar{n})_{|_{\Gamma+}} = ((\bar{\bar{\sigma}}_e \cdot \nabla u_e) \cdot \bar{n})_{|_{\Gamma-}}, \tag{3a}$$

$$\alpha[\![u_i]\!]_{|\Gamma} = ((\bar{\bar{\sigma}}_i \cdot \nabla u_i) \cdot \bar{n})_{|_{\Gamma+}} = ((\bar{\bar{\sigma}}_i \cdot \nabla u_i) \cdot \bar{n})_{|_{\Gamma-}}, \tag{3b}$$

here $\bar{n}$ is the normal vector to Γ oriented from Γ^- towards Γ^+. The coefficient α is a positive constant homogeneous to a surface conductance. It takes into account the fact that the treated region has a higher resistance due to RFA than the healthy tissue. This parameter α is crucial because it is responsible for whether or not the transmembrane potential wave can overcome the Γ interface. In particular, when $\alpha = 0$, it is equivalent to a complete decoupling of the two domains $\mathcal{D}^{\mathrm{LA}}$ and $\mathcal{D}^{\mathrm{PV}}$, resulting in a perfect isolation of the pulmonary vein.

In the asymptotic regime $\alpha \gg 1$, the potentials u_e and u_i become asymptotically continuous. The Kedem-Katchalsky can then be seen as a penalty term which weakly enforces the continuity of the potential through the interface Γ [1]. In the following numerical section, we consider different values of α showing different levels of pulmonary vein isolation.

2.4 Pulsed Field Ablation

It has been experimentally observed that PFA preserves the tissue scaffold and targets the myocardium through the nonthermal, irreversible electroporation process. After PFA, the mechanical properties (stiffness, elasticity...) of the cardiac tissue are preserved, whereas the electrical functionalities of the ablation area are altered [2]. Most likely, PFA leads to local death of cardiomyocytes, which are then replaced by nonexcitable fibroblasts.

Based on these considerations, we propose to model the electrical effect of PFA by a continuity of both the extracellular potential u_e and the extracellular normal flux $(\bar{\bar{\sigma}}_e \cdot \nabla u_e) \cdot \bar{n}$, while assuming that the intracellular potential of $\mathcal{D}^{\mathrm{LA}}$ is isolated from the intracellular potential of $\mathcal{D}^{\mathrm{PV}}$, thus using a homogeneous Neumann boundary condition for u_i. In other words the transmission conditions describing the effect of PFA read

$$[\![u_e]\!]_{|\Gamma} = 0, \ [\![(\bar{\bar{\sigma}}_e \cdot \nabla u_e) \cdot \bar{n}]\!]_{|\Gamma} = 0, \tag{4a}$$

$$((\bar{\bar{\sigma}}_i \cdot \nabla u_i) \cdot \bar{n})_{|_{\Gamma+}} = ((\bar{\bar{\sigma}}_i \cdot \nabla u_i) \cdot \bar{n})_{|_{\Gamma-}} = 0. \tag{4b}$$

The mathematical justification of these conditions is beyond the scope of the present paper and will be presented in a forthcoming mathematical paper. Roughly speaking, these conditions arise from an asymptotic analysis in which the small parameter is the thickness of the electroporated region and its low intracellular conductivity tensor.

3 Numerical Illustrations

The aim of this section is to compare the transmembrane potential v_m and the extracellular potential u_e satisfying (2) with either the transmission conditions for RFA (3) or the transmission conditions for PFA (4). For the sake of simplicity, we assume that the treated region behaves identically along the interface Γ. It means that the parameter α in (3) is assumed to be constant.

3.1 Mesh, Fibers and Numerical Schemes

The simplified geometry of the left atrium is constructed as an ellipsoid 50mm long and 35mm high (see Fig. 1). Its depth is intersected by a plane corresponding to the position of the mitral valve. Four pulmonary veins – modeled by cutting cones with a mean diameter of 13 mm – are added. The appendage is modeled by an ellipsoid of 10 mm × 5 mm × 5mm. The mesh, presented in Fig. 1 has been generated by `Gmsh` [6]. The surface mesh is composed of 26141 nodes and 51904 triangular elements. The fibers directions – needed to build the tensors (1) – at the endocardium and epicardium are derived according to the literature [8, 13].

The numerical illustrations have been obtained using `FreeFem++` [7], a PDE solver based on finite element method. All the problems are solved with a BDF2 semi-implicit scheme ($\Delta t = 0.01$ ms) to deal with the nonlinear term I_{ion} and with P1 elements. To numerically solve the transmission conditions of RFA with fibrosis, we consider a weak coupling. Indeed, the condition leads to a Neumann condition in which we use the trace of the solution on Γ^- (resp. Γ^+) at the previous time step when solving the solution in PV domain (resp. LA domain). To numerically solve the transmission conditions of PFA, we use a Schwarz-type algorithm in which the penalty parameter is fixed at 2. This value has been chosen very carefully through a mathematical study, following [9]. Mesh, fibers and codes are available here: https://gitlab.inria.fr/snatipol/af-pfa-rfa.

3.2 Numerical Illustrations Before Ablation

To compare the effects of the two ablations considered, a simulation corresponding to atrial fibrillation before ablation is proposed.

Physiologically, the depolarization wave that triggers the heartbeat is initiated in the sinoatrial node in the right atrium. It then propagates to the left atrium via two electric pathways: the fastest leading to Bachmann's bundles (BB) and the second to the fossa ovalis (FO). The BB are modeled as two ellipsoids located at the top of the atrium near the appendage. The FO is located on the right side of the left atrium, on the wall between the right and left atria. They are also shown in Fig. 1. The depolarization wave reaches the FO at $t = 10$ ms after depolarization of the BB corresponding to $t = 0$ ms. To generate the AF synthetically, we use a standard S1-S2 protocol [5]. The S1 stimulus corresponds to the BB and the FO stimulus. The S2 stimulus location is near the left pulmonary inferior vein, see Fig. 1. Pulmonary veins are known to be prone to frequent reentry. The S2 stimulus is triggered at $t = 356$ ms. Parameters were set in [16].

The first column of Fig. 2 shows time snapshots of the transmembrane potential before ablation. One first sees the healthy depolarization occurring during the first 90 ms – see 20, 44, and 70 ms snapshots – followed by the healthy

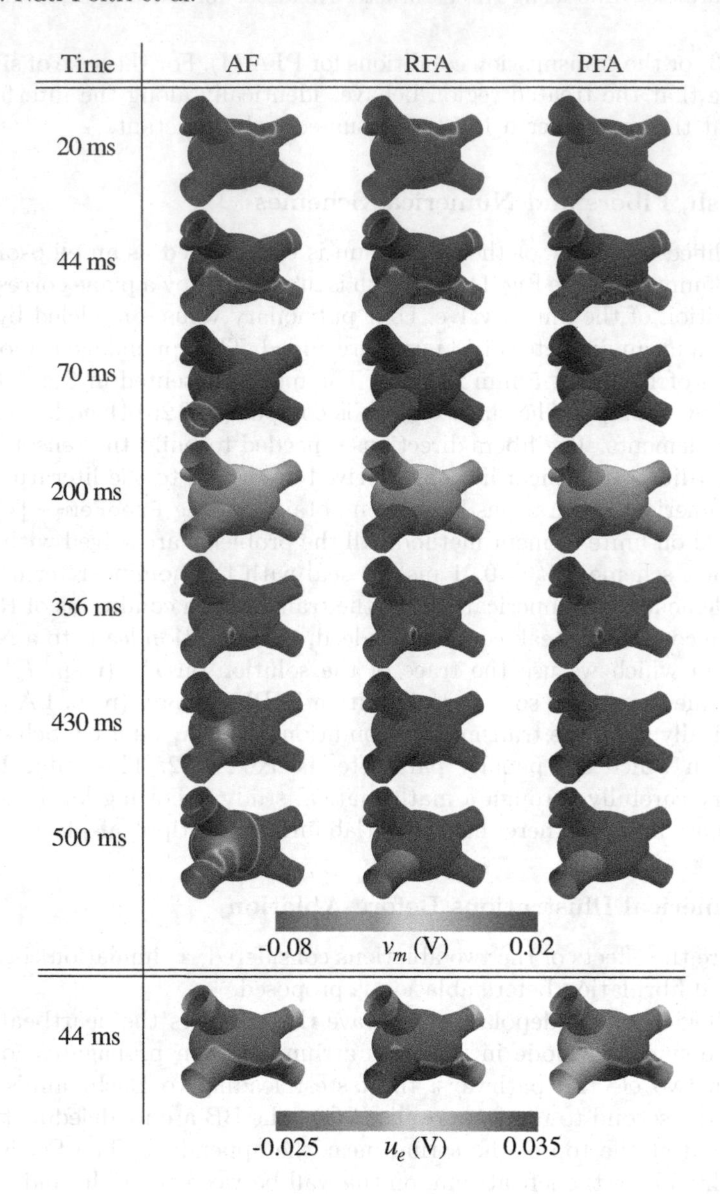

Fig. 2. First column: before ablation. Second column: successful RFA (Eq. (3), $\alpha = 10^{-4}$). Third column: successful PFA. Lines 1 to 7: Snapshots of transmembrane potential v_m. Last line: Snapshot of extracellular potential u_e.

repolarization – see 200 ms snapshot – and by the second stimulus illustrating a pathological area of one of the pulmonary veins – see 356 ms snapshot – that triggers a pathological wave that unfolds in the left atrium – see 430 and 500 ms snapshots.

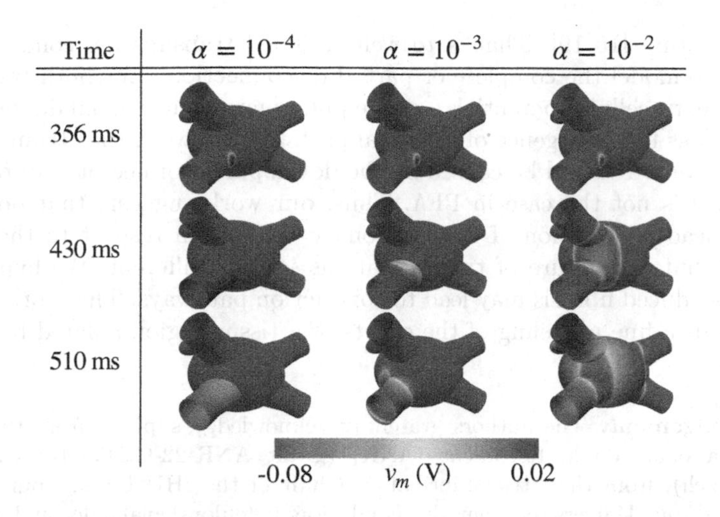

Fig. 3. Snapshots of transmembrane potential v_m in 3 situations of RFA corresponding to transmission conditions (3). First column: $\alpha = 10^{-4}$. Second column: $\alpha = 10^{-3}$. Third column: $\alpha = 10^{-2}$.

3.3 Effects on the Electric Signal

The second column of Fig. 2 gives the transmembrane potential corresponding to successful RFA (Equation (3), $\alpha = 10^{-4}$). Because the two regions are nearly decoupled, the pulmonary vein is well isolated: there is no entry of the wave into the left atrium, see 356, 430, and 500 ms snapshots. Uncoupling is also seen in the evolution of the extracellular potential, shown in the last line of the same figure at time 44ms. The third column of Fig. 2 corresponds to a successful PFA. One can first see that there is no entry of the transmembrane wave, showing the success of the ablation. However and contrary to RFA (second column), the PFA maintains the continuity of the extracellular potential u_e, see the last line of Fig. 2.

The effects of the α parameter are shown in Fig. 3. One can see that as it increases – see the second and the third columns ($\alpha = 10^{-3}$ and $\alpha = 10^{-2}$) – the wave crosses the interface from the pulmonary vein to the left atrium, resulting in a restart of the atrial fibrillation even though the propagation of the wave is delayed. The larger the value of α, the smaller the delay here. The increase in α can be viewed as the emergence of electrical pathways between $\mathcal{D}^{\mathrm{LA}}$ and $\mathcal{D}^{\mathrm{PV}}$.

4 Conclusion

In this work, we propose a mathematical modeling of the cardiac electric signal of a cardiac domain containing a region ablated by RFA or PFA. It consists in determining the transmission conditions of the very classical bidomain model at the interface of the ablated area. Our goal was to propose a mathematical explanation for the lower recurrence of AF after PFA compared with RFA as reported in the literature [15,19]. Thanks to well-designed transmission conditions, we were able to model the complete or partial disconnection – for both transmembrane and extracellular potentials – of the pulmonary vein. Partial disconnection can be seen as the emergence of electrical pathways between the atrium and the pulmonary vein. It could be caused by the development of necrotic fibrosis after RFA, which is not the case in PFA. Thus, our work suggests that both RFA and PFA lead to isolation of the pulmonary veins with respect to the electrical signal, but the nature of these isolations is very different. We hypothesize that RFA-induced fibrosis may lead to conduction pathways. This work is a first step towards a fine modeling of the effects of a tissue region ablated by PFA in cardiology.

Acknowledgement. The authors gratefully acknowledge support from the French Agence Nationale de la Recherche (ANR) (grant ANR-22-CE45-0014-01, project MIRE4VTach), from the Atrial Fibrillation Chair of the IHU Liryc, from the Fondation Bordeaux Université, from the Fondation Lefoulon-Delalande, and from the French Federation of Cardiology - Grands projets - 2022 (project DIELECTRIC).

References

1. Babuška, I.: The finite element method with penalty. Math. Comput. **27**(122), 221–228 (1973)
2. Bulvik, B.E., et al.: Irreversible electroporation versus radiofrequency ablation: a comparison of local and systemic effects in a small-animal model. Radiology **280**(2), 413–424 (2016)
3. Caluori, G., et al.: AC pulsed field ablation is feasible and safe in atrial and ventricular settings: a proof-of-concept chronic animal study. Front. Bioeng. Biotechnol. **8**, 552357 (2020)
4. Chapelle, D., Collin, A., Gerbeau, J.F.: A surface-based electrophysiology model relying on asymptotic analysis and motivated by cardiac atria modeling. Math. Models Methods Appl. Sci. **23**(14), 2749–2776 (2013)
5. Franzone, P.C., Pavarino, L.F., Scacchi, S.: Mathematical Cardiac Electrophysiology, vol. 13. Springer, Heidelberg (2014). https://doi.org/10.1007/978-3-319-04801-7
6. Geuzaine, C., Remacle, J.F.: Gmsh: a 3-D finite element mesh generator with built-in pre-and post-processing facilities. Int. J. Numer. Meth. Eng. **79**(11), 1309–1331 (2009)
7. Hecht, F., Pironneau, O., Le Hyaric, A., Ohtsuka, K.: FreeFEM++ manual (2005)
8. Ho, S.Y., Anderson, R.H., Sánchez-Quintana, D.: Atrial structure and fibres: morphologic bases of atrial conduction. Cardiovasc. Res. **54**(2), 325–336 (2002)
9. Hubert, F.B.F.: Méthodes de décomposition de domaine de type schwarz (2014)
10. Joseph, J., Rajappan, K.: Radiofrequency ablation of cardiac arrhythmias: past, present and future. QJM Int. J. Med. **105**(4), 303–314 (2012)
11. Kedem, O., Katchalsky, A.: A physical interpretation of the phenomenological coefficients of membrane permeability. J. Gen. Physiol. **45**(1), 143–179 (1961)
12. Koruth, J., et al.: Preclinical evaluation of pulsed field ablation: electrophysiological and histological assessment of thoracic vein isolation. Circ. Arrhythmia Electrophysiol. **12**(12), e007781 (2019)

13. Krueger, M.W., et al.: Modeling atrial fiber orientation in patient-specific geometries: a semi-automatic rule-based approach. In: Metaxas, D.N., Axel, L. (eds.) FIMH 2011. LNCS, vol. 6666, pp. 223–232. Springer, Heidelberg (2011). https://doi.org/10.1007/978-3-642-21028-0_28
14. Mitchell, C.C., Schaeffer, D.G.: A two-current model for the dynamics of cardiac membrane. Bull. Math. Biol. **65**(5), 767–793 (2003)
15. Reddy, V.Y., et al.: Pulsed field ablation of paroxysmal atrial fibrillation: 1-year outcomes of IMPULSE, PEFCAT, and PEFCAT II. Clin. Electrophysiol. **7**(5), 614–627 (2021)
16. Schenone, E., Collin, A., Gerbeau, J.F.: Numerical simulation of electrocardiograms for full cardiac cycles in healthy and pathological conditions. Int. J. Numer. Methods Biomed. Eng. **32**(5), e02744 (2016)
17. Skanes, A.C., Klein, G., Krahn, A., Yee, R.: Cryoablation: potentials and pitfalls. J. Cardiovasc. Electrophysiol. **15**, S28–S34 (2004)
18. Tung, L.: A bi-domain model for describing ischemic myocardial dc potentials. Ph.D. thesis, Massachusetts Institute of Technology (1978)
19. Wittkampf, F.H., Nakagawa, H.: RF catheter ablation: lessons on lesions. Pacing Clin. Electrophysiol. **29**(11), 1285–1297 (2006)

Modeling Cardiac Stimulation by a Pacemaker, with Accurate Tissue-Electrode Interface

Valentin Pannetier[1], Michael Leguèbe[1,2], Yves Coudière[1](✉), Richard Walton[3], Philippe Dhiver[4], Delphine Feuerstein[4], and Diego Amaro[4]

[1] Univ. Bordeaux, CNRS, Inria, Bordeaux INP, IMB, UMR 5251, IHU Liryc, 33400 Talence, France
valentin.pannetier@math.u-bordeaux.fr
[2] Inria, 33400 Talence, France
[3] Univ. Bordeaux, Inserm, CRCTB, U 1045, IHU Liryc, 33000 Bordeaux, France
[4] Microport CRM, Clamart, France

Abstract. In this paper we model a cardiac pacemaker placed in a bath with a cardiac excitable tissue. We take into account electrochemical phenomena observed at the electrodes during pacing by using equivalent circuits, whose parameters are calibrated with respect to bench tests data. The complete model consists of a pacemaker model coupled to a re-scaled cardiac ionic model through these circuits. It is compared with ex-vivo experimental data of stimulation threshold detection. We perform an additional study of the influence of the scaling parameters, that can help matching experimental results.

1 Introduction

An implantable pacemaker aims to restore a cardiac beat when the intrinsic conduction system fails. It sends energy to the heart in the form of a voltage pulse for a certain duration via pacing leads implanted inside the heart. The device is programmed to deliver enough energy to trigger a cardiac depolarization (which is called *capture*). For battery saving reason, the energy must be just above the cut-off threshold between capturing and non-capturing regions of the voltage – duration plane, known as the Lapicque curve [2]. We aim to reproduce by computer modeling and simulation the experimental tests run to identify this curve. This would ultimately enable manufacturers to try out and optimize the delivered energy of several leads with different electrode shapes in-silico, instead of prototyping, manufacturing and testing leads experimentally.

We propose to couple a computational model of cardiac excitable cells (tissue) in an electrolyte (blood) to a model of the pacemaker's circuitry, through interfaces located at the metal electrodes of the leads. Modeling these interfaces by themselves is an extensive field of research, because many complex biophysical phenomena occur when an electrode delivers current into an electrolyte, as compiled in [3], and because a large part of the energy is dissipated by the contacts. It is then crucial to model correctly these bio-electrode interfaces, in order to

O. Bernard et al. (Eds.): FIMH 2023, LNCS 13958, pp. 194–203, 2023.
https://doi.org/10.1007/978-3-031-35302-4_20

reproduce quantitatively the threshold voltages. Cardiac simulations generally consider only the myocardium, and apply an artificial, irrealistic stimulation. Instead, we introduce boundary conditions that include impedance of the contacts, and act as current source of the cardiac model, following the approach of Somersalo *et al.* [4].

In this paper, Sects. 2, 3, and 4, describe the complete modeling approach, model calibration with respect to experimental data, and comparison of computed Lapicque curves to experimental ones.

2 Pacemaker Model Coupled to Cardiac Ionic Equations

Pacemaker Circuit. We model the electronics of the pacemaker with the circuit shown in Fig. 1 (left), which appeared to be standard among manufacturers. We focus on the short time interval dedicated to pacing, ignoring the sensing ability of the device, as it is not linked directly to the triggering of a cardiac action potential. A tank capacitor C_{pulse} ($\sim$ 10 μF) is charged (not modelled here), and then the actual pacing follows the three steps below (see Fig. 2).

Pulse. When the device triggers a stimulation, the tank capacitor acts as source of current. In the meantime, a secondary capacitance C_{ocd} is charged. The duration of this step is programmed by the clinician, and can range from 0.25 to 2 ms. The amplitude of the stimulation, proportional to the charge of the tank capacitor, is also programmable, typically between 0.25 V and 2 V.

Switch. There is a short transition phase (in the order of hundreds of μs) during which the two switches are open (see Fig. 1, left, black part).

Output Capacitor Discharge (OCD). The capacitor C_{pulse} is isolated from the circuit, and C_{ocd} discharges. This step lasts for 13 ms and is intended to discharge the equivalent capacity due to the polarization of the electrodes that occurs during the pulse (see below).

After the OCD, the stimulating part of the device is switched off, and is replaced by a sensing circuit until the next stimulation. In the meantime, the tank capacitor C_{pulse} is recharged. During the pulse and OCD steps, the circuit can be simplified into an equivalent RC series circuit. In the following, we simply denote by R and C the equivalent resistance and capacitance of this simplified circuit, specifically, $\frac{1}{C} = \frac{1}{C_{\text{pulse}}} + \frac{1}{C_{\text{ocd}}}$ during the pulse and $\frac{1}{C} = \frac{1}{C_{\text{ocd}}}$ during OCD, while $R = R_{\text{pulse}} + R_{\text{gnd}}$ during the pulse and $R = R_{\text{gnd}}$ during OCD.

Modeling the Bio-Electrode Contacts. Multiple electrochemical interactions occur in the vicinity of the surface of an electrode placed in an electrolyte when a current is applied, as described extensively in [3]. These interactions are modeled by equivalent electric circuits, called contact models. There exists a large collection of contact models, as reviewed in [3]. We choose to model the tip and ring contacts by a resistance in parallel with a capacitance (parallel R-C, see Fig. 1, green part). This model is a good compromise that can reproduce experimental measures while relying on a limited number of parameters that

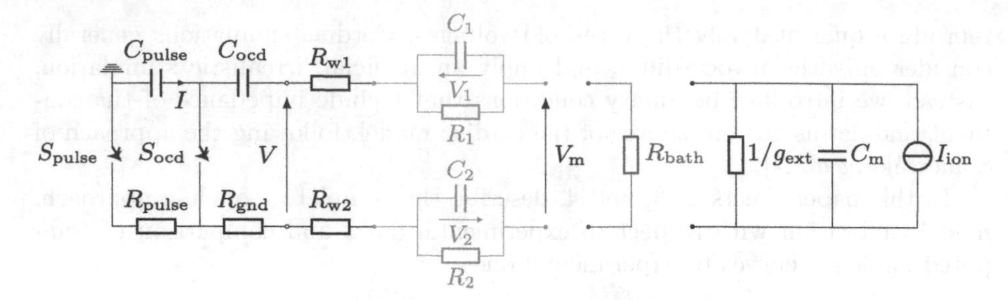

Fig. 1. Equivalent circuit of a pacemaker (black, left) coupled to either a resistive medium (blue, middle), or a single cell model (red, right) via connecting dots. On the left circuit, S_{pulse} is closed and S_{ocd} is open during the pulse step, S_{pulse} and S_{ocd} are open during switch step, S_{pulse} is open and S_{ocd} is closed during the OCD step. The resistances R_{pulse} and R_{gnd} are located within the canister of the pacemaker, whereas the resistances R_{w1} and R_{w2} are equivalent resistances for the wires between the canister and the leads.

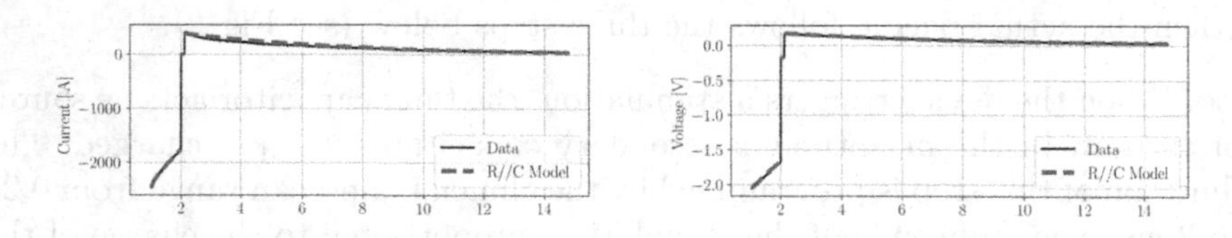

Fig. 2. Current (left) and voltage (right) delivered by a pacemaker lead in a saline solution (black solid line), between the tip and ring electrodes. Red dashed lines are the result of the calibration (Sect. 3). The voltage and duration set on the device were 2 V and 1 ms. (Color figure online)

can, in consequence, be identified. For instance, a simple resistive-only model of contact cannot reproduce the charging behavior of the bio-electrode contact, either during the pacing step nor during the switch step. On the other hand, the parameters of a complex Cole-Cole model cannot be identified from our data.

We state the equations of the contacts and pacemaker model, during pulse and OCD, as follows:

$$\frac{\mathrm{d}I}{\mathrm{d}t} + \frac{1}{\tau} I = \frac{1}{R_{\text{tot}}} \left(\frac{V_1}{\tau_1} + \frac{V_2}{\tau_2} \right), \tag{1a}$$

$$\frac{\mathrm{d}V_i}{\mathrm{d}t} + \frac{V_i}{\tau_i} = \frac{I}{C_i}, \quad i = 1, 2 \tag{1b}$$

where the current I, and voltages V_2 and V_1 are such as represented on Fig. 1. The other constants are $\tau_1 = R_1 C_1$, $\tau_2 = R_2 C_2$, $\tau = R_{\text{tot}} C_{\text{tot}}$, with $R_{\text{tot}} = R + R_{\text{w1}} + R_{\text{w2}} + R_{\text{bath}}$ and $\frac{1}{C_{\text{tot}}} = \frac{1}{C} + \frac{1}{C_1} + \frac{1}{C_2}$. Experimentally, the voltage

$V = V_1 + V_2 + (R_{\text{bath}} + R_{\text{w1}} + R_{\text{w2}})I$ is measured, where R_{bath} is the equivalent resistances of the saline solution.

In absence of applied current, the two contacts naturally discharge. Hence, we assume that $V_1(t = 0) = V_2(t = 0) = 0$. Meanwhile, the main capacitor is charged to its nominal capacity, corresponding to a voltage V_{stim}. In consequence, a current $I(t = 0) = -V_{\text{stim}}/R_{\text{tot}}$ initially flows out of it. This voltage V_{stim} is referred to as *the amplitude of stimulation.*

Pacemaker and Cardiac Tissue – Surrogate Model. To account for the excitability of the cardiac tissue, the resistance R_{bath} is not sufficient. It is replaced by a cardiac cell ionic model. The ionic model is adapted to account for the propagation of the electric current in the extracellular medium, of equivalent conductance g_{e}. The membrane is modeled by a capacitance C_{m} in parallel with the total ionic current I_{ion}. The intracellular path of conduction through gap junctions between cells is not taken into account in this simplification. We use the Beeler-Reuter ionic model [1], and we denote the transmembrane voltage by V_m.

As the Beeler-Reuter model is written in units per cm^2 of cell membrane, we introduce a scaling parameter S, representing a given surface of cell membrane. We assume that the total current I spreads uniformly on this surface. We also introduce the total conductance g_{e} of the extracardiac medium between the electrodes. Assuming that a volume of 1 mm^3 is relevant to this approximation, we obtain orders of magnitude of 100 cm^2 for S and 10^{-2} mS for g_{e}. However, the excited medium consists of blood and cardiac tissue, which is highly heterogeneous. Therefore, these estimates can only be seen as a starting point for any calibration procedure, and the current model as a surrogate to a complete 3D bidomain model. The coupled 0D model reads as follow

$$\frac{\mathrm{d}I}{\mathrm{d}t} + \frac{1}{\tau}I = \frac{1}{R_{\text{tot}}}\left(\frac{V_1}{\tau_1} + \frac{V_2}{\tau_2} + \frac{I_{\text{ion}}(h, V_{\text{m}})}{C_{\text{m}}} + \frac{V_{\text{m}}}{\tau_{\text{m}}}\right), \tag{2a}$$

$$\frac{\mathrm{d}V_i}{\mathrm{d}t} + \frac{V_i}{\tau_i} = \frac{I}{C_i}, \quad i = 1, 2 \tag{2b}$$

$$\frac{\mathrm{d}V_{\text{m}}}{\mathrm{d}t} + \frac{I_{\text{ion}}(h, V_{\text{m}})}{C_{\text{m}}} + \frac{V_{\text{m}}}{\tau_{\text{m}}} = \frac{I}{SC_{\text{m}}}, \tag{2c}$$

$$\frac{\mathrm{d}h}{\mathrm{d}t} + g(h, V_{\text{m}}) = 0, \tag{2d}$$

with τ, τ_1, τ_2 as above, but with $R_{\text{tot}} = R + R_{\text{w1}} + R_{\text{w2}}$ and $\frac{1}{C_{\text{tot}}} = \frac{1}{C} + \frac{1}{C_1} + \frac{1}{C_2} + \frac{1}{SC_{\text{m}}}$, and $\tau_{\text{m}} := \frac{SC_{\text{m}}}{g_{\text{e}}}$. The ionic current I_{ion}, state variables h and evolution function g are given by the Beeler-Reuter model. The voltage measured in experiments is now $V = V_1 + V_2 + (R_{\text{w1}} + R_{\text{w2}})I + V_{\text{m}}$. In addition to the parameters S and g_{e}, we need to determine the parameters R_1, τ_1 and R_2, τ_2 that characterize each electrode.

3 Calibration from Bench Experiments

In order to evaluate the accuracy of the parallel R-C model for the electrode polarization, we measure the current I and the voltage V delivered by pacemaker leads directly in a saline solution, without cardiac tissue (Fig. 3, left). We used a Microport CRM Borea DR pacemaker, with a Vega lead in a solution with measured conductivity 2.85 mS cm^{-1}, to obtain 9 datasets of current and voltage (I_{meas} and V_{meas}). The datasets differ exclusively by the amplitude (1, 2 or 4 V) and duration (0.25, 0.5 or 1 ms) of the delivered pulse.

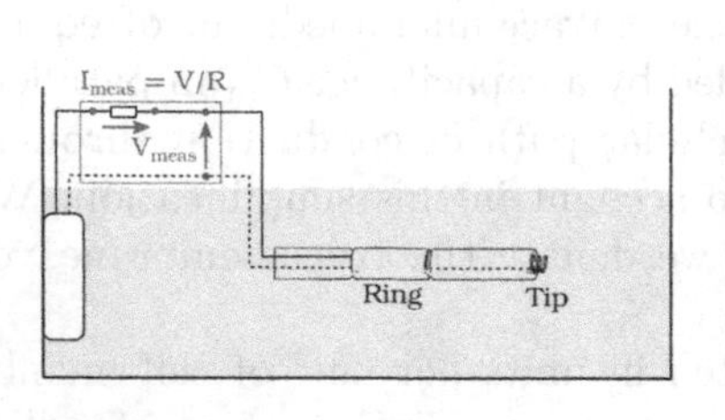

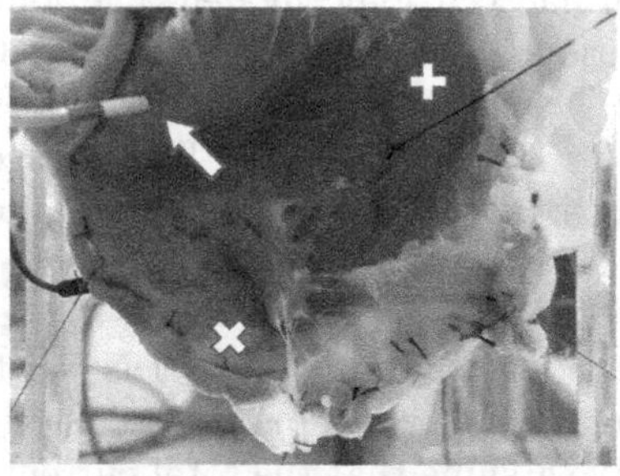

Fig. 3. Left: sketch of the bench test used to calibrate the electrode contact model. Right: right ventricle endocardial view of the tissue wedge preparation with the pacemaker electrode implanted in the septum (pointed by arrow). Other implantations sites were located in the apex ($\times$) and at the base of the ventricle ($+$).

The experimental setup can be modelled by the circuit shown in Figs. 1 (left and middle), where we used the equivalent circuit from the previous section for the pacemaker, and R_{bath} denotes the resistance of the saline medium. The corresponding differential equations system (1) has 5 parameters to be identified: R_{bath} and the properties of the contacts R_1, R_2, τ_1 and τ_2. The parameter R_{bath} was deduced from the measured initial conditions, as the electrodes are considered depolarized: $R_{\text{bath}} = \frac{V_{\text{meas}}(t=0)}{I_{\text{meas}}(t=0)} - R_{\text{w2}} - R_{\text{w1}}$. The four remaining parameters $\theta = (R_1, R_2, \tau_1, \tau_2)$ were calibrated using the following cost function:

$$J(\theta) = \sum_{\text{s}} \omega_{\text{s}} \mathbb{1}_{\text{s}}(t) \left(\frac{\|V(\theta) - V_{\text{meas}}\|_{\ell^2}}{\|V_{\text{meas}}\|_{\ell^2}} + \frac{\|I(\theta) - I_{\text{meas}}\|_{\ell^2}}{\|I_{\text{meas}}\|_{\ell^2}} \right),$$

where ω is a vector of three weights of sum 1 that allows to give more importance on a specific step, s $\in$ {pulse, switch, OCD}. The function $\mathbb{1}_{\text{s}}$ restricts the computation of the relative difference of the ℓ^2 norms of voltage and current on a specific step, V_{meas} and I_{meas} are the experimental data of the voltage and current, and $V(\theta)$ and $I(\theta)$ are the numerical solutions of the voltage and current, respectively. We used the SciPy implementation of the L-BFGS-B algorithm to minimize the cost function J within a subset of $\mathbb{R}^4$ for θ. Each evaluation of J requires to solve system (1) that is linear in I, V_1 and V_2. We solve it semi-analytically using eigenvalue decomposition at each time step in a dedicated Python code. The convexity of J was not established. Hence, we first optimize the logarithm of each parameter ($R \in [1, 10^5]$ Ω for the resistances, and $\tau \in [10^{-6}, 1]$ s for the characteristic times), in order to fix a correct order of magnitude, and then fit the values of the parameters.

First, the model was calibrated using the 9 data files separately. For each file, and for a given choice of Ω, the L-BFGS-B algorithm was started at a value of the parameters θ for which the cost function $J(\theta)$ is the lowest among 10 000 random samples of θ. Then, we minimized J using the 9 files all at once, with the average value of the previously found parameters as starting point. We set $\omega = (0.45, 0.45, 0.1)$, so as to emphasize the steps in which the information can be more easily recovered.

The results of the fits are illustrated on Fig. 4, for both voltage and current. We found that one of the couples of parameters is well determined: $R_2 = 26.0 \pm 1.77$ Ω, $\tau_2 = 0.0548 \pm 6.93\ 10^{-3}$ ms ($C_2 = 2.12 \pm 0.34$ μF). However, we found values of R_1 with much larger variability across the files: between 2 and 6 kΩ, with average $\sim$ 4 kΩ, which is one order of magnitude larger than the total impedance measured by the device (940 Ω). We believe that it is due to the over-simplification of our contact model, leading to this equivalent resistance having no real physical meaning. We tried to use more complex models to reproduce the data during the OCD step, to overcome this limitation of the simple R-C parallel model. However, we faced problems of identifiability of the parameters during the calibration (*ie*, finding different local minima for repeated computations).

Anyway, the currents produced by the parallel R-C model are close to the data, and consequently, we consider this model to be accurate enough as a current source for the cardiac model.

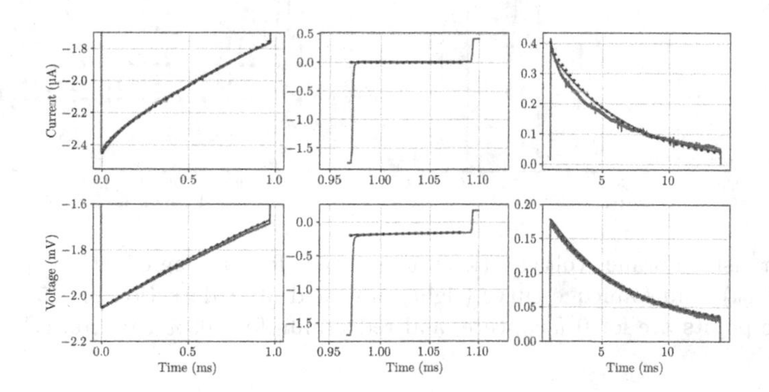

Fig. 4. Fitted currents and voltages when calibrating parameters $\theta = (R_1, R_2, \tau_1, \tau_2)$. Gray solid lines are one experimental dataset, blue dotted lines are computed with θ calibrated with this dataset, and red solid lines from the calibration of the 9 datasets together. (Color figure online)

4 Numerical Results, Comparison with Animal Experiments

Simulation of Single Stimulations, Computation of Lapicque Curves. With all parameters set, we computed the solutions of Eqs. (2) using the same software. An example of simulated transmembrane voltage is given on Fig. 5 (left), for 5 stimulations of 428.5 mV and 1 ms, with $S = 50\ \text{cm}^2$ and $g_e = 0.01$ mS at 90 bpm. This set of parameters results in several captures and non-captures, due to variables of the Beeler-Reuter model that do not return to their original state between each pulse. This shows that a single stimulation with these parameters is not sufficient to trigger a complete depolarization.

From transmembrane voltages, we extract the ratio of successful captures (60% in the illustrated case). We can compute this ratio for several amplitudes V_{stim} and durations d of stimulation, following a 2D dichotomy algorithm in the Lapicque space (Fig. 5 right). The algorithm detects when the capture ratio exceeds 50%, and returns a collection of intervals that should contain the Lapicque curve. We then compute the rheobase V_{rh} (V) and chronaxie T_{ch} (S) of Lapcique's law $V_{\text{stim}}(d) = V_{\text{rh}}(1 + T_{\text{ch}}/d)$ that best match the midpoints of these intervals. Our model seems in agreement with this law, for the set of parameters used to generate Fig. 5.

We then computed several Lapicque curves, for other values of S and g_e, and compared the results with experimental data described in the next paragraph.

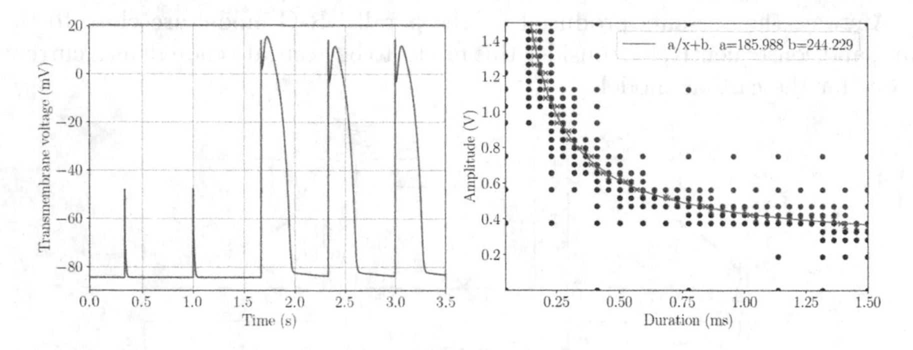

Fig. 5. Transmembrane voltage (left) computed for a single set of parameters $(V_{\text{stim}}, d, S, g_e)$; and Lapicque curve (right) obtained after detection of a 50% capture ratio. Blue points are for 0% capture, and red points for 100% capture. (Color figure online)

Ex-vivo Measurements. Isolated coronary-perfused ventricular wedge preparations were carried out on two sheep hearts, as described in [5]. Perfusion leaks along cut surfaces were carefully occluded for homogeneous perfusion resistance across the preparation. The wedge preparation was then stretched on to a frame to immobilize the tissue into a bath of saline solution (Fig. 3, right).

To study the MicroPort lead (contacts characteristics and voltages applied to the tissue), three lead implantation regions were identified (RV apex, RV septum and RV basal sites) and the lead was implanted in the heart by fully deploying the screw electrode perpendicularly to the tissue's surface. Electrocardiography (ECG) and electrograms (EGM) measured with the lead were simultaneously recorded through an independent acquisition system (PowerLab, ADInstruments). The EGMs were also measured with the pacemaker.

We tested a range of pulse durations and stimulus amplitudes to localize the stimulation threshold. Due to the limited number of steps in stimulation amplitude and duration of the pacemaker, the measured data points cover only sparsely the Lapicque plane. Nevertheless, these data can later be used for fitting the computational model responses to lead stimuli.

Comparison Between Experiment and Simulation. From experimental data, we identified, for each pacing pulse duration, the minimum pulse amplitude that captured. Below this pulse amplitude, any stimulation fails to capture the tissue. We could therefore define a region, delimited on the top by the lowest capturing amplitude, and on the bottom by the highest non-capturing amplitude, in which the Lapicque curve should be found (Fig. 6).

We show from the 0D model that the parameter S has great influence, since it determines the amount of current seen by square centimeter of membrane. On the contrary, we found very little influence of the parameter g_e on the output (not shown). We also studied the influence of parameters of the bio-electrode contact, by varying the resistance R_1. This is justified by the fact that the tip electrode is not any more in contact with a saline solution only, but with the cardiac tissue too (Fig. 6 right).

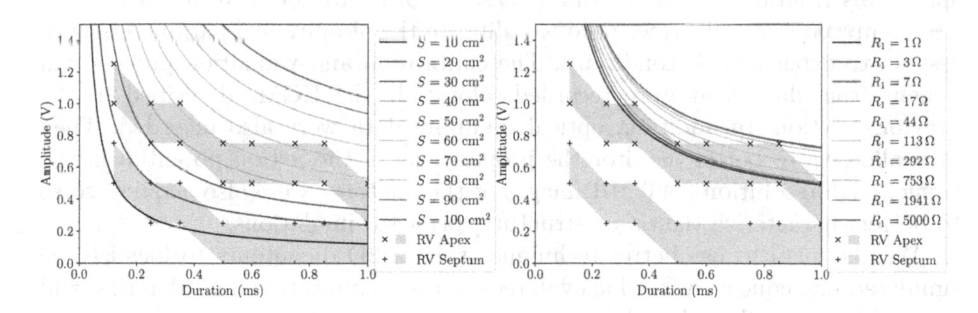

Fig. 6. Comparison of Lapicque curves (solid lines) computed with different values of S (left) and R_1 (right). The shaded regions indicate where the experimental capture threshold is located, for different stimulation sites on the tissue wedge.

The fits of Lapicque law $V_{\mathrm{stim}} = V_{\mathrm{rh}}(1 + T_{\mathrm{ch}}/d)$ do not fall within these regions for all durations. This is expected because the parameters S ad g_e of the 0D model cannot be completely characterized, as surrogates of the geometry and electric conductivity coefficients of the real tissue. Additionally, these parameters may depend on the amplitude of the stimulation, as the excited volume of tissue may change. As a consequence, further calibration work is required, especially to determine the parameter S which plays a crucial role in this 0D model.

5 Conclusion

In this paper, an innovative model is proposed, which couples a pacemaker electrical circuit to equations modeling cardiac action potential, accounting for the complex physics at the bio-electrode interfaces. Animal experiments were conducted which provide information on the electrical function of a cardiac tissue sample stimulated by a commercial pacemaker. It allows us to compare Lapicque curves obtained with the numerical model with experimental capture data.

It is noticeable that the literature provides no model for coupling a pacemaker to a cardiac tissue, because all studies focus either on characterizing in-depth the physics of the bio-electrode contact, or on characterizing the onset of cardiac activation in the tissue sample (virtual electrodes, make and break activations, etc.). To our knowledge, this work provides a first possible model that links the energy delivered by the pacemaker to the onset of activation, trying to evaluate with a model the real distribution of energy between all the concerned elements, pacemaker circuit, bio-electrical contact, and cardiac tissue. The model proposed in this article includes many simplifications, for example it assumes that the contact properties calibrated from bench experiments do not change when the contact is made with cardiac tissue. However, we prefered not to use a more descriptive model that could not be validated against the experimental data at hand.

We are currently deriving and implementing a more reliable 3D model, with bidomain equations for the cardiac tissue in blood. This model will characterize more accurately the delivery of current into cardiac cells, accounting for the spatial distribution of current. In any case, in order to derive such a 3D model and computational solver, we need to calibrate the bio-physical parameters with respect to experimental conditions. The calibration and validation process will benefit from data that were recorded, alongside the EGMs described in the previous section. In addition, optical mapping data were also recorded. They may allow us to better monitor the propagation of the action potential in the tissue. High resolution 9.4T MR images of the anatomy were also acquire, so as to better characterize the tissue structure, prior to simulations.

We also plan to use better techniques than a 2D dichotomy to localize the simulated Lapicque curves, which will reduce the computational burden that will arise when using 3D bidomain equations.

Acknowledgements. This work was supported by the H2020 EU SimCardioTest project (Digital transformation in Health and Care SC1-DTH-06-2020; grant agreement number 101016496). This study received financial support from the French Government as part of the "Investments of the Future" program managed by the National Research Agency (ANR), Grant reference ANR-10-IAHU-04. Experiments presented in this paper were partially carried out using the PlaFRIM experimental testbed, supported by Inria, CNRS (LABRI and IMB), Université de Bordeaux, Bordeaux INP and Conseil Régional d'Aquitaine.

References

1. Beeler, G.W., Reuter, H.: Reconstruction of the action potential of ventricular myocardial fibres. J. Physiol. **268**(1), 177–210 (1977)
2. Blair, H.: On the intensity-time relations for stimulation by electric currents. I. J. Gen. Physiol. **15**(6), 709–729 (1932)
3. Grimnes, S., Martinsen, O.G.: Bioimpedance and Bioelectricity Basics, 3rd edn. Academic Press, Oxford (2015)
4. Somersalo, E., Cheney, M., Isaacson, D.: Existence and uniqueness for electrode models for electric current computed tomography. SIAM J. Appl. Math. **52**(4), 1023–1040 (1992)
5. Walton, R.D., et al.: Compartmentalized structure of the moderator band provides a unique substrate for macroreentrant ventricular tachycardia. Circ.: Arrhythmia Electrophysiol. **11**(8), e005913 (2018)

Simulated Excitation Patterns in the Atria and Their Corresponding Electrograms

Joshua Steyer(✉), Lourdes Patricia Martínez Diaz, Laura Anna Unger, and Axel Loewe

Institute of Biomedical Engineering, Karlsruhe Institute of Technology (KIT), Karlsruhe, Germany
joshua.steyer@kit.edu

Abstract. Gaining an insight into atrial excitation dynamics is crucial for a thorough understanding of the mechanisms that underlie rhythm disturbances and their appropriate treatment, e.g. via substrate-based ablation therapy guided by electroanatomical mapping. Methods based on simulations can be helpful to understand electrogram genesis and morphology caused by different excitation patterns. State-of-the art *in silico* approaches studied these phenomena only on simplified geometries, such as 2D patches, which neither considered the local curvature of these patches nor heterogeneity between different atrial regions. In this study, we calculate unipolar and bipolar electrograms derived from a clinically inspired multielectrode array in a realistic atrial geometry, which was obtained from magnetic resonance imaging. The array is placed on the endocardium on six different basic excitation patterns. Most of the quantitative features of clinically measured electrograms for these phenomena could be reproduced and thus mechanistically underpinned. Future studies using even finer meshes and more sophisticated methods to calculate electrograms may shed more light also on the genesis of fractionation.

Keywords: Cardiac Electrogram Genesis · Arrhythmogenic Excitation Patterns · Atrial Modelling

1 Introduction

Electroanatomical mapping (EAM) is a helpful method to understand cardiac excitation dynamics [1]. Being the drivers of arrhythmias, a thorough understanding of abnormal excitation patterns, usually via EAM, is vital for a successful treatment, for example with ablation therapy [2,3]. Whilst *in vivo* studies are restricted to the specific rhythm disturbance present in the patient at the time of the mapping, *in silico* models provide a way to simulate different arrhythmogenic patterns and to study them systematically in terms of how they influence both the electrogram (EGM) amplitude and morphology [4]. Insights obtained from such computational models may thus help to enhance clinical mapping systems in terms of their set up as well as verification.

The original version of this chapter was revised: the text in acknowledgement section was missing. This was corrected. The correction to this chapter is available at https://doi.org/10.1007/978-3-031-35302-4_73

O. Bernard et al. (Eds.): FIMH 2023, LNCS 13958, pp. 204–212, 2023.
https://doi.org/10.1007/978-3-031-35302-4_21

So far, computational studies on the influence of cardiac excitation dynamics on EGMs were limited to simple 2D or 3D patches [5,6] or realistic geometries which, however, did not incorporate important properties, such as differing wall thickness [7]. In contrast, we consider varying electrical properties in different tissue regions, the heterogeneity of the wall thickness and a realistic curvature by using a model derived from patient data. The EGMs are obtained from a multielectrode array (MEA) that mimics the clinically established Advisor™ HD Grid mapping catheter (Abbott, Abbott Park, IL, USA). In total, the impact of six different clinically relevant basic excitation patterns generated on this realistic geometry on unipolar and exemplarily shown bipolar EGMs (uEGMs and biEGMs, respectively) are studied.

2 Methods

We used a volumetric right atrial model obtained from magnetic resonance imaging [8], consisting of approximately 2.26×10^6 tetrahedra with an average edge length of approximately 0.64 mm and its fibre orientation was calculated via a semi-automatic rule-based algorithm [9]. To account for electric heterogeneity of the atria, we imposed conductivities and anisotropy values for the crista terminalis, pectinate muscles and Bachmann's bundle that differ from the rest of the atria [10,11]. For the simulations, we used the electrophysiology simulator openCARP [12,13], while the visualisation of the excitation wave dynamics on the geometry was done with the application meshalyzer [14]. Electrical propagation inside the cardiac tissue was modelled using the monodomain approach, which assumes equal anisotropy ratios between the intracellular and extracellular space. The Courtemanche-Ramirez-Nattel model [15] was used to represent membrane dynamics and $dt_{\text{int}} = 10\,\mu\text{s}$ as integration time step, and thus, a sampling rate for the EGMs of around 100 kHz. To characterise the excitation dynamics inside the atrial geometry, we modelled the bidirectionally steerable Advisor™ HD Grid mapping catheter, which has an inter-electrode distance of 4 mm along each spline and a distance of $3\,\text{mm} + 2.5\,\text{F} = 3.8\overline{3}\,\text{mm}$ between two splines as a 4×4 MEA of point-like electrodes with an equidistant spacing of $h_x = h_y = 4\,\text{mm}$. Electrodes are named according to Fig. 1. This 2D MEA was then placed above the region where the excitation dynamics were to be investigated. For each of the 16 electrodes, a uEGM was calculated via the infinite volume conductor method [4]:

$$\phi_{\text{e}} = \frac{1}{4\pi\sigma_{\text{b}}} \int_V \frac{\sigma_{\text{i}} \cdot \vec{\nabla} V_{\text{m}}}{||\vec{x} - \vec{x}_{\text{src}}||} \text{d}V, \tag{1}$$

where $\sigma_{\text{b}} = 1\,\text{S/m}$ [16] denotes the isotropic conductivity of the volume conductor, σ_{i} the intracellular conductivity and the denominator represents the Euclidean distance between the source location $\vec{x}_{\text{src}}$ and the electrode location $\vec{x}$. This method is the most simple approach of a monodomain realisation combined with an independent forward calculation of extracellular potentials, ϕ_{e} [4]. In total, we investigated six different basic excitation patterns and their EGMs.

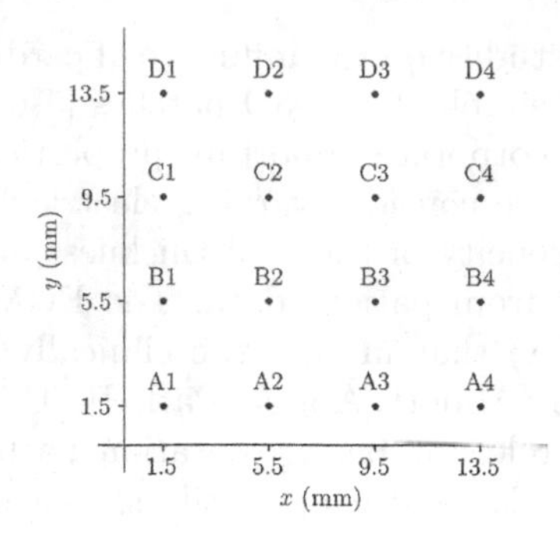

Fig. 1. Nomenclature of the electrodes the MEA is composed of.

Each excitation pattern was initialised by depolarising the sinus node. We then modified this healthy control setup to reproduce arrhythmogenic excitation patterns observed during electroanatomical mapping [17–19]. We analysed an unperturbed wave starting at the sinus node (SIN), wave collision with a line of block (LOB), wave propagation through a slow conduction zone (SCZ), a wave pivoting around a non-conducting block (PIV), wave propagation through a gap in a lesion (GAP) and collision of two wavefronts (COL). Non-conducting regions, i.e. for the LOB, PIV and GAP patterns, were realised by setting conductivities in all directions to 10^{-7} S/m, while for the SCZ zone, we set the conductivity in all directions to 0.1 S/m. The second wave the SIN wave collides with in the COL case is initialised by depolarising a small region in the right pulmonary vein to mimic an ectopic beat [20].

3 Results

We show the SIN excitation pattern and corresponding EGMs acquired by the virtual MEA placed in the right atrium above the pectinate muscles in Fig. 2. For each pattern shown here, we highlighted the tricuspid valve ring (TVR) and the right atrial appendage (RAA), as well as the upper left and lower right electrode name (D1 and A4, respectively) for the sake of orientation. Furthermore, arrows indicate the wave propagation direction and manipulated regions are highlighted in white. In general, each uEGM exhibits a biphasic and approximately symmetric potential. However, there are differences in the amplitudes and uEGM sharpness. In the LOB case (see Fig. 3), the position of the MEA was the same as in the SIN case. Except for the COL pattern, this is also the case for the other patterns. The uEGMs derived from the electrodes B1, C2, D3 are the ones of highest interest, as they are closest to where the wave collides with the block and they show a dominant positive deflection. uEGMs derived from electrodes behind the LOB, on the other hand, exhibit double potentials of different extents.

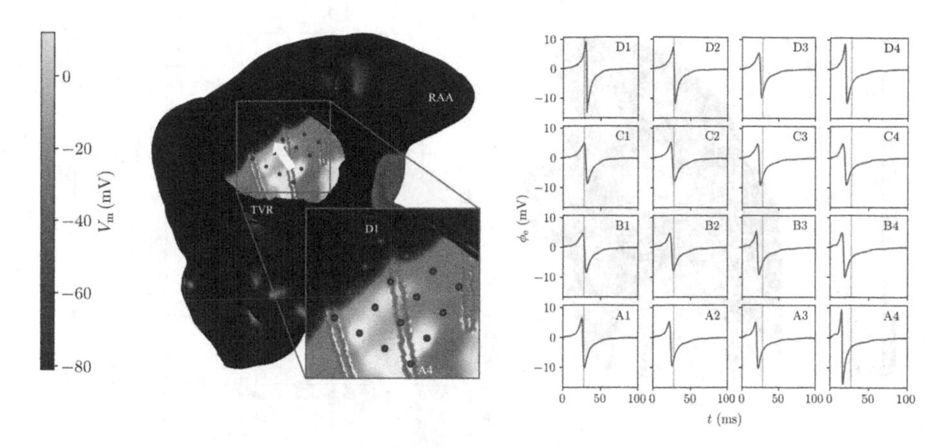

Fig. 2. Left: Wave propagating over the atrial geometry 28 ms after the sinus node was depolarised (**SIN** case). The MEA, whose electrodes are highlighted in bronze, is placed at the right atrial endocardium and can be seen through the TVR. Right: Corresponding uEGMs. (Color figure online)

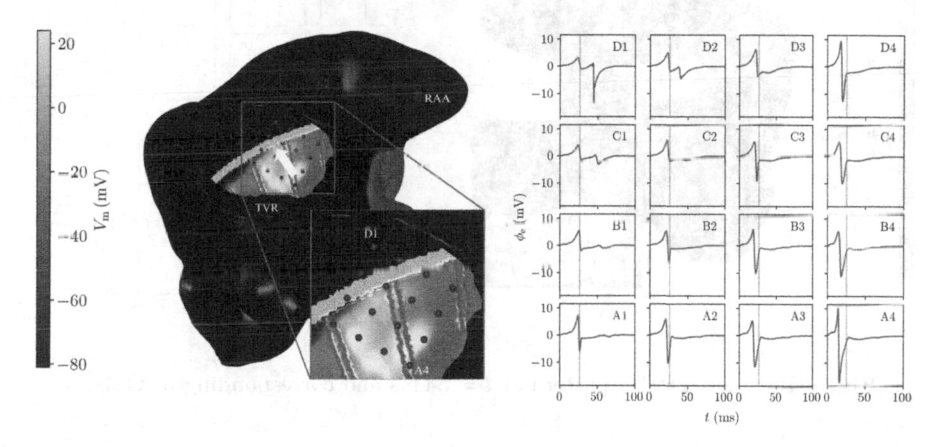

Fig. 3. Wave colliding with a line of block (**LOB**) at t = 26 ms and corresponding uEGMs.

Similar to the LOB case, we severely reduced the conductivity within a region, however, still allowing for (slow) propagation over this slow-conduction zone. The corresponding dynamics and uEGMs are shown in Fig. 4.

The electrodes placed above the SCZ region (approximately all B and C ones) are the ones of most interest. These uEGMs show a, in parts, severe reduction in amplitude when compared to the SIN case and we can observe small double and triple deflections. In the PIV case, the wave initiated from the sinus node had to circumvent a line of block, see Fig. 5. We can observe several double potentials for the uEGMS from electrodes placed behind the LOB. An extension of the PIV case is the GAP case, in which the wave propagates through a gap

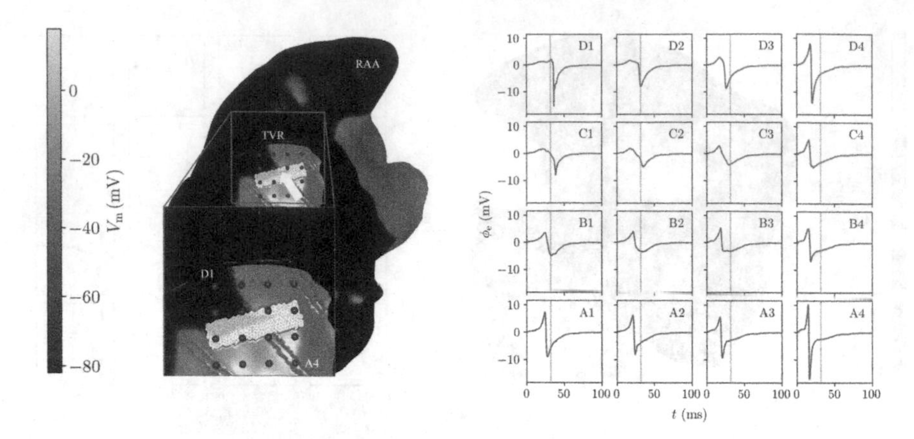

Fig. 4. Slow conduction zone (**SCZ**) at $t = 32$ ms and corresponding uEGMs.

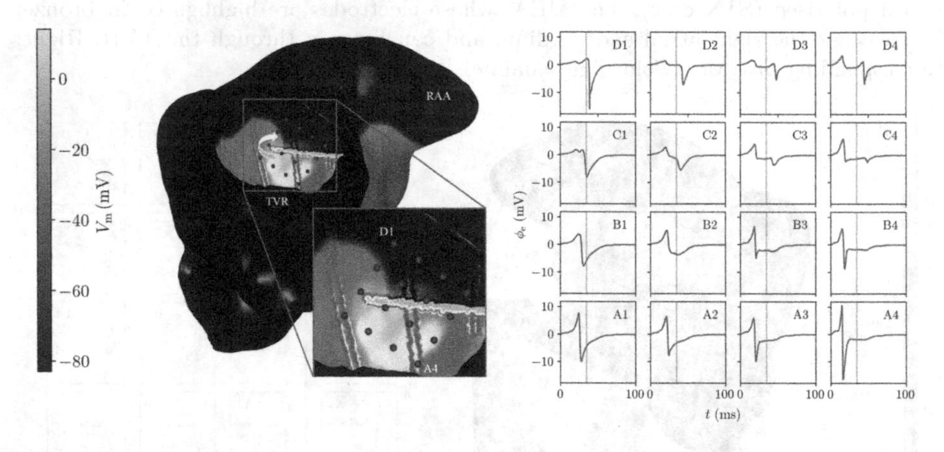

Fig. 5. Pivot site (**PIV**) pattern at $t = 34$ ms and corresponding uEGMs.

inside a LOB. The corresponding dynamics and uEGMs are shown in Fig. 6. The electrode placed right above the gap (B1) yields a biphasic uEGM of reduced amplitude and a small deflection between both extrema. Almost all uEGMs obtained from electrodes placed behind the gap have a dominating minimum, while the maximum is smaller. In the COL case (Fig. 7), the MEA was placed at another position in order to ensure that we actually measure two frontally colliding waves. Compared to the biphasic uEGMs from the SIN activation, the minimum of uEGMs obtained along the collision line are severely reduced.

Finally, for each of the six patterns, we exemplarily show biEGMs derived from uEGMs close to the regions of interest (Fig. 8). While the SIN, SCZ, GAP and PIV cases show a triphasic biEGM and the COL case a biphasic one, the LOB biEGM shows multiple potentials.

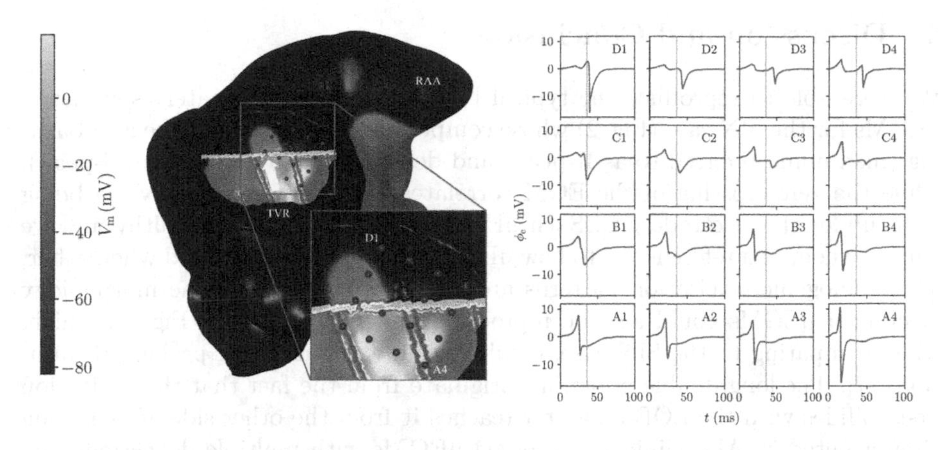

Fig. 6. GAP pattern at $t = 28$ ms and corresponding uEGMs. The gap has the size of a few mesh elements.

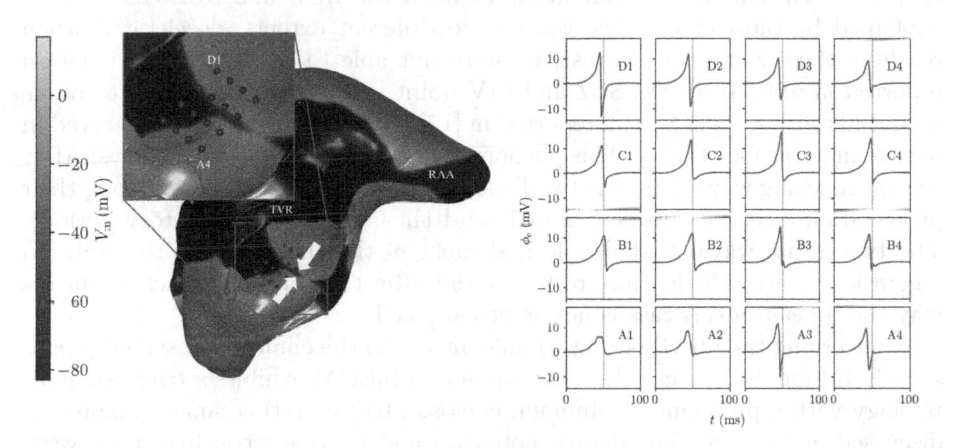

Fig. 7. Two colliding wavefronts (**COL**) at $t = 50$ ms and corresponding uEGMs.

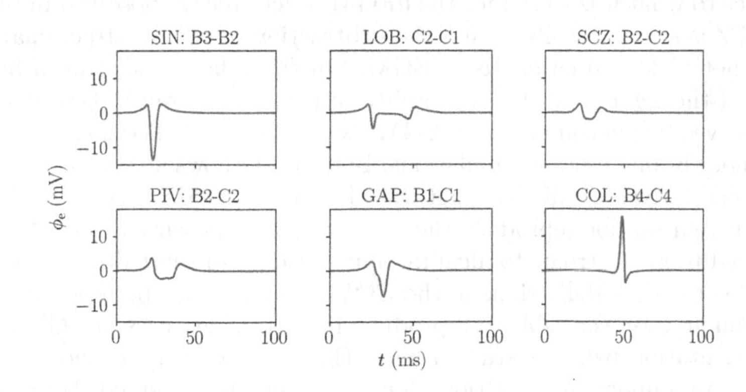

Fig. 8. BiEGMs obtained from the subtraction of uEGMs of two neighbouring electrodes of the MEA.

4 Discussion and Conclusion

We were able to reproduce the typical biphasic and approximately symmetric uEGMs for the SIN case (Fig. 2) whose components correspond to the approaching (maximum, referred to as R-peak) and departing wave (minimum, S-peak), while the zero-crossing of the EGM correlates with the excitation wave being right under the electrode [18]. Such uEGMs can be considered healthy and we will use them as a reference as to how uEGM morphology is changed when other, arrhythmogenic, excitation patterns are investigated instead. The morphology of clinical uEGMs could also be reproduced in the LOB case (Fig. 3), which, when comparing to the SIN case, exhibit double potentials of prolonged duration [17]. The long double potentials originate from the fact that the excitation first collides with the LOB and later reaches it from the other side after having circumvented it. Also clinically observed uEGMs with multiple deflections and of reduced amplitude in the SCZ case [17] could be reproduced in the simulations (Fig. 4). The lowered amplitude of both the LOB and SCZ case can be explained by the fact that the wave is, to different extents, decelerated when reaching these zones. In contrast, we were not able to reproduce fractionation reported in uEGMs for the SCZ and PIV point [17]. Furthermore, short double potentials with a small S-peak reported in [17] for the COL case were observed in our simulations (see Fig. 7). This phenomenon can be explained as follows: After the approaching waves (causing two R-peaks) have collided, all nodes along their propagation direction are depolarised (and thus within the refractory period). The two initial waves that depolarised most of the tissue within the collision region leave rather little space to be excited after the collision, which in turn is why the S-peak in this case is not as pronounced.

Concerning the biEGMs (Fig. 8), in contrast to the clinically observed biphasic biEGM for the SIN case [19], the simulated biEGM exhibits a triphasic morphology with a pronounced minimum enclosed by two rather small maxima, as discussed e.g. in [18]. The double potential and an isoelectric line in between observed clinically for the LOB case [19] were reproduced in our simulations. In accordance to what was shown for the uEGM, fractionation observed in biEGMs for the SCZ is also not visible, since the subtraction of two non-fractionated signals will not yield a fractionated biEGM. On the other hand, the rather long duration of the signal and its low amplitude is well reproduced here [19]. Clinically observed single-component biEGMs with multiple deflections for the PIV case are not obtained in our simulations. Instead, it shows a triphasic potential, which is qualitatively similar to what was shown for the SIN case. Also the GAP biEGMs herein do not reproduce the narrow-spaced potentials with fractionation derived from electroanatomical mapping [19]. Also short duration biEGMs with double or triple deflections in the COL case could not be reproduced.

In summary, we were able to reproduce important features in EGMs for the different excitation patterns studied here. However, several morphological phenomena, most importantly fractionation, could not be observed. Usually, fractionation is associated with electrical propagation in fibrotic cardiac tissue [4]. During electroanatomical mapping it has been reported in uEGMs and biEGMs for SCZ and PIV cases and in biEGMs in GAP scenarios [17,19]. In order to reproduce this important property of EGMs taken from arrhythmogenic excitation propagation, we suggest to use models with a finer mesh resolution and microscopic heterogeneity of conduction properties for future studies [21,22]. In the best case, these meshes should also consider sub-cellular resolution instead of homogenising the substrate over several biological cells [23], which in turn will require high-performance computing methods. In addition, the infinite volume conductor method for EGM calculation is the most simplistic method and more elaborated approaches, e.g. a pseudo-bidomain formulation [16] that additionally considers the spatial extension of the electrodes [22] will have to be taken into account. Finally, different catheter orientations and potential deformations (e.g. when pushing against the endocardium) can be considered for high-fidelity simulations of uEGMs and biEGMs.

Acknowledgements. This project has received funding from the European High-Performance Computing Joint Undertaking EuroHPC (JU) under grant agreement No 955495. The JU receives support from the European Union's Horizon 2020 research and innovation programme and France, Italy, Germany, Austria, Norway, Switzerland.

References

1. Kim, Y.-H., et al.: 2019 APHRS expert consensus statement on three-dimensional mapping systems for tachycardia developed in collaboration with HRS, EHRA, and LAHRS. J. Arrhythmia **36**(2), 215–270 (2020)
2. Hong, K., Borges, J., Glover, B.: Catheter ablation for the management of atrial fibrillation: current technical perspectives. Open Heart **7**, e001207 (2020)
3. Greenspoin, A.J., Hsu, S.S., Datorre, S.: Successful radiofrequency catheter ablation of sustained ventricular tachycardia postmyocardial infarction in man guided by a multielectrode "basket" catheter. J. Cardiovasc. Electrophysiol. **8**(5), 565–570 (1997)
4. Sánchez, J., Loewe, A.: A review of healthy and fibrotic myocardium microstructure modeling and corresponding intracardiac electrograms. Front. Physiol. **13** (2022)
5. Reich, C., Oesterlein, T., Rottmann, M., Seemann, G., Dössel, O.: Classification of cardiac excitation patterns during atrial fibrillation. Curr. Direct. Biomed. Eng. **2**(1), 161–166 (2016)
6. Pollnow, S., Greiner, J., Oesterlein, T., Wülfers, E., Loewe, A., Dössel, O.: Mini electrodes on ablation catheters: valuable addition or redundant information? Insights from a computational study. Comput. Math. Methods Med. **2017** (2017)
7. Hwang, M., Kim, J., Lim, B., Song, J.-S., Joung, B., Shim, E.: Multiple factors influence the morphology of the bipolar electrogram: an in silico modeling study. PLoS Comput. Biol. **15**, 1–13 (2019)
8. Krüger, M.W., et al.: Personalization of atrial anatomy and electrophysiology as a basis for clinical modeling of radio-frequency ablation of atrial fibrillation. IEEE Trans. Med. Imaging **32**, 73–84 (2013)
9. Wachter, A., Loewe, A., Krueger, M.W., Dössel, O., Seemann, G.: Mesh structure-independent modeling of patient-specific atrial fiber orientation. Curr. Direct. Biomed. Eng. **1**, 409–412 (2015)

10. Loewe, A., Krueger, M.W., Platonov, P.G., Holmqvist, F., Dössel, O., Seemann, G.: Left and right atrial contribution to the p-wave in realistic computational models. In: van Assen, H., Bovendeerd, P., Delhaas, T. (eds.) FIMH 2015. LNCS, vol. 9126, pp. 439–447. Springer, Cham (2015). https://doi.org/10.1007/978-3-319-20309-6_50
11. Loewe, A., Krueger, M.W., Holmqvist, F., Dössel, O., Seemann, G., Platonov, P.G.: Influence of the earliest right atrial activation site and its proximity to interatrial connections on P-wave morphology. EP Eur. **18**, iv35–iv43 (2016)
12. Plank, G., et al.: The openCARP simulation environment for cardiac electrophysiology. Comput. Methods Program. Biomed. **208**, 106223 (2021)
13. openCARP consortium, C., et al.: OpenCARP (v12.0) (2022). https://doi.org/10.35097/874
14. Vigmond, E., de Francesco, G., Neic, A., Huang, Y.-L.C., Loewe, A.: meshalyzer (v4.1) (2023). https://doi.org/10.35097/881
15. Courtemanche, M., Ramirez, R.J., Nattel, S.: Ionic mechanisms underlying human atrial action potential properties: insights from a mathematical model. Am. J. Physiol.-Heart Circulatory Physiol. **275**(1), H301–H321 (1998)
16. Bishop, M.J., Plank, G.: Bidomain ECG simulations using an augmented monodomain model for the cardiac source. IEEE Trans. Biomed. Eng. **58**(8), 2297–2307 (2011)
17. Konings, K.T., Smeets, J.L., Penn, O.C., Wellens, H.J., Allessie, M.A.: Configuration of unipolar atrial electrograms during electrically induced atrial fibrillation in humans. Circulation **95**(5), 1231–1241 (1997)
18. de Bakker, J.M.: Electrogram recording and analyzing techniques to optimize selection of target sites for ablation of cardiac arrhythmias. Pacing Clin. Electrophysiol. **42**, 1503–1516 (2019)
19. Frontera, A., et al.: Electrogram signature of specific activation patterns: analysis of atrial tachycardias at high-density endocardial mapping. Heart Rhythm **15**(1), 28–37 (2018)
20. Haïssaguerre, M., et al.: Spontaneous initiation of atrial fibrillation by ectopic beats originating in the pulmonary veins. N. Engl. J. Med. **339**, 659–666 (1998)
21. Vigmond, E., Pashaei, A., Amraoui, S., Cochet, H., Hassaguerre, M.: Percolation as a mechanism to explain atrial fractionated electrograms and reentry in a fibrosis model based on imaging data. Heart Rhythm **13**, 1536–1543 (2016)
22. de Sa, D.D.C., et al.: Electrogram fractionation. Circul. Arrhythmia Electrophysiol. **4**(6), 909–916 (2011)
23. Tveito, A., Mardal, K., Rognes, M.: Modeling Excitable Tissue: The EMI Framework. Simula Springer Briefs on Computing, Springer, Heidelberg (2020). https://doi.org/10.1007/978-3-030-61157-6

Deep Learning-Based Emulation of Human Cardiac Activation Sequences

Ambre Bertrand[1(✉)], Julia Camps[1], Vicente Grau[2], and Blanca Rodriguez[1]

[1] Department of Computer Science, University of Oxford, Oxford, UK
ambre.bertrand@reuben.ox.ac.uk, blanca.rodriguez@cs.ox.ac.uk
[2] Institute of Biomedical Engineering, University of Oxford, Oxford, UK

Abstract. The vision of digital twins for precision cardiology is to combine expert knowledge and data of patients' cardiac pathophysiology with advanced computational methods, in order to generate accurate, personalised treatment strategies. When studying cardiac electrophysiology, the twinning pipeline commonly requires a large amount of simulations, e.g. when exploring parameter spaces for personalisation or when scaling up to large cohorts of virtual patients in Big Data studies. In these cases, state-of-the-art methods are computationally expensive, even when applying relatively fast algorithms such as the Eikonal model. In this work, we investigate the performance of a U-Net-based model for electrical excitation throughout the human ventricles. The approach provides the advantage of reducing the input parameter space by representing anatomical and electrophysiological properties of the heart in a standardised three-dimensional space. Results demonstrate the ability of the model to emulate the Eikonal simulation scheme and predict cardiac activation time maps with average accuracy of 4.7 ms RMSE and an improved performance at point of prediction, yielding results up to 500 times faster. This new method provides promising results for personalised simulations of cardiac propagation in large cohorts of human heart models.

Keywords: deep learning · emulation · cardiac electrophysiology

1 Introduction

Early diagnosis of cardiovascular disease (CVD) is crucial to ensure that patients can be monitored, and for tailored treatment strategies to be implemented, in order to avoid long-term complications. By leveraging the information provided by routine clinical testing modalities such as electrocardiogram (ECG) and cine magnetic resonance (CMR) imaging, it is possible to create accurate models to represent and predict the heart's behaviour *in silico*, and to generate an extensive wealth of synthetic information that mimics real patient data whilst avoiding the need to conduct invasive and resource-consuming clinical procedures [1, 2].

Recent advances in computational methods have exploited the synergy between mechanistic and statistical models to produce a "digital twin" of the heart [3–5]. It is now possible to simulate electrical propagation through the human heart using image

O. Bernard et al. (Eds.): FIMH 2023, LNCS 13958, pp. 213–222, 2023.
https://doi.org/10.1007/978-3-031-35302-4_22

anatomically-accurate models with detailed electrophysiological properties [6, 7]. The monodomain and bidomain models, based on reaction-diffusion, are often used for multiscale simulations [8], but their main drawback is the high computational costs incurred. The Eikonal model is used as an alternative, as it is both fast and accurate [6, 9]. However, personalisation of cardiac properties to clinical data can require hundreds of thousands of simulations, for example scaling to large cohorts of patients and performing parameter inference [3,[10–13]. Inference is an ill-posed problem; small changes in the input can result in drastic changes in the solution. As such, a large number of simulations is required to represent the parameter space and calibrate the model. Building a simplified cardiac personalisation model based on patient-specific electrophysiological properties could enable faster parameter inference, deployment in real-time applications, analysis of large patient datasets, as well as alleviating the requirement of high-performance computing (HPC) infrastructure.

Deep learning (DL) models can process large quantities of information and combine different input modalities in order to extract underlying patterns in the data; identify correlations between different features; and emulate the dynamics of the system studied. Certain studies have used DL-based approaches to build emulators of cardiac mechanics and electrophysiology, to replace and speed up complex biophysical simulations. These methods include physics-informed neural networks (PINNs) for cardiac activation mapping [14]; a physics-based DL framework to predict cardiac electrophysiology dynamics [15]; Gaussian Processes (GPs) to analyse cardiac re-entries [16]; graph convolutional networks (GCNs) to emulate cardiac mechanics [17] and interpolate electrophysiological properties [18]; and mesh neural networks for surrogate modelling of computational fluid dynamics (CFD) for human arteries [19]. U-Net, a popular DL framework, has been successfully applied to many medical imaging applications to date, such as automatic location and segmentation of the left atrium in CMR images [20]. Reduced order models (ROMs) have also shown promise in improving the computational efficiency of cardiac modelling by reducing the dimensionality of the problem, using methods such as randomized proper orthogonal decomposition and poly-affine transformations [21, 22].

In this work, we propose a new emulation approach for cardiac activation sequences, providing a data-driven alternative to current electrophysiological models of cardiac propagation (Fig. 1). We design a convolutional neural network based on 3D U-Net and adapt it specifically for voxel-wise regression. Our model is trained to learn the ventricular activation sequence, given a specific configuration of input parameters. This approach enables faster and more efficient model deployment, with the potential to generate large amounts of activation sequence predictions on-the-fly in a very short timeframe. Our emulator should thus be able to enable the fast generation of personalised cardiac simulations and deep information extraction on large patient cohorts.

2 Methods

2.1 Dataset

The data used in this work originates from a group of 100 patients in the UK Biobank, consisting of 51 females (mean age 54 ± 8 years, mean BMI 26.4 ± 4.4 kg/m^2) and 49 males (mean age 54 ± 7 years, mean BMI 27.9 ± 4.4 kg/m^2). Patients are healthy with no known heart condition. The processing pipeline developed in [23] enables the reconstruction of 3D patient-specific meshes of the heart, based on 2D CMR slices. This four-step pipeline consists of automatic 2D slice selection, deep learning-based segmentation to extract the heart contours, contour alignment in 3D space to correct for patient motion, and finally mesh smoothing. The pipeline was applied to yield one mesh per patient in our virtual study, providing an accurate representation of each subject's cardiac anatomy. Each mesh contains n nodes ($n_{mean} = 142512$, $n_{min} = 75163$, $n_{max} = 382689$ across 100 patients) constituting its shape in 3D.

2.2 Data Representation

In order to use this data in a standard convolution-based framework, the mesh is converted from a point cloud to a sparse, structured grid. Each grid is a three-dimensional array containing $64^3 = 262{,}144$ cubic voxels, with a voxel resolution of 3 mm. This resolution is chosen to match the minimum cardiac wall thickness while retaining a computationally sensible model size–a compromise between special resolution, computational expense and memory size.

Each unique training case consists of one mesh coupled with one point of earliest activation or root node location – physiologically, this would correspond to earliest activation by the Purkinje system. Voxels in the input grid are assigned categorical

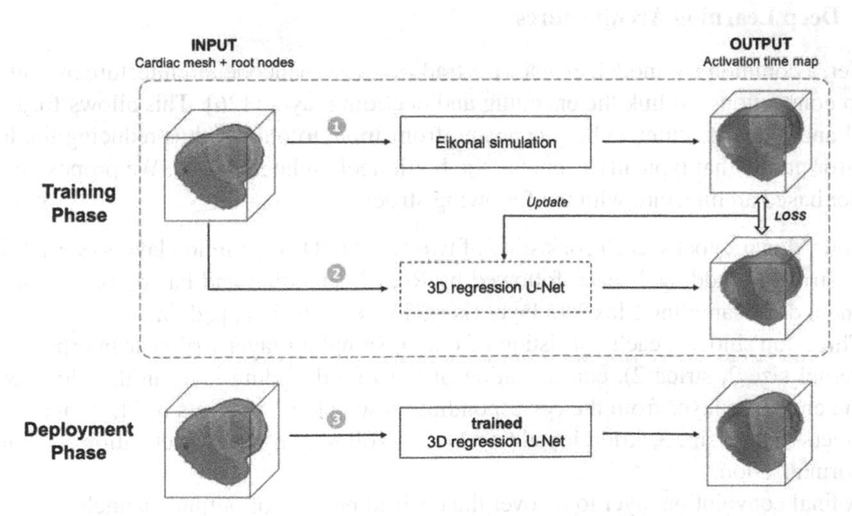

Fig. 1. Overview of the proposed emulation pipeline: (1) generating ground truth data, (2) training the deep learning model, (3) obtaining final activation time map predictions.

values (0, 1, 2). The grid is initialised with voxel values of 0 to represent empty space, then partially filled with 1s to represent the tissue mesh. This mesh is built by scaling down point cloud coordinates to grid coordinates based on the desired resolution, then translating the mesh to the centre of the grid by aligning the midpoints of the grid and scaled mesh coordinates. For each case, the root node voxel is located using a nearest neighbour approach and assigned a value of 2. For every input training case, the output of the model is a prediction of an activation time map in the same 3D space. Voxels in the output grids contain either empty space (value = 0), or a continuous activation time value ranging between 0 and 350 ms approximately.

2.3 Training Data Generation

The ground truth data for our deep learning model is generated using the graph-based Eikonal model [6]. This Eikonal formulation simulates the activation sequence by representing the electrical wave propagation in the heart as a 'shortest path' problem, which is solved by traversing the mesh using a multi-destination extension of Dijkstra's algorithm [24]. We use this model on the original 3D meshes to simulate one activation sequence per training case. The resulting activation time map is converted to a voxel grid, to match the representation described in Sect. 2.2.

The physiological parameters that are relevant to our problem include root node locations and tissue conduction velocity. A subset of all possible root node locations is selected at random amongst all endocardial nodes available for each mesh. Three conduction velocities parameterise the cardiac activation times: speed in the direction normal to the myocardium fibres, sparse endocardial speed, and dense endocardial speed. Typical physiological values range from 30 to 180 cm/s but for the purpose of this simplified model we assume isotropic conduction velocities of 50 cm/s [25].

2.4 Deep Learning Architectures

U-Net, a common DL model, extends the traditional autoencoder architecture by adding 'skip connections' to link the encoding and decoding layers [26]. This allows for both local and global features to be propagated from input to output, thus reducing the loss of information that typically occurs at the bottleneck in latent space. We propose a 3D U-Net based architecture with the following structure:

1. Four 'down' blocks, each consisting of two sets of 3D convolution layers (kernel size 3, stride 1, padding 1) each followed by ReLU activation and batch normalisation, and a downsampling MaxPool layer (kernel size 2, stride 2, padding 0)
2. Three 'up' blocks, each consisting of one upsampling layer (trilinear interpolation, kernel size 2, stride 2), concatenation of the first decoding layer in the block with the encoding layer from the corresponding down block, two sets of 3D convolution layers (kernel size 3, stride 1, padding 1) each followed by ReLU activation and batch normalisation.
3. A final convolution layer to recover the desired number of output channels.

4. A final linear activation layer.

The model has a total of 5,306,081 parameters. A non-linear activation function (e.g. a sigmoid) is typically used in the final layer for classification tasks. The choice of linear activation here is due to the nature of our problem being a multi-output regression, with a single continuous prediction at each voxel, as opposed to a classification problem.

We also propose a fully connected Multi Layer Perceptron (MLP) to use as a baseline DL model. The MLP consists of five layers: a first layer of input size 64^3, output size 128; three hidden layers of input and output size 128; and a final layer of input size 128 and output size 64^3 to recover the original dimensions of the input data. This model has 67,420,672 parameters in total. The outputs of both models are passed through a binary mask representing the original cardiac anatomy, such that final predictions are filtered and obtained for tissue voxels only.

2.5 Training

We select the root mean squared error (RMSE) as loss function during the training phase. Due to class imbalance in favour of background voxels, a naïve loss metric could cause the model to predict voxels as background (i.e. not assign an activation time) in order to obtain small errors and high accuracy at the expense of a lower positive predictive value. To overcome this, the loss function is "masked" such that it considers tissue voxels only and ignores empty space, and the model is not rewarded for achieving zero error on background voxels. We train both architectures with a batch size of 4, the U-Net for 100 epochs and learning rate of 1e−5, and the MLP for 1000 epochs and learning rate of 1e−3. Pre-trained model weights are obtained from the uniform Xavier initialisation [27]. Biases are initialised as zeros. The optimiser used is Adam [28].

2.6 Cross-Validation

All models are evaluated using a k-fold cross-validation scheme, to minimise bias in the final evaluation metrics reported. The training set is divided into $k = 5$ folds of equal size; the model is trained on 4 folds and the last fold is held out as a test set. This process is repeated k times until each fold has been used independently for testing. Final metrics are obtained by a taking an average of the results across all folds.

To quantify the effects of training data on the prediction error, we repeat cross-validation on the same anatomy, this time omitting an increasing number of folds at each test to reduce training data. Results are presented in Sect. 3.2.

2.7 Implementation and Software

We used Python v3.8 (Pytorch v1.4) for the implementation and a machine with an Intel Xeon E5–2690 v3 at 2.60 GHz (12 cores, 64 GB RAM) CPU and an NVIDIA Tesla P100 16 GB GPU with 5704 nodes for deployment and training.

3 Results

The Eikonal simulation model is used to generate ground truth activation sequences for 1185 unique cases, using a single cardiac anatomy and a collection of randomly sampled root node locations. Models are trained to predict activation times for each case using cross-validation as described in Sect. 2.6. Figure 2 illustrates two sample cases for activation time map predictions based on two distinct root node locations.

We evaluate our model based on two criteria, namely performance time and accuracy based on target-prediction error. We report mean absolute error (MAE), RMSE, and mean absolute percentage error (MAPE $= \frac{1}{n}\sum_{t=1}^{n}|\frac{y_t-\hat{y}_t}{y_t}|$ where y_t is target value and $\hat{y}_t$ are predictions). RMSE accounts for extreme error values; it is a reliable estimator of the standard deviation of the error distribution assuming a distribution mean of zero. MAPE offers a more interpretable and scale-independent metric to quantify the difference between target and prediction values. To avoid division by zero in MAPE when considering target values of 0, we scale our data linearly by adding a value of 1 to every voxel, both in target and prediction grids.

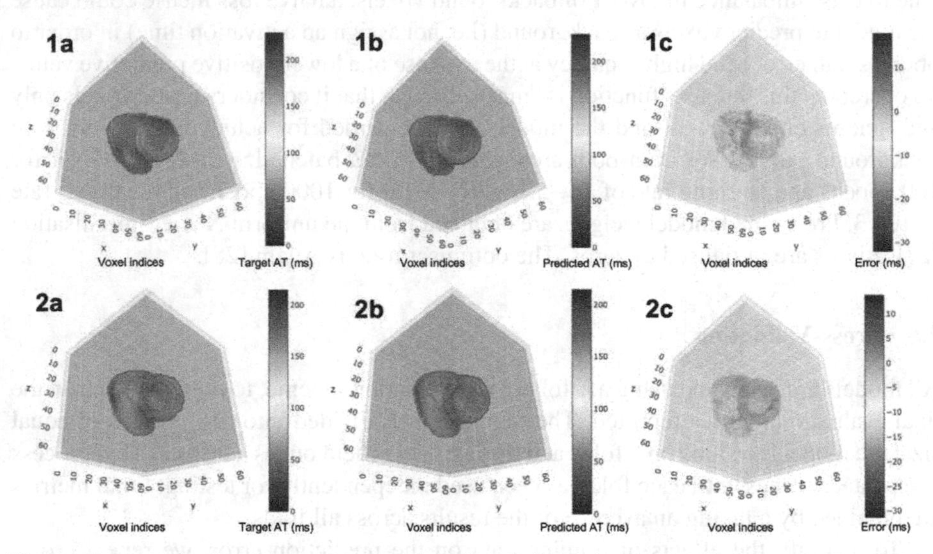

Fig. 2. Target activation time maps using the Eikonal simulation (a), predictions using our U-Net model (b), and residual error (c) for two sample validation cases, with the root node located in the right ventricle (1) and in the septum (2).

3.1 Performance

Table 1 presents the differences in performance for the Eikonal simulation, U-Net and MLP models. Training times are reported for the two DL models, based on one fold of cross-validation using the full training set. Prediction times for a single case are reported for all three models. We observe a significant improvement in prediction time for the trained DL models, with activation sequences being predicted up to 500 times as fast as an equivalent prediction using the Eikonal simulation.

Table 1. Model performance including training times and single case prediction times.

Model	Training time (1185 cases)	Prediction time (1 case)
Eikonal simulation	-	~ 20 s
U-Net	5 h 32 min	0.04 s
MLP	7 h 59 min	0.04 s

3.2 Error Quantification

To assess the effect of data size on model prediction error, we omit n fold(s) and perform cross-validation on k-n folds of the original training set. We repeat this experiment for n = 1, 2, 3 resulting in 948, 711, and 474 training samples respectively. Evaluation metrics are averaged across all iterations of each experiment and visualised in Fig. 3, including results for the full training dataset of 1185 samples. The smallest prediction error is obtained when using the full training set for both models, with 3.8 ± 4.0 ms MAE, 4.7 ms RMSE, 6.0% MAPE for U-Net, and 24.9 ± 35.2 ms MAE, 37.0 ms RMSE, 28.3% MAPE for MLP. Results are discussed in the following section.

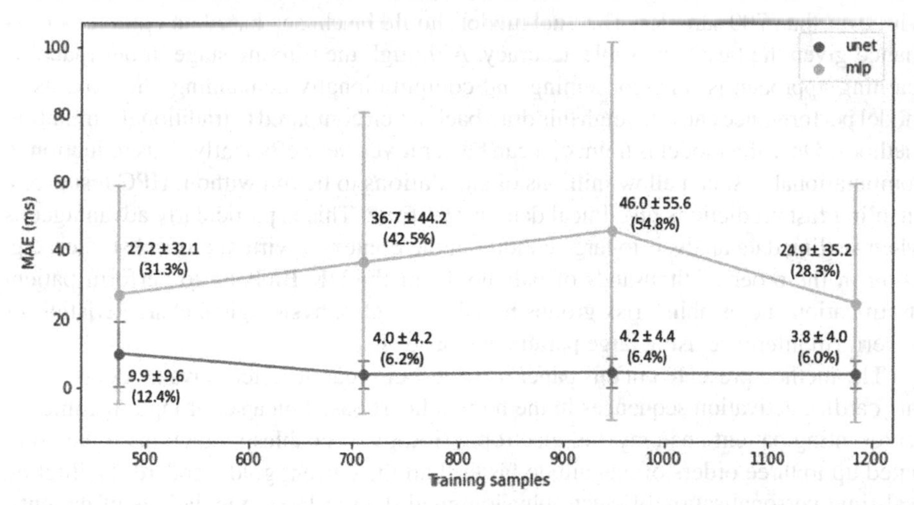

Fig. 3. Model evaluation metrics for different training set sizes. Values at each point indicate MAE ± std and MAPE.

4 Discussion and Conclusion

In this paper, we demonstrate two key benefits of using a deep learning model for emulation of activation sequences in human hearts. Given enough training data, our model is able to:

- learn the electrophysiological propagation pattern in a cardiac mesh and replicate activation time map within 4.7 ms RMSE (6.0%) of ground truth, and
- drastically improve model deployment times, yielding simulations up to 500 times faster compared to the Eikonal model at the point of prediction.

When trained on a single mesh, the U-Net model is able to predict cardiac activation times within 6.0% of the ground truth Eikonal simulation, which represents the gold standard for our problem. The significance of this result will depend on the future clinical applications of this method. Different extents of prediction accuracy and robustness may be required, for example, to recover specific segments of the ECG or to explore the variability of the activation sequence in patient populations that exhibit different clinical characteristics. One main advantage of the deep learning model compared to a physical simulation is that it does not require any pre-existing knowledge of the physics underlying the problem. In our case the physics is known, however, a similar data-driven machine learning approach could be applied to other problems where no mathematical model is available to represent the physical phenomenon of interest.

The training times reported in Sect. 3.1 scale linearly with training data size. Error metrics reported in Sect. 3.2 show that the validation error for MLP predictions increases with number of training samples, suggesting that the model may be overfitting. The U-Net error is an order of magnitude smaller and stabilises at around 4ms for models trained with over than 700 samples. The final model should be chosen based on optimal performance given the best achievable accuracy. Although the training stage in our machine learning approach is time-consuming and computationally demanding, the benefits in model performance can outweigh this drawback when compared to traditional simulation methods. Once the model is trained, it can be deployed very efficiently. This reduction in computational cost can allow millions of simulations to be run without HPC resources, enabling fast predictions on clinical data in real-time. This is particularly advantageous when scaling data analysis to large cohorts in the context of virtual clinical studies, e.g. using in the order of thousands of patients from the UK Biobank to perform patient stratification and establish risk groups based on electrophysiological characteristics, or to perform inference using large parameter spaces [3].

The method presented in this paper provides an accurate, alternative model to simulate cardiac activation sequences in the human heart based on a set of input parameters representing patient anatomy and electrophysiology. It enables predictions to be computed up to three orders of magnitude faster than the current gold standard, facilitating real-time personalisation of electrophysiological data in large populations of patients. Our model could easily scale to cases with multiple activation sites. Individual activation sequence maps should be obtained for each root node, and subsequently aggregated using voxel-wise minimum across all maps to obtain the final result. In order to be robust to heterogeneities in different patient cohorts, and to provide a more complete personalisation approach, the model may benefit from a more flexible representation of physiological parameters such as conduction velocities and cardiac fibres. Improving our model's parameter space would enable it to represent substrates for arrythmia e.g. scar regions, low conduction velocities. Achieving this would require a larger and more heterogeneous training dataset that accurately reflects these population differences. Testing the generalisation of the method to unseen cardiac anatomies, is another key step to further establish the reliability of the model.

Ethical Considerations and Acknowledgements. This virtual study was carried out using computer simulations which did not require ethical approval. This research has been conducted using the UK Biobank Resource under Application Number 40161. The authors express no conflict of interest. This work was funded by an Engineering and Physical Sciences Research Council doctoral award, a Wellcome Trust Fellowship in Basic Biomedical Sciences (214290/Z/18/Z), the CompBioMed2 Centre of Excellence in Computational Biomedicine (European Commission Horizon 2020 research and innovation programme, grant agreement No. 823712). The computation costs were incurred through a PRACE ICEI project (icp019), which provided access to Piz Daint at the Swiss National Supercomputing Centre, Switzerland. For the purpose of open access, the author has applied a Creative Commons Attribution (CC BY) public copyright license to any Author Accepted Manuscript (AAM) version arising from this submission.

References

1. Topol, E.: High-performance medicine: the convergence of human and artificial intelligence. Nat. Med. **25**, 44–56 (2019)
2. Beetz, M.: Multi-domain variational autoencoders for combined modelling of MRI-based biventricular anatomy and ECG-based cardiac electrophysiology. Front. Physiol. **13**, 886723 (2022)
3. Camps, J.: Inference of ventricular activation properties from non-invasive electrocardiography. Med. Image Anal. **73**, 102143 (2021)
4. Alber, M.: Integrating machine learning and multiscale modeling - perspectives, challenges, and opportunities in the biological, biomedical, and behavioral sciences. NPJ Digital Med. **2**, 15 (2019)
5. Corral-Acero, J.: The "Digital Twin" to enable the vision of precision cardiology. Eur. Heart J. **41**(48), 4556–4564 (2020)
6. Wallman, M., Smith, N., Rodriguez, B.: Estimation of activation times in cardiac tissue using graph based methods. In: Metaxas, D. N., Axel, L. (eds.) FIMH 2011. LNCS, vol. 6666, pp. 71–79. Springer, Heidelberg (2011). https://doi.org/10.1007/978-3-642-21028-0_9
7. Sermesant, M., Coudière, Y., Moreau-Villéger, V., Rhode, K.S., Hill, D.L.G., Razavi, R.S.: A Fast-Marching Approach to Cardiac Electrophysiology Simulation for XMR Interventional Imaging. In: Duncan, J.S., Gerig, G. (eds.) Medical Image Computing and Computer-Assisted Intervention – MICCAI 2005. MICCAI 2005. Lecture Notes in Computer Science, vol 3750. Springer, Heidelberg (2005). https://doi.org/10.1007/11566489_75
8. Tung, L.:A bi-domain model for describing ischemic myocardial d-c potentials. Dept Electr Eng Comput Sci MIT, Cambridge, MA (1978)
9. Plank, G.: Efficient computation of electrograms and ECGs in human whole heart simulations using a reaction-eikonal model. J. Comput. Phys. **1**, 191–211 (2017)
10. Giffard-Roisin, S., et al.: Sparse Bayesian non-linear regression for multiple onsets estimation in non-invasive cardiac electrophysiology. In: Pop, M., Wright, G.A. (eds.) FIMH 2017. LNCS, vol. 10263, pp. 230–238. Springer, Cham (2017). https://doi.org/10.1007/978-3-319-59448-4_22

11. McCarthy, A.: Variational Inference over Non-differentiable Cardiac Simulators using Bayesian Optimization. In: NeurIPS (2017)
12. Coveney, S.: Bayesian calibration of electrophysiology models using restitution curve emulators. Front. Physiol. **12**, 1120 (2021)
13. Niederer, S.: Simulating human cardiac electrophysiology on clinical time-scales. Front. Physiol. **2**, 14 (2011)
14. Costabal, F.: Physics-informed neural networks for cardiac activation mapping. Front. Phys. **8**, 42 (2020)
15. Kashtanova, V.: APHYN-EP: physics-based deep learning framework to learn and forecast cardiac electrophysiology dynamics. In: Statistical Atlases and Computational Models of the Heart. Regular and CMRxMotion Challenge Papers, STACOM (2022)
16. Lawson, K.: Slow recovery of excitability increases ventricular fibrillation risk as identified by emulation. Frontiers Physiol. **9**, 1114 (2018)
17. Dalton, D.: Emulation of cardiac mechanics using Graph Neural Networks. Comput. Methods Appl. Mech. Eng. **401**, 115645 (2022)
18. Meister, F.: Extrapolation of ventricular activation times from sparse electroanatomical data using graph convolutional neural networks. Front. Physiol. **12**, 1724 (2019)
19. Suk, J.: Mesh convolutional neural networks for wall shear stress estimation in 3D artery models. In: Statistical Atlases and Computational Models of the Heart. Multi-Disease, Multi-View, and Multi-Center Right Ventricular Segmentation in Cardiac MRI Challenge, STACOM (2021)
20. Jia, S., et al.: Automatically segmenting the left atrium from cardiac images using successive 3D U-nets and a contour loss. In: Pop, M., et al. (eds.) STACOM 2018. LNCS, vol. 11395, pp. 221–229. Springer, Cham (2019). https://doi.org/10.1007/978-3-030-12029-0_24
21. Fresca, S.: POD-enhanced deep learning-based reduced order models for the real-time simulation of cardiac electrophysiology in the left atrium. Front. Physiol. **12**, 1431 (2021)
22. Desrues, G.: Towards hyper-reduction of cardiac models using poly-affine transformations In: Multi-Sequence CMR Segmentation, CRT-EPiggy and LV Full Quantification Challenges, STACOM (2020)
23. Banerjee, A.: A completely automated pipeline for 3D reconstruction of human heart from 2D cine magnetic resonance slices. Philos Trans R. Soc. **379**, 20200257 (2021)
24. Djikstra, E.: A note on two problems in connexion with graphs. Numer Math **1**, 269–271 (1959)
25. Taggart, P.: Inhomogeneous transmural conduction during early ischaemia in patients with coronary artery disease. J. Mol. Cellular Cardiol. **32**(4), 621–630 (2000)
26. Çiçek, Ö., Abdulkadir, A., Lienkamp, S.S., Brox, T., Ronneberger, O.: 3D U-Net: learning dense volumetric segmentation from sparse annotation. In: Ourselin, S., Joskowicz, L., Sabuncu, M.R., Unal, G., Wells, W. (eds.) MICCAI 2016. LNCS, vol. 9901, pp. 424–432. Springer, Cham (2016). https://doi.org/10.1007/978-3-319-46723-8_49
27. Bengio, Y.: Understanding the difficulty of training deep feedforward neural networks. In: Proceedings of the 13th International Conference on Artificial Intelligence and Statistics, pp. 249–256 (2010)
28. Kingma, D.: Adam: a method for stochastic optimization. In: ICLR (2014)

Influence of Myocardial Infarction on QRS Properties: A Simulation Study

Lei Li[1(✉)], Julia Camps[2], Zhinuo Wang[2], Abhirup Banerjee[1,3], Blanca Rodriguez[2], and Vicente Grau[1]

[1] Department of Engineering Science, University of Oxford, Oxford, UK
lei.li@eng.ox.ac.uk

[2] Department of Computer Science, University of Oxford, Oxford, UK
julia.camps@cs.ox.ac.uk

[3] Division of Cardiovascular Medicine, Radcliffe Department of Medicine, University of Oxford, Oxford, UK

Abstract. The interplay between structural and electrical changes in the heart after myocardial infarction (MI) plays a key role in the initiation and maintenance of arrhythmia. The anatomical and electrophysiological properties of scar, border zone, and normal myocardium modify the electrocardiographic morphology, which is routinely analysed in clinical settings. However, the influence of various MI properties on the QRS is not intuitively predictable. In this work, we have systematically investigated the effects of 17 post-MI scenarios, varying the location, size, transmural extent, and conductive level of scarring and border zone area, on the forward-calculated QRS. Additionally, we have compared the contributions of different QRS score criteria for quantifying post-MI pathophysiology. The propagation of electrical activity in the ventricles is simulated via a Eikonal model on a unified coordinate system. The analysis has been performed on 49 subjects, and the results imply that the QRS is capable of identifying MI, suggesting the feasibility of inversely reconstructing infarct regions from QRS. There exist sensitivity variations of different QRS criteria for identifying 17 MI scenarios, which is informative for solving the inverse problem.

Keywords: Myocardial Infarction · Sensitivity Analysis · Simulation · Cardiac Digital Twin

1 Introduction

Myocardial infarction (MI) is a major cause of mortality and disability worldwide [12,30]. Assessment of myocardial viability is essential in the diagnosis and treatment management for patients suffering from MI. In particular, the position, size, and shape of the scarring region and the border zone could provide important information for the selection of patients and delivery of therapies for MI.

L. Li and J. Camps—Two authors contribute equally.

O. Bernard et al. (Eds.): FIMH 2023, LNCS 13958, pp. 223–232, 2023.
https://doi.org/10.1007/978-3-031-35302-4_23

The electrocardiogram (ECG) is one of the most commonly used clinical diagnostic tools for MI [33]. It can provide useful information about the heart rhythm and reveal abnormalities related to the conduction system [15]. For example, ST-segment elevation and T-wave inversion are widely investigated indicators of cardiac remodeling associated with different stages of MI [11]. In contrast, the QRS patterns has received less attention when analyzing ECG abnormalities associated with MI. It is not yet fully clear how QRS abnormalities reflect MI characteristics, with some previous papers reporting conflicting results [27,32].

In-silico computer ECG simulations offer a powerful tool for mechanistic investigations on the MI characteristics [22,32]. For example, Arevalo et al. constructed a cardiac computational model, where simulations of the electrical activity were executed for arrhythmia risk stratification of MI patients [1]. Wang et al. developed a multi-scale cardiac modeling and ECG simulation framework for mechanistic investigations into the pathophysiological ECG and mechanical behavior post-MI [32]. Costa et al. employed a computational ventricular model of porcine MI to investigate the impact of model anatomy, MI morphology, and EP personalization strategies on simulated ECGs. Que et al. designed a multi-scale heart-torso computational model to simulate pathological 12-lead post-MI ECGs with various topographies and extents for ECG data augmentation [24].

In this work, we investigate the association between QRS abnormalities and MI characteristics in a unified coordinate system. In this preliminary study, we only investigate QRS morphology rather than the complete ECG cycle, as the QRS simulation is quite efficient compared to the whole cycle. For each subject, we examine 17 MI scenarios, summarize their effects on the simulated QRS, and identify the scenarios with the most significant alterations in the QRS morphology. This study highlights the potential of QRS to improve the identification and localization of MI and further facilitate patient-specific clinical decision-making. It also demonstrates the feasibility of developing a cardiac "digital twin" deep computational model for the inference of MI by solving an inverse problem. The computational model provides an integrated perspective for each individual that incorporates the features from multi-modality data on cardiac systems. To the best of our knowledge, this is the first sensitivity analysis of QRS complex for quantifying the MI characteristic variation in cardiac electrical activities.

2 Methodology

2.1 Anatomical Model Construction

To obtain anatomical information, we generate a subject-specific 3D biventricular tetrahedral mesh from multi-view cardiac magnetic resonance (CMR) images for each subject, using the method outlined in [2]. We employ the cobiveco coordinate reference system for mesh representation to ensure a symmetric, consistent and intuitive biventricular coordinate system across various geometries [25]. The cobiveco coordinate is represented by (tm, ab, rt), where tm, ab, and rt refer to transmural, apicobasal, and rotational coordinates, respectively. Figure 1 presents the cobiveco coordinate system.

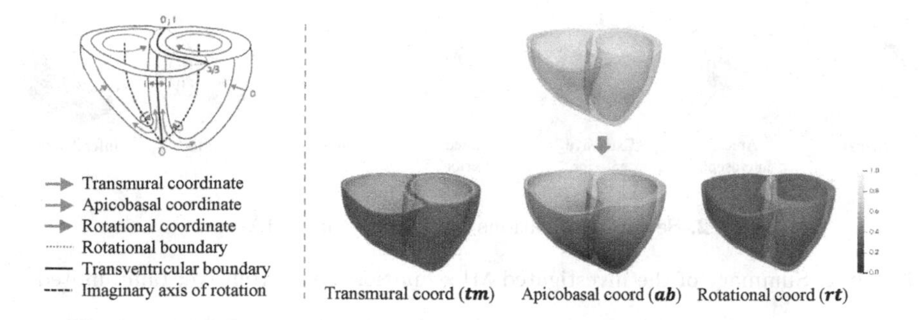

Fig. 1. Consistent biventricular coordinates in the cobiveco system.

We use ellipses with radii tm_r, ab_r, and rt_r to represent infarct regions in the myocardium, represented as

$$\frac{(tm_i - tm_0)^2}{tm_r{}^2} + \frac{(ab_i - ab_0)^2}{ab_r{}^2} + \frac{(rt_i - rt_0)^2}{rt_r{}^2} \leq 1, \tag{1}$$

where (tm_0, ab_0, rt_0) is the center coordinate of the infarct region. To study the effects of MI location at a population level, we employ the American Heart Association (AHA) 17-segment model [14] and consistently select the infarct areas and the transmural extent via cobiveco.

2.2 Electrophysiological Simulation

Cardiac electrophysiology is simulated via an efficient orthotropic Eikonal model [4,31] that incorporates a human-based Purkinje system into the formulation of the root node (RN) activation times. The simulation is performed over the generated cobiveco mesh in Sect. 2.1 and can be defined as,

$$\begin{cases} \sqrt{\nabla^T t \mathcal{V}^2 \nabla t} = 1, \\ t(\Gamma_0) = pk(\Gamma_0) - \min(pk(\Gamma_0)), \end{cases} \tag{2}$$

where $\mathcal{V}$ are the orthogonal conduction velocities (CVs) of fibre, sheet (transmural), and sheet-normal directions, t is the time at which the activation wavefront reaches each point in the mesh, Γ_0 is the set of locations (*i.e.*, RNs) in the endocardium, and pk is a Purkinje-tree delay function from the His-bundle to every point in the mesh. Thus, the earliest activation time at the RNs is defined as their delay from the His-bundle through the Purkinje tree normalized by the earliest activation. The QRS can be calculated from the activation time map via a pseudo-ECG equation [10] for a 1D cable source with constant conductivity at a given electrode location (x', y', z'), as

$$\phi_e(x', y', z') = \frac{a^2 \sigma_i}{4\sigma_e} \int -\nabla V_m \cdot \left[\nabla \frac{1}{r}\right] dx\, dy\, dz, \tag{3}$$

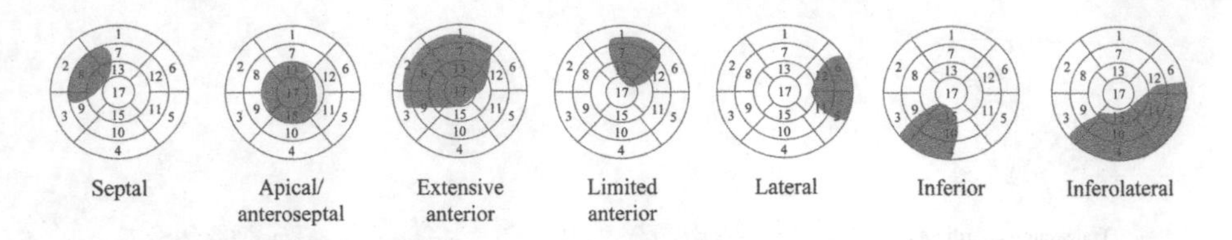

Fig. 2. Seven MI locations on 17-segment AHA-map.

Table 1. Summary of the investigated MI scenarios. Ext: extensive; Lim: limited.

Location	Septal	Apical	Ext anterior	Lim anterior	Lateral	Inferior	Inferolateral
Transmural extent	✓	✓	✓	✓	✓	✓	✓
Size					✓		
CV in MI region					✓		

where V_m is the transmembrane potential, ∇V_m is its spatial gradient, r is the Euclidean distance from a given point (x, y, z) to the electrode location, a is a constant that depends on the fiber radius, and σ_i and σ_e are the intracellular and extracellular conductivities, respectively. QRS is obtained by considering this integral throughout the ventricular activation sequence period. For the measurement of electrode locations, we utilize the automated 3D torso reconstruction pipeline from the CMR images [26]. Note that the pseudo-ECG method can efficiently produce normalized ECG signals with a comparable level of morphological information as the bidomain simulation [19] (Table 1).

For simulation, we consider electrophysiological heterogeneities in the infarct regions, including *seven locations (see Fig. 2), two transmural extents (transmural and subendocardial MI), two different sizes, and two different sets of slower CVs in the infarct areas* [18]. Note that for the comparison of different MI sizes and CV decreasing extents, we only report on lateral MI as an illustrative case. Therefore, for each subject we simulate 17 heterogeneous MI scenarios and one normal ECG as the baseline. Figure 3 provides examples of generated MI heterogeneity scenarios. We vary the CVs of infarct and healthy myocardial areas during its simulation, as slower CVs have been observed in the infarcted human heart [8]. Conduction pathways for electrical propagation in the infarct regions might exist, as observed in clinical data [28]. Therefore, we set the CVs of scarring and border zone areas to 10% and 50% (another CV set: 5% and 25%) of the values in healthy myocardium, respectively.

2.3 Univariate Sensitivity Analysis

In the sensitivity analysis, we introduce a global QRS measure, dynamic time warping (DTW), to calculate the dissimilarity of QRS with different lengths [4]. Moreover, we investigate four local QRS criteria, corresponding to QRS abnormalities of MI reported in the literature, namely, *QRS duration prolongation* [6], *pathological Q-waves* [9], *fragmented QRS (fQRS)* [7], and *poor R wave progression (PRWP)* [13]. An example of each QRS abnormality is illustrated in Fig. 4.

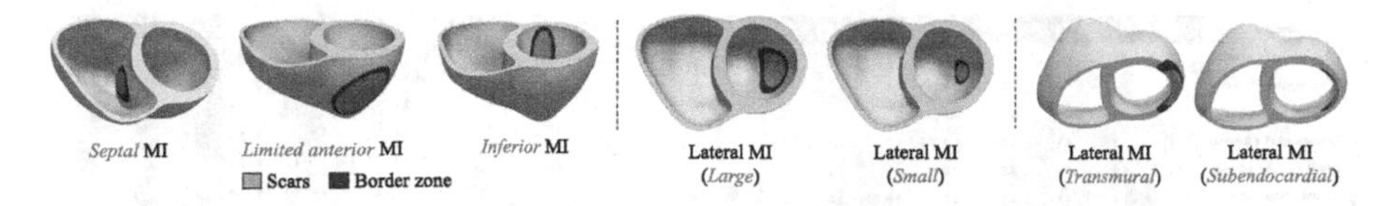

Fig. 3. Illustration of several MI scenarios, including different MI locations, sizes, and transmural extents.

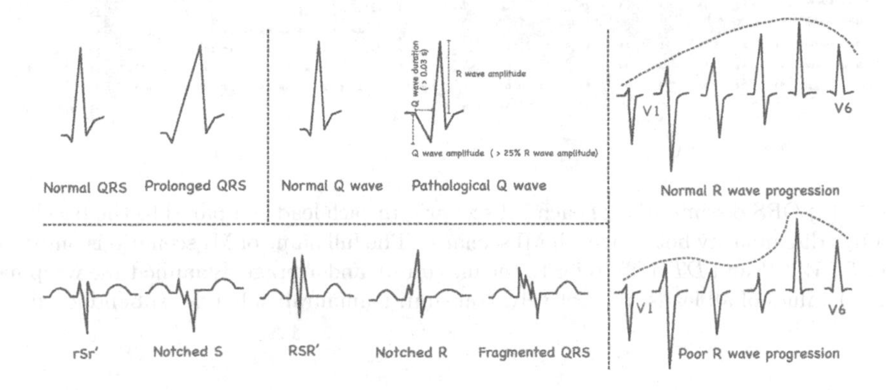

Fig. 4. Sketch map of normal QRS and MI-related QRS abnormalities.

The QRS duration is the time interval between the beginning of the Q wave and the end of the S wave. Pathological Q waves are described as the presence of Q wave with duration $\geq$0.03 s and/or amplitude $\geq$25% of R-wave amplitude [9]. fQRS is defined as the number of additional spikes within the QRS complex [7]. PRWP refers to the absence of the normal increase in amplitude of the R wave in the precordial leads when advancing from lead V1 to V6 [13]. In the literature, different definitions of PRWP exist [17]. In this work, we employ criteria including R wave amplitude of 2 mm or less in the lead V3/ V4 and the presence of reversed R-wave progression such as R of V5 < R of V6 or R of V2 < R of V1, or any combination of these.

3 Experiments and Results

3.1 Data Acquisition and Activation Property Configuration

We collect 49 subjects with paired ECGs and CMR images, including cine short-axis, two- and four-chamber long-axis, localizer, and scout slices, from the UK Biobank study [3]. The locations of root nodes are set to seven fixed homologous locations to allow comparisons [5]. Specifically, four left ventricular (LV) earliest activation sites (LV mid-septum, LV basal-anterior paraseptal, and two LV mid-posterior) and three in the right ventricle (RV), namely, RV mid-septum and two RV free wall, are selected as root nodes. The CVs along the fiber, sheet, sheet-normal, and sparse/dense endocardial directions are set to 65 cm/s, 48 cm/s, 51 cm/s, and 100/150 cm/s, respectively, in agreement with velocities reported for human healthy ventricular myocardium in [21,29].

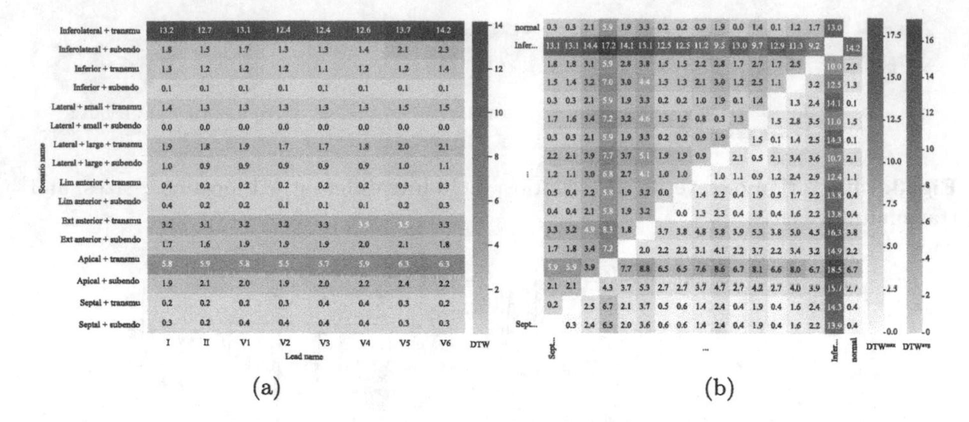

Fig. 5. (a) QRS dissimilarity of each MI scenario in each lead compared to the baseline; (b) QRS dissimilarity between each MI scenario. The full name of MI scenario is omitted here. DTW^{max} and DTW^{avg} refer to the maximum and average dynamic time warping (DTW) values of all leads, respectively. transmu: tranmural; subendo: subendocardial.

3.2 Results

QRS Differences Depending on MI Characteristics. To investigate the sensitivity of QRS on the 17 MI scenarios, we compare the dissimilarity of each of these with the baseline as well as the dissimilarity between them, as shown in Fig. 5. It is clear that there exist significant morphological changes in the post-MI QRS compared to the normal QRS, especially for inferolateral, extensive anterior, and apical MIs. However, differences from healthy QRS are highly reduced, as expected, when we reduce the size of the lateral MI or its transmurality. In addition, there is a significant variation in the QRS of lateral MI among different subjects. As Fig. 6(a–b) shows, the QRS of lateral MI can range from substantially different to almost identical to baseline. The extent of transmurality has evident effects on QRS morphology at each infarct location: as expected, transmural scars tend to present more evident morphological changes in the QRS than subendocardial ones. Even for the septal scars, in which transmural and subendocardial QRS dissimilarities are the smallest ($DTW^{max} = 0.2$ and $DTW^{avg} = 0.3$), one still can observe their morphology difference (see Fig. 6(c)). Nevertheless, differences in QRS between infarct locations appear to be larger than those depending on the extent of transmurality, suggesting that the QRS has higher sensitivity for localizing MI than predicting its transmural extent. The major QRS morphological variation for different degrees of CV reduction setting appears to be the QRS duration, which is not unexpected. However, according to our limited test for this purpose, we get particularly unusual QRS simulation results when we significantly reduce the CVs in the MI regions. Therefore, the CV configuration of MI areas during simulation is still an open question that demands more exploration in the future.

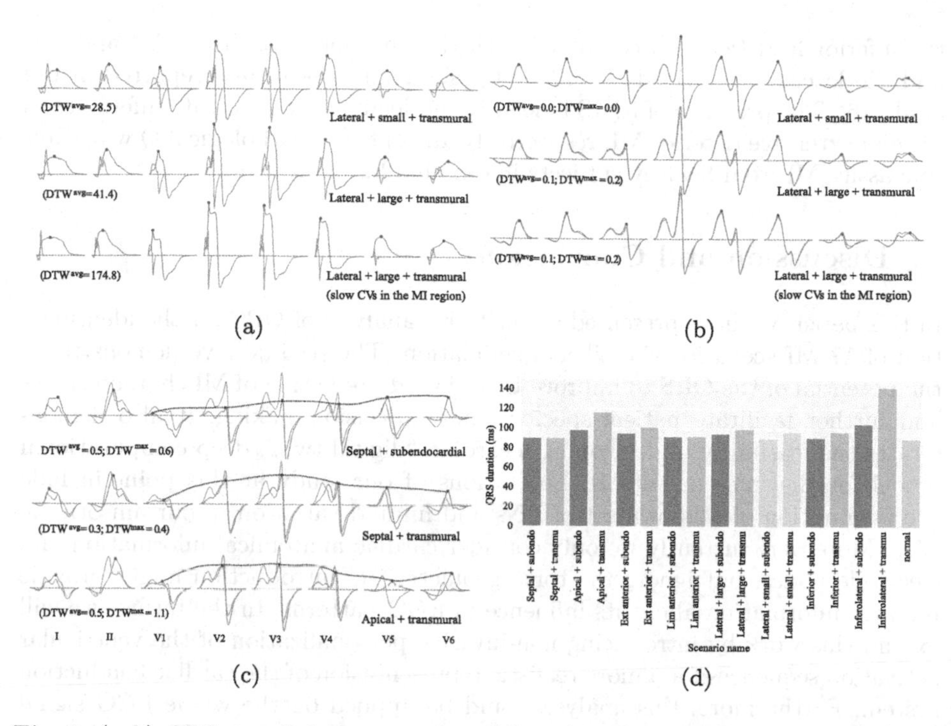

Fig. 6. (a–b) QRS morphology examples of lateral MIs with different sizes and CV setting. Here, MI and baseline QRS are labeled in red and grey, respectively; (c) QRS morphology difference among transmural and subendocardial septal MIs and poor R wave progression examples occurs in apical and septal MIs. The R wave progression is labeled with a black dashed line; (d) QRS duration of MI and baseline. (Color figure online)

The sensitivity of different QRS leads for detecting infarct location is varied. As Fig. 5(a) shows, most infarct locations are represented on the QRS by leads I, V5, and V6, whereas septal MI is represented by leads V1–V4 and V3–V4 for subendocardial and transmural ones, respectively. This result is generally consistent with those reported in clinical practice [23]. In general, larger scars tend to result in QRS changes appearing in more leads.

Sensitivity of Different QRS Criteria for MI Classification The changes in QRS morphology for different MI scenarios are reflected in various perspectives. Here, we introduce several QRS criteria and compare the contribution of each of these for infarct detection. Apical, extensive anterior, and inferolateral MI tend to present prolongation of the QRS duration, as Fig. 6(d) shows. PRWP mainly occurred in extensive anterior, septal, and apical MIs, similar as reported in the literature [13,20]. Specifically, the R wave amplitude in the septal MI is sometimes flattened, while the R wave of V6 tends to be larger than the R of V5 in the apical MI, as Fig. 6(c) shows. The prevalence of fQRS IS more common in the inferior lead (lead II) compared with the anterior leads (leads V3 and V4) and the lateral leads (leads V5 and V6), similar to the results reported in Liu et al. [16]. The presence of fQRS in lead II and leads V3-V4 indicate inferolateral MI and extensive anterior MI, respectively. In contrast, pathological Q wave fails to classify MI from healthy subjects in our simulation system.

4 Discussion and Conclusion

In this paper, we have presented a sensitivity analysis of QRS for the identification of 17 MI scenarios via Eikonal simulation. The results have demonstrated the potential of the QRS to improve ECG-based prediction of MI characteristics and further facilitate patient-specific clinical decision-making. It also demonstrates the feasibility of developing a cardiac "digital twin" deep computational model for the inference of MI. Limitations of our study at this point include the assumption of a known set of RNs and limited variation in our anisotropic CVs. Moreover, currently we only consider cardiac anatomical information and electrode nodes, but ignore the torso geometry. Its introduction might provide relevant information about its influence in ECG patterns. In the future, we will extend this work by introducing non-invasive personalization of the ventricular activation sequences for a more realistic representation of the cardiac conduction system. Furthermore, this analysis could be applied on the whole ECG signal instead of only QRS, necessitating large computational costs. The results can be further validated from relevant clinical outcomes in the ECGs of real MI patients. Consequently, the developed models and techniques will enable further research in cardiac digital twins.

Acknowledgement. This research has been conducted using the UK Biobank Resource under Application Number '40161'. The authors express no conflict of interest. This work was funded by the CompBioMed 2 Centre of Excellence in Computational Biomedicine (European Commission Horizon 2020 research and innovation programme, grant agreement No. 823712). L. Li was partially supported by the SJTU 2021 Outstanding Doctoral Graduate Development Scholarship. A. Banerjee is a Royal Society University Research Fellow and is supported by the Royal Society Grant No. URF\R1\221314. The work of A. Banerjee and V. Grau was partially supported by the British Heart Foundation (BHF) Project under Grant PG/20/21/35082.

References

1. Arevalo, H.J., et al.: Arrhythmia risk stratification of patients after myocardial infarction using personalized heart models. Nat. Commun. **7**(1), 11437 (2016)
2. Banerjee, A., et al.: A completely automated pipeline for 3D reconstruction of human heart from 2D cine magnetic resonance slices. Phil. Trans. R. Soc. A **379**(2212), 20200257 (2021)
3. Bycroft, C., et al.: The UK Biobank resource with deep phenotyping and genomic data. Nature **562**(7726), 203–209 (2018)

4. Camps, J., et al.: Inference of ventricular activation properties from non-invasive electrocardiography. Med. Image Anal. **73**, 102143 (2021)
5. Cardone-Noott, L., Bueno-Orovio, A., Mincholé, A., Zemzemi, N., Rodriguez, B.: Human ventricular activation sequence and the simulation of the electrocardiographic QRS complex and its variability in healthy and intraventricular block conditions. EP Eur. **18**(suppl_4), iv4–iv15 (2016)
6. Cupa, J., et al.: Diagnostic and prognostic value of QRS duration and QTc interval in patients with suspected myocardial infarction. Cardiol. J. **25**(5), 601–610 (2018)
7. Das, M.K., Khan, B., Jacob, S., Kumar, A., Mahenthiran, J.: Significance of a fragmented QRS complex versus a Q wave in patients with coronary artery disease. Circulation **113**(21), 2495–2501 (2006)
8. De Bakker, J., et al.: Slow conduction in the infarcted human heart. 'Zigzag' course of activation. Circulation **88**(3), 915–926 (1993)
9. Delewi, R., et al.: Pathological Q waves in myocardial infarction in patients treated by primary PCI. JACC Cardiovasc. Imaging **6**(3), 324–331 (2013)
10. Gima, K., Rudy, Y.: Ionic current basis of electrocardiographic waveforms: a model study. Circ. Res. **90**(8), 889–896 (2002)
11. Hanna, E.B., Glancy, D.L.: ST-segment depression and T-wave inversion: classification, differential diagnosis, and caveats. Clevel. Clin. J. Med. **78**(6), 404 (2011)
12. John, R.M., et al.: Ventricular arrhythmias and sudden cardiac death. Lancet **380**(9852), 1520–1529 (2012)
13. Kurisu, S., et al.: Poor R-wave progression and myocardial infarct size after anterior myocardial infarction in the coronary intervention era. IJC Heart Vasculature **7**, 106–109 (2015)
14. Lang, R.M., et al.: Recommendations for cardiac chamber quantification by echocardiography in adults: an update from the American Society of Echocardiography and the European Association of Cardiovascular Imaging. Eur. Heart J.-Cardiovasc. Imaging **16**(3), 233–271 (2015)
15. Li, L., Camps, J., Banerjee, A., Beetz, M., Rodriguez, B., Grau, V.: Deep computational model for the inference of ventricular activation properties. In: Camara, C., et al. (eds.) STACOM 2022. LNCS, vol. 13593, pp. 369–380. Springer, Cham (2023). https://doi.org/10.1007/978-3-031-23443-9_34
16. Liu, P., Wu, J., Wang, L., Han, D., Sun, C., Sun, J.: The prevalence of fragmented QRS and its relationship with left ventricular systolic function in chronic kidney disease. J. Int. Med. Res. **48**(4), 0300060519890792 (2020)
17. MacKenzie, R.: Poor R-wave progression. J. Insur. Med. **37**(1), 58–62 (2005)
18. Martinez-Navarro, H., Mincholé, A., Bueno-Orovio, A., Rodriguez, B.: High arrhythmic risk in antero-septal acute myocardial ischemia is explained by increased transmural reentry occurrence. Sci. Rep. **9**(1), 1–12 (2019)
19. Mincholé, A., Zacur, E., Ariga, R., Grau, V., Rodriguez, B.: MRI-based computational torso/biventricular multiscale models to investigate the impact of anatomical variability on the ECG QRS complex. Front. Physiol. **10**, 1103 (2019)
20. Mittal, S., Srivastava, P.: Differentiation of poor R wave progression of old anteroseptal myocardial infarction from that due to emphysema. Int. J. Cardiol. **13**(1), 92–94 (1986)
21. Myerburg, R.J., Gelband, H., Nilsson, K., Castellanos, A., Morales, A.R., Bassett, A.L.: The role of canine superficial ventricular muscle fibers in endocardial impulse distribution. Circ. Res. **42**(1), 27–35 (1978)
22. Neic, A., et al.: Efficient computation of electrograms and ECGs in human whole heart simulations using a reaction-eikonal model. J. Comput. Phys. **346**, 191–211 (2017)

23. Nikus, K., Birnbaum, Y., Eskola, M., Sclarovsky, S., Zhong-Qun, Z., Pahlm, O.: Updated electrocardiographic classification of acute coronary syndromes. Curr. Cardiol. Rev. **10**(3), 229–236 (2014)
24. Que, W., Han, C., Zhao, X., Shi, L.: An ECG generative model of myocardial infarction. Comput. Methods Programs Biomed. **225**, 107062 (2022)
25. Schuler, S., Pilia, N., Potyagaylo, D., Loewe, A.: Cobiveco: Consistent biventricular coordinates for precise and intuitive description of position in the heart-with matlab implementation. Med. Image Anal. **74**, 102247 (2021)
26. Smith, H.J., Banerjee, A., Choudhury, R.P., Grau, V.: Automated torso contour extraction from clinical cardiac MR slices for 3D torso reconstruction. In: 44th Annual International Conference of the IEEE Engineering in Medicine & Biology Society (EMBC), pp. 3809–3813. IEEE (2022)
27. Strauss, D.G., Selvester, R.H.: The QRS complex-a biomarker that "images" the heart: QRS scores to quantify myocardial scar in the presence of normal and abnormal ventricular conduction. J. Electrocardiol. **42**(1), 85 (2009)
28. Strauss, D.G., et al.: ECG quantification of myocardial scar in cardiomyopathy patients with or without conduction defects: correlation with cardiac magnetic resonance and arrhythmogenesis. Circul. Arrhythmia Electrophysiol. **1**(5), 327–336 (2008)
29. Taggart, P., Sutton, P.M., Opthof, T., Coronel, R., Trimlett, R., Pugsley, W., Kallis, P.: Inhomogeneous transmural conduction during early ischaemia in patients with coronary artery disease. J. Mol. Cell. Cardiol. **32**(4), 621–630 (2000)
30. Thygesen, K., Alpert, J.S., Jaffe, A.S., Chaitman, B.R., Bax, J.J., Morrow, D.A., White, H.D.: Fourth universal definition of myocardial infarction (2018). Eur. Heart J. **40**(3), 237–269 (2019)
31. Wallman, M., Smith, N.P., Rodriguez, B.: A comparative study of graph-based, eikonal, and monodomain simulations for the estimation of cardiac activation times. IEEE Trans. Biomed. Eng. **59**(6), 1739–1748 (2012)
32. Wang, Z.J., Santiago, A., et al.: Human biventricular electromechanical simulations on the progression of electrocardiographic and mechanical abnormalities in post-myocardial infarction. EP Eur. **23**, i143–i152 (2021)
33. Zimetbaum, P.J., Josephson, M.E.: Use of the electrocardiogram in acute myocardial infarction. N. Engl. J. Med. **348**(10), 933–940 (2003)

Image and Shape Analysis

Effect of Spatial and Temporal Resolution on the Accuracy of Motion Tracking Using 2D and 3D Cine Cardiac Magnetic Resonance Imaging Data

Kateřina Škardová[1], Tarique Hussain[2], Martin Genet[3,4], and Radomír Chabiniok[2(✉)]

[1] Department of Mathematics, Faculty of Nuclear Sciences and Physical Engineering, Czech Technical University in Prague, Prague, Czech Republic

[2] Division of Pediatric Cardiology, Department of Pediatrics, UT Southwestern Medical Center, Dallas, TX, USA

radomir.chabiniok@utsouthwestern.edu

[3] Solid Mechanics Laboratory (LMS), École Polytechnique, Institut Polytechnique de Paris, CNRS, Palaiseau, France

[4] Inria, Palaiseau, France

Abstract. In this paper, we investigate the effect of spatial and temporal resolution of cardiac MRI cine images on the extracted left ventricle motion. A previously developed and validated finite-element-based image registration method was used for the motion extraction. The study is performed on three subjects, for which the standard 2D cine stack (SA) in short axis orientation and 3D cine MRI series were acquired. The set of acquired image series was augmented by artificially constructed SA-like cine series created from the 3D cine images. Image series with several combinations of spatial and temporal resolution were evaluated for each subject. The study showed a strong correlation between the slice thickness and the accuracy of extracted displacement in the longitudinal direction. The effect of a lower temporal resolution was shown to be less significant. This could prove useful to update current guidelines for cardiac MRI acquisitions.

Keywords: Cardiac magnetic resonance imaging · 3D cine MRI · left ventricular motion · image registration · equilibrated warping

1 Introduction

Cardiac magnetic resonance imaging (MRI) allows to assess the movement of the heart in vivo non-invasively. The analysis of motion patterns and derived quantities, such as the global measure of ejection fraction or local myocardial displacements or strains (typically in directions corresponding to the local coordinate system – radial, circumferential and longitudinal), is invaluable in the

O. Bernard et al. (Eds.): FIMH 2023, LNCS 13958, pp. 235–244, 2023.
https://doi.org/10.1007/978-3-031-35302-4_24

evaluation of the function of the ventricle. The quality of the image data – resolution, signal-to-noise ratio, etc. – may affect the accuracy of the quantities extracted from the image series, however.

In this paper, we intend to evaluate the difference in the tracked left ventricle (LV) motion that occurs when either a stack of 2D cine images with a through-plane resolution lower than the in-plane resolution is used (currently, standard clinical approach) or when a novel 3D cine image sequence providing isotropic spatial resolution is employed. We use a previously developed motion extraction method [4], already validated against open-access databases [5,9] as well as in silico data [1], and applied to multiple clinical projects (e.g., [2]). The study is performed on three subjects, for which the standard 2D cine stack in short axis orientation (SA cine) and in addition a new 3D cine MRI series [6] were acquired. In order to get a better insight into the relationship between the image resolution and the accuracy of the extracted quantities, additional SA-like cine series were constructed from the 3D cine images with various slice thicknesses (both thinner and thicker than the originally acquired 2D SA cine stack) and temporal resolutions.

2 Methods

2.1 Image Acquisition

The computational study uses 3D and SA cine data from three asymptomatic patients with known repaired congenital heart disease and normal volumetric analysis. The data collections were performed under the ethical approval of Institutional Review Board of the UT Southwestern Medical Center, Dallas (STU 032016-009). The IRBs waived the need for a consent to use the anonymized retrospective data. The parameters of both types of cine images are shown in Table 1. The scan time of free-breathing 3D cine sequence was 5.9 ± 2.7 min [6] which is comparable to the acquisition time of the SA cine series using parallel imaging acceleration factor 2.

2.2 Construction of SA-Like Cine Images

The 3D cine images were used to generate SA-like images by resampling a subdomain of the 3D cine image. The cine SA images of slice thickness l mm constructed from the 3D cine will be denoted by cSA_l. These images were constructed with the same position and orientation in the real space, as were acquired the SA images.

First, the 3D cine image was resampled to the voxel size $1 \times 1 \times 1$ mm^3 using 3rd-order spline interpolation (using Python library SciPy). In the construction process, a voxel grid in the orientation of the acquired SA cine image and voxel spacing $1 \times 1 \times 1$ mm^3 was generated. Subsequently, the image intensities of the 3D cine image were projected onto this grid. The resolution of the constructed SA cine image was then decreased along the long axis to obtain the target spacing

Table 1. Parameters of the SA and 3D cine image series.

	Subject A	Subject B	Subject C
SA cine image series			
Time frames	30	30	30
Field of view [mm]	280 × 280 × 88	300 × 300 × 90	330 × 330 × 110
Spacing between slices [mm]	8	10	10
Original pixel dimensions [mm]	1.79 × 1.93	1.8 × 2.38	1.79 × 2.15
Resampled pixel dimensions [mm]	1.09 × 1.09	0.93 × 0.93	1.03 × 1.03
3D cine image series			
Time frames	30	30	30
Field of view [mm]	512 × 512 × 102	512 × 512 × 150	512 × 512 × 150
Spacing between slices [mm]	1.19	1.19	1.19
Original pixel dimensions [mm]	2 × 2.39	2 × 2.39	2 × 2.39
Resampled pixel dimensions [mm]	1.18 × 1.18	1.18 × 1.18	1.18 × 1.18

by averaging the signals from the voxels within the given reconstructed 2D slice (i.e., a slice of thickness n mm was obtained by averaging the image values in n 1-mm slices). The slice thicknesses used in the constructed SA cine for each subject are shown in Table 2.

Table 2. Slice thickness of the constructed SA cine. The original slice thickness of the acquired SA cine is in bold.

	Slice thickness [mm]					
Subject A	4	**8**	10	12		
Subject B	5	8	**10**	12	15	18
Subject C	5	8	**10**	12	15	18

2.3 Image Series with Lower Temporal Resolution

The original 30-time-frame image series were used to generate series with 1/2 and 1/3 temporal resolution, using the scheme shown in Fig. 1. In the series with the original temporal resolution, i-th image is assumed to be acquired over the time interval $(t_i - \Delta t, t_i + \Delta t)$. The length of the acquisition time interval increases to $4\Delta t$ in the case of series with 1/2 temporal resolution, and to $6\Delta t$ in the case of series with 1/3 temporal resolution. The images in the series with the lower temporal resolution are generated by averaging the corresponding images from the original series while assuming the periodicity of the original data. The image series with 1/2 and 1/3 temporal resolution are generated for the 3D cine and two constructed SA cine series for all three subjects.

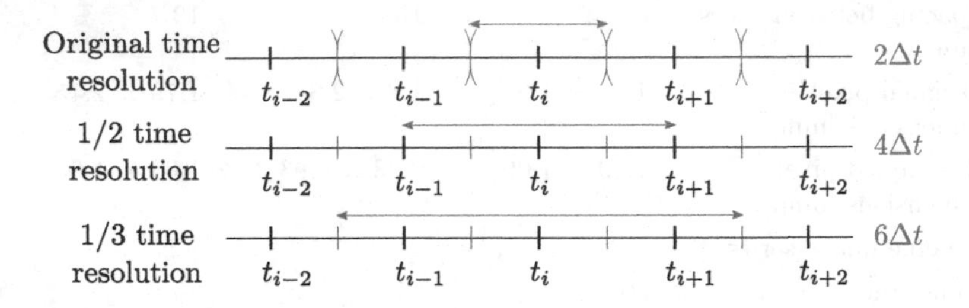

Fig. 1. The scheme of temporal averaging.

2.4 Segmentation of the Left Ventricle

The left ventricle (LV) is manually segmented from the first frame of the 3D cine corresponding to the end-diastole using the image processing framework MeVisLab[1]. The segmented surface meshes of the LV, exported from MeVisLab, are further remeshed in the finite element mesh generator GMSH[2] and subsequently used to generate volume meshes following the method described in [3]. In order to improve the tracking of boundary points of the LV, a layer of elements is added to the surface of the mesh using GMSH (see [4] for details). These elements are then removed before analyzing the deformation fields. The same mesh is used for all variants of cine images of each subject.

2.5 Finite Element-Based Image Registration with Mechanical Regularization

All image series are resampled to uniform resolution $1 \times 1 \times 1$ mm^3 by the spline interpolation before the motion tracking to ensure the correct integration of image similarity terms discussed below for all image series. The motion is extracted from the image series using a finite element-based image registration method, described in detail in [4]. In image registration, we generally want to determine the transformation between two images I_0 and I_i of the object $\mathcal{B}$ – in our case the LV – acquired at times t_0 and t_i. The images are represented by image functions $I_0 : \Omega \to \mathbb{R}$ and $I : \Omega \to \mathbb{R}$, where Ω is the domain of the image. This problem of finding the transformation between the two images is ill-posed, and therefore we use a variational formulation suitable for the inclusion

[1] https://www.mevislab.de.
[2] https://gmsh.info.

of regularization terms. The transformation, represented by mapping ϕ_i is then found as a minimizer of the following functional:

$$E(\phi_i) = E_{\text{image}}(\phi_i) + E_{\text{reg}}(\phi_i), \quad (1)$$

which combines the image similarity term $E_{\text{image}}(\phi_i)$ and the regularization term $E_{\text{reg}}(\phi_i)$. The image term evaluates the difference of image intensities of image I_0 and I_i at corresponding points:

$$E_{\text{image}}(\phi_i) = \frac{1}{2} \int_{\Omega_0} \left(I_i(\phi_i(\vec{X})) - I_0(\vec{X}) \right)^2 \mathrm{d}\vec{X}, \quad (2)$$

where $\Omega_0 \subset \Omega$ is the domain occupied by the tracked object at the t_0. Alternatively, the transformation can be represented by displacement field U_i such that $\phi_i(\vec{X}) = \vec{X} + U_i(\vec{X})$.

The selection of the regularization term may have a large impact on the extracted displacement. When imposing too strong assumptions, such as incompressibility of the tracked object or smoothness of mapping ϕ_i, the actual motion captured in the image series can be suppressed. In this work, we use a regularization term proposed in [4], that penalizes deviation from the solution of a hyperelastic body in equilibrium with arbitrary boundary loads.

The mappings ϕ_i are determined as a solution of (1), for each frame i of the series. Problem (1) is solved using the Newton method with gradient-free golden section line search. In the first frame ($i = 0$), the mapping $\phi_0 = \mathbb{I}$ is prescribed. The estimation of each subsequent ϕ_i, $i > 0$ is initialized by the previously computed mapping ϕ_{i-1}. The numerical solution is described in detail in [4], with the implementation of the solver freely available[3].

3 Results

In this section, we compare displacement fields extracted from different image series. In order to evaluate the difference between displacement fields U^{ref} and U, the following functions will be used:

- mean signed error in the radial direction (defined analogically for circumferential and longitudinal direction)

$$\text{MSE}_\text{r}(t) = \frac{1}{|\Omega_0|} \int_{\Omega_0} \left(U_r^{\text{ref}}(\vec{X}, t) - U_r(\vec{X}, t) \right) \mathrm{d}\vec{X} \quad (3)$$

- root-mean-square error in the radial direction (defined analogically for circumferential and longitudinal direction)

$$\text{RMSE}_\text{r} = \sqrt{\frac{1}{T} \int_0^T \frac{1}{|\Omega_0|} \int_{\Omega_0} |U^{\text{ref}_r}(\vec{X}, t) - U_r(\vec{X}, t)|^2 \mathrm{d}\vec{X} \mathrm{d}t} \quad (4)$$

[3] https://gitlab.inria.fr/mgenet/dolfin_warp.

– normalized root-mean-square error

$$\mathrm{NRMSE} = \frac{\sqrt{\frac{1}{T}\int_0^T \frac{1}{|\Omega_0|}\int_{\Omega_0} \|U^{\mathrm{ref}}(\vec{X},t) - U(\vec{X},t)\|^2 \mathrm{d}\vec{X}\mathrm{d}t}}{\sqrt{\frac{1}{T}\int_0^T \frac{1}{|\Omega_0|}\int_{\Omega_0} \|U^{\mathrm{ref}}(\vec{X},t)\|^2 \mathrm{d}\vec{X}\mathrm{d}t}} \tag{5}$$

3.1 The Effect of Slice Thickness

In this section, we compare the accuracy of the extracted displacements obtained from image series with different slice thicknesses. Due to the absence of a ground truth, the displacement field extracted from the 3D cine image series will be used as reference value U^{ref}.

The dependence of $\mathrm{RMSE}_{\mathrm{r}}$, $\mathrm{RMSE}_{\mathrm{c}}$ and $\mathrm{RMSE}_{\mathrm{l}}$ on the slice thickness for all three subjects is shown in Fig. 2. The figure shows the trend of an increasing error with an increasing slice thickness. $\mathrm{RMSE}_{\mathrm{l}}$ is larger and increases steeper than both $\mathrm{RMSE}_{\mathrm{r}}$ and $\mathrm{RMSE}_{\mathrm{c}}$ for the same values of slice thickness in all three subjects.

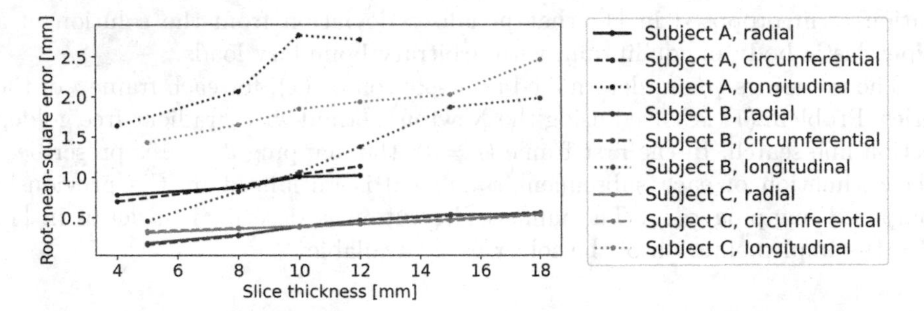

Fig. 2. The effect of the slice thickness on root-mean-square errors $\mathrm{RMSE}_{\mathrm{r}}$, $\mathrm{RMSE}_{\mathrm{c}}$ and $\mathrm{RMSE}_{\mathrm{l}}$.

Table 3 shows the values of NRMSE for the cSA cine image series as well as the original SA image series tSA_8 and tSA_{10}. The error in displacement obtained from the original SA cine is larger than in the constructed image with the same slice.

Table 3. Normalized root-mean-square errors NRMSE [%] evaluated for the cSA and SA cine image series.

	cSA_4	cSA_8	SA_8	cSA_{10}	cSA_{12}		
Subject A	45.07	55.77	55.31	72.11	71.98		
	cSA_5	cSA_8	cSA_{10}	SA_{10}	cSA_{12}	cSA_{15}	cSA_{18}
Subject B	13.22	23.21	30.41	45.86	38.82	51.23	54.18
Subject C	32.60	37.54	41.80	46.35	43.98	46.28	55.94

In the scale of Fig. 2, the values of RMSE in the radial and circumferential directions for Subject B and C appear to be almost identical. However, the temporal distributions of the errors differ, as can be seen in the plots of MSE for Subject C in Fig. 3. The radial displacement is overestimated only in the systole, while the circumferential displacement is alternately underestimated and overestimated throughout the cardiac cycle. In the longitudinal direction, the sign of the mean error is also changing.

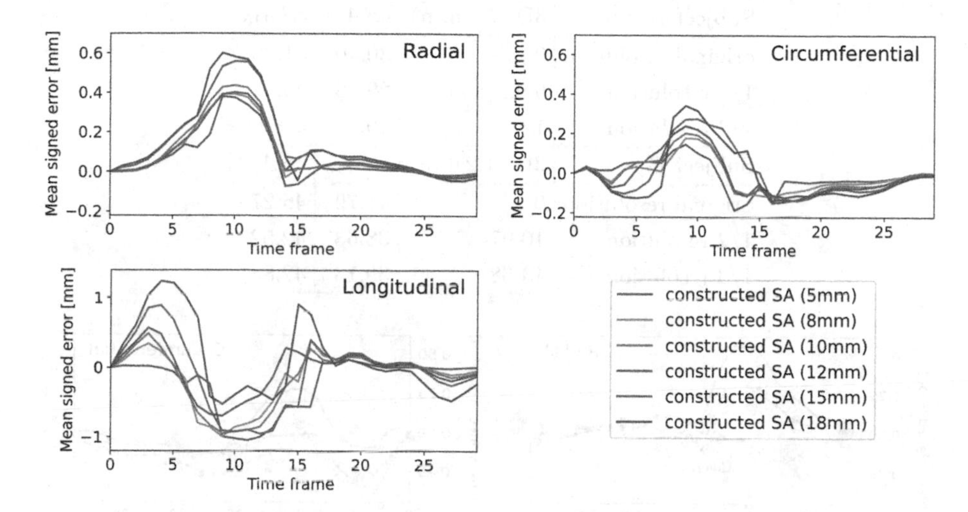

Fig. 3. Mean signed errors MSE_r, MSE_c and MSE_l extracted from cine series with different spatial resolutions for subject C.

3.2 The Effect of Temporal Resolution

In this section, we compare the accuracy of the displacement extracted from image series with the lower temporal resolution to the displacement field U^{3D}. We remark that the time interval T in the definition of NRMSE (5) is independent of the temporal resolution. The time integrals are evaluated using the trapezoidal rule with $\Delta t = 1$, $\Delta t = 2$, and $\Delta t = 3$ for the original, 1/2, and 1/3 temporal resolution, respectively.

The values of NMRSE in Table 4 show that in the case of 3D cine, a decrease in temporal resolution leads to an increase in displacement error. This does not hold for the cSA images with a lower spatial resolution, where the accuracy of the extracted displacement field does not increase. In most cases, the change in NMRSE caused by the reduction in spatial resolution at fixed temporal resolution is larger than the change caused by the reduction in temporal resolution at fixed spatial resolution. The plots of MSE for Subject C are shown in Fig. 4.

Table 4. Normalized root-mean-square errors NRMSE [%] evaluated for different temporal resolutions in the original 3D cine (voxel size 1.2 mm) and constructed SA cine of selected slice thicknesses 8–15 mm.

Subject A	3D (1.2 mm)	cSA_8	cSA_{12}
original resolution	0	55.76	71.98
1/2 resolution	10.97	51.89	70.30
1/3 resolution	12.96	50.88	67.77
Subject B	**3D (1.2 mm)**	cSA_{10}	cSA_{15}
original resolution	0	30.40	51.23
1/2 resolution	9.92	29.38	49.81
1/3 resolution	12.27	29.34	49.70
Subject C	**3D (1.2 mm)**	cSA_{10}	cSA_{15}
original resolution	0	41.79	46.27
1/2 resolution	10.97	38.63	42.52
1/3 resolution	13.68	39.13	42.83

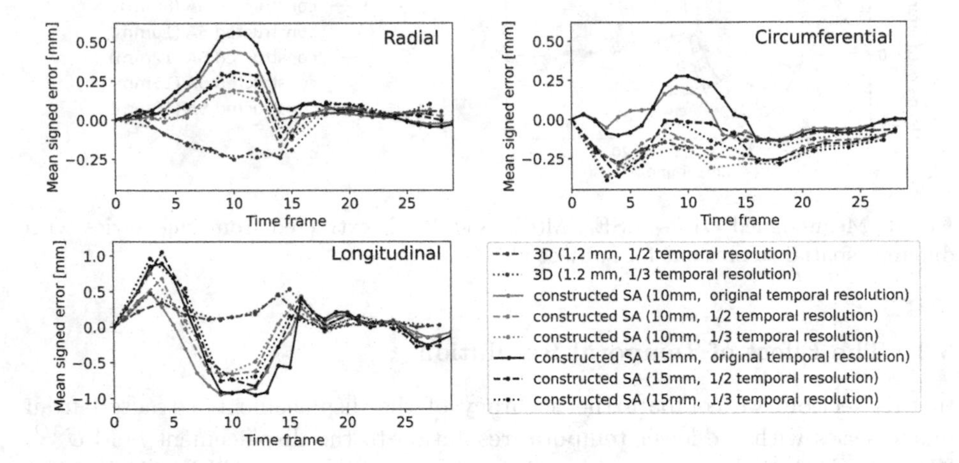

Fig. 4. Mean signed errors MSE_r, MSE_c and MSE_l extracted from cine series with different spatial and temporal resolutions for subject C.

4 Discussion and Conclusion

We investigated the dependence between spatial and temporal resolution of MR cine images and the accuracy of LV motion extracted from the images. A previously developed and validated motion tracking method was used. The study was performed using an acquired series of 3D and SA cine images and SA-like constructed cine image series, which allowed to extend the set of combinations of spatial and temporal resolution.

Given the limitation of working directly with image data, the required spatial and temporal resolution in the constructed image series was obtained by appropriately averaging the original images. The averaging of sub-slices was done to simulate the 2D acquisition with an optimal ramp-like slice selection gradient. The temporal averaging was designed as an image-space approximation of the segmented filling of k-space for each time frame.

The displacements extracted from the image series with various spatial and temporal resolutions were compared with the displacement obtained from the 3D cine with the original temporal resolution. For all subjects, the displacement error was largest in the longitudinal direction. Moreover, the error in the longitudinal direction was also affected by the increase in slice thickness. The effect of a reduced temporal resolution on the studied quantities proved to be smaller than the effect of slice thickness. The effect on other features of the LV motion, such as the strain or strain rate should be investigated in the future.

The trend in the dependency of displacement accuracy on the spatial and temporal resolution was consistent in all three subjects. However, the RMSE, as well as the NRMSE, were higher for Subject A than for Subjects B and C. This is possibly due to a lower ratio between the dimensions of the ventricle and the voxel size, which may result in more voxels being influenced by the partial volume effect.

The present study provides a good insight into the use of 3D cine in routine clinical acquisitions. We have shown that no information is lost compared to standard multiple 2D cine acquisition and a higher quality local motion pattern can possibly be extracted. A similar approach using constructed images can also be used to quantify expected acquisition errors with high undersampling (in space or time), e.g., in clinical applications requiring real time acquisition to capture complex physiology [7,8].

Acknowledgements. This work was supported by the Ministry of Education, Youth and Sports of the Czech Republic under the OP RDE grants number CZ2.11/0/0/16_019/0000765 and by the Ministry of Health of the Czech Republic project No. NV19-08-00071. This work was also supported by the Inria-UTSW Associated Team TOFMOD.

References

1. Berberoğlu, E., Stoeck, C.T., Moireau, P., Kozerke, S., Genet, M.: In-silico study of accuracy and precision of left-ventricular strain quantification from 3D tagged MRI. PLOS One **16**(11) (2021). https://doi.org/10.1371/journal.pone.0258965, https://dx.plos.org/10.1371/journal.pone.0258965
2. Castellanos, D.A., et al.: Left ventricular torsion obtained using equilibrated warping in patients with repaired tetralogy of fallot. Pediatr. Cardiol. **42**(6), 1275–1283 (2021). https://doi.org/10.1007/s00246-021-02608-y
3. Genet, M., et al.: Distribution of normal human left ventricular myofiber stress at end diastole and end systole: a target for in silico design of heart failure treatments. J. Appl. Physiol. **117**, 142–152 (2014). https://doi.org/10.1152/japplphysiol.00255.2014, http://www.ncbi.nlm.nih.gov/pubmed/24876359

4. Genet, M., Stoeck, C.T., Von Deuster, C., Lee, L.C., Kozerke, S.: Equilibrated warping: finite element image registration with finite strain equilibrium gap regularization. Med. Image Anal. **50**, 1–22 (2018)
5. Lee, L.C., Genet, M.: Validation of equilibrated warping—image registration with mechanical regularization—on 3D ultrasound images. In: Coudière, Y., Ozenne, V., Vigmond, E., Zemzemi, N. (eds.) FIMH 2019. LNCS, vol. 11504, pp. 334–341. Springer, Cham (2019). https://doi.org/10.1007/978-3-030-21949-9_36
6. Moghari, M.H., Barthur, A., Amaral, M.E., Geva, T., Powell, A.J.: Free-breathing whole-heart 3D cine magnetic resonance imaging with prospective respiratory motion compensation. Magn. Reson. Med. **80**(1), 181–189 (2018)
7. Ruijsink, B., et al.: Synergy in the heart: RV systolic function plays a key role in optimizing LV performance during exercise. Am. J. Physiol.-Heart Circulatory Physiol. **319**(3), H642–H650 (2020)
8. Ruijsink, B., et al.: Dobutamine stress testing in patients with Fontan circulation augmented by biomechanical modeling. PLoS One **15**(2), e0229015 (2020)
9. Tobon-Gomez, C., et al.: Benchmarking framework for myocardial tracking and deformation algorithms: an open access database. Med. Image Anal. **17**(6), 632–648 (2013)

Extraction of Volumetric Indices from Echocardiography: Which Deep Learning Solution for Clinical Use?

Hang Jung Ling[1(✉)], Nathan Painchaud[1,2], Pierre-Yves Courand[1,3,4], Pierre-Marc Jodoin[2], Damien Garcia[1], and Olivier Bernard[1]

[1] Univ Lyon, INSA-Lyon, Université Claude Bernard Lyon 1, UJM-Saint Etienne, CNRS, Inserm, CREATIS UMR 5220, U1294, 69621 Lyon, France
hang-jung.ling@insa-lyon.fr

[2] Department of Computer Science, University of Sherbrooke, Sherbrooke, QC, Canada

[3] Cardiology Department, Hôpital Croix-Rousse, Hospices Civils de Lyon, Lyon, France

[4] Cardiology Department, Hôpital Lyon Sud, Hospices Civils de Lyon, Lyon, France

Abstract. Deep learning-based methods have spearheaded the automatic analysis of echocardiographic images, taking advantage of the publication of multiple open access datasets annotated by experts (CAMUS being one of the largest public databases). However, these models are still considered unreliable by clinicians due to unresolved issues concerning i) the temporal consistency of their predictions, and ii) their ability to generalize across datasets. In this context, we propose a comprehensive comparison between the current best performing methods in medical/echocardiographic image segmentation, with a particular focus on temporal consistency and cross-dataset aspects. We introduce a new private dataset, named CARDINAL, of apical two-chamber and apical four-chamber sequences, with reference segmentation over the full cardiac cycle. We show that the proposed 3D nnU-Net outperforms alternative 2D and recurrent segmentation methods. We also report that the best models trained on CARDINAL, when tested on CAMUS without any fine-tuning, still manage to perform competitively with respect to prior methods. Overall, the experimental results suggest that with sufficient training data, 3D nnU-Net could become the first automated tool to finally meet the standards of an everyday clinical device.

Keywords: Ultrasound · cardiac segmentation · temporal segmentation · deep learning · CNN

1 Introduction

Echocardiographic imaging has undergone major advances in recent years thanks to artificial intelligence, especially the deep learning (DL) paradigm. In par-

Supplementary Information The online version contains supplementary material available at https://doi.org/10.1007/978-3-031-35302-4_25.

O. Bernard et al. (Eds.): FIMH 2023, LNCS 13958, pp. 245–254, 2023.
https://doi.org/10.1007/978-3-031-35302-4_25

ticular, the automated extraction of clinical indices from the segmentation of cardiac structures has been the subject of intense research leading to major breakthroughs. A key component of these advances has been the publication of open access annotated datasets, including CETUS (45 patients, 3D images annotated at End-Diastole - ED and End-Systole - ES) [1], CAMUS (500 patients, 2D images annotated at ED and ES in apical two-chamber - A2C - and apical four-chamber - A4C - views) [5], EchoNet-Dynamic (10,036 patients, 2D sub-sampled images annotated at ED and ES in A4C view) [7], HMC-QU (109 patients, 2D sequences annotated in A4C view) [2] and TED (98 patients from the CAMUS dataset, 2D sequences annotated in A4C view) [8].

These datasets allowed effective and fair comparisons of methods, whether they are generic image segmentation models [5,6] or were specifically designed to process echocardiographic images [9,11]. Thus, the performance of current state-of-the-art (SOTA) methods on the CAMUS dataset confirmed the dominance of the DL-based methods, which finally achieved inter- and intra-observer variability for most of the geometric (Dice score, Hausdorff distance, mean absolute distance - MAD) and clinical metrics (ejection fraction - EF, volumes at ED/ES).

Although these results are extremely promising and represent a crucial step towards the automation of echocardiographic image analysis, they are not sufficient to justify confidence in fully automated methods in a clinical context. Indeed, two crucial challenges on the path to the practical application of these algorithms remain understudied in the field: i) the frame-by-frame temporal consistency of the predictions, and ii) the generalization of the methods across datasets. Based on this observation, we propose the following contributions:

1. We study the performance of two generic architectures based on common temporal data processing techniques on 2D echocardiography sequences, and compare them to current SOTAs in the same field;
2. We present a new private dataset called CARDINAL (240 patients, 2D sequences annotated in A4C and A2C views), and report the performance impact of training our methods exclusively on CARDINAL and testing on CAMUS.

2 Benchmarked Methods

CAMUS is currently the only dataset where an evaluation platform has been established to effectively compare the performance of segmentation methods[1]. We therefore relied on this dataset to select the methods we retained in this study.

2.1 2D DL Methods

The currently best performing method on the CAMUS dataset exploits the nnU-Net formalism [6]. This model is based on the U-Net architecture and implements several successful DL tricks, such as a patch-wise approach to preserve image

[1] https://www.creatis.insa-lyon.fr/Challenge/camus/results.html.

resolution, data augmentation during both training and inference to enforce generalization and automatic hyperparameter search of the U-Net architecture to increase accuracy. Note that the 2D version uses only one U-Net model [4].

Recently, Sfakianakis *et al.* [9] developed a DL solution called GUDU based on three key aspects. First, they proposed to use data augmentations tailored to ultrasound acquisition, i.e. variation of the contrast between the myocardial tissue and the left ventricular (LV) cavity, random rotation from the origin of the sectorial shape to mimic different probe positioning, and perspective transformations to simulate probe twisting. Inspired by ensemble models, the authors also trained 5 U-Nets with different architectures and averaged their outputs during inference in order to compute the final prediction. Finally, a new loss function was proposed that takes into account the relative position of the cardiac structures with respect to each other. The authors demonstrated the usefulness and complementarity of each contribution in an ablation study.

2.2 2D+t DL Methods

Despite the success of 2D methods in producing accurate segmentations for individual echocardiographic frames, they often fail to maintain temporal consistency between frames [8]. Prior to the publication of HMC-QU and TED, there were no publicly available datasets to train and compare methods that incorporate the temporal dimension in 2D+time echocardiography. This explains why so few papers have focused on this topic.

Due to the lack of 2D+time annotated datasets, Wei *et al.* [11] proposed a method to leverage the limited ED/ES annotated frames and propagate them to unannotated frames. This was achieved by training a 3D U-Net designed to predict both the deformation fields between each pair of consecutive frames and the segmentation masks at each frame of the sequence. The deformation fields are used to propagate the ED/ES reference annotations forward and backward in time through the sequence. The corresponding propagated masks are then used as targets for self-supervised segmentation of the entire sequence. This encourages the model to learn consistent temporal dynamics to find the best match between the predicted segmentation masks and the propagated annotations.

More recently, Smistad *et al.* [10] and Hu *et al.* [3] added convolutional long-term memory blocks to each layer of the encoder of a 2D U-Net (this type of model is hereafter referred to as U-Net LSTM). Thus, instead of processing a single frame, these methods take a series of frames as input and store the extracted features over time to produce the final segmentation of the entire sequence. Results show that such a strategy tends to reduce segmentation shifts from one frame to another.

By their very nature, echocardiographic sequences exhibit regular properties along the time axis. Therefore, it seems logical to consider 2D ultrasound sequences as complete volumes containing coherent 3D shapes and to extract 3D features using 3D convolutional layers to promote temporal consistency. Thus, in this paper, we propose to train a 3D nnU-Net to segment the complete cardiac sequences in a single run. We hypothesize that this model will inherently learn temporal consistency while maintaining a high level of segmentation accuracy.

3 Experimental Setup

3.1 CARDINAL Dataset

Acquisition Protocol: The proposed dataset consists of clinical examinations of 240 patients, acquired at the University Hospital of Lyon (Croix-Rousse Lyon Sud, France) under the regulation of the local ethics committee of the hospital. The complete dataset was acquired with GE ultrasound scanners. For each patient, 2D A4C and A2C view sequences were exported from the EchoPAC analysis software. Each exported sequence corresponds to a set of B-mode images expressed in polar coordinates. The same interpolation procedure as used for the CAMUS dataset was applied to express all sequences in Cartesian coordinates with a single grid resolution of $0.31\,\text{mm}^2$. Each sequence in the CARDINAL dataset corresponds to a complete cardiac cycle defined as the interval between peaks of maximal LV cavity surface area.

Reference Annotations: To tackle the total number of frames to be annotated, an experienced observer first delineated the different contours using semi-automatic tools to ensure temporal consistency of the segmented shapes. Each corresponding output was then checked/corrected by two other experienced observers. To identify the ED/ES frames in the sequence, the ED frames correspond (by definition) to the beginning and end of each sequence, and the ES frame corresponds to the frame where the LV cavity surface is smallest.

3.2 Implemented DL Methods

For a fair comparison, we implemented the 2D nnU-Net, U-Net LSTM, and 3D nnU-Net described in Sect. 2 using the same Python library called ASCENT[2]. These models shared the following training hyperparameters: batch size of 2, SGD optimizer with a learning rate of 0.01 coupled with a polynomial decay scheduler, and 1000 training epochs. The 2D and 3D nnU-Net used a patch-wise approach to avoid resizing the images, thus preserving the native image resolution. To train the U-Net LSTM, the input images were resized to 256×256 and 24 consecutive frames were randomly selected and fed to the model to produce the corresponding segmentations. For inference, the sliding window approach with a Gaussian importance map was used. The prediction was given by the average of the *softmax* probabilities of all windows. To improve segmentation accuracy, the final prediction was obtained by averaging the predictions of the original and mirrored images along different axes. More implementation details for each model can be found in Table 1.

4 Results

We evaluate the methods described in Sect. 2 using three types of measures to get a complete picture of their performance in terms of segmentation accuracy (Table

[2] https://github.com/creatis-myriad/ASCENT.

Table 1. Details of the implementation of the three methods evaluated in this study. *Lowest resolution*: Size of the lowest resolution of feature maps in pixels. *Optimization scheme*: Optimizer + initial learning rate + learning rate scheduler used. *Training duration (hours)*: number of hours required to train each model for 1000 epochs. The configurations shared between models are only shown once in their respective rows.

Configurations	3D nnU-Net	2D nnU-Net	U-Net LSTM
Patch size (pixels)	$320 \times 256 \times 24$	640×512	$256 \times 256 \times 24$
Batch size		2	
Nb. feature maps		$32 \downarrow 480 \uparrow 32$	
Lowest resolution	$10 \times 8 \times 6$	5×4	$8 \times 8 \times 24$
Downsampling scheme		Stride pooling	
Upsampling scheme		Deconvolution	
Normalization scheme		Instance normalization	
Optimization scheme		SGD + 0.01 + polynomial decay	
Loss function		Cross entropy + Dice	
Number of parameters	41.3 M	30.4 M	49 M
Training duration (hours)	22.8	8	69.5

2), extraction of clinical indices (Table 3) and temporal consistency (Table 4). In each of these tables, we group the methods according to the datasets on which they were trained and tested (CARDINAL is abbreviated as *CL* and CAMUS is abbreviated as *CS*) to make it easier to observe the change in performance when generalizing to a new dataset.

4.1 Geometric and Clinical Accuracy

Table 2 shows the segmentation accuracy computed from the CARDINAL and CAMUS datasets for the 5 algorithms described in Sect. 2. The values in bold correspond to the best scores for each metric for a given training/test dataset setup. From the results on the CARDINAL dataset (CL/CL case), we can see that the 3D nnU-Net has the best segmentation scores for all metrics, for both ED and ES. It is also interesting to note that the two temporal consistency methods (3D nnU-Net and U-Net LSTM) produce better results than the 2D nnU-Net method. This can be explained by the fact that the reference segmentation has regular properties along the temporal axis due to the annotation process. Methods that integrate the temporal dimension into their architecture are therefore more likely to produce segmentation results that are closer to the manual references.

It is worth mentioning that methods trained and tested on the same dataset (sections CL/CL and CS/CS in Table 2) get overall better results up to $1.7x$ for the Hausdorff and MAD metrics. One reason for such an improvement is the larger amount of annotated training images for CARDINAL (18,793 images from 190 training/validation patients, reference frames for the full cardiac cycle in A2C and A4C views) than for CAMUS (1,800 images from 450 training/validation patients, reference frames at ED and ES in A2C and A4C views).

Table 3 reports the clinical metrics for the 5 methods. As in Table 2, the methods enforcing temporal consistency gets the best results on CARDINAL, especially for the ejection fraction for which temporal consistency is essential (mean correlation score of 0.917). Furthermore, the best models trained on CARDINAL or CAMUS produce similar results for volume estimation (average correlation of 0.978), revealing a limit reached by these approaches, certainly due to the resolution of the imaging systems.

Table 2. LV segmentation accuracy of the benchmarked methods, on different subsets of frames. The columns *All*, *ED* and *ES* indicate results averaged over all frames, only ED frames, and only ES frames, respectively. Since CAMUS only provides annotation for ED/ES frames, results over *all* frames are not available when testing on it.

Methods	Train/test	Dice			Hausdorff (mm)			MAD (mm)		
		All	ED	ES	All	ED	ES	All	ED	ES
3D nnU-Net	CL/CL	**.969**	**.968**	**.960**	**2.3**	**2.7**	**2.5**	**0.7**	**0.8**	**0.7**
2D nnU-Net		.957	.961	.942	2.9	3.1	3.1	0.9	1.0	1.1
U-Net LSTM		.964	.964	.956	2.5	2.8	2.6	0.8	0.9	0.8
3D nnU-Net	CL/CS	-	**.939**	**.926**	-	5.2	**4.6**	-	**1.6**	**1.5**
2D nnU-Net		-	.934	.921	-	**4.9**	4.6	-	1.8	1.6
U-Net LSTM		-	.925	.903	-	6.0	5.8	-	2.1	2.1
2D nnU-Net	CS/CS	-	**.952**	**.935**	-	**4.3**	**4.2**	-	**1.3**	**1.3**
CLAS		-	.947	.929	-	4.6	4.6	-	1.4	1.4
GUDU		-	.946	.929	-	4.7	4.7	-	1.4	1.4

(CL:CARDINAL, CS:CAMUS)

4.2 Integration of Temporal Consistency

Table 4 allows a better investigation of the temporal performance of the methods by providing additional information on the number/percentage of frames considered temporally inconsistent w.r.t. their neighboring frames. As expected, the methods incorporating temporal persistence produced fewer temporal errors. Looking at the number of sequences with at least one temporally inconsistent frame, the 3D nnU-Net clearly outperforms U-Net LSTM, with only 4 inconsistent sequences over 100 compared to 98 sequences for U-Net LSTM. This result illustrates the greater ability of features computed from 3D convolutional layers to extract relevant spatio-temporal information. The few remaining temporal errors for the 3D nnU-Net are more an indication that the metrics we

Table 3. Clinical metrics of the benchmarked methods. The ED/ES volumes were computed from both the predicted and reference masks using Simpson's biplane method. *Corr.*: Correlation between the ejection fraction (EF) derived from the predicted/reference segmentation. *MAE*: Mean Absolute Error between the EF derived from predicted/reference segmentation.

Methods	Train/test	EF		Volume ED		Volume ES	
		Corr.	MAE (%)	Corr.	MAE (ml)	Corr.	MAE (ml)
3D nnU-Net	CL/CL	.913	2.9	**.978**	**3.3**	**.974**	**2.7**
2D nnU-Net		.850	3.8	.967	4.4	.957	3.2
U-Net LSTM		**.922**	**2.7**	.973	3.4	.969	2.8
3D nnU-Net	CL/CS	**.869**	**5.3**	**.974**	**9.6**	**.976**	**4.9**
2D nnU-Net		.810	7.0	.970	12.8	.959	6.2
U-Net LSTM		.822	11.1	.879	15.9	.903	8.2
2D nnU-Net	CS/CS	.857	4.7	**.977**	**5.9**	**.987**	**4.0**
CLAS		**.926**	**4.0**	.958	7.7	.979	4.4
GUDU		.897	**4.0**	**.977**	6.7	.981	4.6

used are (overly) strict on the temporal smoothness. Indeed, the 3D nnU-Net temporal "inconsistencies" appear invisible to the expert eye. As a qualitative evaluation, Fig. 1 illustrates in detail the temporal consistency of each of our own method on one patient from the CARDINAL test set. To complement this, we also provide in the supplementary material examples of temporally consistent and inconsistent segmentation results obtained by the 3D nnU-Net method for the CARDINAL and CAMUS datasets.

4.3 Generalization Across Datasets

The ability to generalize across datasets is crucially important to gauge the capacity of a method to properly analyze data affected by a distributional shift. To this end, the models trained on CARDINAL were also evaluated on the CAMUS test set without any fine-tuning. The results are reported in the "CL/CS" sections of Tables 2, 3 and 4. Among the methods evaluated, 3D nnU-Net is the undisputed best. It even produces competitive geometric and clinical scores compared with SOTA methods trained directly on CAMUS. Thanks to the integration of temporal consistency, the 3D nnU-Net trained on CARDINAL also produces one of the best correlation scores for the EF calculated on the CAMUS dataset, even when compared to SOTA methods trained directly on CAMUS. In view of these results, and considering that the annotation process between the two databases was not identical and was carried out by different experts (which inevitably introduces a bias during the learning phase), the generalization capacity of the 3D nnU-Net model seems remarkable.

Table 4. Temporal consistency of the benchmarked methods, as defined in [8]. *Nb of seq. w/ err.*: number of sequences (out of the 100 testing sequences) where at least one frame is temporally inconsistent. *% of frames w/ err.*: percentage of frames that are inconsistent in the sequences with at least one temporally inconsistent frame. *Err. to thresh. ratio*: average ratio between the measure used to identify temporal inconsistencies and the threshold for temporal inconsistencies. A lower value indicates "smoother" temporal segmentations.

Methods	Train/test	Nb of seq. w/ err.	% of frames w/ err.	Err. to thresh. ratio
3D nnU-Net	CL/CL	**4**	**4**	**.045**
2D nnU-Net		100	30	.210
U-Net LSTM		98	13	.110
3D nnU-Net	CL/CS	**28**	**12**	**.095**
2D nnU-Net		85	21	.162
U-Net LSTM		83	16	.114

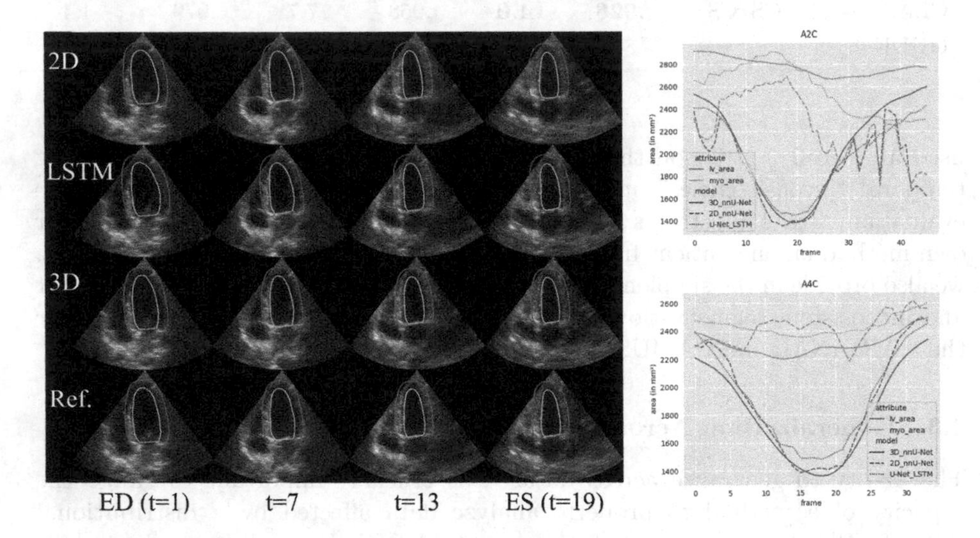

Fig. 1. Visualization of the temporal consistency of the segmentations on one patient from the CARDINAL test set. (left) Frames sampled between ED and ES, with segmentation masks from our own methods + reference. (right) Curves of the LV and myocardium surfaces w.r.t. frame in the sequence.

5 Conclusion

We evaluated the ability of different methods to accurately segment echocardiographic images, with a focus on temporal consistency and cross-dataset generalization. To this end, we introduced a new private database called CARDINAL, with annotations from an expert on the full cardiac cycle for each sequence. The results show that 3D nnU-Net and U-Net LSTM produce the best geometric and clinical scores on the CARDINAL dataset due to the integration of temporal persistence. Regarding the temporal consistency metrics, 3D nnU-Net performed significantly better than U-Net LSTM with only four sequences (instead of 98) out of 100 having at least one image that was temporally inconsistent. As far as cross-dataset generalization is concerned, 3D nnU-Net is also the best performing method. When trained on CARDINAL and tested on CAMUS, it achieved comparable geometric and clinical scores to the best methods both trained and tested on CAMUS. All these results clearly show that 3D nnU-Net is a serious candidate to become the first automated tool to meet the requirements of routine clinical examinations.

Acknowledgment. The authors gratefully acknowledge financial support of the MEGA doctoral school (ED 162), the NSERC Canada Graduate Scholarships-Doctoral Program, the FRQNT Doctoral Scholarships Program, the French National Research Agency (ANR) through the "ORCHID" project (ANR-22-CE45-0029-01), and the LABEX PRIMES (ANR-11-LABX-0063) of Université de Lyon, within the program "Investissements d'Avenir" operated by the French ANR. The authors also thank GENCI-IDRIS for providing access to HPC resources (Grant 2022-[AD010313603]).

References

1. Bernard, O., et al.: Challenge on Endocardial Three-dimensional Ultrasound Segmentation (CETUS). In: MICCAI CETUS, pp. 1–8 (2014)
2. Degerli, A., et al.: Early detection of myocardial infarction in low-quality echocardiography. IEEE Access **9**, 34442–34453 (2021)
3. Hu, J., Smistad, E., Salte, I.M., Dalen, H., Lovstakken, L.: Exploiting temporal information in echocardiography for improved image segmentation. In: 2022 IEEE International Ultrasonics Symposium (IUS) (2022)
4. Isensee, F., Jaeger, P.F., Kohl, S.A.A., Petersen, J., Maier-Hein, K.H.: nnU-net: a self-configuring method for deep learning-based biomedical image segmentation. Nat. Methods **18**(2), 203–211 (2021)
5. Leclerc, S., et al.: Deep learning for segmentation using an open large-scale dataset in 2D echocardiography. IEEE Trans. Med. Imaging **38**(9), 2198–2210 (2019)
6. Ling, H.J., Garcia, D., Bernard, O.: Reaching intra-observer variability in 2-D echocardiographic image segmentation with a simple U-Net architecture. In: 2022 IEEE International Ultrasonics Symposium (IUS) (2022). https://hal.science/hal-03979523
7. Ouyang, D., et al.: Video-based AI for beat-to-beat assessment of cardiac function. Nature **580**(7802), 252–256 (2020)
8. Painchaud, N., Duchateau, N., Bernard, O., Jodoin, P.M.: Echocardiography segmentation with enforced temporal consistency. IEEE Trans. Med. Imaging **41**(10), 2867–2878 (2022)
9. Sfakianakis, C., Simantiris, G., Tziritas, G.: GUDU: geometrically-constrained Ultrasound Data augmentation in U-Net for echocardiography semantic segmentation. Biomed. Signal Process. Control **82**, 104557 (2023)

10. Smistad, E., Salte, I.M., Dalen, H., Lovstakken, L.: Real-time temporal coherent left ventricle segmentation using convolutional LSTMs. In: 2021 IEEE International Ultrasonics Symposium (IUS) (2021)
11. Wei, H., et al.: Temporal-consistent segmentation of echocardiography with co-learning from appearance and shape. In: Martel, A.L., et al. (eds.) MICCAI 2020. LNCS, vol. 12262, pp. 623–632. Springer, Cham (2020). https://doi.org/10.1007/978-3-030-59713-9_60

Whole Heart 3D Shape Reconstruction from Sparse Views: Leveraging Cardiac Computed Tomography for Cardiovascular Magnetic Resonance

Hao Xu[1], Marica Muffoletto[1], Steven A. Niederer[1], Steven E. Williams[1,2], Michelle C. Williams[2], and Alistair A. Young[1(✉)]

[1] Department of Bioengineering, King's College London, London, UK
alistair.young@kcl.ac.uk

[2] University/BHF Centre for Cardiovascular Science, University of Edinburgh, Edinburgh, Scotland

Abstract. The 3D shape of the atria and ventricles is important for studying the mechanisms of disease processes. Common imaging methods such as cardiovascular magnetic resonance (CMR) often acquire a limited number of short and long axis slices. We trained a label completion U-Net (LC-U-Net) to automatically predict 3D shapes for the ventricles, atria, and valves from standard CMR views. We used dense 3D segmentations from a large coronary computed tomography (CCTA) study and tested the method using simulated short and long axis CMR slices. Slice position errors were included as well as breath-hold misalignments (motion artefacts) to mimic actual CMR scans. The method outperformed a previous state of the art method and showed good robustness to different levels of motion artefacts. Explicit motion correction at inference time marginally improved performance. Dice for all chambers was > 90% and mean surface distance was $\leq$ 1.0 mm, even for the right and left atria which had only two and one slice(s) labelled respectively. All four valve positions were accurately reconstructed with mean surface distance $\leq$ 2.0 mm, even for the pulmonary valve which was not included in any of the simulated CMR views. In conclusion, this method can reconstruct 3D shape in all chambers and valves of the heart despite limited information for the atria and valves.

Keywords: Shape Reconstruction · CMR · Whole Heart · 3D

1 Introduction

Three-dimensional (3D) heart shape adapts in response to sub-clinical disease and risk factors [7] and has been shown to predict future adverse events [8]. However, common imaging methods such as cardiovascular magnetic resonance (CMR) acquire a limited number of slices, and reconstruction of 3D shape from sparse information may improve the utility of these studies.

O. Bernard et al. (Eds.): FIMH 2023, LNCS 13958, pp. 255–264, 2023.
https://doi.org/10.1007/978-3-031-35302-4_26

Previous methods for the reconstruction of 3D heart shape from CMR slices have included volume super-resolution [9, 11], point-to-mesh prediction [1], label inpainting [12], and shape deformation [3, 5]. Methods have focused on one or two ventricles, using inputs in the form of regular grids or point clouds. Graph convolution networks typically output one mesh per network, and reconstructions of bi-ventricular shape are performed considering ventricular myocardium as one structure [3] or predicting ventricular cavities and myocardium using different networks [1], either ignoring important structures or introducing potential overlapping between different structures. Convolutional Neural Networks can output label maps with different channels and prevent overlap between structures. Anatomically-constrained methods [9] typically take only the short axis (SAX) CMR slices to simplify the network architecture, either as a set of 2D networks or one 3D network, but often require the number of slices and the distance between adjacent slices to be fixed across the dataset, making it difficult to apply the pre-trained network to different datasets. Fully 3D networks [11, 12] are capable of handling CMR slices with a large variations of slice numbers and orientations, i.e., long axis (LAX) slices could also be included and the reconstruction of more accurate basal/apical structures including both atria are possible.

Supervised-learning methods typically require paired sparse and dense data to train the model. However, such pairs are usually not available and sparse data is typically simulated from dense data. Dense data often comes from dense image segmentations [9, 11] and statistical shape atlases [1, 12]. The correction of motion artifacts improves the plausibility and accuracy of the reconstructed heart shape [7] and is usually integrated to the training stage for supervised-learning methods [1, 9, 12], or explicitly applied for unsupervised-learning methods [11].

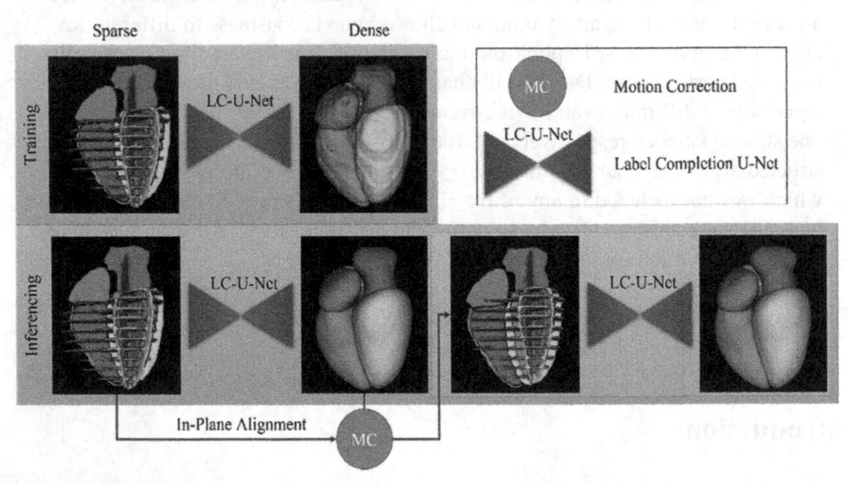

Fig. 1. Method overview. Top: Training protocol. Bottom left: Inferencing protocol. Bottom right: Optional refinement with explicit motion correction. The CCTA ground truth dense segmentation is shown in the top row

In this paper, we address the challenge of dense whole heart shape reconstruction from sparse CMR segmentations. Extending [12], we provide the following improvements: 1) Our method accurately estimates the structure of the whole heart including left ventricle (LV), LV myocardium (LVM), right ventricle (RV), RV myocardium (RVM), left atrium (LA), right atrium (RA), ascending aorta (AA) and pulmonary artery (PA), by taking both LAX and SAX slices. 2) Our method also reconstructs detailed 3D valve representations including mitral valve, aortic valve, tricuspid valve, and pulmonary valve. 3) We test on multiple levels of motion artifacts with two types of motion correction (implicit and explicit) and show the method is robust to different amounts of motion. 4) we directly compare the method with a state-of-the-art super-resolution approach (SR-Heart) [11].

2 Method

In this work, we trained and validated the networks using 1700 dense whole heart segmentations obtained from CCTA images [13], focusing on the chamber blood pools and endocardial surfaces, excluding pulmonary veins and LA appendage from the output labels. The cohort consisted of patients with suspected coronary artery disease who participated in the Scottish COmputed Tomography of the HEART (SCOT-HEART) trial [10]. The overview of our method is shown in Fig. 1. CMR sparse segmentations were simulated by sampling appropriate views from the 3D segmentations.

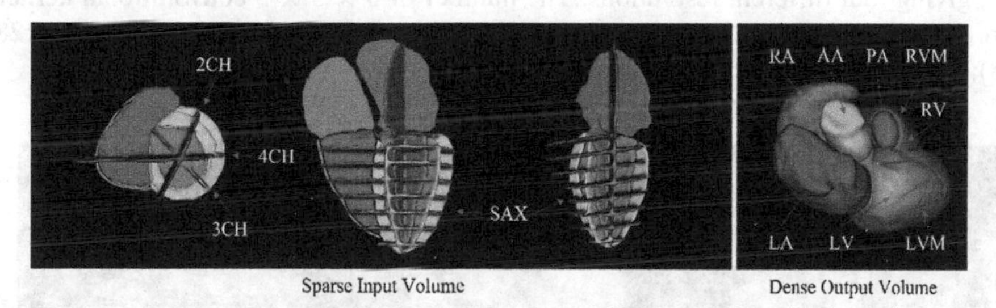

Fig. 2. Sparse input and dense output volumes. LV: left ventricle, LVM: LV myocardium, RV: right ventricle, RVM: RV myocardium, LA: left atrium, RA: right atrium, AA: ascending aorta, PA: pulmonary artery, SAX: short axis, 2CH/3CH/4CH: two/three/four chamber

2.1 Cardiac Coordinate System

To generate spatially consistent input data we defined a cardiac coordinate system using the simulated CMR four-chamber (4CH) and two-chamber (2CH) views. The axes were: 1. Apex to base (intersection of the two views, i.e., the long axis of the LV), 2. Left to right (perpendicular to the long axis within 4CH view), and 3. Cross product of the previous two. We defined the 4CH as the plane passing though the apex of the LV and the centroids of the mitral and tricuspid valves, and the 2CH as the plane passing through the LV apex, the centroid of the mitral valve and normal to the RV, giving the long axis as the intersecting line of the two. The three-chamber (3CH) view was defined as the plane passing through the apical point and the centroids of the mitral and aortic valves. The SAX slices were then defined with a stack of planes perpendicular to the long axis between the centroid of the mitral valve and the apex, with 10 mm slice spacing. The number of SAX slices were first matched to the length of the ventricle, given an allowance of random motion along the long axis of < 10 mm. To mimic CMR slice planning errors, we introduced a series of slice augmentations including a random rotation offset for 2CH and 3CH views around the long axis, and rotation offsets around the other two axes for the SAX views, following a uniform distribution within $\pm$ 5 degrees. Breath-hold inconsistency (motion artefacts) were also introduced as explained below.

2.2 Label Completion U-Net (LC-U-Net)

We used a variation of 3D U-Net [4] for label completion, with both the input and output being 3D grid volumes of size $160 \times 160 \times 160$ voxels. The input volume was a sparse label map consisting of 7 labels: background (0), LV (1), LVM (2), RV (3), RVM (4), LA (5), and RA (6), and the output volume was a dense label map consisting of 9 labels: background (0), LV (1), LVM (2), RV (3), RVM (4), LA (5), RA (6), AA (7), and PA (8). Details of the volume preparation are discussed in the Sect. 2.3 and 2.4, and visualisations of the volumes are shown in Fig. 2.

The network had three max-pooling and deconvolution pairs with a stride of $2 \times 2 \times 2$, giving four different resolutions. The number of $3 \times 3 \times 3$ convolutional kernels were (16, 32), (32, 64), (64, 128), (128, 256) for the encoder and bottle neck, and (128, 128), (64, 64), (32, 9) for the decoder.

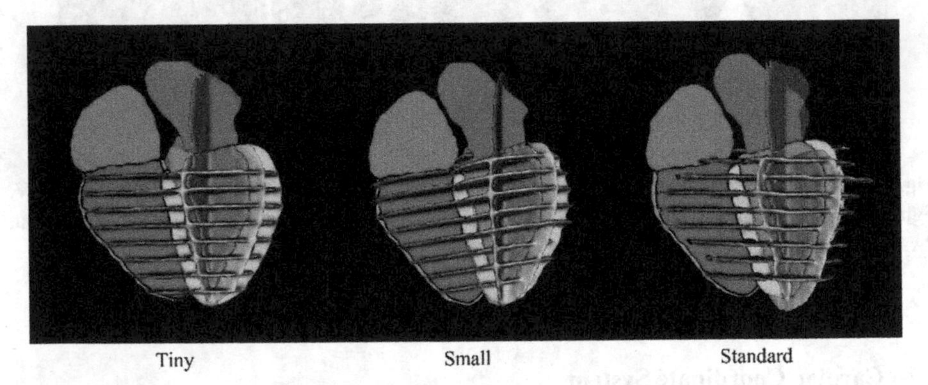

Fig. 3. Motion simulation examples

2.3 Sparse Input Volume

The size of the input and output volumes was determined by a bounding box centred at the centroid of the heart and 20% larger than the largest dimension of the CMR segmentations registered to the cardiac coordinate system. Note this gives a variable slice spacing in the resized volume. The sparse input volume was constructed using 2CH (LV, LVM, LA), 3CH (LV and LVM), 4CH (all labels) and SAX slices (all labels). All voxels were given a 0 (background) label, except for those within 1 voxel of a 2D labelled pixel, which were assigned the corresponding label, giving a slice thickness of 2 voxels.

2.4 Motion Simulation and CMR Segmentation Simulation

Motion artifact was simulated by a 3D translation of the heart. From previous studies [7] we found an average absolute 2D shift of 4.4 ± 5.7 mm. We therefore set three different levels of Gaussian standard deviation for offsets in each direction: tiny (0.5 mm), small (2.0 mm), and standard (3.5 mm). Examples of different levels are shown in Fig. 3. We applied the uniaxial motion in all three axes. For each slice, one set of motions was applied to the reference shape and its intersection with the slice plane was considered as the simulated segmentation of the slice.

For SAX slices, CMR segmentations typically confuse PA and RV labels in the basal slice, whilst the most apical and basal slices might not be segmented at all [2]. We therefore randomly removed the most apical and basal SAX slices (probability 0.5), and labelled PA as RV in the input SAX slice (if present).

2.5 Explicit Motion Correction

To apply explicit motion correction, we first produced the prediction of LC-U-Net and aligned the sparse input slices to the predicted label maps with rigid in-plane transformation and generated a 'motion-corrected' sparse input volume. We then simply applied the LC-U-Net a second time to the updated input to get the final prediction. Examples of our explicit motion correction is shown in Fig. 1.

2.6 Experiments

We randomly split the CCTA dataset into 1400 training cases, 100 validation cases and 200 testing cases. For each training and validating case, we prepared 10 augmentations of LAX and SAX slices, and for each epoch a random combination of individual LAX slices and the SAX stack were chosen to enlarge the variation of the training dataset. We prepared one set of slice planning errors to the testing cases, but with three different levels of motion artefacts. The LC-U-Net was trained with the standard level of motion simulation using cross-entropy loss and tested on all three levels of motion artefacts.

For the SR-Heart β-VAE model, we retrained the network (available at https://github.com/shuowang26/SRHeart) using the dense segmentation from the same training and validation cases. We trained 9 β-VAEs with values of β = 1e-3, 1e0 to 1e6, 1e9 for the combination of total cross-entropy similarity loss and the LK divergence, and the best overall performing network had β = 1e-3. The testing was performed using latent space searching with all three levels of motion simulation, with two types of loss described in [11]: SAX only (SR-Heart SAX) and SAX with 4CH view (SR-Heart SAX + 4CH).

We used Adam [6] optimiser with a learning rate of 1e-3 for all the trainings.

3 Results and Discussion

3.1 Motion Simulation Analysis

For our LC-U-Net, the overall Dice similarity for cavities and myocardium (6 labels) were 0.893, 0.893, and 0.888 for tiny, small, and standard levels respectively, and we did not see any performance drop when testing on smaller motion artifacts. Similar trend was found for both SR-Heart SAX and SR-Heart SAX + 4CH, with Dice similarities of (0.774, 0.751, 0.716), and (0.709, 0.658, 0.590) for tiny, small and standard levels. There was a performance drop for SR-Heart SAX + 4CH, and therefore we kept SR-Heart SAX as the optimal representation of the SR-Heart method.

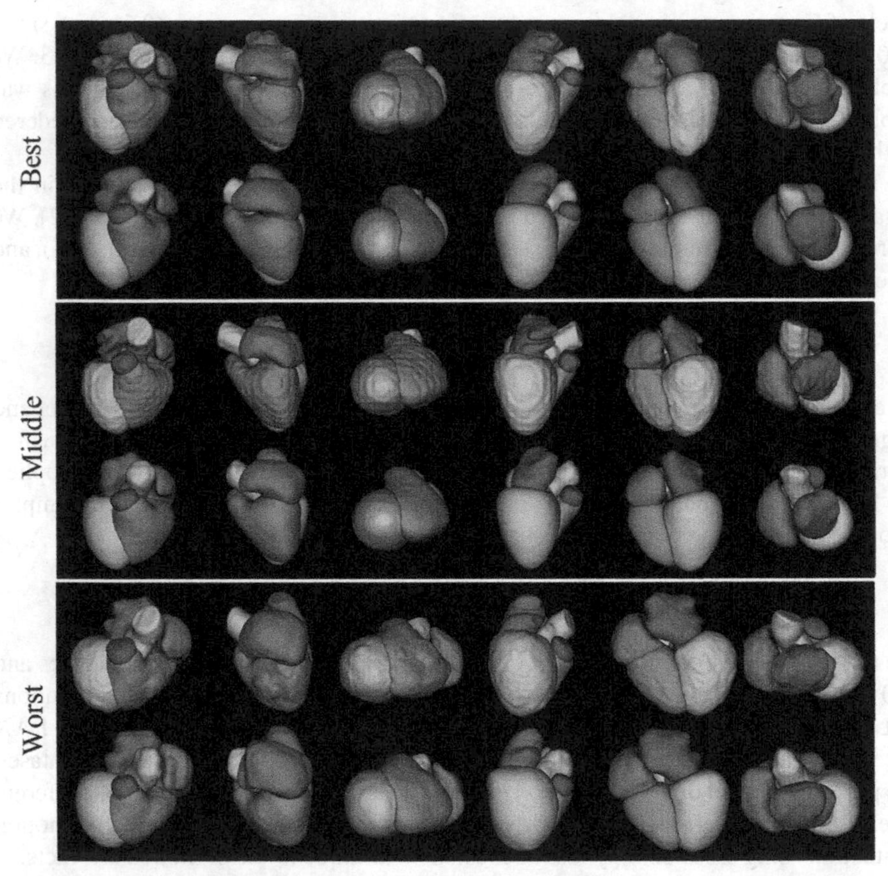

Fig. 4. Reconstruction visualisations of the best, middle, and worst cases of our method. Reference shapes are on the top and the reconstructions are at the bottom

3.2 CCTA Testing Quantitative Analysis

We evaluated the reconstruction accuracy using 3D Dice similarity, Hausdorff distance and mean surface distance between predictions and the reference, and the results are shown in Table 1. The reconstruction of LV and RV were more accurate compared to the atria due to the rich information provided by the SAX slices. The RA was the most challenging cavity but achieved good accuracy with just one cross-section available in the 4CH view. The RVM Dice score was lower than other segments due to its small thickness, however, the HD and MSD accuracy were still acceptable. The LC-U-Net and LC-U-Net MC were both significantly better than SR-Heart ($p < 0.001$ for all). LC-U-Net MC had smaller HD and MSE than LC-U-Net ($p < 0.001$) except for RA HD ($p > 0.05$).

Table 1. CCTA test set accuracy (mean ± standard deviation). LC-U-Net (ours) LC-U-Net MC (our + explicit motion correction), SR-Heart [11]. HD: Hausdorff distance; MSD mean surface distance

		LV	LVM	RV	RVM	LA	RA
Dice	LC-U-Net	0.96 ± 0.01	0.91 ± 0.02	0.95 ± 0.01	0.63 ± 0.08	0.93 ± 0.02	0.92 ± 0.02
	LC-U-Net MC	0.96 ± 0.01	0.92 ± 0.02	0.95 ± 0.01	0.65 ± 0.09	0.93 ± 0.02	0.92 ± 0.02
	SR-Heart	0.88 ± 0.04	0.75 ± 0.08	0.86 ± 0.05	0.29 ± 0.11	0.77 ± 0.06	0.76 ± 0.08
HD (mm)	LC-U-Net	4.2 ± 1.2	4.4 ± 1.7	4.7 ± 1.6	6.8 ± 3.0	5.8 ± 1.6	6.9 ± 2.4
	LC-U-Net MC	4.0 ± 1.2	4.2 ± 1.6	4.4 ± 1.4	6.4 ± 2.8	5.6 ± 1.6	6.9 ± 2.4
	SR-Heart	18.4 ± 11.3	18.8 ± 7.4	15.8 ± 9.3	23.8 ± 7.7	23.6 ± 14.9	43.3 ± 11.8
MSD (mm)	LC-U-Net	0.6 ± 0.2	0.6 ± 0.2	0.7 ± 0.2	0.6 ± 0.2	0.9 ± 0.3	1.1 ± 0.4
	LC-U-Net MC	0.6 ± 0.2	0.5 ± 0.2	0.6 ± 0.2	0.5 ± 0.2	0.8 ± 0.3	1.0 ± 0.4
	SR-Heart	1.8 ± 0.7	1.6 ± 0.5	2.1 ± 0.8	2.2 ± 1.0	3.7 ± 1.2	3.9 ± 1.4

Table 2. CCTA test data valve accuracy (mean + standard deviation)

		Mitral	Aortic	Tricuspid	Pulmonary
HD (mm)	LC-U-Net	3.7 ± 1.2	3.4 ± 0.9	3.9 ± 1.4	4.5 ± 2.0
	LC-U-Net MC	3.7 ± 1.1	3.4 ± 0.9	3.8 ± 1.4	4.3 ± 1.7
	SR-Heart	19.4 ± 12.8	17.0 ± 10.5	26.1 ± 13.1	39.7 ± 14.8
MSD (mm)	LC-U-Net	0.8 ± 0.4	1.2 ± 0.5	1.0 ± 0.5	1.9 ± 1.4
	LC-U-Net MC	0.8 ± 0.3	1.2 ± 0.5	0.9 ± 0.4	1.8 ± 1.3
	SR-Heart	3.1 ± 1.7	3.6 ± 2.2	2.9 ± 1.9	10.2 ± 3.7

We extracted 3D representations of the valve annuli as the contacting voxels of the relevant labels, for example the mitral valve is a two-layer voxel representation merged from LV and LA. The LC-U-Net and LC-U-Net MC were similar, and both significantly better than SR-Heart ($p < 0.001$ for all). We calculated the Hausdorff distance and mean surface distance to evaluate the accuracy of the reconstructed valves (Table 2). Mitral and tricuspid valves had the lowest MSD values as SAX slices were informative as well as the LAX slices. The aortic valve reconstruction was mainly based on the outflow of LV from the 3CH view and achieved compatible results to the mitral and tricuspid valve.

The pulmonary valve was the most challenging valve to reconstruct, considering there was no direct information of its position in the LAX slices and the SAX PA label was mixed with the RV label in the sparse input volume, making it more difficult to predict.

The absolute error and signed error for volumes and masses of the predictions given by the learning-based methods and standard methods (slice summation for LV and RV, area length methods for RA and LA) are shown in Table 3. The LC-U-Net and LC-U-Net-MC were similar, and both had significantly better absolute error than SR-Heart ($p < 0.001$ for all) and standard methods ($p < 0.001$ for all).

Table 3. CCTA test volume and mass accuracy (mean ± standard deviation). STD: standard methods. Reference: dense segmentation voxel summation. AE: absolute error; SE: signed error (Reference-estimate)

		LV/ml	LVM/g	RV/ml	RVM/g	LA/ml	RA/ml
AE	LC-U-Net	2.1 ± 1.5	2.0 ± 1.6	4.1 ± 3.1	2.4 ± 1.8	3.1 ± 2.8	4.9 ± 4.2
	LC-U-Net MC	2.0 ± 1.7	1.9 ± 1.8	3.4 ± 2.4	2.5 ± 2.1	2.9 ± 2.5	5.0 ± 4.4
	SR-Heart	6.2 ± 5.1	7.7 ± 6.8	14.8 ± 9.9	5.3 ± 3.4	25.3 ± 17	21.9 ± 16
	STD	10.1 ± 7.5	10.1 ± 7.7	10.5 ± 8.4	4.0 ± 2.9	18.9 ± 28	33.9 ± 20
SE	LC-U-Net	0.4 ± 2.6	-0.1 ± 2.6	-2.7 ± 4.4	-1.4 ± 2.6	-1.0 ± 4.0	1.3 ± 6.3
	LC-U-Net MC	0.3 ± 2.6	0.9 ± 2.5	-1.6 ± 3.8	-2.3 ± 2.3	-1.3 ± 3.6	2.4 ± 6.3
	SR-Heart	1.2 ± 7.9	-5.1 ± 8.9	-13.0 ± 12.2	4.2 ± 4.8	-24.0 ± 19	-18.1 ± 21
	STD	8.0 ± 9.7	8.6 ± 9.4	-0.8 ± 13.4	3.6 ± 3.3	-4.1 ± 33.5	10.3 ± 38.1
Reference Values		138.3 ± 32	150.4 ± 35	163.5 ± 39	33.9 ± 5	83.7 ± 20	94.5 ± 24

3.3 CCTA Testing Qualitative Analysis

We evaluated our method visually by comparing the predicted dense volume to the reference volume. Examples of the best 5%, middle 5%, and worst 5% based on the average Dice score of the 6 labels are shown in Fig. 4. The worst-case example had a visually different shape, with larger atria oriented to the right and the LV having a larger curvature. The differences between the examples of the middle and best cases were not clearly visible. The shape and orientation (after transformation into cardiac coordinate system) appeared to be important for reconstruction accuracy.

4 Conclusions

In this study, we developed a deep learning method for reconstructing the whole heart shape from CMR image positions and showed higher accuracy compared with the state-of-the-art voxel-based super-resolution approach. We also evaluated the accuracy of the predicted mass and volumes, and the LC-U-Net showed better results compared to the previous method and standard clinical approaches. Our method showed high robustness against different levels of motion artefacts and can incorporate an explicit motion correction stage to further improve the results. We also included all key segments of the heart in our label maps to produce 3D representations of the valves, and the accuracy of the valve reconstruction was very promising. The most challenging task was the reconstruction of RA since only one slice (4CH) contained RA information. To accurately reconstruct the RA shape our method successfully pulled global information from other slices to give a mean Dice score of 92%.

The 3D U-Net provided whole heart reconstruction from CMR slices with good performance on both cavities and valves, considering multi-level motion artefacts. A limitation of the work is that only simulated CMR segmentations from CCTA were used for both development and evaluation of the method, and the performance of the method for different frames in cardiac cycle was not evaluated. The intention of using the cardiac coordinate system was to normalise the shape and orientation of the hearts and to improve the reconstruction accuracy, and irregular shapes and orientations were identified in the worst cases. In future work, we plan to evaluate our method using real CMR data, and a better normalisation approach to improve the results of the worst performing cases.

Acknowledgements. This research was supported by the Innovate UK (104691) London Medical Imaging & Artificial Intelligence Centre for Value Based Healthcare, the USA National Institutes of Health R01HL121754, and core funding from the Wellcome/EPSRC Centre for Medical Engineering [WT203148/Z/16/Z]. SCOT-HEART was funded by The Chief Scientist Office of the Scottish Government Health and Social Care Directorates (CZH/4/588), with supplementary awards from Edinburgh and Lothian's Health Foundation Trust and the Heart Diseases Research Fund. MCW was supported by the British Heart Foundation FS/ICRF/20/26002 and CH/09/002. SEW is supported by the British Heart Foundation (FS/20/26/34952). The authors acknowledge the support of the British Heart Foundation Centre for Research Excellence Award III (RE/18/5/34216).

References

1. Marcel, B., Banerjee, A., Grau, V.: Biventricular surface reconstruction from cine MRI contours using point completion networks. In: 2021 IEEE 18th International Symposium on Biomedical Imaging (ISBI), pp. 105–109 (2021)
2. Vincent, Chen, et al.: Effect of age and sex on fully automated deep learning assessment of left ventricular function, volumes, and contours in cardiac magnetic resonance imaging. The International Journal of Cardiovascular Imaging **37**, 3539–3547 (2021)
3. Xiang, Chen, et al.: Shape registration with learned deformations for 3D shape reconstruction from sparse and incomplete point clouds. Medical Image Analysis **74**, 102228 (2021)
4. Özgün, Ç., et al.: 3D U-Net: learning dense volumetric segmentation from sparse annotation. Medical Image Computing and Computer-Assisted Intervention. Lecture Notes in Computer Science **9901**, 424–432 (2016)
5. Thomas, Joyce, et al.: Rapid inference of personalised left-ventricular meshes by deformation-based differentiable mesh voxelization. Medical Image Analysis **79**, 102445 (2022)
6. Kingma, D.P., Ba, J.: Adam: A method for stochastic optimization. arXiv preprint arXiv: 1412.6980 (2014)
7. Mauger, C., et al.: Right ventricular shape and function: cardiovascular magnetic resonance reference morphology and biventricular risk factor morphometrics in UK Biobank. Journal of Cardiovascular Magnetic Resonance **21**, 1–13 (2019)

8. Mauger, C.A., et al.: Multi-ethnic study of atherosclerosis: relationship between left ventricular shape at cardiac MRI and 10-year outcomes. Radiology **306**, e220122 (2022)
9. Oktay, O., et al.: Anatomically constrained neural networks (ACNNs): application to cardiac image enhancement and segmentation. IEEE Transactions on Medical Imaging **37**, 384–395 (2017)
10. SCOT-Heart Investigators: Coronary CT angiography and 5-year risk of myocardial infarction. New England Journal of Medicine **379**, 924–933 (2018)
11. Wang, S., et al.: Joint motion correction and super resolution for cardiac segmentation via latent optimisation. In: International Conference on Medical Image Computing and Computer-Assisted Intervention. Lecture Notes in Computer Science **12903**, 14–24 (2021)
12. Xu, H., et al.: Ventricle surface reconstruction from cardiac MR slices using deep learning. Functional Imaging and Modeling of the Heart. Lecture Notes in Computer Science **11504**, 342–351 (2019)
13. Xu, H., et al.: Whole Heart Anatomical Refinement from CCTA Using Extrapolation and Parcellation. Functional Imaging and Modeling of the Heart. Lecture Notes in Computer Science **12738**, 63–70 (2021)

Comparison of CNN Fusion Strategies for Left Ventricle Segmentation from Multi-modal MRI

Cylia Ouadah[1], Azadeh Hadadi[2,3], Alain Lalande[1,4], and Sarah Leclerc[1](✉)

[1] Medical Image Processing Team, Institute of Molecular Chemistry of the University of Burgundy, ICMUB UMR CNRS 6302, University of Burgundy, Dijon, France
sarah.leclerc@u-bourgogne.fr
[2] Arts et Metiers Institute of Technology, LISPEN, HESAM Université, UB, Chalon-sur-Saône, France
[3] Institute for Information Management in Engineering, Karlsruhe Institute of Technology, Karlsruhe, Germany
[4] Medical Imaging Department, University Hospital of Dijon, Dijon, France

Abstract. Delayed enhancement magnetic resonance imaging (DE-MRI) is the gold standard to evaluate the state of the heart after myocardial infarction (MI). To measure the relative extent of MI and help assess the myocardium tissue viability, automatic segmentation of the myocardial border is required. In this work, we focus on the use of combined information from both kinetic MRI (CINE) and delayed enhancement MRI (DE) modalities for left ventricle segmentation, and its impact on DE-MRI. To do so, we introduce a newly constructed dataset, CINEDE, that contains MRI volumes of 124 patients for both modalities. In total, five different strategies are investigated, whose architectures are all U-Net based. In comparison with single modality segmentation, it appeared intermediate fusion architectures were more robust and more precise for the myocardium segmentation, with an average dice score for MI cases of 0.82 compared to 0.78 for single modality. We also observed that prior localization and registration always helped further improving the segmentation in multi-modal fusion.

Keywords: Cardiac imaging · Segmentation · MRI · Fusion · Convolutional Neural Networks

1 Introduction

1.1 Clinical Context

Myocardial infarction (MI) is the most common manifestation of ischemic heart disease (IHD). Colloquially called "heart attack", it can be defined as the death of the myocardial cell secondary to prolonged lack of oxygen supply (ischemia) [1]. MRI can provide information about the cardiovascular system structure and

O. Bernard et al. (Eds.): FIMH 2023, LNCS 13958, pp. 265–273, 2023.
https://doi.org/10.1007/978-3-031-35302-4_27

function [2], and in clinical routine the analysis of MRI images helps in the diagnosis, treatment and monitoring of several heart conditions including MI.

In particular, the left ventricle (LV) segmentation allows to compute cardiac parameters like systolic and diastolic volumes, ejection fraction, wall thickness, and myocardium mass when using the Kinetic magnetic resonance (CINE-MRI) modality [3], and most importantly the quantification of the MI extent from the Delayed enhancement MRI also known as late gadolinium MRI (DE-MRI) modality [4]. Accordingly, the precision of the myocardial borders detection is a determining factor to estimate the extension of MI.

1.2 Objective

Segmentation of myocardial borders is more challenging from DE-MRI than CINE-MRI because i) MI results in the appearance of a hyper signal similar to the one of the cardiac cavity, and ii) the intensity associated to healthy myocardium is hardly differentiable from that of surrounding structures such as the lung or the liver. In this work, we study the possibility of improving the segmentation of the myocardium on DE-MRI using the information of paired CINE and DE images acquired during the same cardiac exam. We focus on five fusion strategies that can be applied when using two image modalities. Comparing the outcome with the single modality performance allowed us to evaluate whether and how much the CINE information could be useful in the segmentation task of DE-MRI. Additionally, since we know that both CINE and DE segmentation are complementary in providing helpful measurements, we explored the design of multi-task approaches that provide both modality segmentations simultaneously.

2 Background

2.1 DE Segmentation

A first international challenge organized in 2012 was dedicated to DE segmentation [5], during which several automatic and semi-automatic traditional image processing and machine learning methods were proposed. Later with the EMIDEC challenge [4], more up-to-date automatic methods were introduced to the task, relying mainly on deep neural networks. In medical imaging, the most frequently used deep learning models for semantic segmentation are encoder-decoder-based architectures. The reference model for this task is the U-Net introduced in 2015 [6]. Several novel architectures that were introduced in DE segmentation challenges are based on the generic U-Net architecture, introducing additional blocks or other slight modifications [4]. For instance, attention blocks [7] are now commonly used in medical imaging segmentation to focus on the most significant features at the skip connections stage. Brahim et al. [8] proposed a novel architecture for myocardium segmentation that incorporates attention modules into U-Net. The best method from the EMIDEC challenge of 2021 [9], consisting of a cascade of two U-Nets (2D and 3D), reached an average Dice score of 87,9% and 13.01 mm for the Hausdorff distance, which is promising but remains unsatisfactory for a clinical application [4].

2.2 Deep Learning-Based Multi-modal Segmentation

Multi-modal segmentation architectures were more widely addressed in scene segmentation. Zhou et al. [10] summarized in their work the different deep learning-based fusion approaches in scene segmentation using different datasets and modalities. The first attempt of deep multi-modal fusion was a simple input fusion that consists of concatenation into multiple channels [11]. Another classical fusion category is late fusion that consists of concatenating the information at a final (or almost final) stage. Intermediate fusion and hybrid fusion represent more recent multi-modal segmentation approaches. Recent work on multi-modal fusion tackles the problem of fusion by introducing a self-supervised modal adaptation (SSMA) module [12] to dynamically adapts the fusion of multi-scale representations. This latter approach performed the best on the Cityscapes dataset [13] and ranked top 5 on other databases. As reported results tend to differ depending on the dataset and the task, it is difficult to conclude globally about optimal fusion architectures in medical imaging, however intermediate fusion is often successful. Guo's work [14] for instance showed that performing the fusion at the layers' stage performs better than the mono-modality and the output fusion strategy when using MRI, CT scans, and PET images for the segmentation of lesions in soft tissue sarcomas.

3 Materials and Methods

3.1 CINEDE Dataset

The CINEDE dataset contains the cardiac exams of 124 patients, gathered from 2020 to 2022 at the University Hospital of Dijon, France using two MRI scanners of different magnetic field strength (1.5 T - Siemens Area and 3.0 T - Siemens Trio Tim, Siemens Healthineers, Germany). It contains 124 CINE MRI volumes and the corresponding DE-MRI volumes, as well as the ground truth contours for the left ventricle cavity and surrounding myocardium (i.e. expert annotations) on both modalities. The CINEDE dataset is to our knowledge the first dataset containing both the delayed-enhanced and cine cardiac MRI annotated volumes that were acquired during a same exam. Infarcted areas are not associated to a specific label, contrary to the EMIDEC dataset, so as to learn to retrieve the full myocardium even on pathological cases. Each MRI volume is composed of several slices (6 to 11) acquired from a short-axis view of the left ventricle, from the base to the apex. In total, the number of 2D slices in the CINEDE dataset is 984 for each modality.

The CINEDE dataset is particularly heterogeneous because the volumes come from patients suffering from various cardiac pathologies such as myocarditis, dilated cardiomyopathy, hypertrophic cardiomyopathy, myocardial infarction, and some other rare pathologies. Since DE-MRI is acquired 10 min after the injection of the contrast agent (gadolinium based), the traces of this contrast agent are visible only in pathological zones. Therefore, unlike in CINE MRI, MI is translated by a hyper signal in the DE modality. This tends to make myocardium segmentation a harder task, as intensities are no longer homogeneous throughout the myocardium and because the MI appears similar to the cardiac cavity.

3.2 Data Preparation

Figure 1 summarizes the data preparation stages, which include image pre-processing, LV localization, and registration between the two modalities.

Normalization. Several pre-processing steps were applied to introduce more homogeneity in the dataset. First, image intensities were normalized to range between 0 and 255. Then, the resolution of the images was adjusted to be the same for both modalities and for all images, at the median value for the DE images (1.87 mm). Padding was applied to obtain the same size for each pair of images. Finally, we applied contrast-limited adaptive histogram equalization [15] with a clip limit of 2 and a neighborhood of 8×8.

Localization and Registration. The two modalities show shifts in the localization of LV because CINE and DE are acquired during two different breathholds. To address this problem, image registration is applied, where the moving image is the CINE modality and the fixed image is the DE modality. As the resolutions were previously even, a simple translation transformation is expected to align the LV structures on the two modalities. A region of interest (ROI) detection is first applied to help focus the registration on the left ventricle region and guide the transformation. In our study, the LV detection was obtained by fine-tuning two pre-trained Mask R-CNN [16] on our dataset, one for each modality. At inference time, only the regression network localization results were considered to obtain the LV regions from the resulting bounding boxes with an added 10% margin. After ROI detection, a pixel-based matching registration step is applied, which consists in finding the best transformation matrix (translation along x and y in our case) to align the ROI of CINE to the ROI of DE. Mattes mutual information and mean squared error were tried as metrics, both with a linear interpolator. Experiments established the latter as the best choice. For the optimization, a regular gradient descent optimizer was employed. The impact of the registration process is discussed in the results and discussion parts.

3.3 Fusion Approaches

We conducted experiments using several fusion approaches, simple well-known approaches (input and output fusions), and three more complex architectures in which the information fusion is performed at intermediate levels.

Input Fusion. Input fusion or FIUNet, is the simplest and most intuitive fusion approach, as the architecture consists of a simple concatenation of CINE and DE images at the input level of a classic 2D U-Net architecture.

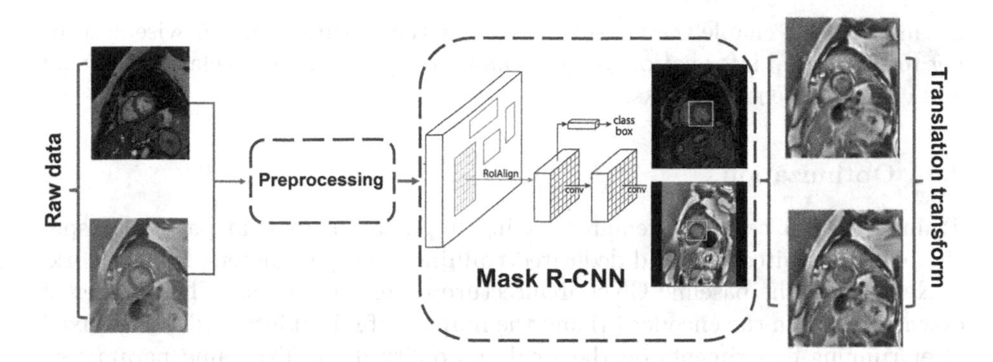

Fig. 1. Data preparation steps

Output Fusion. Two separate 2D U-Nets are trained each on one modality, with the feature maps from both modalities being concatenated at the decision level, i.e. just before the last convolution layer. We refer to this architecture by FOUNet.

Layer Fusion - LFUNet. The Layers' fusion U-Net method (LFUNet) is inspired by FuseNet [17], which is considered as an early fusion scheme since the information is fused at the features extraction level. The network consists of two identical encoders, one for the CINE modality and the other for the DE modality. Unlike FuseNet, in which the feature maps from both modalities are added to an encoder at each resolution, the feature maps resulting from the CINE encoder are concatenated to those of the DE one.

Intermediate Fusion - DualUnet. Inspired by [18], the fusion in DualUnet is performed at the level of a fusion block introduced along the skip connections. At each resolution, the two output feature maps of size N × X × Y, where N is the number of convolutional filters and (X, Y) the feature maps size, are stacked, adding an extra dimension resulting in a shape of N × 2 × X v Y. The fusion is then performed using a 2 × 1 × 1 3D convolution that produces an output shape of N × 1 × X × Y. This output is squeezed to recover the original feature maps shape. The decoder part of the network is kept identical to the original U-Net.

Self Supervised Module Adaptation fusion - SSMAUNet. In SSMAUNet, the fusion is based on the self-supervised model adaptation fusion [12]. The architecture of this model is composed of two identical U-Net encoders and one decoder. The fusion is made at the skip connections level and at the input of the decoder by using the SSMA module, which aims to model the correlation between the two modality-specific feature maps. It is composed of 2 convolutions of the stacked maps, the first one followed by a ReLu activation function and the second by a sigmoid activation to scale the dynamic range of activations between 0 and 1. The resulting output of this path is then multiplied with the original stacked feature maps to enable the network to weight the features element-wise according to the spatial information and the channel depth, similar to classical spatial and channel attention blocks.

3.4 Optimization

Training neural networks requires setting hyperparameters, in particular specific model architectures and dedicated training hyperparameters. In this work, U-Net acts as the baseline CNN architecture of all the models. The number of downsampling in the encoder (4) and the number of initial filters (64) were fixed after running experiments on the single modality data. The same parameters were then extended to the other architectures in order to maintain a unique baseline model that allows to observe the effect of fusion choices. Instead of performing exhaustive search for each architectures, several trials were run in order to decide on an ensemble of common training hyper-parameters for all models (optimizer, loss, learning rate). The final training setting was established based on the validation performance from a single data split, prior and independent to the cross-validation experiment.

We used in fine the FocalDice loss with the adaptive moment estimation (ADAM) optimizer associated to a small learning rate of 10^{-4}, and a weight decay of 10^{-5}. A batch size of 16 was kept for the simple U-Net and the input fusion, whereas a batch size of 8 was used for the rest of the approaches, as the reached performance was slightly better. All the models were trained until convergence, without a predefined limitation of epochs, using early stopping monitoring the training over the validation loss with a patience set to 10.

4 Results

4.1 Cross-Validation

We conducted 5 folds cross-validation in order to compare the fusion approaches. The first experiment consists in establishing the baseline using the conventional U-Net (single modality setting). Table 1 shows the Dice score for the cavity and the myocardium, and the Hausdorff distances (HD) computed for the endocardium and epicardium borders. Table 2 displays the results of all fusion architectures obtained from the registered images. All models were trained with the same loss and optimizer.

4.2 Registration Effect

The effect of the registration step can be observed by comparing Table 2 and Table 3. The same 5 folds cross-validation was performed without the localization and registration to observe how dependent fusion architectures are of this preprocessing step.

Table 1. Dice and HD on DE for a single modality U-Net.

	Dice	HD
LV (endo)	0.93 ± 0.05	6.65 ± 7.68
MYO (epi)	0.83 ± 0.07	8.12 ± 9.38

Table 2. Dice and HD results on DE, obtained from the registered modalities

		FIUNet	FOUNet	LFUNet	DualUNet	SSMAUNet
Dice	LV	0.93 ± 0.04	0.92 ± 0.04	0.93 ± 0.03	**0.95 ± 0.03**	**0.95 ± 0.03**
	MYO	0.81 ± 0.07	0.77 ± 0.06	0.83 ± 0.05	**0.85 ± 0.04**	0.85 ± 0.05
HD	Endo	6.19 ± 4.25	6.95 ± 6.00	6.33 ± 8.22	**4.44 ± 1.77**	4.88 ± 2.57
	Epi	7.12 ± 4.09	8.24 ± 5.52	8.42 ± 9.92	5.59 ± 2.54	**5.49 ± 2.41**

Table 3. Dice and HD results on DE, obtained from the unregistered modalities

		FIUNet	FOUNet	LFUNet	DualUNet	SSMAUNet
Dice	LV	0.92 ± 0.04	0.66 ± 0.20	0.93 ± 0.04	0.93 ± 0.03	**0.94 ± 0.03**
	MYO	0.79 ± 0.08	0.62 ± 0.09	0.81 ± 0.07	0.83 ± 0.06	**0.83 ± 0.05**
HD	Endo	6.12 ± 4.02	19.21 ± 11.94	6.07 ± 5.62	6.21 ± 6.79	**4.93 ± 2.33**
	Epi	9.40 ± 8.87	21.42 ± 11.03	7.83 ± 6.96	6.98 ± 5.88	**6.90 ± 5.51**

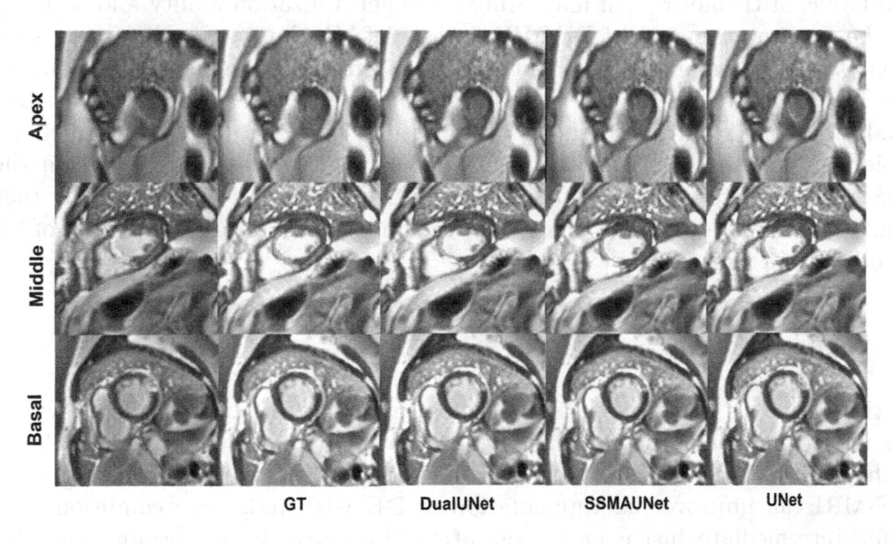

Fig. 2. Visual results at three levels of the heart, comparing the two best methods and the baseline to the ground truth on a MI case

5 Discussion

Both simple fusion strategies (i.e. input and output fusions) lead to worse performance than single modality, implying the addition of information from CINE to improve DE segmentation is not straightforward. Among all fusion methods, output fusion (FOUNet) gave the worst results, followed by input fusion (FIUNet). We therefore established that, on our application at least, the performance of output fusion was lesser than the one of input fusion, contradicting the study in [10]. This contradiction may come from a higher sensitivity of misalignment between the two modalities for simple fusion strategies. Indeed, the efficiency of both FIUNet and FIOUNet showed a significant improvement when training and providing the networks with registered data compared to non-registered images. This was especially the case for FOUNet, improving the epicardium HD from an average of 21.42 mm to 8.24 mm. It should be noted that the registration step had a positive impact for all fusion methods (not juste simple fusion strategies) but its impact was a lot less significant for more complex fusion strategies, leading to a decrease from 6.98 mm to 5.59 mm for DualUNet and from 6.90 mm to 5.49 mm for SSMAUNet for the epicardium HD.

The fusion at the encoder level LFUNet and the intermediate fusion models, DualUNet and SSMAUNet, all outperformed the simple fusion strategies and the baseline. This is in agreement with the work previously done in [14], which demonstrated that a layer-level fusion was more effective and robust than input and output fusions. The best fusion scheme was the intermediate fusion, represented by the models DualUNet and SSMAUNet. Both architectures achieved very close results with an average Dice score on the myocardium of 0.85 and an epicardium HD of 5.59 mm and 5.49 mm, respectively. These results confirm that the CINE-MRI information may bring more generalization ability and stability to the segmentation task of DE-MRI. In case of MI, the average dice score went from 0.78 (single modality U-Net) to 0.82 (bothintermediate fusion models), and the HD from 9.73 to 6.68 mm for DualUNet and 7.02 mm for SSMAUNet. Our analysis of the results also shows that most of the time (8/12), intermediate fusion approaches helped in the segmentation of hard cases (observed on the 10% worst results of single modality U-Net). From this, we can conclude that multi-modality fusion tends to help myocardium segmentation in presence of MI (see Fig. 2 for some visual examples).

6 Conclusion

In this paper, we compared several fusion strategies in encoder-decoder networks for the segmentation of the left ventricle cavity and myocardium in Delayed Enhancement MRI. Our results show that coupling CINE-MRI information to DE-MRI can improve the segmentation of DE-MRI under two conditions, i.e. using intermediate fusion and registration. Interestingly, performing intermediate fusion rather than early or later fusion seems to alleviate the need for registration, however not entirely nullifying its benefits.

References

1. Hashmi, S., et al.: Acute myocardial infarction and myocardial ischemia-reperfusion injury: a comparison. Int. J. Clin. Exp. Pathol. **8**, 8786–96 (2015)
2. Hundley, W.G., et al.: ACCF/ACR/AHA/NASCI/SCMR 2010 expert consensus document on cardiovascular magnetic resonance. Circulation **121**, 2462–2508 (2010). https://doi.org/10.1161/CIR.0b013e3181d44a8f
3. Bernard, O., et al.: Deep learning techniques for automatic MRI cardiac multi-structures segmentation and diagnosis: is the problem solved? IEEE Trans. Med. Imaging **37**(11), 2514–2525 (2018)
4. Lalande, A., et al.: Deep learning methods for automatic evaluation of delayed enhancement-MRI. The results of the EMIDEC challenge. Med. Imaging Anal. **79**, 102428 (2022). https://doi.org/10.1016/j.media.2022.102428
5. Karim, R., et al.: Evaluation of state-of-the-art segmentation algorithms for left ventricle infarct from late gadolinium enhancement MR images. Med. Image Anal. **30**, 95–107 (2016). https://doi.org/10.1016/j.media.2016.01.004
6. Ronneberger, O., Fischer, P., Brox, T.: U-net: convolutional networks for biomedical image segmentation. In: Navab, N., Hornegger, J., Wells, W.M., Frangi, A.F. (eds.) MICCAI 2015. LNCS, vol. 9351, pp. 234–241. Springer, Cham (2015). https://doi.org/10.1007/978-3-319-24574-4_28
7. Oktay, O., et al.: Attention U-net: learning where to look for the pancreas. In: Medical Imaging with Deep Learning (MIDL) Conference (2018)
8. Brahim, K., et al.: A 3D network based shape prior for automatic myocardial disease segmentation in delayed-enhancement MRI. IRBM **42**, 424–434 (2021). https://doi.org/10.1016/j.irbm.2021.02.005
9. Zhang, Y.: Cascaded convolutional neural network for automatic myocardial infarction segmentation from delayed-enhancement cardiac MRI. In: Puyol Anton, E., et al. (eds.) STACOM 2020. LNCS, vol. 12592, pp. 328–333. Springer, Cham (2021). https://doi.org/10.1007/978-3-030-68107-4_33
10. Zhou, T., et al.: A review: deep learning for medical image segmentation using multi-modality fusion. Array (2019). https://doi.org/10.1016/j.array.2019.100004
11. Couprie, C., et al.: Indoor semantic segmentation using depth information. In: ICLR Conference (2013)
12. Valada, A., et al.: Self-supervised model adaptation for multimodal semantic segmentation. Int. J. Comput. Vision **128**(5), 1239–1285 (2019). https://doi.org/10.1007/s11263-019-01188-y
13. Cityscapes dataset. https://www.cityscapes-dataset.com/
14. Guo, Z., et al.: Deep learning-based image segmentation on multimodal medical imaging. IEEE Trans. Radiat. Plasma Med. Sci. **3**, 162–169 (2019). https://doi.org/10.1109/TRPMS.2018.2890359
15. Xue, Y., et al.: Contrast limited adaptive histogram equalization. In: Graphics Gems (1994)
16. He, K., et al.: Mask R-CNN. In: IEEE International Conference on Computer Vision (ICCV) (2017)
17. Hazirbas, C., Ma, L., Domokos, C., Cremers, D.: FuseNet: incorporating depth into semantic segmentation via fusion-based CNN architecture. In: Lai, S.-H., Lepetit, V., Nishino, K., Sato, Y. (eds.) ACCV 2016. LNCS, vol. 10111, pp. 213–228. Springer, Cham (2017). https://doi.org/10.1007/978-3-319-54181-5_14
18. Xue, Y., et al.: Multi-path 2.5 dimensional convolutional neural network system for segmenting stroke lesions in brain MRI images. Neuroimage: Clin. **25**, 102118 (2020)

Long Axis Cardiac MRI Segmentation Using Anatomically-Guided UNets and Transfer Learning

Andre Von Zuben[1(✉)], Emily Whitt[2], Felipe A. C. Viana[1], and Luigi E. Perotti[1]

[1] Department of Mechanical and Aerospace Engineering, University of Central Florida, Orlando, FL, USA
avzuben@knights.ucf.edu

[2] Burnett School of Biomedical Sciences, University of Central Florida, Orlando, FL, USA

Abstract. In this work we present a machine learning model to segment long axis magnetic resonance images of the left ventricle (LV) and address the challenges encountered when, in doing so, a small training dataset is used. Our approach is based on a heart locator and an anatomically guided UNet model in which the UNet is followed by a B-Spline head to condition training. The model is developed using transfer learning, which enabled the training and testing of the proposed strategy from a small swine dataset. The segmented LV cavity and myocardium in the long axis view show good agreement with ground truth segmentations at different cardiac phases based on the Dice similarity coefficient. In addition the model provides a measure of segmentations' uncertainty, which can then be incorporated while developing LV computational models and indices of cardiac performance based on the segmented images. Finally, several challenges related to long axis, as opposed to short axis, image segmentation are highlighted, including proposed solutions.

Keywords: Cardiac image segmentation · Long axis MRI · Machine learning · Anatomically-guided UNet

1 Introduction

Subject specific computational cardiology has the potential to significantly improve diagnosis, prognosis, and therapy planning for patients affected by cardiac diseases. One of the current challenges in deploying computational models to the clinic consists in quickly generating accurate anatomical models from imaging data. In recent years, machine learning (ML) has made it possible to automatically process imaging data. For example, in the context of cardiac magnetic resonance imaging (cMRI), ML models have been extensively used to segment cine MRI data to compute indices of cardiac function.

O. Bernard et al. (Eds.): FIMH 2023, LNCS 13958, pp. 274–282, 2023.
https://doi.org/10.1007/978-3-031-35302-4_28

Currently, most of the ML effort has been directed toward segmenting the left ventricle (LV) myocardium (LVM) and cavity (LVC) from slices acquired in a short axis view. However, there could be significant benefits from building LV models based on images acquired in the long axis view. For example, in contrast to short axis based models, the base and the apex of the LV are clearly identified in the long axis views. Furthermore, fewer long axis versus short axis slices may be needed to build a full LV model, potentially shortening the acquisition time.

In this work we propose a ML model to segment the LVM and LVC from long axis (LA) images. To enable the development of this new model based on a significantly smaller and swine dataset, we apply transfer learning [9] to our previous model [13], which was developed to segment short axis (SA) images and applied to the ACDC [1] human dataset. Therefore, the current study applies transfer learning across species (human to swine) and between different image views (short axis to long axis). In the following, after describing the key features of our algorithm, we present representative results and the new challenges encountered when segmenting long axis images.

2 Methods

2.1 Long Axis Cine MRI Dataset

We tested the proposed ML segmentation model using MRI data acquired in nine (N = 9) healthy swine subjects. All animal experiments were approved by the UCLA Institutional Animal Care and Use Committee (ARC protocol # 2015-124). Subjects were imaged using a 3T MRI scanner (Prisma, Siemens) and, among other data, balanced steady state free precession (bSSFP) 2D cine MRI data was acquired in short and long axis views. Here, we focus on images acquired along 6 long axis planes approximately 30° apart. The in plane spatial image resolution is $1.18 \times 1.18\,\text{mm}^2$. For each subject and long axis location, three cardiac phases (the beginning, mid, and end of systole) are selected and segmented. In total, our dataset consists of 162 unique images. In each segmentation, the LVM and LVC are identified. Data augmentation via elastic deformation is included similarly to our previous work [12]. Via data augmentation, we generated ten variations for each image, leading to a total of 1782 images including the original images.

2.2 Deep Learning Segmentation Strategy

Model Architecture. The proposed pipeline illustrated in Fig. 1 is composed of three neural networks designed to perform the following tasks:

1. Localization of the LVC: the LVC-Locator consists in a traditional UNet.
2. Anatomically-guided feature extraction: two anatomically-guided deep neural networks (UNet$_{AG}$) generate the contours of the LVC and LVM, i.e., the LV endocardial and epicardial walls.

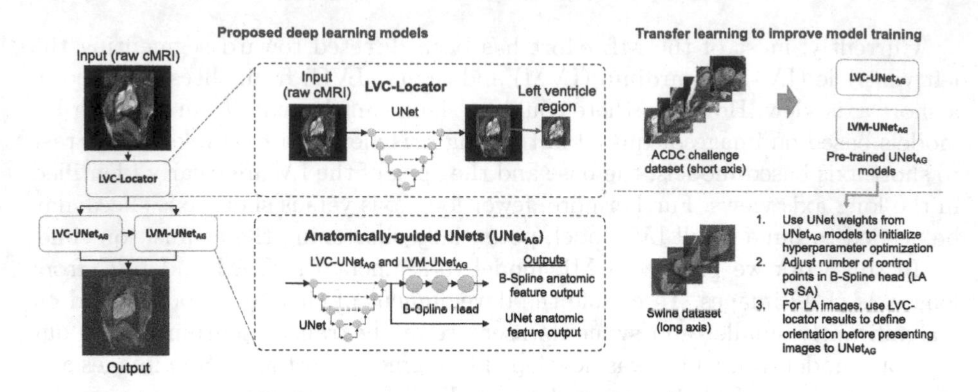

Fig. 1. Proposed pipeline of deep learning models for LV long axis segmentation.

The LVC Locator Network is a UNet [6,11] that receives a 352×352 pixels image as input and outputs each pixel's probability to be part of the heart. In training, a combination of the Dice similarity coefficient (DSC) loss and the binary cross-entropy loss is used as the loss function. The process to locate the region of the image containing the LV includes identifying the pixels with the highest and lowest horizontal and vertical locations, calculating their midpoint, and extracting a 144×144-pixel region centered at that midpoint. Before extracting the region of interest, the image is rotated using the singular value decomposition of the UNet output to align the LV long axis along the vertical or horizontal directions. This step simplifies the implementation of the B-Spline layers in the UNet$_{AG}$.

*The Anatomically-Guided UNet (*UNet$_{AG}$*) model* is a specialized implementation of the UNet proposed in [13] in which the UNet is followed by a B-Spline head. In our UNet implementation, the encoder contains four convolutional blocks (2D convolutional layers and batch normalization) and the decoder contains the four respective transposed convolutional blocks. The B-Spline head is composed of three layers: a contour detection, a B-Spline, and a perimeter-filling layer. As further detailed in [13], the B-Spline head conditions the training of the UNet by returning a smooth and constrained region with high probability of containing the region of interest. The LVM-UNet$_{AG}$ receives the LVC-UNet$_{AG}$ output as an additional channel, enhancing the overall left ventricle contours generated by the pipeline and leading to a robust wall thickness estimation. These characteristics, combined with the smooth and continuous nature of the B-Spline head (i.e., the B-Spline penalizes non-physiological protrusions, jaggedness, and discontinuities), provide lower segmentation uncertainty at the endocardial and epicardial walls. Furthermore, to increase robustness during training and prediction, four copies of the same image are passed to the UNet$_{AG}$ (each copy corresponds to the original image rotated by 0°, 90°, 180°, and 270°): the probabilities computed for each image are then averaged to compute the final segmentation.

Transfer Learning (TL). TL is adopted to implement the UNet_{AG} such that:

- The UNet portion has the same architecture presented in [13] and the weights generated during training with the ACDC dataset [1] will be used as initial guess for the training with the swine dataset. This approach will lead to faster convergence during training and robustness with respect to the reduced dataset.
- The B-Spline head for the long axis will have 45 control points, as opposed to 20 found in [13]. The larger number of control points will allow the UNet_{AG} to conform to the LA geometry, which departs from the circular-like SA geometry.

The choice of 45 B-Spline control points is motivated to promote smoothness of the segmentation while enabling flexibility to match the long axis LV geometry. We conducted studies using 20, 45, 90, 180, and 360 control points and observed only a moderate increase in DSC, which reaches a plateau at ≈45 control points. Additionally, a lower number of control points limits memory usage.

Ablation Study. Given the size of the swine dataset, a 3-fold cross-validation strategy was used to assess the quality of the resulting models. The data was split into 3 groups, each containing the images for 3 subjects: two groups are used for training and the remaining one for validation. Using this approach, we perform an ablation study by varying the initial weights used in training and the utilization of the B-Spline head:

- *Case #1:* UNet_{AG} that uses the UNet transferred from [13] to initialize the training and B-Spline head with 45 control points.
- *Case #2:* UNet_{AG} that uses the UNet transferred from [13] to initialize the training but **without** the B-Spline head.
- *Case #3:* UNet_{AG} with UNet trained from initially randomized weights (Glorot Uniform initialization function) and B-Spline head with 45 control points.
- *Case #4:* UNet_{AG} with UNet trained from initially randomized weights (Glorot Uniform initialization function) **without** the B-Spline head.

3 Results

First, we summarize in Table 1 the model size and computational cost associated with training each network of the pipeline illustrated in Fig. 1. In our study, both the model and the data fit well in the GPU memory. The LVC-Locator is a straightforward segmentation model with low accuracy requirement (other than the LVC rough segmentation and the localization of the region of interest); therefore, we do not report further results on it.

Figure 2 shows the convergence of the loss function throughout training of the LVC-UNet_{AG} for each one of the cases detailed in the "Ablation study" subsection. Similar results were observed for the LVM-UNet_{AG} (data not shown).

Table 1. Trainable parameters and training computational cost. All networks were trained on a GPU NVIDIA Tesla P100 with 16 GB of memory.

Model	Trainable parameters	Time per epoch [s]
LVC-Locator	17,660,694	66
LVC-UNet$_{AG}$	17,660,694	72
LVM-UNet$_{AG}$	17,661,126	114

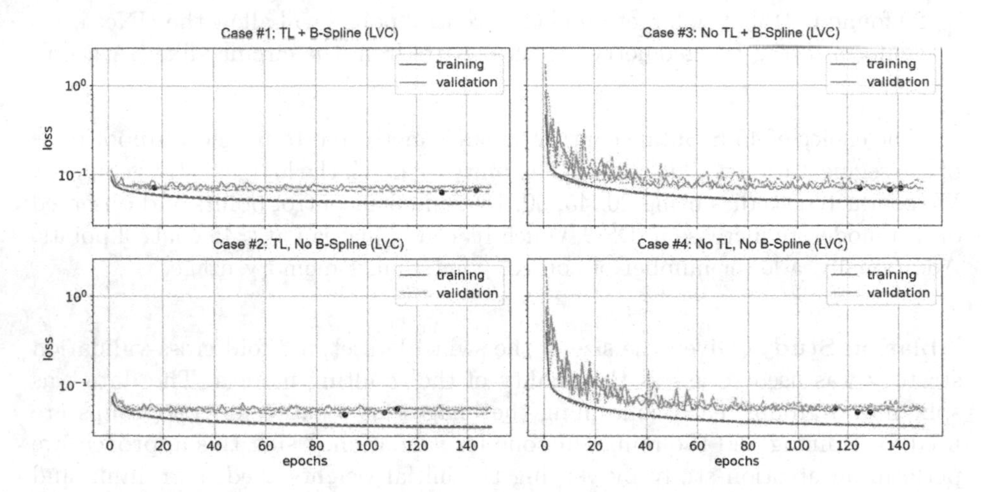

Fig. 2. Loss vs. epochs for LVC-UNet$_{AG}$ models during training and validation. Black circular markers correspond to the lowest validation loss within 150 epochs.

Next, in Table 2, we report the cross-validation performance of the model in terms of DSC for segmentations obtained at the beginning, mid, and end of systole for the swine dataset. The obtained DSC values are comparable with, although slightly lower than, the results reported for SA image segmentation computed from the large ACDC human dataset [1] in our previous work [13].

Figure 3 illustrates the image segmentations obtained with the proposed UNet$_{AG}$ transferred from [13] and using the B-Spline head (case #1). The right and left panels present results for end diastole (ED), mid systole (MS), and end systole (ES) for two subjects in the validation fold. These cases represent different levels of complexity due to contrast between the myocardium and the background, the presence of image artifacts, and different views of the ventricles, including the prominence of the right ventricle.

We conclude by illustrating two challenges specific to LA images' segmentation (Fig. 4). First, the LV myocardium prediction may be closed at the base (Fig. 4, top). Second, the left and right ventricles may be connected due to low contrast or artifacts in the image close to the basal plane (Fig. 4, bottom).

Table 2. Comparison of DSC values obtained in the ablation study and values obtained for SA segmentation using our previous model [13] and the large human ACDC dataset. TL: transfer learning; ED: end diastole; MS: mid systole; ES: end systole.

Method	LVC			LVM		
	ED	MS	ES	ED	MS	ES
TL + B-Spline	0.924	0.903	0.866	0.821	0.824	0.815
TL − No B-Spline	0.926	0.907	0.867	0.825	0.828	0.817
No TL + B-Spline	0.921	0.906	0.858	0.801	0.818	0.799
No TL − No B-Spline	0.923	0.910	0.859	0.801	0.814	0.788
ACDC Dataset (**short axis only**)	0.951	–	0.859	0.866	–	0.874

4 Discussion

In the ablation study, we proposed investigating the performance of transferring the models from [13] as well as the benefits of the B-Spline head. From the analysis of the cross-validation study reported in Tables 1 and 2 as well as Fig. 2, we conclude:

- While the cost-per-epoch is the same for all cases, transferring the models from [13] helps significantly, as the optimization starts at a significantly lower loss function value. We opted for training all the tested cases for the same number of epochs (150) for a fairer comparison. Within 150 epochs, we chose the lowest validation loss (marked with black dots in Fig. 2) to select the weights of the models. We also note that the final validation loss obtained when TL is implemented is lower than the final validation loss obtained without TL, although this difference is small. In practice, other approaches requiring fewer epochs to converge may be used. For example, a different strategy consists in terminating the optimization procedure when the validation loss stabilizes (low noise). Such alternative strategy would benefit even more from the use of transfer learning as, in this case, the validation loss noise is low from the start.
- Given that the original model [13] was trained on the much larger ACDC dataset **with** the B-Spline head, it is difficult to isolate the benefit of the B-Spline head only when using TL. Indeed, the transferred model could have benefited from the B-Spline head in both cases #1 and #2, although case #2 does not include the B-Spline head during the fine tuning using the swine dataset. In addition, the models trained without TL were based only on the small swine dataset, making it difficult to conclusively analyze the role of the B-Spline head on convergence from scratch. However, we highlight that the main benefit of the B-Spline head is not in the speed of training (although it is beneficial), but consists in regularizing the UNet output and in generating smoother contours.

Given the practical advantage during training and the DSC results shown in Table 2, we proceeded to report the predictions' results of the UNet_{AG} trans-

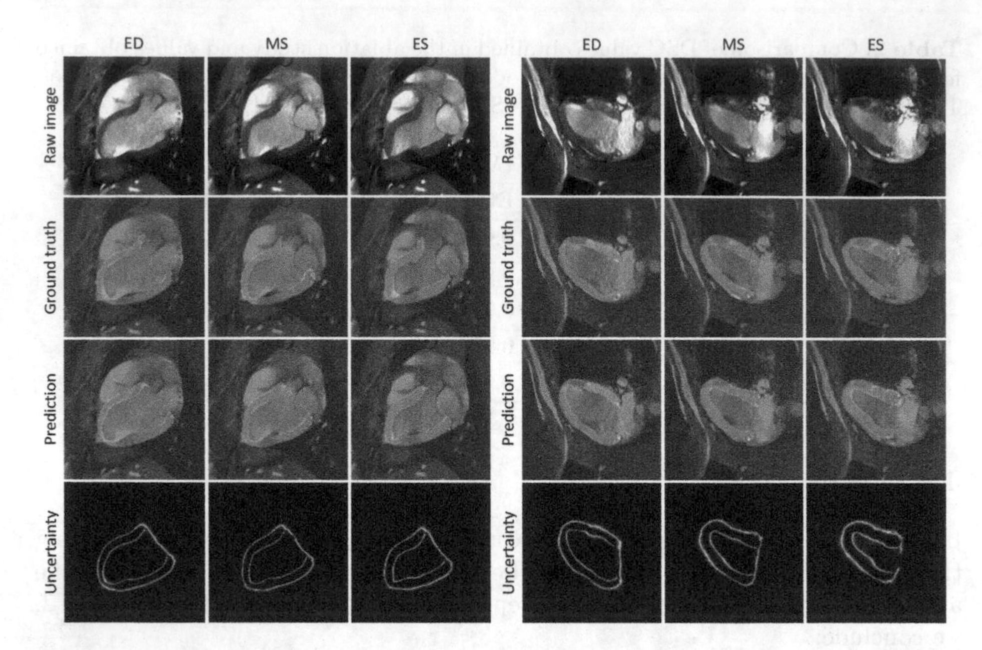

Fig. 3. Raw image, ground truth, prediction, and uncertainty results obtained in case #1. In the ground truth and prediction images, LVM and LVC are shown in green and blue, respectively. The prediction row is obtained by selecting the 50th percentiles of the UNet_{AG} probabilities. Uncertainty results illustrate the 95% prediction interval. (Color figure online)

ferred from [13] and using the B-Spline head (case #1). Overall, the obtained segmentations agree well with the ground truth regions (Fig. 3) at the beginning, mid, and end of systole. The LVM regions are smooth and anatomically correct, with reduced jaggedness due to image resolution and artifacts. The LVC predictions also present good agreement with the ground truth segmentations. The LVM and LVC predictions are paired with their uncertainty estimates (Fig. 3, bottom). Overall, the uncertainty estimates present small bands (few pixels across), which indicates prediction robustness with respect to the dataset. Larger uncertainties are present in regions with lower contrast and image artifacts, for example close to the apex and base of the LV.

Our proposed pipeline starts with an LVC-Locator network used to identify the region of interest (ROI). Different techniques were recently introduced for ROI isolation based on attention mechanisms [2,4,7], bounding boxes [3,10], or spatiotemporal statistical analysis [5]. Although these methods are both robust and lightweight, their outputs do not carry enough information for our model. Instead, we chose to employ a UNet to compute a rough preliminary segmentation of the LVC so that we can find the ROI and, at the same time, utilize the preliminary segmentation to rotate the input images and align the LV long axis along the vertical or horizontal direction.

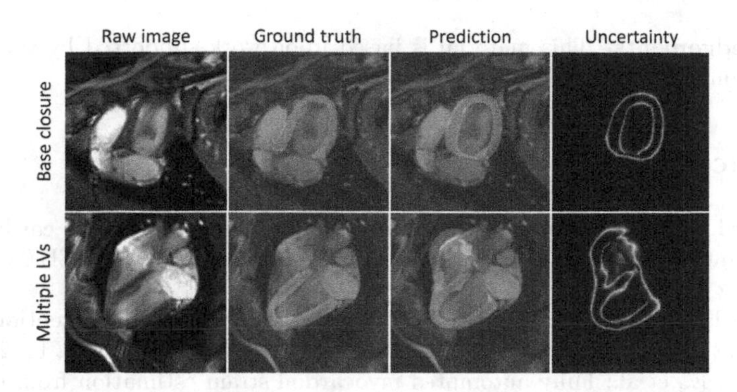

Fig. 4. Examples of specific challenges found during long axis segmentation. Top row: the LV myocardium prediction is incorrectly closed at the base. Bottom row: the left and right ventricles are incorrectly connected due to low image contrast and artifacts near the LV base.

Subsequently, the anatomically-guided deep learning models segment the LVC and LVM. The current cascade approach (the LVM-UNet_{AG} is informed by the LVC-UNet_{AG}) and the use of the B-Spline head lead to lower uncertainty regarding the location of the epicardial and endocardial outlines. An alternative approach consists in connecting two B-Spline heads to the same UNet backbone. Although losing the cascade approach would likely increase the uncertainty of epicardial and endocardial walls, this negative effect may be compensated, at least in part, by the presence of the two B-Spline heads, making this another approach to consider.

The proposed methodology also presents limitations. The first open challenge regards the possible closure of the LV base in the prediction (Fig. 4, top). A strategy to correct this problem consists in repeating the training while including a classifier to identify if the shape of the predicted LVM is physiologically correct (e.g., it has a 'U' versus an 'O' shape). Another strategy to reinforce a physiologically correct LV anatomy could follow the work presented in [8]. Additionally, one could segment the LVM first and then use the LVM segmentation as starting point to estimate the LVC. As this strategy reverses the current order (the current pipeline uses the LVC segmentation as an additional input channel for the LVM-UNet_{AG}), transfer learning could not be directly applied. However, predicting the LVC from the LVM could eliminate the over-prediction at the LVM base. A second challenge with our current approach consists in occasionally predicting multiple LV chambers (Fig. 4, bottom). As multiple chambers are usually predicted only in one cardiac phase, presenting the UNet_{AG} with images from different phases (e.g., ED, MS, and ES) could benefit the predictions. However, this solution may not be feasible with small datasets as multiple images could not be segmented independently, effectively reducing further the size of the dataset. Several of these approaches are currently being implemented to improve the proposed model for LA MRI segmentation.

Acknowledgements. This material is based upon work supported by the National Science Foundation under Grant Number 2205043.

References

1. Bernard, O., et al.: Deep learning techniques for automatic MRI cardiac multi-structures segmentation and diagnosis: is the problem solved? IEEE Trans. Med. Imaging **37**(11), 2514–2525 (2018)
2. Cheng, H., Lu, J., Luo, M., Liu, W., Zhang, K.: PTANet: triple attention network for point cloud semantic segmentation. Eng. Appl. Artif. Intell. **102**, 104239 (2021)
3. Ferdian, E., et al.: Fully automated myocardial strain estimation from cardiovascular MRI-tagged images using a deep learning framework in the UK Biobank. Radiol. Cardiothorac. Imaging **2**(1), e190032 (2020)
4. Islam, M., Vibashan, V.S., Jose, V.J.M., Wijethilake, N., Utkarsh, U., Ren, H.: Brain tumor segmentation and survival prediction using 3D attention UNet. In: Crimi, A., Bakas, S. (eds.) BrainLes 2019. LNCS, vol. 11992, pp. 262–272. Springer, Cham (2020). https://doi.org/10.1007/978-3-030-46640-4_25
5. Khened, M., Kollerathu, V.A., Krishnamurthi, G.: Fully convolutional multi-scale residual DenseNets for cardiac segmentation and automated cardiac diagnosis using ensemble of classifiers. Med. Image Anal. **51**, 21–45 (2019)
6. Litjens, G., et al.: A survey on deep learning in medical image analysis. Med. Image Anal. **42**, 60–88 (2017)
7. Oktay, O., et al.: Attention U-Net: learning where to look for the pancreas. arXiv preprint arXiv:1804.03999 (2018)
8. Popescu, D.M., et al.: Anatomically informed deep learning on contrast-enhanced cardiac magnetic resonance imaging for scar segmentation and clinical feature extraction. Cardiovasc. Digit. Health J. **3**(1), 2–13 (2022)
9. Raghu, M., Zhang, C., Kleinberg, J., Bengio, S.: Transfusion: understanding transfer learning for medical imaging. In: Wallach, H., Larochelle, H., Beygelzimer, A., d' Alché-Buc, F., Fox, E., Garnett, R. (eds.) Advances in Neural Information Processing Systems. vol. 32. Curran Associates, Inc. (2019)
10. Redmon, J., Divvala, S., Girshick, R., Farhadi, A.: You only look once: unified, real-time object detection. In: Proceedings of the IEEE Conference on Computer Vision and Pattern Recognition, pp. 779–788 (2016)
11. Ronneberger, O., Fischer, P., Brox, T.: U-Net: convolutional networks for biomedical image segmentation. In: Navab, N., Hornegger, J., Wells, W.M., Frangi, A.F. (eds.) MICCAI 2015. LNCS, vol. 9351, pp. 234–241. Springer, Cham (2015). https://doi.org/10.1007/978-3-319-24574-4_28
12. Von Zuben, A., Heckman, K., Viana, F.A.C., Perotti, L.E.: A multi-step machine learning approach for short axis MR images segmentation. In: Ennis, D.B., Perotti, L.E., Wang, V.Y. (eds.) FIMH 2021. LNCS, vol. 12738, pp. 122–133. Springer, Cham (2021). https://doi.org/10.1007/978-3-030-78710-3_13
13. Zuben, A.V., Perotti, L.E., Viana, F.A.C.: Anatomically-guided deep learning for left ventricle geometry generation with uncertainty quantification based on short-axis MR images. Eng. Appl. Artif. Intell. **121**, 106012 (2023)

Deep Active Learning for Left Ventricle Segmentation in Echocardiography

Eman Alajrami[1(✉)], Preshen Naidoo[1], Jevgeni Jevsikov[1], Elisabeth Lane[2], Jamie Pordoy[1], Nasim Dadashi Serej[1], Neda Azarmehr[1], Fateme Dinmohammadi[1], Matthew J. Shun-shin[2], Darrel P. Francis[2], and Massoud Zolgharni[1,2]

[1] School of Computing and Engineering, University of West London, London, UK
alajrami.eman@gmail.com

[2] National Heart and Lung Institute, Imperial College London, London, UK

Abstract. The training of advanced deep learning algorithms for medical image interpretation requires precisely annotated datasets, which is laborious and expensive. Therefore, this research investigates state-of-the-art active learning methods for utilising limited annotations when performing automated left ventricle segmentation in echocardiography. Our experiments reveal that the performance of different sampling strategies varies between datasets from the same domain. Further, an optimised method for representativeness sampling is introduced, combining images from feature-based outliers to the most representative samples for label acquisition. The proposed method significantly outperforms the current literature and demonstrates convergence with minimal annotations. We demonstrate that careful selection of images can reduce the number of images needed to be annotated by up to 70%. This research can therefore present a cost-effective approach to handling datasets with limited expert annotations in echocardiography.

Keywords: Echocardiography · Deep learning · Active learning

1 Introduction

Echocardiography (Echo) is one of the preferred modalities for evaluating Cardiovascular diseases [19]. Manual diagnosis is performed by trained clinicians in a time-consuming and error-prone process due to intra- and inter-observer variability. Therefore, accurate automated echo interpretation is highly desirable to improve the quality and reproducibility of crucial diagnostic measurements while freeing up valuable clinicians' time. Research has focused on the application of state-of-the-art deep learning (DL) models for automated medical image analysis [12]. One of the most popular networks is U-Net [17], introduced for medical image segmentation; it is fast, outperforms alternative architectures, and is commonly applied to left ventricle (LV) segmentation tasks [1]. Despite significant advancements, DL models require large datasets with accompanying high-quality labels for training. However, image annotation is costly and

O. Bernard et al. (Eds.): FIMH 2023, LNCS 13958, pp. 283–291, 2023.
https://doi.org/10.1007/978-3-031-35302-4_29

time-consuming, particularly in the medical domain [11]. Deep active learning (DeepAL) is a promising approach for maximising the utility of limited annotations. DeepAL techniques reduce the cost and time of annotating the entire dataset and improve DL model accuracy [11,16]. The most common DeepAL approaches are uncertainty sampling [5], representativeness and diversity sampling [14], and hybrid methods [9]. Uncertainty sampling is the favoured method for DeepAL, during which the model's predictions are used to get an uncertainty score for each image [3]. For instance, in image segmentation, the least confident (pixel-wise) samples are assumed to be the most valuable samples to annotate [6,16]. Similarly, Monte Carlo dropout (MCD) at inference time is proposed as a Bayesian approximation to represent model uncertainty [4]. This approach was applied for image classification [5] using uncertainty policies such as Bayesian active learning with disagreement (BALD) [6,8]. An additional AL annotation policy is representativeness sampling which aims to reduce the annotation cost by selecting diverse samples representing the entire dataset [13,14]. To date, most DeepAL research has focussed on classification tasks. However, a small number of published studies tackle image segmentation [3]. Therefore, this study investigates the efficiency of AL strategies for cardiac image segmentation, proposes a new method for uncertainty sampling, and optimises representativeness sampling.

2 Datasets and Methodology

2.1 Patient Datasets

Two echo datasets are used for LV segmentation tasks. **Unity**: our own dataset which was retrieved from Imperial College Healthcare NHS Trust's echocardiogram database. The images were obtained using ultrasound equipment from GE and Philips manufacturers, and the acquisition process was ethically approved by the Health Regulatory Agency (Integrated Research Application System identifier 243023). It contains 2800 images sampled from different time-points in the cardiac cycle and labelled by a pool of experts using an in-house online labelling platform (https://unityimaging.net). After cleaning, 2094 images of this dataset were used for model developments (i.e., training and validation). **CAMUS**: a publicly available dataset containing 450 images. These 450 images are obtained from 450 patients, and one apical four-chamber (A4C) end-diastolic (ED) annotated image is selected from each patient [10]. Both datasets are split into 70% for training, 15% for validation and 15% for testing.

2.2 Network Architecture

The U-Net architecture with MCD is used for LV segmentation (see Fig. 1). It is customised by adding a dropout layer after each encoder and decoder block with a dropout probability of 0.1, and a dropout layer with a probability of 0.25 is added at the centre. The graphic was created with the PlotNeuralNet tool (https://github.com/HarisIqbal88/PlotNeuralNet).

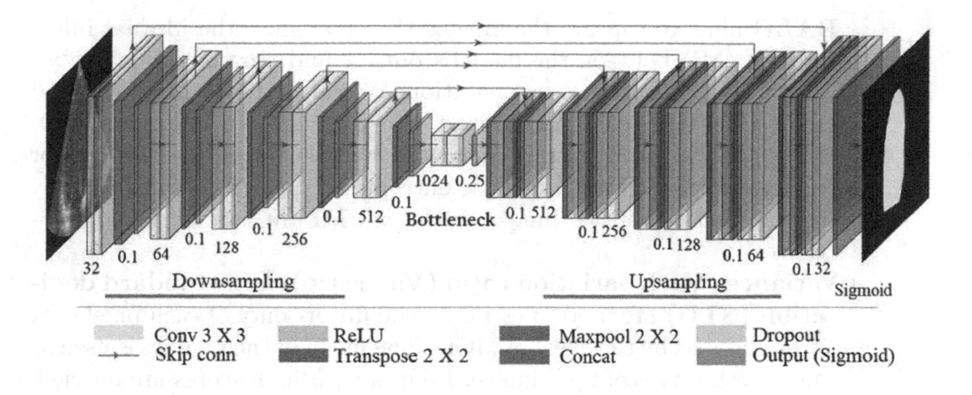

Fig. 1. A Customised U-Net architecture

2.3 Methodology

A standard AL methodology is applied throughout, encompassing four steps: training the U-Net model on the initial annotated data (L), calculating the model's uncertainty or representativeness scores on the unlabelled pool (U), selecting the top-ranked batch of images (K) to obtain their labels from oracle, adding them to L and removing them from U, and finally retraining the model on the updated L. These steps are repeated until the optimal number of AL iterations is reached.

Random sampling and a variety of different selective sampling approaches were used for selecting the next batch of images from the unlabelled pool:

- **Random** the baseline technique which randomly selects the next batch of images from the pool of dataset.
- **Uncertainty scoring**
  - **Classification Uncertainty (Pixel-wise)** known as the least confidence, in which the closer the pixel probabilities are to 0.5, the more uncertain the model is.
  - **Predictive entropy** known as Shannon Entropy [5], is adopted to our segmentation model for binary pixel classification. We derived the probabilities for both pixel classes to calculate the entropy score for each pixel in an image and sum the pixels' entropies. Images with maximum entropy are then selected for annotation.
  - **Ensembles techniques** use MCD to activate the dropped neurons at inference time to have ensembles of the DL model, so different predictions of each image in U are obtained over T runs with T= 100. Next, uncertainty measures are used, including:
    - ∗ **MCD-Entropy** by which we predict every image in U over T times and average the predictions to calculate the entropy of mean predictions. Images with high scores are selected for annotations.

  * **BALD** aims to capture the images that maximise the Mutual Information (MI) between the model's output and parameters [8], so we adopt BALD to binary segmentation. We calculate Entropy1 (H) and Entropy2 (E) over T predictions; the first one is similar to MCD-Entropy; the second one is computed by calculating the entropy for each prediction, averaging these entropies. Then, MI is calculated by subtracting E from H. Images with high MI are selected for annotations.
  * **Variance (Var), variation ratio (Var-ratio), and standard deviation (STD)** are used to estimate the uncertainty of each pixel over T predictions by the effect of MCD. The mean of their values is used as an uncertainty score per image. Images with high scores are queried.
  * **Coefficient of variation (Coef-var)** is proposed as a new measure of uncertainty with MCD ensembles to be added to the selection policies in the heuristics [2]. It is calculated for each pixel over T predictions. The mean of pixels' Coef-var will be computed, and samples with maximum Coef-var will be labelled.
- **Representativeness sampling (RS)** is applied in which samples from U represent others while outliers for L are selected for labelling [13]. First, we extracted the features of the images in L and U using two different methods: VGG16 [18] and Gray Level Run Length Matrix (GLRM) [15]. Then, cosine similarity is used to measure the samples' similarity [7]. For each instance x in U, we calculate the maximum similarity score between x and the remaining samples and the maximum similarity score between x and L data points. Then, for every x, the representativeness score is calculated as follows:
  Representativeness-score (x) = max-sim(x, U) - max-sim (x, L) [13]. The unlabelled images are ranked in descending order by representativeness score, and the top K samples are selected for annotations.
  A drawback of the RS method is that it only focuses on selecting highly representative samples in U, leaving others with lower similarities out of selection.
- **Optimised Representativeness sampling (ORS)** is proposed to overcome the mentioned drawback of RS by including feature-based outliers, images with lower similarities with other U samples and not seen previously by the model, to K samples for annotation. Training feature-based outliers helps with domain adaption [13]. Therefore, we divided U into two lists: list 1, the most representative images, and list 2, feature-based outliers. We followed the same steps in the previous section to get the most representative images sorted in descending order. For feature-based outliers, we picked the images with similarities to U samples below a threshold, the mean similarity of U, which can be tuned. Then, we computed their representativeness score and ranked them in ascending order. Finally, to get the K samples, we used both lists as follows:

K1 = 0.75% of K from list 1, K2 = 0.25% of K from list 2, and then K = K1 + K2. This split can be tuned; we tried different splits and achieved optimal results with this approach.

2.4 Implementation and Training Settings

Tensorflow and Keras frameworks are used for the development of DL models. The training was conducted using an Nvidia RTX3090 GPU. The U-Net was trained using binary cross-entropy loss and ADAM optimiser with a learning rate of 0.0001 for 200 epochs. Images(and corresponding ground-truth) were resized to 512×512, and a fixed batch size of 8 was applied. For the CAMUS dataset, we selected 10% (35 images) of the initial training data as L, and U was the remaining 90%. For the Unity dataset, we chose 4% (82 images) as the initial L, and the remaining was the U dataset. For each dataset, all experiments using different approaches for selective sampling started with the same model, trained on the initial labels in L. The K samples in the next batch were 5% and 1% of the total CAMUS and Unity datasets, respectively.

The size selection of the initially labelled data and the K samples are decided based on various experiments. We achieved better performance with these percentages of initial training and sampling size.

2.5 Evaluation Metrics

We evaluated the model after every AL iteration using the Dice-Coefficient (DC) metric on the testing dataset [6]. Then the mean of Dice scores of all images is calculated to present the model's accuracy. Each AL selection strategy is trained three times, and the average of the DC at each AL iteration is computed.

3 Results and Discussions

3.1 Uncertainty Sampling Results

We trained the whole annotated datasets using the U-Net model described in Sect. 2.2 to evaluate its maximum performance on the fully labelled data and compared it to the performance of AL methods. Using 350 images to train the CAMUS dataset, our model achieved a mean DC of 0.941 on the testing dataset; however, on the Unity dataset, we trained the whole annotated data, and it approached a mean DC of 0.913 on the testing dataset.

Based on our experiments, the predictive entropy is the best uncertainty strategy compared to others on both datasets at the initial AL stages, while Coef-Var performs well on the Unity dataset (see Fig. 2). The predictive entropy achieved 97.7% of the maximum accuracy achievable using the entire CAMUS dataset using only 25% of the annotations, while other approaches required approximately 35% or more to reach similar performance. This would mean a reduction of 10% in the cost of labelling images. Coef-Var achieved 98% of

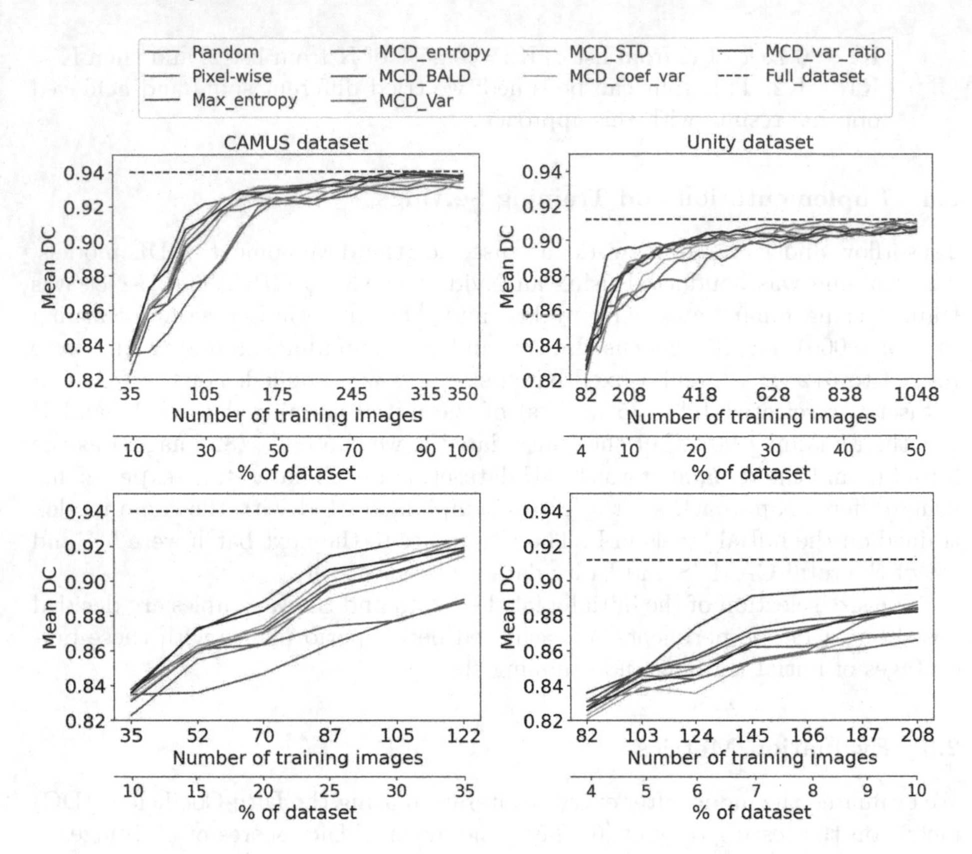

Fig. 2. Performance profiles for various uncertainty selection strategies at each active learning iteration; lower panel shows a magnified version of early stages presented at the upper panel. Black broken lines indicate the maximum achievable accuracy if the entire dataset and annotations are used for training.

maximum performance using 40% of the labels in the CAMUS dataset, outperforming random, pixel-wise, and MCD-Entropy methods.

For the Unity dataset, the predictive entropy significantly outperformed all other methods from the early stages of AL achieving 98.3% and 98.6% of the maximum performance using only 7% and 14% of the labels, respectively. After using 27% of the annotations, pixel-wise selection approach almost converged to the best performance, outperforming all other uncertainty techniques.

3.2 Representativeness Sampling Results

Figure 3 illustrates performance plots for representativeness sampling strategies. Evidently, our proposed optimised method (ORS) outperformed the existing approaches (RS) for both datasets, improving the performance particularly at early stages. This is likely due to selection of images with various distributions in the dataset.

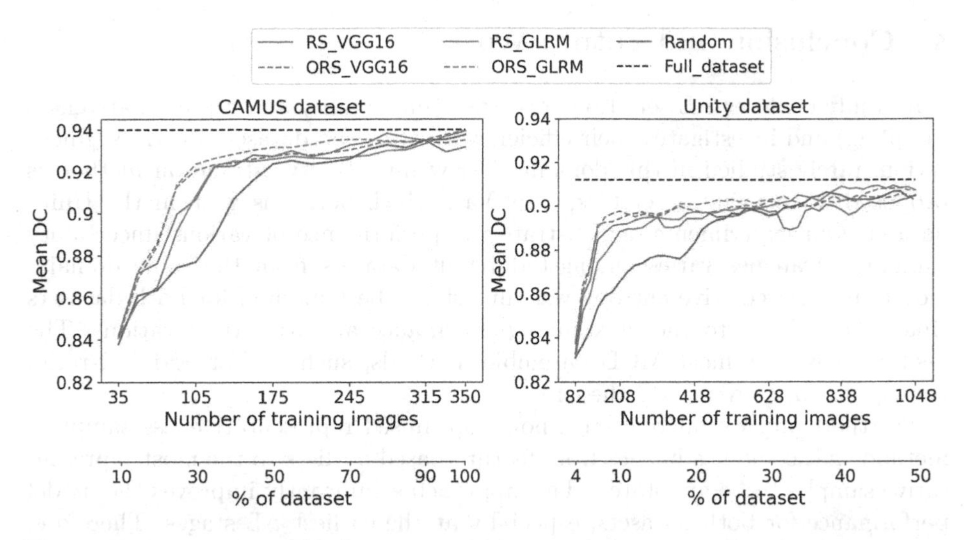

Fig. 3. Performance profiles for various representativeness sampling strategies versus optimised. Black broken lines indicate the maximum achievable accuracy if the entire dataset and annotations are used for training.

Our proposed ORS method achieved 98% of the maximum achievable performance, using only 25% of labels in the CAMUS dataset, outperforming the RS method, which required 35% of the annotations to approach the same performance level. In addition, when GLRM is used, our ORS method achieved almost the same full dataset performance using 60% of annotations, while RS method needed 90% of labels to reach that performance. However, with VGG16, the ORS method outperformed the RS method at the early AL stages until 30% of labels; after that, both methods performed almost similarly.

For the Unity dataset, the proposed ORS method achieved 98.6% of the performance of the entire dataset using only 9% of annotations with GLRM, and it converged to that performance after using 15% of labels reducing the labelling effort significantly by 24% compared to the RS method, which required 33% of annotations to approach that performance. The proposed method, ORS, using VGG16, achieved 98.5% of the whole dataset performance with only 12% of the annotations.

The RS using the GLRM reached its peak performance once 40% of the Unity dataset had been utilized. However, in the case of AL, the primary objective is to attain maximum performance during the early stages of the iterative learning process, as this facilitates the deployment of the model earlier without the need for a huge amount of annotations. Consequently, ORS methods were successful in achieving this objective during the initial phases of AL. Generally, most methods converged to the maximum performance of the whole dataset before training on 50% of the Unity dataset due to its size, and the AL selection methods focus on obtaining the most informative images for training.

4 Conclusion and Future Work

This study explores AL selection strategies (uncertainty and representativeness sampling) and investigates their efficiency on two echo datasets for LV segmentation, rarely studied in this domain. A new uncertainty estimation method is proposed when using ensembles, Coef-Var, which performs well on the Unity dataset. Our experiments demonstrate the performance of various uncertainty sampling strategies varies amongst different datasets from the same domain. Additionally, predictive entropy was almost the best method for both datasets since it converged to the maximum performance at early AL iterations. The results show that most MCD ensemble methods, such as Var and Var-ratio, usually do not converge till the end.

Furthermore, we introduced a novel optimised representativeness sampling method that combines images from feature-based outliers to the most representative samples to be annotated. This approach significantly improves the model performance for both datasets, especially at the earliest AL stages. Therefore, we recommend this policy in AL's early iterations and when it is expensive to get large number of annotations. In future work, we will combine our optimised representativeness method with uncertainty techniques.

Acknowledgement. This work was supported in part by the British Heart Foundation, UK (Grant no. RG/F/22/110059). E Alajrami is supported by the Vice Chancellor's Scholarship at the University of West London. In addition, we would like to thank the Schlumberger Foundation (Faculty for the Future) and the Funds for Women Graduates (FfWG) for their funding and support for E Alajrami. FfWG provide grants, bursaries and fellowships to women graduates to help with living expenses.

References

1. Azarmehr, N., Ye, X., Sacchi, S., Howard, J.P., Francis, D.P., Zolgharni, M.: Segmentation of left ventricle in 2D echocardiography using deep learning. In: Zheng, Y., Williams, B.M., Chen, K. (eds.) MIUA 2019. CCIS, vol. 1065, pp. 497–504. Springer, Cham (2020). https://doi.org/10.1007/978-3-030-39343-4_43
2. Bindu, K., Morusupalli, R., Dey, N., Rao, C.: Coefficient of Variation and Machine Learning Applications. CRC Press (2019). https://doi.org/10.1201/9780429296185
3. Budd, S., Robinson, E.C., Kainz, B.: A survey on active learning and human-in-the-loop deep learning for medical image analysis. Med. Image Anal. **71**, 102062 (2021). https://doi.org/10.1016/j.media.2021.102062
4. Gal, Y., Ghahramani, Z.: Dropout as a Bayesian approximation: representing model uncertainty in deep learning. In: Proceedings of the 33rd International Conference on Machine Learning (2015)
5. Gal, Y., Islam, R., Ghahramani, Z.: Deep Bayesian active learning with image data. arXiv (2017). https://doi.org/10.48550/arXiv.1703.02910
6. Gorriz, M., Carlier, A., Faure, E., Giro-i Nieto, X.: Cost-effective active learning for melanoma segmentation. arXiv (2017). https://doi.org/10.48550/arXiv.1711.09168

7. Gupta, V., Sachdeva, S., Dohare, N.: Deep similarity learning for disease prediction. In: Piuri, V., Raj, S., Genovese, A., Srivastava, R. (eds.) Trends in Deep Learning Methodologies, pp. 183–206. Hybrid Computational Intelligence for Pattern Analysis, Academic Press (2021). https://doi.org/10.1016/B978-0-12-822226-3.00008-8
8. Houlsby, N., Huszár, F., Ghahramani, Z., Lengyel, M.: Bayesian active learning for classification and preference learning. arXiv (2011). https://doi.org/10.48550/arXiv.1112.5745
9. Huang, S.J., Jin, R., Zhou, Z.H.: Active learning by querying informative and representative examples. IEEE Trans. Pattern Anal. Mach. Intell. **36**(10), 1936–1949 (2014). https://doi.org/10.1109/TPAMI.2014.2307881
10. Leclerc, S., et al.: Deep learning for segmentation using an open large-scale dataset in 2D echocardiography. IEEE Trans. Med. Imaging **38**(9), 2198–2210 (2019)
11. Liu, X., Song, L., Liu, S., Zhang, Y.: A review of deep-learning-based medical image segmentation methods. Sustainability **13**(3), 1224 (2021). https://doi.org/10.3390/su13031224
12. Lundervold, A., Lundervold, A.: An overview of deep learning in medical imaging focusing on MRI. Zeitschrift für Medizinische Physik **29** (2018). https://doi.org/10.1016/j.zemedi.2018.11.002
13. Monarch, R.M.: Human-in-the-loop machine learning: active learning and annotation for human-centered AI. Manning (2021)
14. Nguyen, H., Smeulders, A.: Active learning using pre-clustering. In: ICML (2004). https://doi.org/10.1145/1015330.1015349
15. Ozturk, S., Akdemir, B.: Application of feature extraction and classification methods for histopathological image using GLCM, LBP, LBGLCM, GLRLM and SFTA. Procedia Comput. Sci. **132**, 40–46 (2018)
16. Pengzhen, R., Xiao, Y., Chang, X., Huang, P.Y., Chen, X., Wang, X.: A survey of deep active learning. arXiv (2020). https://doi.org/10.48550/arXiv.2009.00236
17. Ronneberger, O., Fischer, P., Brox, T.: U-Net: convolutional networks for biomedical image segmentation. In: Navab, N., Hornegger, J., Wells, W.M., Frangi, A.F. (eds.) MICCAI 2015. LNCS, vol. 9351, pp. 234–241. Springer, Cham (2015). https://doi.org/10.1007/978-3-319-24574-4_28
18. Simonyan, K., Zisserman, A.: Very deep convolutional networks for large-scale image recognition. CoRR abs/1409.1556 (2014). http://arxiv.org/abs/1409.1556
19. Wang, S., Hu, P.: Deep learning for automated echocardiogram analysis. J. Stud. Res. **11** (2023). https://doi.org/10.47611/jsrhs.v11i3.3085

Right Ventricular Volume Prediction by Feature Tokenizer Transformer-Based Regression of 2D Echocardiography Small-Scale Tabular Data

Tuan A. Bohoran[1(✉)], Polydoros N. Kampaktsis[2], Laura McLaughlin[2], Jay Leb[2], Serafeim Moustakidis[3], Gerry P. McCann[4], and Archontis Giannakidis[1]

[1] School of Science and Technology, Nottingham Trent University, Nottingham, UK
tuan.bohoran@ntu.ac.uk

[2] Division of Cardiology, Columbia University Irving Medical Center, New York, NY, USA

[3] AiDEAS, Tallinn, Estonia

[4] Department of Cardiovascular Sciences, University of Leicester and the NIHR Leicester Biomedical Research Centre, Glenfield Hospital, Leicester, UK

Abstract. Quantitative evaluation of right ventricular (RV) volumes is of paramount importance in many cardiovascular conditions and is best performed by cardiovascular magnetic resonance imaging (CMR). However, CMR scanners are scarce, costly, and lack portability. Two-dimensional transthoracic echocardiography (2DE) allows for widely available, low cost and bedside evaluation of RV size and function. 2DE-based quantitative RV analysis is nevertheless restricted by the lack of accurate models of the complex RV shape. In this paper, we propose to calculate the RV end-diastolic (ED) and end-systolic (ES) volume by using an attention-based deep learning (DL) model on tabular data. Morphological measurements (areas) from eight standardized 2DE views are used as input to the regression model along with age, cardiac phase and gender information. The proposed architecture comprises a feature tokenizer module to transform all features (categorical and numerical) to embeddings, before applying a stack of Transformer layers. Our pipeline is trained and tested on 50 ED and 50 ES RV volumes (100 in total). The predicted volumes are compared to reference CMR values. Our method achieved impressive performance (R^2=0.975) on this relatively small-scale dataset, while it outperformed other state-of-the-art methods. The RV function evaluation using tabular Transformers shows promise. This work questions the superiority of tree-based ensemble models over DL-based solutions for tabular data in the context of functional imaging of the heart. Our pipeline is also appealing as it may allow building multi-modal cardiovascular frameworks, where only one part of the data is tabular, and other parts include images and text data.

T. A. Bohoran and P. N. Kampaktsis—Authors contributed equally.

O. Bernard et al. (Eds.): FIMH 2023, LNCS 13958, pp. 292–300, 2023.
https://doi.org/10.1007/978-3-031-35302-4_30

Keywords: Ultrasound · Right ventricular volume · Regression · Transformers · Tabular data

1 Introduction

Assessment of right ventricular (RV) volumes via non-invasive imaging is the cornerstone of RV size and function evaluation, which in turn is required for the diagnosis and management of numerous cardiovascular diseases [1]. Cardiac magnetic resonance imaging (CMR) is considered the gold standard method for quantifying RV volumes [2]. However, in UK there are only 6.1 magnetic resonance imaging scanners per million people, while at the same time over 1 in 10 people are affected by heart and circulatory diseases alone [3,4]. As a result, at present hospitals urgently need to deal with a huge backlog of scans for heart conditions, and the number of cardiac patients waiting for CMR imaging studies has been ballooning [5]. The long waits for imaging (>6 weeks) could lead to more deaths or other complications which would escalate the pressure on hospitals. In addition, it is costly to replace the aging CMR scanners. There is an urgent need to find new and more cost-effective methods for the accurate measurement of RV volumes.

An alternative and commonly used method for RV evaluation is transthoracic two-dimensional echocardiography (2DE). Compared to CMR, 2DE is by far more widely available in clinical practice and has minimal to no risk. In 2018, 1731 CMR scans were performed per million people in UK, as opposed to ~45K 2DE tests per million people [3,6]. In addition, 2DE offers rapid inexpensive acquisition, and its equipment is highly portable. In fact, hand-held 2DE devices permit imaging in any healthcare setting, including emergency bedside monitoring. However, the major drawback of 2DE-based estimation of RV volumes currently is low accuracy due to the lack of accurate geometric models to approximate the complex RV shape [7]. Three-dimensional echocardiography (3DE) could theoretically bridge this gap. Nevertheless, it requires special transducers, hardware and software, advanced sonographic training and it's often limited by poor image quality [8]. Furthermore, it is also associated with a higher financial and computational cost, when compared to 2DE.

This work proposes a novel deep learning (DL) pipeline for accurately predicting RV volumes by relying on conventional 2D morphological (area) measurements from eight standardized 2DE views and also on gender, cardiac phase, and age information. The main component of our pipeline is a tabular feature tokenizer Transformer [9]. It is yet unexplored how tabular attention-based DL models would perform in the context of functional imaging of the heart. Our pipeline is trained and tested on a relatively small-scale dataset, comprising 50 end-diastolic (ED) and 50 end-systolic (ES) RV volumes (100 in total). The RV volumes, predicted by our pipeline, are compared to reference values obtained from CMR. Our method is also compared to other state-of-the-art (SOTA) methods for tabular data-based regression.

2 Materials and Methods

2.1 Study Population

The study population was a retrospective cohort of 50 adult patients for which 2DE and CMR were obtained within 30 days from each other as part of routine cardiac care. The 30-day interval between echocardiography and CMR has been previously used in clinical research [13]. We excluded patients with prior cardiac surgery of the right ventricle, tricuspid or pulmonic valve repair or replacement, cyanotic congenital heart disease, L-transposition of the great arteries, more than small pericardial effusion, cardiac tamponade, atrial fibrillation at the time of either study, severe RV dysfunction or inadequate imaging quality. The study was approved by the Columbia University Irving Medical Center Institutional Review Board and the Nottingham Trent University Ethics Committee.

2.2 2D Ultrasound and CMR Acquisition

Echocardiographic images were obtained using 4 types of Philips Ultrasound systems, namely iE33 x-Matrix, Epic CVx, Epic 7C and Affiniti 70C. Images were obtained by experienced sonographers and according to the standard institutional scanning protocol of the Columbia University Irving Medical Center Adult Echocardiography Laboratory. CMR images were obtained by 1.5T or 3T GE HealthCare scanners, namely 3T-Signa Premier, 1.5T Discovery (750 W/450 W) and 1.5T Twin HDxT. Standard cine images were obtained for the quantitative evaluation of the right ventricle.

2.3 Data Annotation and Data Set

The right ventricular endocardial-myocardial interface was manually traced in end-systole and end-diastole by a single cardiologist (PNK) with expertise in cardiovascular ultrasound for the following 8 standardised echocardiographic views: parasternal long axis (PLAX), right ventricular inflow (RV Inflow), parasternal short axis at the level of the aortic valve (PSAX AV), basal (PSAX Base), mid (PSAX mid) and apical left ventricular segments (PSAX Distal), four chamber (Four C) and subcostal views (Sub C). The cardiologist was blinded to the CMR results. Short-axis stack cine CMR was used to calculate ED and ES volumes in a semi-automated way. This analysis was performed by experienced cardiac radiologists. Echocardiographic annotations were performed using the Siemens Syngo Workstation software, whereas CMR RV volumes were derived using the Circle Cardiovascular Imaging cvi42 software. The 8 area measurements described above along with the patient age were the numerical input variables of our model, whereas the gender and cardiac phase information served as the categorical input variables. For each patient, the CMR-derived ED and ES RV volumes were recorded, making a total of 100 data points.

2.4 Neural Network Architecture

The proposed pipeline employed the method introduced in [9], but all design choices were optimised to our problem. It constitutes an adaptation of the Transformer architecture to 2DE tabular (both categorical and numerical) data. Figure 1 portrays the inner workings of the proposed pipeline. In essence, all categorical and numerical inputs are tokenized and then forwarded to cascaded Transformer layers.

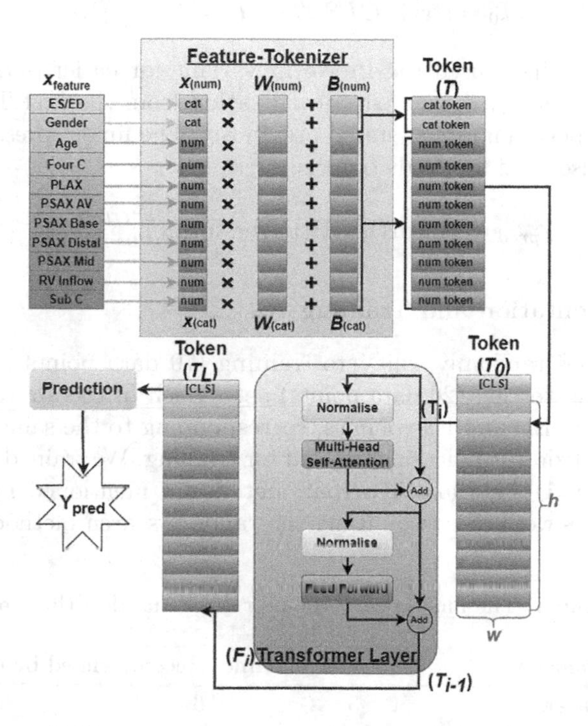

Fig. 1. The proposed DL model. $X_{feature}$ is the input. Token (T) is the input embedding by the Feature Tokenizer. T_0 is the [CLS] appended embedded token, where [CLS] is a special symbol added in front of every input example. F_i is the ith Transformer layer. T_L is the output product of the cascaded Transformers.

The **Feature-Tokenizer** module transforms inputs $X_{feature}$ to embeddings denoted by Token ($T \in \mathbb{R}^{h \times w}$). The embedding for a feature x_i is obtained by:

$$T_i = B_i + f_i(x_i) \in \mathbb{R}^w \qquad f_i : \mathbb{X}_i \rightarrow \mathbb{R}^w \tag{1}$$

where B_i is the ith feature bias, $f_{i(num)}$ is the element-wise multiplication of $x_{i(num)}$ (numerical features) with vector $W_{i(num)} \in \mathbb{R}^w$, whereas $f_{i(cat)}$ is a lookup table with $W_{i(cat)} \in \mathbb{R}^{S_i \times w}$ for $x_{i(cat)}$ (categorical features). The whole process is described by:

$$T_{i(num)} = B_{i(num)} + x_{i(num)} \cdot W_{i(num)} \in \mathbb{R}^w, \tag{2}$$

$$T_{i(cat)} = B_{i(cat)} + e_i^T W_{i(cat)} \in \mathbb{R}^w, \quad (3)$$

$$\text{Token}(T) = \text{stack}[T_{1(cat)}, T_{2(cat)}, T_{1(num)}, T_{2(num)}, ..., T_{9(num)}] \in \mathbb{R}^{w \times h}, \quad (4)$$

where e_i^T is the one-hot vector for the corresponding categorical variable.

The embeddings T are transformed into a classification token in Transformer layers as described in [14]. Then the [CLS] token is appended to T and L Transformer layers F_1, ..., F_L giving:

$$T_0 = \text{stack}[[CLS], T] \quad T_i = F_i(T_{i-1}). \quad (5)$$

Following [9], we have used the PreNorm variant for easier optimisation [15]. In addition, we discarded the first normalization from the first Transformer to achieve a good performance, as stated previously. The final representation of the [CLS] token (also used for prediction) is:

$$Y_{pred} = \text{Linear}(\text{ReLU}(\text{LayerNorm}(T_L^{[CLS]}))). \quad (6)$$

2.5 Implementation and Training

The data set was randomly split into training (70 data points), validation (10 data points) and testing (20 data points) sets. Each of the above sets involved equal numbers of ED and ES volumes, corresponding to the same patients. The mean squared error loss function was used for training. We trained for 500 epochs with batch size=1. The final hyperparameters are mentioned in Table 1. The hyperparameters were chosen by using the random search method.

Table 1. The final hyperparameter values used in this work.

Parameter	Our pipeline	Recommended by [9]
#Layers	3	6
Feature embedding size	16	512
Residual Dropout	0.3	0.2
Attention Dropout	0.3	0.5
FFN Dropout	0.3	0.5
FFN Factor	4/3	4/3
Learning Rate	0.01	LogUniform[3e-5,3e-4]
Weight Decay	0	LogUniform[3e-6,3e-3]
Optimizer	Adamax	AdamW

2.6 Model Evaluation

Our method is compared to other SOTA methods for tabular data, both shallow tree-based ensemble models (such as XGBoost [10] and CatBoost [11]) and attention-based deep architectures (such as the Tab-Transformer [12]). All SOTA methods were implemented according to the recommended values in [9]. To evaluate the predicted ES and ED volumes predicted by each method relative to the reference (CMR), we employed the R^2 coefficient and the absolute percentage error (APE). Apart from volume prediction, the four methods are also compared in terms of right ventricular ejection fraction (RVEF) values, which is a clinical metric used to describe the percentage of blood leaving the right ventricle each time it contracts. Finally, all predicted volumes and calculated EF values are plotted against the reference (CMR) values for all four methods.

3 Results

Table 2 details the R^2 and APE values for predicting RV ED and ES volumes. Also given are the *p*-values obtained from the Wilcoxon signed-rank test. The proposed method achieved a higher R^2 score and a lower APE when compared to the SOTA methods. All the *p*-values are less than 0.05. The cross-validation set performance is comparable to the test set performance. Table 3 lists the APE in RVEF, which was derived from the predicted volumes. The proposed method retains the best performance among all the methods. Figure 2 displays the predictions for RV volumes (a) and RVEF values (b) for the proposed and SOTA methods, versus the ground truth. According to those plots, our predictions are closer to the CMR reference values than the other SOTA methods.

Table 2. Quantitative comparison of the predicted RV volumes between the proposed and SOTA methods. APE stands for absolute percentage error. SD is the standard deviation. The *p*-values were obtained from the Wilcoxon signed-rank test ($\alpha = 0.05$).

Method	R^2 Score	APE (%) mean (±SD)	*p*-value
Proposed	**0.975**	**5.46 (±4.87)**	–
TabTransformer [12]	0.784	21.05 (±16.74)	2.1×10^{-5}
CatBoost [11]	0.797	15.13 (±10.81)	6.3×10^{-6}
XGBoost [10]	0.600	15.90 (±18.62)	7.3×10^{-6}

Table 3. Quantitative comparison of the RVEF between the proposed and SOTA methods. All abbreviations are the same as in Table 2.

Method	APE in EF (%) mean (±SD)	*p*-value
Proposed	**5.80 (±3.91)**	–
TabTransformer [12]	11.29 (±6.11)	0.004
CatBoost [11]	13.48 (±5.31)	0.048
XGBoost [10]	21.93 (±20.63)	0.013

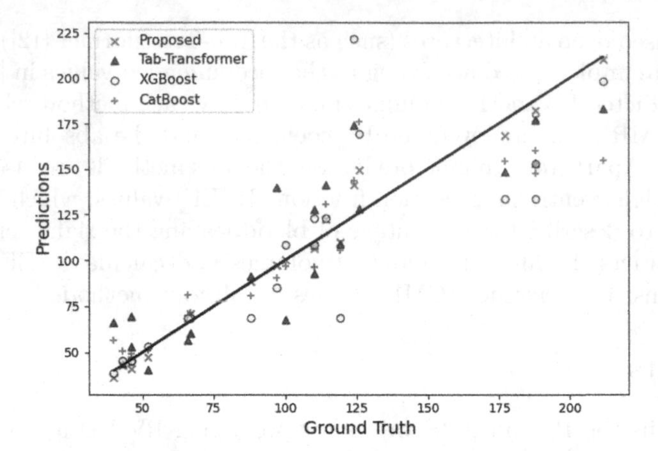

(a) Ground truth Vs Predicted RV volumes for the four methods.

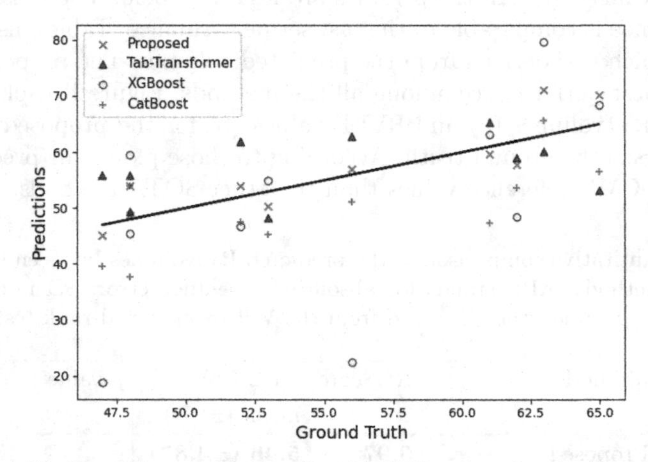

(b) Ground truth Vs Predicted RVEF values for the four methods.

Fig. 2. Visual comparison of model predictions against CMR (ground truth).

4 Discussion and Conclusions

This paper dealt with the problem of quantifying RV ED and ES volumes by using attention-based models on tabular 2DE data. Morphological measurements (areas) from 8 standardised 2DE views were used as input to the model along with age, cardiac phase, and gender information. Our method achieved impressive performance ($R^2 = 0.975$) on a relatively small-scale dataset, while it outperformed other SOTA methods. This work questions the superiority of tree-based ensemble models over DL-based solutions for tabular data in the context of functional imaging of the heart.

The proposed architecture comprised a feature tokenizer module to transform all features (both categorical and numerical) to embeddings, before applying a stack of Transformer layers. Transformers are the key modules in most SOTA architectures for natural language processing and computer vision. However, their performance on tabular data-based evaluation of the cardiac function had not been explored. Our pipeline is also appealing as it may allow building multi-modal pipelines for cardiovascular imaging problems, where only one part of the data is tabular, and other parts include images and text data.

Further studies are needed with a higher number of patients and a wider RV volume and RVEF range. Future work will involve adjusting our architecture to enable interpretability. We will investigate transferable tabular Transformers towards determining the most informative 2DE views. The proposed method can be combined with DL approaches for RV segmentation and ES/ED frame detection towards developing a fully-automated end-to-end pipeline for RVEF estimation.

Funding Information. Tuan Aqeel Bohoran is funded by the European Union's Horizon 2020 reasearch and innovation programme under the Marie Sklodowska-Curie grant agreement No 801604.

Data Availibility Statement. The link to the GitHub repository containing the code of the image analysis pipeline is: https://github.com/tuanaqeelbohoran/RV-Volume-Prediction.git.

Ethics Approval and Consent to Participate. Each study was approved by the relevant research and ethics services and written informed consent was obtained from all subjects prior to participation.

Competing Interests. The authors declare that they have no known competing financial interests or personal relationships that could have appeared to influence the work reported in this paper.

References

1. Haddad, F., Hunt, S., Rosenthal, D., Murphy, D.: Right ventricular function in cardiovascular disease, part I: anatomy, physiology, aging, and functional assessment of the right ventricle. Circulation **117**, 1436–1448 (2008)
2. Grothues, F., Moon, J., Bellenger, N., Smith, G., Klein, H., Pennell, D.: Interstudy reproducibility of right ventricular volumes, function, and mass with cardiovascular magnetic resonance. Am. Heart J. **147**, 218–223 (2004)
3. Keenan, N., et al.: Regional variation in cardiovascular magnetic resonance service delivery across the UK. Heart **107**, 1974–1979 (2021)
4. Treibel, T., et al.: United Kingdom standards for non-invasive cardiac imaging: recommendations from the Imaging Council of the British Cardiovascular Society. Heart **108**, e7–e7 (2022)

5. Marsh, H.: English hospitals in urgent need of more scanners and staff to deal with backlog. https://www.theguardian.com/society/2020/jul/14/english-nhs-hospitals-in-urgent-need-of-more-scanners-and-staff-to-deal-with-backlog,0. Accessed 10 Feb 2023
6. Tilling, L., González Gómez, A., Gros Otero, J., Becher, H.: Performing a comprehensive echocardiogram study: audit of adherence to the British Society of echocardiography minimum dataset guidelines. Br. J. Cardiol. **15**, 151–154 (2008)
7. Jenkins, C., Chan, J., Bricknell, K., Strudwick, M., Marwick, T.: Reproducibility of right ventricular volumes and ejection fraction using real-time three-dimensional echocardiography: comparison with cardiac MRI. Chest **131**, 1844–1851 (2007)
8. Fernández-Golfín, C., Zamorano, J.: Three-dimensional echocardiography and right ventricular function: the beauty and the beast? Circ. Cardiovasc. Imaging **10**, e006099 (2017)
9. Gorishniy, Y., Rubachev, I., Khrulkov, V., Babenko, A.: Revisiting deep learning models for tabular data. NeurIPS **34**, 18932–18943 (2021)
10. Chen, T., Guestrin, C.: XGBoost. In: Proceedings of the 22nd ACM SIGKDD International Conference on Knowledge Discovery and Data Mining (2016)
11. Prokhorenkova, L., Gusev, G., Vorobev, A., Dorogush, A., Gulin, A.: CatBoost: unbiased boosting with categorical features. NeurIPS **31** (2018)
12. Huang, X., Khetan, A., Cvitkovic, M., Karnin, Z.: TabTransformer: Tabular Data Modeling Using Contextual Embeddings (2020)
13. Kochav, J., et al.: Novel echocardiographic algorithm for right ventricular mass quantification: cardiovascular magnetic resonance and clinical prognosis validation. J. Am. Soc. Echocardiogr. **34**, 839-850.e1 (2021)
14. Devlin, J., Chang, M.W., Lee, K., Toutanova, K.: BERT: pre-training of deep bidirectional transformers for language understanding. arXiv preprint arXiv:1810.04805 (2018)
15. Wang, Q., et al.: Learning deep transformer models for machine translation. arXiv preprint arXiv:1906.01787 (2019)

Detection of Aortic Cusp Landmarks in Computed Tomography Images with Deep Learning

Luka Škrlj[1], Matija Jelenc[2], and Tomaž Vrtovec[1(✉)]

[1] Faculty of Electrical Engineering, University of Ljubljana, Ljubljana, Slovenia
tomaz.vrtovec@fe.uni-lj.si

[2] Department of Cardiovascular Surgery, University Medical Center Ljubljana, Ljubljana, Slovenia

Abstract. To perform aortic valve morphology for the assessment of the valvular heart disease in cardiovascular medicine, an accurate identification of specific anatomical points, i.e. landmarks, which define the aortic cusps, is required. In this study, we investigate the application of a deep learning framework, namely the spatial configuration network, for aortic cusp landmark detection in 120 contrast-enhanced end-diastolic coronary computed tomography images of normal patients. By performing three-fold cross-validation experiments, we obtained a mean detection error of 1.45 ± 0.82 mm for six landmarks located at the nadirs and commissures of the aortic valve sinuses, which dropped to 1.15 ± 0.62 mm when landmarks were detected in images that were cropped around the aortic valve by applying atlas-based segmentation. The obtained accuracy is comparable to existing methods, however, additional improvements in the form of image pre- or post-processing, or by applying advanced methodological concepts, may improve the landmark detection performance.

Keywords: Landmark detection · Deep learning · Aortic valve · Aortic cusp morphometry

1 Introduction

As the prevalence of the valvular heart disease is growing worldwide due to the improved survival rate and the aging population, its clinical management has become a focus of cardiovascular medicine [6]. Modern surgical treatment of both congenitally malformed and dysfunctional aortic valves, which holds the potential of an improved surgical planning, repair and outcome, requires a thorough morphological comprehension of the complex three-dimensional (3D) anatomy of the aortic root and valve [13]. Although the 3D preoperative assessment is based mostly on echocardiography [4], cardiac computed tomography (CT) is emerging as a complementary modality because of an acceptable radiation exposure and, in particular, high image resolution, which allows for an accurate and reliable morphological assessment [13].

O. Bernard et al. (Eds.): FIMH 2023, LNCS 13958, pp. 301–309, 2023.
https://doi.org/10.1007/978-3-031-35302-4_31

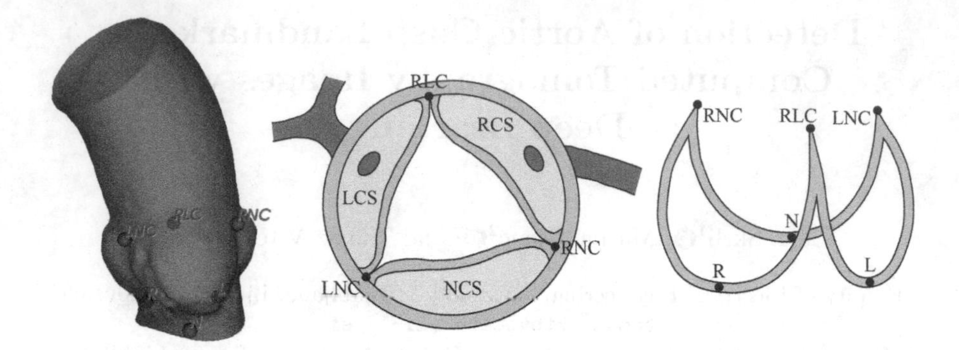

Fig. 1. The simplified anatomy of the aortic valve, shown as a three-dimensional (3D) rendering (*left*), an axial cross-section (*middle*), and a 3D representation of aortic cusps (*right*). RCS, LCS and NCS: right-, left- and non-coronary sinus; R, L and N: nadir of the RCS, LCS and NCS; RLC, RNC and LNC: right-/left-coronary, right-/non-coronary and left-/non-coronary commissure.

The starting point of aortic valve morphology is an accurate identification of specific anatomical points, i.e. landmarks, that define the aortic cusps, such as the nadirs and commissures of the sinuses (Fig. 1). However, manual identification of these landmarks from CT images can be difficult and time-consuming for physicians. Unfortunately, locally similar structures often present problems due to the ambiguity of landmarks, which make it difficult to achieve a low landmark detection error and robustness against landmark mis-identification. To address these problems, landmark detection based on automated medical image analysis has the potential to provide a more accurate and reliable morphological assessment. In the past decade, the advancements in deep learning for automated medical image analysis have impacted also the field of cardiovascular medicine [8]. Especially convolutional neural networks (CNNs) as one of the most representative deep learning models have been extensively used in many aspects of medical image analysis [14], and have been as such applied also for automated aortic valve landmark detection [2,9], resulting in an improved performance over conventional approaches [1,12].

In this study, we will address the application of a CNN-based framework [10] for the task of landmark detection in coronary CT images, in detail, for the detection of specific landmarks on the aortic valve, i.e. the nadirs and commissures of the sinuses, which represent the starting point for the aortic valve morphology. In addition, we will also observe the impact on the resulting landmark detection accuracy when landmark detection is performed in original images and in images that were first cropped according to aortic valve segmentation masks.

2 Methods

To detect the aortic valve landmarks, we make use of a CNN-based framework known as the spatial configuration network [10], which is based on representing each landmark with a probability distribution map, i.e. a heatmap, and forming a network of spatial configurations among landmarks (Fig. 2).

2.1 Landmark Heatmap Generation and Regression

First, we create a 3D heatmap g_i for each i-th landmark in each image that follows the Gaussian distribution and models the probability of that landmark to be located at its current position [10]:

$$g_i(x, \sigma_i) = \frac{\gamma}{(\sigma_i \sqrt{2\pi})^3} \exp\left(- \frac{\|\boldsymbol{x} - \hat{\boldsymbol{x}}_i\|^2}{2\sigma_i^2} \right), \tag{1}$$

where $\boldsymbol{x} = (x, y, z)$ and $\hat{\boldsymbol{x}}_i = (\hat{x}_i, \hat{y}_i, \hat{z}_i)$ are the arbitrary and reference 3D coordinates of the i-th landmark, respectively, and $\|\cdot\|$ is the L^2 norm (i.e. the Euclidean distance). Parameter σ_i is the corresponding standard deviation of the Gaussian function causing the heatmap pixels near the target landmark to have high values that smoothly but rapidly decrease with the increasing distance from that landmark, while parameter γ is a scaling factor used to avoid numerical instabilities during training that may be caused by very small values of the Gaussian function. Next, we train a U-Net [11], a standard encoder-decoder network used in many biomedical image analysis applications, to generate heatmaps, taking σ_i as a parameter. In addition to heatmaps, we also learn the optimal σ_i, which defines regions wide enough to avoid incorrect predictions but still small enough to avoid loss of information due to smoothing. The network learns to simultaneously regress all heatmaps by minimizing the loss L between the predicted and

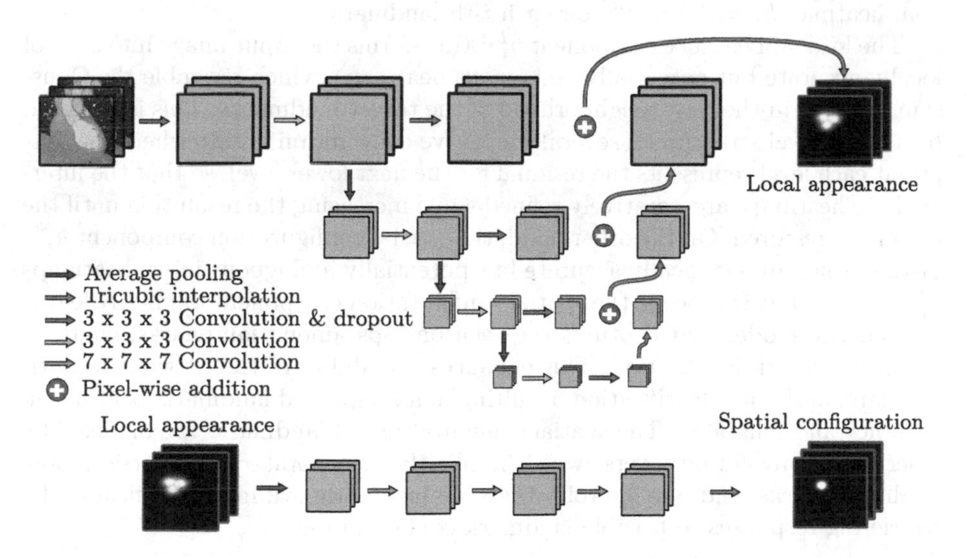

Fig. 2. The architecture of the applied spatial configuration network [10], consisting of the local appearance and spatial configuration components, which are joined with pixel-wise multiplication to obtain the final heatmaps.

reference heatmaps h_i and g_i, respectively:

$$L(\boldsymbol{w}, \boldsymbol{\sigma}) = \sum_{i=1}^{N} \|h_i(\boldsymbol{x}, \boldsymbol{w}) - g_i(\boldsymbol{x}, \sigma_i)\|^2 + \alpha \|\boldsymbol{\sigma}\|^2 + \lambda \|\boldsymbol{w}\|^2, \tag{2}$$

where N is the number of landmarks, $\boldsymbol{\sigma} = \{\sigma_i\}$ is the vector of Gaussian standard deviations for all landmarks (1) and $\boldsymbol{w}$ are the network weights, while α and λ respectively control their corresponding contributions. Vector $\boldsymbol{\sigma}$ as a trainable parameter is included in two terms. On one hand, the distance $\|h_i(\cdot) - g_i(\cdot)\|$ between predicted heatmaps h_i and reference heatmaps g_i drives $\boldsymbol{\sigma}$ towards infinity. On the other hand, the penalizing term with $\boldsymbol{\alpha}$ acts in the opposite direction. These conflicting contributions may result in a cancellation of effects, allowing $\boldsymbol{\sigma}$ to be efficiently utilized as a trainable parameter of the model. During inference, the predicted coordinates $\boldsymbol{x}_i' = (x_i', y_i', z_i')$ for the i-th landmark are obtained from the location of the maximal value of h_i.

2.2 Landmark Spatial Configuration

The proposed spatial configuration network [10] uses two interacting components to achieve a low landmark detection error with limited training datasets. The first component h_i^{LA}, which models the local appearance of landmarks, is responsible for providing locally accurate but potentially ambiguous predictions. The second component h_i^{SC}, which models the spatial configuration of landmarks, aims to improve the robustness against landmark mis-identification. The combination of these two components is achieved by pixel-wise multiplication that forms the final heatmap $h_i = h_i^{LA} \odot h_i^{SC}$ for each i-th landmark.

The local appearance component h_i^{LA} transforms the input image into a set of locally accurate but potentially ambiguous heatmaps, which resemble the Gaussian function in the close neighborhood of the target landmarks. This is achieved by a multi-level structure of several successive convolution layers, where the output at each level represents the residual for the next lower level, so that the intermediate heatmaps are iteratively refined while increasing the resolution until the original is restored. On the other hand, the spatial configuration component h_i^{SC} distinguishes among locally accurate but potentially ambiguous input heatmaps h_i^{LA}. Through this process, the spatial configuration component implicitly learns a geometric model that captures the relationships among landmarks, enabling robust landmark localization. This enhances the ability of the network to mitigate landmark mis-identification, resulting in an improved landmark localization accuracy and reliability. The spatial configurations of landmarks are modeled by a series of convolution layers, which implicitly incorporate a geometric model of all landmarks, and provide robustness against landmark mis-identification by restricting responses to feasible landmark configurations.

3 Experiments and Results

3.1 Images

In this study, we used a dataset of 120 contrast-enhanced end-diastolic coronary CT images of patients with healthy aortic valves that were provided by the Department of the Cardiovascular Surgery at University Medical Center Ljubljana, Slovenia. All images were acquired by the *Somatom Force* CT scanner (Siemens Healthineers, Erlangen, Germany), with the volume size of $512 \times 512 \times 229 - 574$ pixels and corresponding pixel size of $0.4 \times 0.4 \times 0.6\,\text{mm}^3$. By using *Mimics* software (Materialise NV, Leuven, Belgium), an experienced cardiovascular surgeon manually annotated $n = 6$ anatomical landmarks of the aortic cusps in each image (Fig. 1), i.e. the nadirs of the right-, left- and non-coronary sinus (R, L and N, respectively), and the left-/right-coronary, right-/non-coronary and left-/non-coronary commissure (RLC, RNC and LNC, respectively), which served as reference landmark locations for our study.

3.2 Implementation Details

We used the publicly available implementation[1] of the applied spatial configuration network [10]. The local appearance component is implemented with four levels, each consisting of three consecutive $3 \times 3 \times 3$ convolution layers with 128 outputs, while a $2 \times 2 \times 2$ average pooling after the second convolution layer is used to create the lower level. To improve generalizability, a dropout of 0.5 after the first convolution layer is included at each level. Local appearance heatmaps are generated from a convolution layer of $3 \times 3 \times 3$, with the number of outputs equal to the number of landmarks. The spatial configuration component is computed at 1/4 of the input resolution, and consists of three consecutive $7 \times 7 \times 7$ convolution layers with 128 outputs and an additional $7 \times 7 \times 7$ convolution layer with the number of outputs equal to the number of landmarks. The obtained outputs are scaled back to the input resolution by tricubic interpolation to generate the final heatmaps. Each intermediate convolution layer has a leaky rectified linear unit activation function with a negative slope of 0.1 to facilitate convergence during training. The weights $\boldsymbol{w}$ (2) are initialized by a Gaussian distribution with $\sigma = 0.001$, and the remaining parameters as $\gamma = 1000$ (1), $\alpha = 1000$ (2) and $\lambda = 0.0005$ (2). The loss function (2) is minimized by the Nesterov accelerated gradient with 10^{-6} learning rate and 0.99 momentum.

To test the impact of image segmentation on landmark detection accuracy, we devised a relatively simple yet efficient framework for aortic valve segmentation from original coronary CT images, i.e. an atlas-based segmentation. All images were first non-rigidly registered (i.e. with B-spline functions) to a selected reference image by using *SimpleElastix*[2], an extension of the open-source image registration toolbox *elastix* [7], to obtain an atlas with corresponding geometrical transformations that best aligned such image pairs. The aortic valve was then

[1] https://github.com/christianpayer/MedicalDataAugmentationTool-HeatmapRegression.
[2] https://simpleelastix.github.io.

Table 1. The aortic cusp landmark detection error (mean ± standard deviation), shown separately for original coronary computed tomography images and for cropped images, obtained by atlas-based segmentation of the aortic valve from original images.

Landmark name	Label	Detection error (mm)	
		Original	Cropped
Nadir of the right-coronary sinus	R	1.39 ± 0.61	1.06 ± 0.56
Nadir of the left-coronary sinus	L	1.32 ± 0.62	0.95 ± 0.45
Nadir of the non-coronary sinus	N	1.27 ± 0.83	1.18 ± 0.66
Right-/left-coronary commissure	RLC	1.50 ± 0.62	1.09 ± 0.52
Right-/non-coronary commissure	RNC	1.82 ± 1.11	1.48 ± 0.95
Left-/non-coronary commissure	LNC	1.31 ± 0.87	1.14 ± 0.62
All landmarks		$\mathbf{1.45 \pm 0.82}$	$\mathbf{1.15 \pm 0.62}$

manually segmented by pixel-wise annotation in the constructed atlas, resulting in the reference segmentation mask. The atlas was propagated from the reference image to all other images by using the inverse geometrical transformations, so that segmentation masks were obtained in all images. Finally, cropping with a fixed margin of 15 pixels around the obtained segmentation masks was applied, forming a new dataset of cropped CT images.

3.3 Experiments

We performed three-fold cross-validation experiments, meaning that our dataset of 120 images was randomly split into three folds of 40 images. In each of the three separate experiments, two folds (80 images or 66.6%) were used for training, and one fold (40 images or 33.3%) was used for testing. The images were rescaled to the size of $96 \times 96 \times 128$ pixels with the corresponding pixel size of $1 \times 1 \times 1\,\text{mm}^3$, and their intensities were normalized to the range of $[-1, 1]$. Image augmentation was performed by random intensity multiplication and shift of $[0.75, 1.25]$ and $[-0.25, 0.25]$, respectively, and by random translation, rotation and scaling of $[-25, 25]$ pixels, $[-15°, 15°]$ and $[0.5, 1.5]$, respectively. The applied deep learning framework was then trained for 80,000 epochs. The landmark detection error was evaluated by the Euclidean distance $d_i = \|\boldsymbol{x}'_i - \hat{\boldsymbol{x}}_i\|$ between the predicted 3D coordinates $\boldsymbol{x}'_i$ and reference 3D coordinates $\hat{\boldsymbol{x}}_i$ of each i-th landmark. The same experiments were performed on original as well as on cropped CT images.

3.4 Results

By applying the described spatial configuration network [10] to detect $n = 6$ aortic cusp landmarks in the devised original dataset of 120 coronary CT images, we obtained a mean landmark detection error of 1.45 ± 0.82 mm across the three-fold cross-validation experiments. When the same experiments were applied to the dataset of cropped CT images, obtained by atlas-based segmentation of the aortic valve, the resulting landmark detection error was 1.15 ± 0.62 mm. For specific landmarks, the obtained detection errors are listed in Table 1, while an example for a selected coronary CT image is shown in Fig. 3.

4 Discussion

We applied an existing landmark detection framework for detecting aortic cusp landmarks in coronary CT images. By optimizing the components of local appearance and spatial configuration simultaneously, the problem of landmark detection was split into two simpler sub-problems that can be modeled by using a relatively small amount of training data [10]. As a result, we achieved relatively small landmark detection errors for the nadirs and commissures of the sinuses that represent the basis of the aortic cusp morphology.

The obtained mean landmark detection error is comparable to existing approaches. Al et al. [1] applied colonial walk, a regression tree-based machine learning but not deep learning algorithm, with a two-phase optimized search space learning model on coronary CT images of 71 patients with significant or mild valvular calcification, and achieved a 2.05 ± 1.15 mm detection error for nadirs and commissures. By applying deep learning approaches, the accuracy and precision of landmark detection was improved. Noothout et al. [9] applied a coarse-to-fine CNN-based approach, which resulted in a median detection error of 1.87 mm for nadirs and commissures in 672 coronary CT images. Aoyama

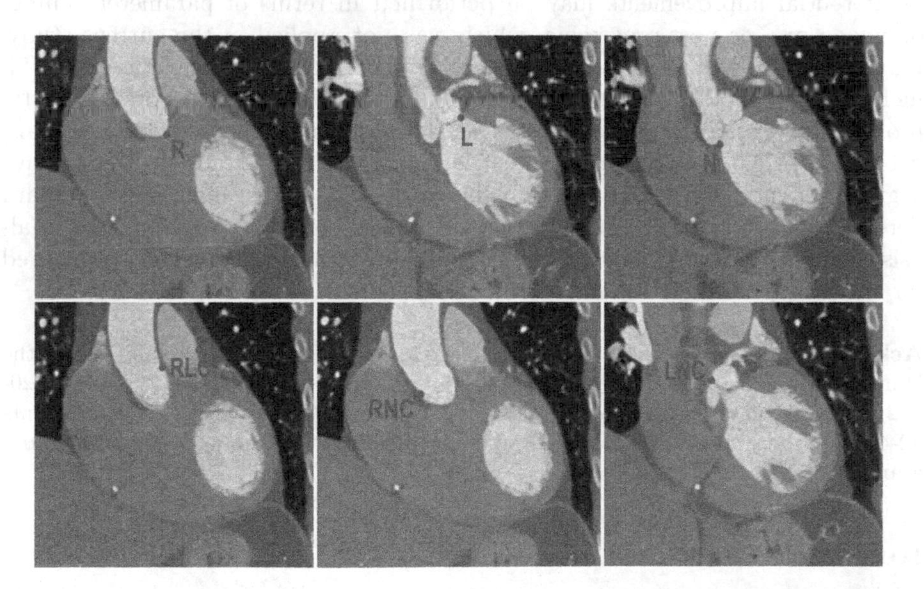

Fig. 3. An example of landmark detection results for a selected original computed tomography image, shown in corresponding coronal cross-sections. R, L and N: nadir of the right-, left- and non-coronary sinus; RLC, RNC and LNC: right-/left-coronary, right-/non-coronary and left-/non-coronary commissure.

et al. [2] applied the same spatial configuration network in a cascaded manner, where they first performed coarse landmark detection that was used to segment the aortic root, followed by a more accurate detection of aortic cusp landmarks in 138 coronary CT images of normal patients and patients with severe calcification, with the resulting detection error of 1.57 ± 1.34 mm for nadirs and commissures. In the study that proposed the original spatial configuration network, Payer et al. [10] reported a 0.66 ± 0.74 mm detection error for 37 landmarks in 895 two-dimensional hand X-rays, a 0.84 ± 0.62 mm detection error for 28 landmarks in 60 3D magnetic resonance (MR) hand images, and a 6.2 ± 9.9 mm and 2.9 ± 4.4 mm detection error for 26 landmarks in 224 and 60, respectively, 3D CT spine images. In our study, we obtained a mean landmark detection error of 1.45 ± 0.82 mm for six landmarks in 120 3D coronary CT images, which dropped to 1.15 ± 0.62 mm when landmark detection was applied to cropped images, obtained by prior atlas-based segmentation of the aortic valve from original images. This indicates that image preprocessing in the form of segmentation of the region of interest is beneficial for the applied spatial configuration network, especially since the aortic valve itself occupies a relatively small portion of the whole coronary CT image, but further investigations are needed to support and generalize such conclusions. Nevertheless, when compared to the results of Aoyama et al. [2], our results are slightly better but there were no pathological cases (e.g. calcification) in our dataset. On the other hand, the results reported by Payer et al. [10], especially those for 3D MR hand images, indicate that there may still be room for improvement when using the proposed spatial configuration network.

Potential improvements may be performed in terms of parameter tuning, or image pre- or post-processing, which was not applied in the current study. Our future work will be therefore focused on the segmentation of the aorta [5], inclusion of pathological images with, for example, dilated aortic roots (i.e. aortic root aneurysms), and application of other mechanisms that may aid in landmark detection. In particular, reinforcement learning [15] has already proved to have a great potential in various medical image analysis tasks, including landmark detection [3]. In addition, we plan to focus on the complete morphometric analysis of the complex 3D anatomy of the aortic root and valve, which is required for the management of valvular heart disease in cardiovascular medicine.

Acknowledgements. The study was approved by the Ethics Committee of the University Medical Center Ljubljana, Slovenia, under 0120-133/2021/3 and 0120-312/2022/3, and supported by the Slovenian Research Agency (ARRS) under grants J2-4453 and P2-0232, and by the University Medical Center Ljubljana, Slovenia, under grant 20190174.

References

1. Al, W.A., Jung, H.Y., Yun, I.D., Jang, Y., Park, H.B., Chang, H.J.: Automatic aortic valve landmark localization in coronary CT angiography using colonial walk. PLoS ONE **13**(7), e0200317 (2018). https://doi.org/10.1371/journal.pone.0200317

2. Aoyama, G., et al.: Automatic aortic valve cusps segmentation from CT images based on the cascading multiple deep neural networks. J. Imaging **8**(1), 11 (2022). https://doi.org/10.3390/jimaging8010011
3. Bekkouch, I.E.I., Maksudov, B., Kiselev, S., Mustafaev, T., Vrtovec, T., Ibragimov, B.: Multi-landmark environment analysis with reinforcement learning for pelvic abnormality detection and quantification. Med. Image Anal. **78**, 102417 (2022). https://doi.org/10.1016/j.media.2022.102417
4. Calleja, A., et al.: Automated quantitative 3-dimensional modeling of the aortic valve and root by 3-dimensional transesophageal echocardiography in normals, aortic regurgitation, and aortic stenosis. Circ. Cardiovasc. Imaging **6**(1), 99–108 (2013). https://doi.org/10.1161/CIRCIMAGING.112.976993
5. Chen, C., et al.: Deep learning for cardiac image segmentation: a review. Front. Cardiovasc. Med. **7**, 25 (2020). https://doi.org/10.3389/fcvm.2020.00025
6. Coffey, S., et al.: Global epidemiology of valvular heart disease. Nat. Rev. Cardiol. **18**, 853–864 (2021). https://doi.org/10.1038/s41569-021-00570-z
7. Klein, S., Staring, M., Murphy, K., Viergever, M.A., Pluim, J.P.W.: elastix: a toolbox for intensity-based medical image registration. IEEE Trans. Med. Imaging **29**(1), 196–205 (2010). https://doi.org/10.1109/TMI.2009.2035616
8. Krittanawong, C., et al.: Deep learning for cardiovascular medicine: a practical primer. Eur. Heart J. **40**(25), 2058–2073 (2019). https://doi.org/10.1093/eurheartj/ehz056
9. Noothout, J.M.H., et al.: Deep learning-based regression and classification for automatic landmark localization in medical images. IEEE Trans. Med. Imaging **39**(12), 4011–4022 (2020). https://doi.org/10.1109/TMI.2020.3009002
10. Payer, C., Štern, D., Bischof, H., Urschler, M.: Integrating spatial configuration into heatmap regression based CNNs for landmark localization. Med. Image Anal. **54**, 207–219 (2019). https://doi.org/10.1016/j.media.2019.03.007
11. Ronneberger, O., Fischer, P., Brox, T.: U-Net: convolutional networks for biomedical image segmentation. In: Navab, N., Hornegger, J., Wells, W.M., Frangi, A.F. (eds.) MICCAI 2015. LNCS, vol. 9351, pp. 234–241. Springer, Cham (2015). https://doi.org/10.1007/978-3-319-24574-4_28
12. Tahoces, P.G., et al.: Automatic detection of anatomical landmarks of the aorta in CTA images. Med. Biol. Eng. Comput. **58**(5), 903–919 (2020). https://doi.org/10.1007/s11517-019-02110-x
13. Tretter, J.T., et al.: Understanding the aortic root using computed tomographic assessment: a potential pathway to improved customized surgical repair. Circ. Cardiovasc. Imaging **14**(11), e013134 (2021). https://doi.org/10.1161/CIRCIMAGING.121.013134
14. Yu, H., Yang, L.T., Zhang, Q., Armstrong, D., Deen, M.J.: Convolutional neural networks for medical image analysis: state-of-the art, comparisons, improvement and perspectives. Neurocomputing **444**, 92–110 (2021). https://doi.org/10.1016/j.neucom.2020.04.157
15. Zhou, S.K., Le, H.N., Luu, K., Nguyen, H.V., Ayache, N.: Deep reinforcement learning in medical imaging: a literature review. Med. Image Anal. **73**, 102193 (2021). https://doi.org/10.1016/j.media.2021.102193

Automatic Detection of Coil Position in the Chest X-ray Images for Assessing the Risks of Lead Extraction Procedures

YingLiang Ma[1(✉)], Vishal S. Mehta[2], C. Aldo Rinaldi[2], Pengpeng Hu[3], Steven Niederer[4], and Reza Razavi[4]

[1] School of Computing Sciences, University of East Anglia, Norwich, UK
yingliang.ma@uea.ac.uk
[2] Department of Cardiology, Guy's and St. Thomas' Hospitals NHS Foundation Trust, London, UK
[3] Centre for Computational Science and Mathematical Modelling, Coventry University, Coventry, UK
[4] School of Imaging Sciences and Biomedical Engineering, King's College London, London, UK

Abstract. The lead extraction procedures are for the patients who already have pacemaker implanted and leads need to be replaced. The procedure is a high-risk procedure and it could lead to major complications or even procedure-related death. Recently, an Electra Registry Outcome Score (EROS) was designed to create a risk assessment tool using the data about personal health records and an accuracy of 0.70 was achieved. In this paper, we hypothesized that a coil inside the superior vena cava (SVC) is a very important risk factor. By integrating it into the risk assessment model, the accuracy can be further improved. Therefore, an automatic detection method was developed to localize the positions of coils in the X-ray images. It was based on a U-Net convolutional network. To determine the coil position relative to the SVC position inside the chest X-ray image, the heart region was first detected by using a modified VGG16 model. Then, the bounding box of the SVC can be estimated based on the heart anatomy. Finally, a XGBoost classifier was trained on the data about personal health records and the risk factor about the coil position. An accuracy of 0.85 was achieved.

Keywords: Deep learning · Wire detection · Risk assessment

1 Introduction

The number of patients with implantable pacemaker has grown exponentially in recent decades. Transvenous lead extraction (TLE) is the primary treatment for complications related to the implanted pacemaker. The complications include lead fracture, device failure, lead erosion and infection [1]. Although the success rate of TLE procedure remain high, the procedure is sometimes complex and leads to severe complications and even procedure-related death. The studies in major European centers have reported

O. Bernard et al. (Eds.): FIMH 2023, LNCS 13958, pp. 310–319, 2023.
https://doi.org/10.1007/978-3-031-35302-4_32

a 1.7% rate of major complications including deaths [2]. Therefore, it is essential to assess the risk of the TLE procedure for individual patients to reduce the mortality rate. Sidhu et al. [3] has proposed an Electra Registry Outcome Score (EROS) to create a risk assessment tool using a number of variables about personal health data and an accuracy of 0.70 was achieved. Recently, machine learning (ML) based approaches have been proposed to improve the accuracy. Vishal et al. [4] has trained ML models and tested them on a ELECTRa database [3] and achieved an accuracy of 0.74. The results are marginally improvement from EROS.

We hypothesized that adding additional information extracted from a plain chest X-ray image will further improve the accuracy. The coil inside the superior vena cava (SVC) is a very important risk factor as the common reason for the major complication inside the SVC is the fibrotic tissue around the lead or the coil. It does more likely happen to the coil which has a large surface area. The fibrotic tissue likely causes mechanical damages such as tear during the procedure [5]. To find out the coil position, the plain chest X-ray images were used and they are normally acquired before the procedure. Robust computer vision algorithms were designed, which are based on deep learning techniques for image segmentation. The computer vision algorithms were able to automatically compute the positions of the coils in the X-ray image and detect whether majority of the coil is inside the SVC or not. The main contribution of this manuscript is the methodology of the development of a deep learning framework using the combination of an additional feature from images and other clinical data to achieve higher accuracy to detect high-risk patients. The proposed approach is not limited to the TLE procedure using the chest X-ray image and it could apply to assess the risk of other cardiac interventional procedures using X-ray or CT images.

2 The Detection of Coils in X-ray Images

In the chest X-ray images from patients undergoing the TLE procedure, there are often coils mixing with leads. The leads are thin metal wires. On the other hand, the coils have a large surface area of electrodes and they often conduct strong currents to depolarize the heart. They can be recognized on x-ray images as focal areas of wire thickening. In order to robustly detect the position of all coils inside the X-ray image, a two-step approach was proposed. Firstly, both leads and coils were segmented by using a U-Net convolutional network [6] and this acted as the coarse region segmentation. Then, the masked images created from the coarse region segmentation were feed to the second U-net model to extract the exact location of the coils.

3 The Detection of Both Leads and Coils

To segment both leads and coils from X-ray images, a U-Net convolutional network was trained and tested by using a database of 737 chest X-ray images from 737 patients.

Creating Manual Segmentations of Leads and Coils. The manual segmentation of leads and coils in X-ray images is very time-consuming. To speed up, a vessel enhancement filter [7] was used to extract all wire-like objects and the result image was automatically binarized by an adaptive binarization method: Otsu's method [8]. Then an experienced clinician manually removed the objects which are not leads or coils. In addition, the pacemaker and nearby wire segments were also removed as they are not useful for finding the coils. Figure 1 presented the process of manual segmentation.

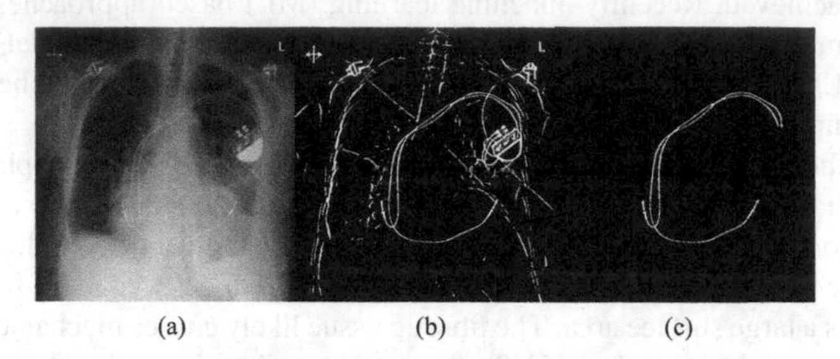

(a) (b) (c)

Fig. 1. Manual segmentation of leads. (a) The original image. (b) Image after applying the vessel enhancement filter. (c) The final result of manual segmentation

The U-Net was trained to behave like a selective wire-enhancement filter and it is able to extract high-contrast leads and coils and ignore other wire-like objects such as surface ECG leads, rib bone shadows, the border of the cardiac silhouette and pacemaker.

Training and Results. The ratio for train-test split is 70:30. The loss function for the U-Net model is the dice similarity coefficient (DSC) and it is defined as:

$$DSC = \frac{2|X \cap^{Y}|}{|X| + |Y|} \tag{1}$$

The optimizer is Adam. The U-Net convolutional network was trained using 516 images. Images with the resolution of 512 × 512 were directly input into the network without any down-sampling as the leads might not be visible in the low-resolution images. The U-Net model was implemented using TensorFlow API and training was carried out using an apple MacBook pro (GPU: Radeon Pro 560 and CPU: 2.9 GHz quad-core Intel Core i7). The training process took about 8 h and it was accelerated using GPU. The lengthy training process is because the high-resolution images (512 × 512) were used for training. The trained U-Net was tested on remaining 221 images. An accuracy of 0.69 ± 0.10 was achieved for the segmentation of leads and coils and it was measured in DSC against the ground truth. Although the U-Net is not always able to detect the completed length of leads, it is sufficient for our next task: extracting coil position. The recall of coils is 1.0 and it means that all coils within X-ray images have been successfully detected as the coils are high-contrast objects and relatively easy to be detected. Two examples of segmentation results are presented in Fig. 2.

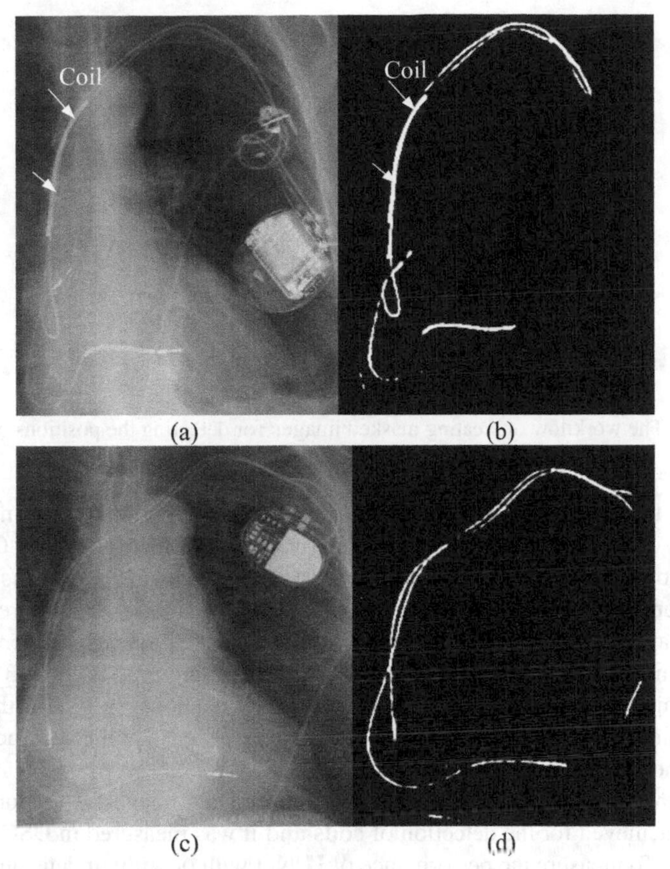

Fig. 2. The results of detecting leads and coils. (a)(c) The original image. (b)(d) The segmentation results output by the U-Net. (a) The coils are clearly within the detected segmentations

4 The Detection of Coils

To localize the exact location of the coils within the X-ray images, we use the results of detected leads and coils as the initial data input. Although the previous results already include the coils, the location of the coil relative to the location of SVC has to be determined as it is important to assess the risk of TLE procedures. [9] reported that the fibrotic tissue surrounding the coil is one of leading factors contributing to the mechanical damages such as tear during the procedure. To detect only the coil, the similar U-Net model is used but the input image is different. The input image is created by applying an image mask onto the original X-ray image and the mask is generated from the binary image after applying an image dilation operation with the kernel size of 10×10. The Fig. 3 illustrate the process of creating masked images.

In order to reduce the distraction from the thin lead wires, the input images were down-sampled to 256×256 and they are used as input data for the U-Net model. The U-Net is the same as the previous one except the dimension of input and output data. As the low-resolution images reduced the computation cost of model training, we were

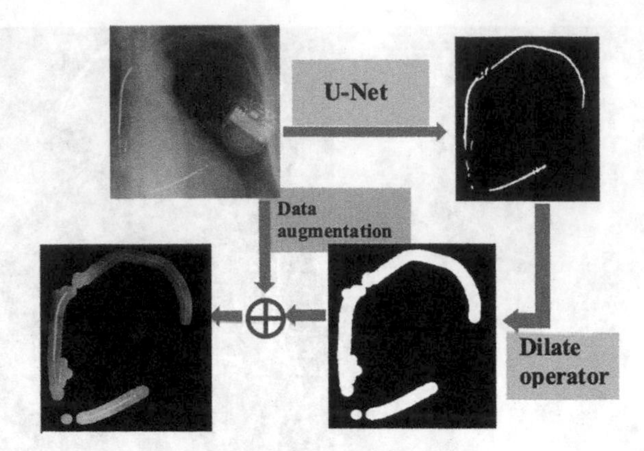

Fig. 3. The workflow of creating masked images for detecting the positions of coils

able to apply the data augmentation technique to increase the number of training images. The contrast of the 516 images were reduced by a random factor between 0.6 to 0.9 to create additional 1,032 training images. The reason for reducing contrast is to improve the performance of coil detection for low-contrast X-ray images. Therefore, the image masks were applied onto 1,548 training images and masked images were created. They are used as input images for the U-Net model. The training time is about 3 h using the same PC which trained the previous U-Net model. As shown in Fig. 4, the accuracy of model training using data augmentation increases faster than the accuracy of model training without it.

Finally, the trained U-Net was tested on remaining 221 images. An accuracy of 0.87 ± 0.10 was achieved for the detection of coils and it was measured in DSC against the ground truth. To measure the performance of U-Net with or without data augmentation, key metrics such as accuracy, precision, recall and F1-score were calculated and presented in Table 1. A true positive detection is defined as at least 75% length of the target coil object was detected. A false positive detection is defined as the other wire object was detected as the target coil object. The examples of successful and unsuccessfully detections are presented in Fig. 5.

Table 1. Key metrics for model performance with or without data augmentation

Data augmentation	Accuracy	Precision	Recall	F1 score
without	0.87	0.88	0.91	0.89
with	0.98	0.98	0.99	0.98

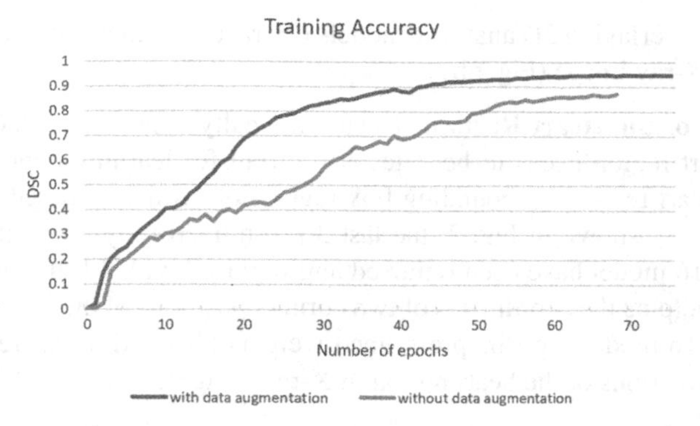

Fig. 4. Accuracy by epochs of the U-Net training in terms of the DSC

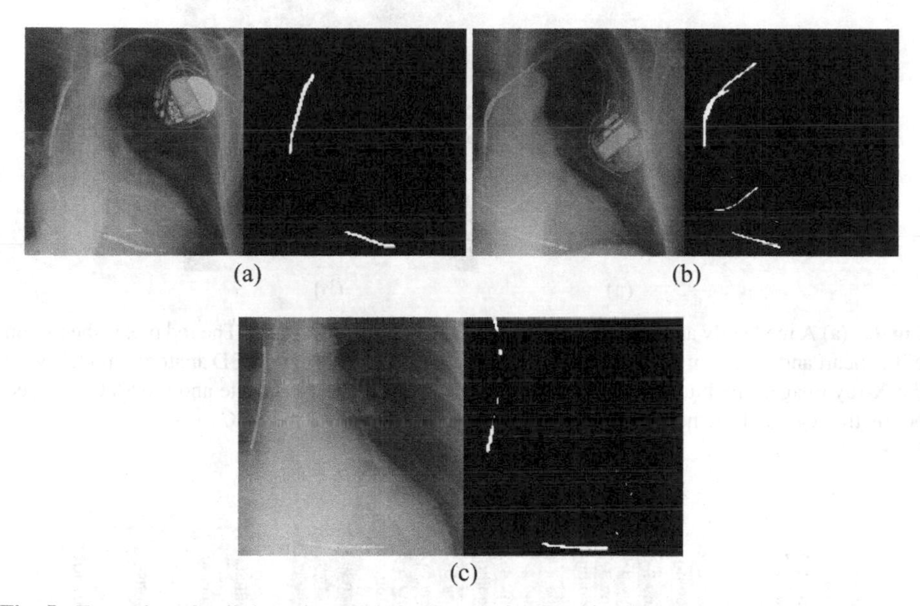

Fig. 5. Examples of coil detections. (a) (b) Successful detections. (c) A false negative detection which only part of coil (less than 75%) has been detected

5 The Detection of the Approximate Location of SVC

SVC is a major blood vessel inside the heart and it provides an important pathway for inserting the pacing leads into the right atrium and other heart chambers. The SVC is the most common location requiring surgical repair as the result of a major complication after the TLE procedure. It is not possible to detect the exact location of SVC in the X-ray image as the SVC is not visible in the image unless a contrast agent is injected. However, it is possible to estimate the location based on the heart anatomy. As show in Fig. 6a, the location of the SVC (green box) is the top left corner of the heart region (red box). The height of the green box is approximately half the height of the heart region. The width of the green box is approximately one third of the heart region. The method was verified by overlaying 3D anatomy models (extracted from pre-procedure CT scan) with the chest X-ray image (Fig. 6b).

The Detection of the Heart Region. To automatically located the SVC in the X-ray image, the heart region needs to be detected. A transfer learning approach was used to detect the heart region via bounding box regression, which is based on a modified VGG16 model. As shown in Fig. 7, the last 3 pre-trained fully-connected layers of a standard VGG16 model have been removed and replaced with 4 full-connected layers. The last layer outputs the coordinates of two corners positions of the bounding box. The modified VGG16 model uses the pre-trained weights [10] and it was re-trained using the manual annotations of the heart region in X-ray images.

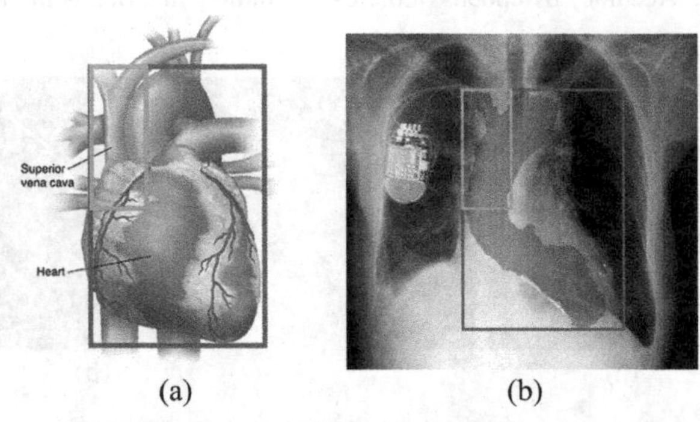

Fig. 6. (a) A medically accurate illustration of the anatomy of the heart. The red box is the region of the heart and the green box is the location of the SVC. (b) Overlaying 3D anatomy models with the X-ray image. The blue shadow is the 3D model of aorta, left ventricle and the SVC. The red box is the region of the heart and the green box is the location of the SVC

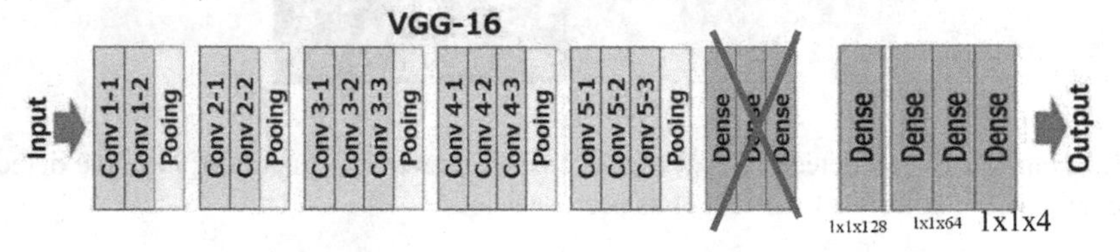

Fig. 7. The architecture of the modified VGG16 model

The Accuracy of Detection. The accuracy of heart region detection is measured by the Intersection Over Union (IOU). IOU in this application is defined as:

$$IOU = \frac{|B_p \cap B_{gt}|}{|B_p \cup B_{gt}|} = \frac{|B_p \cap B_{gt}|}{|B_p| + |B_{gt}| - |B_p \cap B_{gt}|} \quad (2)$$

where B_p is the predicted bounding box or the detected bounding box and B_{gt} is the ground truth bounding box. A correct detection is that IOU is larger than or equal to the preset threshold. The preset threshold in this application is set to both 50% and 75%, as the detection of heart region has to be accurate to ensure that the location SVC is relatively accurate. An example of IOU was presented in Fig. 8. The precisions and recalls of the detect method were given in Table 2.

Table 2. The precisions and recalls for assessing the accuracy of heart region detection

IOU Threshold	50%	75%
Precision	1.0	0.818
Recall	1.0	1.0

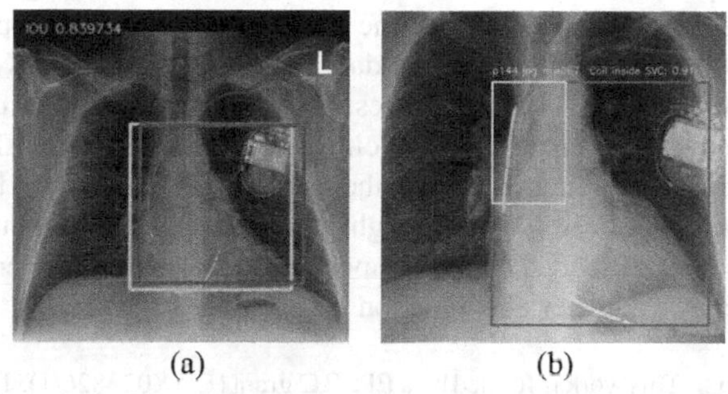

Fig. 8. (a) The results of IOU. The green box is the ground truth and the red box is the detected bounding box. (b) The coil position related to the SVC. The green box is the bounding box of the SVC and the red box is the bounding box of the heart region. The yellow lines are the centerlines of the detected coils

6 Machine-Learning Model for Risk Assessment

Once the coil position was determined, it was compared with the bounding box of the SVC. If 50% of the coil length is within the SVC, the coil will be labelled as "inside". Otherwise, the coil will be labelled as "outside". The coil position label combined with personal health records were feed into a machine-learning model to predict the TLE procedure risk. The XGBoost classifier [11] was chosen to predict the risk of TLE procedures as it uses sequentially-built shallow decision trees to provide accurate results and a highly-scalable training method that avoids overfitting. Furthermore, the XGBoost classifier has been used to diagnose chronic kidney disease [12] and detect credit card fraud [13] because it works well with imbalanced datasets and binary classifications. There is a total 737 sets of data from 737 TLE clinical cases and the outcome from 18 cases (2.4%) are a major complication or procedure related death. To balance the data, Adaptive Synthetic (ADASYN) sampling techniques [14] were applied to the data and generated additional 701 sets of data which was labelled as the cases of a major complication or procedure related death. The balanced data were split into two groups.

70% of data were used for training and 30% of data were used for testing. The learning rate was set to 0.02 and the number of gradient boosted trees was set to 200. The maximum depth of a tree, the minimum child weight and the subsample ratio of columns for each tree were set to 5, 1 and 0.75, respectively. The balanced accuracy of 0.85 was achieved, which was tested on the balanced test dataset.

7 Conclusions and Future Work

This paper presents a novel deep-learning framework for predicting the risk of the TLE procedures. Robust computer vision algorithms were developed to extract the position of coils in the chest X-ray images. By comparing the coil positions with the estimated location of the SVC, a new risk-factor variable was created. By adding this risk-factor variable with the personal health data, we were able to achieve a higher classification accuracy for detecting high-risk cases in the TLE procedures. Our approach only use plain chest X-ray images as the additional data source and the chest X-ray images are routinely acquired before the TLE procedures. Therefore, our approach does not require additional data and will not change current clinical practices for the TLE procedures.

Additional geometric features such as the number of overlapping leads inside the SVC, the angulation of a lead inside the right ventricle and the angulation of the lead near the entry point of the SVC could be important for further improvement of the risk modelling and achieve a higher classification accuracy.

Acknowledgements. This work is funded by a EPSRC grant (EP/X023826/1). The study was also supported by the Wellcome/EPSRC Centre for Medical Engineering (WT203148/Z/16/Z) and the National Institute for Health Research (NIHR) Biomedical Research Centre based at Guy's and St Thomas' NHS Foundation Trust and King's College London. The views expressed are those of the author(s) and not necessarily those of the NHS, the NIHR or the Department of Health.

References

1. van Erven, L., et al.: Attitude towards redundant leads and the practice of lead extractions: a European survey. Europace **12**(2), 275–276 (2010)
2. Bongiorni, M.G., et al.: The European Lead Extraction ConTRolled (ELECTRa) study: A European Heart Rhythm Association (EHRA) Registry of Transvenous Lead Extraction Outcomes. European Heart Journal **38**(40), 2995–3005 (2017)
3. Sidhu, B.S., et al.: Risk stratification of patients undergoing transvenous lead extraction with the ELECTRa Registry Outcome Score (EROS): an ESC EHRA EORP European lead extraction ConTRolled ELECTRa registry analysis. Europace **23**(9), 1462–1471 (2021)
4. Mehta, V.S., et al.: Machine learning–derived major adverse event prediction of patients undergoing transvenous lead extraction: Using the ESC EHRA EORP European lead extraction ConTRolled ELECTRa registry. Heart Rhythm **19**(6), 885–893 (2022)
5. Tułecki, Ł, et al.: A Study of Major and Minor Complications of 1500 Transvenous Lead Extraction Procedures Performed with Optimal Safety at Two High-Volume Referral Centers. International journal of environmental research and public health **18**(19), 10416–29 (2021)

6. Ronneberger, O., et al.: U-Net: Convolutional Networks for Biomedical Image Segmentation. In: Proc. Int. Conf. Medical Image Computer Assisted Intervention (MICCAI), LNCS, vol 9351, pp. 234–241. Springer (2015)
7. Frangi, A.F., et al.: Multiscale vessel enhancement filtering. In: Proc. Int. Conf. Medical Image Computer Assisted Intervention (MICCAI) (1998)
8. Otsu, N.: A threshold selection method from gray-level histograms. IEEE Transactions on Systems, Man and Cybernetics **9**(1), 62–66 (1979)
9. Tułecki, Ł, et al.: Analysis of Risk Factors for Major Complications of 1500 Transvenous Lead Extraction Procedures with Especial Attention to Tricuspid Valve Damage. Int. J. Environ. Res. Pub. Heal. **18**(17), 9100–9113 (2021)
10. Li, J., et al.: Transfer Learning Performance Analysis for VGG16 in Hurricane Damage Building Classification. In: 2nd International Conference on Big Data & Artificial Intelligence & Software Engineering (ICBASE), pp 177–184 (2021)
11. Chen, T., et al.: XGBoost: A Scalable Tree Boosting System. In: Proceedings of the 22nd ACM SIGKDD International Conference on Knowledge Discovery and Data Mining (KDD '16), pp 785–794 (2016)
12. Ogunleye, A., et al.: XGBoost model for chronic kidney disease diagnosis. IEEE/ACM Transactions on Computational Biology and Bioinformatics **17**(6), 2131–2140 (2020)
13. Abdulghani, A.Q., et al.: Credit card fraud detection using XGBoost algorithm. In: 14th International Conference on Developments in eSystems Engineering, pp. 487–492 (2021)

Cardiac MRI Tagline Extraction Based on Diffeomorphic Active Contour Algorithm

Ruiyi Zhang[1], Jinchi Wei[1], Dnyanesh Tipre[1], Robert G. Weiss[2], Laurent Younes[1], and Siamak Ardekani[1](✉)

[1] The Center for Imaging Science, Johns Hopkins University, Baltimore, MD 21218, USA
sardekani@jhu.edu

[2] The Department of Medicine, Section of Cardiology, Johns Hopkins Medical Institutions, Baltimore, MD 21218, USA

Abstract. Cardiac tagged Magnetic Resonance Imaging (MRI) is readily available in clinical and preclinical scanners and can be used to quantitatively characterize cardiac motion. Because tag lines fade towards the end of cardiac cycle, accurately extracting their location over the full cardiac cycle is particularly challenging. In this work, we present a geometric flow approach to the segmentation of tag lines in cardiac MRI. This algorithm is based on the diffeomorphic active contour algorithm that evolves straight lines placed on the tagged MR images to match taglines. We evaluated the accuracy of the algorithm using the Hausdorff distance, with errors of 1.62 ($\pm$0.21 SD) pixel units for horizontal tag lines and 1.79 ($\pm$0.62 SD) pixel units for vertical ones generated from long axis cardiac images. Our algorithm performed well in the presence of noise or large deformations.

Keywords: Cardiac tagged MRI · Diffeomorphic active contours

1 Introduction

The use of small rodents models of human cardiovascular diseases, facilitated by genetic modification, surgical or pharmaceutical interventions, can provide valuable information with regards to underlying molecular mechanisms and can assist in the development and evaluation of novel therapeutic approaches. The value of these models relies on the accurate characterization of cardiac structure and function in response to disease. Accordingly, *in-vivo* non-invasive imaging techniques that can be used to perform morphological and functional analysis of rapidly moving murine heart over the course of the disease is of utmost importance. Cardiovascular Magnetic Resonance Imaging (CMR) provides excellent combination of spatial and temporal resolution to perform longitudinal analysis of murine heart [7].

Over the years, many CMR-based techniques such as spatial modulation of magnetization (SPAMM), displacement encoding with stimulated echoes

O. Bernard et al. (Eds.): FIMH 2023, LNCS 13958, pp. 320–328, 2023.
https://doi.org/10.1007/978-3-031-35302-4_33

(DENSE), or strain encoded magnetic resonance (SENC), each with its own advantages and disadvantages have been developed [13]. Among these techniques, SPAMM is typically most readily available MR sequence in the preclinical scanners.

The SPAMM technique uses slice selective radio-frequency saturation pulses to place stripes or grids (called tags) on the CMR images and makes it possible to characterize the cardiac motion non-invasively as tags moving along during the cardiac cycle [2,3]. Many model-based methods are developed using the tagged MRI to analyze myocardial strain in details [5,9,11,12], including tracking material points in 3D [4,5,8,10].

The Harmonic Phase (HARP) analysis technique [10] is a method of frequency domain analysis, while the active contours (snakes) technique [6] works on the spatial domain of images. Diffeomorphic active contours (DAC) were introduced in [1] as a variant ensuring that contours evolution remains diffeomorphic, preventing, in particular, the occurrence of intersections. In cardiac tag MRI, due to degradation of the image with time, phase aliasing occurs and sensitivity to noise increases, which reduces the accuracy of motion estimation using frequency domain analysis. The new method for tag tracking proposed in this paper is the combination of both methods in which we apply the active contour idea onto the phase images after frequency-filtered on the Fourier domain.

2 Methods

2.1 Imaging Protocol

In-vivo heart images of adult male wild type mice were acquired using Bruker NMR/MRI spectrometer equipped with a 11.7T magnet (Bruker Biospin, Germany). SPAMM tagged MRI was collected (15 frames, echo time (TE) = 1.7852 ms, repetition time (TR) varied according to the heart rate, slice thickness = 0.8 mm, in-plane resolution was 0.1307 × 0.1307 mm^2, flip angle = 10°, NEX = 15, tag spacing = 0.2 mm, tag distance = 0.75 mm) at both short and long axis (3 slices) which resulted in images that contained orthogonally oriented tags. The animal protocol was approved by the Institutional Animal Care and Use Committee of the Johns Hopkins University.

For the remaining of this section, we present a diffeomorphic-active-contour-based (DAC-based) tagline extraction method. The basic idea is first to enhance the tagline on the MRI image and then to evolve a straight line to detect the tagline on that enhanced image. We start this section with an introduction of the Gabor filters bank used for enhancing the original MRI images.

2.2 Image Preprocessing

A Gabor Filter parameterized by $(u_0, v_0, \sigma_x, \sigma_y, \theta)$ computes Fourier coefficients at horizontal and vertical frequencies u_0 and v_0, over elliptical windows with x, y width given by σ_x and σ_y, rotated by θ. It is defined as

$$h(x, y) = g(x', y') \exp[-i2\pi(u_0 x + v_0 y)] \tag{1}$$

where (u_0, v_0) is the center frequency, $\begin{bmatrix} x' \\ y' \end{bmatrix} = \begin{bmatrix} \cos(\theta) & \sin(\theta) \\ -\sin(\theta) & \cos(\theta) \end{bmatrix} \begin{bmatrix} x \\ y \end{bmatrix}$ and

$$g(x', y') = \frac{1}{2\pi\sigma_x\sigma_y} \exp\left(-\frac{(x'/\sigma_x)^2 + (y'/\sigma_y)^2}{2}\right) \tag{2}$$

is a Gaussian filter with the spatial standard deviations $\sigma_{x'}, \sigma_{y'}$. When applied to tagged MRI images such as the one described in Fig. 1, it gets its largest magnitude at frequencies close to the spacing between tag lines. When computed at such frequencies, the phase map of the transform provides images such as I_{ph} and I_{pv} in Fig. 1, which exhibit sharp discontinuities at tag lines. Those lines can then detected using standard edge detection methods, as described in Algorithm 1, which also takes as input a segmented mask of the epicardium.

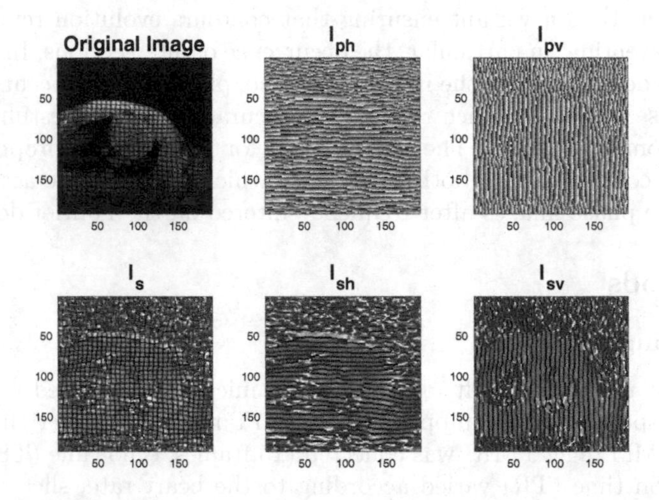

Fig. 1. Image Sharpening to enhance the horizontal and vertical taglines. I_{ph} and I_{pv} are the 2D phase map generated by Gabor filters bank at frequencies adjusted for optimal sharpening. Taglines are extracted based on images I_{sh} and I_{sv} (described in Algorithm 1) filtered with the mask and the Gaussian filter.

Algorithm 1. *Image preprocessing [10]. (See Fig. 1 and Fig. 2)*

1. *Let I_{ph} and I_{pv} denote the 2D phase map generated by Gabor filters bank which enhance the horizontal taglines and the vertical taglines on I respectively.*
2. *Let $I_{phy} = \frac{\partial I_{ph}}{\partial y}$, $I_{pvx} = \frac{\partial I_{pv}}{\partial x}$, $I_s = \frac{1}{I_{phy}^2 + I_{pvx}^2}$, $I_{sh} = \frac{1}{1+I_{phy}^2}$, $I_{sv} = \frac{1}{1+I_{pvx}^2}$.*
3. *Filter I_s, I_{sh}, I_{sv} with the mask which is the dilated epicardium to get I_{sm}, I_{shm}, I_{svm}.*
4. *Apply Gaussian filter to I_{sm}, I_{shm}, I_{svm} to get $I_{smg}, I_{shmg}, I_{svmg}$. For simplicity, we still use the same notation of I_s, I_{sh}, I_{sv} to denote $I_{smg}, I_{shmg}, I_{svmg}$.*
5. *Find the boundary of the mask. Let m denotes the mask. $m_x = \frac{\partial m}{\partial x}$, $m_y = \frac{\partial m}{\partial y}$, Boundary $= m_x^2 + m_y^2 > 0$.*

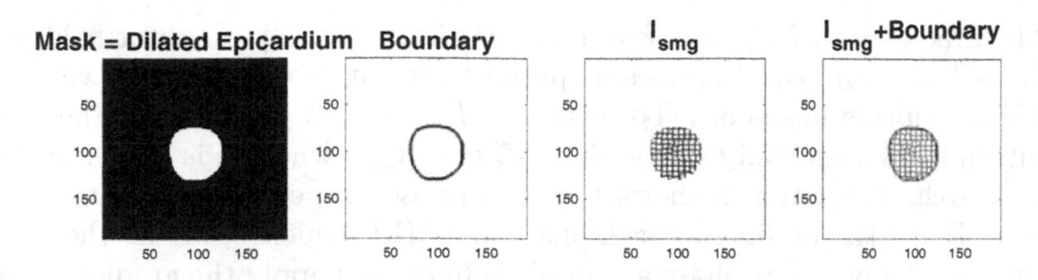

Fig. 2. Image Filtering with the mask and the Gaussian filter. Boundary $= m_x^2 + m_y^2 > 0$. I_{smg} is obtained by filtering I_s with the mask and the Gaussian filter. Without confusion, I_{smg} will still be noted as I_s later on.

2.3 Tagline Extraction Based on Diffeomorphic Active Contours

The DAC model evolves a curve to detect objects in a given image subject to some constraints. To extract a tagline on the preprocessed image, we evolve a family of initial lines toward the taglines. Let f be the preprocessed image and $l_{0k} : [0,1] \to \mathbb{R}^2$ be the initial lines, for $k = 1, \ldots, H$. The DAC model minimizes $\sum_{k=1}^{m} E(l_k; l_{0,k})$ over families of lines $l_k : [0,1] \to \mathbb{R}^2$, where

$$E(l; l_0) = \int_0^1 \frac{\|l'(t) - l_0'(t)\|^2}{\|l_0'(t)\|} dt + \lambda_1 \frac{1}{\|l\|} \int_0^1 f(l(t)) \|l'(t)\| dt + \lambda_2 [B(l(0)) + B(l(1))] \quad (3)$$

where $\|l\|$ denotes the length of l and prime denotes derivatives in t. The first term measures the movement from l_0 to l compared with the location of l_0. The second term measures the gray scale of l on f. The third term measures the greyscale of tag line endpoint coordinates on the image mask boundary (Fig. 2). It is a penalty term added to ensure that the endpoints of l stay on the boundary of the segmented mask, preventing l from shrinking. If there is no such penalty, the minimization of the energy will force the line to shrink to a point. Since the boundary has the smallest greyscale of the whole image, the endpoints of the lines staying on the boundary will be the best choice for the energy to stay as small as possible.

The gradient of $E(l)$ is needed in the minimization, and is given by:

$$\begin{aligned}
\nabla E(l; l_0) = &-2 \left(\frac{l'(t) - l_0'(t)}{\|l_0'(t)\|} \right)' + \frac{\lambda_1}{\|l\|} \left(\nabla f(l(t)) \|l'(t)\| - \left((f - \bar{f}) \frac{l'(t)}{\|l'(t)\|} \right)' \right) \\
&+ \left(\lambda_2 \nabla B(l(0)) - 2 \frac{(l'(0) - l_0'(0))}{|l_0'(0)|} - \frac{\lambda_1}{\|l\|} \left((f(l(0)) - \bar{f}) \frac{l'(0)}{\|l'(0)\|} \right) \right) \delta_{t=0} \\
&+ \left(\lambda_2 \nabla B(l(1)) + 2 \frac{(l'(1) - l_0'(1))}{|l_0'(1)|} + \frac{\lambda_1}{\|l\|} \left((f(l(1)) - \bar{f}) \frac{l'(1)}{\|l'(1)\|} \right) \right) \delta_{t=1}.
\end{aligned}$$

with $\bar{f} = \int_0^1 f(l(t)) \|l'(t)\| dt / \|l\|$.

The gradients ($\nabla E(l_k, l_{0k}), k = 1, \ldots, H$) form a family of vector fields along the lines $l_1, \ldots, l_H$. DAC applies a spatial Gaussian kernel to these vectors all together, resulting in smoothed directions $\bar{\nabla} E(l_k, l_{0k})$, $k = 1, \ldots, H$. The descent algorithm is then applied as: $l_k \mapsto l_k - \alpha \bar{\nabla} E(l_k; l_{k0})$, where α is estimated using a line search. Algorithm 2 shows how to impose proper straight lines on f at phase 1. To get the taglines at each phase in a MRI sequence, we use the taglines obtained at the previous phase as initial taglines, and apply the gradient descent algorithm. Results are shown in Fig. 3.

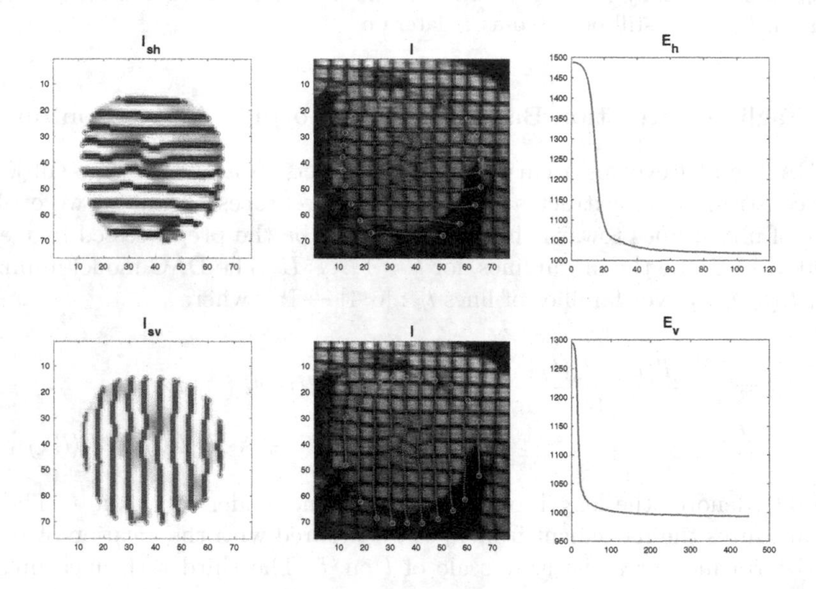

Fig. 3. Tagline Extraction at phase = 1.

Algorithm 2. *Estimate proper straight horizontal lines on f as initial horizontal taglines at phase 1.*

1. *Visually estimate the gap, δ pixels, between adjacent horizontal taglines and use it to place a set of straight horizontal lines on f.*
2. *Move the set of straight horizontal lines up or down by $k \in \mathbb{N}$ pixels to best fit the dark lines on f, $k = 0, 1, \cdots, \lceil \delta \rceil$. Suppose there are H lines in the set, and we find the best location of the set of straight lines by computing:* $\hat{k} = \text{argmin}_{k \in \{0,1,\cdots,\lceil \delta \rceil\}} \sum_{i=1}^{H} \int_{l_i} f ds.$
3. *Crop the segments outside the mask.*

To evaluate the performance of our tag extraction algorithm, we computed the Hausdorff distance (the maximum distance across all points in a set to their closest point in the other set) using four different experiments: 1) we used a single long axis image (15 cardiac phases) to compare the Hausdorff distance between automated and manually traced tag lines, 2) to evaluate noise sensitivity, we repeated the process after adding Rician noise with two different standard deviations ($\sigma = 0.05$ and $\sigma = 0.1$) to the long axis image, 3) to evaluate the ability of the algorithm to identify large deformations, we applied our algorithm using every 2, 3 and 5 slices, and 4) we calculated Hausdorff distance using 5 different long axis images from 3 different animals, two of which had MRI scans 56 days apart. Selecting the long axis allowed us to evaluate our algorithm using a larger number of horizontal tags, as well as longer vertical tags (Fig. 4).

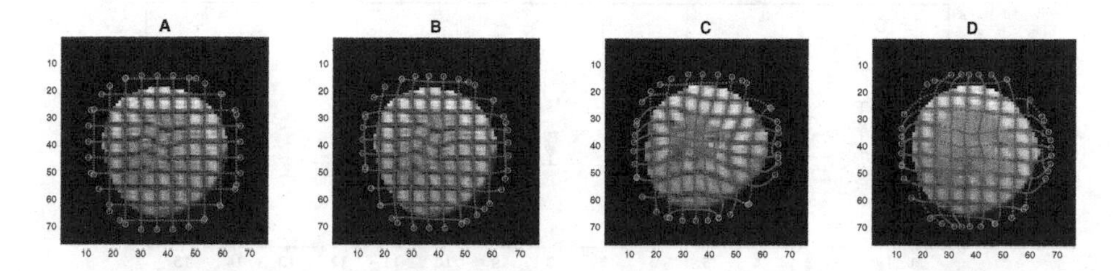

Fig. 4. Tagline-Extraction on the epicardium. A: Initial taglines at phase = 1. B: Extracted taglines at phase = 1. C: Extracted taglines at phase = 8. D: Extracted taglines at phase = 15.

3 Results

Figure 5 shows the comparison of the DAC-based taglines and the manually-extracted taglines. Visual assessment indicates good agreement between the

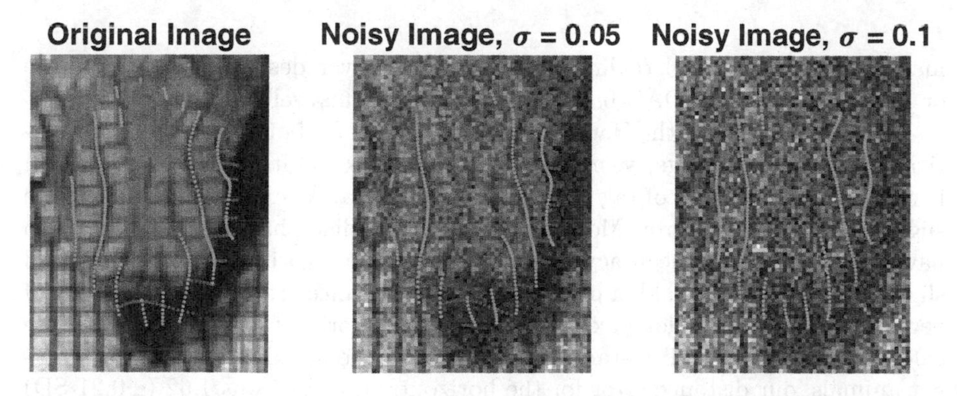

Fig. 5. Robustness of the DAC-based tagline extraction method to added noise. Extracted taglines are superimposed on long axis cardiac image on an original image (left), images with added Rician noise with $\sigma = 0.05$ (middle), and $\sigma = 0.1$ (right), respectively.

Table 1. Evaluations of Tagline Extraction

Tag Orientation	DAC (original image)	DAC added noise ($\sigma = 0.05$)	DAC added noise ($\sigma = 0.1$)	DAC (skip one phase)	DAC (skip two phases)	DAC (skip four phases)
Horizontal	1.78 (0.22)	2.07 (0.65)	2.54 (0.58)	1.73 (0.14)	1.79 (0.1)	1.74 (0.07)
Vertical	2.24 (0.31)	2.59 (0.68)	2.54 (0.52)	2.08 (0.24)	2.31 (0.38)	2.47 (0.63)

– The unit of Hausdorff distance is *pixels*. Smaller Hausdorff distances mean better alignment between ground truth and automatically extracted contours. Numbers in parenthesis represent standard deviation.

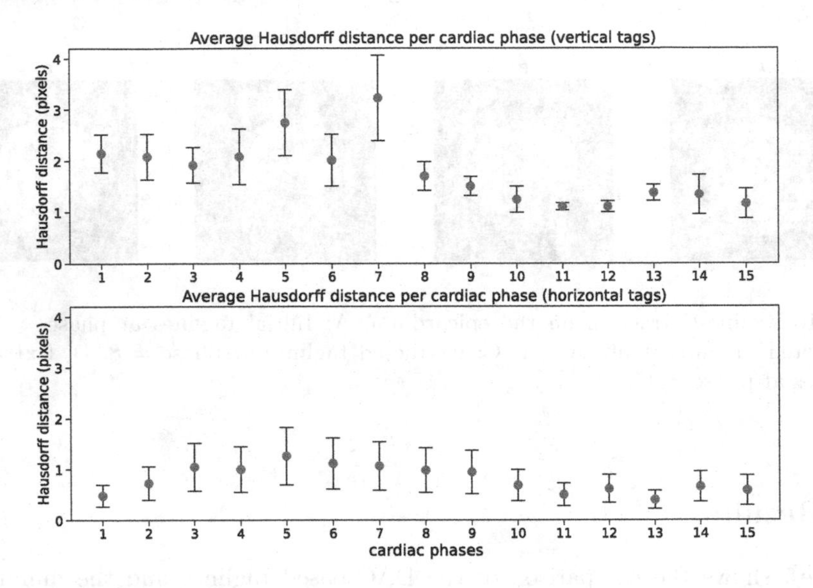

Fig. 6. Average Hausdorff distance error per cardiac phase for vertical (top panel) and horizontal (bottom panel) tag lines. Bars represent standard error.

manual and DAC-based tagline extraction. Moreover despite the added noise or larger deformation, DAC-based algorithm performs well.

Table 1 summarizes the Hausdorff distance error for both horizontal and vertical tag lines. On average, we noticed around two pixel units distance error which is roughly the thickness of tag lines in our MRI data. As expected, adding noise slightly increased the error. Moreover, reducing cardiac phases does not seem to have any effect on the accuracy of our tag detection algorithm. We also noticed slight difference (about half a pixel unit) larger distance error in segmentation of vertical tag lines in the long axis image versus the horizontal tag lines. When we calculated the Hausdorff distance using 5 distinct long axis images from different animals, our distance error for the horizontal tag lines was 1.62 ($\pm$0.21 SD) pixels and for the vertical tag lines was 1.79 ($\pm$0.62 SD). Figure 6 illustrates the trajectory of Hausdorff error across different cardiac phases for both vertical and horizontal tag lines computed from the 5 long axis slices. For the horizontal tag lines, the error is within one or two pixels, while the error is slightly higher for the vertical taglines. More interestingly, the Hausdorff distance error is not increasing for the last few cardiac phases, instead it is increases during the mid to end systole cardiac phases (phases 4–8).

4 Discussion and Conclusion

In this work, we used the DAC algorithm to extract tag lines in cardiac MRI. Both visual assessment and quantitative analysis indicate good agreement between our algorithm and the ground truth (manually extracted tag lines). Our algorithm is relatively insensitive to the added noise and can accommodate larger deformation between one slice to another. Using long axis images, we noticed slightly larger average distance errors for vertical tag lines (aligned with the long axis of heart) compared to the horizontal tag lines. We also noticed that during the systolic cardiac phases there is slightly a higher mismatch between the automated and manually extracted tag lines. Both can be attributed to the signal loss that typically occurs in the apical region of the heart due to magnetic field inhomogeneity in high magnetic fields (see Fig. 5 for an example of null signal represented as a large dark area in the bottom of left panel masking the apex of myocardium). In long axis images we have fewer vertical tag lines than horizontal. These vertical tag lines are passing through the apical region with signal loss make it more difficult to identify those part of taglines. Moreover, the extent of signal loss is increased during the systolic phase exaggerating the problem for those phases.

Acknowledgement. This work was supported by several grants from the National Institutes of Health (HL130292, HL61912, HL63030).

References

1. Arrate, F., Ratnanather, J.T., Younes, L.: Diffeomorphic active contours. SIAM J. Imaging Sci. **3**(2), 176–198 (2010)
2. Axel, L., Dougherty, L.: Improved method of spatial modulation of magnetization (SPAMM) for MRI of heart wall motion. Radiology **172**, 349–350 (1989)
3. Axel, L., Dougherty, L.: MR imaging of motion with spatial modulation of magnetization. Radiology **171**(3), 841–845 (1989)
4. Chen, T., Chung, S., Axel, L.: Automated tag tracking using Gabor filter bank, robust point matching, and deformable models. In: Sachse, F.B., Seemann, G. (eds.) FIMH 2007. LNCS, vol. 4466, pp. 22–31. Springer, Heidelberg (2007). https://doi.org/10.1007/978-3-540-72907-5_3
5. Denney, T.S., Jr., McVeigh, E.R.: Model-free reconstruction of three-dimensional myocardial strain from planar tagged MR images. J. Magn. Reson. Imaging **7**(5), 799–810 (1997)
6. Kass, M., Witkin, A., Terzopoulos, D.: Snakes: active contour models. Int. J. Comput. Vis. **1**(4), 321–331 (1988)

7. Ku, M.-C., Huelnhagen, T., Niendorf, T., Pohlmann, A.: Cardiac MRI in small animals. In: García-Martín, M.L., López-Larrubia, P. (eds.) Preclinical MRI. MMB, vol. 1718, pp. 269–284. Springer, New York (2018). https://doi.org/10.1007/978-1-4939-7531-0_16
8. McVeigh, E.R.: MRI of myocardial function: motion tracking techniques. Magn. Reson. Imaging **14**(2), 137–150 (1996)
9. Moore, C.C., O'Dell, W.G., McVeigh, E.R., Zerhouni, E.A.: Calculation of three-dimensional left ventricular strains from biplanar tagged MR images. J. Magn. Reson. Imaging **2**(2), 165–175 (1992)
10. Osman, N.F., Kerwin, W.S., McVeigh, E.R., Prince, J.L.: Cardiac motion tracking using CINE harmonic phase (HARP) magnetic resonance imaging. Magn. Reson. Med. **42**(6), 1048–1060 (1999)
11. Park, J., Metaxas, D., Axel, L.: Analysis of left ventricular wall motion based on volumetric deformable models and MRI-SPAMM. Center for Human Modeling and Simulation 99 (1996)
12. Young, A.A., Axel, L.: Three-dimensional motion and deformation of the heart wall: estimation with spatial modulation of magnetization-a model-based approach. Radiology **185**(1), 241–247 (1992)
13. Zhang, X., Alexander, R.V., Yuan, J., Ding, Y.: Computational analysis of cardiac contractile function. Curr. Cardiol. Rep. **24**(12), 1983–1994 (2022). https://doi.org/10.1007/s11886-022-01814-1

Weighted Tissue Thickness

Nicolas Cedilnik(✉) and Jean-Marc Peyrat

inHEART, Pessac, France
nicolas.cedilnik@inheart.fr
https://inheartmedical.com

Abstract. Measuring the thickness of a tissue can provide valuable clinical information; anatomical structures segmented on medical images can include sub-structures ("inclusions") corresponding to a different biological tissue. This article presents a method, based on partial differential equations, to measure the thickness of one specific tissue in this particular configuration.

After describing the mathematical formulation of our "weighted thickness" definition, we show on synthetic geometries in one, and two dimensions that it outputs the expected results. We then present three possible applications of our method on cardiac imaging data: measuring the muscular thickness of a ventricle with fat infiltration; measuring the thickness of an infarct scar; visualising the transmural extent of an infarct scar.

Keywords: imaging · thickness · infarct · fat

1 Introduction

Measuring the thickness of an anatomical structure is a common task for radiologists. More specifically, in the field of cardiology, it is often needed to measure the thickness of the myocardial muscle, as it convey information about its health status, which has been shown to be related to its electrophysiological properties ([2,5]).

The visible myocardium segmented on medical images can have inclusions of non-muscle tissue such as fat infiltration, calcifications, or fibrosis. It can happen that we are interested in measuring the thickness of the actual muscle fibers, ignoring these inclusions (Fig. 1). In this article we describe a method to perform this task, and show that it can reciproqually be used to measure the thickness of the inclusion.

2 Methodology

Our method is an extension of the method described by Yezzi and Prince ([6]).

O. Bernard et al. (Eds.): FIMH 2023, LNCS 13958, pp. 329–337, 2023.
https://doi.org/10.1007/978-3-031-35302-4_34

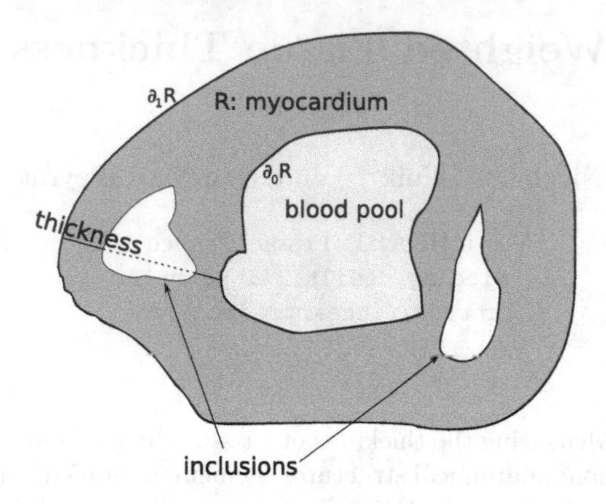

Fig. 1. Illustration of the problem addressed by our method. We want to measure the thickness of R, excluding the thickness of the inclusions, i.e. the length of the thickness line, exclusing the dotted part. $\partial_0 R$: inner boundary. $\partial_1 R$: outer boundary.

2.1 Correspondence Trajectories

Briefly, Yezzi and Prince define thickness at each point x of the tissue region R as the total arclength of a unique curve, passing through x. This curve originates on the the inner boundary of the tissue region $\partial_0 R$, and terminates on its outer boundary $\partial_1 R$ (see Fig. 1)

Such curves can be obtained by solving the Laplace equation over R:

$$\Delta u = 0 \tag{1}$$

with the Dirichlet boundary conditions:

$$u(\partial_0 R) = 0 \text{ and } u(\partial_1 R) = 1 \tag{2}$$

u defines a scalar field in R and is used to define curves within R with the tangent field $\overrightarrow{T}$:

$$\overrightarrow{T} = \frac{\nabla u}{||\nabla u||} \tag{3}$$

For didactic purposes, in Fig. 1 the curve is represented as a straight dotted line ("thickness"), but $\overrightarrow{T}$ actually defines curvy trajectories for complex shapes of R.

2.2 Arclength Computation

Yezzi and Prince ([6]) define the length functions L_0, where $L_0(x)$ gives the arclength of the correspondence trajectory between $\partial_0 R$ and x (reciprocally, L_1(x) between $\partial_1 R$ and x).

$$\nabla L_0 \cdot \overrightarrow{T} = 1, \text{ with } L_0(\partial_0 R) = 0 \tag{4}$$

$$-\nabla L_1 \cdot \overrightarrow{T} = 1, \text{ with } L_1(\partial_1 R) = 0 \tag{5}$$

Thickness at every point x is then defined by summing these length functions:

$$W(x) = L_0(x) + L_1(x) \tag{6}$$

2.3 Weighted Arclength

To account for the inclusion of tissue which thickness we want to exclude, we propose to replace the right term (constant) of eq. (4) and eq. (5) by a function f varying over the tissue region.

$$\nabla L_0 \cdot \overrightarrow{T} = f(x) \tag{7}$$

$$-\nabla L_1 \cdot \overrightarrow{T} = f(x) \tag{8}$$

We call f the arclength weight function and it has the following properties:

- If x is fully occupied by the tissue region we want to measure, $f(x) = 1$
- If x is fully occupied by an inclusion we want to exclude from the thickness measure, $f(x) = 0$

$f(x)$ can also take any value in the $[0, 1]$ interval. This can be useful to account for partial volume effects by using a transfer function between the intensity (eg, Hounsfield Units) of the medical image and f.

2.4 Ray-Tracing as a Possible alternative

Another approach to achieve comparable measurements is to use ray-tracing. In such framework, one needs to define straight lines emanating from one surface and measure the length of the segment between the two surfaces $\partial_0 R$ and $\partial_1 R$. Typically, these lines would be normal to the surface of the considered anatomical structure at the point where they emanate. The measured distance (ie, the tissue thickness) could be similarly weighted with the weight function f.

However, our methodology presents several advantages over this approach:

- It operates directly on voxels and it does not require to define a surface, something usually achieved by converting voxel data to triangular meshes. While there are several algorithms and tools to achieve this conversion, arbitrary choices must be made to control, among other things, the smoothness of the surface; and these choices, influencing the directions of the rays, can have drastic effects on the thickness values.
- The results with our method are thickness values for all voxels of the segmented structure, and not on a single surface. This can be leveraged for further analysis, or parameterizing fast electrophysiological models based on rectangular-grid data ([1,4]).

- There is no need to define out of which surface these trajectories have to be calculated. Since normals have different directions depending on whether they are defined from the inner or outer surface, this would result in visually different thickness maps depending on the considered surface. With our method, and as a consequence of both the definition of the correspondence trajectories and the voxel nature of the output, one can see topologically similar thicknesses on the epicardial, endocardial, or even mid-wall surface.

3 Implementation

3.1 Algorithm

It has been shown ([6]) that computing eq. (4) and eq. (5) over R, in 3D amounts to solving (equations (8) and (9) of [6])

$$L_0[i,j,k] = \frac{f(x) + |T_x|\, L_0[i \mp 1, j, k] + |T_y|\, L_0[i, j, \mp 1, k] + |T_z|\, L_0[i, j, k \mp 1]}{|T_x| + |T_y| + |T_z|}$$

$$L_1[i,j,k] = \frac{f(x) + |T_x|\, L_1[i \pm 1, j, k] + |T_y|\, L_1[i, j, \pm 1, k] + |T_z|\, L_1[i, j, k \pm 1]}{|T_x| + |T_y| + |T_z|}$$

NB: in [6], $f(x)$ is a constant equal to 1.

3.2 Numerical Solving

Results can be obtained by the same methods described by Yezzi et Prince ([6]). Iterative approaches until convergence are possible, but a fast-marching-like algorithm is more efficient (computational resource-wise), especially in a single-threaded context, since it requires a single pass over R, in theory. However we want to note here, than unlike [6] describe, our experiments showed that this fast-marching-like approach does not reach convergence in a single pass. We propose an iterative/fast-marchine-like combined approach where the traversal order of the first pass is memorized for subsequent passes, reaching convergence more rapidly than a naive iterative approach.[1]

3.3 Results on Toy Geometries

One Dimension. We conducted experiments in one dimension to verify that the results were those expected. We set correspondence trajectories to straight horizontal lines and set a 4-element wide "hole" along these trajectories where the weight function values ranged from 0 to 1 (Fig. 2).

[1] An informal benchmark with a real left ventricular wall geometry of 363268 voxels (voxel spacing = 0.8 mm), on an AMD Ryzen 9 3950X CPU, single-threaded, took 11.2 ± 0.2 s for the iterative approach vs 7.5 ± 0.2 s for the semi-ordered approach over 10 runs.

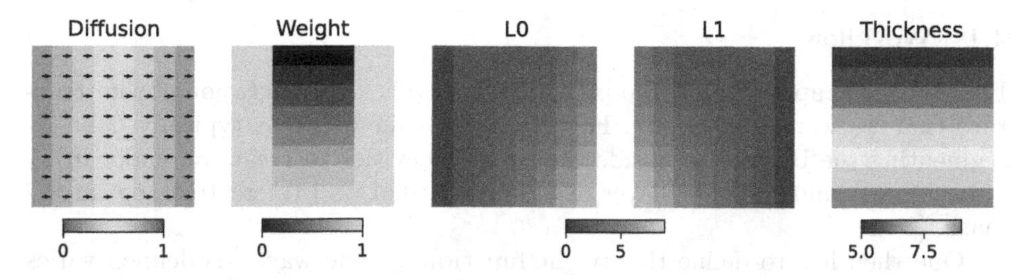

Fig. 2. Toy example in 1D. From top to bottom, each row has an inclusion with a increasing weight. From left to right: scalar field u ("diffusion") from eq. (1) and tangent field $\overrightarrow{T}$ (eq. (3)); weight function f; weighted arclengths $L0$ and $L1$ (eq. (4), eq. (5)) and the resulting weighted thickness.

As expected, the resulting *weighted thickness* $W_p(x)$ equals to the number of elements n in R along the horizontal line when the weight function $f(x)$ is set to 1 along the line, corresponding to the non-weighted thickness computation. When $f(x_{\text{holes}}) = 0$, $W_p(x) = n - n_{\text{holes}}$. For $0 < f(x_{\text{holes}}) < 1$, $n < W_p(x) < -n_{\text{holes}}$.

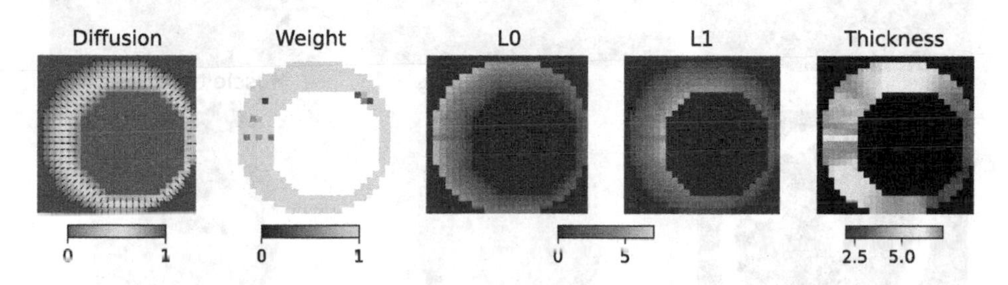

Fig. 3. Toy example in 2D. Refer to the legend of Fig. 2 for details.

Two Dimensions. For illustrative purposes, before showing applications on clinical data in 3D, we also present a 2 dimensional toy geometry (Fig. 3). It resembles a radiological "short-axis view" of a human heart, in which the myocardium presents inclusion of pixels where $f(x) < 1$. As can be seen on Fig. 3 (rightmost picture), the parts of the (toy) myocardium on the left where such inclusions are present present a smaller *weighted thickness* values compared to the surrounding, "non-weighted" regions.

4 Applications

In this section we present possible applications of our method on clinical data.

4.1 Workflow

For all clinical applications, the preliminary step is a segmentation of the structure that we want to measure. For the myocardium, this is typically done by segmenting the blood pool inside the cardiac cavities to create an endocardial binary mask, and the outer layer of the myocardial wall to create an epicardial mask. [3]

One then has to define the weight function f. The way it is defined varies depending on the application, as shown in the following sections.

4.2 Muscle-Only Thickness

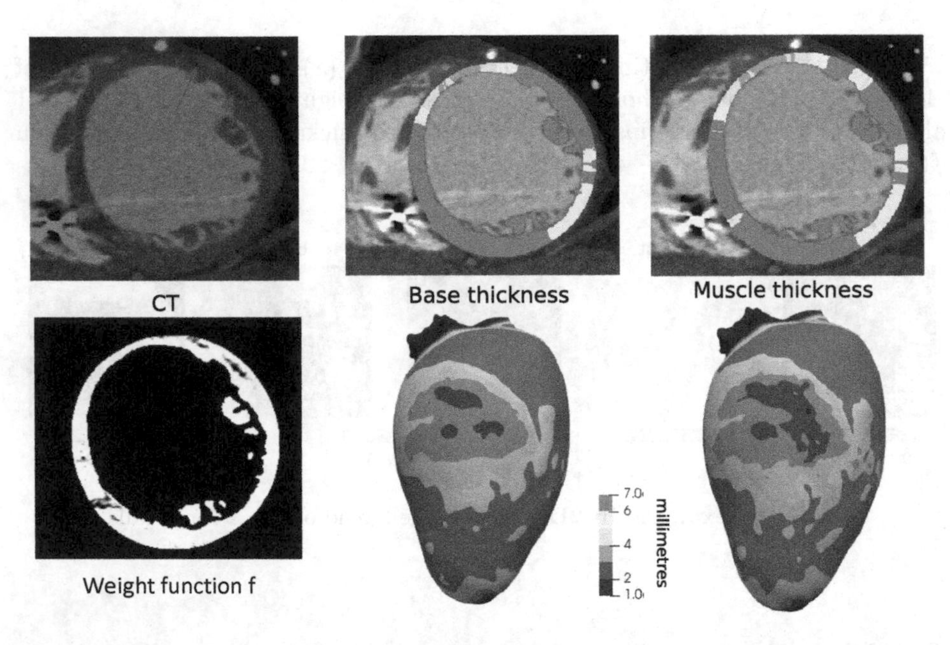

Fig. 4. Myocardium-only thickness. From left to right: (top) CT image showing fat inclusions in the left myocardium (red arrows); (bottom) result of $f(x)$; non-weighted thickness; weighted thickness showing more severe thinning. The thicknesses are shown both on a 2D slice (top) and projected on a surfacic mesh of the ventricular wall.

Fat and calcification inclusions in the myocardium are common. Evaluating the thickness of the muscle fibers excluding these inclusions is the most straightforward application of our method.

The Hounsfield units (HU) of muscle and fat are close, especially in the context of iodine-injected CT. After segmentation, we performed a histogram analysis of HU values within the myocardium to determine the peak value P, and standard deviation δ. We define a transfer function to set $f(x)$ depending on the HU such that:

- for $I(x) > P - \delta$, $f(x) = 1$
- for $I(x) < P - 2\delta$, $f(x) = 0$
- for $2\delta < I(x) < \delta$, $f(x)$ is linearly interpolated

An example result is shown on Fig. 4.

4.3 Scar Thickness

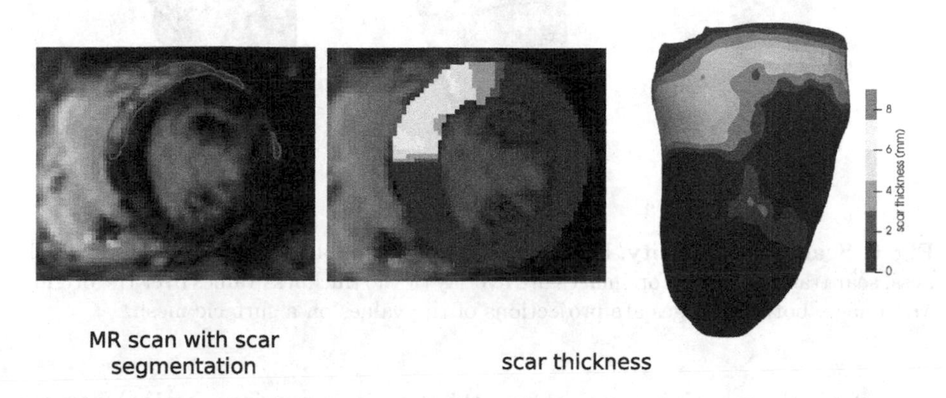

Fig. 5. Scar thickness. From left to right: MR image with late gadolinium enhancement typical of scar tissue (outlined in red); scar thickness visualise overlayed on a MR slice; scar thickness projected on a surfacic mesh

Our method can also be used to quantify the local thickness of an infarct scar. To achieve this, after segmentation of the endocardial and epicardial masks, one has to segment the scar mask (how to obtain such segmentation is outside the scope of this article).

The weight function f is then defined such that $f(x) = 0$ in the myocardium and $f(x) = 1$ in the scar. This can be used to visually assess the thickness of the fibrosis, as shown on Fig. 5.

4.4 Scar Transmurality

Instead of the absolute thickness of the scar, in millimeters, our method can also be used to visualise the local proportion of fibrosis in the myocardial wall, a ratio we call *scar transmurality*. A value of 1 would mean that an area of the myocardial wall is pure fibrosis and reciproqually, a value of 0 must be interpreted as the absence of fibrosis in this area, ie. no ischemic scar.

To obtain such measure, a possible approach is to:

1. Compute the non-weighted thickness of the wall, ie, setting $f(x) = 1$ all over the myocardial wall.

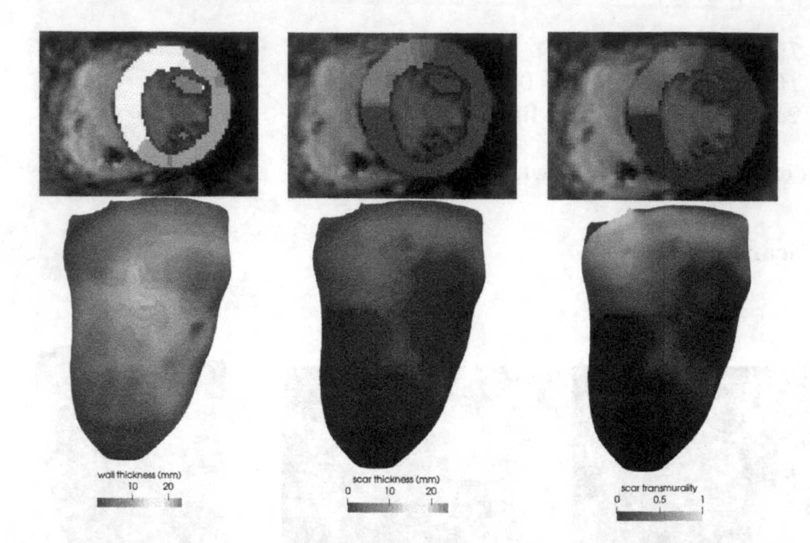

Fig. 6. Scar transmurality. From left to right, total, total wall thickness, scar thickness, scar transmurality. Top images are overlay of the thickness values over the original MR image, bottom images are projections of the values on a surfacic mesh.

2. Perform the voxel-wise ratio of scar thickness (see previous section) over the total thickness.

The transmural extent, *i.e.*, the scar *transmurality* is defined for every voxel of the myocardial wall. See Fig. 6 for an example result.

5 Conclusion

In this article, we defined a measure we named **weighted thickness**, suited to measure the thickness of a tissue on medical images, when it presents inclusions of a different tissue that we want to exclude from the measure. It does so by giving voxels different weights depending on a weight function that must be defined accordingly to the considered application. Like other partial differential equation-based thickness measures, it outputs a scalar map of the same dimensions as the input segmentation, in which each pixel (or voxel) has a thickness value, coherent from the inner to the outer boundary of the measured structure.

The main limitation of our article is that we do not evaluate the clinical relevance of the weighted thickness measure. In particular, the transfer function described in 4.2 has been arbitratily defined and would require tuning and validation for the implied application, i.e. defining the health status of the myocardial on CT images. Our future work will focus on such validation.

References

1. Cedilnik, N., Duchateau, J., Dubois, R., Jaïs, P., Cochet, H., Sermesant, M.: VT scan: towards an efficient pipeline from computed tomography images to ventricular tachycardia ablation. In: Pop, M., Wright, G.A. (eds.) FIMH 2017. LNCS, vol. 10263, pp. 271–279. Springer, Cham (2017). https://doi.org/10.1007/978-3-319-59448-4_26
2. Ghannam, M., et al.: Correlation between computer tomography-derived scar topography and critical ablation sites in postinfarction ventricular tachycardia **29**(3), 438–445. https://doi.org/10.1111/jce.13441. https://onlinelibrary.wiley.com/doi/abs/10.1111/jce.13441
3. Komatsu, Y., et al.: Regional myocardial wall thinning at multidetector computed tomography correlates to arrhythmogenic substrate in postinfarction ventricular tachycardia: assessment of structural and electrical substrate **6**(2), 342–350. https://doi.org/10.1161/CIRCEP.112.000191. http://circep.ahajournals.org/cgi/doi/10.1161/CIRCEP.112.000191
4. Rapaka, S., et al.: LBM-EP: Lattice-Boltzmann method for fast cardiac electrophysiology simulation from 3D images. In: Ayache, N., Delingette, H., Golland, P., Mori, K. (eds.) MICCAI 2012. LNCS, vol. 7511, pp. 33–40. Springer, Heidelberg (2012). https://doi.org/10.1007/978-3-642-33418-4_5
5. Takigawa, M., et al.: Detailed comparison between the wall thickness and voltages in chronic myocardial infarction **30**(2), 195–204. https://doi.org/10.1111/jce.13767
6. Yezzi, A., Prince, J.: An Eulerian PDE approach for computing tissue thickness **22**(10), 1332–1339. https://doi.org/10.1109/TMI.2003.817775. http://ieeexplore.ieee.org/document/1233930/

Strainger Things: Discrete Differential Geometry for Transporting Right Ventricular Deformation Across Meshes

Gabriel Bernardino[1,2(✉)], Thomas Dargent[2,3], Oscar Camara[1], and Nicolas Duchateau[2,4]

[1] BCN Medtech, Department of Information and Communication Technologies, Universitat Pompeu Fabra, Barcelona, Spain
gabriel.bernardino@upf.edu
[2] Univ Lyon, Université Claude Bernard Lyon 1, INSA-Lyon, CNRS, Inserm, CREATIS UMR 5220, U1294, 69621 Lyon, France
[3] Laboratoire de Météorologie Dynamique (UMR8539), École Polytechnique, IPSL, CNRS, Palaiseau, France
[4] Institut Universitaire de France (IUF), Paris, France

Abstract. Cardiac dynamics have been a focus of image analysis, and their statistical models are used in a wide range of applications: generating synthetic datasets, derivation of specific biomarkers of pathologies, or atlas-based motion estimation. Current representations of dynamics, mainly based on displacements, often overlook the physiological basis of cardiac contraction. We propose to use local strain as a more accurate representation, and demonstrate this on 3D echocardiography surface meshes of the right ventricle. Our methodology, based on a differential geometry algorithm, deforms the surface mesh according to a preimposed strain field. This approach allows for a clearer disentanglement between cardiac geometry and dynamics, better differentiating deformation changes than those due to changes in cardiac morphology. The methodology is demonstrated in two toy examples: transporting deformation from one individual to another; and simulating the effects of a pathology on a healthy patient, namely an akinetic right ventricular outflow tract.

Keywords: Right ventricular strain · cardiac function · discrete differential geometry · surface mesh reconstruction

1 Introduction

The heart ventricles are made up of myocytes, which are muscle cells that contract during systole to generate force, thereby raising the pressure inside the chamber and ejecting blood. The shape and function of the heart are closely linked, and any change in shape can affect how the heart works, even if each individual myocyte is still working properly [14]. This is crucial to understand

O. Bernard et al. (Eds.): FIMH 2023, LNCS 13958, pp. 338–346, 2023.
https://doi.org/10.1007/978-3-031-35302-4_35

because some illnesses primarily alter the myocardium shape (such as hypertrophy) and, subsequently, affect the cardiac function, while others primarily alter the conditions under which the myocardium works (such as infarction), leading to altered shape as well as impaired function.

Finding ways to represent cardiac function that are independent to its shape is a significant area of research with various applications. These include identifying functional abnormalities in populations with diverse shape variations, creating synthetic populations of shapes with different functional patterns, and transferring function between different individuals or imaging modalities (for example, using an atlas to estimate the function of a patient for whom only computed tomography (CT) images are available).

Researchers in medical image analysis have proposed various methods for better considering independent components of shape and deformation when analyzing the cardiac function, such as spatio-temporal statistical shape models [1,9,11]. These approaches describe the pointwise displacement of every point of the mesh. More modern approaches use parallel transport to transport displacement fields among two meshes while accounting for the non-linear geometry of the mesh spaces [7,10]. However, this strategy does not consider the physiological basis of cardiac mechanics: it ignores that the resulting pointwise displacement is the aggregation of the local contraction and relaxation of individual myocytes, and thus movement at a certain point is affected by the contraction of many myocytes, which will pull the shape. Therefore, strain-based indices are often used in the clinical community as they are size-independent and better comparable between individuals, allowing for better identification of abnormalities [13]. Strain-based analysis has been performed by authors, but these representations cannot be mapped back to the original space, generating synthetic representative elements for visualisation [5].

While there are standardized strain representations for the left ventricle (LV), which is divided into segments [3], analyzing strain in the right ventricle (RV) is more challenging due to its thin walls and difficulty in accurately imaging it with magnetic resonance imaging (MRI) or ultrasound [12]. This complicates the calculation of strain as the RV is represented as a surface and requires the use of differential geometry, which is the field of mathematics that studies smooth surfaces. Additionally, the RV has higher anatomical variability and irregularity [8] compared to the regular ellipsoid shape of the LV, making standardization of segments more difficult.

In the present work, we propose a strain-based methodology to represent deformation across the cardiac cycle and analyse the cardiac function across a population. In particular, the developed strain representation can be imposed to a given surface mesh at end-diastole (ED), to generate realistic dynamics. This methodology is based on differential geometry, where we solve the inverse problem of strain computation in a local anatomical system of reference, to construct a mesh satisfying the desired strain. To allow a direct transportation and comparison of strain between different individuals, strain is expressed in a local anatomical system of coordinates, consisting of the longitudinal and circumferential coordinates. We demonstrate the utility of this representation on two toy examples: to transport deformation between different subjects, and how the strain of a healthy individual can be modified to model a pathology.

2 Methodology

2.1 Data

To illustrate the methodology, we used patient-specific triangular surface meshes of the RV obtained from a commercial software (4D RV-Function 2.0, TomTec Imaging Systems GmbH, Germany) for segmentation and tracking along 3D echocardiographic sequences. The meshes of different individuals are in approximate point-to-point correspondence thanks to the model-based segmentation.

We tested the methodology on the data from two male controls. They were selected from a small control cohort for being the subjects with the largest global longitudinal and circumferential strain, respectively. We focused on the meshes at end-systole (ES) and ED.

2.2 Discrete Differential Geometry

Discrete differential geometry studies 2D surface meshes embedded in the 3D space. In discrete differential geometry, a triangular surface is characterized (modulo translations and rotations) by its first and second fundamental forms [4], as in classical differential geometry. The first fundamental form is defined at every triangle and consists of the length of each of its sides (this uniquely defines a triangle, modulo translations and rotations). The second fundamental form is defined at each edge, and corresponds to the dihedral angle, which is the angle between the normals of the incident triangles. It is worth noting that not all combinations of dihedral angles and lengths correspond to a valid mesh, but they require to satisfy integrability conditions, which are the discrete equivalent to the Codazzi-Mainardi equations (also known as Gauss-Codazzi) [15]. Informally, they correspond to ensuring that for every possible loop traversing triangles, the sum of the traversed dihedral angles must be a multiple of 2π and is that the common edge of two triangles must have the same length in both triangles.

At each cell of the mesh, we define a local system of coordinates that corresponds to the longitudinal and circumferential directions. These directions are estimated using a heat diffusion method, where the longitudinal direction is defined at each triangle as the gradient of the stationary solution to the heat equation from the apex to the pulmonary and tricuspid valves, as explained in [6]. These directions are orthogonal and defined in the tangent space of the mesh, which in a discrete mesh corresponds to the plane in which the cell is contained.

2.3 Deformation Representation

Given an ED mesh and its corresponding ES mesh in point-to-point correspondence, we can compute all parameters needed to represent its ED-ES deformation. The different steps of these computations can be seen in Fig. 1. Basically,

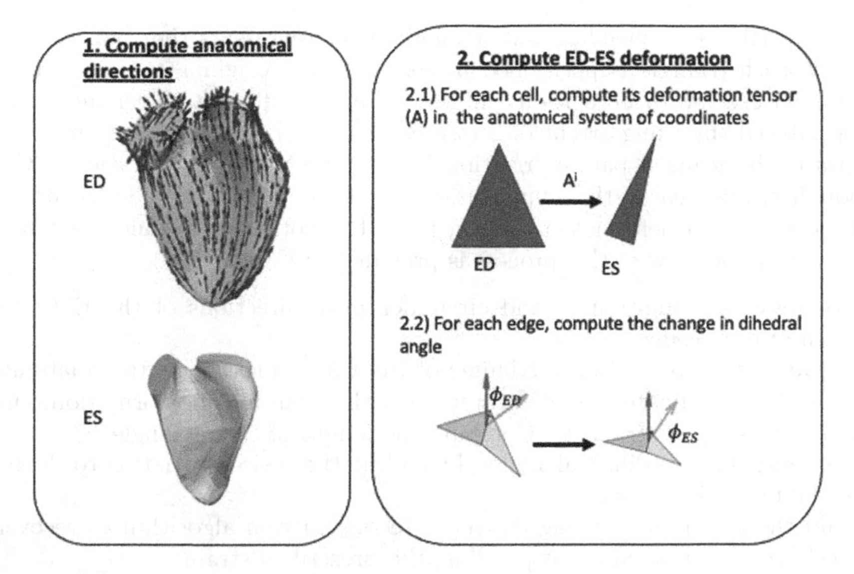

Fig. 1. Computation of the strain representation from an input pair of end-diastole (ED) and end-systole (ES) meshes. First, the anatomical directions are computed at the ED surface (the longitudinal direction is represented as red lines in the mesh, panel 1). Then, we compute the deformation tensors for each triangle, and express them in the anatomical system of coordinates at ED (panel 2.1). Finally, the variation of dihedral angles is estimated (panel 2.2). (Color figure online)

for each i-th cell, we need to compute the deformation tensor $\mathbf{A}^i$, which is the linear mapping from the undeformed cell to the deformed cell. We express this tensor in the local anatomical system of coordinates, leading to a 2×2 matrix. Since we use the Lagrangian formulation, this is defined in the system of coordinates of the ED. The second part of the deformation consists of the changes of dihedral angles, which are represented for each edge, as the difference of the angles at ED minus the ones at ES.

2.4 Strain Transport Algorithm

Here we describe the method to impose a strain (expressed as defined in Sect. 2.3) from a given source individual to a new target individual for which only the ED mesh is required. This strain can come from an acquisition where both ED or ES are available (that might correspond to another individual, but also to the same individual in a different modality), but it may also come from an atlas, or be a purely synthetic strain.

A linear mesh reconstruction has been proposed [15], which given the desired edge lengths and dihedral angles, finds a mesh (with a fixed topology) that satisfies as much as possible these constraints in the L_2 manner. This approach works by optimizing the 3D coordinates of each node, and defining a dummy matrix variable **F**, defined at each triangle, that represents the rotation defining how each triangle is positioned in space. In the original approach, **F** was not forced exactly to be a rotation, which allowed efficient solver algorithms, but produced shrinking artefacts. In an extension [2], **F** was constrained to be a rotation by using a parametrization in the Lie Algebra of rotations SO(3), although this increased the computational cost. The steps of the algorithm to transport a strain deformation from a mesh to another are defined as follows, and a schematic view of this process is provided in Fig. 2. :

1. Compute the longitudinal and circumferential directions of the ED target mesh at each cell.
2. Express each point of each triangle of the target mesh using its anatomical system of coordinates and deform it using the anatomical deformation tensor of the prescribed strain $\mathbf{A}^i$. Compute the length of each triangle.
3. Compute the new dihedral angles, by adding the desired variation to the ones of the target ED mesh.
4. Find the ES mesh by using the mesh reconstruction algorithm to recover a mesh satisfying, as much as possible, the prescribed strain.

Note that a full match to the prescribed strain is not possible, since the desired lengths and dihedral angles computed in steps (1)–(3) above might not exactly satisfy the Codazzi-Mainardi equations (note that they depend on both the target mesh, and the desired strain). This is because, even if we can locally deform every triangle/edge to satisfy the desired new lengths and dihedral angles, imposed strain needs to be globally consistent, as expressed in the Codazzi-Mainardi equations.This implies that the reconstruction step performed in step (4) above can be seen as a projection to the space of dihedral angles and edge lengths that exactly satisfy the equations.

2.5 Experiment Design

We assessed the utility of the proposed methodology in two toy examples:

1. We first examined how to synthetically simulate the impact of pathology on strains. A specific region was selected to become akinetic and undergo no deformation. We selected the right ventricular outflow tract, which can be impacted by diseases such as pulmonary hypertension or arhythmogenic right ventricular dysplasia. The procedure was as follows: we calculated the deformation (consisting of deformation tensors and dihedral angle variations) using the original ED and ES data, and manually selected a region near the outflow tract. We then removed the kinetics from that region by setting the deformation tensors to the identity matrix and the dihedral angle variation to 0. Finally, this modified deformation was imposed to the ED mesh.
2. We tested the capacity of our method to impose a real deformation to a new geometry. We selected the mesh with the largest longitudinal strain as the target, to which we imposed the strain from the mesh with the largest circumferential strain. The evaluation metrics consisted of the residual of the longitudinal and circumferential strains.

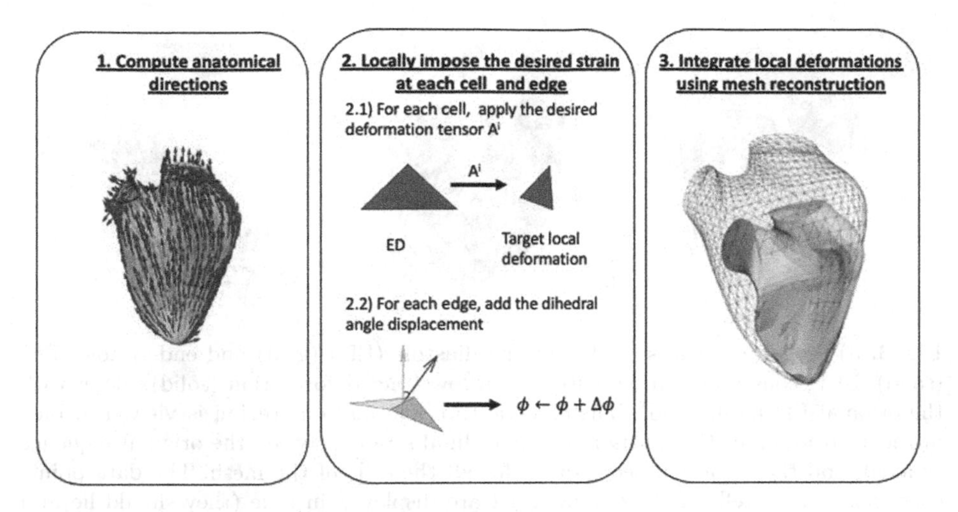

Fig. 2. Algorithm to impose an input strain (composed of the deformation tensors $\mathbf{A}^i$ at each cell i, and the variation in dihedric angles $\Delta\phi$ from end-diastole (ED) to end-systole (ES)).

2.6 Code Availability

The Python code is publicly available as a library in the github (https://github.com/gbernardino/rvmep) for non-commercial academic usage.

3 Results

3.1 Synthetic Deformation

As described above, we first estimated the deformation of a real sequence, synthetically removed deformation at the outflow tract, and applied it to the original ED mesh. This can be seen in Fig. 3. We can also observe that the outflow tract has low deformation, both quantitatively (Fig. 3.c, blue points), and visually, since it has the bloated form characteristic of ED, instead of a more squeezed shape which would be expected at non-pathological ES. As expected, this region has high displacements, as the rest of the ventricle is contracting and pulling it.

3.2 Inter-Individual Transport

We also assessed the ability of our method to correctly transport the real ED–ES deformation of a given individual to a new target ED mesh, both qualitatively and quantitatively, by comparing the original target ES mesh to the one reconstructed after transporting the source deformation. This can be seen in Fig. 4. We observe that the resulting mesh undergoes large circumferential motion, especially near the apical region (marked with arrows), as for the reference. This is visible in the global circumferential strain (GCS), which increased from 16% in

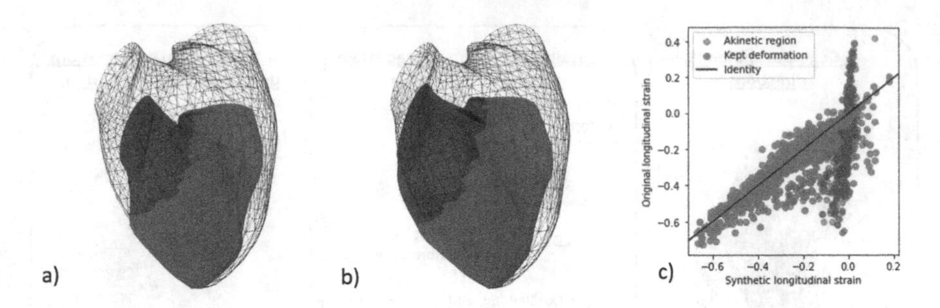

Fig. 3. a) Original surfaces meshes at end-diastole (ED) (grid) and end-systole (ES) (solid). b) Reconstructed mesh with no outflow tract deformation (solid), along with the original ED mesh (grid). The akinetic (no deformation) region is viewed in blue in the solid mesh. c) Comparison of longitudinal strain between the original sequence (x-axis), and the reconstructed (y-axis) for all the cells of the mesh. The data points corresponding to cells of the outflow tract are displayed in blue (they should lie in a vertical line at 0), and the rest are displayed in red (they should have the same value as those of the original deformation). (Color figure online)

the original sequence to 22% in the transported sequence, with the GCS of the target strain being 20%. The mean absolute difference between the desired and obtained longitudinal and circumferential strain (averaged over all cells) was 4% (±6%) and 3% (±3%), respectively. In comparison, the standard deviation of the strain across all triangles was 19%. This mismatch is mainly due to the fact that for a deformation to be exactly applicable to a surface, its elements need to jointly satisfy the Codazzi-Mainardi equations.

4 Discussion

Our strain-based methodology was able to transport a given deformation (either synthetic, or from another individual) to the mesh of another individual. We obtained very promising results in our preliminary experiments. Perfect reconstruction was not obtained, as it is not possible to exactly impose a strain field and recover a valid mesh: the strain field must satisfy the discrete Codazzi-Mainardi conditions.

Contrary to previous methods [1,7,9,11], we use strain to model cardiac dynamics. While this is very well established in the clinical community, and has strong physiological reasons, the engineering community has focused on modelling displacement, and afterwards computing strain. This approach has issues, as displacement has a high dependence on the shape as well as the strain. Our strain-based method allows modelling easily how some pathologies would affect the deformation of an individual, such as a region becoming akinetic. This would be more difficult to model in a displacement-based approach, as the akinetic region would still experience displacement due to pulling from the other contracting areas. The strain representation of the cardiac dynamics is more decou-

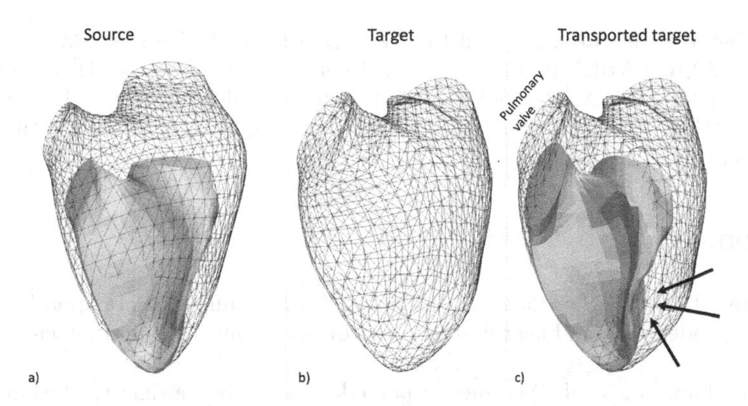

Fig. 4. a): Source end-diastole (ED) (grid) and end-systole (ES) (solid) meshes. b): target mesh at ED (grid). c): resulting ES mesh (solid) after inter-individual transport of the source to the target. Note the circumferential contraction at the apical region (indicated by arrows), which presents a more squeezed shape at ES as in the source ES. No self-intersection artefacts where observed.

pled from the cardiac shape than displacement, since it represents the actual process of myocyte contraction.

Our method has assumptions: it relies on point-to-point correspondences and assumes that the computed anatomical directions are consistent among different individuals. For the former, we think its effect is lower than for displacement-based methods, as strain is less dependent to shape. Regarding the latter, while it can be a problem for pathologies such as some congenital heart diseases, where anatomy suffers large abnormalities, we believe it represents a more appropriate assumption compared to displacement-based strategies.

Limitations of our methods is that local strain data is noisier than displacement, being its spatial derivative, and the fact that our method is not able to capture any rigid-body motion. The latter is not a big issue, since in most cases, the RV apex is fixed due to pericardial constraints, and rigid motion of the heart is mostly due to breathing artefacts.

5 Conclusion

We proposed a discrete differential geometry method to deform a surface mesh using an imposed deformation field, based on a mesh reconstruction algorithm. The method defines strain in a local anatomical system of coordinates, allowing transportation of strain across different individuals. In the future, we would like to apply the method to the left atria and show its usefulness for multimodality: there are modalities with higher temporal resolution (such as 2D echocardiography), from which local strain can be estimated and transported to a higher spatial resolution modality, such as MRI or CT.

Acknowledgements. The authors acknowledge the support from the European Union - NextGenerationEU, Ministry of Universities and Recovery, Transformation and Resilience Plan, through a call from Pompeu Fabra University (Barcelona), and the French ANR (LABEX PRIMES of Univ. Lyon [ANR-11-LABX-0063] and the JCJC project "MIC-MAC" [ANR-19-CE45-0005]). They are also grateful to E. Saloux and A. Hodzic (CHU Caen, France) for providing the imaging data related to the studied population.

References

1. Adams, J., Khan, N., Morris, A., Elhabian, S.: Learning spatiotemporal statistical shape models for non-linear dynamic anatomies. Front. Bioeng. Biotechnol. **11**, 83 (2023)
2. Bernardino, G., et al.: Volumetric parcellation of the cardiac right ventricle for regional geometric and functional assessment. Med. Image Anal. **71**, 102044 (2021)
3. Cerqueira, M.D., et al.: Standardized myocardial segmentation and nomenclature for tomographic imaging of the heart. J. Cardiovasc. Magn. Reson. (2002)
4. Crane, K., Wardetzky, M.: A glimpse into discrete differential geometry. Not. Am. Math. Soc. **64**(10), 1153–1159 (2017)
5. Di Folco, M., Moceri, P., Clarysse, P., Duchateau, N.: Characterizing interactions between cardiac shape and deformation by non-linear manifold learning. Med. Image Anal. **75**, 102278 (2022)
6. Doste, R., et al.: A rule-based method to model myocardial fiber orientation in cardiac biventricular geometries with outflow tracts. Int. J. Numer. Methods Biomed. Eng. **35**(4), e3185 (2019)
7. Guigui, N., Moceri, P., Sermesant, M., Pennec, X.: Cardiac motion modeling with parallel transport and shape splines. In: Proceedings - International Symposium on Biomedical Imaging, pp. 1394–1397 (2021)
8. Haddad, F., Doyle, R., Murphy, D.J., Hunt, S.A.: Right ventricular function in cardiovascular disease, Part II. Circulation **117**(13), 1717–1731 (2008)
9. Hoogendoorn, C., Sukno, F.M., Ordás, S., Frangi, A.F.: Bilinear models for spatio-temporal point distribution analysis: application to extrapolation of left ventricular, biventricular and whole heart cardiac dynamics. Int. J. Comput. Vis. **85**, 237–252 (2009)
10. Piras, P., et al.: Transporting deformations of face emotions in the shape spaces: a comparison of different approaches. J. Math. Imaging Vis. **63**(7), 875–893 (2021). https://doi.org/10.1007/s10851-021-01030-6
11. Puyol-Antón, E., et al.: A multimodal spatiotemporal cardiac motion atlas from MR and ultrasound data. Med. Image Anal. **40**, 96–110 (2017)
12. Shiota, T.: 3D echocardiography: evaluation of the right ventricle. Curr. Opin. Cardiol. **24**(5), 410–414 (2009)
13. Smiseth, O.A., Torp, H., Opdahl, A., Haugaa, K.H., Urheim, S.: Myocardial strain imaging: how useful is it in clinical decision making? Eur. Heart J. **37**(15), 1196–1207 (2016)
14. Stokke, T.M., et al.: Geometry as a confounder when assessing ventricular systolic function: comparison between ejection fraction and strain. J. Am. Coll. Cardiol. **70**(8), 942–954 (2017)
15. Wang, Y., Liu, B., Tong, Y.: Linear surface reconstruction from discrete fundamental forms on triangle meshes. Comput. Graph. Forum **31**(8), 2277–2287 (2012)

Shape Morphing and Slice Shift Correction in Congenital Heart Defect Model Generation

Puck Pentenga[1], Ashley Stroh[2], Wouter van Genuchten[3], Wim A. Helbing[3], and Mathias Peirlinck[1(✉)]

[1] Delft University of Technology, Delft, The Netherlands
mplab-3me@tudelft.nl
[2] Dassault Systemes Simulia Corp., Johnston, RI, USA
[3] Erasmus MC-Sophia Children's Hospital, Rotterdam, The Netherlands

Abstract. Computational heart modeling is a promising approach for improving the prognosis of patients born with congenital heart defects. To create accurate physics-based digital cardiac twins of this population, it is crucial to accurately represent the highly diverse and unique subject-specific heart geometry. In young pediatric patients, this is a challenging endeavor given the lack of high-spatial-resolution imaging data and the risk of slice misalignment. In this study, we set up a multistep shape morphing and slice correction approach to accommodate these challenges and establish a population of biventricular heart models for a variety of healthy, Fallot, and Fontan pediatric patients.

Keywords: Shape Morphing · Congenital Heart Defects · Cardiac digital twins

1 Introduction

Congenital heart defects (CHDs) are one of the most common birth defects, affecting approximately 1% of newborns worldwide. Within this population, approximately 40% of the patients require one or more surgeries during their lifetime. As surgical treatment is seldom curative, many patients with CHD suffer from complications later in life, the most common being heart failure [9]. Heart failure leads to severe debilitating symptoms, drastically reducing the patient's quality of life. Furthermore, it is the most common cause of death in CHD patients worldwide.

Computational heart models offer a promising platform to improve the long-term outlook for this challenging patient population [12]. These models integrate imaging and diagnostic data with physiological and physical principles to provide detailed insights into cardiac function [13,16]. Given the large variability and complexity of CHD anatomical configurations, a computational analysis of CHD cardiac function requires a patient-specific approach [4,8,15,19].

Setting up patient-specific heart models involves geometric reconstruction and meshing based on cardiac imaging data [17]. Within a CHD population,

O. Bernard et al. (Eds.): FIMH 2023, LNCS 13958, pp. 347–355, 2023.
https://doi.org/10.1007/978-3-031-35302-4_36

this is a challenging and cumbersome task. Avoiding harmful radiation in children, magnetic resonance imaging (MRI) is the most suitable technique [2,6]. Unfortunately, MRI in younger children poses significant challenges due to noncompliance with breath-holding and faster heart and respiratory rates [11]. This limits the spatiotemporal resolution of clinically available cardiac MR images in this young population. Deducing detailed three-dimensional models from these images is often time consuming and inconsistent [21]. Shape morphing can be an interesting technique to overcome some of these challenges. This is a technique to convert one three-dimensional geometrical model into another through global and local geometric interpolation techniques. Doing so, we can reconstruct one geometry from another without losing the original topology. Preserving the topology is highly favorable for quantifying individual variations in heart anatomy between patients and for mapping routinely unavailable geometrical details, such as the cardiac myofiber architecture or the Purkinje networks, from one model to another. In this study, we set up a shape morphing framework to construct subject-specific CHD geometries and explore the intrinsic opportunities and challenges this framework entails.

2 Methods

2.1 Cardiac Imaging Data Collection and Segmentation

Anonymized MRI data of seven pediatric patients was collected at the Erasmus University Medical Center. The image data set comprised two healthy female patients aged 8 and 9 years, two male patients with repaired tetralogy of Fallot aged 7 and 8 years, and three male single ventricle patients with Fontan physiology aged 5, 8 and 19. On average, our imaging dataset had a spatial resolution of 1.8–2.0 mm × 1.8–2.1 mm, a temporal resolution of 27–32 ms repetition time 3.41–3.75 ms, echo time 1.31–1.62 ms, flip angle 45 deg, and slice thickness 8 mm and an interslice gap of 1–2 mm. Left and right ventricular contours in the short-axis slices were semiautomatically segmented by a clinician in Medis (Medical Imaging Systems, Leiden, Netherlands), see Fig. 1 - left. The left and right ventricular top contours were segmented on the short axis slices containing the mitral and tricuspid valve respectively.

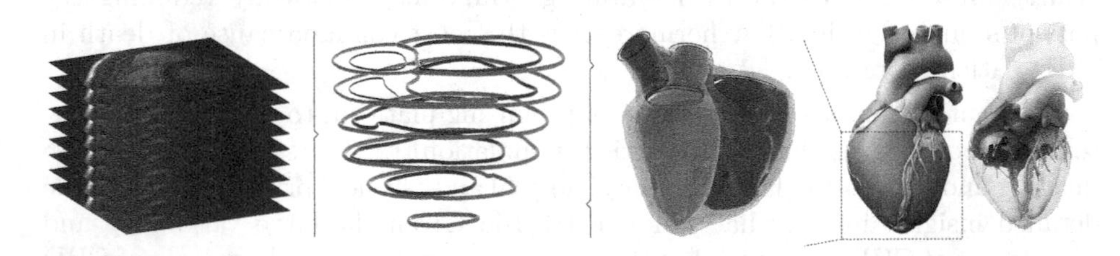

Fig. 1. Inputs to the shape morphing workflow: (left) cardiac image data collection and segmentation contours - (right) high-resolution biventricular CAD geometry deduced from an average 21year-old Caucasian male [24].

2.2 High-Resolution Baseline Model

For our baseline geometry, we make use of a three-dimensional model of the heart of a healthy, 21 year-old, 50th percentile Caucasian U.S. male, created by the Zygote Media Group [24]. This model was deduced from high-resolution MRI data consisting of 0.75 mm thick slices. We imported this geometry in the 3DEXPERIENCE platform (Dassault Systemes, Rhode Island, USA) and constructed three NURBS multipatch surfaces of the left ventricular endocardium, the right ventricular endocardium, and the biventricular epicardium, respectively, with the Zygote geometry as a reference (see Fig. 1 - right). The left ventricular geometry was further morphed to another patient-specific MRI scan.

2.3 Shape Morphing Workflow

We followed a four-step approach to morph the high resolution baseline geometry to the subject-specific contours: global scaling, global alignment, parametric morphing, and manual morphing (see Fig. 2). In the *global scaling* step, we matched the volumetric dimensions of the left and right ventricles to those of the segmented contours. We updated the dimensions by adjusting the affinity dimensions of the endocardium surfaces, which automatically rescaled the epicardial surface. In the *global alignment* step, we semi-automatically found the best translation and rotation parameters to align the globally scaled reference geometry with the endo- and epicardial constraints. The *parametric morphing* step involved making localized adjustments to the endo- and epicardial surfaces. We adjusted the LV and RV long axis length, septal long axis angle, anterior/posterior long axis angle parameters, and ventricle diameter parameters. In the final *manual morphing* step, we selected control nodes on the LV endocardium, RV endocardium, and epicardium surfaces and made additional localized adjustments to each NURBS surface. Throughout the entire shape morphing workflow, we conducted interference checks to ensure that no surfaces intersected.

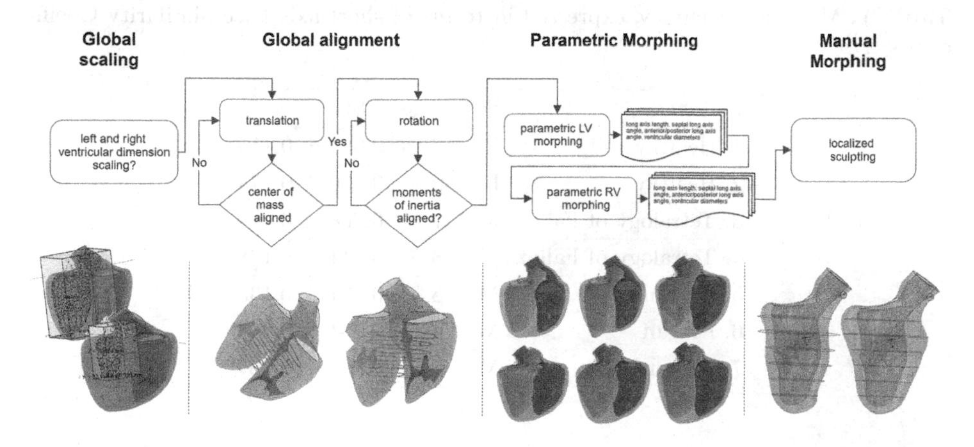

Fig. 2. Shape morphing workflow including a global scaling, global alignment, parametric morphing and manual morphing step.

2.4 Slice Shift Correction

MR imaging can result in both short- and long-axis slice misalignment, also known as slice shift, due to inconsistent breath-holding and/or movement. To address this issue prior to morphing, we proposed two correction approaches. In our first approach, we assumed that the left ventricular *papillary muscles* was an *anatomical straight line* landmark. Following this assumption, we translated all short-axis MR segmentation contours within their respective plane such that the segmented papillary muscles were aligned. In our second approach, we assumed the *center of mass* of each *LV endocardial contour* to be *aligned.* We translated all short-axis MR segmentation contours with respect to their respective LV endocardial centers of mass.

For the patient in our cohort with the slice shift, we qualitatively and quantitatively compared long-axis MR-based segmentation with long-axis slices made from our resulting short-axis-image-morphed geometry. More specifically, we computed the Dice Similarity Coefficient (DSC), which evaluates the overlap between the ground truth (long-axis MR image contours) and long-axis slices of the morphed geometry on a pixel-by-pixel basis [23].

3 Results

3.1 Morphing Accuracy

The final result of the morphing workflow for one of the patients is shown in Fig. 3. For each of the patients, we computed DSC scores that quantified the local match between the clinically segmented short-axis contours and short-axis slices of the final morphed geometries. The respective resulting DSC scores are shown in Table 1. We report DSC scores ranging between 0.830 (best case - patient 7) and 0.681 (worst case - patient 6).

Table 1. Morphing accuracy, expressed in terms of short-axis Dice Similarity Coefficients (DSC).

Patient	Sex	Age	DSC
1. Healthy	F	8	0.691 ± 0.155
2. Healthy	F	9	0.750 ± 0.073
3. Tetralogy of Fallot	M	7	0.768 ± 0.040
4. Tetralogy of Fallot	M	8	0.711 ± 0.138
5. Fontan	M	5	0.801 ± 0.138
6. Fontan	M	8	0.681 ± 0.204
7. Fontan	M	19	0.830 ± 0.037

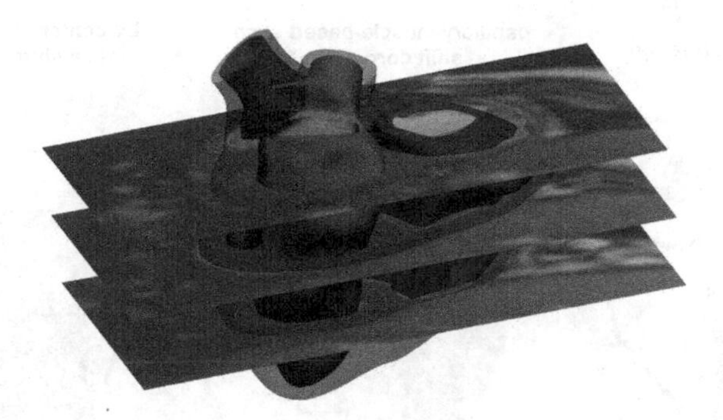

Fig. 3. Final result of shape morphing workflow: a patient-specific Tetralogy of Fallot heart CAD geometry (patient 3).

3.2 Slice Shift Correction

Figure 4 showcases the original slice shift present in Patient 2, and the two different approaches we followed to correct for this misalignment. It can be seen that both approaches tend to restore a more natural ellipsoidal shape for the left ventricle after correction. More specifically, the papillary muscle-based and endocardial center of mass correction approaches amounted to slice shifts of 13.937 ± 3.11 mm and 10.123 ± 4.26 mm respectively.

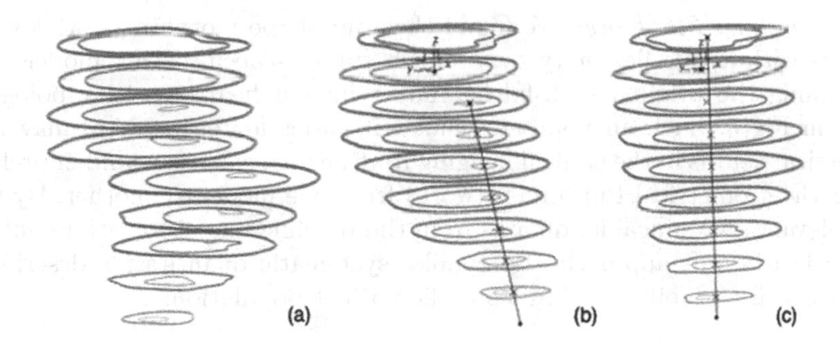

Fig. 4. Slice shift misalignment comparison of correction methods in LV of Patient 2 (healthy heart). Comparison of original data with two correction methods using the center-of-gravity points. (a) Original contours (b) Alignment through papillary muscle (c) Alignment through LV endocardial center of masses.

A qualitative and quantitative validation of this slice shifting approach can be found in Fig. 5. As can be seen, the original model suffered from the slice shift with a long-axis based DSC = 0.747. Both the papillary muscle-based (DSC = 0.916) and LV endocardial center of mass-based (DSC = 0.877) improved the fit to the collected long-axis image slice.

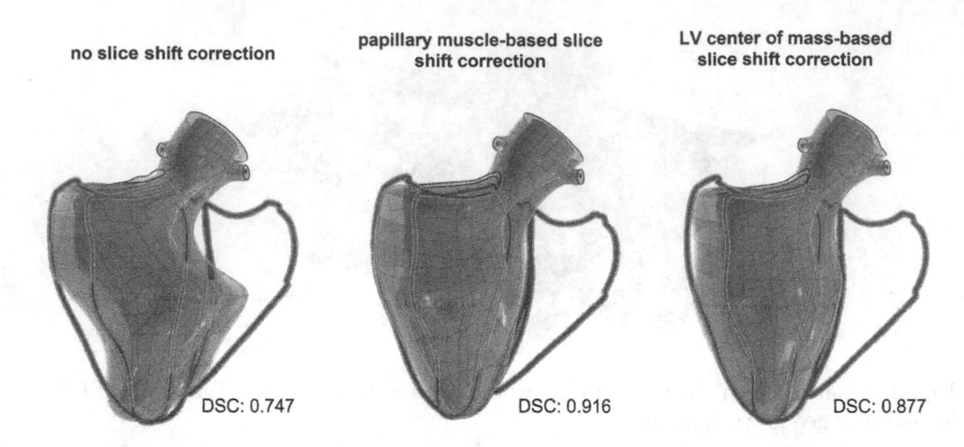

Fig. 5. Qualitative and quantitative validation of our slice shift correction approaches.

4 Discussion

We established and validated a shape morphing framework to develop subject-specific heart models for pediatric patients with complex CHD heart anatomies. Subsequently, we tested two approaches to correct slice shifts, a known problem in pediatric magnetic resonance imaging, both qualitatively and quantitatively.

Shape Morphing Flexibility. We found that our shape morphing workflow provided us with ample flexibility to establish subject-specific heart models while maintaining the original high-fidelity adult human heart model topology, as shown in Fig. 3. Such an approach makes it easier for us to map unavailable geometrical details in the clinical imaging routine (e.g. cardiac myofiber architecture or the geometrical Purkinje network) from one model to another. By mapping relevant anatomical landmarks from the baseline model to various subject-specific hearts, our approach also enables systematic quantitative descriptions of anatomical variability within a specific patient population.

Accuracy. Our local geometric accuracy studies, shown in Table 1, show an average Dice Similarity Coefficient ranging from 0.681 to 0.830. This demonstrates that our approach generated geometric models with good spatial agreement (DSC $>$ 0.70 [23]) with the segmented MR images. Current state-of-the-art deep learning cardiac image segmentation approaches have reached biventricular segmentation (differentiating tissue and blood volumes) Dice scores ranging between 0.780 and 0.950 [5]. However, these approaches worked with vast amounts of ground-truth image segmentations for healthy adults to train these networks. For the more challenging young CHD populations, we did not find any works reporting Dice scores with respect to deep learning based myocardial tissue segmentation accuracy. For relatively easier segmentation of left and right ventricular blood volume, state-of-the-art approaches achieved Dice scores ranging between 0.537 and 0.906 in a young CHD population [7].

Slice Shift Correction. We demonstrated the validity of two strategies to correct for slice shift misalignment in the original short-axis image stack. Based on our analysis of the healthy patient with the greatest slice shift, we found that the papillary-muscle based re-centering approach provided slightly better results than the LV center of mass re-centering approach. However, it is important to note that this conclusion is limited by the absence of long-axis image slices for the other patients, which prevented us from conducting a systematic evaluation across the entire pediatric CHD patient population. If these long-axis slices were available, a multi-view loss objective function could be incorporated to automate the slice shift correction [20], and quantify the validity of our two approaches. Working only with short-axis image slices, another potential solution would be to incorporate an additional anatomic landmark, such as the spine, into the imaged region of interest, and use this landmark for realignment. In our case, such a landmark unfortunately fell outside the imaged region of interest.

Manual Work. Our proof-of-concept framework involves a substantial amount of manual morphing work, where we iteratively tune global dimensions (global scaling), update the translation and rotation vector (global alignment), modify the LV and RV shape parameters (parametric morphing), and finally locally sculpt the endocardial and epicardial surfaces (manual morphing). To a trained fellow, these steps easily take a few hours per subject-specific heart. As such, we aim to automate the global alignment, scaling and parametric morphing steps in our framework. With this goal in mind, we envision our approach can greatly benefit from – but also provide greater flexibility to – statistical shape modeling techniques [3,10,18,22]. On the one hand, statistical shape modeling techniques can automate some steps in our shape morphing framework [14]. On the other hand, the flexibility of our final manual morphing allows us to create more subject-specific heart models that are potentially not represented by statistical shape models trained on limited pediatric CHD imaging datasets [1]. Additionally, we currently started from a healthy human heart as a high-fidelity starting point. In the future, we aim to collect CHD-specific medical imaging data with a higher spatial resolution that allows the development of CHD phenotype-specific high-fidelity heart models to start the shape morphing process from.

Note

This chapter is a summary of the MSc graduation project conducted by Puck Pentenga (AY 2022–2023) at Delft University of Technology. A more detailed description of this work and the developed population of CHD-image-informed biventricular CAD model geometry files can be found on the TU Delft MSc Thesis repository.

References

1. Albà, X., et al.: Reusability of statistical shape models for the segmentation of severely abnormal hearts. In: Camara, O., Mansi, T., Pop, M., Rhode, K., Sermesant, M., Young, A. (eds.) STACOM 2014. LNCS, vol. 8896, pp. 257–264. Springer, Cham (2015). https://doi.org/10.1007/978-3-319-14678-2_27
2. Babu-Narayan, S.V., Giannakoulas, G., Valente, A.M., Li, W., Gatzoulis, M.A.: Imaging of congenital heart disease in adults. Eur. Heart J. **37**(15), 1182–1195 (2015). https://doi.org/10.1093/eurheartj/ehv519
3. Bai, W., et al.: A bi-ventricular cardiac atlas built from 1000+ high resolution MR images of healthy subjects and an analysis of shape and motion. Med. Image Anal. **26**(1), 133–145 (2015). https://doi.org/10.1016/j.media.2015.08.009
4. Biglino, G., Capelli, C., Bruse, J., Bosi, G.M., Taylor, A.M., Schievano, S.: Computational modelling for congenital heart disease: how far are we from clinical translation? Heart **103**(2), 98–103 (2016). https://doi.org/10.1136/heartjnl-2016-310423
5. Chen, C., et al.: Deep learning for cardiac image segmentation: a review. Front. Cardiovasc. Med. **7** (2020). https://doi.org/10.3389/fcvm.2020.00025
6. Gilbert, K., Cowan, B.R., Suinesiaputra, A., Occleshaw, C., Young, A.A.: Rapid D-affine biventricular cardiac function with polar prediction. In: Golland, P., Hata, N., Barillot, C., Hornegger, J., Howe, R. (eds.) MICCAI 2014. LNCS, vol. 8674, pp. 546–553. Springer, Cham (2014). https://doi.org/10.1007/978-3-319-10470-6_68
7. Karimi-Bidhendi, S., Arafati, A., Cheng, A.L., Wu, Y., Kheradvar, A., Jafarkhani, H.: Fully-automated deep-learning segmentation of pediatric cardiovascular magnetic resonance of patients with complex congenital heart diseases. J. Cardiovasc. Magn. Resonance **22**(1) (2020). https://doi.org/10.1186/s12968-020-00678-0
8. Levine, S., Battisti, T., Butz, B., D'Souza, K., Costabal, F.S., Peirlinck, M.: Dassault systèmes' living heart project. In: Modelling Congenital Heart Disease, pp. 245–259. Springer International Publishing, Cham (2022). https://doi.org/10.1007/978-3-030-88892-3_25
9. van der Linde, D., et al.: Birth prevalence of congenital heart disease worldwide. J. Am. Coll. Cardiol. **58**(21), 2241–2247 (2011). https://doi.org/10.1016/j.jacc.2011.08.025
10. Marciniak, M., et al.: A three-dimensional atlas of child's cardiac anatomy and the unique morphological alterations associated with obesity. Eur. Heart J. Cardiovasc. Imaging **23**(12), 1645–1653 (2021). https://doi.org/10.1093/ehjci/jeab271
11. Mitchell, F.M.: Cardiovascular magnetic resonance: diagnostic utility and specific considerations in the pediatric population. World J. Clin. Pediatr. **5**(1), 1 (2016). https://doi.org/10.5409/wjcp.v5.i1.1
12. Naci, H., et al.: Impact of predictive medicine on therapeutic decision making: a randomized controlled trial in congenital heart disease. NPJ Digit. Med. **2**(1) (2019). https://doi.org/10.1038/s41746-019-0085-1
13. Niederer, S.A., Lumens, J., Trayanova, N.A.: Computational models in cardiology. Nat. Rev. Cardiol. **16**(2), 100–111 (2018). https://doi.org/10.1038/s41569-018-0104-y
14. Ordas, S., Oubel, E., Leta, R., Carreras, F., Frangi, A.F.: A statistical shape model of the heart and its application to model-based segmentation. In: Manduca, A., Hu, X.P. (eds.) SPIE Proceedings. SPIE, March 2007. https://doi.org/10.1117/12.708879

15. Peirlinck, M., et al.: Flow optimization in the reconstructed hypoplastic aortic arch. In: Proceedings of the 5th International Conference on Engineering Frontiers in Pediatric and Congenital Heart Disease, pp. 76–78 (2016)
16. Peirlinck, M., et al.: Precision medicine in human heart modeling. Biomech. Model. Mechanobiol. **20**(3), 803–831 (2021). https://doi.org/10.1007/s10237-021-01421-z
17. Peirlinck, M., et al.: Kinematic boundary conditions substantially impact in silico ventricular function. Int. J. Numer. Methods Biomed. Eng. **35**(1), e3151 (2018). https://doi.org/10.1002/cnm.3151
18. Rodero, C., et al.: Linking statistical shape models and simulated function in the healthy adult human heart. PLOS Comput. Biol. **17**(4), e1008851 (2021). https://doi.org/10.1371/journal.pcbi.1008851
19. Tikenoğulları, O.Z., Peirlinck, M., Chubb, H., Dubin, A.M., Kuhl, E., Marsden, A.L.: Effects of cardiac growth on electrical dyssynchrony in the single ventricle patient. Comput. Methods Biomech. Biomed. Eng. (2023). https://doi.org/10.1080/10255842.2023.2222203
20. Wang, S., et al.: Joint motion correction and super resolution for cardiac segmentation via latent optimisation. In: de Bruijne, M., et al. (eds.) MICCAI 2021. LNCS, vol. 12903, pp. 14–24. Springer, Cham (2021). https://doi.org/10.1007/978-3-030-87199-4_2
21. Weissmann, J., Charles, C.J., Richards, A.M., Yap, C.H., Marom, G.: Cardiac mesh morphing method for finite element modeling of heart failure with preserved ejection fraction. J. Mech. Behav. Biomed. Mater. **126**, 104937 (2022). https://doi.org/10.1016/j.jmbbm.2021.104937
22. Young, A.A., Frangi, A.F.: Computational cardiac atlases: from patient to population and back. Exp. Physiol. **94**(5), 578–596 (2009). https://doi.org/10.1113/expphysiol.2008.044081
23. Zou, K.H., et al.: Statistical validation of image segmentation quality based on a spatial overlap index1. Acad. Radiol. **11**(2), 178–189 (2004). https://doi.org/10.1016/s1076-6332(03)00671-8
24. Zygote Media Group Inc: Zygote solid 3d human anatomy - generation ii - development report. Technical report (2014). https://www.zygote.com

Implicit Neural Representations for Modeling of Abdominal Aortic Aneurysm Progression

Dieuwertje Alblas[1(✉)], Marieke Hofman[1], Christoph Brune[1], Kak Khee Yeung[2,3], and Jelmer M. Wolterink[1]

[1] Department of Applied Mathematics, Technical Medical Centre, University of Twente, Enschede, The Netherlands
d.alblas@utwente.nl

[2] Department of Surgery, Amsterdam UMC location Vrije Universiteit Amsterdam, Amsterdam, The Netherlands

[3] Amsterdam Cardiovascular Sciences, Microcirculation, Amsterdam, The Netherlands

Abstract. Abdominal aortic aneurysms (AAAs) are progressive dilatations of the abdominal aorta that, if left untreated, can rupture with lethal consequences. Imaging-based patient monitoring is required to select patients eligible for surgical repair. In this work, we present a model based on implicit neural representations (INRs) to model AAA progression. We represent the AAA wall over time as the zero-level set of a signed distance function (SDF), estimated by a multilayer perception that operates on space and time. We optimize this INR using automatically extracted segmentation masks in longitudinal CT data. This network is conditioned on spatiotemporal coordinates and represents the AAA surface at any desired resolution at any moment in time. Using regularization on spatial and temporal gradients of the SDF, we ensure proper interpolation of the AAA shape. We demonstrate the network's ability to produce AAA interpolations with average surface distances ranging between 0.72 and 2.52 mm from images acquired at highly irregular intervals. The results indicate that our model can accurately interpolate AAA shapes over time, with potential clinical value for a more personalised assessment of AAA progression.

Keywords: Abdominal aortic aneurysm · Implicit neural representation · Deep learning · Aneurysm progression

1 Introduction

Abdominal aortic aneurysms (AAAs) are progressive local dilatations of the abdominal aorta of at least 30 mm that most frequently occur below the renal

Supplementary Information The online version contains supplementary material available at https://doi.org/10.1007/978-3-031-35302-4_37.

O. Bernard et al. (Eds.): FIMH 2023, LNCS 13958, pp. 356–365, 2023.
https://doi.org/10.1007/978-3-031-35302-4_37

arteries. AAAs are mostly asymptomatic, but rupture of an AAA has a mortality rate of 70–80% [3]. To avert rupture, patients can undergo elective repair via either open surgery or an endovascular procedure. Patients become eligible for surgical repair if the diameter of the AAA exceeds a threshold (5.5 cm in men, 5.0 cm in women) or if the AAA diameter has increased more than 1 cm in a year [16].

Prior to elective repair, patients are monitored via periodic outpatient clinic visits and imaging with ultrasound or CT. Although these longitudinal images are primarily used to measure the diameter of the aneurysm, they contain a wealth of information that may be leveraged to better model AAA progression in individual patients [6]. Detailed insight into personalised AAA progression has the potential to aid the physician in clinical decision-making by filling in the gaps in surveillance data. Previous efforts to model the progression of AAAs based on longitudinal imaging include models based on Gaussian processes that represent an underlying deformation field [4], Markov chains [21], deep belief networks [9], or CNNs operating on the surface of the AAA [10].

Recently, implicit neural representations (INRs) have gained traction as natural representations for signals on a spatial or spatiotemporal domain [20]. INRs are multilayer perceptrons that take continuous coordinates as input and output the value of the signal or function at that point [14]. INRs are attractive representation models as derivatives of the signal can be analytically computed using automatic differentiation. In medical imaging, INRs have been used for, e.g., sparse-view CT reconstruction [13,15] and image registration [19]. Moreover, INRs can be used to accurately represent shapes [12], which has led to applications in cell shape synthesis [18] statistical shape modeling [2,11] or surface fitting based on point cloud annotations [1].

In this work, we propose to use INRs with a time coordinate to represent a longitudinal 3D AAA model of a patient and investigate to what extent such a model can be used to *interpolate* and *extrapolate* the AAA surface in time.

2 Methods

We represent the evolving AAA surface as the zero level set of its temporal signed distance function (SDF). We parametrize this function by a neural network $f(\boldsymbol{x}, t; \theta)$, with weights θ.

2.1 Signed Distance Function

A surface can be implicitly represented by the zero level set of its signed distance function. We consider a manifold evolving over time, that we represent by a temporal SDF: $SDF(\boldsymbol{x}, t) : \mathbb{R}^3 \times \mathbb{R} \mapsto \mathbb{R}$. The value of the $SDF(\boldsymbol{x}, t)$ represents the minimum distance to the surface at location $\boldsymbol{x}$ at time t. The temporal SDF of an evolving 2D manifold $\mathcal{M}$ embedded in $\mathbb{R}^3 \times \mathbb{R}$ is defined as:

$$SDF_{\mathcal{M}}(\boldsymbol{x}, t) = \begin{cases} -d(\boldsymbol{x}, \mathcal{M}) & \boldsymbol{x} \text{ inside } \mathcal{M} \text{ at time } t \\ 0 & \boldsymbol{x} \text{ on } \mathcal{M} \text{ at time } t \\ d(\boldsymbol{x}, \mathcal{M}) & \boldsymbol{x} \text{ outside } \mathcal{M} \text{ at time } t. \end{cases} \quad (1)$$

The zero-level set of this function thus describes the manifold $\mathcal{M}$. Moreover, the signed distance function is a solution to the Eikonal equation at each instance in time: $||\nabla_{\boldsymbol{x}} SDF(\boldsymbol{x}, t)|| = 1, \forall \boldsymbol{x}, t$.

2.2 Implicit Neural Representations

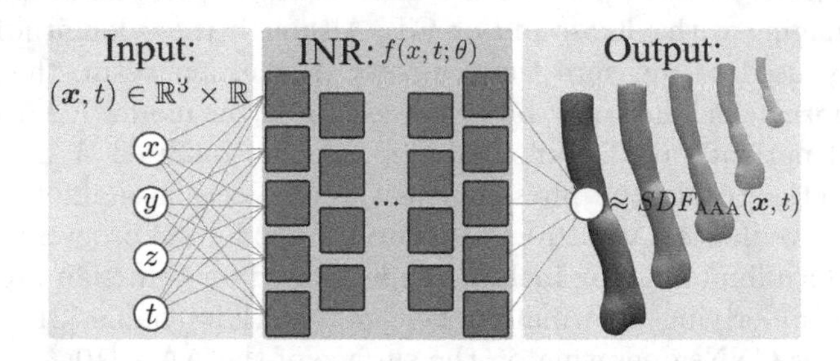

Fig. 1. Schematic representation of our INR, taking spatiotemporal coordinates $(\boldsymbol{x}, t)$ as an input, outputting $SDF(\boldsymbol{x}, t)$ of the AAA surface. Note that a single INR represents the complete evolving AAA of a patient.

In previous work, it has been shown that an SDF of a manifold can be represented by a neural network [1,7,12,18]. Similarly, we embed the remodeling of the AAA over time in an implicit neural representation (INR). We use 4D coordinates from the spatiotemporal domain $\Omega := [-1, 1]^3 \times [-1, 1]$ as input to the network $f(\boldsymbol{x}, t; \theta)$. The output node of our INR approximates the SDF value at the input coordinate. Figure 1 shows a schematic overview of our INR.

We aim to reconstruct $SDF_{\mathrm{AAA}}(\boldsymbol{x}, t)$ given a sequence of point clouds of the AAA surface, representing the aneurysm shape of a single patient over J scans: $\{\mathcal{X}_j\}_{j=1,\dots,J}$, where $\mathcal{X}_{j,\cdot} \subset [-1, 1]^3$. We denote individual points on the jth AAA surface $\boldsymbol{x}_i^j$. To optimise the INR, we sample points on and off the AAA surface at multiple instances in time.

The loss function we use to optimize the INR consists of two terms: a term $\mathcal{L}_{\text{data}_j}$ at each time point t_j where we have ground-truth scan data, and a term $\mathcal{L}_{\text{reg}}$ that regularises the SDF at times the surface is unknown.

$$\mathcal{L}(\theta) = \sum_{1 \le j \le J} \mathcal{L}_{\text{data}_j}(\theta) + \mathcal{L}_{\text{reg}}(\theta), \tag{2}$$

$$\begin{aligned} \mathcal{L}_{\text{data}_j}(\theta) = \frac{1}{N_j} \sum_{1 \le i \le N_j} |f(\boldsymbol{x}_i^j, t_j; \theta)| + \lambda_1 \mathbb{E}(||\nabla_{\boldsymbol{x}} f(\boldsymbol{x}, t_j; \theta)|| - 1)^2 \\ + \lambda_2 \mathbb{E}(|\nabla_t f(\boldsymbol{x}, t_j; \theta)|) \end{aligned} \tag{3}$$

$$\mathcal{L}_{\text{reg}} = \lambda_3 \mathbb{E}\left(||\nabla_{\boldsymbol{x}} f(\boldsymbol{x}, \tilde{t}; \theta)|| - 1\right)^2 + \lambda_4 \mathbb{E}\left(|\nabla_t f(\boldsymbol{x}, t; \theta)|\right). \tag{4}$$

The first term of $\mathcal{L}_{\text{data}_j}$ was introduced in [7]. It ensures $SDF(\boldsymbol{x}_i^j, t_j) = 0$ for all points $\boldsymbol{x}_i^j$ in pointcloud $\mathcal{X}_j$, i.e. that points that are known to be on the AAA

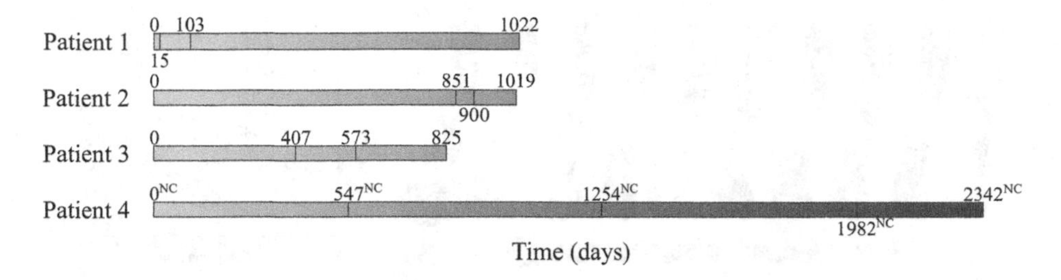

Fig. 2. Timeline of the CT scans of the four patients with longitudinal data, showing scan instances in days. Non-contrast scans are indicated with ^{NC}.

surface are indeed on the zero level set of the SDF. The remaining terms in both parts of the loss function regularise the INR's spatial and temporal gradient. As these terms do not depend on pointcloud data, we evaluate them both at times t_j as well as times data is unavailable. Regularising the norm of the spatial gradient was also introduced in [7] and enforces the INR to be a solution to the Eikonal equation. We evaluate this term at time t_j in $\mathcal{L}_{\mathrm{data}_j}$, and at an arbitrary time point $\tilde{t}$ in $\mathcal{L}_{\mathrm{reg}}$. The temporal regularisation term is introduced in this work to restrict temporal changes of the INR. These are evaluated at time t_j and at multiple arbitrary time points in $\mathcal{L}_{\mathrm{data}_j}$ and $\mathcal{L}_{\mathrm{reg}}$ respectively.

2.3 Data

We retrospectively included longitudinal CT data of four patients scanned at Amsterdam AMC (Amsterdam, The Netherlands) between 2011 and 2020. Three patients were scanned four times, and one patient was scanned five times (Fig. 2). Scan dates were shifted so that the first scan date of each patient became day 0. Patient 1 was scanned three times between day 0 and day 103, followed by a gap of almost three years. The first follow-up image of Patient 2 was after 851 days, after which two additional follow-up images were acquired relatively soon. Patient 3 was scanned more regularly. The follow-up for Patient 4 is the longest, with over 75 months of follow-up. CT scans were a mixture of non-contrast and contrast-enhanced images.

We obtained automatic segmentations of the AAA and vertebra in each of these patients. All CT scans were processed using TotalSegmentator [17], a Python library based on nn-UNet [8] that segments >100 structures in 3D CT images. This library segmented the vertebra with good accuracy in both non-contrast and contrast-enhanced images and the AAA with high accuracy in all non-contrast images. However, segmentation of the AAA in contrast-enhanced images was unsatisfactory. Instead, we used an in-house dataset of 80 contrast-enhanced CT images of AAA patients with annotations of the AAA ranging between the top of the T12 vertebra and the iliac bifurcation to train an additional nn-UNet model. This model achieved a mean Dice similarity coefficient of 0.90 on a separate test set consisting of 13 contrast-enhanced CT scans.

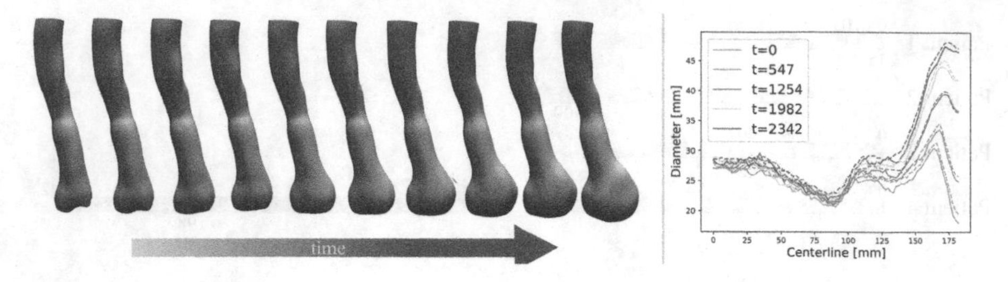

Fig. 3. An optimised INR can be used to extract shape interpolations at an arbitrary number of time points, here we show results for Patient 4. *Left:* We show extracted shapes at ten regularly spaced intervals in time. *Right:* Diameter plots along the centerlines of the aorta, comparing the ground-truth segmentation mask (solid) to the surface fitted by the network at five time points where reference CT scans are available (dashed).

2.4 Preprocessing

In order to evaluate local changes in shape over time, all shapes should be aligned in the same coordinate system. For this, we used rigid registration in ITK on the vertebra segmentations [4]. Subsequently, the surface of each aorta was extracted from the mask and represented as a point set. This resulted in aligned pointcloud representations of the AAA surface for each scan. Finally, before serving as input to the network, the spatial coordinates of the pointclouds of each patient were jointly normalized to the $[-1, 1]^3$ domain. Similarly, the time scale of each patient was normalized to the $[-1, 1]$ interval.

3 Experiments and Results

In all cases, we used an MLP with six fully connected layers containing 256 nodes with Softplus ($\beta = 100$) activations and a final node representing the estimated SDF of the AAA surface. Like [1,7], we used a skip connection, connecting the input to the third hidden layer. The regularization coefficients were set to $\lambda_1 = \lambda_2 = \lambda_3 = \lambda_4 = 0.1$. We used an Adam optimizer with a learning rate of 0.0001 to train our network for 25,000 epochs on an NVIDIA Quadro RTX 6000 GPU. The batch sizes depended on the size of point clouds and ranged between 2877 and 6027.

3.1 Interpolation and Extrapolation

For each patient, we first optimised a single INR (Fig. 1) based on point clouds from all available scans. Because the spatiotemporal input coordinates to the INR are continuous, we can retrieve a shape at any point in time at any resolution. We visualize this in Fig. 3(*left*), where we show ten AAA shapes of Patient 4 at regularly spaced intervals. In Fig. 3(*right*) we compare the diameters along the AAA centerlines of the ground-truth segmentation masks to the

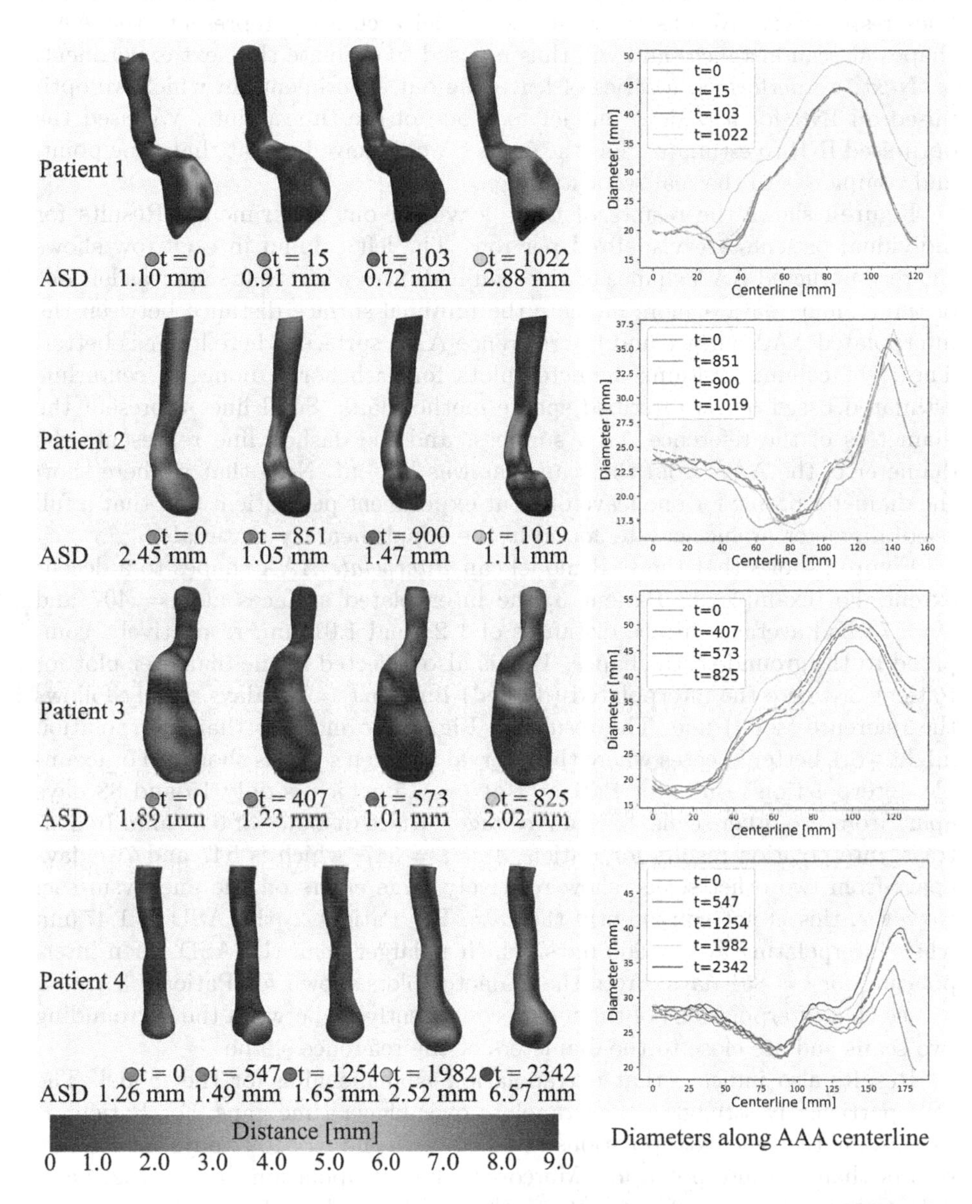

Fig. 4. *Left:* Inter- and extrapolated AAA shapes for each scan. Surface colors indicate distances to reference shapes, averages are indicated below each AAA. Colored dots indicate the corresponding diameter profile in the graph on the right. *Right:* Diameter profiles along each aorta. Solid lines represent reference diameters, dashed lines show interpolated or extrapolated diameters. A complete set of diameter plots can be found in the supplementary materials.

AAA surfaces reconstructed by the network, represented by solid and dashed lines respectively. We observe that the model accurately represents the AAA shapes at scan instances and will thus be used to evaluate the next experiment.

Next, we performed a series of leave-one-out experiments in which we optimised an INR for a patient but left out one of the time points. We used the optimised INR to estimate what the surface would have been at that time point, and compare it to the real reconstruction.

Figure 4 shows the results of these leave-one-out experiments. Results for individual patients are visualized per row. The left column in each row shows the reconstructed AAA shapes at the scan instances when that scan was left out of the training data. Colors indicate the minimal surface distance between the interpolated AAA surface and the reference AAA surface, where lower is better. The right column contains diameter plots for each aorta along its centerline estimated based on an inscribed sphere method [5,9]. Solid lines represent the diameters of the reference AAA surfaces, and the dashed line represents the diameter of the AAA from the scan that was left out. Note that we here show the diameter profile for one leave-one-out experiment per patient and that a full set of diameter profiles can be found in the supplementary materials.

Figure 4 shows that the INR model can *interpolate* AAA shapes to a decent extent. For example, in Patient 3, the interpolated surfaces at $t = 407$ and $t = 573$ had average surface distances of 1.23 and 1.01 mm, respectively, compared to the ground-truth shapes. This is also reflected in the diameter plot for Patient 3, where the interpolated (dashed) line for $t = 407$ days closely follows the reference (solid) line. The results in Fig. 4 also indicate that interpolation might work better in cases where the interval between scans is shorter. For example, interpolation results for Patient 1 at $t = 15$, which is only 15 and 88 days apart from two other scans, have an average surface distance of 0.91 mm. In contrast, interpolation results for Patient 4 at $t = 547$, which is 547 and 707 days apart from two other scans, show relatively large errors on the aneurysm sac. However, this is not consistently the case. For Patient 2, the ASD is 1.47 mm when interpolating at $t = 900$ days, which is larger than the ASD when interpolating for $t = 851$ days. From the diameter plots shown for Patients 3 and 4, we see that interpolations of the model consistently lie between the surrounding two scans and are close to the diameters of the reference shape.

Results also indicate that extrapolation is challenging for the model. The INR particularly struggles to extrapolate over bigger time gaps. For Patient 1, we observe that the extrapolations at $t = 0$ days and $t = 1022$ days have worse results than the interpolations. Moreover, the extrapolation at $t = 1022$ days differs more from the reference shape than at $t = 0$ days due to the difference in time gaps. The diameter profiles for Patient 1 and Patient 2 reveal that the model tends to reconstruct the surface of the last known shape. We hypothesize that this might be due to the temporal regularization term in Eq. 4.

Finally, Fig. 4 indicates that our INR model reacts strongly to small misalignments of the original AAA shapes. Following [4], we register AAA shapes based on segmentation masks of the vertebrae, but this alignment might lead to small local shifts of the AAA. For example, the result for Patient 2, $t = 0$ in Fig. 4 shows errors on the healthy part of the aorta, an area that, in principle, should not show growth over time.

4 Discussion and Conclusion

In this work, we have obtained a personalised model for AAA progression, based on longitudinal CT data. We combine fully automatic state-of-the-art image segmentation methods, registration, and shape modeling with implicit neural representations and adequate regularisation terms to build personalised models of an evolving anatomical structure. In experiments with four longitudinally scanned AAA patients, we have demonstrated how the model represents the evolving shape of an AAA over time. This may impact patient monitoring and treatment; accurate knowledge about the progression of an AAA allows the physician to personalise surveillance and time intervention better based on AAA diameter and growth rate [16].

One appealing aspect of our approach is the continuity of the implicit neural representations. This allows us to reconstruct an AAA mesh at any point in time, at any desired resolution. We have here modeled shape changes over multiple years with sparse and irregularly spaced shape data. Modeling this change through linear interpolation of alternative surface representations, such as meshes or point clouds, would require point-to-point correspondence, a challenging problem that we here circumvent. Moreover, since our network relies on pointcloud data, it is agnostic to imaging modality. This is important for longitudinal studies of AAAs, where imaging modalities such as MRI and 3D US are increasingly used. All these scans can be incorporated into this framework as long as we can extract AAA surfaces. Furthermore, because we represent an evolving shape in space and time in a differentiable neural network, we can add any gradient-based regularisation term to the loss function. We have here included an Eikonal term and temporal regularization, but this framework could be further extended.

Lastly, we found that our model is sensitive to errors in the initial alignment of AAA shapes. Although we have followed [4] in registering based on the location of the vertebrae, better results can likely be achieved by registering based on other landmarks, such as the renal arteries and iliac bifurcation.

One limitation of the current approach is the relatively limited test set of four longitudinally scanned patients, which we aim to increase in future work. By increasing the data set and combining the here proposed model-driven with a data-driven approach [4,9,21], we might address current limitations of our model in extrapolation. Moreover, our current method is based on AAA morphology, but can be extended with biomarkers related to AAA growth and rupture [6]. Furthermore, additional optimization constraints could more properly model the pathophysiology of aneurysms. Whereas our temporal regularization term now aimed to minimise the gradient of the SDF, in future work, we could optimise this gradient within biologically plausible growth rates. This kind of regularization could also be obtained in a data-driven way, by learning a generalisable model from a larger set of patients with longitudinal data. Finally, there is evidence that intraluminal thrombus shape plays a key role in AAA remodeling [21], and it might be beneficial to explicitly represent thrombus in our INR [1].

In conclusion, we have shown that INRs are promising tools in modeling AAA evolution. In future work, this flexible model could be extended with biologically plausible regularization terms and hemodynamic parameters.

Acknowledgements. Jelmer M. Wolterink was supported by the NWO domain Applied and Engineering Sciences VENI grant (18192).

References

1. Alblas, D., Brune, C., Yeung, K.K., Wolterink, J.M.: Going off-grid: continuous implicit neural representations for 3D vascular modeling. In: Statistical Atlases and Computational Models of the Heart. Regular and CMRxMotion Challenge Papers. STACOM 2022. LNCS, vol. 13593, pp. 79–90. Springer, Cham (2022). https://doi.org/10.1007/978-3-031-23443-9_8
2. Amiranashvili, T., Lüdke, D., Li, H.B., Menze, B., Zachow, S.: Learning shape reconstruction from sparse measurements with neural implicit functions. In: International Conference on Medical Imaging with Deep Learning, pp. 22–34. PMLR (2022)
3. Brewster, D.C., Cronenwett, J.L., Hallett, J.W., Jr., Johnston, K.W., Krupski, W.C., Matsumura, J.S.: Guidelines for the treatment of abdominal aortic aneurysms: report of a subcommittee of the joint council of the American association for vascular surgery and society for vascular surgery. J. Vasc. Surg. **37**(5), 1106–1117 (2003)
4. Do, H.N., et al.: Prediction of abdominal aortic aneurysm growth using dynamical Gaussian process implicit surface. IEEE Trans. Biomed. Eng. **66**(3), 609–622 (2018)
5. Gharahi, H., Zambrano, B., Lim, C.Y., Choi, J., Lee, W., Baek, S.: On growth measurements of abdominal aortic aneurysms using maximally inscribed spheres. Med. Eng. Phys. **37**(7), 683–691 (2015)
6. Groeneveld, M.E., et al.: Systematic review of circulating, biomechanical, and genetic markers for the prediction of abdominal aortic aneurysm growth and rupture. J. Am. Heart Assoc. **7**(13), e007791 (2018)
7. Gropp, A., Yariv, L., Haim, N., Atzmon, M., Lipman, Y.: Implicit geometric regularization for learning shapes. In: International Conference on Machine Learning, pp. 3789–3799. PMLR (2020)
8. Isensee, F., Jaeger, P.F., Kohl, S.A., Petersen, J., Maier-Hein, K.H.: nnU-Net: a self-configuring method for deep learning-based biomedical image segmentation. Nat. Methods **18**(2), 203–211 (2021)
9. Jiang, Z., Do, H.N., Choi, J., Lee, W., Baek, S.: A deep learning approach to predict abdominal aortic aneurysm expansion using longitudinal data. Front. Phys. **7**, 235 (2020)
10. Kim, S., et al.: Deep learning on multiphysical features and hemodynamic modeling for abdominal aortic aneurysm growth prediction. IEEE Trans. Med. Imaging **42**, 196–208 (2023)

11. Lüdke, D., Amiranashvili, T., Ambellan, F., Ezhov, I., Menze, B.H., Zachow, S.: Landmark-free statistical shape modeling via neural flow deformations. In: Wang, L., Dou, Q., Fletcher, P.T., Speidel, S., Li, S. (eds.) Medical Image Computing and Computer Assisted Intervention – MICCAI 2022. MICCAI 2022. LNCS, vol. 13432, pp. 453–463. Springer, Cham (2022). https://doi.org/10.1007/978-3-031-16434-7_44
12. Park, J.J., Florence, P., Straub, J., Newcombe, R., Lovegrove, S.: Deepsdf: learning continuous signed distance functions for shape representation. In: 2019 IEEE/CVF Conference on Computer Vision and Pattern Recognition (CVPR), pp. 165–174. IEEE Computer Society (2019)
13. Shen, L., Pauly, J., Xing, L.: NeRP: implicit neural representation learning with prior embedding for sparsely sampled image reconstruction. IEEE Transactions on Neural Networks and Learning Systems (2022)
14. Sitzmann, V., Martel, J., Bergman, A., Lindell, D., Wetzstein, G.: Implicit neural representations with periodic activation functions. In: NeurIPS (2020)
15. Sun, Y., Liu, J., Xie, M., Wohlberg, B., Kamilov, U.S.: Coil: coordinate-based internal learning for tomographic imaging. IEEE Trans. Comput. Imaging **7**, 1400–1412 (2021)
16. Wanhainen, A., et al.: Editor's choice-european society for vascular surgery (ESVS) 2019 clinical practice guidelines on the management of abdominal aorto-iliac artery aneurysms. Eur. J. Vasc. Endovasc. Surg. **57**(1), 8–93 (2019)
17. Wasserthal, J., Meyer, M., Breit, H.C., Cyriac, J., Yang, S., Segeroth, M.: TotalSegmentator: robust segmentation of 104 anatomical structures in CT images. arXiv preprint arXiv:2208.05868 (2022)
18. Wiesner, D., Suk, J., Dummer, S., Svoboda, D., Wolterink, J.M.: Implicit neural representations for generative modeling of living cell shapes. In: Wang, L., Dou, Q., Fletcher, P.T., Speidel, S., Li, S. (eds.) Medical Image Computing and Computer Assisted Intervention – MICCAI 2022. MICCAI 2022. LNCS, vol. 13434, pp. 58–67. Springer, Cham (2022). https://doi.org/10.1007/978-3-031-16440-8_6
19. Wolterink, J.M., Zwienenberg, J.C., Brune, C.: Implicit neural representations for deformable image registration. In: International Conference on Medical Imaging with Deep Learning, pp. 1349–1359. PMLR (2022)
20. Xie, Y., et al.: Neural fields in visual computing and beyond. Comput. Graph. Forum **41**(2), 641–676 (2022)
21. Zhang, L., Zambrano, B.A., Choi, J., Lee, W., Baek, S., Lim, C.Y.: Intraluminal thrombus effect on the progression of abdominal aortic aneurysms by using a multistate continuous-time Markov chain model. J. Int. Med. Res. **48**(11), 0300060520968449 (2020)

Prototype of a Cardiac MRI Simulator for the Training of Supervised Neural Networks

Marta Varela[1(✉)] and Anil A. Bharath[2]

[1] National Heart and Lung Institute, Imperial College London, London, UK
marta.varela@imperial.ac.uk

[2] Bioengineering Department, Imperial College London, London, UK
a.bharath@imperial.ac.uk

Abstract. Supervised deep learning methods typically rely on large datasets for training. Ethical and practical considerations usually make it difficult to access large amounts of healthcare data, such as medical images, with known task-specific ground truth. This hampers the development of adequate, unbiased and robust deep learning methods for clinical tasks.

Magnetic Resonance Images (MRI) are the result of several complex physical and engineering processes and the generation of synthetic MR images provides a formidable challenge. Here, we present the first results of ongoing work to create a generator for large synthetic cardiac MR image datasets. As an application for the simulator, we show how the synthetic images can be used to help train a supervised neural network that estimates the volume of the left ventricular myocardium directly from cardiac MR images.

Despite its current limitations, our generator may in the future help address the current shortage of labelled cardiac MRI needed for the development of supervised deep learning tools. It is likely to also find applications in the development of image reconstruction methods and tools to improve robustness, verification and interpretability of deep networks in this setting.

Keywords: Cardiac MRI · MRI Simulator · Synthetic Cardiac Images · Training of Supervised Neural Networks · Cardiac Volume Estimation

1 Introduction

Recent developments in machine learning (ML) have improved rapid reconstruction techniques, enabled automatic image analysis and enhanced the interpretation of medical images in general and cardiac Magnetic Resonance Imaging

---

Supplementary Information The online version contains supplementary material available at https://doi.org/10.1007/978-3-031-35302-4_38.

O. Bernard et al. (Eds.): FIMH 2023, LNCS 13958, pp. 366–374, 2023.
https://doi.org/10.1007/978-3-031-35302-4_38

(MRI) in particular [7]. Progress in this area has nevertheless been restricted by shortages in large anonymised curated datasets, and difficulties in obtaining high-quality ground truth annotations for supervised learning tasks [13]. These problems are compounded by the under-representation of patients from minority backgrounds and those with infrequent anatomical variations or rare diseases. Moreover, deploying ML models to novel imaging sequences is often delayed until a large numbers of similarly parameterised images have been accrued.

The creation of large datasets of synthetic images whose properties follow prescribed statistical distributions could help alleviate some of the current constraints in the deployment of supervised neural networks (NNs) for medical imaging tasks. The physical processes underlying the nuclear magnetic resonance (NMR) of water protons in biological tissue are complex, as are the interactions of the protons' magnetization vectors with the MRI magnets and other engineering equipment. As such, simulating MRI acquisition necessarily involves simplifications and trade-offs between accuracy and speed. Several simulators with varying degrees of complexity and focusing on different anatomical regions have been proposed [1,2,6,11,14]. So far none of these has allowed the automatic generation of large sets of MR images with controlled variation in anatomical or imaging parameters suitable for the training of NNs.

Aims. We present a generator of synthetic cardiac MR images, designed to allow the creation of large imaging datasets with controlled parameter variations. As an application, we show how the synthetic data can be used to train a network that estimates left ventricular myocardial volume (LVMV).

2 Methods

2.1 MRI Simulator

We created a Python 3.8-based modular simulator of cardiac MRI. The simulator takes the following information as independent inputs:

1. Cardiac phantom (described below);
2. Scanner characteristics. We assumed simulated image acquisition at a $B_0 = 1.5$ T, with a spatially uniform B_1 field and perfect gradient coils with an infinite slew rate. We also assumed a single receive coil with a spatially uniform sensitivity.
3. List of MR sequence parameters similar the ones inputted by radiographers at the time of scanning. We used an axial 3D gradient echo sequence, with: an echo time of 40–60 ms, a repetition time of 3000 ms, an excitation flip angle of $10 - 20°$ and a 80 MHz bandwidth. We used a Cartesian sampling scheme with a linear phase encode order, RF spoiling and no parallel imaging capabilities.

We solved the Bloch equations with the CPU-based forward Euler single-compartment solver initially proposed by Liu *et al.* [8] and a fast Fourier transform for image reconstruction. Spatial motion of the excited protons during the image acquisition process (caused by blood flow, cardiac and respiratory cycles, diffusion or patient motion) was not simulated.

2.2 Cardiac MRI Phantom

Digital cardiac phantoms are a necessary component of cardiac medical image simulators. There are several atlases and mesh-based models of the human heart [16], but realistic cardiac MR images rely on accurate representations of the entire thoracic anatomy. Digital models that rely on non-uniform rational B-splines (NURBS) are particularly flexible and efficient and have been employed in computed tomography and nuclear medicine image simulators in the past. **blue**The different approaches employed to generate computational human phantoms are carefully reviewed elsewhere [5].

We used the XCAT phantom [9,10], which is a detailed NURBS-based representation of human anatomy originally based on the segmentations of the high-resolution Visible Human Male and Female images. We took the thoracic region of the XCAT phantom adjusted to 50th-percentile organ volume values [10] as our baseline anatomical representation. Taking advantage of NURBS's flexibility in representing structures with varying morphology and volume, we independently varied the following anatomical parameters: heart dimensions (varied independently in the FH, LR and AP directions); LV radius; apico-basal length; LV thickness; LV thickness close to the apex. All these parameters followed independent uniform distributions, so that their value varied between 80% and 120% of the original value in the reference XCAT phantom. For each instance, we randomly chose between the male or female phantom representation. Clinical experts visually inspected the generated phantom anatomy instances to ensure that they were anatomically plausible (i.e., that they could be segmentations of realistic torso anatomies). We also confirmed that the LVMV range in the generated phantoms was similar to the LVMV in the image patients (see below). This was the only quantitative test of anatomical plausibility we performed on the simulated images.

We then voxelised the phantom representations into images with dimensions $256 \times 256 \times 15$ with a $1.0 \times 1.0 \times 6.0\,\text{mm}^3$-resolution. Before voxelisation, we randomly introduced translations of <2 cm and rotations of <6° along any axis to the phantom in order to model the variability of body position within the scanner. For each of the 17 segmented tissues in the phantom (see Supplementary Table 1), we created Gaussian distributions of water proton NMR relaxation time constants (T_1, T_2, T_2^*). The distributions were centred on literature values at 1.5 T for each NMR parameter and their standard deviation was set to 10% of the corresponding literature value. An exception was made for arterial and venous blood, whose standard deviation was set to 30% to simplistically model the greater variability in signal due to blood flow. We neglected variations in magnetic susceptibility across the body. Each voxel in the simulated image had NMR parameters randomly chosen from the multivariate Gaussian distribution corresponding to its tissue type (Fig. 1).

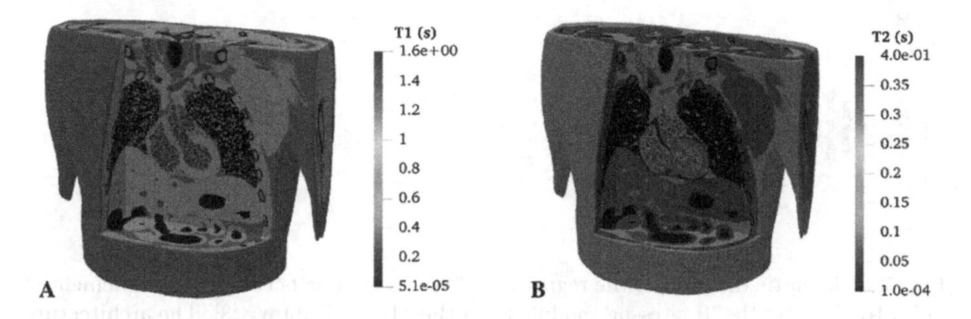

Fig. 1. Cross section of the T_1 (A) and T_2 (B) maps for the XCAT phantom. The values assigned to each voxel follow a normal distribution around the literature values of the respective tissue. The standard deviation is 30% for blood and 10% for all other tissues. Lumen and cortical bones were treated like air and assigned proton densities close to zero. A similar approach was followed when assigning T_2^* values.

2.3 Left Ventricular Myocardial Volume (LVMV) Estimation

To demonstrate the potential application of the cardiac MRI simulator as a generative model for deep learning applications, we create an *in silico* dataset of 500 T_2-weighted axial stacks of cardiac images, with variations in phantom anatomy and MR sequence parameters detailed above. We train a regression NN to automatically estimate the volume of the LV myocardium (VLVM) in three different experimental setups - see Table 1. The patient data was part of an ethically approved retrospective study. It was acquired in a 1.5T Siemens scanner, using a T_2 prepared multi slice gradient echo sequence (TE/TR: 1.4/357 ms, FA: 80°, resolution: $1.4 \times 1.4 \times 6\,\text{mm}^3$). Ground truth VLVMs are calculated using an existing segmentation CNN [4] applied to the patient images.

Table 1. Number of simulated and patient images used in each of the experiments, as part of the training and test sets.

Experiment	# Training Images	# Test Images
A	400 (simulated)	100 (simulated)
B	393 (patient)	99 (patient)
C	393 (patient) + 500 (simulated)	99 (patient)

We trained the CNN shown in Fig. 2 for the regression task using a mean square error loss function. We trained for 150 epochs, using Adam optimisation with a 10^{-4} learning rate, 10^{-4} weight decay, batch size of 2 and 0.1 dropout.

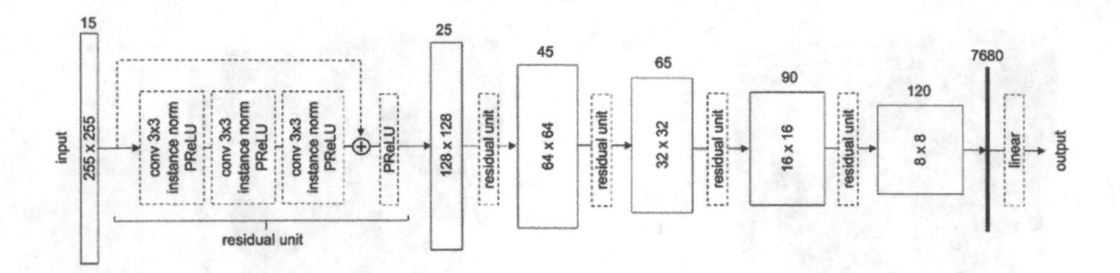

Fig. 2. Schematic diagram of the regression CNN used. The network was implemented in PyTorch using the 'Regressor' module from the MONAI library [3]. The architecture consists of 5 residual units which downsample the inputs by a factor of 2 via strided convolutions. Each residual unit consists of three convolutional layers (kernel size 3×3) followed by instance normalisation and a PReLU activation function. Skip connections are employed to bypass the convolutions. The network ends with a fully connected layer resizing the output from the residual blocks to a single value to which a linear activation function is applied.

3 Results

MRI Simulator. The simulator was able to generate realistic sets of cardiac MR images according to the visual assessment of experts. All represented anatomical structures preserved their smooth, non-overlapping boundaries. A representative example is shown in Fig. 3A. The simulation of each 3D cardiac image took approximately 2 h on a single CPU. By distributing the image generation process across 64 CPUs, we were able to simulate sets of 500 cardiac MR images in less than 16 h. Despite differences in contrast and anatomy, the simulated images compare well with the cropped patient images depicted in Fig. 3B, showing similar cardiac structures in an equivalent anatomical context.

3.1 Left Ventricular Myocardial Volume (LVMV) Estimation

With the introduced variations in cardiac anatomy, we were able to simulate images with varying realistic cardiac sizes and morphology. VLVM in the simulated images was $102 \pm 19\,\text{cm}^3$ (range: 55–163 cm^3), compared to 123 ± 60 cm^3 (range: 43–242 cm^3) from the segmentations of the patient data.

The regression CNN was able to estimate LVMV, as shown in Fig. 4. LVMV estimates on simulated data were accurate and precise (Fig. 4a), when compared to LVMV estimates performed on patient images (Figs. 4b and c). Enhancing the training dataset with simulated data (Fig. 4c) led to a small decrease in the dispersion of the estimates (root mean square error (RMSE) of 39.0 cm^3 instead of 45.1 cm^3, whilst introducing a small drop in accuracy (best fit slope of 0.94 ± 0.03 vs 0.97 ± 0.03). The comparisons are performed against the data shown in Fig. 4b, where no simulated images were used for training.

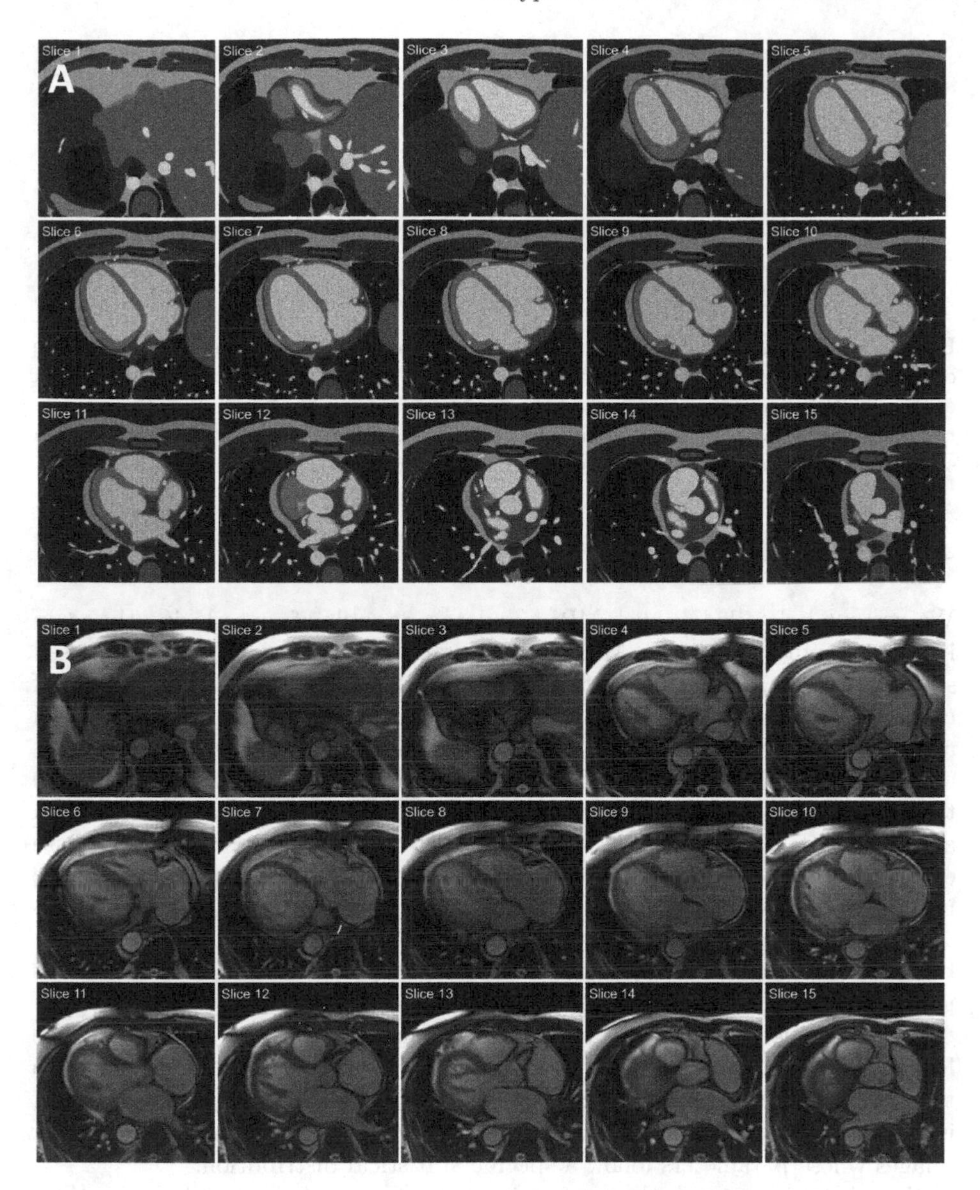

Fig. 3. Examples of stacks of axial cardiac images. A: Images simulated using the proposed framework. B: Patient images acquired in a 1.5T scanner.

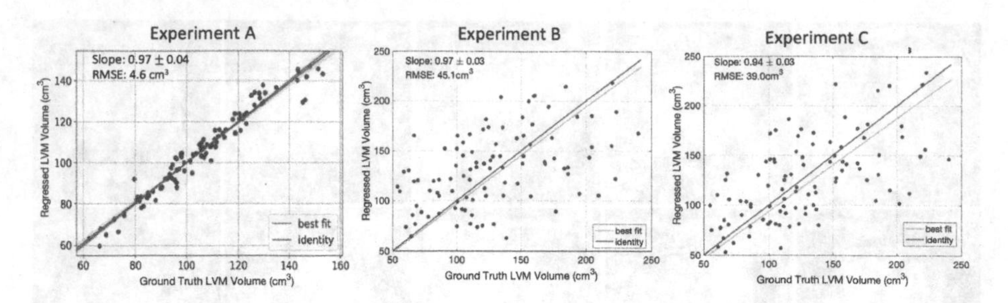

Fig. 4. Scatter plots on test data of LVMV estimation. From left to right, we depict the outcomes of experiments A, B and C respectively (see Table 1 for further details). In each case, the identity y = x line and the best fit line are also shown. We also indicate the values of the slope of the best fit line (going through the origin) and the root mean square error of the volume estimates in each experiment.

4 Conclusions

We present a detailed cardiac MRI simulator, capable of reproducing the main imaging features of cardiac MRI with a high anatomical detail. The proposed simulator is designed to allow the rapid and efficient generation of batches of cardiac MR images, with controlled variation in some parameters (anatomical and/or related to MR imaging). This makes the simulator ideal for the development and testing of machine learning applications. Suitable tasks include the training of supervised NNs which rely on large amounts of training data, such as NNs for image analysis tasks, as the LVMV estimation task presented here exemplifies. It is also well suited to train NNs for image reconstruction or image manipulation tasks, for which ground truth information is often not readily available. Finally, it offers a platform to study the robustness of existing NNs to controlled training distribution shifts or adversarial perturbations, and to address potential imbalances in existing datasets.

Others have also seen the potential of simulated MR images for machine learning. For example, Xanthis *et al.* [15] have generated synthetic MR images to train a neural network for a cardiac segmentation task. Their approach, however, is not specifically catered towards allowing the rapid parallel generation of MRI images whose parameters follow a specific statistical distribution.

Despite its potential, the current implementation of the simulator is still a work in progress and suffers from some limitations that restrict the realism of the images it produces. For example, the simulator currently does not include: cardiac and breathing motion; the effects of water diffusion or blood flow; interactions between different proton pools; partial volume effects; inhomogeneities in B_0 or B_1 fields; variations in magnetic susceptibility or realistic transmit and receive coil sensitivities. These limitations explain the relative lack of success we achieved when enhancing the training of our LVMV regression network with simulated data (see Fig. 4). The LVMV regression network performed well on the synthetic data where tissue intensities are relatively flat and well defined, but poorly on real data, with its more complex intensity distribution. In this instance, adding simulated data to the training pool did not help the network's performance, but this is likely to change when more receive coil sensitivities are included in the simulation process.

Future work will test the usefulness of the proposed cardiac MRI simulator in other datasets and tasks. We will test the proposed LVMV estimation CNN on open access cardiac MRI databases with LV segmentations delineated by experts, such as the short-axis images available from the 2011 LV Segmentation Challenge [12]. The proposed simulator is also well suited for training networks for other tasks, such as the identification of anatomical structures in different MRI slices. It is currently being tested for this purpose.

The XCAT phantom setup we used allows for several variations in anatomy and properties, from variations in heart dimensions to alterations in LV thickness, which mimic both physiological variability and disease-induced remodelling. It currently does not allow changes in the anatomy of the other thoracic and abdominal organs present in the images or in the orientation of the heart relative to its surroundings, although patient data shows a large degree of variation in this latter parameter. The current version of XCAT phantom also does not permit the straightforward inclusion of focal pathology, such as myocardial scarring.

Despite its current shortcomings, we believe that the presented cardiac MRI simulator is a useful and attractive platform for the generation of large datasets of synthetic cardiac MR images in a controlled and simple manner. We expect that in the future these synthetic datasets will be used to train NNs and improve their robustness and explainability.

Acknowledgments. This work was supported by the British Heart Foundation Centre of Research Excellence at Imperial College London (RE/18/4/34215). The authors would like to thank Abhishek Roy, Tommy Chen and Krithika Balaji for their contributions.

References

1. Benoit-Cattin, H., Collewet, G., Belaroussi, B., Saint-Jalmes, H., Odet, C.: The SIMRI project: a versatile and interactive MRI simulator. J. Magn. Reson. **173**(1), 97–115 (2005). https://doi.org/10.1016/j.jmr.2004.09.027
2. Bittoun, J., Taquin, J., Sauzade, M.: A computer algorithm for the simulation of any nuclear magnetic resonance (NMR) imaging method. Magn. Reson. Imaging **2**(2), 113–120 (1984). https://doi.org/10.1016/0730-725X(84)90065-1
3. Cardoso, M.J., et al.: MONAI: an open-source framework for deep learning in healthcare (2022). https://doi.org/10.48550/arxiv.2211.02701, https://arxiv.org/abs/2211.02701v1
4. Howard, J.P., et al.: Automated analysis and detection of abnormalities in transaxial anatomical cardiovascular magnetic resonance images: a proof of concept study with potential to optimize image acquisition. Int. J. Cardiovasc. Imaging **37**(3), 1033–1042 (2020). https://doi.org/10.1007/s10554-020-02050-w

5. Kainz, W., et al.: Advances in computational human phantoms and their applications in biomedical engineering - a topical review (2019). https://doi.org/10.1109/TRPMS.2018.2883437
6. Kose, R., Kose, K.: BlochSolver: a GPU-optimized fast 3D MRI simulator for experimentally compatible pulse sequences. J. Magn. Reson. **281**, 51–65 (2017). https://doi.org/10.1016/j.jmr.2017.05.007
7. Litjens, G., et al.: State-of-the-art deep learning in cardiovascular image analysis. JACC: Cardiovasc. Imaging **12**(8P1), 1549–1565 (2019). https://doi.org/10.1016/J.JCMG.2019.06.009, https://www.jacc.org/doi/10.1016/j.jcmg.2019.06.009
8. Liu, F., Velikina, J.V., Block, W.F., Kijowski, R., Samsonov, A.A.: Fast realistic MRI simulations based on generalized multi-pool exchange tissue model. IEEE Trans. Med. Imaging **36**(2), 527–537 (2017). https://doi.org/10.1109/TMI.2016.2620961
9. Segars, W.P., Sturgeon, G., Mendonca, S., Grimes, J., Tsui, B.M.: 4D XCAT phantom for multimodality imaging research. Med. Phys. **37**(9), 4902–4915 (2010). https://doi.org/10.1118/1.3480985
10. Segars, W.P., Tsui, B.M., Cai, J., Yin, F.F., Fung, G.S., Samei, E.: Application of the 4-D XCAT phantoms in biomedical imaging and beyond. IEEE Trans. Med. Imaging **37**(3), 680–692 (2018). https://doi.org/10.1109/TMI.2017.2738448
11. Stoecker, T., Vahedipour, K., Pflugfelder, D., Shah, N.J.: High-performance computing MRI simulations. Magn. Reson. Med. **64**(1), 186–193 (2010). https://doi.org/10.1002/mrm.22406, https://onlinelibrary.wiley.com/doi/10.1002/mrm.22406
12. Suinesiaputra, A., et al.: A collaborative resource to build consensus for automated left ventricular segmentation of cardiac MR images. Med. Image Anal. **18**(1), 50–62 (2014). https://doi.org/10.1016/J.MEDIA.2013.09.001
13. Varoquaux, G., Cheplygina, V.: Machine learning for medical imaging: methodological failures and recommendations for the future (2022). https://doi.org/10.1038/s41746-022-00592-y
14. Xanthis, C.G., Aletras, A.H.: coreMRI: a high-performance, publicly available MR simulation platform on the cloud. PLOS One **14**(5), e0216594 (2019). https://doi.org/10.1371/journal.pone.0216594, https://dx.plos.org/10.1371/journal.pone.0216594
15. Xanthis, C.G., Filos, D., Haris, K., Aletras, A.H.: Simulator-generated training datasets as an alternative to using patient data for machine learning: an example in myocardial segmentation with MRI. Comput. Methods Program. Biomed. **198**, 105817 (2021). https://doi.org/10.1016/j.cmpb.2020.105817
16. Young, A.A., Frangi, A.F.: Computational cardiac atlases: from patient to population and back. Exp. Physiol. **94**(5), 578–596 (2009). https://doi.org/10.1113/expphysiol.2008.044081

Deformable Image Registration Using Vision Transformers for Cardiac Motion Estimation from Cine Cardiac MRI Images

Roshan Reddy Upendra[1(✉)], Richard Simon[2], Suzanne M. Shontz[3,4,5], and Cristian A. Linte[1,2]

[1] Center for Imaging Science, Rochester Institute of Technology, Rochester, NY, USA
ru6928@rit.edu

[2] Biomedical Engineering, Rochester Institute of Technology, Rochester, NY, USA

[3] Electrical Engineering and Computer Science, University of Kansas, Lawrence, KS, USA

[4] Bioengineering Program, University of Kansas, Lawrence, KS, USA

[5] Institute for Information Sciences, University of Kansas, Lawrence, KS, USA

Abstract. Accurate cardiac motion estimation is a crucial step in assessing the kinematic and contractile properties of the cardiac chambers, thereby directly quantifying the regional cardiac function, which plays an important role in understanding myocardial diseases and planning their treatment. Since the cine cardiac magnetic resonance imaging (MRI) provides dynamic, high-resolution 3D images of the heart that depict cardiac motion throughout the cardiac cycle, cardiac motion can be estimated by finding the optical flow representation between the consecutive 3D volumes from a 4D cine cardiac MRI dataset, thereby formulating it as an image registration problem. Therefore, we propose a hybrid convolutional neural network (CNN) and Vision Transformer (ViT) architecture for deformable image registration of 3D cine cardiac MRI images for consistent cardiac motion estimation. We compare the image registration results of our proposed method with those of the VoxelMorph CNN model and conventional B-spline free form deformation (FFD) non-rigid image registration algorithm. We conduct all our experiments on the open-source Automated Cardiac Diagnosis Challenge (ACDC) dataset. Our experiments show that the deformable image registration results obtained using the proposed method outperform the CNN model and the traditional FFD image registration method.

Keywords: Vision Transformer · Cardiac MRI · Cardiac Motion Estimation · Medical Image Registration · Deep Learning

1 Introduction

The assessment of regional myocardial function such as myocardial wall deformation, strain, torsion and wall thickness, plays a crucial role in understanding,

O. Bernard et al. (Eds.): FIMH 2023, LNCS 13958, pp. 375–383, 2023.
https://doi.org/10.1007/978-3-031-35302-4_39

diagnosis, risk stratification and planning treatment of several myocardial disorders. Therefore, accurate cardiac motion estimation is an important step in assessing the kinematic and contractile properties of the myocardium, thereby, quantifying dynamic regional heart function.

Cine cardiac MRI provides high-resolution, dynamic 3D images of the cardiac chambers, which depict cardiac motion throughout the cardiac cycle. Thus, cardiac motion estimation can be formulated as an image registration problem, which involves finding an optical flow representation between the consecutive 3D frames of a 4D cine cardiac MRI dataset [19].

In the past decade, deep learning models have gained increased popularity in medical image registration [6]. A number of researchers leveraged these deep learning-based 4D deformable registration methods to estimate cardiac motion from cine cardiac MRI images [11,12,21]. In our earlier work [16], we presented a deep learning-based 4D deformable registration method for cardiac motion estimation from cine cardiac MRI dataset by leveraging the VoxelMorph framework [1]. Additionally, we demonstrated the application of the VoxelMorph-based cardiac motion estimation method to build dynamic patient-specific left ventricle (LV) myocardial models across subjects with different pathologies [17]. Although the convolutional neural network (CNN)-based cardiac motion estimation presented in our previous work [16,17] showed promising performance, the CNN-based approaches usually exhibit limitations in modeling explicit long-range spatial relations due to the limited receptive fields of convolution operations [3]. Therefore, the large variations in shape and size of the cardiac chambers can affect the registration performance of the CNN-based cardiac motion estimation methods.

In recent years, self-attention-based architectures (Transformer-based), due to their great success in sequence-to-sequence prediction in natural language processing have gained increasing interests in computer vision tasks [5], including medical image segmentation [3] and registration [4]. These current research studies show that fusing the self-attention mechanism with the CNN models overcome the limitation of the convolution operation in learning global semantic information, which is critical for the image registration task in cardiac motion estimation from the cine cardiac MRI images.

In this work, we propose a hybrid CNN-ViT architecture (Fig. 1) for consistent cardiac motion estimation from 4D cine cardiac MRI images. Here, we leverage the VIT-V-Net [4] to register the moving and fixed frame of the cardiac MRI volumes. We evaluate the proposed method by training the models on the ACDC dataset [2].

2 Methodology

2.1 Cardiac MRI Dataset

In this study, we use the 2017 ACDC dataset [2], consisting of short-axis cine cardiac MRI images from 150 subjects, divided into five equally-distributed subgroups: normal, dilated cardiomyopathy (DCM), hypertrophic cardiomyopathy

(HCM), prior myocardial infarction (MINF) and abnormal right ventricle (RV). These images were acquired as part of clinical diagnostic exams conducted on two different MRI scanners of 1.5 T and 3.0 T magnetic strength. These series of short axis MRI slices cover the LV from base to apex with a through-plane resolution of 5 mm to 10 mm and a spatial resolution of 1.37 mm^2/pixel to 1.68 mm^2/pixel.

2.2 Vision Transformer-Based Deformable Image Registration

Here, the aim of the deep learning model is to find an optical flow representation between a sequence of image volume pairs $\{(I_{ED}, I_{ED+t})\}_{t=1,2,3,...,N_T-1}$ where I_{ED} is the image volume frame at end-diastole (ED) and N_T is the total number of frames for a particular subject. That is, for the given image volume pair (I_{ED}, I_{ED+t}), the deep learning model should predict a differentiable transformation function ϕ to warp the moving image volume I_{ED}, to produce a warped image volume $I_{ED} \circ \phi$. The similarity loss is computed between the fixed image volume I_{ED+t} and warped image volume $I_{ED} \circ \phi$, and this loss is used to back-propagate the deep learning network.

In this work, we employ the ViT-V-Net [4] architecture to estimate the differentiable optical flow representation (to ensure smoothness of the displacement field from ED to ES) between the image volume pairs (I_{ED}, I_{ED+t}). The direct application of ViT to the full-resolution cine cardiac MRI volume increases computational complexity. Similarly, splitting the image volume into 3D patches is not ideal, as it leads to the model not learning the local context information across the spatial and depth dimensions for volumetric registration [20]. Therefore, in ViT-V-Net [4], instead of feeding the whole high-resolution image volumes to the ViT, the image volumes are first encoded to low-resolution and high-level feature representations using a CNN encoder. Next, the high-level 3D context features are split into patches, which are then mapped onto a latent space using a trainable linear projection, i.e., patch embedding. These patch embeddings are added to the learnable position embeddings to retain the positional information of the patches, which are then fed into the ViT. The ViT consists of multiple alternating layers of multihead self-attention (MSA) and multi-layer

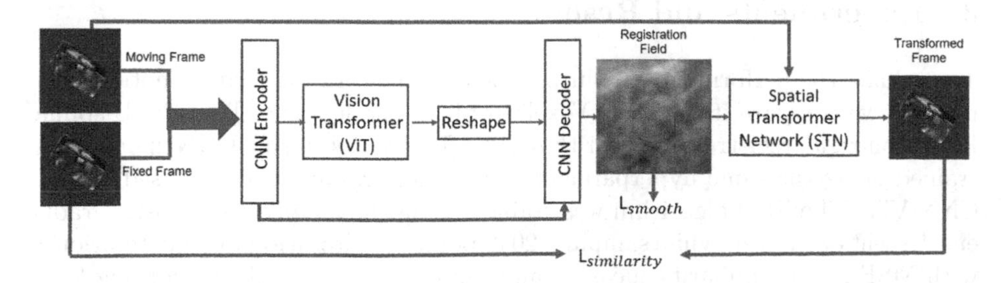

Fig. 1. Overview of the proposed hybrid CNN-ViT architecture for cardiac motion estimation.

perceptron (MLP) blocks. Finally, the output of ViT is fed into a CNN decoder to output the deformation field ϕ [4,5,18]. Also, long skip-connections are used between the V-Net [10] style encoder-decoder architecture. This deformation field is fed to the spatial transformer network (STN) [7] along with the moving image volume to produce a warped image volume (Fig. 1).

The loss function used to optimize the network described above is given by:

$$L = L_{\text{similarity}} + \lambda L_{\text{smooth}}, \tag{1}$$

where $L_{\text{similarity}}$ is the mean squared error (MSE) between the fixed image volume I_{ED+t} and the warped image volume $I_{ED} \circ \phi$. The smoothing loss L_{smooth} is the diffusion regularizer used in [4], on the spatial gradients of the deformation field ϕ, and λ is the regularization parameter.

2.3 Network Training

In order to rectify the inherent slice misalignments that occur during the cine cardiac MRI image acquisition, we train a variant of the U-Net model [13] to segment the cardiac chambers such as LV blood-pool, LV myocardium and RV blood-pool from 2D cine cardiac MRI images. We identify the centroid of these predicted segmentation maps as the LV blood-pool centers and stack the 2D MRI slices collinearly for all the frames of the cardiac cycle, resulting in slice misalignment corrected 3D images. These slice misalignment corrected 3D images were used to train all the registration algorithms reported in this work.

As mentioned earlier, we aim to find the optical flow representation between image pairs $\{(I_{ED}, I_{ED+t})\}_{t=1,2,3,\ldots,N_T-1}$. In order to do this, we employ 110 of the available 150 cardiac MRI dataset for training, 10 for validation and 30 for testing. The data-split in this work is consistent with our earlier work that involves VoxelMorph-based cardiac motion estimation [16,17], for comparison. We train our networks using an Adam optimizer with a learning rate of 10^{-4}, reduced by half every 10th epoch for 50 epochs. Furthermore, all the deep learning models in this work were trained on a machine equipped with a NVIDIA RTX 2080 Ti GPU.

3 Experiments and Results

To evaluate the performance of our proposed framework for cardiac motion estimation, we compare it with the VoxelMorph [1] model, as well as the B-spline FFD non-rigid registration algorithm [14]. The VoxelMorph CNN model was trained using the same hyperparameters used for training the proposed hybrid CNN-ViT. The FFD algorithm was trained using the adaptive stochastic gradient descent optimizer, while sampling 2048 points per iteration for 500 iterations, with MSE as the similarity measure and binding energy as the smoothing loss. This FFD-based non-rigid image registration algorithm was implemented using SimpleElastix [8,9] on an Intel(R) Core(TM) i9-9900K CPU.

Table 1. Summary of registration evaluation on the test set (30 subjects): Unregistered (post slice misalignment correction), B-spline free form deformation (FFD), VoxelMorph and the proposed ViT-V-Net. Mean Dice score (std-dev) and Hausdorff distance (HD) for LV blood-pool (LV), LV myocardium (MC) and RV blood-pool (RV), for both "gold" and "silver" standard comparisons. Statistically significant differences between the registration metrics of VoxelMorph and ViT-V-Net registration were evaluated using the Student t-test and are reported using * for $p < 0.05$ and ** for $p < 0.01$. The best evaluation metrics achieved are labeled in **bold**.

		Dice (%)			HD (mm)		
		LV	MC	RV	LV	MC	RV
ED to ES frames: Gold standard	Unregistered	87.30 (3.20)	69.15 (2.99)	70.18 (3.85)	7.22 (1.64)	8.93 (2.72)	11.85 (2.47)
	FFD	88.94 (2.42)	74.93 (2.12)	73.38 (3.22)	6.35 (3.61)	8.87 (2.42)	11.89 (1.94)
	VoxelMorph	92.17 (4.21)	79.39 (3.22)	77.58 (1.30)	5.59 (1.21)	8.05 (2.94)	11.75 (2.11)
	ViT-V-Net	**93.31 (2.10)**	**82.24** (2.14)**	**81.27** (1.30)**	**5.11 (1.11)**	**6.50* (1.83)**	**9.02* (1.73)**
ED to all frames: Silver standard	Unregistered	81.29 (4.93)	80.15 (3.64)	77.32 (3.91)	3.13 (2.44)	6.08 (2.91)	8.61 (3.12)
	FFD	84.34 (2.34)	82.57 (1.03)	78.23 (4.10)	3.01 (1.03)	6.11 (2.89)	8.75 (1.41)
	VoxelMorph	**94.67 (5.96)**	84.08 (4.32)	82.73 (3.76)	**2.51 (1.31)**	6.07 (2.79)	8.96 (3.47)
	ViT-V-Net	93.67 (4.61)	**88.53** (3.78)**	**83.92* (3.17)**	2.66 (1.42)	**5.02* (0.93)**	**7.83* (1.08)**

We evaluate the performance of all our models by warping the segmentation map of the ED frame to the end-systole (ES) frame using the estimated registration field, and computing the Dice score and Hausdorff distance (HD) between ground truth ES segmentation map of the cardiac chambers, namely LV blood-pool, LV myocardium and RV blood-pool, and warped segmentation map of the ED frame. Since the segmentation maps of the ED and ES frame are manually annotated by experts, we refer to this comparison as the "gold" standard comparison.

Additionally, we warp the segmentation map of the ED frame to all the subsequent frames of the cardiac cycle, and compute the evaluation metrics between the warped segmentation map of the ED frame and segmentation maps predicted by the U-Net model (as described in Sect. 2.3). Since the segmentation maps used here were generated using techniques that were previously validated against expert annotations, we refer to it as "silver" standard comparison. These results are shown in Table 1.

In Table 1, we show that our proposed method achieved a mean Dice score of 85.67% and a mean HD of 6.87 *mm* for our "gold" standard comparison, and a mean Dice score of 88.71% and a mean HD of 5.17 *mm* for our "silver" standard comparison.

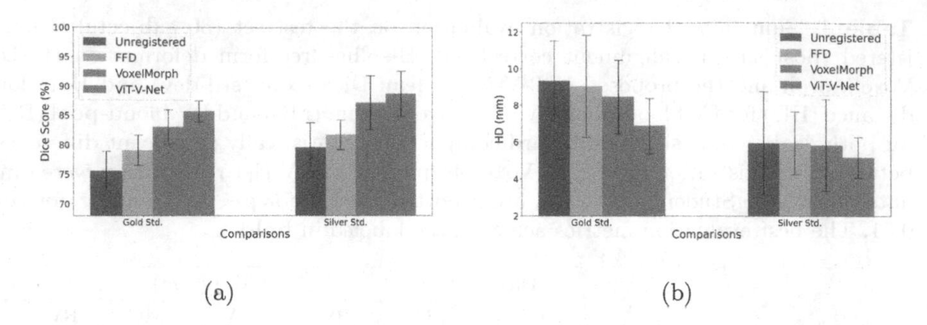

Fig. 2. "Gold" and "silver" comparison of (a) mean Dice score, and (b) mean HD values before registration (post slice misalignment correction), B-spline free form deformation (FFD) registration, VoxelMorph and the proposed ViT-V-Net, on the test set (30 subjects).

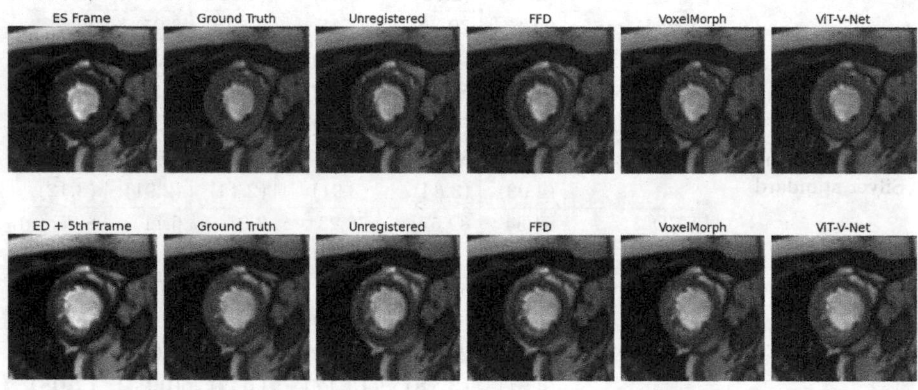

Fig. 3. Panel 1-1: End-systole (ES) frame; Panel 1-2: ground truth segmentation map of the cardiac chambers overlaid on the slice; Panel 1-3: segmentation map of the end-diastole (ED) frame overlaid on the ES frame without registration (Dice: 77.29%, HD: 9.09 mm); Panel 1-4: post registration contours using FFD algorithm (Dice: 79.02%, HD: 8.99 mm); Panel 1-5: post registration contours using VoxelMorph model (Dice: 83.11%, HD: 7.01 mm); Panel 1-6: post registration contours using ViT-V-Net framework (Dice: 85.57%, HD: 6.82 mm). Panel 2-1: ED + 5th frame; Panel 2-2: U-Net predicted segmentation map of the cardiac chambers overlaid on the slice; Panel 2-3: segmentation map of the end-diastole (ED) frame overlaid on the ED + 5th frame without registration (Dice: 78.07%, HD: 5.55 mm); Panel 2-4: post registration contours using FFD algorithm (Dice: 81.03%, HD: 4.92 mm); Panel 2-5: post registration contours using VoxelMorph model (Dice: 85.03%, HD: 3.97.01 mm); Panel 2-6: post registration contours using ViT-V-Net framework (Dice: 86.84%, HD: 3.28 mm). The Dice score and HD reported here are the average of the LV blood-pool, LV myocardium and RV blood-pool registration results.

We can observe that the proposed hybrid CNN-ViT model outperforms the CNN-only VoxelMorph model, as well as the FFD registration method (Fig. 2). In Fig. 3, we show an example of the cardiac chamber contours propagated using

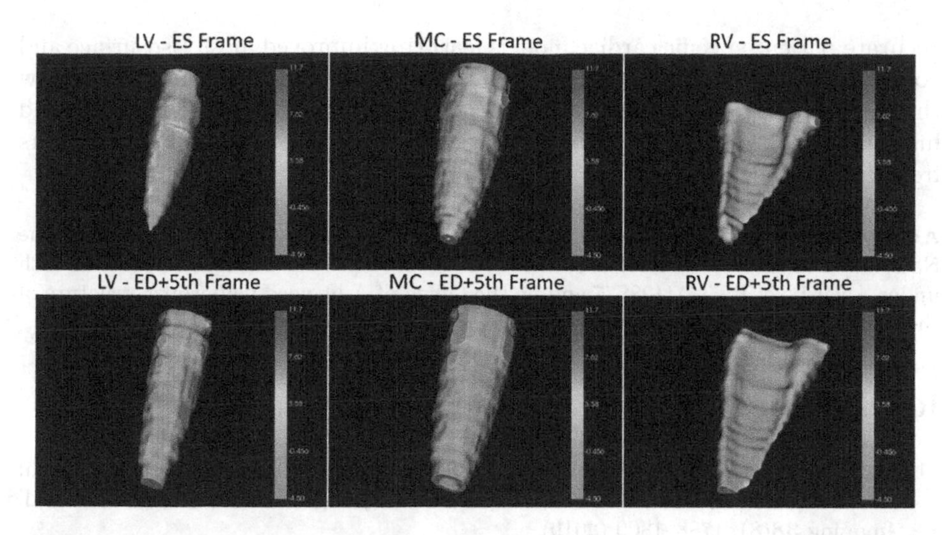

Fig. 4. Model-to-model distance between the isosurface mesh generated from VoxelMorph and ViT-V-Net propagation at end-systole (ES) frame (*top row*) and end-diastole (ED)+5th frame (*bottom row*) for (*left to right*) left ventricle blood-pool (LV), left ventricle myocardium (MC) and right ventricle blood-pool (RV)

the registration methods from the ED frame to the ES frame, as well as the ED + 5th frame. Additionally, in Fig. 4, we show an example of the model-to-model distance between the isosurface meshes of the cardiac chambers propagated using VoxelMorph framework and the proposed hybrid CNN-ViT framework from the ED frame to the ES frame, as well as the ED + 5th frame. Here, we can observe that the two sets of isosurface meshes are in close agreement with each other.

4 Discussion and Conclusion

In this paper, we present a hybrid CNN-ViT deformable image registration method for consistent cardiac motion estimation from 3D cine cardiac MRI images. To the best of our knowledge, this is the first study to investigate the usage of ViT for cardiac motion estimation. In addition to the local context information learnt by the CNN encoder-decoder layers, the ViT encodes global context information by treating the CNN-encoded features as sequences.

We evaluate the performance of the proposed hybrid CNN-ViT framework by comparing it with the VoxelMorph framework, which is essentially a CNN encoder-decoder architecture without the ViT. We observe that the proposed hybrid framework outperforms the VoxelMorph framework for cardiac motion estimation from cine cardiac MRI images (Fig. 2).

In our earlier work [15,17], we showed that the VoxelMorph framework can be used to build patient-specific LV myocardial and RV models, respectively. However, thanks to the improved registration accuracy of the proposed method compared to the VoxelMorph model, this work will enable us to generate more accurate patient-specific cardiac models featuring improved mesh (isosurface and volumetric) quality. As such, as part of our future work, we will demonstrate how the cardiac motion estimated using this proposed method may be used to build high quality, deformable patient-specific geometric models of cardiac chambers from cine cardiac MRI.

Acknowledgment. Research reported in this publication was supported by the National Institute of General Medical Sciences of the National Institutes of Health under Award No. R35GM128877 and by the Office of Advanced Cyberinfrastructure of the National Science Foundation under Award No. 1808530 and Award No. 1808553.

References

1. Balakrishnan, G., Zhao, A., Sabuncu, M.R., Guttag, J., Dalca, A.V.: VoxelMorph: a learning framework for deformable medical image registration. IEEE Trans. Med. Imaging **38**(8), 1788–1800 (2019)
2. Bernard, O., et al.: Deep learning techniques for automatic MRI cardiac multi-structures segmentation and diagnosis: is the problem solved? IEEE Trans. Med. Imaging **37**(11), 2514–2525 (2018)
3. Chen, J., et al.: TransUnet: transformers make strong encoders for medical image segmentation. arXiv preprint arXiv:2102.04306 (2021)
4. Chen, J., He, Y., Frey, E.C., Li, Y., Du, Y.: ViT-V-net: vision transformer for unsupervised volumetric medical image registration. arXiv preprint arXiv:2104.06468 (2021)
5. Dosovitskiy, A., et al.: An image is worth 16×16 words: transformers for image recognition at scale. arXiv preprint arXiv:2010.11929 (2020)
6. Haskins, G., Kruger, U., Yan, P.: Deep learning in medical image registration: a survey. Mach. Vision Appl. **31**(1), 1–18 (2020)
7. Jaderberg, M., Simonyan, K., Zisserman, A., et al.: Spatial transformer networks. Adv. Neural Inf. Process. Syst. **28** (2015)
8. Klein, S., Staring, M., Murphy, K., Viergever, M.A., Pluim, J.P.: Elastix: a toolbox for intensity-based medical image registration. IEEE Trans. Med. Imaging **29**(1), 196–205 (2009)
9. Marstal, K., Berendsen, F., Staring, M., Klein, S.: SimpleElastix: a user-friendly, multi-lingual library for medical image registration. In: Proceedings of the IEEE Conference on Computer Vision and Pattern Recognition Workshops, pp. 134–142 (2016)
10. Milletari, F., Navab, N., Ahmadi, S.A.: V-net: fully convolutional neural networks for volumetric medical image segmentation. In: 2016 Fourth International Conference on 3D Vision (3DV), pp. 565–571. IEEE (2016)
11. Qin, C., et al.: Joint motion estimation and segmentation from undersampled cardiac MR image. In: Knoll, F., Maier, A., Rueckert, D. (eds.) MLMIR 2018. LNCS, vol. 11074, pp. 55–63. Springer, Cham (2018). https://doi.org/10.1007/978-3-030-00129-2_7
12. Qiu, H., Qin, C., Le Folgoc, L., Hou, B., Schlemper, J., Rueckert, D.: Deep learning for cardiac motion estimation: supervised vs. unsupervised training. In: Pop, M., et al. (eds.) STACOM 2019. LNCS, vol. 12009, pp. 186–194. Springer, Cham (2020). https://doi.org/10.1007/978-3-030-39074-7_20

13. Ronneberger, O., Fischer, P., Brox, T.: U-net: convolutional networks for biomedical image segmentation. In: Navab, N., Hornegger, J., Wells, W.M., Frangi, A.F. (eds.) MICCAI 2015. LNCS, vol. 9351, pp. 234–241. Springer, Cham (2015). https://doi.org/10.1007/978-3-319-24574-4_28
14. Rueckert, D., Sonoda, L.I., Hayes, C., Hill, D.L., Leach, M.O., Hawkes, D.J.: Nonrigid registration using free-form deformations: application to breast MR images. IEEE Trans. Med. Imaging **18**(8), 712–721 (1999)
15. Upendra, R.R., et al.: Motion extraction of the right ventricle from 4D cardiac cine MRI using a deep learning-based deformable registration framework. In: 2021 43rd Annual International Conference of the IEEE Engineering in Medicine & Biology Society (EMBC), pp. 3795–3799. IEEE (2021)
16. Upendra, R.R., Wentz, B.J., Shontz, S.M., Linte, C.A.: A convolutional neural network-based deformable image registration method for cardiac motion estimation from cine cardiac MR images. In: 2020 Computing in Cardiology, pp. 1–4. IEEE (2020)
17. Upendra, R.R., Wentz, B.J., Simon, R., Shontz, S.M., Linte, C.A.: CNN-based cardiac motion extraction to generate deformable geometric left ventricle myocardial models from cine MRI. In: Ennis, D.B., Perotti, L.E., Wang, V.Y. (eds.) FIMH 2021. LNCS, vol. 12738, pp. 253–263. Springer, Cham (2021). https://doi.org/10.1007/978-3-030-78710-3_25
18. Vaswani, A., et al.: Attention is all you need. Adv. Neural Inf. Process. Syst. **30** (2017)
19. Wang, H., Amini, A.A.: Cardiac motion and deformation recovery from MRI: a review. IEEE Trans. Med. Imaging **31**(2), 487–503 (2011)
20. Wang, W., Chen, C., Ding, M., Yu, H., Zha, S., Li, J.: TransBTS: multimodal brain tumor segmentation using transformer. In: de Bruijne, M., et al. (eds.) MICCAI 2021. LNCS, vol. 12901, pp. 109–119. Springer, Cham (2021). https://doi.org/10.1007/978-3-030-87193-2_11
21. Zhang, X., You, C., Ahn, S., Zhuang, J., Staib, L., Duncan, J.: Learning correspondences of cardiac motion from images using biomechanics-informed modeling. arXiv preprint arXiv:2209.00726 (2022)

Unsupervised Polyaffine Transformation Learning for Echocardiography Motion Estimation

Yingyu Yang[✉] and Maxime Sermesant

Université Côte d'Azur, Inria Epione Team, Sophia Antipolis, Valbonne, France
{yingyu.yang,maxime.sermesant}@inria.fr

Abstract. Echocardiography plays an important role in the diagnosis of cardiac dysfunction. In particular, motion estimation in echocardiography is a challenging task since ultrasound images suffer largely from low signal-to-noise ratio and out-of-view problems. Current deep learning-based models for cardiac motion estimation in the literature estimate the dense motion field with spatial regularization. However, the underlying spatial regularization can only cover a very small region in the neighborhood, which is not enough for a smooth and realistic motion field for the myocardium in echocardiography. In order to improve the performance and quality with deep learning networks, we propose applying polyaffine transformation for motion estimation, which intrinsically regularizes the myocardium motion to be polyaffine. Our thorough experiments demonstrate that the proposed method not only presents better evaluation metrics on the registration of cardiac structures but also shows great potential in abnormal wall motion detection.

Keywords: Motion Estimation · Echocardiography · Polyaffine Transformation

1 Introduction

Echocardiography is one of the most widely used modalities for cardiac dysfunction diagnosis. It's radiation-free, non-invasive, and real-time, making echocardiography very suitable for portable analysis, such as myocardial motion evaluation. However, ultrasound images generally have poorer quality than other modalities, such as MRI and CT, due to their low signal-to-noise ratio. Additionally, limitations in acquisitions and patient variability may cause the myocardium to occasionally be outside the imaging field-of-view. These burdens make motion estimation of the myocardium in echocardiography a very challenging task.

Traditional methods for motion estimation in echocardiography, such as block matching [3,5] or optical flow [3,6,11], are typically time-consuming and not suitable for real-time or portable analysis. However, recent advances in deep learning (DL) have improved both the time efficiency and tracking performance of

Supplementary Information The online version contains supplementary material available at https://doi.org/10.1007/978-3-031-35302-4_40.

O. Bernard et al. (Eds.): FIMH 2023, LNCS 13958, pp. 384–393, 2023.
https://doi.org/10.1007/978-3-031-35302-4_40

motion estimation algorithms. DL-based models can generally be classified into two types: optical flow-based models and registration-based models. Optical flow-based DL models estimate dense displacement fields between consecutive image frames and typically require ground truth displacement for supervised training [10,17]. Since obtaining ground truth displacement from real-world echocardiography images can be difficult, synthetic echocardiography datasets are often used for supervised training of these models [2,10]. Registration-based DL models typically register all other frames to end-diastole either in an unsupervised manner or with weak supervision using segmentation masks, enabling them to work on real-world datasets [1,25]. U-net-like architectures are often used as the core design of such methods [1,25]. Many other studies on cardiac MRI motion estimation have also utilized the registration approach with different temporal smoothness strategies [13,21]. Recently, researchers have also incorporated biomechanical modeling knowledge into deep learning networks with the aim of improving the generalizability of motion estimation performance [22,26].

The motion of the myocardium is not arbitrary, and various spatial regularization techniques have been explored in the literature, such as the divergence penalty [15] for enforcing incompressibility, the rigidity penalty for smoothness [24], and the elastic strain energy for mechanical correctness [19], among others. Many deep learning models have incorporated these regularization techniques into their work [1,8,26]. However, these regularization techniques are usually based on first- or second-order derivatives of the displacement vector field, which are applied at the pixel level in discrete implementation and are insufficient for studying the myocardium in echocardiography. The high noise-to-signal ratio in echocardiography degrades the quality of the estimated motion field, despite the pixel-wise constraints in deep learning methods. Other methods tackle the smoothness issue at the transformation level, for instance using regionally affine deformations [16].

In this work, we introduce a deep learning-based motion model leveraging such approach and tailored for 2D echocardiography motion estimation, called the PolyAffine Motion model (PAM). It incorporates global-level smoothness regularization for the myocardium through polyaffine motion fusion. Our contributions include:

- Proposing a comprehensive pipeline for using PAM in a weakly-supervised manner on a large public echocardiography dataset, demonstrating the effectiveness of our approach for unsupervised learning of myocardial motion.
- Conducting extensive experiments on various real-world datasets to show that PAM outperforms other popular deep learning-based motion estimation methods, thereby highlighting its potential for clinical applications.

2 Methodology

2.1 Polyaffine Motion Model

Given source image I_S and target image I_T, we would like to estimate a dense motion field $\mathcal{T}_{S \leftarrow T} : \mathrm{R}^2 \rightarrow \mathrm{R}^2$ from target image I_T to source image I_S such that

$$I_S(\mathcal{T}_{S \leftarrow T}(\cdot)) = I_T(\cdot). \quad (1)$$

Inspired by the polyaffine motion fusion framework proposed by Arsigny et al. [4], we adapted the motion estimation module from [23] to develop our proposed method for cardiac motion estimation in echocardiography.

The Polyaffine motion model consists of two steps. Firstly, the motion of the left ventricle is approximated through the sparse motion of several key points. An encoder-decoder network is used to output the location of key points and their local affine mapping for both I_S and I_T separately. Secondly, from the sparse motion, we obtain the final dense motion field through polyaffine motion fusion. The proposed method is illustrated in Fig. 1.

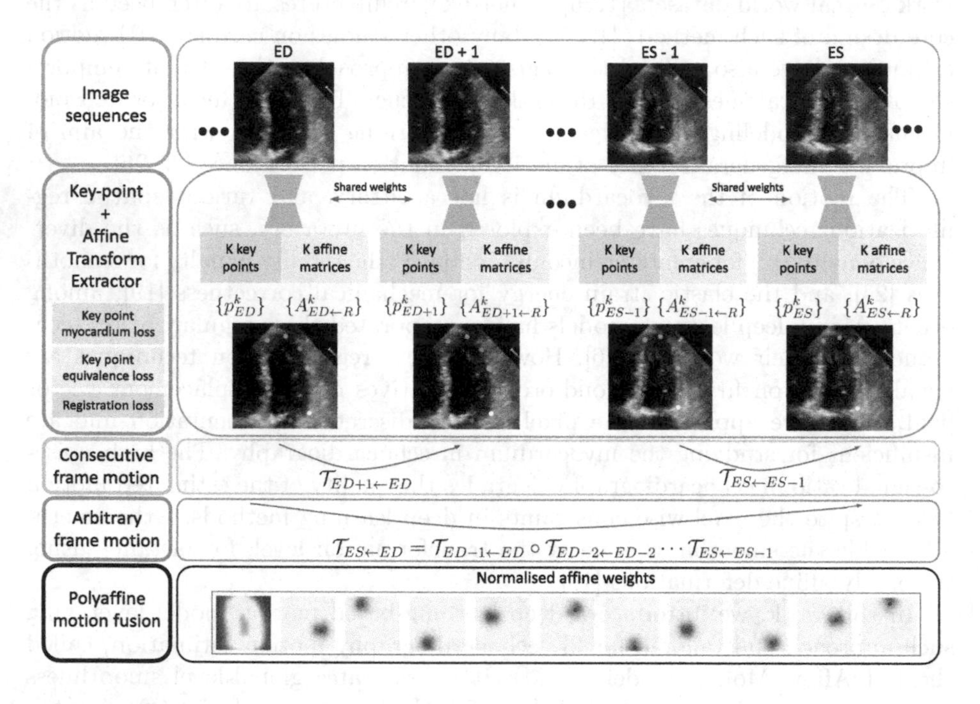

Fig. 1. Method overview.

Key Point and Affine Transformation Estimation. We adopt the encoder-decoder architecture for key point extraction as presented in [23] and provide a brief review of the method from the point of view of affine transformation. In order to process each image separately, we assume there exists an abstract reference frame R. Given an image X, the encoder-decoder network outputs the estimated key points $p^k_X, k = 1, 2, ...K$ as well as the corresponding linear mappings $A^k_{X\leftarrow R} \in \mathcal{R}^{2\times 2}, k = 1, 2, ...K$. The local affine transformation from target image I_T to source image I_S is then computed using the following equation

$$\mathcal{T}^k_{S\leftarrow T}(z) = \underbrace{\bar{A}^k_{S\leftarrow T} \cdot z}_{\text{Linear mapping}} + \underbrace{(p^k_S - \bar{A}^k_{S\leftarrow T} \cdot p^k_T)}_{\text{Translation}}, \tag{2}$$

where $z \in \mathrm{R}^2$ represents the coordinate in the target image, and $\bar{A}^k_{S\leftarrow T} = A^k_{S\leftarrow R}(A^k_{T\leftarrow R})^{-1}$. In order to capture motion around the myocardium, we introduce a myocardium-related key-point prior (Sect. 3.2) and associated loss functions (Sect. 2.2). This encourages the network to output key points close to the myocardium, in contrast to [23], which employed a self-supervised approach for learning key-point positions.

Polyaffine Motion Fusion. Once we obtain the local affine transformation, the dense motion field is computed through direct polyaffine motion fusion. For each local affine motion, a spatial weight $W_k(p^k_T, \sigma^2)$ controls its influenced region. It is a 2D isotropic Gaussian distribution centered at key point p^k_T with variance σ^2. W_0 represents the weight for the background region and the left ventricle cavity area. It is computed as follows:

$$W_0 = \mathbf{RELU}(1 - \sum_{k=1}^{K} W_k(p^k_T, \sigma^2)). \tag{3}$$

We normalise all weights into $\bar{W}_k = \frac{W_k}{\sum_{k=0}^{K} W_k}$, where $k = 0, 1, 2...K$. An example is shown in Fig. 1. We apply them to compute the polyaffine dense motion

$$\mathcal{T}_{S\leftarrow T}(z) = \bar{W}_0 z + \sum_{k=1}^{K} \bar{W}_k(p^k_T)\mathcal{T}^k_{S\leftarrow T}(z). \tag{4}$$

The original first-order motion model (FOMM) [23] utilised a second encoder-decoder network to estimate the normalised weights for each local affine transformation, which can not guarantee the center of weights close to the corresponding key point, thereby unstable for myocardial motion estimation.

Sequence Motion Estimation. Considering a sequence of image frames $I_1, I_2, ...I_N$, the assumption of abstract reference frame enables a fast way to obtain the dense motion field between any arbitrary pair of frames in the sequence. Without loss of generality, we assume I_{ED} is the frame at end-diastole and I_{ES} is the frame at end-systole and that I_{ED} is ahead of I_{ES} in time. The local affine transformation from I_{ED} to I_{ES} can be calculated by composition:

$$\begin{aligned} \mathcal{T}^k_{ES\leftarrow ED}(z) &= \mathcal{T}^k_{ED+1\leftarrow ED} \circ \mathcal{T}^k_{ED+2\leftarrow ED+1} \cdots \mathcal{T}^k_{ES\leftarrow ES-1} \\ &= \bar{A}^k_{ES\leftarrow ED} \cdot z + (p^k_{ES} - \bar{A}^k_{ES\leftarrow ED} \cdot p^k_{ED}), \end{aligned} \tag{5}$$

where $\bar{A}^k_{ES\leftarrow ED} = A^k_{ES\leftarrow R}(A^k_{ED\leftarrow R})^{-1}$. The final dense motion is computed by combining the local motion, as described in Eq. 4.

2.2 Loss Functions

The model is trained end-to-end using a combination of loss functions, which can be grouped into the following subsets for sequence motion estimation.

Keypoint Myocardium Losses. We propose two loss functions to enforce the position of key points near the myocardium region. The first loss, denoted by $\mathcal{L}_{\text{kp\_prior}}$, penalizes the L_2 norm between the estimated key-point position and a prior position. We obtain the prior key-point position from the available training masks, as described in Sect. 3.2. The second loss, denoted by $\mathcal{L}_{\text{kp\_myo\_ED}}$, constrains key points at end-diastole (ED) to reside within the myocardium region by penalizing the distance between each key point and the myocardium.

$$\mathcal{L}_{\text{kp\_myo\_ED}} = -\sum_{k=1}^{K} H(p^k, \sigma_H^2) * (\text{Mask}_{\text{myo\_ED}} - 0.5), \tag{6}$$

where $H(p^k, \sigma_H^2)$ is the isotropic Gaussian heatmap centered at p^k with variance σ_H^2 and $\text{Mask}_{\text{myo\_ED}}$ the binary mask of myocardium at ED.

Keypoint Equivalence Losses. Another two losses $\mathcal{L}_{\text{equi\_kp}}$ and $\mathcal{L}_{\text{equi\_affine}}$ impose an equivalence constraint to the detected key points [23]. They force the model to predict consistent key points and local linear mappings when applying a known transformation to the input image.

Registration Losses. The last four losses regularize the final dense motion field through image similarity and key-point similarity. First, for each pair of consecutive frames in the sequence, we constrain

$$\mathcal{L}_{\text{seq\_im}} = \sum_{seq} NCC(I_T, I_S(\mathcal{T}_{I_S \leftarrow I_T})), \tag{7}$$

$$\mathcal{L}_{\text{seq\_kp}} = \sum_{seq} \frac{1}{K} \sum_{k=1}^{K} |H(p_T^k, \sigma_H^2) - H(p_S^k, \sigma_H^2)(\mathcal{T}_{I_S \leftarrow I_T})|, \tag{8}$$

where NCC represents the normalised cross correlation. Second, in order to force the model to learn temporal motion, we apply a second pair of similarity loss between the image at end-diastole and all frames after the end-diastole frame, denoted as $\mathcal{L}_{\text{ED2any\_im}}$ and $\mathcal{L}_{\text{ED2any\_kp}}$ for image similarity and key-point similarity respectively.

The total loss for the PAM model becomes

$$\begin{aligned}\mathcal{L}_{\text{total}} =& \lambda_1 \mathcal{L}_{\text{kp\_prior}} + \lambda_2 \mathcal{L}_{\text{kp\_myo\_ED}} + \lambda_3 \mathcal{L}_{\text{equi\_kp}} + \lambda_4 \mathcal{L}_{\text{equi\_affine}} \\ &+ \lambda_5 \mathcal{L}_{\text{seq\_im}} + \lambda_6 \mathcal{L}_{\text{seq\_kp}} + \lambda_7 \mathcal{L}_{\text{ED2any\_im}} + \lambda_8 \mathcal{L}_{\text{ED2any\_kp}}.\end{aligned} \tag{9}$$

3 Experiments

3.1 Datasets

Three public datasets of echocardiography are included in our study. EchoNet[1] [18] contains 10030 echocardiography videos of long-axis 4 chamber

[1] https://echonet.github.io/dynamic/.

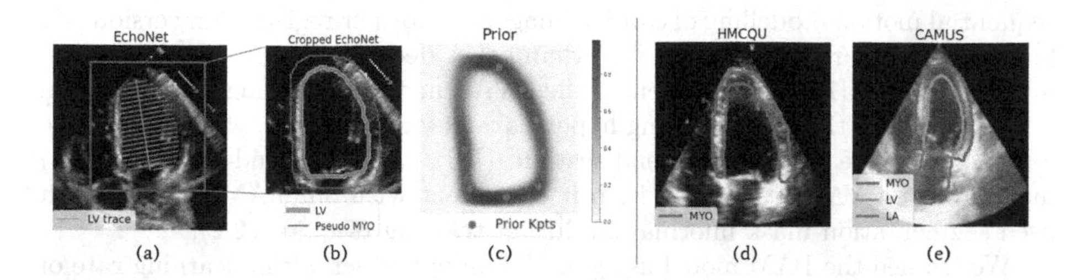

Fig. 2. Dataset overview. (a) EchoNet example and the given annotations of left ventricle tracing. (b) Left ventricle cropping with the generated pseudo myocardium contour (blue). (c) The mean mask from the EchoNet training set and the 10 prior key points. (d) HMC-QU example and the given annotation of the myocardium. (e) CAMUS example and the given annotation of different cardiac structures. (MYO: myocardium, LV: left ventricle, LA: left atrium) (Color figure online)

view. Left-ventricle tracings at end-diastole and end-systole are available (see Fig. 2(a)). CAMUS dataset[2] [14] consists of apical 4-chamber view videos from 450 patients whose left heart segmentation masks are publicly accessible (see Fig. 2(e)). HMC-QU dataset[3] [9] contains 109 4-chamber view echocardiography recordings with the segmentation of the left myocardium along one cardiac cycle (see Fig. 2(d)). All image sequences are cropped around the left ventricle center according to the provided segmentation/tracings (see Fig. 2(b)). In this study, we follow the given data division of EchoNet dataset, with 7465 samples for training, 1288 samples for validation and 1277 samples for testing. CAMUS and HMC-QU datasets are used for evaluation during test phase.

3.2 Dataset Preprocessing

Pseudo Myocardium Mask. There is no ground truth of myocardium mask in EchoNet dataset. To provide guidance for key points, we generate a pseudo myocardium mask for ED and ES by applying a dilation operation using 13×13 and 17×17 structure element to the left ventricle mask at ED and ES frame respectively. The difference between the dilated mask and the original one is regarded as the pseudo myocardium mask (see Fig. 2(b)).

Key-Point Prior at End-Diastole. A mean myocardium mask is computed using all the pseudo myocardium masks at ED from the training set. Then, all the pixels of the mean mask are clustered into 10 groups using the KMeans function from the scikit-learn package [20], where the center of each group is considered as one key-point prior (see Fig. 2(c)).

3.3 Implementation

We compare our proposed model with the conditional variational autoencoder (CVAE) method proposed in [13], which has demonstrated effectiveness in

[2] https://www.creatis.insa-lyon.fr/Challenge/camus/.
[3] https://www.kaggle.com/datasets/aysendegerli/hmcqu-dataset.

sequential motion modelling of cardiac images. Its former registration version [12] has shown more regular motion field than other deep learning methods, including VoxelMorph [7]. We implemented the CVAE method following its description in [13] and used the same training hyper-parameters. However, we added a Dice loss between the transformed end-systolic (ES) mask and end-diastolic (ED) mask during CVAE training to keep it consistent with our PAM model, which used segmentation mask information during training (see Sect. 2.2).

We trained the PAM model using an Adam optimizer with a learning rate of $1e-4$. To determine the optimal hyperparameters, we conducted experiments on a randomly selected subset of 1000 training examples from the EchoNet dataset. The hyperparameters for the loss function were set to $\lambda_1 = 20, \lambda_2 = 0.1, \lambda_3 = 50, \lambda_4 = 50, \lambda_5 = \lambda_6 = \lambda_7 = \lambda_8 = 100$. The variance of the Gaussian heatmap was set to $\sigma^2 = 0.05$ for affine transformation weights (see Eq. 3), and $\sigma_H^2 = 0.005$ for all other scenarios. We trained the model for a maximum of 100 epochs, and applied early stopping if the validation loss did not improve over 10 epochs.

4 Results

We first evaluated the motion estimation accuracy by assessing the registration performance using the available segmentation masks from all the test datasets.

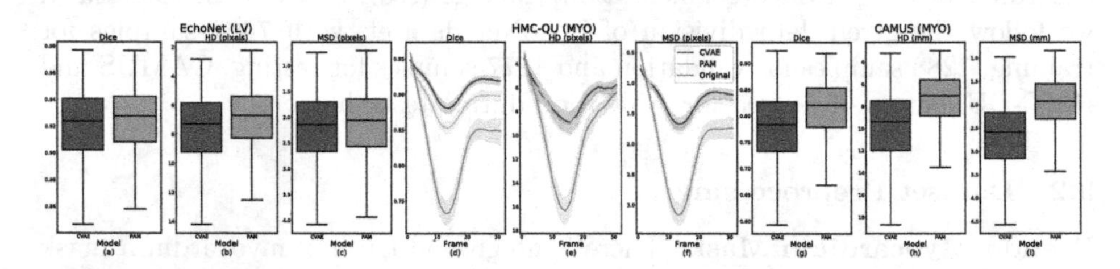

Fig. 3. Evaluation results on test datasets. (a–c) On EchoNet test split using available left ventricle masks of ED/ES. (d–f) On HMC-QU dataset using frame-wise myocardium masks. The original line represents the comparison between the ground truth ED frame and the ground truth mask of each frame. (g–i) On CAMUS 4-chamber view dataset using myocardium masks of ED/ES. (HD: Hausdorff distance, MSD: main surface distance)

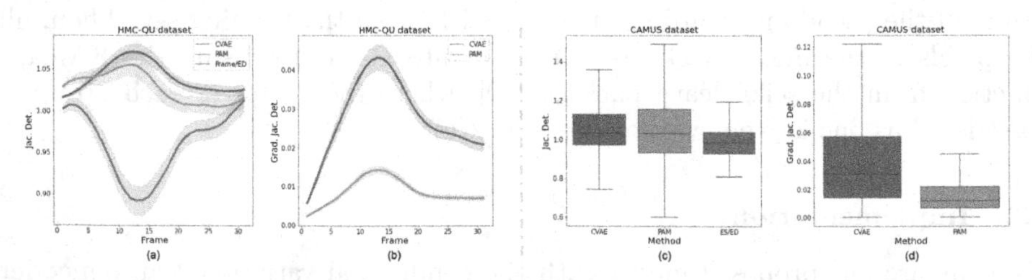

Fig. 4. Evaluation of Jacobian Determinant and its gradient in the myocardium region on the HMC-QU dataset (a–b) and on the CAMUS dataset (c–d). We also present the area change of the myocardium between the considered frame and end-diastole.

For the EchoNet test split, we compared the ground truth left ventricle mask of end-diastole (ED) with that transformed from end-systole (ES). For the CAMUS dataset, the same evaluation was applied to the myocardium mask for ED/ES. For the HMC-QU dataset, the myocardium masks of one cardiac cycle were all transformed to ED using the estimated motion field. To enable group-level statistical analysis along the cardiac cycle, the temporal metric was interpolated to the same length. Our proposed PAM model demonstrated significant outperformance compared to the CVAE model on the test sets of EchoNet and CAMUS datasets. However, it showed slightly lower Dice scores on the HMC-QU dataset, which may be due to the fact that the pseudo-labelling process for generating the ground truth segmentation of the myocardium on the HMC-QU dataset [9] may be not very accurate. Furthermore, the PAM model exhibited more regular deformation fields compared to the baseline model (as shown in Fig. 4) when evaluating the gradient of Jacobian determinant. This smooth displacement field is advantageous for computing dense strain tensor, which is typically constructed using the second-order derivative of the displacement field. Additionally, as listed in Table 1, we demonstrated that our proposition of using myocardium prior and the polyaffine fusion mechanism greatly improved the registration performance, when compared with the original first-order motion model (FOMM) [23]. In particular, our explicit design of fusion weights helps the network to efficiently learn affine transformation locally.

Table 1. Performance comparison of different methods on EchoNet and CAMUS datasets. FOMM: the original model in [23], without prior of any key points (Prior), without explicit design of polyaffine weights (Polyaffine), without registration penalty between ED and any other frames (Sequence). All = FOMM+Prior+Polyaffine+Sequence.

Method	EchoNet-LV			CAMUS-MYO		
	Dice	HD *(pixels)*	MSD *(pixels)*	Dice	HD *(mm)*	MSD *(mm)*
All = PAM (ours)	**0.92**	**7.33**	**2.23**	**0.81**	**7.43**	**1.99**
FOMM+Prior+Polyaffine	0.91	7.53	2.36	0.80	7.95	2.08
FOMM+Prior	0.77	18.36	5.46	0.57	13.85	3.89
FOMM [23]	0.75	22.99	6.07	0.48	17.27	4.41
CVAE [13]	0.91	7.97	2.30	0.77	9.91	2.67

In addition to the good performance from various evaluation metrics, the proposed PAM model has the potential for abnormal wall motion detection. We provide examples of myocardial infarction (MI) from the HMC-QU dataset (see Supplementary video material) and show the abnormal strain values in different segments, highlighting the consistency of our findings with the ground truth diagnosis.

5 Discussion and Conclusion

In this paper, we proposed a polyaffine motion model (PAM) for echocardiography motion estimation. The PAM model demonstrated excellent motion estimation performance on real-world echocardiography datasets and showed good generalization to unseen datasets from other centers. Our explicit design of fusion weights enabled efficient learning of local affine transformation, and the intrinsic polyaffine structure improved the smoothness of the motion field, showing potential for abnormal wall motion detection. In the future, we will focus on integrating temporal regularization for the PAM model and conducting evaluations on synthetic datasets with known ground-truth displacement.

Acknowledgements. This work has been supported by the French government through the National Research Agency (ANR) Investments in the Future with 3IA Côte d'Azur (ANR-19-P3IA-0002) and by Inria PhD funding. The authors are grateful to the OPAL infrastructure from Université Côte d'Azur for providing resources and support.

References

1. Ahn, S.S., Ta, K., Lu, A., Stendahl, J.C., Sinusas, A.J., Duncan, J.S.: Unsupervised motion tracking of left ventricle in echocardiography. In: Medical Imaging 2020: Ultrasonic Imaging and Tomography, vol. 11319, pp. 196–202. SPIE (2020)
2. Alessandrini, M., Chakraborty, B., et al.: Realistic vendor-specific synthetic ultrasound data for quality assurance of 2-D speckle tracking echocardiography: simulation pipeline and open access database. IEEE Trans. Ultrason. Ferroelectr. Freq. Control **65**(3), 411–422 (2017)
3. Alessandrini, M., Heyde, B., et al.: Detailed evaluation of five 3D speckle tracking algorithms using synthetic echocardiographic recordings. IEEE Trans. Med. Imaging **35**(8), 1915–1926 (2016)
4. Arsigny, V., Commowick, O., Ayache, N., Pennec, X.: A fast and log-euclidean polyaffine framework for locally linear registration. J. Math. Imaging Vis. **33**(2), 222–238 (2009)
5. Azarmehr, N., et al.: An optimisation-based iterative approach for speckle tracking echocardiography. Med. Biol. Eng. Comput. **58**(6), 1309–1323 (2020). https://doi.org/10.1007/s11517-020-02142-8
6. Barbosa, D., Heyde, B., Dietenbeck, T., Friboulet, D., D'hooge, J., Bernard, O.: Fast left ventricle tracking in 3D echocardiographic data using anatomical affine optical flow. In: Ourselin, S., Rueckert, D., Smith, N. (eds.) FIMH 2013. LNCS, vol. 7945, pp. 191–199. Springer, Heidelberg (2013). https://doi.org/10.1007/978-3-642-38899-6_23
7. Dalca, A.V., Balakrishnan, G., Guttag, J., Sabuncu, M.R.: Unsupervised learning for fast probabilistic diffeomorphic registration. In: Frangi, A.F., Schnabel, J.A., Davatzikos, C., Alberola-López, C., Fichtinger, G. (eds.) MICCAI 2018. LNCS, vol. 11070, pp. 729–738. Springer, Cham (2018). https://doi.org/10.1007/978-3-030-00928-1_82
8. De Vos, B.D., Berendsen, F.F., Viergever, M.A., Sokooti, H., Staring, M., Išgum, I.: A deep learning framework for unsupervised affine and deformable image registration. Med. Image Anal. **52**, 128–143 (2019)
9. Degerli, A., et al.: Early detection of myocardial infarction in low-quality echocardiography. IEEE Access **9**, 34442–34453 (2021)
10. Evain, E., Sun, Y., Faraz, K., Garcia, D., et al.: Motion estimation by deep learning in 2D echocardiography: synthetic dataset and validation. IEEE Trans. Med. Imaging **41**(8), 1911–1924 (2022)

11. Farnebäck, G.: Two-frame motion estimation based on polynomial expansion. In: Bigun, J., Gustavsson, T. (eds.) SCIA 2003. LNCS, vol. 2749, pp. 363–370. Springer, Heidelberg (2003). https://doi.org/10.1007/3-540-45103-X_50
12. Krebs, J., Delingette, H., Mailhé, B., Ayache, N., Mansi, T.: Learning a probabilistic model for diffeomorphic registration. IEEE Trans. Med. Imaging **38**(9), 2165–2176 (2019)
13. Krebs, J., Mansi, T., Ayache, N., Delingette, H.: Probabilistic motion modeling from medical image sequences: application to cardiac cine-MRI. In: Pop, M., et al. (eds.) STACOM 2019. LNCS, vol. 12009, pp. 176–185. Springer, Cham (2020). https://doi.org/10.1007/978-3-030-39074-7_19
14. Leclerc, S., Smistad, E., et al.: Deep learning for segmentation using an open large-scale dataset in 2D echocardiography. IEEE Trans. Med. Imaging **38**(9), 2198–2210 (2019)
15. Mansi, T., Pennec, X., Sermesant, M., Delingette, H., Ayache, N.: iLogDemons: a demons-based registration algorithm for tracking incompressible elastic biological tissues. Int. J. Comput. Vision **92**, 92–111 (2011)
16. McLeod, K., Sermesant, M., Beerbaum, P., Pennec, X.: Spatio-temporal tensor decomposition of a polyaffine motion model for a better analysis of pathological left ventricular dynamics. IEEE Trans. Med. Imaging **34**(7), 1562–1575 (2015)
17. Østvik, A., Salte, I.M., Smistad, E., Nguyen, T.M., Melichova, D., et al.: Myocardial function imaging in echocardiography using deep learning. IEEE Trans. Med. Imaging **40**(5), 1340–1351 (2021)
18. Ouyang, D., He, B., et al.: Video-based AI for beat-to-beat assessment of cardiac function. Nature **580**(7802), 252–256 (2020)
19. Papademetris, X., Sinusas, A.J., Dione, D.P., Duncan, J.S.: Estimation of 3D left ventricular deformation from echocardiography. Med. Image Anal. **5**(1), 17–28 (2001)
20. Pedregosa, F., et al.: Scikit-learn: machine learning in Python. J. Mach. Learn. Res. **12**, 2825–2830 (2011)
21. Qin, C., et al.: Joint learning of motion estimation and segmentation for cardiac MR image sequences. In: Frangi, A.F., Schnabel, J.A., Davatzikos, C., Alberola-López, C., Fichtinger, G. (eds.) MICCAI 2018. LNCS, vol. 11071, pp. 472–480. Springer, Cham (2018). https://doi.org/10.1007/978-3-030-00934-2_53
22. Qin, C., Wang, S., Chen, C., Qiu, H., Bai, W., Rueckert, D.: Biomechanics-informed neural networks for myocardial motion tracking in MRI. In: Martel, A.L., et al. (eds.) MICCAI 2020. LNCS, vol. 12263, pp. 296–306. Springer, Cham (2020). https://doi.org/10.1007/978-3-030-59716-0_29
23. Siarohin, A., Lathuilière, S., et al.: First order motion model for image animation. In: Advances in Neural Information Processing Systems, vol. 32 (2019)
24. Staring, M., Klein, S., Pluim, J.P.: A rigidity penalty term for nonrigid registration. Med. Phys. **34**(11), 4098–4108 (2007)
25. Ta, K., Ahn, S.S., Lu, A., Stendahl, J.C., Sinusas, A.J., Duncan, J.S.: A semi-supervised joint learning approach to left ventricular segmentation and motion tracking in echocardiography. In: 2020 IEEE 17th International Symposium on Biomedical Imaging (ISBI), pp. 1734–1737. IEEE (2020)
26. Zhang, X., You, C., et al.: Learning correspondences of cardiac motion from images using biomechanics-informed modeling. In: Camara, O., et al. (eds.) Statistical Atlases and Computational Models of the Heart, pp. 13–25. Springer, Cham (2022). https://doi.org/10.1007/978-3-031-23443-9_2

Automated Analysis of Mitral Inflow Doppler Using Deep Neural Networks

Jevgeni Jevsikov[1](✉), Elisabeth S. Lane[2], Eman Alajrami[1], Preshen Naidoo[1], Nasim Dadashi Serej[1], Neda Azarmehr[1], Sama Aleshaiker[1], Catherine C. Stowell[2], Matthew J. Shun-shin[2], Darrel P. Francis[2], and Massoud Zolgharni[1,2]

[1] School of Computing and Engineering, University of West London, London, UK
Jevgeni.Jevsikov@uwl.ac.uk

[2] National Heart and Lung Institute, Imperial College, London, UK

Abstract. Doppler echocardiography is a widely applied modality for the functional assessment of heart valves, such as the mitral valve. Currently, Doppler echocardiography analysis is manually performed by human experts. This process is not only expensive and time-consuming, but often suffers from intra- and inter-observer variability. An automated analysis tool for non-invasive evaluation of cardiac hemodynamic has potential to improve accuracy, patient outcomes, and save valuable resources for health services. Here, a robust algorithm is presented for automatic Doppler Mitral Inflow peak velocity detection utilising state-of-the-art deep learning techniques. The proposed framework consists of a multi-stage convolutional neural network which can process Doppler images spanning arbitrary number of heartbeats, independent from the electrocardiogram signal and any human intervention. Automated measurements are compared to Ground-truth annotations obtained manually by human experts. Results show the proposed model can efficiently detect peak mitral inflow velocity achieving an average F1 score of 0.88 for both E- and A-peaks across the entire test set.

Keywords: Automated Analysis · Deep Learning · Transmitral Doppler Echocardiography

1 Introduction

The commonest method of characterising ventricular filling is assessment of transmitral velocity during the diastole phase, obtained by pulsed-wave Doppler echocardiography, comprising two distinct components: (i) the E-wave (early phase ventricular filling), initiated by active mechanical suction of blood from the atrium by the recoiling and simultaneously relaxing ventricle, and (ii) the A-wave (atrial phase of ventricular filling), caused by left atrial contraction, tops off the ventricle, elevates its pressure and increases its volume.

Currently, the clinicians manually perform velocity measurements on Doppler traces, often resulting in high intra- and inter-observer variability [1]. Moreover, such offline measurements are commonly performed on a single heartbeat, as

O. Bernard et al. (Eds.): FIMH 2023, LNCS 13958, pp. 394–402, 2023.
https://doi.org/10.1007/978-3-031-35302-4_41

opposed to multiple beats and averaging them, due to the time-consuming nature of manual analysis. Automated systems have potential to assist in standardising measurement protocol, thus reducing time and monetary expense acquiring measurements, thereby improving the clinical workflow.

The main aim of this study was to: (i) develop a technique for fully automated measurement of Doppler transmitral velocity profile, and (ii) validate the developed technique aganist the gold-standard manual measurements obtained from the human experts.

We propose a pipeline that is independent from the electrocardiogram (ECG) signal and does not require any pre-processing of the Doppler images. It can accurately detect, isolate, and perform measurements on an arbitrary number of E- and A-wave peaks present in the image, including overlapping waves as well as the singular peak heartbeats.

1.1 Related Works

Jahren et al. [3] explored ways to isolate heartbeat cycles on cardiac spectral Doppler spectrograms without reliance upon an ECG signal. They combined a CNN module to extract local features on an image with an RNN module to connect extracted features temporally. The heartbeat detection model achieved a performance of 97.7% accuracy for true positive detections and 2.5% a false detection rate.

Lane et al. [4] applied a novel Deep Learning technique to localise blood flow velocity peaks on Tissue Doppler Images. The method is ECG-independent and employs a Mask-RCNN model to first isolate heartbeats. Next, beats are classified into complete, where all three Doppler peaks are present or incomplete, where only some peaks are visible. Finally, complete beats are cropped, and S', E' and A' peak velocities are detected using a multi-stage gaussian heatmap regression model.

Zamzmi et al. [10] propose a comprehensive overview of relevant literature in the field of automated mitral inflow velocity measurements, inclusive of their own Deep Learning (DL) framework. In their study, Faster-RCNN is used to isolate the Doppler and ECG regions, which are used to isolate cardiac cycle beats. However, automatic beat segmentation failed in several images due to noisy ECG signal. Finally, E' and A' peak velocities are detected using the peak detection algorithm on the profile curve.

Elwazir et al. [2] apply a differing approach. They train a DL classifier to differentiate between echocardiographic study types. Mitral inflow images are then segmented using a U-net network for the purpose of extracting the envelope profile. ECG tracing was used to separate the beats. The E' and A' peaks were subsequently detected through signal processing of the segmented envelope. A key limitation of the study being that only echo images from normal patients, without abnormalities, were included in the training data.

Park et al. [6] suggested a non-DL approach focussing on triangular shape detection, where multiple triangular-shaped objects are proposed, and a scoring system applied to select the optimal. This paper also addresses the issue of overlapping E' and A' waves.

1.2 Heatmap Regression

Heatmap regression is a popular method used for landmark detection. A gaussian heatmap is a visual representation of the probability of a certain keypoint being present at a given pixel. Wei et al. [9] used heatmap regression in their Convolutional Pose Machine (CPM) framework to predict pose estimates. In their CPM they created a multi-stage CNN which would predict pose landmark estimates. The final output of CPM has N channels, corresponding to the number of predictions for a particular use case.

Stern et al. [8] proposed a U-Net-like [7] architecture to find suture point landmarks during Mitral valve repair surgery. Their network can detect an arbitrary number of landmarks. During post-processing, they threshold the output heatmaps and locate the binarised mass objects' centroids.

Heatmap regression allows the prediction of a variable number of landmarks per image, as each predicted heatmap may contain Gaussian probabilities for all keypoints. Moreover, as a heatmap simply contains probabilities, a threshold can be applied, compared to methods involving direct prediction of Cartesian coordinates.

2 Methodology

2.1 Patient Datasets and Expert Annotations

Development-dataset - A random sample of 700 echocardiographic (echo) examinations of different patients was extracted from Imperial College Healthcare NHS Trust's echocardiogram database. The acquisition of the images was performed by experienced echocardiographers and according to standard protocols. Each image underwent labelling (selecting the E- and A-wave peak coordinates) by one individual from a pool of experts using our in-house online labelling platform (https://unityimaging.net). This dataset was used for model developments (i.e. training and validation).

Test-dataset - An independent database of consecutive studies conducted over 3 working days in 2019 (3 years apart from the date of previous database) was used as the testing dataset. The testing dataset consisted 117 images. Mitral inflow Doppler images in each echo exam were then automatically identified, and each image was labelled by different pool of clinical experts, using the Unity labelling platform.

2.2 Doppler Peak Velocity Annotations

For each image, we created a set of heatmaps that served as the ground truth. There are two heatmaps for each image since symmetric Gaussian distributions were created at all Ground-truth peak coordinates for each peak type (E- and A-wave).

Figure 1 illustrates typical generated heatmaps with Gaussian peaks. The spread of the Gaussian distribution, which is controlled by the standard deviation value, was set to 5.

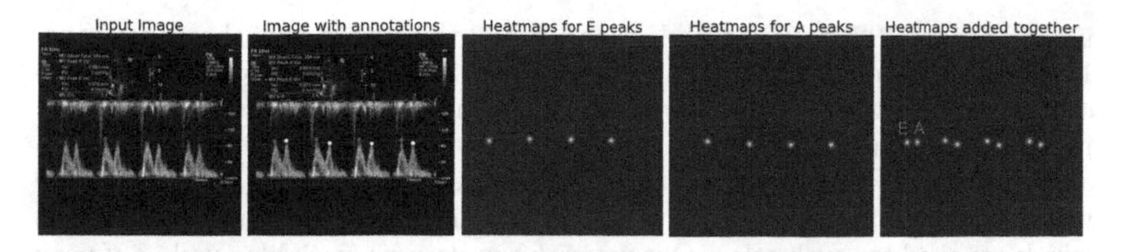

Fig. 1. Example of wave peak annotations (Red - E-wave, White - A-wave) along with its corresponding generated heatmaps which was used as ground-truth. (Color figure online)

2.3 Deep Learning Framework

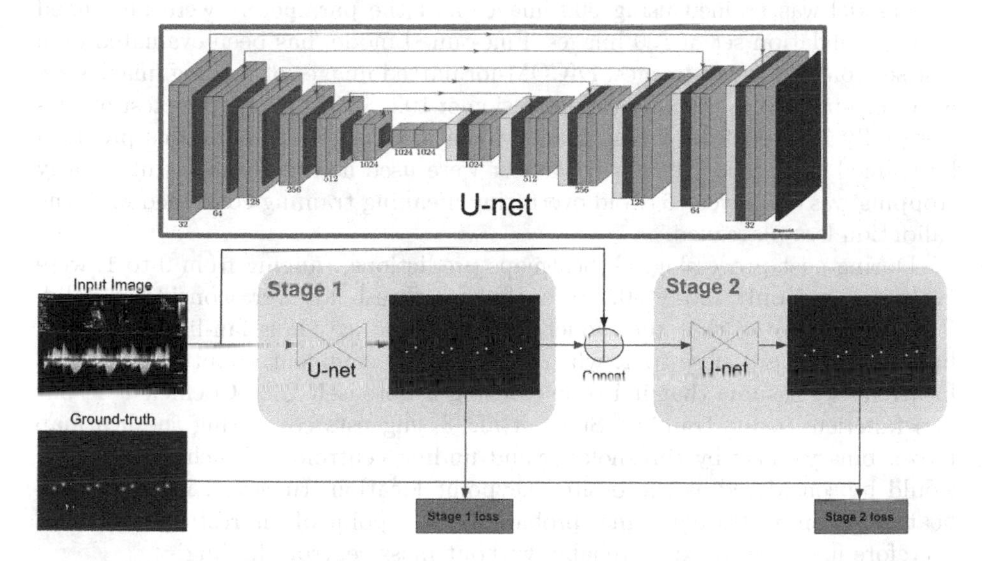

Fig. 2. Illustration of the proposed network. The backbone at each stage is U-NET [7] with the depth of 6. The U-net illustration was created using the PlotNeuralNet tool (https://github.com/HarisIqbal88/PlotNeuralNet)

In Fig. 2 we illustrate the proposed automatic measurement framework, which involves a multi-stage [5,9] DL network, where the output from each stage is concatenated with the original image and passed into the next stage. By passing heatmap estimations to each consecutive stage along with the original image, the network can reassess it's initial predictions. Each stage has its own loss function, which in our work is Binary Cross Entropy + Sørensen Dice Coefficient and are optimised using ADAM optimizer.

We empricially used a 2-stage network, at each stage we implemented a U-Net [7] architecture incorporating an encoder-decoder structure. Each stage takes an input of shape 1024×1024 with N channels and the output is of the same shape, 1024×1024, but with 2 channels, where the first channel contains heatmap estimates for E-wave peaks and the second channel predicts A-wave peaks. The input image convolves through the first stage and outputs a set of heatmap approximations which are concatenated with the original image and used as the input for the second stage. The second stage outputs the final set of heatmaps which are used for evaluation. Our network can detect a variable number of peak locations on image, which is an important aspect, because Doppler Mitral inflow images can contain a variable number of heartbeats.

3 Results and Discussion

3.1 Evaluation and Implementation Details

The model was trained using 600 images and the parameters were fine-tuned using a validation set of 100 images. Fine-tuned model has been evaluated on a test set containing 117 images. DICOM-formatted images of varying image sizes were zero-padded to the uniform dimension of 1024×1024 pixels. The test images were fully isolated from training and validation images. Annotations provided by trained experts for the entire dataset were used as the Ground-truth. Early stopping was employed to avoid overfitting meaning training continued until the validation loss plateaued.

During post-processing, all heatmaps predictions, ranging from 0 to 1, were evaluated and only those >0.7 (a confidence threshold) were considered valid. It should be noted that the model does not produce Gaussian-like heatmaps, but the values are close to 1, when the model is confident about the keypoint location. We assume that it is due to using a Sørensen Dice Coefficient in our loss function during training. Stern et al. [8] suggests converting the heatmap into a binary mask by thresholding and finding centroids of each mass, which would be considered as the desired keypoint location. However, in our model, heatmaps demonstrate a higher probability at a point of interest and we could therefore find local maxima reliably without mass centroid finding.

To get Cartesian coordinates from the predicted heatmaps, all local maxima were identified by a process of iteration. In order to find the matching Ground-truth peak point for each predicted keypoint, a region of 30 pixels wide on either side of the keypoint was considered. The total boundary width would be 60 pixels, which we found was enough to cover an entire wave in most cases. Then, if a Ground-truth keypoint coordinate is found within that boundary, it is considered as True Positive (TP), otherwise it will be counted as False Positive (FP). If for any Ground-truth keypoint no prediction is found, it is considered as False Negative (FN). Table 1 shows the calculated values of Precision, Sensitivity and F1 score, based on the equations below. Some examples of True Positive detections are shown on Fig. 3.

$$Precision = \frac{TP}{TP + FP} \tag{1}$$

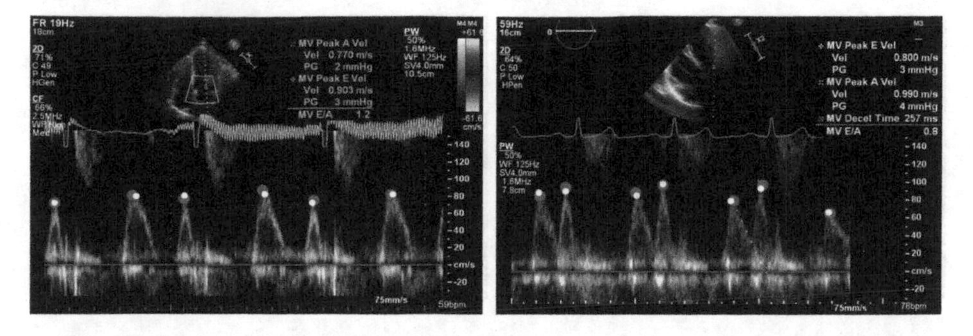

Fig. 3. Examples of True Positive detections, showing predicted E- and A-wave peak coordinates (Red) and Ground-truth E- and A-wave peak annotations (white) (Color figure online)

$$Sensitivity = \frac{TP}{TP + FN} \tag{2}$$

$$F1\ score = \frac{2 * Precision * Sensitivity}{Precision + Sensitivity} \tag{3}$$

Table 1. Model performance for each wave

	E-wave	A-wave
Precision	0.9570	0.9482
Sensitivity	0.8241	0.8120
F1 score	0.8856	0.8748

Our observations show that one of the biggest challenges for our network is dealing with images that have noise artefacts or where the velocities for one or both waves are too low and hardly distinguished from noise. Examples of failing situations are shown on Fig. 4.

3.2 Statistical Analysis

For correctly detected peak points, the automated measurements were compared to the matching Ground-truth values by Bland-Altman analyses. Figure 5 upper row shows such plots for the pool of all wave peaks across all patients. Mean difference for the automated measurements versus the expert measurements was -1.07 cm/s for E-wave and -0.97 cm/s for A-wave peak velocities, respectively. Standard Deviation was ≤ 5.5 cm/s for either of the peak velocities.

The lower row shows the same plots, but for a patient-by-patient analysis, where all heartbeats present in a Doppler image for each patient are averaged.

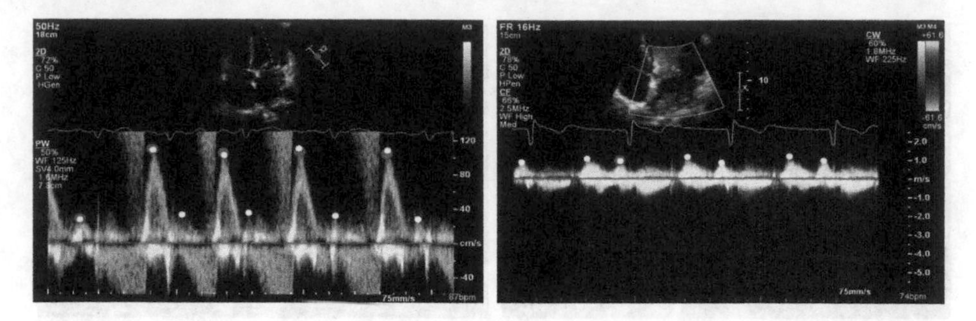

Fig. 4. Examples of missed detection cases (False Negatives), where dots with red colour are model predictions and white dots are Ground-truth annotations. On the left panel, four A-peaks have been missed, and on the right panel, the model has failed to detect most of the peak points, which is likely to be due to the low-quality Doppler image. (Color figure online)

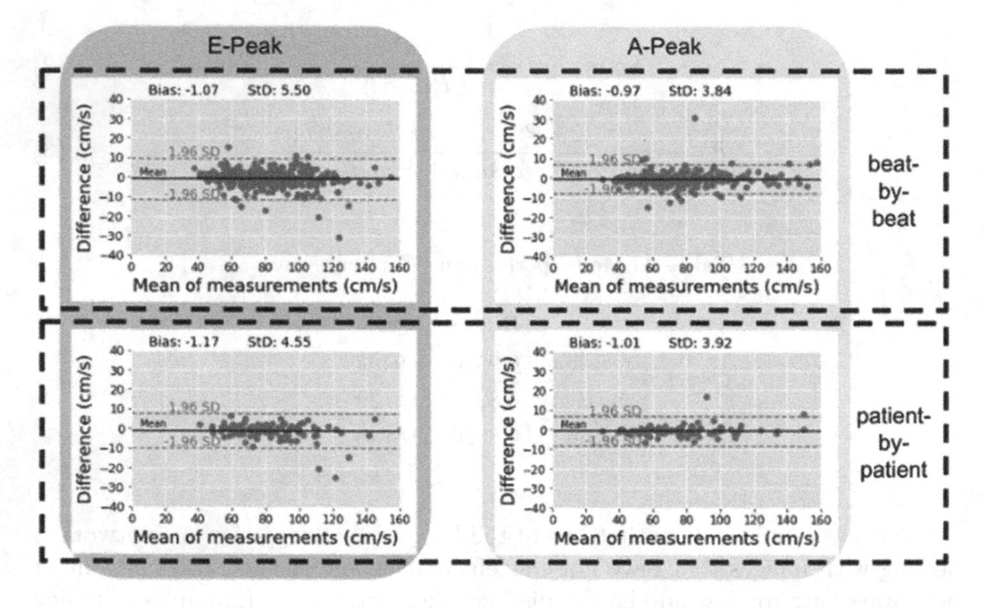

Fig. 5. Bland-Altman plots for patient-by-patient and beat-by-beat analysis

While the mean difference between automated and expert measurements remains at the similar level, the standard deviation of the differences is reduced to $\leq$ 4.55, indicating that beat-averaging for each patient results in more reliable measurements.

4 Conclusion and Future Work

The aim of this study was to develop an automated DL pipeline capable of measuring continuous Doppler Mitral Inflow velocities on images containing an arbitrary number of heartbeats, without using ECG signal. The trained network was evaluated against Ground-truth measurements acquired by clinical experts.

The proposed model accurately calculates velocity peaks on Doppler images. Among all predicted keypoints, the model achieved an average F1 score of 0.8802 for both E- and A-wave peaks. The mean difference of True detections compared to Ground-truth was −1.02 cm/s, and an average Standard deviation was 4.67 cm/s.

As a future work, it would be beneficial for the model to see more examples with noise and artefacts on images. This could be achieved by labelling new samples of images by clinical experts, or using pseudo-labelling.

Acknowledgement. This work was supported in part by the British Heart Foundation, UK (Grant no. RG/F/22/110059). J Jevsikov is supported by the Vice Chancellor's Scholarship at the University of West London. We are grateful to the following experts for their invaluable input in labelling images: Arjun Ghosh, Maysaa Zetani, Mahmoud Tawil, Luxy Ananthan, Camelia Demetrescu, Amar Singh, Sanjeev Bhattacharyya, Joban Sehmi, Kavitha Vimalesvaran, Abdallah Al-Mohammad, Bushra Rana, Tiffany Ng.

References

1. Corriveau, M.M., Johnston, K.W.: Interobserver variability of carotid doppler peak velocity measurements among technologists in an ICAVL-accredited vascular laboratory. J. Vasc. Surg. **39**(4), 735–741 (2004)
2. Elwazir, M.Y., Akkus, Z., Oguz, D., Ye, Z., Oh, J.K.: Fully automated mitral inflow doppler analysis using deep learning. In: 2020 IEEE 20th International Conference on Bioinformatics and Bioengineering (BIBE), pp. 691–696. IEEE (2020)
3. Jahren, T.S., Steen, E.N., Aase, S.A., Solberg, A.H.S.: Estimation of end-diastole in cardiac spectral doppler using deep learning. IEEE Trans. Ultrason. Ferroelectr. Freq. Control **67**(12), 2605–2614 (2020)
4. Lane, E.S., et al.: Automated multi-beat tissue doppler echocardiography analysis using deep neural networks. Med. Biol. Eng. Comput., 1–16 (2023)
5. Newell, A., Yang, K., Deng, J.: Stacked hourglass networks for human pose estimation. In: Leibe, B., Matas, J., Sebe, N., Welling, M. (eds.) ECCV 2016. LNCS, vol. 9912, pp. 483–499. Springer, Cham (2016). https://doi.org/10.1007/978-3-319-46484-8_29
6. Park, J.H., Zhou, S.K., Jackson, J., Comaniciu, D.: Automatic mitral valve inflow measurements from doppler echocardiography. In: Metaxas, D., Axel, L., Fichtinger, G., Székely, G. (eds.) MICCAI 2008. LNCS, vol. 5241, pp. 983–990. Springer, Heidelberg (2008). https://doi.org/10.1007/978-3-540-85988-8_117
7. Ronneberger, O., Fischer, P., Brox, T.: U-Net: convolutional networks for biomedical image segmentation. In: Navab, N., Hornegger, J., Wells, W.M., Frangi, A.F. (eds.) MICCAI 2015. LNCS, vol. 9351, pp. 234–241. Springer, Cham (2015). https://doi.org/10.1007/978-3-319-24574-4_28
8. Stern, A., et al.: Heatmap-based 2D landmark detection with a varying number of landmarks. In: Bildverarbeitung für die Medizin 2021. I, pp. 22–27. Springer, Wiesbaden (2021). https://doi.org/10.1007/978-3-658-33198-6_7

9. Wei, S.E., Ramakrishna, V., Kanade, T., Sheikh, Y.: Convolutional pose machines. In: Proceedings of the IEEE Conference on Computer Vision and Pattern Recognition, pp. 4724–4732 (2016)
10. Zamzmi, G., Hsu, L.Y., Li, W., Sachdev, V., Antani, S.: Fully automated spectral envelope and peak velocity detection from doppler echocardiography images. In: Medical Imaging 2020: Computer-Aided Diagnosis, vol. 11314, pp. 1053–1064. SPIE (2020)

VisHeart: A Visualization and Analysis Tool for Multidimensional Data

Edson A. G. Coutinho[1], Bruno M. Carvalho[1(✉)], Selan R. dos Santos[1], and Leon Axel[2]

[1] Federal University of Rio Grande do Norte, Natal, Brazil
{alyppyo,bruno,selan}@dimap.ufrn.br
[2] New York University, New York, NY, USA
leon.axel@nyulangone.org

Abstract. New MRI techniques enable the reconstruction of multidimensional data sets, including 3D volumes of cardiac image data reconstructed along both cardiac and respiratory phases. However, these new type of data present several challenges in their manipulation, visualization and analysis, and there are no off-the-shelf tools for the interactive visualization of such high-dimensional data sets. Such a tool, in order to be successful, must deal with large amounts of data, producing animations with no lagging and allowing users to interact in real-time with the volumes. Therefore, one must use efficient lossless compression techniques in order to keep the information while allowing the tool to manipulate the data efficiently. In this work, we report on the development of VisHeart, a multi-platform open source tool for the visualization and analysis of multidimensional cardiac data. The experiments reported here culminated in a file format composed of a combination of standard compression techniques, allowing for the compression of the test dataset with a rate of 2.954 in approximately 6 min, while keeping the decompression time to approximately 15 s for a dataset formed by 52 volumes with $192 \times 192 \times 192$ voxels. After the initial decompression, the dataset can be visualized in real-time.

Keywords: Multidimensional data · Cardiovascular MRI · Volume compression · Volume rendering

1 Introduction

New MRI techniques enable the reconstruction of multidimensional data sets, including 3D volumes of cardiac image data reconstructed along both cardiac and respiratory phases [4]. However, there are no available tools for interactive visualization of such high-dimensional data sets, and there are many challenges associated with creating and using such tools [1]. Thus, it will be necessary to develop new visualization tools, in order to be able to fully understand and analyze these new multidimensional data sets.

O. Bernard et al. (Eds.): FIMH 2023, LNCS 13958, pp. 403–411, 2023.
https://doi.org/10.1007/978-3-031-35302-4_42

The data arrangement in this case has five dimensions, where the 3 (three) spatial dimensions are represented by 3D volumes. These volumes are distributed over 2 (two) physiological dimensions - cardiac and respiratory - forming a 2D matrix [4], where lines could, for instance, represent cycles of heartbeats, while columns could represent cycles of inflation and deflation of the lungs.

A visualization tool that deals with this kind of data should provide mechanisms for visualizing it in several different ways, for example, by animating one of the physiological dimensions for a particular cycle phase in the other dimension [1]. The tool may even show more than one of these animations side by side, in order to provide a visual comparison of different cycle phases.

Another requirement for such tool would be an adequate handling of large amounts of data (on the order of a few gigabytes). As this kind of data tends to grow very rapidly in storage requirements, it is necessary to adopt efficient compression standards in order to deal with them.

In addition, an user should be able to manipulate the data without noticing any lags while visualizing them, and the data loading time should be as fast as possible. To address these requirements, efficient compression procedures must be employed.

It is also desirable to be able to readily explore all potentially relevant areas of such large multidimensional data sets. Efficient segmentation procedures must be implemented, in order to isolate organs and structures of interest in reasonable amounts of time. Finally, analysis mechanisms to extract and visualize motion and other temporally varying properties should also be developed, e.g., by producing arrows, glyphs, streamlines or even virtual particle traces [6].

In this work, we report on the development of a visualization and analysis tool for multidimensional cardiac data, called VisHeart. We also present a file format that represents these specific multidimensional data sets, storing relevant information about them and compressing their contents. With this representation, we are currently achieving a compression rate of almost 3 times. We proceed by describing the methods that lead us to this result.

2 Methods

The main objective of this work is to provide a multi-platform visualization and analysis tool capable of efficiently handling these particular multidimensional data sets, by providing all the requirements mentioned previously. Currently under development, VisHeart is a desktop application made in Qt. There are experimental versions for Linux and Windows, and an Android application is also under development. For handling the visualization tasks, VisHeart uses the Visualization Toolkit (VTK). As a very established library in this field, it provides a wide variety of tools for data manipulation and presentation [12].

VisHeart allows the user to visualize and interact with volumetric data in different ways. Figure 1 shows its main window with a visualization in progress.

The content tree can be seen in (1). There, the user can choose to visualize a particular volume contained in the data set. There is also information about the

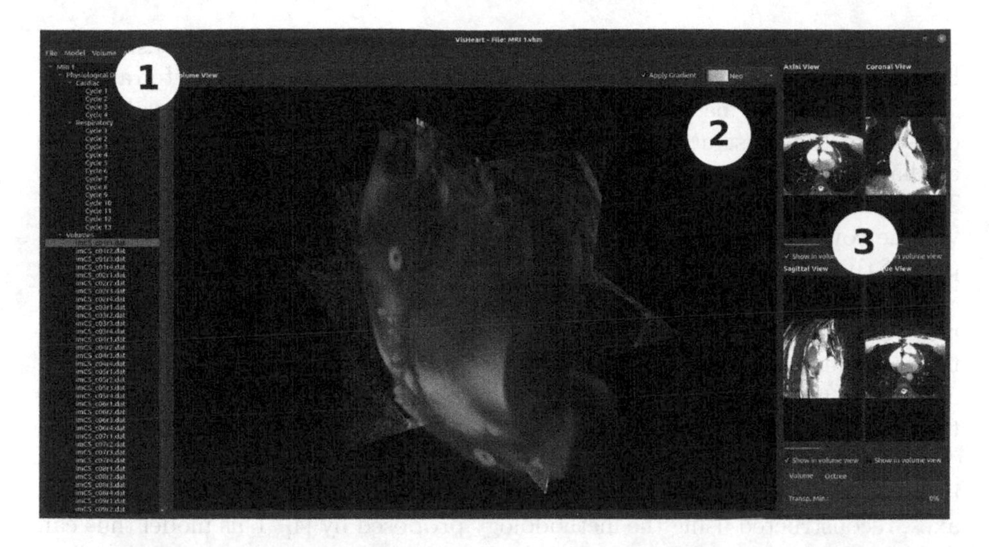

Fig. 1. Screenshot of VisHeart tool under use, showing a full volume rendering. The content tree can be seen in (1). In (2), the main volume view is seen. The reformatted anatomical planes are visible in (3).

cycles registered for each of the physiological dimensions. Choosing one of the cycles will trigger an animation passing throughout all the volumes that make part of it.

In (2), the main volume view is seen. The user may interact directly with the volume presented by rotation, zoom in, zoom out, and translation. It is also possible to change the background color and gradients for volume visualization. The tool allows the creation of custom gradients, as well.

The reformatted anatomical planes are visible in (3). Using the provided sliders, the user can navigate between the slices of the volume in the axial, coronal or sagittal planes. There is also an oblique plane, that can render different reformatted angles of the volume. All of the planes can also be seen in volume view, and have their content animated accordingly when any animation starts. The visualized data can be restricted to ranges of slices, i.e., portions of the volume can be made fully transparent to make it easier to visualize other specific parts.

The user has the option of changing the volume's transparency, add an octree, apply a predefined wavelet function, export slices of anatomical planes as images, save data in specific formats, and so on.

2.1 Use Case

To demonstrate VisHeart's capabilities and the partial results obtained, we will present throughout this paper a use case. We have opted for this approach because these kind of data sets are not easily obtained. The process of data capture and separation in physiological dimensions requires specific equipment that is not widely available. The technique is also brand new and using data from exams requires authorizations, which make these data sets quite rare.

The use case is based on data from a MRI exam that captured volumetric data from the chest area of a patient, synchronized to multiple phases of the cardiac and respiratory cycles (physiologic "axes"), making a net 5D data set. The 3D volume image data were thus created as a 2D matrix of sequences; each volume could be considered as a volumetric snapshot of this body part inside a pair of time intervals.

Wc define as a model a set of such generated volumes, combined with information about how they are distributed over spatial and physiological axes. In this sample, two physiological aspects were captured in different axes: cardiac and respiratory. These axes were entwined, providing spatial volumes that belong to more than one physiologic axis simultaneously.

In this use case, the model contained 52 volumes with 192^3 voxels, arranged in a 2D physiological matrix with cardiac (13 phases) and respiratory (4 phases) axes, reconstructed using the methodology proposed by [4]. This model thus can display the heart in different stages of a heartbeat, and the lungs as they inflate and deflate over time. Figure 2 shows a representation of this model, which has 52 3D reconstructions conceptually organized on a 13 × 4 matrix.

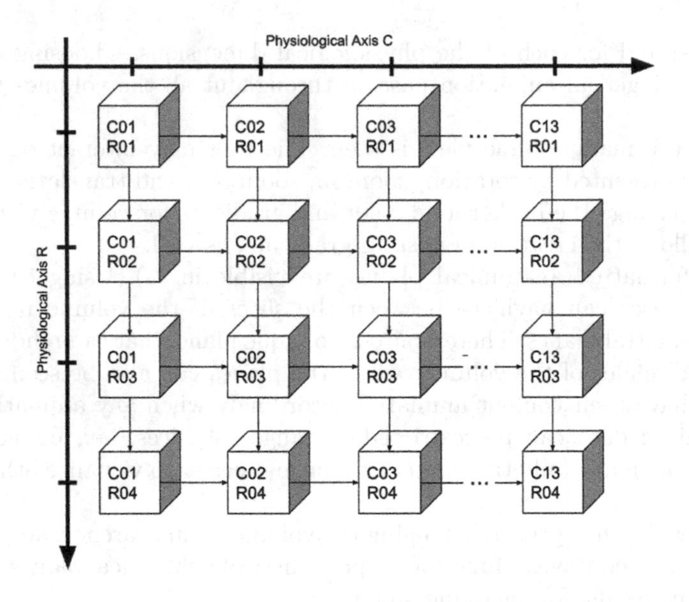

Fig. 2. Use case model with 52 volumes distributed over two physiological axes. The Cardiac (C) axis contains 4 series of 13 volumes, while the Respiratory (R) axis has 13 series of 4 elements. As can be seen, a given volume may correspond to coordinates along more than one physiologic axis at the same time.

2.2 Data

In this study, VisHeart is used to visualize and interact with the data set presented in the use case. VisHeart makes it possible to visualize single volumes or entire models. It accepts data in 3 (three) different formats: **MAT:** MatLab proprietary data format; **VHV:** VisHeart single volume data; and **VHM:** VisHeart model data, where VHV and VHM formats were created specifically for VisHeart, aiming for high fidelity and compression of the data.

The VHV format contains just the original size (in bytes) of the spatial dimensions of a volume, its size after compression, and its own compressed data. The volumes are represented as a sequence of non-negative integers, ranging from zero to 65,535 (16 bits).

The VHM comprises a set of VHV files, compressed individually, and a XML formatted manifest file containing information about the physiological axes and how the represented volumes are distributed over them.

VisHeart can export MAT files in VHV format. The tool for generation of VHM files from single MAT or VHV files is still under development, so these files are currently being manually created. After that, the VHM file can be recognized by VisHeart and new versions can be generated, if desired.

2.3 Compression

To determine a suitable compression methodology for this 5D data, we investigated four different approaches:

Deflate Algorithm. The first step was selecting a widely used algorithm for compression. The ZLIB compression library [5] is well known and adopted in projects that vary from CODECs [3] to Electrocardiograms [10]. It uses the Deflate algorithm [2]—which is a combination of the LZ77 compression [14], a dictionary-base coding technique; and Huffman coding [8], an optimal prefix code technique—as base for lossless compression.

Wavelets. The use of Wavelets [7] allows for the division of the volumetric data into low- and high-frequency components while packing most of the image energy on the low-frequency coefficients. Since we want to employ a lossless compression, the high-frequency coefficients are maintained, but they can be compressed using RLE (run-length encoding).

VisHeart uses the the biorthogonal LeGall-Tabatabai [9] 5/3 wavelet, provided by Blitzwave Library [13]. This wavelet transformation function showed a good balance between compression and processing time.

Difference Coding. This approach is quite simple and direct: one volume is selected as the base of reconstruction and the values of its voxels are not altered in the output. A subtraction is made between the reference volume and its immediate neighbor, and this result is kept instead of the original values of the neighbor. The difference volumes can them be efficiently compressed using an entropy coding technique, such as Arithmetic or Huffman coding [7].

Motion Estimation. Motion Estimation (ME) is performed as one step of the CODECs of the video compression standards H.264 and H.265 [11]. The type of multidimensional data used here has very high spatial and temporal redundancies, potentially allowing high compression gains. The process compares a block of pixels in a reference frame with blocks of pixels from an adjacent frame, computing the motion vector for the most similar block according to a cost function. In this case, we search for the most similar 3D block of voxels in an adjacent 3D reconstruction. Even though the most used cost function, the SAD (Sum of Absolute Differences), is a simple procedure, the process of repeatedly computing many SADs becomes quite expensive in computational terms, specially in a 3D scenario. The block size can be fixed or adaptive, with a higher computational cost associated with larger block sizes.

3 Experiments and Results

We have run some experiments to explore the best combination of strategies for compressing, which have resulted in the VHM file format previously described. In this section, we will present the compression rates obtained during the experiments and the configuration of strategies used to achieve them, in application to the use case defined.

The computer used in the experiments was equipped with an Intel® Core^{TM} 4th Generation i5-4690K (3.50 GHz, 4 Cores, 6 MB Cache) CPU; 24 GB of RAM; and operational system Ubuntu 20.04 LTS.

The original data was represented in a MAT file with 4-byte floating-point values. The first step was converting these data, changing them to a 2-byte integer representation. That was the single lossy step in the process. After that, the data were cleaned, removing part of the external noise inherent to the capture procedure. Even though this may be seen as a lossy step, no data inside the chest volume was affected, so the signal inside the volume of interest was left unaltered.

Table 1 shows the compression rate obtained for each combination of compression strategies adopted.

The values in "Data State" could be Raw, indicating that the data still had noise from the capture process; or Clean.

Column "Compressed" indicates if the data were compressed with the Deflate algorithm. The first configuration was shown just as a base for comparison, as it represents the original data after being converted to 2-byte integer values, the format used in all the other strategies.

The "Subtraction Strategy" indicates if there were any subtraction operations between the volumes contents and, if so, which one was chosen. "Normal" indicates that voxels in equivalent positions were subtracted, while "ME" indicates that a Motion Estimation search in an area for the best match was done. The "Cardiac" or "Respiratory" modifiers show from which axis the neighbor volume was selected.

The "Wavelet" column indicates if a wavelet function was used in the process and the last two-"Size" and "Compression Ratio"-are self-explanatory.

Table 1. Results by strategy (Detailed), with emphasis in the configuration that achieved the best ratio.

#	Data State	Compressed	Subtraction Strategy	Wavelet	Size (MB)	Compression Ratio
1	Raw	No	None	No	736.1	1.000
2	Raw	Yes	None	No	555.6	1.324
3	Raw	Yes	Normal/Cardiac	No	486.0	1.515
4	Clean	Yes	None	No	418.5	1.758
5	Clean	Yes	None	Yes	377.7	1.948
6	Clean	Yes	Normal/Cardiac	No	356.8	2.063
7	Clean	Yes	Normal/Cardiac	Yes	362.6	2.030
8	Clean	Yes	ME/Cardiac	No	357.7	2.058
9	Clean	Yes	ME/Cardiac	Yes	365.6	2.013
10	**Clean**	**Yes**	**ME/Respiratory**	**No**	**249.2**	**2.954**
11	Clean	Yes	ME/Respiratory	Yes	292.8	2.514

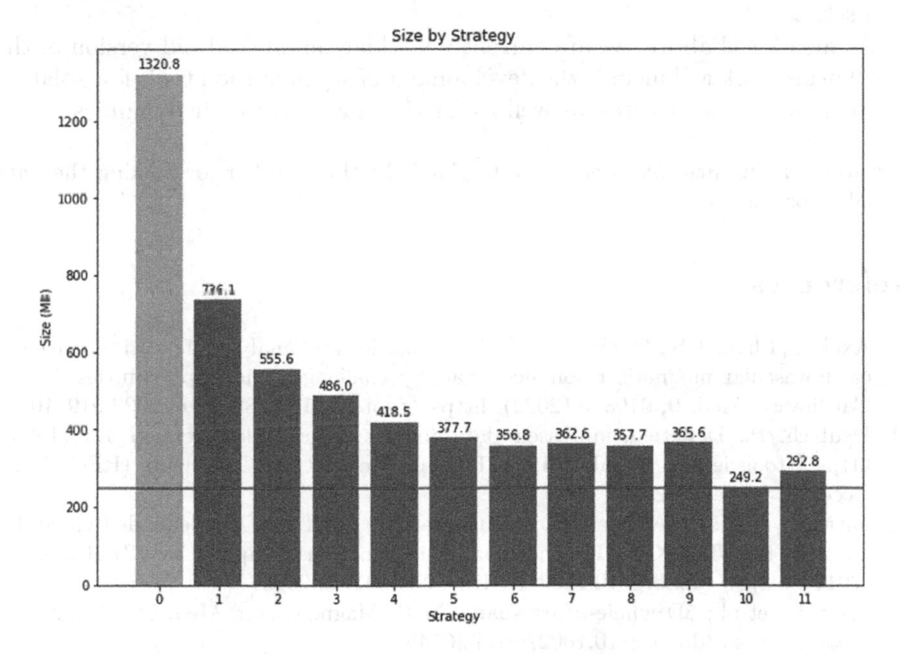

Fig. 3. Total size of VHM model after being processed by different compression strategies. The configuration 0 is the original data, still represented in floating-point values. The red line cuts the bars at the height of the smallest one. (Color figure online)

Figure 3 shows a visualization of the compression rate column from Table 1. Our current largest compression rate was achieved with configuration #10: cleaned data, subtracted after searching for the best matching neighbor, compressed, and without passing through a wavelet function. Using this setup, we achieved a compression ratio of almost 3 times (2.954) the initial configuration (1).

The compression time in this case was 5 min and 59 s while the decompression time was 15.147 s. These very different execution times for the compression and decompression steps are typical of asymmetric compression models and make sense since the compression is performed only once for each data set while the decompression is performed every time a data set is loaded by the tool.

4 Conclusion

In this paper we presented VisHeart, a tool for visualization of multi-dimensional data, composed of volumes with additional physiological dimensions. To efficiently represent these contents, we have created the VHM file format, which uses a combination of compression strategies that allows us to achieve compression rates up to 2.954 times while keeping the compression time to 5 min and 59 s and the decompression in 15.147 s, for a representative 5D cardiac image data set.

As mentioned above, we are currently working on an Android version of the tool. Future work will include the development of segmentation tools for isolating the heart and its structures, as well as for the analysis of their dynamics.

Acknowledgements. We would like to thank Mathias Stuber for making the data available for our use.

References

1. Axel, L., Phan, T.S., Metaxas, D.N.: Visualization and analysis of multidimensional cardiovascular magnetic resonance imaging: challenges and opportunities. Front. Cardiovasc. Med. **9**, 919810 (2022). https://doi.org/10.3389/fcvm.2022.919810
2. Deutsch, P.: Deflate compressed data format specification version 1.3 (1996). https://tools.ietf.org/html/rfc1951. Internet Request for Comments (RFC) 1951. Accessed 09 Feb 2023
3. Farrugia, R.A.: Reversible visible watermarking for H.264/AVC encoded video. In: 2011 IEEE EUROCON - International Conference on Computer as a Tool, pp. 1–4 (2011). https://doi.org/10.1109/EUROCON.2011.5929031
4. Feng, L., et al.: 5D whole-heart sparse MRI. Magn. Reson. Med. **79**(2), 826–838 (2018). https://doi.org/10.1002/mrm.26745
5. Gailly, J., Adler, M.: Zlib compressed data format specification version 3.3 (1996). https://tools.ietf.org/html/rfc1950. Internet Request for Comments (RFC) 1950. Accessed 09 Feb 2023
6. van der Geest, R.J., Garg, P.: Advanced analysis techniques for intra-cardiac flow evaluation from 4D flow MRI. Curr. Radiol. Rep. **4**(7), 1–10 (2016). https://doi.org/10.1007/s40134-016-0167-7
7. Gonzalez, R.C., Woods, R.E.: Digital Image Processing, 4th edn. Pearson, New York (2017)

8. Huffman, D.A.: A method for the construction of minimum-redundancy codes. Proc. IRE **40**(9), 1098–1101 (1952). https://doi.org/10.1109/JRPROC.1952.273898
9. Le Gall, D., Tabatabai, A.: Sub-band coding of digital images using symmetric short kernel filters and arithmetic coding techniques. In: ICASSP-88., International Conference on Acoustics, Speech, and Signal Processing, vol. 2, pp. 761–764 (1988). https://doi.org/10.1109/ICASSP.1988.196696
10. Pulavskyi, A., Krivenko, S., Kryvenko, L.S.: Determination of the signal-to-noise ratio for noisy electrocardiogram using lossless data compression. In: 2019 8th Mediterranean Conference on Embedded Computing (MECO), pp. 1–4 (2019). https://doi.org/10.1109/MECO.2019.8760294
11. Richardson, I.E.G.: H.264 and MPEG-4 Video Compression: Video Coding for Next-Generation Multimedia. Wiley, Hoboken (2003)
12. Schroeder, W., Martin, K., Lorensen, B.: The Visualization Toolkit: An Object-Oriented Approach to 3D Graphics, 4.1 edn. Kitware Inc, New York (2018)
13. Schulz, O.: Blitzwave C++ wavelet library. https://oschulz.github.io/blitzwave/. Accessed 10 Feb 2023
14. Ziv, J., Lempel, A.: A universal algorithm for sequential data compression. IEEE Trans. Inf. Theory **23**(3), 337–343 (1977). https://doi.org/10.1109/TIT.1977.1055714

Generating Short-Axis DENSE Images from 4D XCAT Phantoms: A Proof-of-Concept Study

Hugo Barbaroux[1,2(✉)], Michael Loecher[3], Karl P. Kunze[4], Radhouene Neji[4], Daniel B. Ennis[3,5], Sonia Nielles-Vallespin[2,6], Andrew D. Scott[2,6], and Alistair A. Young[1]

[1] School of Biomedical Engineering and Imaging Sciences, King's College London, London, UK
hugo.barbaroux@kcl.ac.uk
[2] Cardiovascular Magnetic Resonance Unit, The Royal Brompton Hospital (Guy's and St Thomas's NHS Foundation Trust), London, UK
[3] Department of Radiology, Stanford University, Stanford, CA, USA
[4] MR Research Collaborations, Siemens Healthcare Limited, Camberley, UK
[5] Division of Radiology, Veterans Affairs Health Care System, Palo Alto, CA, USA
[6] National Heart and Lung Institute, Imperial College London, London, UK

Abstract. Displacement ENcoding with Stimulated Echoes (DENSE) is a CMR modality that can encode myocardial tissue displacement at a pixel level, enabling the characterization of cardiac disease at early stages. However, we do not currently have a way of evaluating the accuracy of derived results, since the ground truth is unknown. In this study, we developed a proof-of-concept pipeline to generate realistic DENSE images with a known ground truth. We leverage the XCAT tool to create body anatomies, along with associated myocardial tissue displacements, and generate DENSE images with a Bloch simulation based on the time-resolved positions. We generated 6 samples: an apical, a mid, and a basal short-axis slice for both male and female anatomy. We then extracted radial and circumferential strain components using DENSEanalysis, and compared them to the ground-truth strain obtained from the XCAT displacements. While the reproducibility of the strain calculations was similar to the inter-observer variability from previous studies, and the bias in circumferential strain was small (0.03 ± 0.02), the current methods for strain extraction resulted in a bias in radial strain of 0.19 ± 0.19. There is a need to develop better regularization strategies for DENSE analysis, for instance using Deep Learning, and this study provides initial groundwork for obtaining ground-truth strain to evaluate these methods.

Keywords: Cardiac MR · Simulation · DENSE · Strain

1 Introduction

Measurement of myocardial strain is a critical component of a comprehensive assessment of cardiac conditions. In particular, it provides significant added value

O. Bernard et al. (Eds.): FIMH 2023, LNCS 13958, pp. 412–421, 2023.
https://doi.org/10.1007/978-3-031-35302-4_43

over the left-ventricular ejection fraction (LVEF) [19,24], and has the ability to detect disease-related changes in myocardial deformation at an early stage.

Several techniques across different modalities enable the analysis of myocardial deformation. Among them, DENSE (for Displacement ENcoding with Stimulated Echoes, [1,14,27]) is an MRI technique that encodes the tissue displacements in the image phase, providing a regional assessment of functional impairment through the extraction of pixel-wise myocardial strain. DENSE has been shown highly reproducible [3,26], and can provide more accurate displacements at higher resolutions than phase-velocity encoding or tissue-tagging [1,10].

Tools have been developed over the years to process DENSE images and extract myocardial strain information. One widely used solution is the DENSEanalysis Matlab tool [9,22]. However, such tools generally require regularization when calculating displacements to compensate for the noise in the image phase, but there is currently no method to validate the impact of this regularisation on a strain estimate. Another limitation of the current DENSE processing tools is the need for user input. While recent work has been done to automate this process and reduce the need for manual intervention [4,8,12,13], optimizing these methods would benefit from a more realistic ground-truth strain estimate.

One difficulty at present is the lack of realistic synthetic DENSE data with known ground truth strains. Such a dataset would provide accurate ground-truth strain for supervised modeling and studying partial volume effects. MRXCAT [25] is a tool for generating synthetic but realistic cardiac MR images based on XCAT [20], a CT simulation tool that generates torso anatomies. However, MRXCAT does not include DENSE protocols.

In this work, we propose a proof-of-concept pipeline for the generation of synthetic DENSE images, using a combination of XCAT phantoms and a tailored DENSE simulator that could be used in assessing the effects of regularisation in strain analysis or in training Deep Learning models for DENSE analysis. The pipeline provides realistic DENSE images, along with ground-truth strain. We validated our pipeline by conducting a strain analysis with DENSEanalysis.

2 Methods

2.1 XCAT Phantom

XCAT was used in this study to generate short-axis images of 2D+time body anatomies. As a proof-of-concept, we generated 6 slices: a basal, a mid, and an apical slice for the default male and female anatomies. The short-axis plane was obtained using rotations of 115°, 35°, and 240° for the input XCAT rotations parameters in x, y, and z. We generated 30 frames over a 1s cardiac cycle. To obtain high-resolution ground-truth displacements, we used a pixel size of 0.8 mm, forming a square image array of size 360 centered on the left ventricle.

Two XCAT modes were used to generate the data of interest: mode 0, to create anatomical labels for each frame of the cardiac cycle, particularly the first frame which is considered as the resting end-diastolic anatomy; and mode 4, to retrieve the Lagrangian displacement maps. XCAT generates these displacements by fitting 4D NURBS surfaces to displacements obtained from tagged MRI acquired in a normal subject [21].

We extracted pixel-wise Lagrangian strain from the XCAT displacement maps using the means of isoparametric formulation with quadrilateral elements, as described in [17,22]. These high-resolution circumferential and radial pixel strain components were used to validate the DENSE simulation.

2.2 DENSE Simulation

To simulate realistic MR images from the XCAT phantom, a Bloch simulation is performed on the phantom data. Each point in the phantom is considered to be an individual spin with a prescribed T1, T2, and equilibrium magnetization M_0. A conventional DENSE sequence is then simulated using the XCAT displacements for each point to determine the position in time for each spin. The simulated sequence consists of a 90° RF pulse, followed by an encoding gradient with $k_e = 0.1\, cycles/mm$, and then another 90° RF pulse. Separate simulations are performed for different encoding directions, as well as phase cycling the second RF pulse. After the encoding, a cine series of images are simulated using a spoiled gradient sequence with a variable flip angle strategy, as defined below:

$$\theta_i = \arcsin\left(\sin\left(\theta_{i+1}\right) e^{-dt/T1}\right) \tag{1}$$

where i is the frame number from 1 to N, T1 = 850 ms, dt the temporal resolution in ms, and $\theta_N = \frac{\pi}{12}$ (15°). At each imaging time, the transverse magnetization of the spins are then gridded into an image. To simulate intravoxel dephasing, the high-resolution images are then resampled onto a coarser grid with Fourier interpolation and a Hamming window. The final images for each spatial encoding are then obtained after combining the phase-cycled data. Table 1 summarizes the tissue parameters used in the simulations. Parameters have been estimated using previous 3T studies for normal values of in vivo CMR data [6,23], as well as in vivo DENSE images as a reference for signal intensity. The blood T1 used here is much lower than that usually reported for the myocardium at 3T, to model the rapid decay of ventricular blood signal in the first frames of the cardiac cycle observed due to blood washout in vivo DENSE (dark-blood).

Table 1. Tissue parameters for DENSE simulation

	Myocardium	Muscle	Fat	Blood	Liver	Bone	Bone marrow
T1 (ms)	1200	1150	400	500	800	1250	300
T2 (ms)	45	35	100	15	35	35	120
M_0 (relative)	0.4	0.6	0.1	1.0	0.5	0.8	0.7

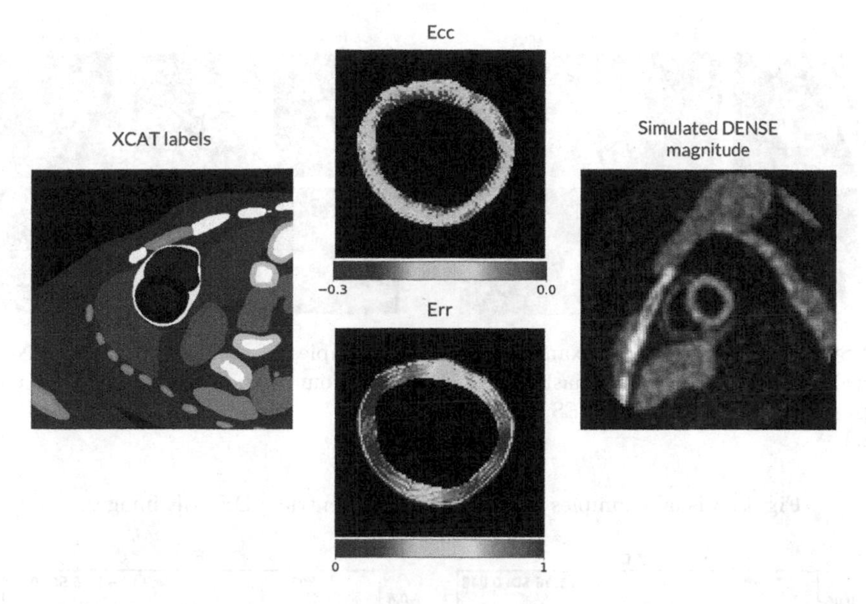

Fig. 1. Simulation pipeline. Left: XCAT anatomy. Middle: high resolution circumferential (top) and radial strain (bottom) maps (unitless) calculated from the XCAT simulation. Right: synthetic DENSE magnitude image (end-systole).

2.3 Validation with DENSEanalysis

We validated the simulated DENSE data by extracting myocardial strain components and comparing them to the strain components extracted from the high spatial resolution XCAT displacements. To this purpose, we processed the images using an accelerated version of the DENSEanalysis [9,22]. In a previous study [5], we engineered this version by eliminating non-essential graphical interfaces and pre-calculating phase unwrapping seed points to facilitate rapid processing for large datasets. The regularization parameters used in the strain calculation were temporal model: 10th-order polynomial and spatial smoothing: 0.5. The high-resolution segmentation obtained from the XCAT anatomy was used to define the LV myocardial segmentation for DENSEanalysis. Analysis was performed using Bland-Altman plots and a comparison of strain-time curves using the XCAT displacement derived strains as the ground truth. Finally, we fitted a second-order polynomial to the pixel strain, after normalizing the radius from endocardium (0) to epicardium (1), in order to analyze the transmural accuracy and precision of the DENSE-derived strain.

3 Results

Ground-truth circumferential and radial strain maps were successfully generated from XCAT displacements, along with simulated DENSE images (see Fig. 1).

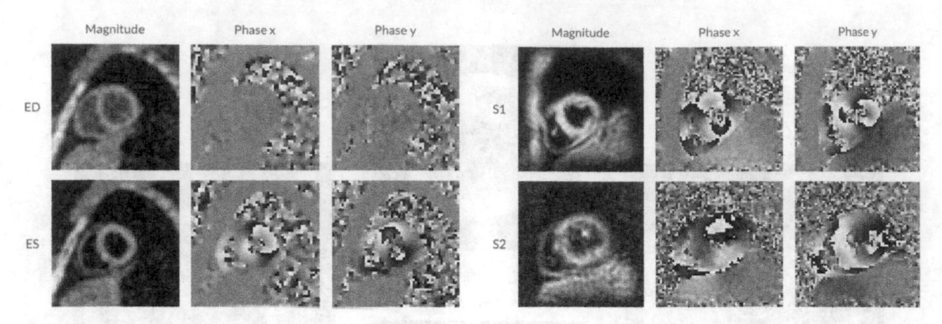

(a) Synthetic DENSE image example (myocardial SNR 100), at end-diastole (ED, top row) and end-systole (ES, bottom row).

(b) Examples of real end-systolic DENSE images, from two different subjects (S1 and S2).

Fig. 2. Visual examples of short-axis mid-ventricle DENSE images.

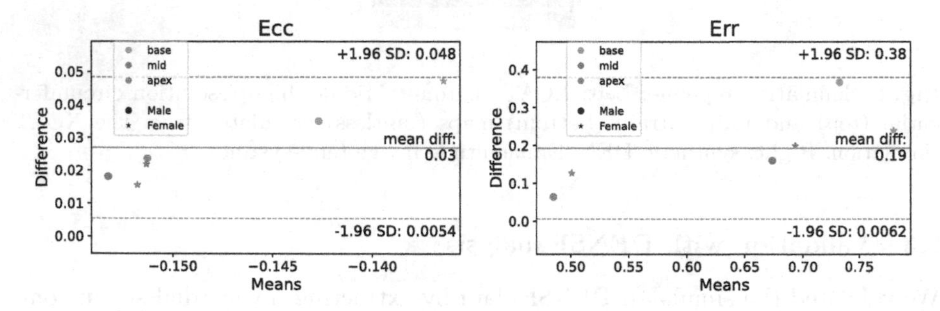

Fig. 3. Bland-Altman plots of average peak strain at end-systole, ground-truth vs DENSE strain (absolute strain differences). Left: Circumferential strain (Ecc). Right: Radial strain (Err).

Simulated DENSE images appear visually similar to real DENSE data acquired using a similar protocol. Figure 2a shows an example of a simulated image, at end-diastole and end-systole. A comparison with real images can be seen in Fig. 2b. Displacements of the myocardium are captured by the phase image, with phase wrapping occurring at end-systole due to the magnitude of the displacements.

From the 6 simulated DENSE datasets that we generated, the Bland-Altman bias of the calculated global peak-systole strain from DENSE vs. ground-truth is 0.03, with limits of agreement from −0.0054 to 0.048 on the circumferential component, and a bias of 0.19 and limits of agreement from 0.0062 to 0.38 on the radial strain. The Bland-Altman plots can be seen in Fig. 3. From the strain-time curves and transmural fitting in Fig. 4 from a mid-ventricle slice, we can confirm that the circumferential strain is well captured, although slightly underestimated at end-diastole and towards the end of the cardiac cycle. The radial strain tends to be underestimated for all frames, and especially at end-systole.

Figure 5 shows a complementary analysis of the average fitting difference and the average variability of the DENSE derived radial strain transmurally across the myocardial wall for 3 different values of the spatial smoothing parameter. We can see that for all SNRs, a higher magnitude of the spatial regularization seems to increase the precision, but also decrease the strain accuracy.

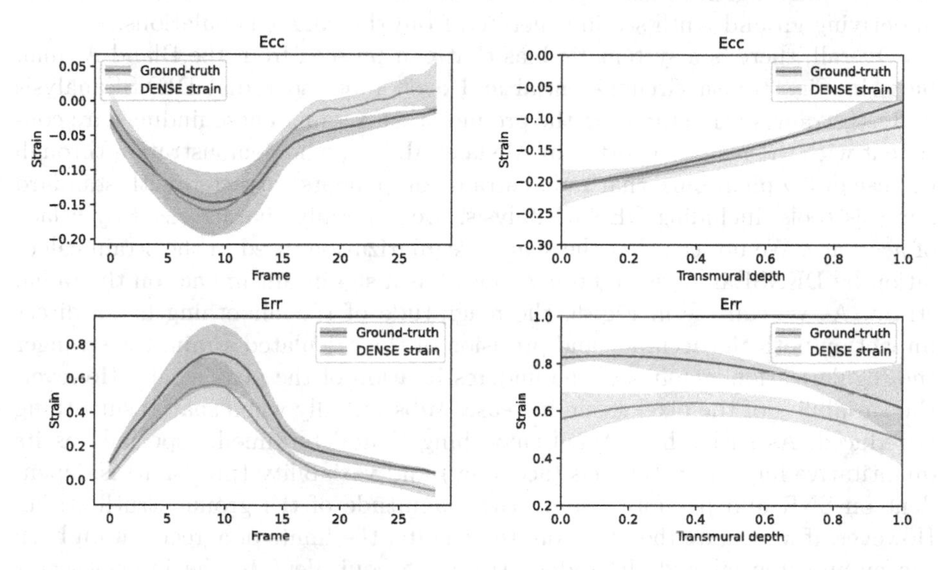

Fig. 4. Strain analysis of ground-truth strain and DENSE strain from a mid-ventricle image. Dark lines: median strain over the myocardium. Shaded area: 25% and 75% quantiles. (Left): Strain-time curves. (Right): 2nd order polynomial fitting of the pixel strain at end-systole, over the normalized radius (0: endocardium, 1: epicardium).

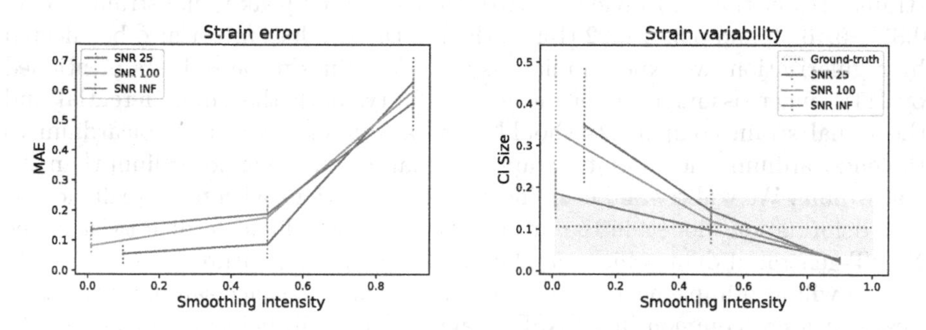

Fig. 5. Impact of SNR and spatial smoothing on Err values from a synthetic DENSE mid short-axis slice. Left: mean (and std error bars) error between DENSEanalysis strain and simulated ground-truth strain on the fitted transmural data. Right: mean variability of the DENSE strain, calculated as the mean CI over the 2nd order regression of the transmural strain. MAE: Mean absolute error. CI: 99% confidence interval.

4 Discussion

We established a novel pipeline for the generation of realistic DENSE images, using XCAT as a basis for both the anatomy and myocardial motion to simulate a DENSE acquisition. We visually validated the resulting images, but also analyzed how the Lagrangian strain extracted from the DENSE data compared to the underlying ground-truth strain generated from the XCAT simulations.

Overall, there is a systematic bias that can be seen from the Bland-Altman plots in Fig. 3 with circumferential and even more so radial DENSEanalysis derived strains over-estimating the ground truth strain. These findings are consistent with the results reported by Mella et al. [16], who demonstrated, through the use of 3D phantoms, that radial strain components extracted with standard analysis tools, including DENSEanalysis, are generally characterized by a lack of accuracy. We noticed that the spatial regularization, used in the strain calculation by DENSEanalysis and other tools, has a significant impact on the radial strain. As we can see in Fig. 5, the magnitude of the smoothing has a direct impact on both the accuracy and precision of the calculated strain: the stronger the regularization, the higher the underestimation of the peak strain. However, the variability of the pixel strain increases substantially when spatial smoothing is reduced. As such, the optimal smoothing is undetermined, especially as its quantitative impact on the bias (accuracy) and variability (precision) is dependent on SNR and possibly also on the magnitude of the ground truth strain. However, if we negate the bias from the results, the limits of agreement on both the circumferential and the radial strain are equivalent to the inter-observer variability found in in-vivo DENSE data reported in a recent study [3].

Our analysis is reliant on the XCAT cardiac motion model. However, as we can see in Fig. 1, the ground-truth radial strain obtained from the XCAT simulation seems to show unexpected circumferential variations and maximum pixel strains greater than 1. Previous in vivo studies showed peak radial strain around 0.35 ± 0.16 [3] and 0.39 ± 0.22 [15]. Although these values might also be affected by regularization, we expect radial pixel strain values to be < 1. As mentioned by [11], under assumptions of incompressibility, both the circumferential and the radial strain components should show a gradient from the epicardium to the endocardium, the absolute strain being larger on the endocardium than the epicardium. We will try and combine XCAT simulations with more realistic cardiac deformation models in the future. We also noted little cardiac twist in the XCAT-generated data, which may be another target for future work.

We will also implement simulations of the spiral k-space sampling trajectories, which are common in DENSE imaging. This will help us understand the impact of specific spiral artifacts, on the strain calculation performance, including off-resonance-related blurring. The dark-blood contrast that is inherent to DENSE images was also simplified in this work by artificially imposing a short T1 for the blood pool. While we acknowledge this limitation, this simplified protocol has a small impact on signal intensity, and correctly generates noisy phase in the blood region, as generally seen on DENSE images acquired in vivo.

Another limitation comes from the lack of heterogeneity within a given organ. Single T1, T2, and M_0 values are used for entire anatomical structures, but this results in an unrealistic appearance of the images. This limitation is also acknowledged in MRXCAT [25]. Previous works in XCAT mitigate this for CT simulations [7], where tissue parameters are estimated at a higher resolution and heterogeneity inside organs was incorporated using data-driven approaches. Similar approaches could be used in this work, for example by using probability distributions for T1, T2, and M_0 to generate tissue parameters at a pixel level.

The possibility to obtain a ground-truth strain will be important for the validation of processing methods, particularly those based on deep learning which generally require a target dataset to learn from. A workshop from FIMH 2021 showed how simple phantom simulations with ground-truth strain could give insights into current methods, leading the way for more research about simulations. As XCAT anatomies can be generated at a high spatial resolution, this opens avenues toward a better analysis of partial volume effects in DENSE-derived strains. Indeed, the spatial resolution of DENSE images is rather limited compared to standard MRI cine data. Partial volume effects have a strong impact around the borders of the myocardium, especially as the displacements are directly estimated from the phase image. This work could ultimately be combined with data-driven modeling, as generative Deep Learning models have been applied for the generation of MR images, improving the realism of tissue signals [2,18]. However, these methods do not currently apply to DENSE as they are, given the lack of control over the physics behind DENSE images, as it is necessary to preserve the relationship between phase and tissue displacements.

5 Conclusion

This study is a promising proof-of-concept, and has several implications for current research in DENSE imaging. Simulations like that presented here could help when learning and modeling partial volume effects in actual DENSE data, but also help achieve super-resolution. Being able to accurately estimate strain is critical in DENSE, and obtaining ground-truth information is necessary to correctly assess the validity of acquisition and processing techniques.

Acknowledgements. This work is funded by EPSRC Centre for Doctoral Training in Smart Medical Imaging (EP/S022104/1), by a Program Grant from the British Heart Foundation (RG/19/1/34160), and Siemens Healthineers. This work was supported by the National Institute of Health (NIH R01-HL131823).

References

1. Aletras, A.H., Ding, S., Balaban, R.S., Wen, H.: DENSE: displacement encoding with stimulated echoes in cardiac functional MRI. J. Magn. Reson. **137**(1), 247–252 (1999). https://doi.org/10.1006/jmre.1998.1676

2. Amirrajab, S., et al.: XCAT-GAN for synthesizing 3D consistent labeled cardiac MR images on anatomically variable XCAT phantoms. In: Martel, A.L., et al. (eds.) MICCAI 2020. LNCS, vol. 12264, pp. 128–137. Springer, Cham (2020). https://doi.org/10.1007/978-3-030-59719-1_13
3. Auger, D.A., et al.: Reproducibility of global and segmental myocardial strain using cine DENSE at 3 T: a multicenter cardiovascular magnetic resonance study in healthy subjects and patients with heart disease. J. Cardiovasc. Magn. Reson. **24**(1), 23 (2022). https://doi.org/10.1186/S12968-022-00851-7
4. Barbaroux, H., et al.: Automated segmentation of long and short axis DENSE cardiovascular magnetic resonance for myocardial strain analysis using spatio-temporal convolutional neural networks. J. Cardiovasc. Magn. Reson. **25**(1), 1–17 (2023). https://doi.org/10.1186/S12968-023-00927-Y
5. Barbaroux, H., Scott, A.D., Young, A.A.: Extending DENSEanalysis for automated and faster DENSE strain processing. In: Proceedings from the 26th Annual Society for Cardiovascular Magnetic Resonance (SCMR) Scientific Sessions, p. 1351770 (2023). https://doi.org/10.13140/RG.2.2.32980.07043
6. Bojorquez, J.Z., Bricq, S., Acquitter, C., Brunotte, F., Walker, P.M., Lalande, A.: What are normal relaxation times of tissues at 3 T? Magn. Reson. Imaging **35**, 69–80 (2017). https://doi.org/10.1016/J.MRI.2016.08.021
7. Bond, J., Frush, D., Samei, E., Segars, W.P.: Simulation of anatomical texture in voxelized XCAT phantoms. In: Medical Imaging 2013: Physics of Medical Imaging, vol. 8668, p. 86680N. SPIE (2013). https://doi.org/10.1117/12.2008422
8. Ghadimi, S., et al.: Fully-automated global and segmental strain analysis of DENSE cardiovascular magnetic resonance using deep learning for segmentation and phase unwrapping. J. Cardiovasc. Magn. Reson. **23**(1), 20 (2021). https://doi.org/10.1186/s12968-021-00712-9
9. Gilliam, A.D., Suever, J.D., et al.: DENSEanalysis (2021). https://github.com/denseanalysis/denseanalysis
10. Ibrahim, E.S.H.: Myocardial tagging by cardiovascular magnetic resonance: evolution of techniques-pulse sequences, analysis algorithms, and applications. J. Cardiovasc. Magn. Reson. **13**(1), 1–40 (2011). https://doi.org/10.1186/1532-429X-13-36
11. Ishizu, T., et al.: Experimental validation of left ventricular transmural strain gradient with echocardiographic two-dimensional speckle tracking imaging. Eur. J. Echocardiogr. **11**(4), 377–385 (2010). https://doi.org/10.1093/EJECHOCARD/JEP221
12. Kar, J., Cohen, M.V., McQuiston, S.A., Poorsala, T., Malozzi, C.M.: Direct left-ventricular global longitudinal strain (GLS) computation with a fully convolutional network. J. Biomech. **130**, 110878 (2022). https://doi.org/10.1016/J.JBIOMECH.2021.110878
13. Kar, J., Cohen, M.V., McQuiston, S.P., Malozzi, C.M.: A deep-learning semantic segmentation approach to fully automated MRI-based left-ventricular deformation analysis in cardiotoxicity. Magn. Reson. Imaging **78**, 127–139 (2021). https://doi.org/10.1016/J.MRI.2021.01.005
14. Kim, D., Gilson, W.D., Kramer, C.M., Epstein, F.H.: Myocardial tissue tracking with two-dimensional cine displacement-encoded MR imaging: development and initial evaluation. Radiology **230**(3), 862–871 (2004). https://doi.org/10.1148/RADIOL.2303021213

15. Lin, K., Meng, L., Collins, J.D., Chowdhary, V., Markl, M., Carr, J.C.: Reproducibility of cine displacement encoding with stimulated echoes (DENSE) in human subjects. Magn. Reson. Imaging **35**, 148–153 (2017). https://doi.org/10.1016/j.mri.2016.08.009
16. Mella, H., Mura, J., Sotelo, J., Uribe, S.: A comprehensive comparison between shortest-path HARP refinement, SinMod, and DENSEanalysis processing tools applied to CSPAMM and DENSE images. Magn. Reson. Imaging **83**, 14–26 (2021). https://doi.org/10.1016/J.MRI.2021.07.001
17. Moaveni, S.: Finite Element Analysis: Theory and Application with ANSYS. Prentice Hall, Hoboken (1999)
18. Muffoletto, M., et al.: Comparison of semi- and un-supervised domain adaptation methods for whole-heart segmentation. In: Oscar C., et al. (eds.) Statistical Atlases and Computational Models of the Heart, Regular and CMRxMotion Challenge Papers, STACOM 2022, vol. 13593 LNCS, pp. 91–100. Springer, Cham (2022). https://doi.org/10.1007/978-3-031-23443-9_9
19. Potter, E., Marwick, T.H.: Assessment of left ventricular function by echocardiography: the case for routinely adding global longitudinal strain to ejection fraction. JACC: Cardiovasc. Imaging **11**(2P1), 260–274 (2018). https://doi.org/10.1016/j.jcmg.2017.11.017
20. Segars, W.P., Mahesh, M., Beck, T.J., Frey, E.C., Tsui, B.M.: Realistic CT simulation using the 4D XCAT phantom. Med. Phys. **35**(8), 3800–3808 (2008). https://doi.org/10.1118/1.2955743
21. Segars, W.P., Lalush, D.S., Frey, E.C., Manocha, D., King, M.A., Tsui, B.M.: Improved dynamic cardiac phantom based on 4D NURBS and tagged MRI. IEEE Trans. Nuclear Sci. **56**(5), 2728–2738 (2009). https://doi.org/10.1109/TNS.2009.2016196
22. Spottiswoode, B.S., et al.: Tracking myocardial motion from cine DENSE images using spatiotemporal phase unwrapping and temporal fitting. IEEE Trans. Med. Imaging **26**(1), 15–30 (2007). https://doi.org/10.1109/TMI.2006.884215
23. Stanisz, G.J., et al.: T1, T2 relaxation and magnetization transfer in tissue at 3T. Magn. Reson. Med. **54**(3), 507–512 (2005). https://doi.org/10.1002/MRM.20605
24. Tops, L.F., Delgado, V., Marsan, N.A., Bax, J.J.: Myocardial strain to detect subtle left ventricular systolic dysfunction. Eur. J. Heart Failure **19**(3), 307–313 (2017). https://doi.org/10.1002/ejhf.694
25. Wissmann, L., Santelli, C., Segars, W.P., Kozerke, S.: MRXCAT: realistic numerical phantoms for cardiovascular magnetic resonance. J. Cardiovasc. Magn. Reson. **16**(1), 1–11 (2014). https://doi.org/10.1186/S12968-014-0063-3
26. Young, A.A., Li, B., Kirton, R.S., Cowan, B.R.: Generalized spatiotemporal myocardial strain analysis for DENSE and SPAMM imaging. Magn. Reson. Med. **67**(6), 1590–1599 (2012). https://doi.org/10.1002/mrm.23142
27. Zhong, X., Spottiswoode, B.S., Meyer, C.H., Kramer, C.M., Epstein, F.H.: Imaging three-dimensional myocardial mechanics using navigator-gated volumetric spiral cine DENSE MRI. Magn. Reson. Med. **64**(4), 1089–1097 (2010). https://doi.org/10.1002/MRM.22503

Cardiovascular Hemodynamics and CFD

Vortex Duration Time to Infer Pulmonary Hypertension: *In-Silico* Emulation and Dependence on Quantification Technique

Malak Sabry[1,2,3](✉), Pablo Lamata[1], Andreas Sigfridsson[3,4], Hamed Keramati[1], Alexander Fyrdahl[3,4], Martin Ugander[3,5], Magdi H. Yacoub[2,6], David Marlevi[3,7], and Adelaide De Vecchi[1]

[1] Department of Biomedical Engineering, King's College London, London, UK
malak.sabry@kcl.ac.uk
[2] Aswan Heart Research Centre, Magdi Yacoub Foundation, Aswan, Egypt
[3] Department of Molecular Medicine and Surgery, Karolinska Institutet, Solna, Sweden
[4] Department of Clinical Physiology, Karolinska University Hospital, Stockholm, Sweden
[5] Kolling Institute, Royal North Shore Hospital, University of Sydney, Sydney, Australia
[6] National Heart and Lung Institute, Imperial College London, London, UK
[7] Institute for Medical Engineering and Science, Massachusetts Institute of Technology, Cambridge, MA, USA

Abstract. The gold-standard for pulmonary hypertension (PH) diagnosis is invasive right-heart catheterisation, a technique ill-suited for general patient screening. Noninvasive markers of PH are thus of direct clinical value. Using 4D Flow Magnetic Resonance Imaging (MRI), a recent empirical correlation has been observed between the duration of a blood vortex in the main pulmonary artery (MPA), and the mean pulmonary arterial pressure. The mechanism underlying this relationship, however, remains unknown. In this context, our aim was to replicate this correlation using computational fluid dynamics simulations. Retrospective MRI data from six subjects with suspected PH were used to generate patient-specific models of the pulmonary arteries. Vortex durations in the MPA were derived from the simulations using four methods: visual assessment, Q-criterion, Lambda2 criterion, and backward flow. Results were compared to reference durations predicted by the empirical correlation. The method most closely reproducing values from the relationship was the backward-flow approach (overall error of 7%, vs. 12–14% for the other methods). *In-silico* simulations are a valid approach to unveil the mechanistic factors driving pulmonary vortex behaviour in future larger modelling cohorts.

Keywords: Pulmonary Hypertension · Vorticity · CFD · 4D Flow MRI

O. Bernard et al. (Eds.): FIMH 2023, LNCS 13958, pp. 425–434, 2023.
https://doi.org/10.1007/978-3-031-35302-4_44

1 Introduction

Pulmonary Hypertension (PH) is a progressive disease characterised by a mean pulmonary arterial pressure (mPAP) greater than 20 mmHg at rest. If left untreated, PH leads to remodelling of the pulmonary vasculature and progressive right ventricular dysfunction with life-threatening consequences. Both the right ventricle (RV) and the pulmonary artery (PA) undergo adverse structural remodelling and stiffening as the disease progresses [1,2]. The gold-standard for diagnosing PH is invasive right-heart catheterisation (RHC) to measure mPAP. As the main symptoms of PH are not unique to the disease, and since RHC has elevated risks and costs, PH is usually diagnosed at a late stage. There is therefore a need for alternative non-invasive markers of PH to support earlier diagnosis and appropriate treatment planning.

PH is accompanied by increased pulmonary vascular resistance, and subsequent changes in ventricular afterload. This results in structural maladaptations such as increased wall stiffness and dilation of the main PA (MPA), and abnormal flow patterns in the PA with formation of vortical and helical flow [3]. Recent studies have shown an empirical relationship between mPAP levels and the persistence of a pathological vortex in the MPA in 145 subjects with suspected PH. The vortex duration was identified by visual assessment of phase-contrast magnetic resonance imaging (4D Flow MRI) [4–6]. However, the fluid mechanical origin of this pathological vortex, and the hemodynamic conditions influencing the observed relationship are still unknown. Visual assessment of vortex duration from 4D Flow data is also a time-consuming and user-dependent process with low reproducibility.

Image-based, personalised computational fluid dynamics (CFD) models have been successfully applied to various cardiovascular diseases to elucidate the mechanisms of complex hemodynamic behaviour [7]. In this context, this work aims to (1) create physiological *in-silico* models of patient-specific pulmonary flow, to recover the empirical relationship observed between clinical mPAP and image-derived vortex duration [4], and (2) utilise these models to evaluate different approaches for automated vortex duration quantification.

2 Methods

2.1 Imaging Protocol

Clinical image data were acquired from patients referred for cardiac MRI exams at the Karolinska University Hospital, Solna, between February and July 2022, with specific study inclusion based on suspicion of elevated mPAP. Six subjects were retrospectively selected, 3 with confirmed PH by manual vortex detection and inferred mPAP >20 mmHg, and 3 without PH (i.e. no vortex detected). All subjects participated under informed consent. Data acquisition and coupled imaging studies had ethical approval from the Swedish Ethical Review Authority (EPM, 2011/1077-31, additional amendment: 2022-01151-02).

CMR images were acquired on a 1.5T scanner (MAGNETON Sola, Siemens Healthcare, Erlangen, Germany), using retrospective ECG gating and phase array receiver coils. Standard breath-hold Cine imaging with steady-state free precession (SSFP) in short- and long-axis cardiac planes were acquired. Phase contrast velocity-encoded time-resolved three-direction flow data (4D Flow MRI) were acquired using a research sequence ($2.0 \times 2.5 \times 2.5\,\text{mm}^3$ interpolated to 2.0 mm isotropic; TR/TE = 58.7/2.6 ms; VENC = 90 cm/s). All sequences were reconstructed into 20 cardiac phases using compressed sensing (CS acceleration R = 7.7).

2.2 Patient-Specific CFD Simulation

Segmentation. The MPA and its two main branches (left and right pulmonary artery, LPA and RPA respectively) were manually segmented from Cine MRI data using 3D Slicer [8] (Fig. 1a). Extensions of length equal to 5 times the vessel diameter were added to the two outlets to ensure computational stability and avoid artifacts from the boundary conditions on the simulated flow.

Meshing. Volumetric meshes of approximately 2 million tetrahedral elements with 3 layers of prismatic elements near the vessel wall were generated for each

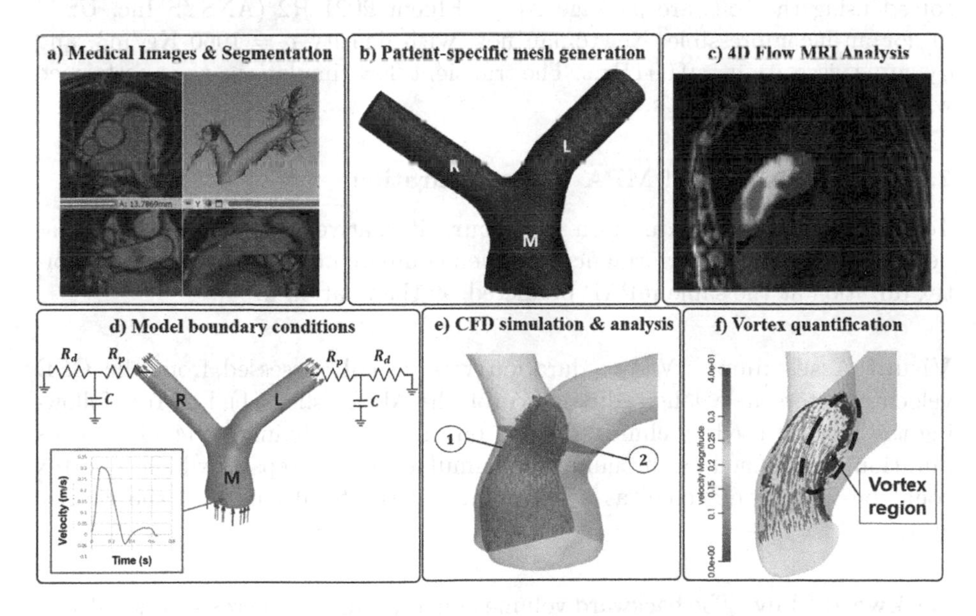

Fig. 1. Patient-specific model construction. a) MR images with lumen segmentation. b) Mesh of the MPA (M), RPA (R), and LPA (L) fluid domains. c) Matching 4D Flow MRI. d) Simulations BCs: velocity inlet, 3-element Windkessel outlets. e) Slices used for post-processing for vortex analysis: (1) longitudinal and (2) cross-sectional slice. f) Velocity vectors and Q-criterion isosurfaces used for vortex quantification.

segmented anatomy using the Ansys Mechanical 2021 R2 (ANSYS Inc, USA) [9] (Fig. 1b). A mesh independence study was carried out by computing mean wall shear stress (WSS) in a plane in the MPA with constant flow by sequentially increasing mesh density by a factor of 1.3. The mesh corresponding to an error between one mesh size and the next below 10% was selected.

Boundary Conditions. As shown in Fig. 1d, the inlet was chosen to be above the pulmonary valve, with inflow boundary condition (BC) derived from through-plane velocities extracted from the corresponding anatomical slice (2D Flow MRI). A 3-element Windkessel model was applied to the two outlets, RPA and LPA, with initial parameters obtained from literature [10]. The three Windkessel parameters (R_d, R_p, and C) were progressively tuned by minimizing the difference between the simulated outlet flow rate and the one derived from the corresponding through-plane flow at the level of the RPA and LPA, respectively, using the least-squares method. The equation used for the Windkessel modelling was:

$$\frac{\partial p}{\partial t} + \frac{p}{R_d C} = \frac{Q}{C}(1 + \frac{R_p}{R_d}) + R_p \frac{\partial Q}{\partial t} \tag{1}$$

Governing Equations and Solver Setup. The Navier-Stokes equations were solved using the software package Ansys Fluent 2021 R2 (ANSYS Inc, USA) [9] for an incompressible, Newtonian flow with density $\rho = 1060$ Kg/m3, and dynamic viscosity $\mu = 0.004$ Pa.s. The transient flow simulations were performed with a time step of 0.001 s.

2.3 Quantification of MPA Vortex Duration

To quantify the vortex duration time four alternative methods were used, as described next. Their performance was then compared to the corresponding vortex duration at the same mPAP predicted by the relation in [4].

Visual Assessment. Vortex duration was visually assessed from the CFD velocity vectors in a longitudinal slice of the MPA (slice (1) Fig. 1e), following the method used by clinical experts on 4D Flow MR images [4,11]. Vortex duration was defined as the number of simulation timesteps in which a vortex could be visually detected, as a percentage of the total length of the cardiac cycle.

Backward Flow. The backward volume flow rate through a cross-sectional slice of the MPA passing through the vortex (see slice (2) in Fig. 1e) was computed over time. The duration of the vortical flow was then defined as the full width at half maximum (FWHM) of the peak backward flow, i.e. the duration between the two instances at 50% of the maximal backward flow of the slice, based on the method presented by [12] as shown in Fig. 2.

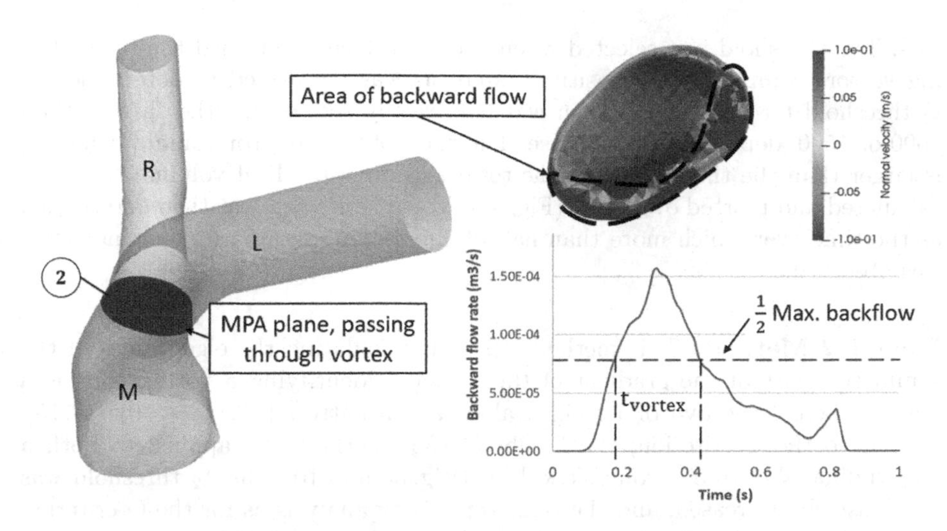

Fig. 2. Backward flow method using slice (2), modified from [12]

Q-Criterion. The Q-criterion is a well-known fluid mechanical method that identifies vortex cores as the locations where the vorticity magnitude is higher than the rate of strain magnitude. A positive Q represents a vorticity-dominating flow field [13]. The choice of threshold allows for the visualisation of smaller or larger vortical structures. To identify an appropriate, subject-specific value for Q and avoid spurious smaller vortices, a reduced region of interest was chosen either by placing a longitudinal slice across mid of the MPA near the posterior wall or by selecting a volumetric region of interest (ROI), as shown in Fig. 3.

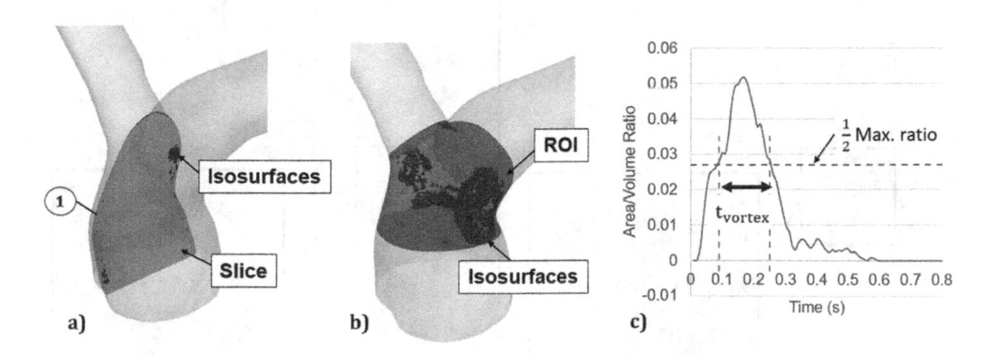

Fig. 3. Q-criterion and λ_2 methods. a) Vortex detection on the longitudinal slice (1). b) Region of interest (ROI) used for vortex detection methods. c) Ratio of elements within the vortex detection threshold, over all elements in ROI.

The appropriate Q threshold was selected by progressively increasing the value from zero and overlaying the resulting Q isosurfaces on the velocity vectors. The threshold was selected when the isosurfaces contained the recirculating vectors completely. A sensitivity analysis was performed to determine the Q threshold for each case, which was empirically set to be either higher than 1000 or 1500, depending on the case. The ratio of the area (or volume) where Q is larger than the threshold over the total slice area (or ROI volume) was then calculated and plotted over time (Fig. 3c). Vortex duration was then determined as the time over which more than half of the maximum area (or volume) ratio was observed.

Lambda2 Method. This method relies on calculating the eigenvalues of the symmetric part of the gradient of the velocity, identifying a vortex core as a region where least two of its eigenvalues are negative, (i.e. $\lambda_2 < 0$) [14,15]. Similar to the Q-criterion, the Lambda2 (λ_2) method was applied to both a longitudinal slice and a volumetric ROI (Fig. 3a and b). The λ_2 threshold was progressively decreased, and the same sensitivity analysis as for the Q-criterion was performed for each case. The threshold was then empirically set to –1500 or –2000, depending on the case. Vortex durations were similarly computed from the area and volume ratios, respectively.

3 Results

3.1 Flow Field

Figure 4 shows the end-systolic velocity vector field at the longitudinal slice in a PH case, and a no PH case (i.e. control case), respectively.

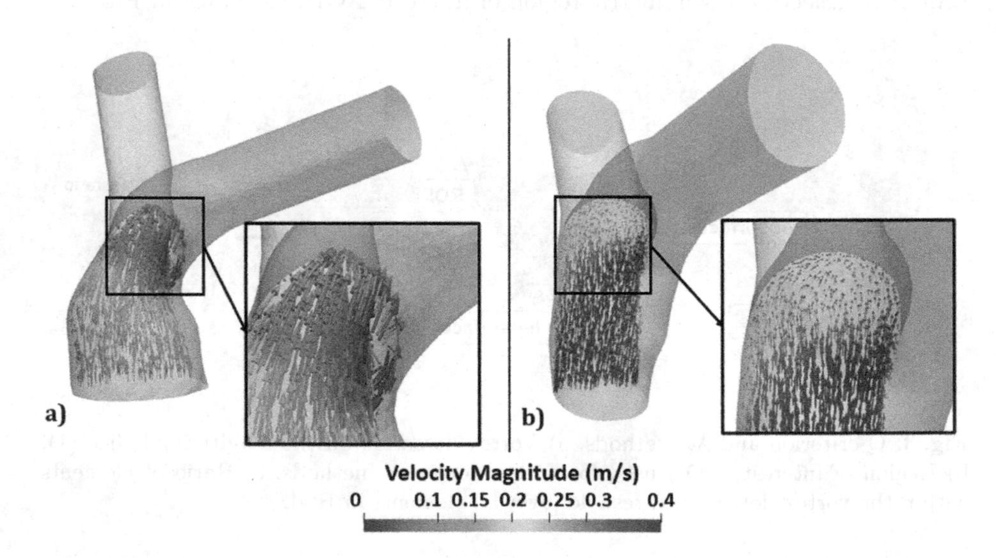

Fig. 4. Velocity vector field on selected slice for a) Subject with PH, with a visible vortex; b) Subject with no PH, with no visible vortex.

There is no strong observable vortex in the control at end-systole, while there appears to be a vortex at the posterior wall of the MPA in the PH patient case, consistent with the pattern reconstructed from 4D Flow data in previous reference work [4].

3.2 Vortex Detection and Durations

The vortex durations given by the different post-processing methods are represented as scattered data points in Fig. 5, while the empirical relationship is shown as a solid line.

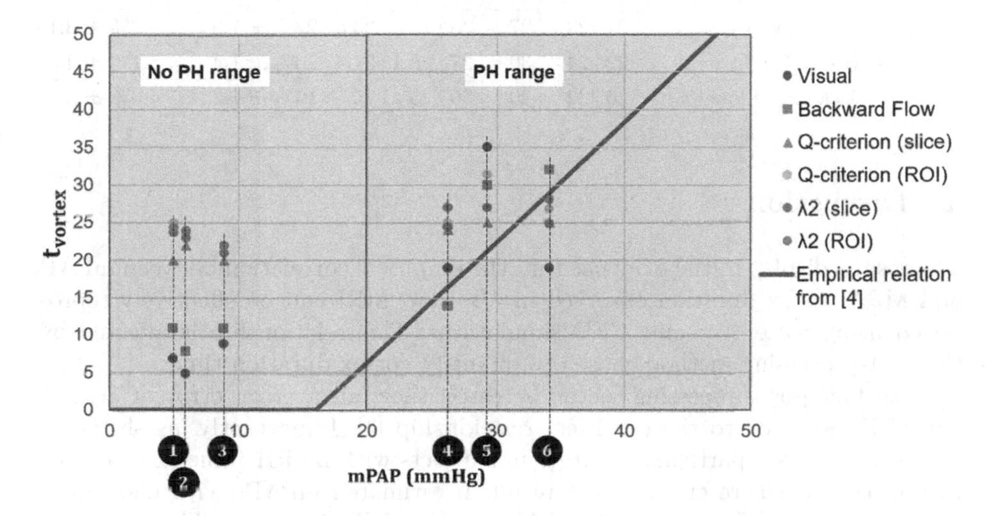

Fig. 5. Correlation between simulated mPAP and vortex duration as a percentage of the cardiac cycle, for geometries 1–6, obtained from the six tested vortex detection methods. The simulation points are superposed on the empirical relationship from [4].

A vortex duration error (VDE) is then obtained in the "PH" and the "No PH" ranges for each post-processing method using Eq. 2 below. It is computed as the average difference between the duration predicted from the empirical relationship [4], $t_{vortex-emp}$, and the one obtained from the post-processing method, $t_{vortex-method}$, at each simulated mPAP, p_i, for n simulations. Values are shown in Table 1.

$$VDE = \frac{\sum_{i=1}^{n}(t_{vortex-emp,i} - t_{vortex-method,i})}{n} \tag{2}$$

Table 1. Estimated vortex duration for cases 1 to 6 using the different proposed vortex detection techniques, and vortex duration errors corresponding to each method. In all cases, values are pooled for controls (no PH) and PH patients, respectively.

	Patient						Vortex duration error [%]	
	1	2	3	4	5	6	Control range	PH range
Simulation mPAP (mmHg)	5	6	9	27	30	35	–	–
$t_{vortex-emp}$ (%)	0	0	0	19	30.4	22.4	-	–
t_{vortex} - Visual from CFD (%)	7	5	9	19	40	19	7% + 2	8.5% ± 6.1
t_{vortex} - Q-criterion (slice) (%)	20	22	25	24	25	20	20% ± 1.7	4.8% ± 2.0
t_{vortex} - Q-criterion (ROI) (%)	22	23.8	31.4	25	26.9	25	23% ± 1.2	4.3% ± 3.2
t_{vortex} - λ_2 (slice) (%)	21	23	27	25	25	23.8	22% ± 1.5	4.3% ± 3.1
t_{vortex} - λ_2 (ROI) (%)	22	24	30	27	28.1	24.4	23% ± 1.4	5.1% ± 4.5
t_{vortex} - Backward Flow (%)	20	22	20	39	25	25	10% ± 1.3	4.3% ± 3.0

4 Discussion

The results display initial evidence that the empirical correlation between mPAP and MPA vortex duration observed in 4D Flow MRI can be effectively reproduced using patient-specific CFD simulations. This relationship is affected by the post-processing method chosen to quantify vortex duration time.

The four post-processing methods tested show an average error of around 4.5–23% compared to the empirical relationship [4]. Importantly, as shown in Table 1, errors are particularly larger in subjects with no PH (blue shaded area in Fig. 5), and where errors might result in estimated mPAPs with discordant diagnostic output (i.e. generating false positive PH diagnosis). The reason to this misclassification could possibly be based on the fact that when the vortex is either weak or virtually non-present, the sensitivity to the choice of vortex detection threshold becomes greater. This is particularly evident for the Q and λ_2-criteria, where the choice of an empirical vortex threshold is a critical step and where the choice of a suboptimal threshold could result in the inclusion of secondary vortices with no physiological significance [15].

Across all the six subjects studied the backward flow method yielded lowest errors out of all evaluated methods. This is a promising finding with respect to semi-automated vortex detection. The visual inspection method (used in previous work [4,11]) has the smallest error in the control range, but the highest error in the PH range (distinct underestimation apparent from visual inspection in subject 6, in Fig. 5). The reason for this is difficult to isolate from a single subject, but could be partially due to the user-dependent nature of a fully manual method. As no method shows a complete lack of vortex in the control range, a key issue to be addressed in future work is to understand whether the vortex detection methods used are capturing a backflow unrelated to the vortex, or whether the vortex is still present in healthy cases but is not detected from the

4D Flow images due to its lower temporal resolution (25 frames per cycle vs. 1000 in CFD simulations).

A few limitations are worth pointing out. First, a small cohort size was utilised, limiting the generalizability of our results. No subjects with severely elevated mPAP (>50 mmHg) could be included, and lower mPAP ranges were also under represented. However, our data should be viewed as a feasibility test of using *in-silico* modelling to recover clinically relevant MPA flows and pulmonary vortex appearance, forming the basis for future mechanistic studies. Second, no direct comparison was performed between vortex durations derived from 4D Flow MRI and corresponding CFD models. Third, the backward flow method, which yielded the lowest errors, was dependent on the choice of slice following an initial visual estimation of the vortex location. An alternative approach using multiple cross-sectional planes was also tested and estimated durations were not different from those obtained with a single slice method. In any case, a fully automated slice selection will be a necessary step to remove user dependency in future developments. Also, the mechanistic reasons behind the correlation between mPAP and vortex duration shown in [4] remain unclear. Previous studies hypothesised that the MPA vortex could be linked to the reflected waves from the downstream pulmonary vasculature [16] or to abnormal vessel dilation and declining RV function [17]. This study therefore lays the ground for performing parametric investigations using patient-specific in-silico modelling to test different mechanistic hypotheses.

In conclusion, we have shown promising initial evidence for the use of patient-specific CFD simulations to model a physiologically realistic correlation between mPAP and MPA vortex duration.

Acknowledgements. We thank Daniel Giese at Siemens Healthcare GmbH, Erlangen, Germany, for providing the 4D flow work-in-progress package. Malak Sabry is supported by a PhD educational grant from Siemens Healthineers and the Magdi Yacoub Foundation.

References

1. Nakaya, T., et al.: Right ventriculo-pulmonary arterial uncoupling and poor outcomes in pulmonary arterial hypertension. Pulm. Circ. **10**, 2045894020957223 (2020)
2. Vanderpool, R.R., et al.: Surfing the right ventricular pressure waveform: methods to assess global, systolic and diastolic RV function from a clinical right heart catheterization. Pulm. Circ. **10**, 1–11 (2020)
3. Schäfer, M., et al.: Differences in pulmonary arterial flow hemodynamics between children and adults with pulmonary arterial hypertension as assessed by 4D-flow CMR studies. Am. J. Physiol.-Heart Circ. Physiol. **316**(5), H1091–H1104 (2019)
4. Reiter, G., et al.: Magnetic resonance derived 3-dimensional blood flow patterns in the main pulmonary artery as a marker of pulmonary hypertension and a measure of elevated mean pulmonary arterial pressure. Circ. Cardiovasc. Imaging **1**(1), 23–30 (2008)

5. Reiter, G., Reiter, U., Kovacs, G., Olschewski, H., Fuchsjäger, M.: Blood flow vortices along the main pulmonary artery measured with MR imaging for diagnosis of pulmonary hypertension. Radiology **275**, 140849 (2014)
6. Kräuter, C., et al.: Automated vortical blood flow-based estimation of mean pulmonary arterial pressure from 4D flow MRI. Magn. Reson. Imaging **88**, 132–141 (2022)
7. Kamada, H., Nakamura, M., Ota, H., Higuchi, S., Takase, K.: Blood flow analysis with computational fluid dynamics and 4D-flow MRI for vascular diseases. J. Cardiol. **S0914–5087**(22), 122–8 (2022)
8. 3D Slicer. https://www.slicer.org/
9. Ansys. https://www.ansys.com/
10. Blanco, P.J., Feijóo, R.A.: A 3D–1D-0D computational model for the entire cardiovascular system. Mecánica Computacional **29**(59), 5887–5911 (2010)
11. Ramos, J.G., et al.: Pulmonary hypertension by catheterization is more accurately detected by cardiovascular magnetic resonance 4D-flow than echocardiography. JACC Cardiovasc. Imaging **16**(4), 558–559 (2023). https://doi.org/10.1016/j.jcmg.2022.09.006. https://www.jacc.org/doi/abs/10.1016/j.jcmg.2022.09.006
12. Kamada, H., et al.: Quantification of vortex flow in pulmonary arteries of patients with chronic thromboembolic pulmonary hypertension. Eur. J. Radiol. **148**, 110142 (2022)
13. Hunt, J.C., Wray, A.A., Moin, P.: Eddies, streams, and convergence zones in turbulent flows. Studying turbulence using numerical simulation databases, 2. In: Proceedings of the 1988 Summer Program (1988)
14. Jeong, J., Hussain, F.: On the identification of a vortex. J. Fluid Mech. **285**, 69–94 (1995)
15. Dong, Y., Yan, Y., Liu, C.: New visualization method for vortex structure in turbulence by lambda2 and vortex filaments. Appl. Math. Model. **40**(1), 500–509 (2016)
16. Mawad, W., Fadnes, S., Løvstakken, L., Henry, M., Mertens, L., Nyrnes, S.A.: Pulmonary hypertension in children is associated with abnormal flow patterns in the main pulmonary artery as demonstrated by blood speckle tracking. CJC Pediatr. Congenit. Heart Disease **1** (2022)
17. Schäfer, M., et al.: Helicity and vorticity of pulmonary arterial flow in patients with pulmonary hypertension: quantitative analysis of flow formations. J. Am. Heart Assoc. **6**(12), e007010 (2017)

Modelling Blood Flow and Biochemical Reactions Underlying Thrombogenesis in Atrial Fibrillation

Ahmed Qureshi[1(✉)], Maximilian Balmus[1,2], Shaheim Ogbomo-Harmitt[1], Dmitry Nechipurenko[2], Fazoil Ataullakhanov[2], Gregory Y. H. Lip[3], Steven E. Williams[1,4], David Nordsletten[1,5], Oleg Aslanidi[1], and Adelaide de Vecchi[1]

[1] School of Biomedical Engineering and Imaging Sciences, King's College London, London, UK
ahmed.qureshi@kcl.ac.uk

[2] Perelman School of Medicine, University of Pennsylvania, Philadelphia, USA

[3] Liverpool Centre for Cardiovascular Science, University of Liverpool and Liverpool Heart and Chest Hospital, Liverpool, UK

[4] Centre for Cardiovascular Science, The University of Edinburgh, Edinburgh, UK

[5] Department of Biomedical Engineering and Cardiac Surgery, University of Michigan, Ann Arbor, US

Abstract. Atrial fibrillation (AF) is a widespread cardiac disease associated with a high risk of thromboembolic stroke. Clinically applicable stroke-risk stratification schemes can be improved with a mechanistic understanding of the underlying thrombogenicity induced by AF – blood stasis, hypercoagulability and endothelial damage – known as Virchow's triad. We propose a coupled biophysical modelling scheme which integrates all aspects of Virchow's triad using computational fluid dynamics (CFD) to represent blood stasis, reaction–diffusion-convection equations for the blood coagulation cascade and the endothelial cell activation potential (ECAP) to quantify endothelial damage. This comprehensive workflow is tested on a 3D patient-specific geometry reproduced from cardiac Cine MRI data. The patient case was tested in both AF and regular sinus rhythm (SR) conditions with two thrombus initiation sites: i) peak ECAP in the LA appendage (LAA) and ii) positioned at the LAA tip, totalling four cases (A-D). Case A (SR and peak ECAP initiation) washed out all thrombogenic proteins after one cardiac cycle showing low risk of thrombus formation. Case D (AF and LAA tip initiation) led to unregulated clot formation, solidification and storage in the LAA. This finding suggests that the solidified thrombus may be ejected from the LAA and travel towards the brain if the patient reverted to SR. This novel pipeline provides a promising tool that can be extended to larger patient cohorts.

Keywords: Atrial Fibrillation · Virchow's triad · Stroke · Computational Fluid Dynamics · Reaction–Diffusion-Convection · Biophysical Modelling · Digital-Twins

O. Bernard et al. (Eds.): FIMH 2023, LNCS 13958, pp. 435–444, 2023.
https://doi.org/10.1007/978-3-031-35302-4_45

1 Introduction

Strokes occur when the brain is starved of oxygen due to the obstruction of cerebrovascular pathways by an embolus travelling in the blood stream [1]. Cardioembolic strokes, particularly from thrombi formed in the left atrium (LA) have severe consequences and carry a high risk of mortality and disability [2]. A common site where greater than 90% of pathological thrombi are known to form is the LA appendage (LAA) [3]. Almost one-third of all strokes are caused by atrial fibrillation (AF), the world's most prevalent cardiac arrythmia [2]. This heightened risk of AF-related stroke can be described by a pathological amalgamation of three prothrombogenic mechanisms outlined in Virchow's triad – blood stasis, hypercoagulability and endothelial dysfunction [4].

In healthy haemostatic conditions, endothelial injury to the vascular lining initiates numerous biochemical reactions in the coagulation cascade resulting in the catalytic conversion of plasma fibrinogen to a viscous fibrin clot by the enzyme thrombin to rapidly heal damage [5]. However, AF promotes Virchow's triad of mechanisms and deviates from normal haemostasis [6]. Disorganised electrical activity in the LA myocardium during AF disrupts the regular sinus rhythm (SR) of the heart, diminishing LA contractility (hypocontractility) leading to stasis of blood within the LA. Velocities of less than 20 cm/s in the LAA are a biomarker for thrombus formation [6, 7]. Hypercoagulability manifests as abnormal blood constituent levels that may further increase propensity for coagulation, such as high plasma fibrinogen, thrombin-anti-thrombin complex and fibrin D-dimer concentrations in blood [8, 9]. The final component of Virchow's triad is endothelial dysfunction which is introduced to the LA as a compensatory effect of the reduced cardiac output during AF, leading to increased volume dilation and myocardial stretching which in turn may result in deposition of fibrosis or rupture of the cardiac endothelial lining [13, 14]. In tandem, these three mechanisms can lead to devastating outcomes for AF patients, yet very few clinically viable approaches are available for the identification of biomarkers of Virchow's triad [3].

This increased propensity for stroke in AF patients has led to the development of risk stratification schemes that rely on comorbidities to predict future stroke risk such as the CHA_2DS_2-VASc score [10]. Such scores are simple, quick to administer and generally effective, however they struggle to provide any quantifiable mechanistic insight to patient stroke risk and remain suboptimal in "low risk" cohorts, prompting a search for novel tools to assess patient-specific thrombogenicity [6].

Clinical imaging procedures provide a powerful tool to capture the presence of thrombi after their formation but fail to provide reliable risk prediction. Image-based biophysical modelling, on the other hand, has the ability to predict the underlying mechanisms of AF-related thrombus formation [11]. Previous models have attempted to capture two components of Virchow's triad, namely computational fluid dynamics (CFD) for blood stasis [12, 13] and metrics of endothelial damage [14, 15], however, comprehensive evaluation of these mechanisms remains to be addressed.

This study proposes a novel modelling pipeline to combine cardiac imaging and biophysical modelling for quantitative assessment of the formation of intracardiac thrombi. Importantly, this tool can be used to predict the risk of thrombus formation prior to *in-vivo* development using advanced mathematical tools.

2 Methodology

The pipeline for full patient-specific modelling of thrombus formation is described in this Section with an overview provided in Fig. 1. Cardiac clinical imaging data was leveraged to derive models of the LA and LV prescribed with patient-specific myocardial motion. Each component of Virchow's triad was assessed through biophysical modelling using the finite-element solver, *CHeart* [16].

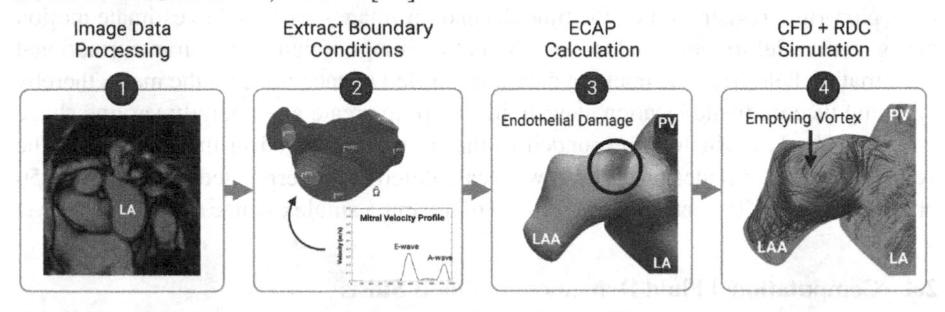

Fig. 1. Workflow for generating patient-specific models and boundary conditions to perform simulations of thrombus formation dynamics

2.1 Clinical Data

Whole heart Cine MRI images were acquired from a study at St Thomas' Hospital by Chubb et al. [17] using a 1.5 T Phillips Ingenia scanner with a balanced steady state free precession (bSSFP) technique at end-expiration breath-hold to ensure motion artefacts from breathing were minimised. The acquired images had an in-plane (axial) spatial resolution of 1.4×1.4 mm^2, slice thickness of 10 mm and temporal resolution of 50 phases per cardiac cycle enabling the entire LA motion to be captured. One AF patient was selected for development and testing of the methodology proposed in this paper. The patient was in SR at the time of imaging with a heart rate of 74 beats per minute and cardiac cycle time of 0.81 s. Moreover, the patient was aged 45, diagnosed with paroxysmal AF three years prior to image acquisition with a CHA_2DS_2-VASc = 0 and no other comorbidities listed.

2.2 Model Generation

High resolution meshes for performing patient-specific LA finite-element simulations were produced from the imaging data. Specifically, semi-automatic segmentation of the LA endocardial surface was performed on the Cine MR images at end-systole using the *Medical Imaging Interaction Toolkit* software package [15]. A 3D interpolation algorithm based on radial basis functions was applied for smoothing and minimisation of discontinuities between slices to form a patient-specific surface mesh of the LA and LV. The pulmonary veins (PVs) and mitral valve (MV) were clipped to define the inflow and outflow boundary planes. Tetrahedral volume meshes were generated using the *Simmodeler* software package (Simmetrix Inc., USA) with 1.12 million linear elements for the LA with average element size of ~ 1 mm. These meshes were produced on the first Cine MR image in the time-series at end-systole, creating an organised structure of trackable nodes and elements.

2.3 Wall Motion Tracking

bSSFP Cine MRI enables the motion tracking of cardiac structures over time in a cyclic sequence. Patient-specific myocardial contractility was mimicked by deforming the static LA and LV meshes generated in Sect. 2.2. The temporally varying Cine MR image data was processed in Eidolon where a feature-based tracking algorithm mapped the motion of structures over time using the Medical Image Registration Toolkit set of routines [18]. Non-rigid registration on the time-dependant images was used to estimate motion and generate deformation fields over the entire image volume. The transformational deformation field from the tracking data was applied to each node on the mesh, thereby updating its coordinate location at each time step, to create a temporally varying shape model of the LA which corresponded to the motion observed in the Cine data. The velocity vectors of nodes on the wall were calculated and interpolated to convert the 50 phases from the Cine data to 1000 steps to represent a single cardiac cycle.

2.4 Computational Fluid Dynamics for Blood Stasis

CFD modelling of blood flow was performed by solving the incompressible Navier–Stokes equations (NSE) in the arbitrary-Eulerian–Lagrangian (ALE) frame of reference for the moving LA domain. Blood density and viscosity were set to average values, $\rho = 1060$ kg/m^3 and $\mu = 3.5 \times 10^{-3}$ Pa s, respectively.

A mitral velocity profile was applied as a Dirichlet condition on the MV outlet in the outward normal direction, with open boundary conditions applied on PV inlets to drive flow through the LA and out of the MV. To generate this MV flow profile, the volumetric flow rate, Q_{MV}, was calculated using the change in volume of the motion-tracked, deforming LV mesh with corresponding MV area. The LV volume curve over the cardiac cycle was extracted and interpolated to match the timestep in subsequent finite-element simulations.

$$Q_{MV} = v_{MV} A_{MV} = \frac{dV_{LV}}{dt} = \begin{cases} 0, & \frac{dV_{LV}}{dt} < 0 \\ \frac{dV_{LV}}{dt}, & \frac{dV_{LV}}{dt} \geq 0 \end{cases} \tag{1}$$

2.5 Endothelial Cell Activation Potential for Endothelial Damage

Clinical assessment of endothelial damage is challenging to perform prior to thrombus formation, especially within the LAA. Regions prone to thrombogenesis can be determined by haemodynamic indices based on the LA wall shear stress ($\boldsymbol{\tau}_w$) such as the oscillatory shear index (OSI), the time averaged wall shear stress (TAWSS), and the endothelial cell activation potential (ECAP) shown in Eqs. (2–4) [19]. The OSI is a unitless scalar with a maximum value of 0.5 and high ECAP values indicate large oscillatory shear flows and low wall shear stresses. We hypothesise that the peak ECAP location in the LAA would be at highest risk of thrombus initiation and was chosen as one of the injury sites in the subsequent simulations.

$$TAWSS = \frac{1}{T} \int_0^T |\boldsymbol{\tau}_W| dt \tag{2}$$

$$OSI = \frac{1}{2}\left(1 - \frac{\left|\int_0^T \tau_W dt\right|}{\int_0^T \tau_W dt}\right) \tag{3}$$

$$ECAP = \frac{OSI}{TAWSS} \tag{4}$$

2.6 Modelling Blood Coagulation for Hypercoagulability

Coagulation of blood was defined by the three clotting proteins described in Sect. 1: thrombin (Th), fibrinogen (Fg) and fibrin (Fn). Each protein was modelled using its respective reaction–diffusion-convection (RDC) Eq. (5–8) to describe the concentration of substances distributed in the LA domain [20].

$$\frac{\partial Th}{\partial t} = D_{Th}\Delta Th - \boldsymbol{u}.\nabla Th + R_{Th} \tag{5}$$

$$R_{Th} = K_1(1 + K_2 Th)[K_3 Th(1 + K_4 Th)] \times \left(1 - \frac{Th}{u_0}\right) - K_5 Th \tag{6}$$

$$\frac{\partial Fg}{\partial t} = D_{Fg}\Delta Fg - u.\nabla Fg + K_{eff}.Fg.Th \tag{7}$$

$$\frac{\partial Fn}{\partial t} = D_{Fn}\Delta Fn - \boldsymbol{u}.\nabla Fn - K_{eff}.Fg.Th \tag{8}$$

The physiological mechanism of thrombus solidification was modelled to capture the transition from concentrations of thrombogenic proteins dispersed in liquid blood to a viscous fibrin gel thrombus at the site of injury. Reproducing this mechanism *in-silico* involved a two-way coupling between the system of RDC equations and NSE. This was achieved by employing the Hill equation to scale viscosity, μ, in the entire domain with parameters of n = 3, K = 0.5, μ_{max} = 2 Pa s (chosen to mimic the viscosity of fibrin gel [21]) with $\phi = Fn/Fg$.

$$\mu = \frac{\mu_{\max}\phi^n}{K^n + \phi^n} \tag{9}$$

2.7 Simulation Set-Up

Application of these tools was performed in *CHeart*. The solution scheme accounts for boundary driven intracardiac blood flow and a system of coupled RDC equations for each of the interconnected protein species. An initial condition for Th concentration was set to 0.09 mmol/m^3 to define the site of endothelial injury and initiate the cascade of coagulation mechanisms. At each solution step, μ was updated using Eq. 9 and fed back into NSE. The time-step was set to 0.81 ms, and the fully coupled RDC-viscosity simulation ran for a total of 12.15 s (15 cardiac cycles).

A simple parametric study was designed to identify the underlying cause of *in-silico* thrombus formation. The original SR patient data was manipulated to induce AF conditions resulting in four cases for investigation, summarised in Table 1. Two thrombus initiation locations were derived, based on the peak ECAP and the tip of the LAA which is the furthest site from the LAA ostium. We propose a metric for the cumulative concentration - defined as the integral of the Fn concentration by time or area under the Fn curve (Fn AUC) – to represent thrombus growth and total Fn generated.

Table 1. Case study design for AF vs. SR and location to initiate thrombogenesis [4]

Case	Patient Data	Initial Location	MV Profile	Initial Fg (mmol/m^3)
A	SR	Peak ECAP	Q_{MV}	4.5
B	SR	LAA Tip	Q_{MV}	4.5
C	AF	Peak ECAP	$0.5 \times Q_{MV}$	9
D	AF	LAA Tip	$0.5 \times Q_{MV}$	9

3 Results

3.1 LAA Morphological Parameters

Morphological assessment of the LAA resulted in a geodesic length of 3.13 cm, tortuosity of 0.066 and volume of 5.35 cm^3. The ECAP distribution is shown in Fig. 2 with peak ECAP near the LAA ostium.

3.2 Case Study Findings

Results for the four Cases listed in Table 1 will focus on the LAA, describing the interplay between Fg and Fn, peak LAA flow velocities and whether the thrombus was washed out. A summary of the results will be presented in Table 2.

Case A: Fn generated at the peak ECAP damage site increases during the first cardiac cycle to 0.19 mmol/m^3 during SR. As blood velocity increases to 17.5 cm/s in the LAA, Fn is immediately washed out after 0.84 s. Fn plateaus at negligible concentrations in the LAA for the remaining simulation. **Case B**: Moving the thrombus initiation site to the LAA tip facilitated Fn generation to a maximum of 3.62 mmol/m^3 after 7.8 s, yet a gradual decrease in Fn can be observed from 6–12 s. **Case C:** Introducing the AF scenario to the peak ECAP location

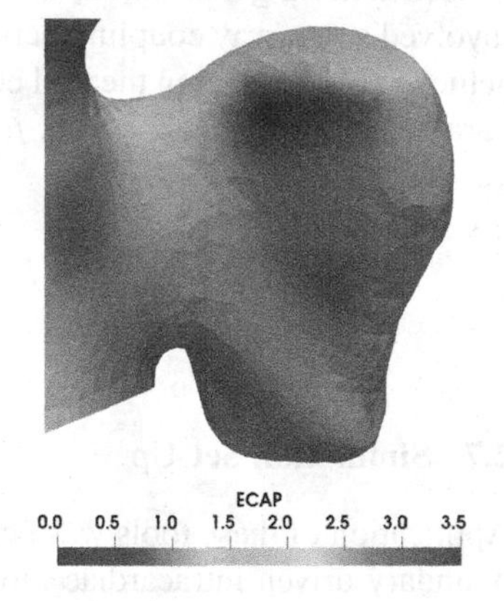

Fig. 2. ECAP distribution on patient LAA in (anterior view). Black circle indicates the peak ECAP location

resulted in lower peak LAA velocities of 13.3 cm/s
(24% lower than SR) at the injury site. However, rapid washout of the Fn concentration still occurred after one cardiac cycle (1.2 s). **Case D:** The combination of AF and

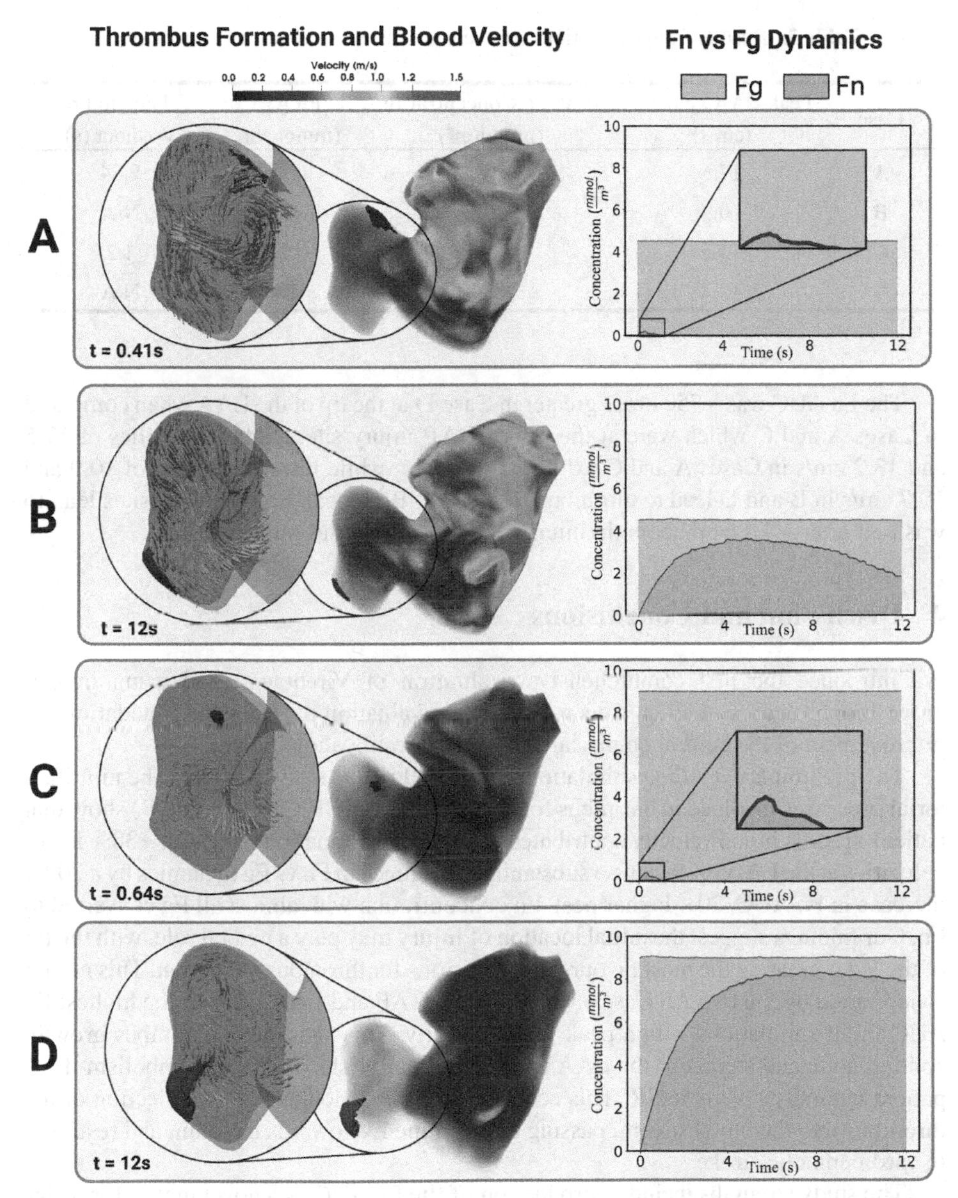

Fig. 3. Snapshots from the coupled RDC-viscosity simulations visualise thrombus formation (red region) and a plot for Fn vs Fg dynamics for Cases A-D (Color figure online)

LAA tip injury resulted in the lowest blood velocity of 10.7 cm/s. Fn generation reached a maximum of 8.34 mmol/m^3 after 4.2 s. Fg vs Fn dynamics indicate the largest Fn generation curve as shown by an Fn AUC value of 106 mmol s/m^3 (Fig. 3).

Table 2. Results from simulations for the four cases. AUC = area under curve

Case	Peak LAA velocity (cm/s)	Peak Fn Concentration (mmol/m^3)	Fn AUC (mmol s/m^3)	Time to Fn washout (s)
A	17.5	0.189	0.141	0.84
B	10.9	3.62	42.4	N/A
C	13.3	0.406	0.297	1.2
D	10.7	8.34	106	N/A

The Fn AUC was ~ 750 times greater in Case D at the tip of the LAA when compared to Cases A and C which were at the peak ECAP injury site. Higher velocities of 17.5 and 13.2 cm/s in Cases A and C exhibited washout, while lower velocities of 10.9 and 10.7 cm/s in B and D lead to thrombus formation. Both peak ECAP injury sites lead to washout after ~ 1 s while both tip injuries resulted in Fn thrombogenesis.

4 Discussion and Conclusions

We introduce the first comprehensive evaluation of Virchow's triad using image-based biophysical modelling. This mechanistic evaluation provides the foundation for improvement of the current empirical stroke risk stratification indices.

The preliminary *in-silico* simulations presented in this study highlight the multifactorial processes involved in thrombus formation. The two SR Cases (A and B) show that patient-specific blood velocity contributed to thrombus formation risk, with ~ 38% lower velocities at the LAA tip causing a substantial difference in Fn vs Fg dynamics by a 200× increase in Fn AUC, 20× higher peak Fn concentration with almost all Fg converted to Fn. Our findings suggest the initial location of injury may play a pivotal role, with the tip of the LAA creating the most favourable conditions for thrombus formation. This notion is supported by the two AF Cases, where Case D (AF and LAA tip) had the highest Fn AUC in all simulations with a peak LAA velocity. This substantial thrombus growth, solidification and storage at the LAA tip may increase risks of thromboembolism if the patient naturally reverts to SR. This could lead to the dislodgement and ejection of the thrombus into the blood stream, passing through the LV towards the brain and resulting in cardioembolic stroke.

The study strengths include introduction of the Fn AUC as a novel metric for quantifying thrombus growth in the LAA. This can further improve patient stratification by incorporating thrombogenic factors such as blood velocity, injury location and initial Fg concentration in the evaluation. The study limitations include using a single patient case and manipulating patient-specific data to mimic AF conditions. This departed from personalised modelling yet allowed us to illustrate the methodology and perform the simple parametric study on the importance of injury site.

This study presents a novel image-based biophysical modelling pipeline integrating Virchow's triad and proposes a new metric for quantifying thrombus growth. Applied to large patient-specific cohorts, this tool can enable clinicians to have a more accurate mechanistic assessment of patient stroke risks.

References

1. Hindricks, G., et al.: 2020 ESC Guidelines for the diagnosis and management of atrial fibrillation developed in collaboration with the European Association for Cardio-Thoracic Surgery (EACTS). Eur Heart J. **42**, 373–498 (2021). https://doi.org/10.1093/eurheartj/ehaa612
2. Freedman, B., Potpara, T.S., Lip, G.Y.H.: Stroke prevention in atrial fibrillation. The Lancet. **388**, 806–817 (2016). https://doi.org/10.1016/S0140-6736(16)31257-0
3. Blackshear, J.L., Odell, J.A.: Appendage obliteration to reduce stroke in cardiac surgical patients with atrial fibrillation (1996)
4. Ding, W.Y., Gupta, D., Lip, G.: Atrial fibrillation and the prothrombotic state: revisiting Virchow's triad in 2020. Heart. heartjnl-2020-316977 (2020). https://doi.org/10.1136/heartjnl-2020-316977
5. Palta, S., Saroa, R., Palta, A.: Overview of the coagulation system. Indian J Anaesth. **58**, 515–23 (2014). https://doi.org/10.4103/0019-5049.144643
6. Fukushima, K., et al.: Correlation between left atrial appendage morphology and flow velocity in patients with paroxysmal atrial fibrillation. Eur Heart J Cardiovasc Imaging. **17**, 59–66 (2016). https://doi.org/10.1093/ehjci/jev117
7. Goldman, M.E., et al.: Pathophysiologic Correlates of Thromboembolism in Nonvalvular Atrial Fibrillation: I. Reduced Flow Velocity in the Left Atrial Appendage (The Stroke Prevention in Atrial Fibrillation [SPAF-III] Study). Journal of the American Society of Echocardiography **12**, 1080–1087 (1999). https://doi.org/10.1016/S0894-7317(99)70105-7
8. Marín, F., Roldán, V., Climent, V.E., Ibáñez, A., García, A., Marco, P., Sogorb, F., Lip, G.Y.H.: Plasma von Willebrand factor, soluble thrombomodulin, and fibrin D-dimer concentrations in acute onset non-rheumatic atrial fibrillation. Heart. **90**, 1162–1166 (2004). https://doi.org/10.1136/HRT.2003.024521
9. Akar, J.G., Jeske, W., Wilber, D.J.: Acute onset human atrial fibrillation is associated with local cardiac platelet activation and endothelial dysfunction. J Am Coll Cardiol. **51**, 1790–1793 (2008). https://doi.org/10.1016/J.JACC.2007.11.083
10. Lane, D.A., Lip, G.Y.H.: Use of the CHA2DS2-VASc and HAS-BLED Scores to aid decision making for thromboprophylaxis in nonvalvular atrial fibrillation. Circulation. **126**, 860–865 (2012). https://doi.org/10.1161/CIRCULATIONAHA.111.060061
11. Qureshi, A., Lip, G., Nordsletten, D.A., Williams, S.E., Aslanidi, O., de Vecchi, A.: Imaging and biophysical modelling of thrombogenic mechanisms in atrial fibrillation and stroke. Front Cardiovasc Med. **9**, 3872. https://doi.org/10.3389/FCVM.2022.1074562
12. Dillon-Murphy, D., et al.: Modeling left atrial flow, energy, blood heating distribution in response to catheter ablation therapy. Front Physiol. **9**, 1757 (2018). https://doi.org/10.3389/fphys.2018.01757
13. Masci, A., et al.: A Patient-Specific Computational Fluid Dynamics Model of the Left Atrium in Atrial Fibrillation: Development and Initial Evaluation 392–400 (2017). https://doi.org/10.1007/978-3-319-59448-4_37
14. Mill, J., et al.: Patient-specific flow simulation analysis to predict device-related thrombosis in left atrial appendage occluders (2021). https://doi.org/10.24875/RECICE.M21000224

15. Morales, X., et al.: Deep Learning Surrogate of Computational Fluid Dynamics for Thrombus Formation Risk in the Left Atrial Appendage. Lecture Notes in Computer Science (including subseries Lecture Notes in Artificial Intelligence and Lecture Notes in Bioinformatics). (2020). https://doi.org/10.1007/978-3-030-39074-7_17
16. Lee, J., et al.: Multiphysics computational modeling in CHeart. SIAM Journal on Scientific Computing. **38**, C150–C178 (2016). https://doi.org/10.1137/15M1014097
17. Chubb, H., et al.: The reproducibility of late gadolinium enhancement cardiovascular magnetic resonance imaging of post-ablation atrial scar: a cross-over study. Journal of Cardiovascular Magnetic Resonance **20**, 21 (2018). https://doi.org/10.1186/s12968-018-0438-y
18. Kerfoot, E., et al.: Eidolon: Visualization and Computational Framework for Multi-Modal Biomedical Data Analysis.
19. di Achille, P., Tellides, G., Figueroa, C.A., Humphrey, J.D.: A haemodynamic predictor of intraluminal thrombus formation in abdominal aortic aneurysms. Proceedings of the Royal Society A: Mathematical, Physical and Engineering Sciences **470**, 20140163 (2014). https://doi.org/10.1098/rspa.2014.0163
20. Ataullakhanov, F.I., Zarnitsyna, V.I., Kondratovich, A.Y., Lobanova, E.S., Sarbash, V.I.: A new class of stopping self-sustained waves: a factor determining the spatial dynamics of blood coagulation. Physics-Uspekhi. **45**, 619–636 (2002). https://doi.org/10.1070/PU2002v045n06ABEH001090
21. Lobanov, A.I., Nikolaev, A.V., Starozhilova, T.K.: Mathematical model of fibrin polymerization. Math. Model. Nat. Phenom. **6**, 55–69 (2011). https://doi.org/10.1051/mmnp/20116705

SE(3) Symmetry Lets Graph Neural Networks Learn Arterial Velocity Estimation from Small Datasets

Julian Suk(✉), Christoph Brune, and Jelmer M. Wolterink

Department of Applied Mathematics & Technical Medical Center,
University of Twente, Enschede, The Netherlands
{j.m.suk,c.brune,j.m.wolterink}@utwente.nl

Abstract. Hemodynamic velocity fields in coronary arteries could be the basis of valuable biomarkers for diagnosis, prognosis and treatment planning in cardiovascular disease. Velocity fields are typically obtained from patient-specific 3D artery models via computational fluid dynamics (CFD). However, CFD simulation requires meticulous setup by experts and is time-intensive, which hinders large-scale acceptance in clinical practice. To address this, we propose graph neural networks (GNN) as an efficient black-box surrogate method to estimate 3D velocity fields mapped to the vertices of tetrahedral meshes of the artery lumen. We train these GNNs on synthetic artery models and CFD-based ground truth velocity fields. Once the GNN is trained, velocity estimates in a new and unseen artery can be obtained with 36-fold speed-up compared to CFD. We demonstrate how to construct an SE(3)-equivariant GNN that is independent of the spatial orientation of the input mesh and show how this reduces the necessary amount of training data compared to a baseline neural network.

Keywords: Geometric deep learning · Graph neural networks · Coronary arteries · Hemodynamics

1 Introduction

Patient-specific, hemodynamic biomarkers have great potential in diagnosis [7], prognosis [4] and treatment planning for patients with cardiovascular disease [8]. Most studies in this field have focussed on hemodynamic biomarkers such as wall shear stress or fractional flow reserve but the post-operative change in hemodynamic velocity, a direct indicator for assessment of coronary artery bypass grafting surgery [1], is also a useful biomarker. Velocity field estimates can also be used in applications like cardiovascular stent design which is an active area of medical research [5]. All these quantities can be accurately estimated by computational fluid dynamics (CFD), based on patient-specific 3D artery models from medical images such as MRI and CT. However, high-fidelity blood flow simulations are computationally intensive.

O. Bernard et al. (Eds.): FIMH 2023, LNCS 13958, pp. 445–454, 2023.
https://doi.org/10.1007/978-3-031-35302-4_46

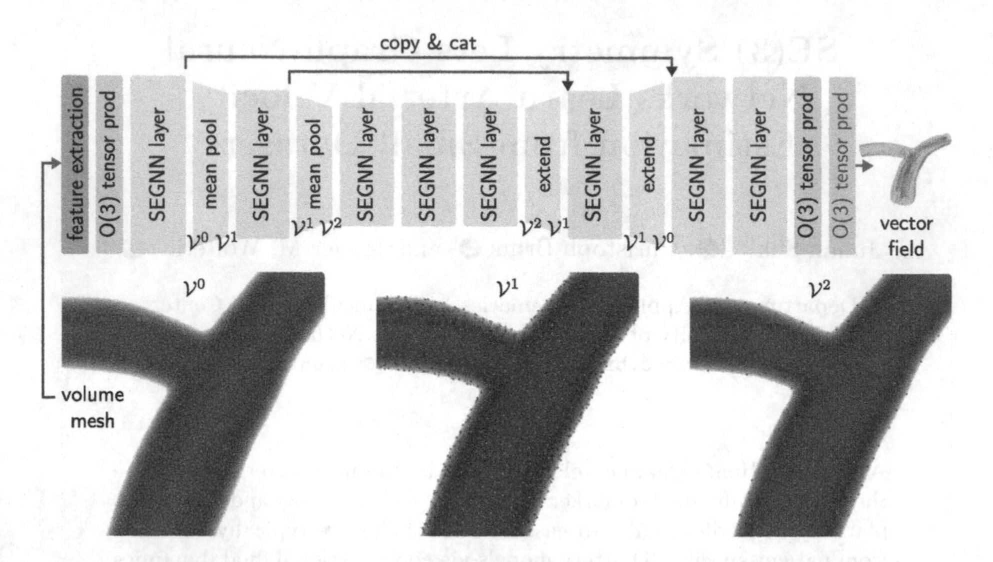

Fig. 1. Steerable E(3)**-equivariant graph neural network** (SEGNN). Inputs are processed by O(3) tensor product layers and SEGNN ResNet layers. We introduce mean pooling and "copy back" extension for coarsening and refinement between $\mathcal{V}^0$, $\mathcal{V}^1$ and $\mathcal{V}^2$. We include skip connections inside the same pooling scales. SEGNN predicts one velocity vector per mesh vertex (visualised via the resulting streamlines).

Recent work has demonstrated the potential for machine learning in cardiovascular biomechanics modelling [3]. In contrast to time-intensive CFD simulations, machine learning estimates can be obtained in a matter of seconds. Estimating hemodynamic scalar and vector fields with deep neural networks, either on the artery wall by projecting it to a 1D or 2D domain and using multilayer perceptrons (MLP) and convolutional neural networks [9,13,14,24], using autoencoders [17,18] or point cloud and mesh-based methods in 3D [10,15,16,25] has been an ongoing area of research. In the above works, neural networks are trained offline on a dataset of results from CFD simulation which typically comprise the velocity field of the blood flow mapped to the vertices of a tetrahedral volume mesh of the artery lumen. However, only in [15,16,18] a complete 3D velocity field is estimated. Other efforts have been made to learn dense velocity fields in arteries with MLPs [2,22], so-called physics-informed neural networks (PINN). However, PINNs are usually limited to fitting flow patterns in a single artery and cannot generalise to other arteries, due to their use of non-local MLPs. We propose to use the generalisation capabilities of graph neural networks (GNN) to learn the relation between artery geometry and velocity field with the ability to generate accurate predictions for new and unseen arteries.

In previous work [25], we have exploited gauge symmetry to estimate hemodynamic fields on the artery *wall*. Here, we aim to estimate volumetric velocity fields within the artery *lumen*. We propose to make use of recent advances in E(3)-equivariant message passing [6] to construct a GNN that is independent of the spatial orientation of the input geometry. Thus, our neural network is able to focus on learning the relation between geometry and hemodynamics without relying on the ambient coordinate system and to make efficient use of available training data. This is important because clinical trials in medical research typically encompass only few patients. We train and validate our GNN on a dataset of synthetic coronary bifurcations with ground truth from CFD and find that the proposed method is both more accurate and more data-efficient than a baseline PointNet++ [21].

2 Data

We consider the estimation of 3D velocity vectors in bifurcating coronary arteries. We synthesized 2,000 arteries after measurement statistics of left main coronary bifurcations [19]. Each artery mesh consists of ca. 175,000 vertices and ca. 1,02 million tetrahedra. Steady-flow CFD was performed with SimVascular to obtain the ground truth, details of which can be found in [25]. Each simulation took ca 15 min. The Reynolds number was low suggesting laminar flow. The dataset comprises 39 GB of simulation data and is made available upon reasonable request.

3 Method

3.1 Group Symmetry

Since hemodynamics in coronary arteries are usually modelled as independent of gravity, they exhibit certain symmetries. rotation or translation of the artery in 3D does not change the flow pattern but only rotates the velocity vectors. This property is formally called equivariance under SE(3) transformation. SE(3) is the (special Euclidean) symmetry group comprised of all rotations and translations in 3D Euclidean space. In contrast, the Euclidean group E(3) consists of rotations, translations and reflections. The orthogonal group O(3) contains all distance-preserving transformations and the special orthogonal group SO(3) all rotations. A group-equivariant learning setup must be composed, from input to output, entirely of group-equivariant operators.

3.2 Input Features

We describe the geometry of an artery by assigning a feature vector to each vertex $\mathcal{V}$ of the input mesh. This feature vector depends on the location of the vertex relative to the artery inlet, wall and outlets used in fluid simulation. Denote the sets of vertices comprising the artery inlet as $\mathcal{P}_{\text{inlet}}$, the artery wall as $\mathcal{P}_{\text{wall}}$ and the artery outlets as $\mathcal{P}_{\text{outlets}}$. We can construct an SE(3)-equivariant descriptor of global geometry for each vertex p^i as

$$\kappa_{\mathcal{P}}(p) := \arg\min_{q \in \mathcal{P}} \|p - q\|_2$$
$$f^i = \left(\kappa_{\mathcal{P}_{\text{inlet}}}(p^i) - p^i,\ \kappa_{\mathcal{P}_{\text{wall}}}(p^i) - p^i,\ \kappa_{\mathcal{P}_{\text{outlets}}}(p^i) - p^i\right) \in \mathbb{R}^9$$
$$i = 1, \ldots, n.$$

We then stack these feature vectors in a vertex feature matrix

$$X = \begin{pmatrix} \text{---} f^1 \text{---} \\ \text{---} f^2 \text{---} \\ \vdots \\ \text{---} f^n \text{---} \end{pmatrix} \in \mathbb{R}^{n \times 9}.$$

Corollary 1. *X is row-wise equivariant under* SE(3) *transformation of $\mathcal{V}$.*

(Proof in Appendix.)

3.3 Steerable E(3)-equivariant GNN

Steerable E(3)-equivariant graph neural networks (SEGNN) [6] represent signals between layers as steerable feature vectors. Steerable feature vectors consist of geometric objects, e.g. Euclidean 3-vectors, which we know how to transform with elements of the symmetry group E(3), e.g. rotations. **SEGNN layers** update the vertex features f^i by message passing

$$m^{ij} = \phi_m(f^i, f^j, \|p^j - p^i\|_2^2, a^{ij})$$
$$f^i \leftarrow \phi_f(f^i, \sum_{j \in \mathcal{N}(i)} m^{ij} a^i)$$

between vertex positions p^i and p^j where a^{ij} are edge attributes, a^i are vertex attributes and ϕ_m and ϕ_f are O(3)-equivariant MLPs. The MLPs are powered by the Clebsch-Gordan O(3)-equivariant **tensor product**. $\mathcal{N}(i)$ denotes the neighbourhood of p^i which we choose so that $|\mathcal{N}(p^i)| \approx 13$. To save memory, we choose the numbers of latent channels so that SEGNN has 20,868 trainable parameters. Note that since SEGNN is E(3)-equivariant, composition with our SE(3)-equivariant input features creates an end-to-end SE(3)-equivariant setup.

Pooling. We introduce pooling to make SEGNN efficiently pass long-range messages within large meshes. As in [25], we sample a hierarchy of vertex subsets $\mathcal{V} = \mathcal{V}_0 \supset \mathcal{V}_1 \supset \mathcal{V}_2$ via farthest point sampling and compute the pooling target for each $p \in \mathcal{V}^i$ as $\kappa_{\mathcal{V}^{i+1}}(p)$. We perform **mean pooling** among all source vertex features towards the target vertex and unpooling by "copy back" **extending** the source feature to its target vertices, unchanged. Figure 1 gives an overview of our neural network.

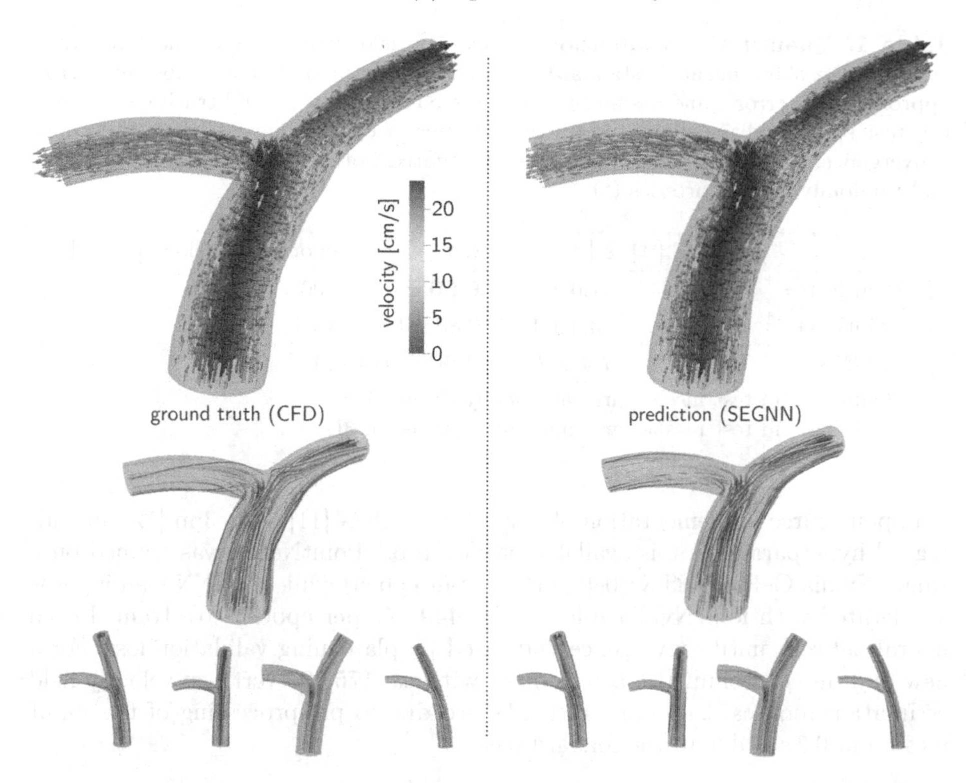

Fig. 2. Velocity field estimation. CFD ground truth (left) and SEGNN prediction (right). On top we show a subset of the velocity vectors and on bottom we visualise the flow via a selection of streamlines evolving from the bifurcation region. We show four additional arteries of the test split.

3.4 PointNet++

As a baseline for comparison, we implement a PointNet++ [21] with five sample & grouping layers using point convolution to pool the input down to a coarse point cloud, followed by five MLP-enhanced interpolation layers unpooling the signal up to its original size. Sampling ratios and grouping radii are chosen so that vertices have thirteen neighbours on average and point convolutions and MLPs are set up so that the overall neural network has 1,029,775 trainable parameters.

4 Experiments

We divided the dataset 80:10:10 into training, validation and test split. PointNet++ and SEGNN were trained with batch size two, L^1 loss and Adam optimiser (learning rate 3e-4). Systematic hyperparameter optimisation was infeasible due to long training times, so we chose hyperparameters by informed trial and error.

Table 1. Quantitative evaluation of velocity field estimation in the bifurcating arteries. We show mean ± standard deviation of normalised mean absolute error, approximation error ε and cosine similarity $\mathfrak{cos}$ compared to ground truth CFD across the test split. Additionally, we state the number of epochs and wallclock time until convergence. PointNet++ was trained and evaluated both on canonically oriented (†) and randomly rotated arteries (‡).

	NMAE [%]	ε [%]	$\mathfrak{cos}$	epochs	wallclock [h:min]
PointNet++ †	1.1 ± 0.6	11.0 ± 7.1	$\mathbf{0.90} \pm 0.02$	500	20:00
PointNet++ ‡	3.4 ± 1.5	33.4 ± 13.4	0.83 ± 0.09	1000	40:00
SEGNN	$\mathbf{0.7} \pm 0.2$	$\mathbf{7.4} \pm 2.2$	$\mathbf{0.90} \pm 0.01$	440	54:54

† training and test meshes are canonically oriented
‡ training and test meshes are randomly rotated in 3D

An open-source implementation of our setup in PyG [11] and e3nn [12], including all hyperparameters, is available on GitHub.[1] PointNet++ was trained on a single Nvidia GeForce RTX 3080 (144.10 s per epoch) while SEGNN training was accelerated with four Nvidia A40 GPUs (449.27 s per epoch). We trained each neural network until convergence, indicated by plateauing validation loss. For a new and unseen volumetric artery mesh with ca. 175,000 vertices, velocity field estimation took ca. 24.5 s of which 24 s are due to pre-processing of the input mesh and 0.5 s is due to the forward pass.

4.1 Quantitative Evaluation Metrics

We evaluate accuracy by normalised mean absolute error (NMAE), approximation error ε and mean cosine similarity $\mathfrak{cos}$. Let $Y^i \in \mathbb{R}^{n\times 3}$ be the GNN's stacked output matrix and $V^i \in \mathbb{R}^{n\times 3}$ the corresponding velocity ground truth matrix for the i-th input mesh of the test split. We define the metrics element-wise as

$$\mathrm{NMAE}_i = \frac{\operatorname*{mean}_j \|(V^i - Y^i)_j\|_2}{\max_k \max_j \|(V^k)_j\|_2}$$

$$\varepsilon_i = \frac{\sum_j \|(V^i - Y^i)_j\|_2^2}{\sum_j \|(V^i)_j\|_2^2}$$

$$\mathfrak{cos}_i = \operatorname*{mean}_j \cos(\sphericalangle(V^i)_j, (Y^i)_j),$$

where $(\cdot)_j$ means taking the j-th row vector. Note that $\cos(\sphericalangle(V^i)_j, (Y^i)_j)$ is 1 if $(V^i)_j$ and $(Y^i)_j$ are proportional, 0 if they are orthogonal and –1 if they are opposing.

[1] github.com/sukjulian/segnn-hemodynamics.

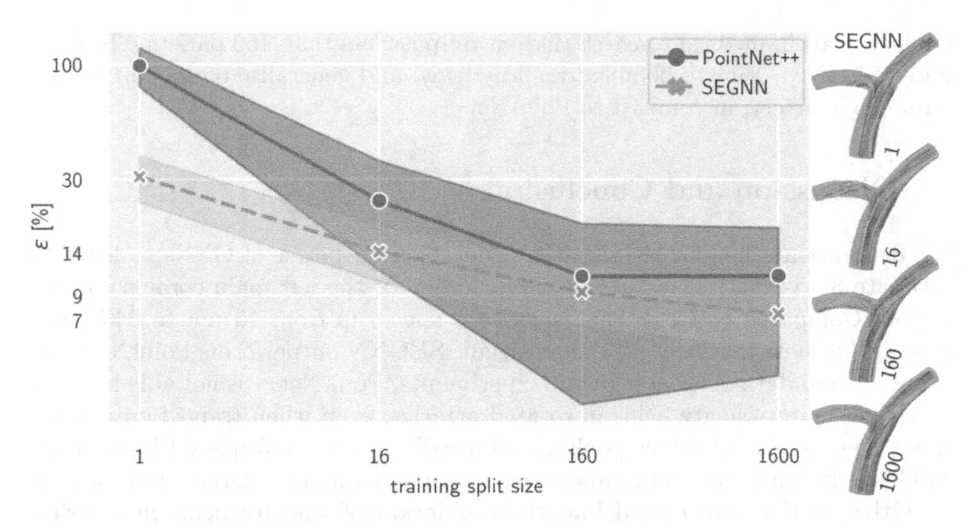

Fig. 3. **ε over training split size** in log-log scale. Data points are the mean approximation errors and the shaded region corresponds to ± one standard deviation. The same test split (200 arteries) is used to evaluate ε for each training split. Streamlines show SEGNN outputs in a test split artery after training on different split sizes.

4.2 Velocity Field Estimation

Figure 2 shows example results of velocity field estimation in unseen arteries. Qualitative comparison of vector field and streamlines suggest good agreement between CFD and SEGNN prediction. Even though streamlines are sensitive to small perturbations, they largely coincide. Table 1 contains quantitative evaluation showing that SEGNN strictly outperforms PointNet++. Both neural networks' estimation has good directional agreement to CFD, indicated by high mean cosine similarity **cos**. We train PointNet++ on canonically oriented samples and evaluate its accuracy for this case, then later randomly rotate the arteries in 3D during further training and testing. In the rotated case, PointNet++ is not able to recover its accuracy of the oriented case with the given input features. Note that since SEGNN is fully equivariant to rotations, its accuracy does not change for rotated input meshes.

4.3 Learning from Small Datasets

Figure 3 shows approximation error ε resulting from PointNet++ and SEGNN trained with different amounts of data in log-log scale. We find that both neural networks can obtain good accuracy when trained on 160 instead of 1,600 meshes. The performance of PointNet++ decreases more rapidly than SEGNN with decreasing training split size. Training PointNet++ on 160 versus 16 meshes doubles its approximation error ε from 10.9% to 23.9% while SEGNN deteriorates 1.5-fold from 9.3% to 14.1%. We conclude that SEGNN is favourable for clinical trials in medical research that encompass less than 160 patients. Intriguingly, SEGNN seems to be able to mildly learn and generalise based on training on a single artery, in contrast to PointNet++.

5 Discussion and Conclusion

We demonstrate how to leverage the generalisation power of GNNs to learn to estimate 3D velocity fields in the artery lumen of the left main coronary bifurcation. Compared to CFD, this leads to a speed-up from 15 min to 24.5 s per artery. We show that our SE(3)-equivariant SEGNN outperforms PointNet++ in accuracy and data-efficiency. In our experiments, PointNet++ is not able to accurately estimate velocity fields in rotated arteries, even when trained on rotated input meshes. We introduce pooling and parallelise over multiple GPUs to make SEGNN efficiently pass long-range messages while scaling to large input meshes.

Other works have looked into the prediction of velocity fields in arteries. Liang et al. [18] proposed a neural network that predicts velocity fields using autoencoder-based shape encoding of the input geometry. Their method requires vertex correspondence between input meshes, achieved trough reparametrisation, to ensure an equal number of mesh vertices with the same nodal connectivity. This can be overly restricting if input geometries differ significantly, e.g. branching artery trees. Li et al. [15,16] proposed an adapted PointNet [20] that combines global and local information to predict velocity fields on arbitrary point clouds. While PointNet is robust to ordering of the input vertices due to the locality of its layers (permutation equivariance), its output implicitly depends on the ambient coordinate system in which its input points are expressed. This may not be a problem if the input geometries are all canonically oriented by construction. However, this is not common in practice and thus PointNet requires canonical alignment ("registration") of the input geometries. Furthermore, the implicit conditioning on the ambient coordinate system misguides the neural network's prediction. Addressing this by data augmentation might waste expressive capacity or may not be possible at all, as our results indicate. In contrast, SEGNN is by design fully equivariant to roto-translations and thus robust to misaligned data. In our experiments, we find that SEGNN is able to estimate velocity fields in new and unseen arteries based on training even with small datasets. We conjecture that due to its SE(3) equivariance, SEGNN avoids conditioning on misleading information like the alignment of the input mesh, which is a key driver of performance.

Clinical datasets with patient-specific artery models are difficult to obtain and typically small. Thus, it is imperative to further increase the data efficiency of our method. To do so, we plan to incorporate knowledge about the flow physics, leading to so-called physics-informed graph networks [23]. A limitation of our study is the reliance on a dataset of synthetic coronary arteries as well as the steady-flow assumption. To address this, we aim to extend our work to real-life patient data and pulsatile flow in the future. Furthermore, we will condition our GNNs on flow boundary conditions like inflow velocity and outlet pressure.

In conclusion, SEGNN is able to learn the relation between artery geometry and hemodynamic velocity in coronary arteries based on training with small datasets while being robust to misalignment of the input meshes through SE(3) equivariance.

Acknowledgements. This work is funded in part by the 4TU Precision Medicine programme supported by High Tech for a Sustainable Future, a framework commissioned by the four Universities of Technology of the Netherlands. Jelmer M. Wolterink was supported by the NWO domain Applied and Engineering Sciences VENI grant (18192). This work made use of the Dutch national e-infrastructure with the support of the SURF Cooperative using grant no. EINF-2675.

We would like to thank Pim de Haan and Johannes Brandstetter for the fruitful discussions about steerable equivariant graph neural networks.

References

1. Amin, S., Werner, R.S., Madsen, P.L., Krasopoulos, G., Taggart, D.P.: Intraoperative bypass graft flow measurement with transit time flowmetry: a clinical assessment. Ann. Thorac. Surg. **106**(2), 532–538 (2018)
2. Arzani, A., Wang, J.X., D'Souza, R.M.: Uncovering near-wall blood flow from sparse data with physics-informed neural networks. Phys. Fluids **33** (2021)
3. Arzani, A., Wang, J.X., Sacks, M., Shadden, S.: Machine learning for cardiovascular biomechanics modeling: challenges and beyond. Ann. Biomed. Eng. **50**(6), 615–627 (2022)
4. Barral, M., et al.: Blood flow and shear stress allow monitoring of progression and prognosis of tumor diseases. Front. Physiol. **12** (2021)
5. Beier, S., et al.: Impact of bifurcation angle and other anatomical characteristics on blood flow - a computational study of non stented and stented coronary arteries. J. Biomech. **49**, 1570–1582 (2016)
6. Brandstetter, J., Hesselink, R., van der Pol, E., Bekkers, E.J., Welling, M.: Geometric and physical quantities improve E(3) equivariant message passing. In: Proceedings of the 10th International Conference on Learning Representations (2022)
7. Candreva, A., et al.: Current and future applications of computational fluid dynamics in coronary artery disease. Rev. Cardiovasc. Med. **23**(11), 377–394 (2022)
8. Chung, B., Cebral, J.R.: CFD for evaluation and treatment planning of aneurysms: review of proposed clinical uses and their challenges. Ann. Biomed. Eng. **43**(1), 122–138 (2015)
9. Ferdian, E., Dubowitz, D.J., Mauger, C.A., Wang, A., Young, A.A.: WSSNet: aortic wall shear stress estimation using deep learning on 4D flow MRI. Front. Cardiovasc. Med. **8** (2022)
10. Ferez, X.M., et al.: Deep learning framework for real-time estimation of in-silico thrombotic risk indices in the left atrial appendage. Front. Physiol. **12** (2021)
11. Fey, M., Lenssen, J.E.: Fast graph representation learning with PyTorch geometric. In: ICLR Workshop on Representation Learning on Graphs and Manifolds (2019)
12. Geiger, M., Smidt, T.: e3nn: Euclidean neural networks (2022)
13. Gharleghi, R., Sowmya, A., Beier, S.: Transient wall shear stress estimation in coronary bifurcations using convolutional neural networks. Comput. Methods Programs Biomed. **225**(21) (2022)

14. Itu, L.M., et al.: A machine learning approach for computation of fractional flow reserve from coronary computed tomography. J. Appl. Physiol. **121**(1), 42–52 (2016)
15. Li, G., et al.: Prediction of cerebral aneurysm hemodynamics with porous-medium models of flow-diverting stents via deep learning. Front. Physiol. **12** (2021)
16. Li, G., et al.: Prediction of 3D cardiovascular hemodynamics before and after coronary artery bypass surgery via deep learning. Commun. Biol. **4**(1) (2021)
17. Liang, L., Liu, M., Martin, C., Sun, W.: A deep learning approach to estimate stress distribution: a fast and accurate surrogate of finite-element analysis. J. R. Soc. Interface **15** (2018)
18. Liang, L., Mao, W., Sun, W.: A feasibility study of deep learning for predicting hemodynamics of human thoracic aorta. J. Biomech. **99** (2020)
19. Medrano-Gracia, P., et al.: A computational atlas of normal coronary artery anatomy. EuroIntervention **12**, 845–854 (2016)
20. Qi, C., Su, H., Mo, K., Guibas, L.J.: PointNet: deep learning on point sets for 3D classification and segmentation. In: IEEE/CVF Conference on Computer Vision and Pattern Recognition (2017)
21. Qi, C., Yi, L., Su, H., Guibas, L.J.: PointNet++: deep hierarchical feature learning on point sets in a metric space. In: Advances in Neural Information Processing Systems, vol. 31 (2017)
22. Raissi, M., Yazdani, A., Karniadakis, G.E.: Hidden fluid mechanics: learning velocity and pressure fields from flow visualizations. Science **367**, 1026–1030 (2020)
23. Shukla, K., Xu, M., Trask, N., Karniadakis, G.E.: Scalable algorithms for physics-informed neural and graph networks. Data-Centric Eng. **3** (2022)
24. Su, B., Zhang, J.M., Zou, H., Ghista, D., Le, T.T., Chin, C.: Generating wall shear stress for coronary artery in real-time using neural networks: feasibility and initial results based on idealized models. Comput. Biol. Med. **126** (2020)
25. Suk, J., de Haan, P., Lippe, P., Brune, C., Wolterink, J.M.: Mesh convolutional neural networks for wall shear stress estimation in 3D artery models. In: MICCAI Workshop on Statistical Atlases and Computational Models of the Heart (2022)

Influence of Anisotropy on Fluid-Structure Interaction Simulations of Image-Based and Generic Mitral Valves

Nariman Khaledian[1(✉)], Pierre-Frédéric Villard[1,2], Peter E. Hammer[3], Douglas P. Perrin[3], and Marie-Odile Berger[1]

[1] Université de Lorraine, CNRS, Inria, LORIA, Nancy, France
nariman.khaledian@inria.fr
[2] Harvard School of Engineering and Applied Sciences, Cambridge, MA, USA
[3] Harvard Medical School, Boston, MA, USA

Abstract. The dynamic behavior of the mitral valve (MV) is highly influenced by the material model used to describe the leaflet motion. Due to the presence of collagen fibers, MV leaflets show an anisotropic behavior. The aim of this study is to investigate the influence of anisotropy on the fluid-structure interaction (FSI) simulation of the MV dynamic closure. The FSI simulation of the MV is performed using an immersed boundary method. Two constitutive models, Holzapfel-Gasser-Ogden for the anisotropic, and third-order Ogden for the isotropic are used. For the anisotropic model, two fiber directions, one that is parallel to the annulus surface and another that follows an arc on the leaflets are considered. In order to take into account the effects of both the chordae structure and leaflet geometry, generic and image-based MV are studied. The quality of the closure is evaluated based on measuring the bulging area, contact map, and flow rate. In both generic and image-based, a significant difference is observed between the anisotropic and isotropic cases. Additionally, the chordae forces during the closure are compared with ex-vivo data of the literature and show good similarities with these results.

Keywords: Mitral valve · Fluid-structure interaction · anisotropy

1 Introduction

The MV ensures the one-way flow of oxygenated blood from the left atrium to the left ventricle. It consists of leaflets that are held in a closed position by chordae. Various pathologies reduce its efficiency of remaining sealed properly and surgery may be necessary to repair the valve. Unfortunately, the outcome of the surgery depends on the surgeon's experience as well as on the patient's data. Computer-based simulations can help to have a predictive treatment.

In [13], an FSI method has been studied to simulate the MV behavior by focusing on perfect closure with no orifice hole. A generic MV geometry and

O. Bernard et al. (Eds.): FIMH 2023, LNCS 13958, pp. 455–464, 2023.
https://doi.org/10.1007/978-3-031-35302-4_47

isotropic hyperelastic material model were used. It was shown that it is possible to have a reliable simulation that can predict if a valve is pathological in a reasonable time. However, the chordae structure, the leaflet geometry, and the microstructure of the leaflet tissue are unique for each patient. We propose here to go toward clinical use and then replicate the accurate behavior of the MV. Two steps are necessary: extracting the image-based patient's valve anatomy and simulating the valve closure with biomechanical parameters based on experimental measurements and more realistic boundary conditions.

In this paper, we focus on those two aspects. The image-based geometry is extracted from the segmentation of a medical image and it leads to various challenges compared to generic geometry. The leaflets have non-regular contours with some noise. Many folds can occur during the closing process. The more realistic mechanical behavior of the valve includes replicating the tissue microstructure that is known to be anisotropic. We study the influence of the anisotropy both on a generic and an image-based valve as well as its influence on the valve shape.

2 State of the Art

Various experimental studies like [14] measured the stress-strain response of the MV leaflet. It is observed that the leaflet tissue shows anisotropic behavior and that anisotropy rates depend on the region where the specimen is taken for the stress-strain test. This can be explained by considering leaflet tissue as a composite of fiber families running through an isotropic base material and assuming that fiber orientation dictates the anisotropy rate. The fiber map can be extracted by using small angle light scattering (SALS) to observe the microstructure of the leaflet [23]. The fiber map is unique for each specimen and obtaining it is time-consuming work and needs resources and expertise in conducting experimental studies. To have a practical simulation, that can be used for any patient-specific case, simplification in modeling different aspects of the MV is inevitable. In the literature, fibers are modeled by implementing the fiber map extracted from observing the micro-structure of the leaflets as described in [2,16,23,25], or simply laying the fibers to be parallel to the annulus as described in [12,27,29]. One attempt to come up with a global fiber map is done by Einstein et al. [7], they used the SALS data reported in [5]. This fiber map is used in [6,21,26] and it follows two rules, in the middle of the leaflet being parallel to the annulus, and close to the commissures being perpendicular to the free edge. The influence of fiber direction in the simulation results is described in [12].

In the literature, various constitutive models are used to model the anisotropic behavior of the MV leaflets. These models can be categorized as coupled [17] and decoupled [10]. May-Newman et al. [17] proposed an anisotropic constitutive model specific for MV leaflets in which the strain energy function consists of a single term describing the material behavior in all directions. In decoupled models, such as Holzapfel-Gasser-Ogden (HGO) constitutive model [10], the strain energy function consists of one isotropic and one anisotropic term which in combination describes the desired anisotropic behavior. In [23] performance of both coupled and decoupled anisotropic models is compared to in vivo measurement on ovine data with shell structure simulations.

Extracting the coefficients of these models can be challenging. Extracting the 3D coefficients of a generalized Fung model is impossible since it requires shear data in all three main directions, whereas only planar biaxial tests are possible. In [1], an inverse finite element study is used to optimize the constitutive model coefficients based on in vitro experimental study. In the case of the HGO model, the strain energy function consists of one linear and one exponential term with four different coefficients. This makes it challenging to fit experimental data and the success of the optimization relies on the initial values of the coefficients.

Depending on the purpose of the study, the geometry used in the literature is generic valve [12], parameterized valve [21], or image-based. In that case, the objective is to work with realistic geometry and to address patient-specific data or pathologies. Since the complete chordae network is not visible under clinical imaging modalities, several studies use generic evenly distributed chordae tendineae [9], adapt existing data to a specific patient [3], or use ex-vivo setup to obtain clean and high-resolution subject-specific MV imaging [28]. Often, the annulus is approximated with an elliptic shape, or cubic Hermitian splines [23] though its actual shape is much more complex [19]. Only limited number of papers address MV simulation with image-based leaflet and chordae geometry as [8] who investigate how chordae structure impacts the MV dynamics and [24] in the context of prolapsus study. Globally, the aim of these studies is to perform comparisons between different configurations but the effectiveness of closure is not the central topic of these studies though it is a fundamental clinical topic.

Our contributions are twofold: Define an FSI model able to handle image-based data and experimental-based tissue properties, and study the added value of this improved model on porcine data. More specifically:

- We first explicitly describe in Sect. 3 the changes in the FSI model which are required to handle image-based valves obtained from image segmentation and specifically: complex and irregular leaflet surfaces, non-elliptic annulus shapes, trees of chordae.
- We study the impact of using isotropic versus anisotropic material on the realism of the simulation. Our study is based on the HGO model which is one of the most widely used anisotropic hyperelastic models. We then study how fiber direction influences the simulation.
- The efficiency of the models is measured with several figures of merits that allow us to evaluate the valve closure: bulging ratio, map of contact, forces applied on the chordae, and flow rate.

3 Method

The method for simulating MV closure is based on the FSI immersed boundary method described in [13] and demonstrated for a generic valve which is implemented in Abaqus explicit solver. We describe in the section the changes which were necessary to adapt this model to image-based data.

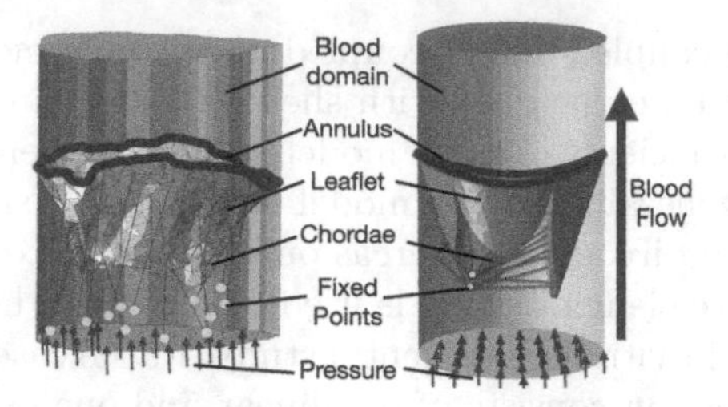

Fig. 1. Case set-up showing the image-based (left) and generic (right) MV mounted in a generalized and elliptic cylinder.

Geometry: Two kinds of MV geometry have been studied. A generic one like in [13] and an image-based one extracted from a 3D CT scan of an explanted porcine heart. In the generic geometry case, the chordae structure is optimal such as the valve always closing. In the case of the geometry extracted from medical data, the leaflets and the chordae have been manually segmented at the open state of the valve under physiologically normal configuration without any subsequent distortion to the segmentation. This porcine valve closes naturally ex-vivo without orifice holes. The chordae structure is represented by a linear elastic two-node beam element with an articulated mechanism to avoid compression load in the chordae. MV is immersed in a tube filled with blood representing the fluid domain. The tube in the generic valve is relatively simple with an elliptic base, but the image-based valve has a more complex annulus shape. The tube base is the shape of the annulus ring to prevent blood pass over the annulus. The tube direction is normal to the annulus plane and is determined thanks to a principal component analysis on the vertices of the annulus ring (see Fig. 1). The numerical domain in the image-based valve has 28 k elements for the leaflets and 34 k elements for the blood. In the case of the generic valve, the leaflets are represented with 40 k elements, and the blood consists of 65 k elements.

Numerical Domain: The tube is represented by a C3D8 element, a hexahedron with one node in each corner, with reduced integration. For the more complex geometry of the image-based valve, the C3D4 element, a tetrahedron with one node in each corner, is used which has better convergence qualities compared to the C3D10 element, a tetrahedron with one node in each corner and an extra node in the middle of each edge, previously used for generic valve with less complex geometry in [13]. The quadratic nature of C3D10 is not suitable when element nodes on complex surfaces are included in a contact model. Therefore, C3D4 is relatively less prone to distortion in such situations and prevents divergence of the simulation caused by element distortion.

In image-based MV geometry, the angle between the leaflet facets is relatively sharper compared to the generic valve. This makes contact behavior more complex by assuming that neighboring facets are constantly going to be engaged in contact. The simulation with the image-based valve diverges when both sides of the leaflet with 1mm thickness are involved in the contact model. Therefore, only the inner surface of the leaflets is included in the contact model. The self-contact between the leaflet surfaces is modeled by using the balanced master-slave technique. The normal forces are computed with a penalty-based formulation, and frictional contact is calculated by Coulomb's friction model [13].

Constitutive Model: To mimic the anisotropic behavior of the MV leaflets, we used the HGO constitutive model [10] which is already used in the literature for modeling the biological tissues. The HGO strain energy function for the incompressible material is as follows:

$$\psi = C_{10}(I_1 - 3) + K_1/2K_2 \sum_{\alpha=1}^{N} [exp(K_2 < E_\alpha >^2) - 1] \qquad (1)$$

In this model, the anisotropic behavior is introduced by families of fibers running through a uniform medium. The C_{10} coefficient in the linear part of the equation defines the characteristics of the uniform medium. K_1 and K_2 define the behavior of tension-only fibers. One family of fibers, $N = 1$, is defined in the simulation. $E_\alpha = \kappa(I_1 - 3) + (1 - 3\kappa)(I_{4(\alpha\alpha)} - 1)$ which I_1 is the first deviatoric strain invariant, $I_{4(\alpha\alpha)}$ is the pseudo-invariant related to the fibers, and κ is the dispersion factor between zero, parallel fiber families, to 1/3, randomly distributed fibers.

The direction of fibers is an important factor influencing the mechanical response of the MV. Experimental studies, for example in [5], confirm that in the central region of the leaflets, fibers are parallel to the annulus. The fiber direction in other regions of the leaflet is specific for each valve, but the general trend is that in the anterior and posterior leaflets, fibers follow an arc from one side of the leaflet to another [18,22], joining one group of chordae endpoints to another. To evaluate the performance of this fiber direction, we consider in this study two fiber directions. One parallel to the annulus surface, and one following an arc by using diffusion to determine the fiber direction in each element (Fig. 2).

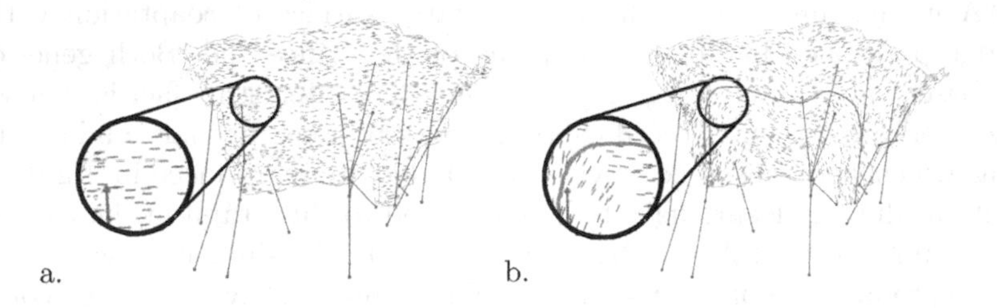

Fig. 2. Fiber map on the anterior leaflet of the image-based valve: (a) with parallel-based fibers (b) with fibers following a curved line with the diffusion process.

Calibration of Material Parameters: HGO coefficients (C_{10}, K_1, K_2) for porcine MV are not reported in the literature. Indeed, Such parameters are available for ovine data in [23] or for porcine data but only for the aortic valve [15]. Therefore, we decided to use stress-strain experimental data from [11] to extract the proper coefficients for our simulation. The optimization process is done with the MCalibration software[1] which is connected to the Abaqus solver to execute the inverse finite element analysis calculations. Depending on the chosen optimization technique, the results turn out to be unstable and sensitive to the initial estimate. We thus use genetic algorithms for minimization and consider several plausible initial values and keep the one that provides the smallest residual.

The parameters computed with this method are the following: $C_{10} = 2272$, $K_1 = 4000, K_2 = 150$, $\kappa = 0.25$. To obtain the isotropic material parameters comparable with the anisotropic one, first, the dispersion coefficient, κ, is set to the maximum, 1/3, which makes the constitutive model act as isotropic. The stress-strain response is then extracted and used as material behavior to obtain the third-order Ogden hyperelastic constitutive model, which results in the following final coefficients for the isotropic mode: $\mu_1 = -481836.3$, $\alpha_1 = -19.4$, $\mu_2 = 484166.6$, $\alpha_2 = -19.3$, $\mu_3 = 742.8$, $\alpha_3 = -25$.

4 Results and Discussion

We compare in this section the simulations obtained with and without considering anisotropy in the case of a generic and a image-based valve, with the same HGO parameters. Results are provided in Fig. 3 top row for the generic valve and in Fig. 3 bottom row, 4, and 5 for the image-based valve. Two fibers models are considered in the anisotropic case (Fig. 2): parallel to the annulus (denoted as anisotropic with parallel-based fibers) and fibers guided by diffusion following one arc outlined by an expert (denoted as anisotropic with arc-based fibers). Both for generic and image-based cases, the use of the isotropic model induces a high bulging of the leaflet (see Fig. 3 top and bottom rows where parts above the annular plane are drawn in red), whereas anisotropy shows decreased bulging. This fact is in good agreement with clinical knowledge which states that "A normal mitral valve has [...] a large surface of coaptation with the free edge positioned low below the plane of the orifice" [4]. Both generic and image-based valves are closing properly. The other interesting fact is that we are able to replicate the imbalanced chordae forces between primary and secondary chordae which were measured ex vivo in [20] and are reproduced in Fig. 4.a. Our simulations during closure have to be compared to this graph in the time interval $[0.1 - 0.2]$ seconds. As can be seen in Fig. 4.b, the simulation with isotropy fails to reproduce the primary and secondary chords behavior. On the contrary, the simulations with anisotropy are closer to the reference and isotropic with arc-based fibers faithfully reproduce the order of magnitude of forces.

The map of contact forces obtained with the anisotropic model is presented in Fig. 5 for isotropic (a), anisotropic with parallel-based fibers (b), and anisotropic with arc-based fibers (c). The cumulative map, i.e. $P(||pressure|| > t)$ is also provided (Fig. 5.d). The map is very irregular and not realistic in the isotropic case because the high bulging produces folds with high contacts near the annulus. With fibers parallel to the annulus, the area of contacts is roughly parallel to

[1] https://polymerfem.com/mcalibration/.

the annulus and shows areas with higher contact forces. When curve-based fibers are used, the amplitude of contacts is higher than in the two other cases as can be seen on the cumulative map. Moreover, in the case of curved-based fibers, the coaptation area is shifted slightly towards the free edge of the leaflet, which more closely resembles the physiological behavior of the MV in its closed state. The flow rates are shown for each case in Fig. 4.e and show that the MV closure is effective for the three cases. These experiments confirm that arc-based fibers allow a more realistic simulation than the other models. Indeed the valve we model is structurally normal and such valves do not exhibit significant bulging above the annulus. The model anisotropic with arc-based fibers best reproduces

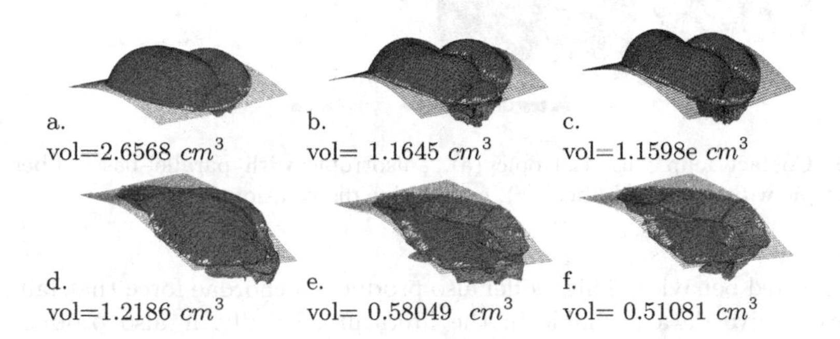

Fig. 3. Top row: bulging on the generic valve with (a) isotropic, (b)anisotropic with parallel-based fibers, (c) anisotropic with arc-based fibers. Bottom row: bulging on the image-based valve with (d) isotropic, (e) anisotropic with parallel-based fibers, (f) anisotropic with arc-based fibers

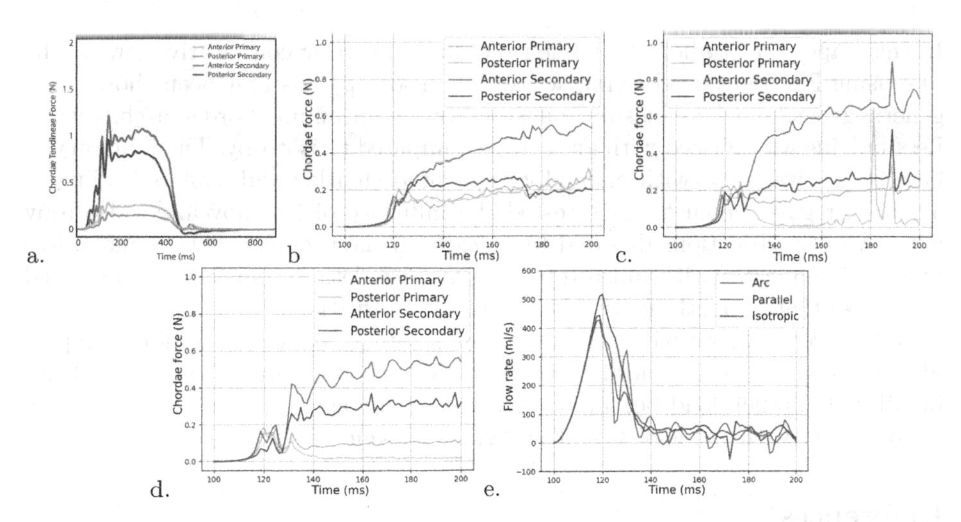

Fig. 4. Chordae force distribution on ex-vivo data from [20] (a), simulation with isotropic (b), anisotropic with parallel-based fibers (c), anisotropic with arc-based fibers (d), Flow rates through time (e).

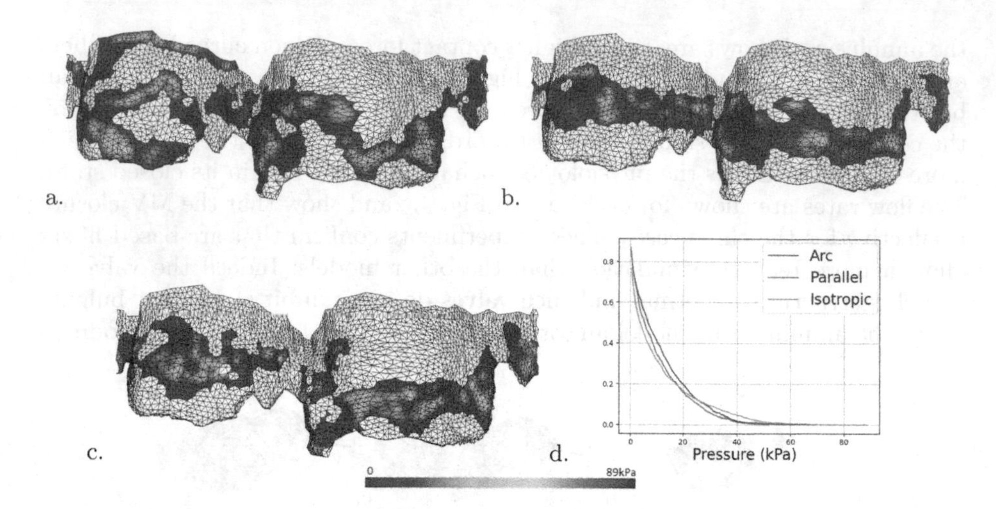

Fig. 5. Contact map with isotropic (a), anisotropic with parallel-based fibers (b), anisotropic with arc-based fibers (c), cumulative distribution (d)

this expected behavior. This model also produces a chordae force that faithfully mimics ex-vivo data available in the literature. Finally, it also produces the strongest contact force which shows the proper functioning of the MV during the closure.

5 Conclusion

In this paper, we studied the influence of an anisotropy constitutive law on the MV using FSI analysis. The impact of the anisotropy has first been shown on a generic geometry where closure is ensured by an optimum chordae architecture. Less bulging was noticed with anisotropy compared to isotropy. Then, an image-based valve was used with manually segmented chordae and leaflets leading to a more irregular geometry. We tested the influence of isotropy and anisotropy with two fiber directions dictated by rules from the literature. Once again, less bulging is observed with anisotropy and a more detailed analysis also showed more cues that correctly match with clinical observations.

In future work, we will focus on a new strategy to extract a more complex fiber map based on the chordae structure. The chordae are known to be dense bundles of parallel, load-bearing fibers, and we can assume that these fibers are in continuity with the leaflets in the regions at which they insert.

References

1. Abbasi, M., Barakat, M.S., Vahidkhah, K., Azadani, A.N.: Characterization of three-dimensional anisotropic heart valve tissue mechanical properties using inverse finite element analysis. J. Mech. Behav. Biomed. Mater. **62**, 33–44 (2016)

2. Amini, R., et al.: On the in vivo deformation of the mitral valve anterior leaflet: effects of annular geometry and referential configuration. Ann. Biomed. Eng. **40**(7), 1455–1467 (2012)
3. Biffi, B., et al.: A workflow for patient-specific fluid-structure interaction analysis of the mitral valve: a proof of concept on a mitral regurgitation case. Med. Eng. Phys. **74**, 153–161 (2019)
4. Carpentier, A., Adams, D.H., Filsoufi, F.: Carpentier's Reconstructive Valve Surgery. Elsevier Health Sciences, Amsterdam (2010)
5. Cochran, R., Kunzelman, K., Chuong, C., Sacks, M., Eberhart, R.: Nondestructive analysis of mitral valve collagen fiber orientation. ASAIO Trans. **37**(3), M447-8 (1991)
6. Einstein, D.R., Kunzelman, K., Reinhall, P., Nicosia, M., Cochran, R.: Haemodynamic determinants of the mitral valve closure sound: a finite element study. Med. Biol. Eng. Comput. **42**(6), 832–846 (2004)
7. Einstein, D.R., Kunzelman, K.S., Reinhall, P.G., Nicosia, M.A., Cochran, R.P.: The relationship of normal and abnormal microstructural proliferation to the mitral valve closure sound. J. Biomech. Eng. **127**(1), 134–147 (2005)
8. Feng, L., et al.: On the chordae structure and dynamic behaviour of the mitral valve. IMA J. Appl. Math. **83**(6), 1066–1091 (2018)
9. Gao, H., Feng, L., Qi, N., Berry, C., Griffith, B.E., Luo, X.: A coupled mitral valve-left ventricle model with fluid-structure interaction. Med. Eng. Phys. **47**, 128–136 (2017)
10. Gasser, T.C., Ogden, R.W., Holzapfel, G.A.: Hyperelastic modelling of arterial layers with distributed collagen fibre orientations. J. R. Soc. Interface **3**(6), 15–35 (2006)
11. Jett, S., et al.: An investigation of the anisotropic mechanical properties and anatomical structure of porcine atrioventricular heart valves. J. Mech. Behav. Biomed. Mater. **87**, 155–171 (2018)
12. Kaiser, A.D., McQueen, D.M., Peskin, C.S.: Modeling the mitral valve. Int. J. Numer. Methods Biomed. Eng. **35**(11), e3240 (2019)
13. Khaledian, N., Villard, P.F., Berger, M.O.: Capturing contact in mitral valve dynamic closure with fluid-structure interaction simulation. Int. J. Comput. Assist. Radiol. Surg. **17**, 1–8 (2022)
14. Laurence, D., et al.: An investigation of regional variations in the biaxial mechanical properties and stress relaxation behaviors of porcine atrioventricular heart valve leaflets. J. Biomech. **83**, 16–27 (2019)
15. Laville, C., Pradille, C., Tillier, Y.: Mechanical characterization and identification of material parameters of porcine aortic valve leaflets. J. Mech. Behav. Biomed. Mater. **112**, 104036 (2020)
16. Lee, C.H., Amini, R., Gorman, R.C., Gorman, J.H., III., Sacks, M.S.: An inverse modeling approach for stress estimation in mitral valve anterior leaflet valvuloplasty for in-vivo valvular biomaterial assessment. J. Biomech. **47**(9), 2055–2063 (2014)
17. May-Newman, K., Yin, F.C.P.: A constitutive law for mitral valve tissue. J. Biomech. Eng. **120**(1), 38–47 (1998)
18. Noack, T., et al.: New concepts for mitral valve imaging. Ann. Cardiothorac. Surg. **2**(6), 787 (2013)
19. Panicheva, D., Villard, P.F., Hammer, P.E., Perrin, D., Berger, M.O.: Automatic extraction of the mitral valve chordae geometry for biomechanical simulation. Int. J. Comput. Assist. Radiol. Surg. **16**, 709–720 (2021)

20. Paulsen, M.J., et al.: Mitral chordae tendineae force profile characterization using a posterior ventricular anchoring neochordal repair model for mitral regurgitation in a three-dimensional-printed ex vivo left heart simulator. Eur. J. Cardiothorac. Surg. **57**(3), 535–544 (2020)
21. Prot, V., Haaverstad, R., Skallerud, B.: Finite element analysis of the mitral apparatus: annulus shape effect and chordal force distribution. Biomech. Model. Mechanobiol. **8**(1), 43–55 (2009)
22. Prot, V., Skallerud, B.: Nonlinear solid finite element analysis of mitral valves with heterogeneous leaflet layers. Comput. Mech. **43**(3), 353–368 (2009)
23. Rausch, M.K., Famaey, N., Shultz, T.O., Bothe, W., Miller, D.C., Kuhl, E.: Mechanics of the mitral valve. Biomech. Model. Mechanobiol. **12**(5), 1053–1071 (2013)
24. Razavi, S.E., Talebi, A.: Comparative modeling of the mitral valve in normal and prolapse conditions. Bioimpacts (2023)
25. Sadeghinia, M.J., Skallerud, B., Holzapfel, G.A., Prot, V.: Biomechanics of mitral valve leaflets: second harmonic generation microscopy, biaxial mechanical tests and tissue modeling. Acta biomaterialia **141**, 244–254 (2022)
26. Skallerud, B., Prot, V., Nordrum, I.: Modeling active muscle contraction in mitral valve leaflets during systole: a first approach. Biomech. Model. Mechanobiol. **10**(1), 11–26 (2011)
27. Stevanella, M., Votta, E., Redaelli, A.: Mitral valve finite element modeling: implications of tissues' nonlinear response and annular motion. J. Biomech. Eng. **131**(12), 121010 (2009)
28. Toma, M., et al.: High-resolution subject-specific mitral valve imaging and modeling: experimental and computational methods. Biomech. Model. Mechanobiol. **15**(6), 1619–1630 (2016). https://doi.org/10.1007/s10237-016-0786-1
29. Votta, E., Caiani, E., Veronesi, F., Soncini, M., Montevecchi, F.M., Redaelli, A.: Mitral valve finite-element modelling from ultrasound data: a pilot study for a new approach to understand mitral function and clinical scenarios. Philos. Trans. R. Soc. A Math. Phys. Eng. Sci. **366**(1879), 3411–3434 (2008)

Computational Modelling of the Cardiovascular System for the Non-invasive Diagnosis of Portal Hypertension

M. Inmaculada Villanueva[1(✉)], Patricia Garcia-Cañadilla[2,3], Oscar Camara[1], Angeles Garcia-Criado[4], Genis Camprecios[5,6], Valeria Perez-Campuzano[5,6], Virgina Hernandez-Gea[5,6], Fanny Turon[5,6], Anna Baiges[5,6], Angela Lopez Sainz[6,7], Juan Carlos García-Pagan[5,6], Bart Bijnens[6,8], and Gabriel Bernardino[1,2,6]

[1] Physense, DTIC, Universitat Pompeu Fabra, Barcelona, Spain
mariainmaculada.villanueva@upf.edu

[2] BCNatal - Barcelona Center for Maternal-Fetal and Neonatal Medicine (Hospital Sant Joan de Déu and Hospital Clínic), Barcelona, Spain

[3] Interdisciplinary Cardiovascular Research Group, Institut de Recerca Sant Joan de Déu, Barcelona, Spain

[4] Radiology, Hospital Clínic, Barcelona, Spain

[5] Barcelona Hepatic Hemodynamic Laboratory. LIVER UNIT. Hospital Clinic. IDIBAPS. CIBEREHD., Barcelona, Spain

[6] August Pi i Sunyer Biomedical Research Institute (IDIBAPS), Barcelona, Spain

[7] Cardiovascular Institute, Hospital Clínic, Barcelona, Spain

[8] ICREA - Catalan Institution for Research and Advanced Studies, Barcelona, Spain

Abstract. Cirrhosis is a prevalent chronic liver disease that causes scarring of the liver, leading to altered mechanics and impaired function. One of its most severe complications is portal hypertension, characterised by an increase in portal vein pressure, associated with abnormal blood flow dynamics. Portal hypertension is usually diagnosed by measuring the hepatic venous pressure gradient (HVPG) through invasive catheterisation. Computational models can help to understand the causes of observed phenomena and assess certain variables that are challenging to measure without invasive procedures. Therefore, the aim of this study was to use a 0D model of the cardiovascular system to study portal hypertension and its haemodynamic effects in the circulation. A sensitivity analysis was conducted to assess the impact of different variables on the model. In addition, the model was personalised based on hepatic Doppler waveforms from two patients, one with elevated HVPG (and cirrhosis) and the other with normal HVPG (and hepatitis). The model-based haemodynamic parameters were compared to the invasive haemodynamic measurements. This study provides insight into how cirrhosis alters haemodynamics and demonstrates the potential of employing computational models of the cardiovascular system to understand haemodynamic changes in individual patients.

O. Bernard et al. (Eds.): FIMH 2023, LNCS 13958, pp. 465–474, 2023.
https://doi.org/10.1007/978-3-031-35302-4_48

Keywords: 0D cardiovascular model · Adult cardiology · Cirrhosis · Doppler ultrasound · Optimisation · Portal hypertension

1 Introduction

Cirrhosis is a chronic and progressive liver disease that involves the scarring of liver tissue, leading to decreased liver function and altered hepatic haemodynamics [1]. It is a major cause of morbidity and mortality worldwide, and the 11th most common cause of death [3]. Its comorbidities include cirrhotic cardiac disease, where the hepatic blood pressure is increased and overloads the right chambers [18].

One of the most severe complications of cirrhosis is portal hypertension [5]. This condition occurs when the blood pressure in the portal vein (PV), which carries blood from the intestines and spleen to the liver, increases. The elevation in pressure produces haemodynamic alterations [13], such as decreased blood pressure, vasodilation of blood vessels, increased cardiac output (CO) and a hyperdynamic circulation. Despite the high CO, there are blunted systolic and diastolic contractile responses to stress, in conjunction with ventricular hypertrophy or chamber dilatation and electrophysiological abnormalities, all contributing to the so called cirrhotic cardiomyopathy. Moreover, the blood flow into the liver is increased, collateral vessels are formed to shunt portal blood to the systemic circulation, and abdominal blood vessels dilate, which can lead to the development of varices and ascites in the esophagus and stomach [7].

The gold standard for diagnosing portal hypertension is measuring the hepatic venous pressure gradient (HVPG) [9,20]. HVPG is defined as the difference between PV and hepatic vein (HV) pressures, and serves as an indirect measure of the portal pressure. It is measured by inserting a catheter into the HV to determine free and wedge hepatic venous pressure, and subtracting both values. However, this procedure is invasive and only available in specialised centers [19]. Haemodynamic changes in the PV have been reported, thus Doppler ultrasound can also aid in diagnosing portal hypertension by evaluating changes in the velocity profile of the liver vasculature [11], although it is typically used in conjunction with other methods.

Computational models have been widely employed in the assessment of the cardiovascular system, the mechanisms of haemodynamic changes and the interactions between its components. To analyse global blood flow and pressure distributions under various physiological conditions, and the relations between different elements of the cardiovascular system, 0D models are frequently used [16]. This approach offers a comprehensive and consistent description, allowing for the assessment of certain variables, such as pressure, that are challenging to measure non-invasively. They also allow for the personalisation to a patient to improve understanding of their diagnosis and treatment [6,15]. Previously published papers modelled the haemodynamic changes observed during partial liver ablation in pigs [4] and simulated the procedure of HVPG measurement to quantitatively investigate its sensitivity and its relation to portal pressure [17]. However, no research has been conducted with personalised models to understand portal hypertension, and its contributors, in specific patients without invasive pressure measurements.

The purpose of this study is to use a 0D model of the cardiovascular system to study portal hypertension in patients with cirrhosis. The model can be personalised based on non-invasive hepatic Doppler waveforms and simulation results compared to invasive measurements. This model could also help understand the mechanisms in which cirrhosis alters haemodynamics in the whole circulation. We demonstrate the potential of this approach by personalising the model to two patients, one with elevated HVPG and the other with normal HVPG.

2 Methodology

2.1 Data Acquisition

Blood velocities in HV, PV and hepatic artery (HA) were acquired via Doppler ultrasound (Canon Aplio i800) for one patient with hepatitis, with normal HVPG, and another patient with cirrhosis, with increased HVPG. The heart rate (HR) of the hepatitis patient varied greatly between different measurements, and thus it was standardised as the average HR for all waveforms. The diameter of HV, PV and HA were also measured on the B-mode images. To measure the HVPG, the difference between free and wedge hepatic venous pressure was determined by introducing a catheter into HV, following international recommendations.

2.2 Lumped Model Description

For the purpose of examining gross changes in haemodynamics, we implemented a closed 0D model of the entire adult cardiovascular system, as shown in Fig. 1. The scheme was based on Audebert et al. [4] with some modifications: (1) for simplicity and as we do not consider any difference between lobes, the liver was treated as a single component; (2) HV and aortic components were included to more accurately reflect the pressure gradient between the liver and the right atrium, and between HA and the aortic valve, respectively; and (3) given the impact of atrial contraction and relaxation to the HV waveform, we used the single fiber ventricular and atrial model described by Arts et al. [2]. The aortic sinuses and systemic arteries were simulated as arterial segments using a combination of hydraulic resistance (R), blood inertia (L) and vessel compliance (C), as implemented in Korakianitis et al. [12]. The lungs, digestive and other organs were represented by three-element vascular bed compartments (RCR); the liver and HV by two-element vascular bed compartments (RC) and HA and PV by a resistance (R_{ha} and R_{pv}). This model was created using OpenCOR and libcellml [8] and then exported to Python. All parameters were manually adjusted to obtain pressures and blood velocities within a physiological range.

The blood pressures and velocities reported in this work for the HV, PV and HA were measured at the entrance of their respective components, marked

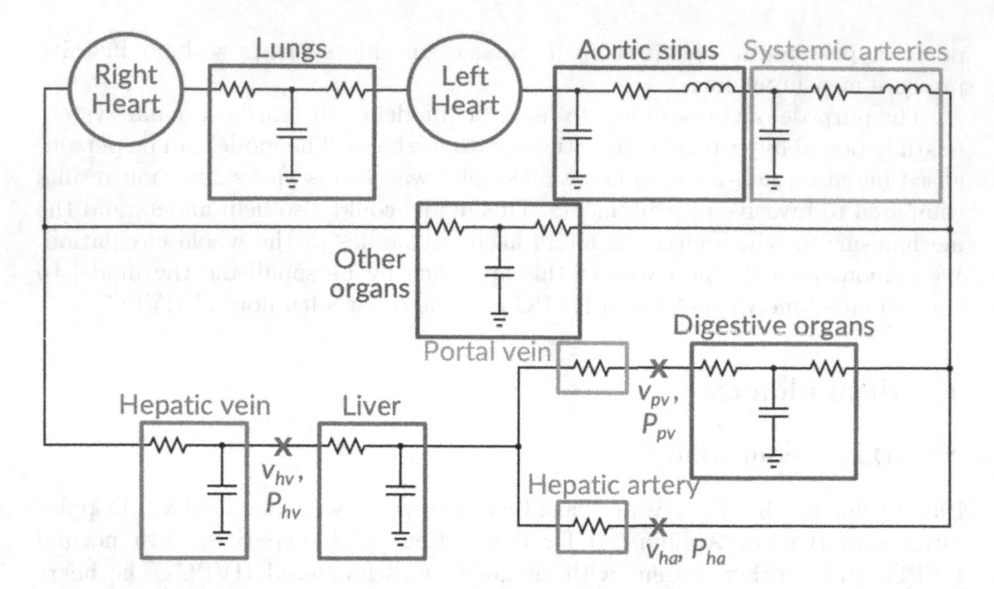

Fig. 1. 0D computational model of the cardiovascular system. Red crosses indicate where the hepatic vein (HV), portal vein (PV) and hepatic artery (HA) blood velocities (v) and pressures (P) are measured. (Color figure online)

in Fig. 1. The blood flow was converted to velocity by assuming the vessels were perfectly circular and that the blood flow in the arteries followed a fully developed laminar pattern, with the maximum velocity defined as two times the average velocity. We also accounted for the fact that humans present three HV and two HA, while our simplified model combined them in a single vein and artery, respectively.

2.3 Sensitivity Analysis

We conducted a sensitivity analysis to evaluate the impact of specific model parameters on simulated velocities in HV, PV, and HA, aortic pressure, CO, and HVPG, and determine the order of magnitude for the optimisation process. Knowing that cirrhosis causes changes in the liver function and vasculature, and that total blood volume affects the CO, a set of 6 model variables were selected for the analysis and increased by a 10% of their original values: (1) initial blood volume in the heart (V_0), (2) R_l, (3) C_l, (4) R_{ha}, (5) R_{pv}, and (6) resistance of other organs (R_{oo}).

2.4 Optimisation

Finally, the model was personalised to one patient with normal HVPG and another with increased HVPG. Based on the previous sensitivity analysis, we estimated those model parameters allowing them to vary only within a physiological range (Table 1).

Table 1. Physiological range in which the different model parameters can vary during the optimisation process and initial values of the model.

	V_0(L)	R_l (mmHg· s/ml)	C_l (ml/mmHg)	R_{ha} (mmHg· s/ml)	R_{pv} (mmHg· s/ml)	R_{oo} (mmHg· s/ml)
Range	1.2–1.6	0.01–10	0.01–10	10–100	0.01–10	0.5–1.5
Initial values	1.25	0.06	7.89	40.00	0.14	1.36

The best model parameters were selected using the Covariance Matrix Adaptation Evolution Strategy (CMA-ES) optimisation method [10]. This technique is a stochastic and derivative-free numerical optimisation algorithm specifically designed for solving non-linear or non-convex continuous problems. The objective function for this study was defined as the weighted sum of the normalised root mean squared error (NRMSE) between model-based and real blood velocity waveforms in the HV, PV and HA, that were manually aligned. Additionally, the minimum and maximum aortic pressure values (80 mmHg and 120 mmHg, respectively), together with the CO, with values of 5 L/min and 5.5 L/min for the hepatitis and cirrhotic patients, respectively, since cirrhosis is expected to result in an increase in CO, were also included in the objective function. All these factors were normalised and assigned a weight of 1, except for the NRMSE of the PV velocity (weight 2), due to its importance in determining portal hypertension, and the minimum and maximum aortic pressure (weight 0.5), to ensure a total contribution of 1 for aortic pressure. The optimal values were selected once a tolerance of 10^{-6} was reached.

3 Results

3.1 Dataset Description

The collected Doppler velocity traces of the hepatitis and cirrhotic patients, measured at the HV, PV and HA for two cardiac cycles, are displayed in Fig. 2. We can observe the typical HV velocity waveform for a normal HVPG: the A wave, corresponding to retrograde flow into the HV due to the right atrium contraction; the S wave, due to atrial filling during ventricular systole; the V wave, a transitional inflection point caused by atrial overfilling; and the D wave, marking the opening of the tricuspid valve. In the cirrhotic patient, on the other hand, this pattern is transformed into an almost constant blood velocity. In addition, there is a decrease in PV velocity and an increase in HA velocity of the patient with cirrhosis with respect to the one with hepatitis.

Table 2 shows the diameter of the vessels and the HR measured for both individuals, used as inputs for the computational model and to convert blood flows to velocities, as well as the reported HVPG.

3.2 Sensitivity Analysis

The sensitivity analysis, whose results are summarised in Table 3, revealed that V_0 and R_{oo} had the greatest impact on the arterial pressure and CO, in addition

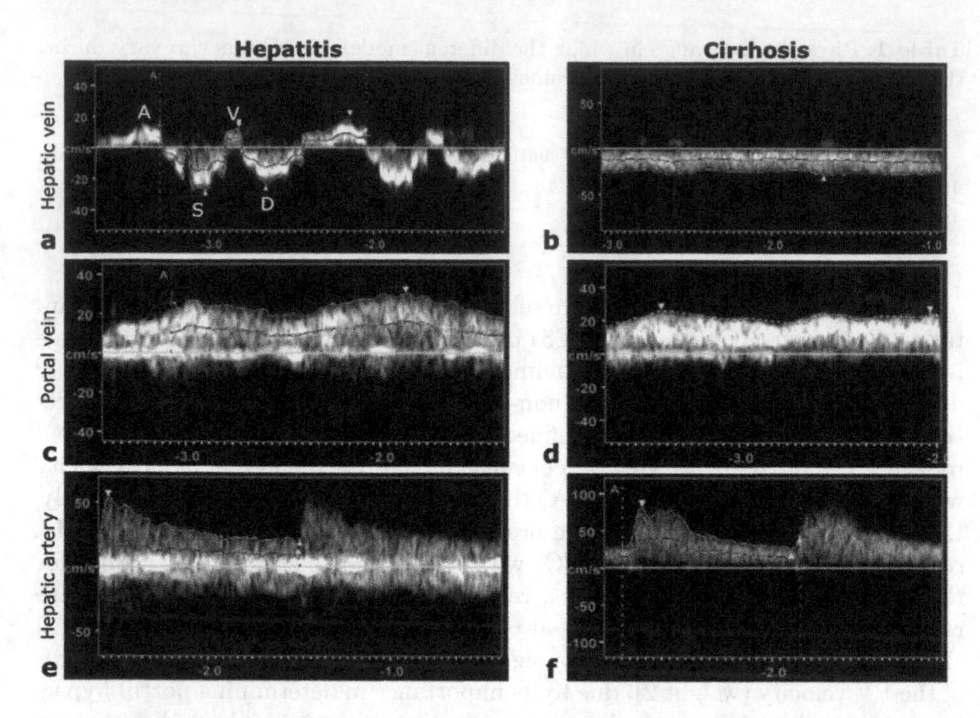

Fig. 2. Hepatic vein (a - b), portal vein (c - d) and hepatic artery (e - f) Doppler velocities measured for the hepatitis and cirrhotic patients. Typical hepatic vein velocity waves (A, S, V, D) are marked on the hepatitis patient's image.

to affecting all other outputs. R_l and C_l had both a slight effect on HVPG, with the first one being responsible for the flatenning of the HV velocity profile, as its maximum value was substantially lowered, while the second one altered its average. Finally, R_{pv} influenced the pressure in PV, but had a limited impact on velocity, and R_{ha} significantly conditioned the velocity of the HA, whereas it had a negligible effect on other outputs.

3.3 Optimisation Results

We personalised the simplified model to represent the patient-specific data. In Table 4, we present the optimal parameters for the patients with preserved and

Table 2. Quantitative measurements obtained from the patients with hepatitis and cirrhosis.

Patient	Diameter (mm)			HR (bpm)	HVPG (mmHg)
	Hepatic vein	Portal vein	Hepatic artery		
Hepatitis	10.2	13.0	2.6	54.5	5.0
Cirrhosis	6.5	10.3	2.7	75.0	26.0

HR: heart rate, HVPG: hepatic venous pressure gradient.

Table 3. Percentage of change in different outputs of the model when increasing the different model parameters by a 10% of their original values.

Variable	HV velocity		PV velocity		HA velocity		Aortic pressure		CO	HVPG
	min	max	min	max	min	max	min	max		
V_0	11.3	36.1	8.6	8.5	8.6	8.1	9.1	8.5	8.5	7.9
R_l	–2.0	–24.9	–0.2	–0.1	–0.2	–0.1	0.0	–0.1	–0.1	4.0
C_l	0.5	–0.2	–0.3	–0.2	–0.3	–0.2	–0.3	–0.2	–0.2	2.3
R_{pv}	–0.1	0.3	–0.2	–0.2	0.0	0.0	0.0	0.0	0.0	6.6
R_{ha}	–0.7	1.6	0.2	0.2	–8.8	-9.0	0.2	0.1	–0.1	0.3
R_{oo}	2.5	–19.1	6.6	4.5	7.3	3.6	6.8	3.4	–1.9	6.9

CO: cardiac output, HVPG: hepatic venous pressure gradient, HV: hepatic vein, PV: portal vein, HA: hepatic artery.

Table 4. Values of parameters obtained for the optimisation of the computational model to the patients with hepatitis and cirrhosis.

	V_0 (L)	R_l (mmHg· s/ml)	C_l (ml/mmHg)	R_{ha} (mmHg· s/ml)	R_{pv} (mmHg· s/ml)	R_{oo} (mmHg· s/ml)
Hepatitis	1.33	0.04	1.65	60.21	0.29	1.34
Cirrhosis	1.36	2.84	0.75	23.23	1.10	1.15

increased HVPG. In the former case, liver compliance and resistance in the HA and other organs were higher, whereas in the latter case, liver and PV resistances were higher, with both patients having similar initial blood volumes.

The root mean squared error between real measures and those obtained from the optimisation is reported in Table 5. As can be observed, for both cases the algorithm successfully fitted the CO and velocity in PV, but it had difficulty adjusting the aortic pressure. Moreover, the hepatitis patient presented the greatest error in HV velocity. Its simulated HVPG was very similar to the real value, indicated by the relatively low error, and still within the acceptable range of less than 5 mmHg. In contrast, the cirrhotic patient had the highest errors in the maximum aortic pressure and HA velocity. Its simulated HVPG indicated portal hypertension, but had a considerable difference compared to the actual measurement.

Regarding HV velocity, displayed in Fig. 3, the model was able to depict the changes in its profile. In the case of the hepatitis patient, the typical pattern with four waves was reproduced, although some peaks were misaligned and the V wave did not reach zero velocity. On the other hand, the cirrhotic patient presented less variation in the velocity.

4 Discussion

We implemented a simplified 0D model of the adult cardiovascular system to study the haemodynamic consequences of portal hypertension, and personalised it to two liver disease patients, one with normal portal pressure and the other

Table 5. Root mean squared error of the simulations obtained for the optimisation of the computational model to the patients with hepatitis and cirrhosis.

Patient	Velocity (cm/s)			Aortic pressure (mmHg)		CO (L/min)	HVPG (mmHg)
	HV	PV	HA	min	max		
Hepatitis	11.06	2.20	4.76	5.31	5.46	0.06	0.12
Cirrhosis	0.71	0.76	8.74	0.21	11.27	0.03	17.43

CO: cardiac output, HVPG: hepatic venous pressure gradient, HV: hepatic vein, PV: portal vein, HA: hepatic artery.

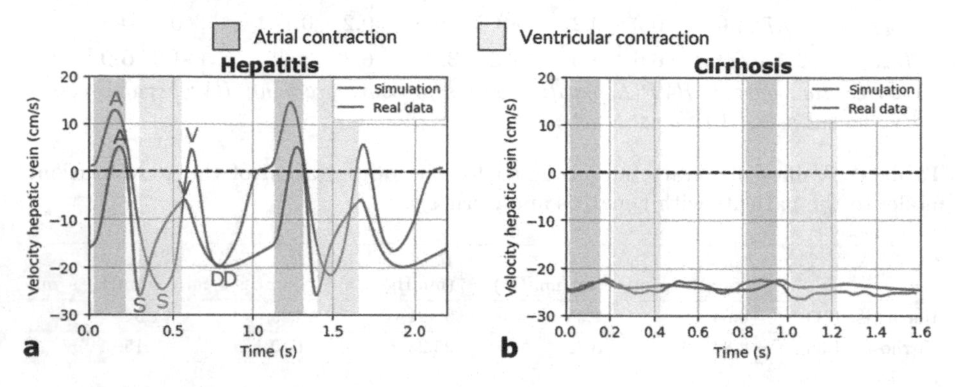

Fig. 3. Simulated blood velocity in the hepatic vein (red) compared to the real traces (blue) for the (a) hepatitis and (b) cirrhotic patients. Atrial and ventricular contraction are depicted in green and yellow, respectively, and the four typical velocity waves (A, S, V, D) are marked on the hepatitis patient's image. (Color figure online)

with portal hypertension. The sensitivity analysis provided insights into the impact of different variables on the outputs of the model. It showed that the parameter that mostly determines the velocity profile in the HV is R_l; that R_{pv} has a great influence on HVPG; and that V_0, which controls the total amount of blood in the cardiovascular system, and R_{oo}, which influences the pressure in the right atrium and regulates the blood flow to the digestive organs and HA, and therefore to the liver, affect all outputs. The results were in agreement with the literature [14].

Moreover, we were able to personalise our model to two patients, one with hepatitis and normal HVPG and the other with cirrhosis, using Doppler velocities in three vessels, their diameters and the HR of the individuals as input data. Our model replicated the HV velocity profile for the hepatitis individual, showcasing the typical 4-wave pattern, and showed its alteration with cirrhosis, which became nearly constant. Liver resistance increased and its capacitance decreased in the cirrhotic patient with respect to the hepatitis patient and to literature values [4], as expected due to the disease. Despite not using the liver vasculature pressures as input, our model was able to estimate a HVPG within a healthy range for the hepatitis patient and an elevated HVPG for the cirrhotic patient, showing the realism of the model and suggesting that model personalisation can help estimating portal pressures, thus avoiding invasive procedures.

While our model demonstrated the ability to capture HV waveform changes and increased HVPG caused by cirrhosis, it had some limitations. First, the HR of the hepatitis patient varied between measurements, requiring rescaling of velocities to its average that could introduce an error to our simulations, whereas a direct use of the velocities would have been more ideal. Secondly, optimisation of the cardiac component would have been desirable, particularly timing and accurate representation of blood velocity through the tricuspid valve, which in turn affects the velocity in HV, but the necessary data were not available. Finally, only data from two patients were provided, whereas having a larger dataset would enhance the validation of our model's capability to non-invasively diagnose portal hypertension.

5 Conclusions

This study provides insight into the mechanisms in which cirrhosis alters haemodynamics and shows the potential of using computational models of the cardiovascular system as a non-invasive tool for studying portal hypertension. In a future work, a comprehensive optimisation including the heart and using a larger dataset should be performed to further validate this approach.

Acknowledgements. This work has been performed under FI 2022 grant number 00237, awarded by the Agency for Management of University and Research Grants (AGAUR), Generalitat de Catalunya. It has been partially funded by European Union-NextGenerationEU, Ministry of Universities and Recovery, Transformation and Resilience Plan, through a call from Pompeu Fabra University (Barcelona).

References

1. Cirrhosis - NHS. https://www.nhs.uk/conditions/cirrhosis/
2. Arts, T., Delhaas, T., Bovendeerd, P., Verbeek, X., Prinzen, F.W.: Adaptation to mechanical load determines shape and properties of heart and circulation: the circadapt model. Am. J. Physiol. Heart Circ. Physiol. **288**, 1943–1954 (2005)
3. Asrani, S.K., Devarbhavi, H., Eaton, J., Kamath, P.S.: Burden of liver diseases in the world. J. Hepatol. **70**, 151–171 (2019)
4. Audebert, C., Bekheit, M., Bucur, P., Vibert, E., Vignon-Clementel, I.E.: Partial hepatectomy hemodynamics changes: Experimental data explained by closed-loop lumped modeling. J. Biomech. **50**, 202–208 (2017)
5. Bosch, J., Garcia-Pagan, J.E., Feu, F., Pizcueta, M.P.: Portal hypertension. In: Prieto, J., Rodés, J., Shafritz, D.A. (eds.) Hepatobiliary Diseases, pp. 429–463. Springer, Berlin, Heidelberg (1992). https://doi.org/10.1007/978-3-642-76802-6_11
6. Garcia-Canadilla, P., et al.: A computational model of the fetal circulation to quantify blood redistribution in intrauterine growth restriction. PLoS Comput. Biol. **10**, 9–11 (2014)

7. Garcia-Tsao, G.: Portal hypertension. Curr. Opin. Gastroenterol. **22**, 254–262 (2006)
8. Garny, A., Hunter, P.J.: OpenCOR: a modular and interoperable approach to computational biology. Front. Physiol. **6** (2015)
9. Groszmann, R.J., Suchat, W.: The hepatic venous pressure gradient: anything worth doing should be done right (2004)
10. Hansen, N., Akimoto, Y., Baudis, P.: Cma-es/pycma: r2.7.0, April 2019
11. Kok, T., Jagt, E.J.V.D., Haagsma, E.B., Bijleveld, C.M.A., Jansen, P.L.M., Boeve, W.J.: The value of doppler ultrasound in cirrhosis and portal hypertension. Scand. J. Gastroenterol. **34**, 82–88 (1999)
12. Korakianitis, T., Shi, Y.: A concentrated parameter model for the human cardiovascular system including heart valve dynamics and atrioventricular interaction. Med. Eng. Phys. **28**, 613–628 (2006)
13. Liu, H., Gaskari, A., Lee, S.S., Gaskari, S.A.: Cardiac and vascular changes in cirrhosis: pathogenic mechanisms. World J. Gastroenterol. **12**, 837–842 (2006)
14. Sala, L., Golse, N., Joosten, A., Vibert, E., Vignon-Clementel, I.: Sensitivity analysis of a mathematical model simulating the post-hepatectomy hemodynamics response. Ann. Biomed. Eng. **51**, 270–289 (2023)
15. Savoji, H., et al.: Cardiovascular disease models: a game changing paradigm in drug discovery and screening. Biomaterials **198**, 3–26 (2019)
16. Shi, Y., Lawford, P., Hose, R.: Review of zero-d and 1-d models of blood flow in the cardiovascular system. Biomed. Eng. Online **10**, 1–38 (2011)
17. Wang, T., Liang, F., Zhou, Z., Shi, L.: A computational model of the hepatic circulation applied to analyze the sensitivity of hepatic venous pressure gradient (HVPG) in liver cirrhosis. J. Biomech. **65**, 23–31 (2017)
18. Xanthopoulos, A., Starling, R.C., Kitai, T., Triposkiadis, F.: Heart failure and liver disease: cardiohepatic interactions. JACC Heart Fail. **7**, 87–97 (2019)
19. Xu, G., Li, F., Mao, Y.: Portal pressure monitoring-state-of-the-art and future perspective. Ann. Transl. Med. **7**, 583–583 (2019)
20. Zipprich, A., Winkler, M., Seufferlein, T., Dollinger, M.M.: Comparison of balloon vs. straight catheter for the measurement of portal hypertension. Aliment. Pharmacol. Ther. **32**, 1351–1356 (2010)

An Image-Based Computational Model of the Newborn Cardiovascular System with Term and Preterm Applications

Robyn W. May[1,2(✉)], Gonzalo D. Maso Talou[1], Finbar Argus[1], Thomas L. Gentles[3], Frank H. Bloomfield[2], and Soroush Safaei[1]

[1] Auckland Bioengineering Institute, 70 Symonds Street, Auckland, New Zealand
{r.may,g.masotalou,finbar.argus,soroush.safaei}@auckland.ac.nz
[2] Liggins Institute, 85 Park Road, Auckland, New Zealand
f.bloomfield@auckland.ac.nz
[3] Starship Hospital, Te Toka Tumai Auckland, 2 Park Road, Auckland, New Zealand
TomG@adhb.govt.nz

Abstract. Computational modelling is a well-established tool for understanding cardiovascular physiology, yet very few models of the neonatal circulation have been developed. Babies born small or early suffer from a range of problems, including cardiovascular instability, and monitoring their haemodynamics in acute care settings remains challenging. We aimed to develop a computational model of the neonatal circulation, customisable for the preterm circulation, where fetal shunts may still be open. Ultrasound imaging of the heart and arterial structure (vessel diameter) and function (Doppler flow) was collected for a term and a preterm baby. A 0D bond graph model of the newborn cardiovascular system was developed which is parameterised using patient-specific arterial measurements and included patent fetal shunts unique to the preterm circulation. This open-source cardiovascular model of the neonatal circulation is readily individualised using image-based anatomical measurements, realistic in its blood pressure and flow predictions, fully conservative for mass and energy and adaptable to the unique circulatory conditions of early life.

Keywords: Computational modelling · haemodynamics · neonatal circulation · preterm birth

1 Background

Computational modelling has well-established utility in the study of cardiovascular haemodynamics for medical research purposes and is increasingly being translated into clinical settings to improve the diagnosis and treatment of cardiovascular diseases [10]. Most cardiovascular models developed so far have been from the adult circulatory system [3,12,13]; however, there are a number of open questions in paediatric medicine that would benefit from the haemodynamic insights that these models can provide [8].

O. Bernard et al. (Eds.): FIMH 2023, LNCS 13958, pp. 475–484, 2023.
https://doi.org/10.1007/978-3-031-35302-4_49

Preterm birth (defined as birth before 37 completed weeks of gestation) and low birth weight deliveries (defined as weight less than 2500 g at birth) are a worldwide burden with up to 30 million babies born globally each year that require specialised care after birth [16]. Babies who are born preterm are at greater risk of a range of short-term morbidities, including cardiac complications such as patent ductus arteriosus (PDA) [14], as well as non-cardiac complications that may nonetheless impact the cardiovascular system. For example, hypotension (low blood pressure) is common in small and preterm babies, and remains controversial for several reasons: how hypotension should be diagnosed, how blood pressure should be monitored, the underlying cause of hypotension and its effect on prognosis, and lack of clear evidence regarding optimal treatment [2,5]. Haemodynamic monitoring of the complex circulatory interactions in neonates, particularly those born preterm, remains challenging [15]. Indirect parameters that are easily monitored, for example blood pressure, do not provide a clear picture of systemic blood flow and end-organ perfusion [15]. A computational model of neonatal circulation could be directly applicable in clinical practice to address this knowledge gap by providing detailed information on local blood pressures and flows. For example, models can be used to combine the information from ultrasound measurements with known physical and haemodynamic laws to make estimations of unobserved blood flows and pressures in regions of interest, e.g. cerebral perfusion. For this to be readily translatable into a clinical setting, it would need to be easily personalised and computationally efficient.

For biological computational models to be robust, they should satisfy the physics of the real-world application. Bond graph methodology is a modelling framework that correctly formulates energy and mass transfer in physical systems, thus ensuring conservation of energy and mass [13]. It is based on two key ideas. The first is based on the recognition that energy and power are the only quantities that are common across all physical systems. It follows that energy transmission can be separated into storage and dissipation components. The first key idea is the definition of *potential* (called "effort" in the bond-graph literature) as the common driving force behind the *flow*. For example, in the fluid mechanics domain used for cardiovascular haemodynamics, potential u is energy density or pressure ($J.m^{-3}$, equivalent to Pascals) and the flow v is volumetric flow rate ($m^3.s^{-1}$). The product of potential and flow is then power (in units of Joules per second). The second key concept is the definition of junctions. The system is examined to find points of (1) common potential (blood pressure) - defined as a *0-junction*, where mass balance is applied; and (2) common blood flow - defined as a *1-junction*, where potential balance is applied. From these bond graph formulations, the mathematical formulations of the computational model can be derived, ensuring that they are fully conservative. The extraordinary utility of this approach is that multiple physical domains (e.g. fluid dynamics for blood flow, electromagnetics for the electrical formulation of 0D modelling, chemical domain if gas and nutrient exchange is included) can all be included in the same mathematical formulation and remain fully conservative. Safaei et al. developed a library of bond graph elements that represent different configurations of blood vessel segments for cardiovascular computational modelling, depending on whether potential or flow are prescribed as boundary conditions at their inlets and outlets [13]. Argus et al. extended from this model to develop a software for automatic generation and calibration of arbitrary complexity circulatory system models [1].

Most neonatal models developed to date - with the notable exception of [11] - have been simplified and none have been readily customisable to the individual. In this paper, we present an open-source cardiovascular model of the neonatal circulation developed to address this gap, which is complex enough to be individualised using anatomical measurements obtained by ultrasound, while also being simple enough for real-time computational simulation. It will also be a novel application of bond graph methodology to lumped parameter neonatal cardiovascular modelling, thus ensuring a fully conservative formulation for mass and energy. Finally, the neonatal model is customisable for the preterm circulation where fetal shunts may still be patent, a less explored field and could be expanded to study both the earlier fetal and later infant cardiovascular physiology.

2 Methods

2.1 Data Collection

Cardiovascular anatomical and functional data was collected at Auckland City Hospital, Te Toka Tumai Auckland (formerly Auckland District Health Board). The inclusion criteria for the term neonate was birth at or after 37^{+0} weeks gestation and for the late preterm neonate was birth at or after 34^{+0} and before 37^{+0} weeks gestation. The exclusion criterion was any known medical conditions or cardiovascular abnormalities. Both babies underwent an ultrasound examination within 48 h of birth. All ultrasound scans were performed without sedation using a Philips Epiq ultrasound machine. All imaging sequences were recorded and stored digitally as moving images, with measurements taken retrospectively. From the recordings, arterial diameter measurements were taken in systole from longitudinal arterial images. Doppler measurements of the blood velocity profiles were measured and recorded for at least the duration of at least 3 heartbeats. Cardiac parameters were measured according to the recommendation of the American Society of Echocardiography for targeted neonatal echocardiography [9]. Blood pressure was measured on a lower limb using an automated sphygmomanometer with a neonatal cuff. While the baby was resting quietly or sleeping, heart rate and oxygen saturation was measured using an electronic pulse oximeter on either foot with heart rate and oxygen saturation recorded every 15 s over 150 s and averaged out. Ethical approval for this data collection was obtained from the Northern A Health and Disability Ethics Committee (20/NTA/187).

2.2 Computational Modelling

The neonatal model was adapted from an existing 0D adult bond graph model of systemic circulation [13] and is composed of the following bond graph segments:

- Thirteen arterial segments which are parameterised using measurable anatomical data (vessel length and radius): aortic root, ascending aorta, aortic arch (three segments), descending aorta, abdominal aorta, main pulmonary, brachiocephalic, subclavian and common carotid arteries (indicated in red in Fig. 1).
- Seven arterial segments and peripheral vascular beds for which anatomical data are unavailable: "arms" (including all arteries from the axillary artery onwards and the peripheral vascular bed in the upper limb), "legs" (including all arteries from the common iliac artery onwards and the peripheral vascular bed in the lower limb), "head" (including all arteries from the common carotid onwards and the cerebral vascular bed) and the splanchnic bed (indicated in blue in Fig. 1).
- Two venous compartments - SVC representing all venous drainage from the upper body and IVC from the lower body.
- Two pulmonary compartments representing arterial and venous pulmonary vasculature.
- A four-chamber time-varying elastance heart model.
- To characterise the preterm circulation, a ductus arteriosus segment connects the aortic arch to the main pulmonary artery and the foramen ovale is modelled as an orifice between the right and left atria (indicated in orange in Fig. 1).

Model Parameterisation and Personalisation. Modelling inputs and outputs are shown in Fig. 2. The following were used to parameterise the model:

- The main patient specific data used to parameterise the model are the arterial radii obtained from ultrasound imaging. These, as well as the patient specific heart rate, were used to parameterise the model directly (all patient specific modelling inputs are shown in Table 2).
- Parameters that could not be measured directly were estimated using a genetic algorithm (freely available from https://github.com/FinbarArgus/circulatory_autogen) [1]. For these models, estimated parameters were: total blood volume; terminal resistances for the head, arm, leg and abdomen modules; compliances of the pulmonary, SVC and IVC venous modules; Young's modulus of the arterial segments; and amplitude, baseline and timing values of the time-varying elastance function of the left ventricle. The patient-specific data used as inputs to the parameter estimation algorithm were: blood pressure measurements (Table 3) and a single digitised Doppler flow waveform from the abdominal aorta (Fig. 3). Further details of the estimation algorithm, including the objective function, can be found in [1].

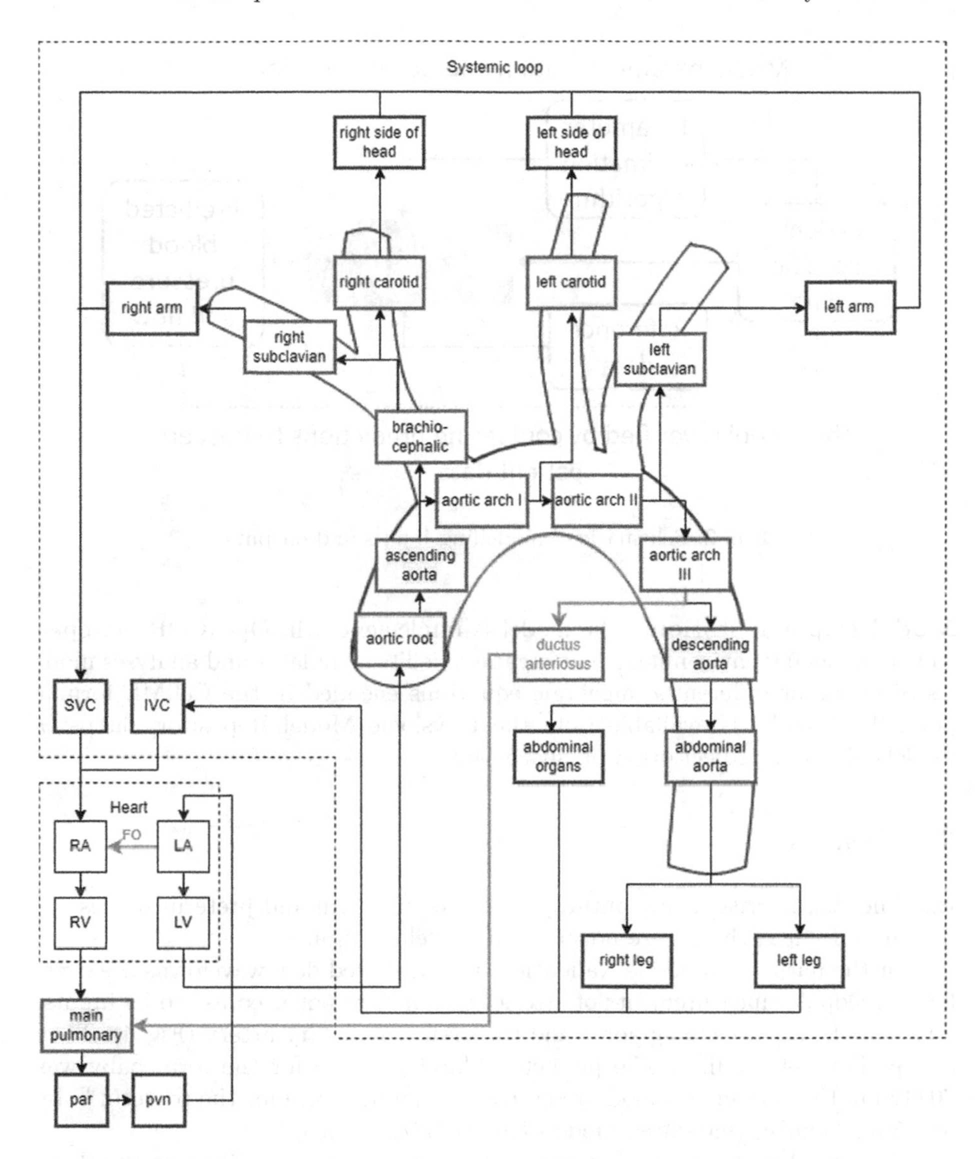

Fig. 1. Schematic of closed-loop model of simplified neonatal circulation. Modules in red are parameterised using patient-specific arterial measurements. Unknown parameter values for modules in blue are estimated using a genetic algorithm. Modules and connections unique to the preterm circulation in orange. FO: foramen ovale; IVC: inferior vena cava; LA: left atrium; LV: left ventricle; par: pulmonary arteries; pvn: pulmonary veins; RA: right atrium; RV: right ventricle; SVC: superior vena cava (Color figure online)

– Where no patient-specific measurements were available for the previous two steps, values were taken from the literature (e.g. blood viscosity, arterial length).

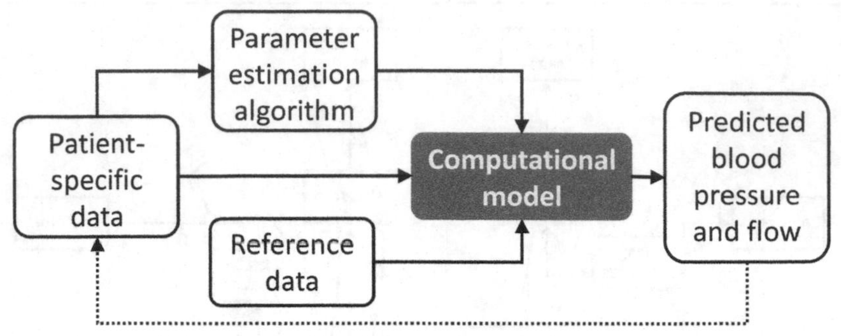

Fig. 2. Schematic of modelling inputs and outputs

Model Implementation. The model is implemented in OpenCOR, an open source modelling environment that organises, edits, simulates and analyses models of ODEs or differential algebraic equations encoded in the CellML format [4,6]. This model is available from the Physiome Model Repository https://models.physiomeproject.org/workspace/662.

3 Results

Baseline characteristics and anthropometry of the term and preterm babies are shown in Table 1; these were not used as modelling inputs.

For the purposes of model validation, the predicted flow waveforms are overlaid on Doppler measurements of arteries which were not used as model inputs, for example, the ascending aorta and the main pulmonary artery (Fig. 4). They are qualitatively similar. The predicted blood pressure for the term baby was 70/42 mmHg (measured blood pressure 81/48 mmHg) and for the preterm baby was 50/24 mmHg (measured blood pressure 57/27 mmHg).

We were also able to predict parameters that are not readily measurable in clinical practice. For example, the fitted pulmonary compliance for the preterm baby was lower than that of the term baby, as would be expected [7]; however, further validation is required before these could be clinically useful.

Table 1. Participant baseline characteristics and anthropometry

	Term	Preterm
Gestational age at birth (weeks+days)	38+6	35+0
Birth weight (g)	3580	2680
Weight-for-age z-score	0.72	1.43
Birth length (cm)	53	49
Length-for-age z-score	1.53	2.37

Table 2. Participant measurements used as direct model inputs

	Term	Preterm
Aortic root diameter (cm)	0.93	0.81
Ascending aorta diameter (cm)	0.95	0.77
Proximal transverse aortic arch diameter (cm)	0.67	0.59
Distal transverse aortic arch diameter (cm)	0.67	0.56
Abdominal aorta diameter (cm)	0.75	0.55
Main pulmonary artery diameter (cm)	0.96	0.89
Brachiocephalic artery diameter (cm)	0.47	0.35
Right subclavian artery diameter (cm)	0.33	0.14
Right common carotid artery diameter (cm)	0.23	0.29
Patent ductus arteriosus	no	yes
Ductus arteriosus diameter (cm)		0.07
Patent foramen ovale	yes	yes
Foramen ovale area (cm^2)	0.11	0.013
Heart rate (bpm)	110	137

Table 3. Participant measurements used as parameter estimation targets

	Term	Preterm
Blood pressure (mmHg)	81/48	57/27
Mean blood pressure (mmHg)	61	39

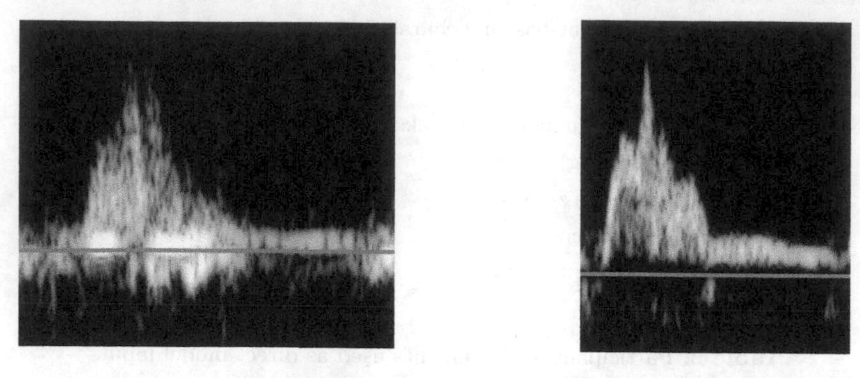

Fig. 3. Term (left) and preterm (right) abdominal aorta Doppler waveforms were targets for the genetic algorithm optimisation process to estimate unknown parameters.

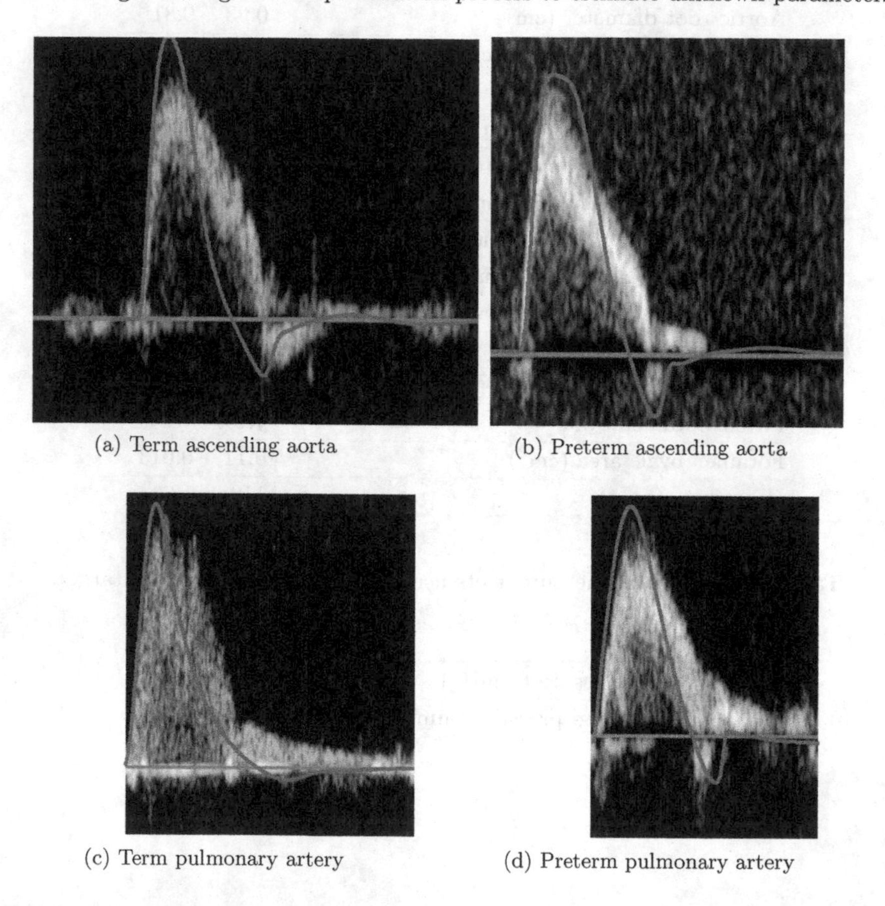

(a) Term ascending aorta

(b) Preterm ascending aorta

(c) Term pulmonary artery

(d) Preterm pulmonary artery

Fig. 4. Modelling results: Doppler waveforms compared to model predicted flows.

4 Discussion

This open-source cardiovascular model of the neonatal circulation is:

- readily individualised using image-based anatomical measurements. In future work we will explore the allometric relationships between body size and cardiovascular structure. If the model could be paramterised using easily acquired body size measurements, its clinical utility would be greatly improved.
- realistic in its blood pressure and flow predictions for both a term and a preterm test case. Further model validation will be done to ensure outputs are quantitatively similar.
- fully conservative for mass and energy owing to the use of the bond graph modelling approach. A strength of this approach is that this model could also be extended to other domains, e.g. including gas exchange.
- adaptable to the unique circulatory conditions of early life. In future, this model will be applied to neonatal data collected at birth and within the first few weeks of life.

These models meet the requirements for computational efficiency for translation to clinical settings. The computational model simulates 50 s of flow in 7 s. For these models, the genetic algorithm used for parameter estimation took 52 min on an hpc with 32 processors; however, in future work, parameter estimation will be performed for half of the term and preterm data to get population estimates for unknown parameters and these population estimates will be verified using the remainder of the data, thus removing this step from the workflow for a more computationally efficient modelling workflow.

5 Conclusion

This paper is a proof-of-concept for the personalisation of a computational model of the cardiovascular system to the term and preterm circulations. In future work, this model will be applied to a larger dataset of 15 term and 10 preterm babies at two time points: shortly after birth and 3–6 weeks later. This large dataset will allow for model verification and validation. By combining both the clinical dataset and the modelling results that provide additional insights into haemodynamics, we will be able to explore the effect of gestational age on early cardiovascular development.

Acknowledgements. RWM thanks the Auckland Medical Research Foundation for their support through a Doctoral Scholarship. SS acknowledges the financial support provided by the Aotearoa Foundation.

References

1. Argus, F., Zhao, D., Gamage, T.P.B., Nash, M.P., Talou, G.D.M.: Automated model calibration with parallel MCMC: applications for a cardiovascular system model. Front. Physiol. **13** (2022). https://doi.org/10.3389/fphys.2022.1018134

2. Barrington, K.J., Janaillac, M.: Treating hypotension in extremely preterm infants. The pressure is mounting. Arch. Disease Child. Fetal Neonatal Edit. **101**(3), F188–F189 (2016). https://doi.org/10.1136/archdischild-2015-309814
3. Blanco, P.J., Watanabe, S.M., Passos, M.A.R.F., Lemos, P.A., Feijóo, R.A.: An anatomically detailed arterial network model for one-dimensional computational hemodynamics. IEEE Trans. Biomed. Eng. **62**(2), 736–753 (2015). https://doi.org/10.1109/TBME.2014.2364522
4. Cuellar, A.A., Lloyd, C.M., Nielsen, P.F., Bullivant, D.P., Nickerson, D.P., Hunter, P.J.: An overview of CellML 1.1, a biological model description language. Simulation **79**(12), 740–747 (2003). https://doi.org/10.1177/0037549703040939
5. Dempsey, E.M.: What should we do about low blood pressure in preterm infants. Neonatology **111**(4), 402–407 (2017). https://doi.org/10.1159/000460603
6. Garny, A., Hunter, P.J.: OpenCOR: a modular and interoperable approach to computational biology. Front. Physiol. **6**(FEB), 26 (2015). https://doi.org/10.3389/fphys.2015.00026
7. Hjalmarson, O., Sandberg, K.: Abnormal lung function in healthy preterm infants. Am. J. Respir. Crit. Care Med. **165**(1), 83–87 (2002). https://doi.org/10.1164/ajrccm.165.1.2107093
8. May, R.W., et al.: From fetus to neonate: a review of cardiovascular modeling in early life. WIREs Mechanisms of Disease, p. e1608 (2023). https://doi.org/10.1002/wsbm.1608
9. Mertens, L., et al.: Targeted neonatal echocardiography in the neonatal intensive care unit: practice guidelines and recommendations for training: Writing group of the American Society of Echocardiography (ASE) in collaboration with the European Association of Echocardiograph. Eur. J. Echocardiogr. **12**(10), 715–736 (2011). https://doi.org/10.1093/ejechocard/jer181
10. Morris, P.D., et al.: Computational fluid dynamics modelling in cardiovascular medicine. Heart **102**(1), 18–28 (2016). https://doi.org/10.1136/heartjnl-2015-308044
11. Mynard, J.: Computer modelling and wave intensity analysis of perinatal cardiovascular function and dysfunction. Doctoral dissertation, University of Melbourne (2011). http://hdl.handle.net/11343/36318
12. Mynard, J.P., Smolich, J.J.: One-dimensional haemodynamic modeling and wave dynamics in the entire adult circulation. Ann. Biomed. Eng. **43**(6), 1443–1460 (2015). https://doi.org/10.1007/s10439-015-1313-8
13. Safaei, S., Blanco, P.J., Müller, L.O., Hellevik, L.R., Hunter, P.J.: Bond graph model of cerebral circulation: toward clinically feasible systemic blood flow simulations. Front. Physiol. **9**(MAR), 1–15 (2018). https://doi.org/10.3389/fphys.2018.00148
14. Schneider, D.J., Moore, J.W.: Patent ductus arteriosus. Circulation **114**(17), 1873–1882 (2006). https://doi.org/10.1161/CIRCULATIONAHA.105.592063
15. Vrancken, S.L., van Heijst, A.F., de Boode, W.P.: Neonatal hemodynamics: from developmental physiology to comprehensive monitoring. Front. Pediatr. **6**, 1 (2018). https://doi.org/10.3389/fped.2018.00087
16. World Health Organisation: Survive and thrive: transforming care for every small and sick newborn. Technical report, Geneva (2019)

Impact of Blood Rheological Strategies on the Optimization of Patient-Specific LAAO Configurations for Thrombus Assessment

Carlos Albors[1(✉)], Andy L. Olivares[1], Xavier Iriart[2], Hubert Cochet[2], Jordi Mill[1], and Oscar Camara[1]

[1] Sensing in Physiology and Biomedicine (PhySense), Department of Information and Communication Technologies, Universitat Pompeu Fabra, 08018 Barcelona, Spain
carlos.albors@upf.edu

[2] Hôspital Haut-Lévêque, Bordeaux, France

Abstract. Left atrial appendage occlusion devices (LAAO) are a feasible alternative for non-valvular atrial fibrillation (AF) patients at high risk of thromboembolic stroke and contraindication to antithrombotic therapies. However, optimal LAAO device configurations (i.e., size, type, location) remain unstandardized due to the large anatomical variability of the left atrial appendage (LAA) morphology, leading to a 4–6% incidence of device-related thrombus (DRT). In-silico simulations can be used to estimate the risk of DRT and identify the critical parameters, such as suboptimal device positioning. However, simulation outcomes depend a lot on a series of modelling assumptions such as blood behaviour. Therefore, in this work, we present fluid simulations results computed on two patient-specific LA geometries, using two different commercially available LAAO devices, located in two positions: 1) mimicking the real post-LAAO intervention configuration; and 2) an improved one better covering the pulmonary ridge for DRT prevention. Different blood modeling strategies were also tested. The results show flow recirculations at low velocities with significant platelet accumulation in LAA-deep device positioning uncovering the pulmonary ridge, potentially leading to thrombus formation. In addition, assuming Newtonian blood behaviour may result in an overestimation of DRT risk.

Keywords: Left atrial appendage occlusion · Blood behaviour · Platelet adhesion · Device related thrombus · In-silico fluid simulations

1 Introduction

Atrial fibrillation (AF) is a life-threatening condition with stroke prevention as a cornerstone of its clinical management. Among AF patients, >90% AF-related strokes develop in the ear-shaped and highly trabeculated cavity called the left atrial appendage (LAA) [4]. Although anticoagulation therapy is the first-line treatment, the non-negligible number (>50%) of non-valvular AF patients with

O. Bernard et al. (Eds.): FIMH 2023, LNCS 13958, pp. 485–494, 2023.
https://doi.org/10.1007/978-3-031-35302-4_50

anticoagulants contraindications, along with the increasing incidence of the AF population, creates the unmet need to reduce the risk of stroke [9]. In the past decade, left atrial appendage occlusion (LAAO) has emanated with a potential role for stroke prevention. However, an effective LAAO device implantation requires a high level of expertise from clinicians, as the shape and size of the LAA can greatly vary. Therefore, device settings (e.g., design, size, position) must be accordingly tailored to avoid unusual events at follow-up such as device embolization, peri-device leaks, or device-related thrombus (DRT), due to sub-optimal device characteristics selection. Of these, thrombus formation on device surface, DRT, is the major concern of LAAO with a documented incidence of 4–6% [2]. The DRT mechanism appears to be a multifactorial process enrolling abnormal hemodynamic patterns related to the clinical patient characteristics (e.g., age, comorbidities), postoperative complications (e.g., incomplete device endothelialization, effusion), or deep-device-LAA implantation; recently correlated as a DRT-independent predictor [3,5].

Transesophageal echocardiography (TEE) is the most commonly used imaging technique for LA/LAA morphological and hemodynamic (Doppler echocardiography) analysis, but the spatial and time resolution limitations may hinder the hemodynamic interpretations or measurements for device selection and consequently, DRT assessment. Cardiac computed tomography (CCT) modality or pre-planning tools (e.g., FEOps HeartGuide [17], 3mensio [10], VIDAA platform [1]) can provide accurate morphological estimations and recommend better device settings, albeit disregarding the importance of blood flow patterns. Alternatively, in-silico flow simulations are an emerging powerful concept for the prevention of DRT that can help understand the morphology of LA/LAA and the complex characterization of blood flow. Multiple computational fluid dynamics (CFD) studies (see [15] for a recent literature review) have been published on the hemodynamics representation of the LA in patients with AF [6,7,14], including thrombus models with discrete phase methods (DPM) [23], and more recently, rheological characterization of blood with non-Newtonian approaches [8]. However, only a few investigations have incorporated the occlusion procedure. Some studies have focused on blood flow behavior in several device configurations [1]. Others have also added particles to mimic platelet adhesion to the surface of the device [19], and more advanced ones explore the mechanical properties of the devices for optimal in-silico representation of deployment [26].

Accordingly, the main goal of this work was to identify the impact of device configuration (type, size, position) in DRT risk post-LAAO intervention on two patient-specific geometries with DRT in follow-up scans with the resulting in-silico haemodynamic analysis. The most common commercialized devices (Amplatzer Amulet from Abbott Vascular, USA; and Watchman from Boston Scientific, USA) were virtually implanted in the real post-LAAO device settings and other proposed improved configurations for DRT prevention. Rheological blood variations and platelet concentrations were also tested.

2 Material and Methods

2.1 Clinical Data

Patient-specific 3D left atria geometries were extracted in Slicer 4.11[1] from the manually obtained binary mask segmentations of the retrospective pre- and post-occlusion computed tomography (CT) images provided by ANONYMOUS. The six-month follow-up CT scan showed the presence of a device-related thrombus in both of the cases analysed. Cardiac CT studies were performed on a 64-slice dual source CT system (Siemens Definition, Siemens Medical Systems, Forchheim, Germany) and then reconstructed into isotropic voxel sizes (0.37–0.5 mm range; 512 × 512 × [270–403] slices) after approval from the ethical committee and informed consent of the patient.

2.2 Occluder Device Configurations and 3D Model Generation

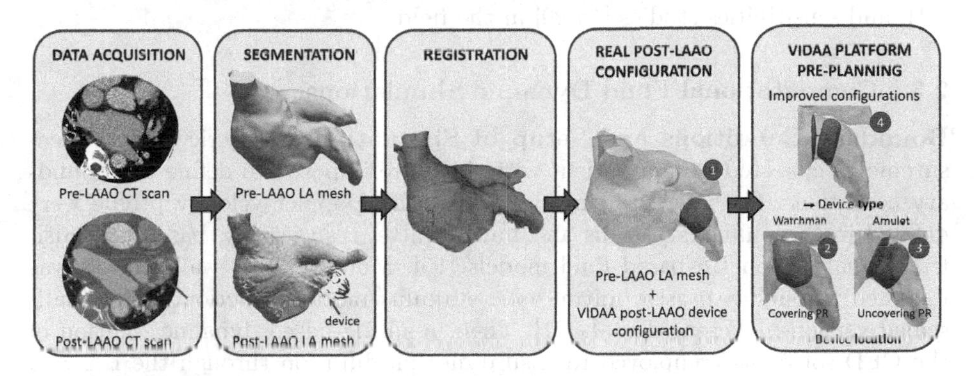

Fig. 1. Modeling pipeline of several device deployments in each of the patients. LAAO: Left atrial appendage occlusion; CT: computed tomography; PR:pulmonary ridge.

Two device positioning were defined for each patient. First, the real location of the LAAO device had to be extracted. To achieve that, apart from the pre- and post-CT LA geometries, the LAAO device was manually segmented in the post-CT images (Fig. 1 segmentation). Then, a fiducial registration from Meshlab v2021-07[2] was employed between pre- and post-LAAO meshes to place the CT segmented device in the pre-oclusion mesh (Fig. 1 registration). Finally, the segmented device and the pre-CT LA mesh were together uploaded into the web-based VIDAA platform [1] to simulate the device deployment with computer-aided design (CAD) models of the most used occluders; the plug type Watchman and the pacifier type Amplatzer Amulet. The device segmentation allows the identification of the device's size, type, and location from the post-CT images, which is then used for the selection of the CAD model of the device to deploy in

[1] https://www.slicer.org/.
[2] http://www.meshlab.net.

the pre-CT LA geometry. In both analysed patients, clinicians ended up implanting the devices quite deep in the LAA, i.e., uncovering the pulmonary ridge (lateral fold formed by the coalescence of the LAA and the Left Superior Pulmonary Vein (LSPV); therefore, a second device positioning, closer to the ostium (interface between LA main cavity and the LAA) and covering the pulmonary ridge, was also evaluated. For this second configuration, the recommended LAAO device size was estimated within the VIDAA platform based on anatomical measurements following the guidelines provided by device manufacturers [13]. Both device designs (e.g., plug and pacifier devices) were implanted in each patient to test their performance. For Patient 1, both plug type device configurations (the real and the second proposed configuration) were mimicked with the pacifier type device. In Patient 2, the paficier was the starting device. The result was four configurations per patient (see Fig. 1).

To solve the fluid domain, tetrahedral volumetric meshes of 12×10^5 elements were generated in Gmsh 4.8.4[3] with the LAAO devices deployed and LA geometry downloaded from VIDAA. Mesh resolution follows the request proposed by [11], and sensitivities studies [15,20] in the field.

2.3 Computational Fluid Dynamic Simulations

Boundary Conditions and Setup of Simulations: Generic clinical measurements collected from a patient with AF were imposed to define the boundary conditions. Catheter pressure data and echo Doppler velocity profile were defined in the pulmonary veins and mitral valve, respectively, based on sensitivity analyses on LA-based fluid models [15]. Moreover, LA wall motion was included through a passive mitral valve annulus motion function from a half-scaled version of Veronesi et al. [21]. Then, a spring-based dynamic solution of the CFD solver was employed to ensure motion diffusion through the LA wall geometry. The model setup also included three boundary layers of 0.0001 mm thickness. Two cardiac cycles were simulated in 176 steps per beat with a time-step of 0.01 s, according to the patients heart rate (HR). In-silico studies were performed within the fluid flow solver Ansys Fluent 2022 (ANSYS Inc, USA).

Modelling Strategies for Rheological Variations: The different approaches to test the blood rheological scenarios were the following: i) the assumption of blood as a homogeneous and incompressible Newtonian fluid [6,16] with constant 1060 kg kg/m3 density and 0.0035 Pa Pa/s viscosity; and ii) a Carreau model defining a non-Newtonian approach [7], where the viscosity is a function dependent on the shear rate. The dynamic viscosity behavior in Carreau's model is described by the following equation:

$$\eta = \eta_\infty + (\eta_0 - \eta_\infty)[1 + \gamma^2 \lambda^2]^{(n-1)/2} \tag{1}$$

where λ as time constant, n the power-law index, η_0 the zero shear viscosity and η_∞, the infinite shear one. The values, $\mu_0 = 0.056$ Pa$\cdot$ s, $\mu_\infty = 0.0035$ Pa

[3] http://gmsh.info.

Pa· s, $\lambda = 1.902$ s, n = 0.3568, were implemented from [25] to model the blood conditions.

Platelet Adhesion Model: Similar to Planas et al. [19], a first approximation of a thrombus model based on a discrete phase modeling (DPM) was combined with the continuous phase (blood flow). The DPM was coupled within the CFD solver to interact with the fluid flow and complement the DRT risk estimation. A given number of particles representing clusters of platelets were injected through the pulmonary veins during the initial 10 time steps of each beat of the cardiac cycles. The number of platelets per injection $n_{plat} = c_p \cdot V_{LA}$, particle diameter d_p, LA volume (V_{LA}), and total flow rate Q, were calculated accordingly for each patient. To do so, values such as particle density ($\rho_p = 1550$ kg m^{-3}), molecular viscosity (μ_p), surface tension (σ_p), and blood platelet concentration ($c_p = 2 \cdot 10^8$ mL^{-1}) were assumed under physiological conditions[4].

$$m_p \frac{d\mathbf{u}_p}{dt} = m_p \frac{\mathbf{u} - \mathbf{u}_p}{\tau_r} + \mathbf{F}, \tag{2}$$

where $\mathbf{u}$ represents the fluid velocity, m_p the particle mass, $\mathbf{u}_p$ the particle velocity, ρ the fluid density and the term $m_p (\mathbf{u} - \mathbf{u}_p) / \tau_r$, the drag force. The term, $\mathbf{F}$, describe the Saffman's lift force [12]. The inclusion of the DPM interaction in the CFD solver reported a slight difference in the LAA ostium average velocities in comparison to the cases only with the continuous domain solution. Nevertheless, due to the personalized characteristics of the particles, the influence of DPM on flow behavior was negligible. The platelets adhesion model defines the worst-case DRT scenario, in which all particles touching the LAA under certain conditions are adhered by the wall-film constraint imposed on the entire endothelial LAA wall. In platelets, the assumption of no division after their collision has been made, so there is no particle splashing. The O'Rourke [18] separation model with an angle of 90° is also taken into account. The number and location of platelets were used to estimate the risk of thrombus formation.

2.4 Hemodynamic Indices

Average velocities, the simulated blood flow patterns, and the endothelial cell activation potential (ECAP) were assessed on the device surface at key instants in the cardiac cycle. The δ *viscosity* metric ($\delta_{visc.} = nonNewt._{visc.} - Newt._{visc.}$) was incorporated to define the viscosity differences between non-Newtonian and Newtonian regimes. Additionally, the number of wall-adhered platelets was computed at the end of the simulation. All indices were calculated on the second cardiac beat to avoid convergence problems.

3 Results

The device configurations with the covered pulmonary ridge shown in Fig. 3 (third and fifth row) provided higher velocities without re-circulations patterns

[4] https://bionumbers.hms.harvard.edu/search.aspx.

in the surrounding areas near the device surface. The presence of mean velocities below 0.2 m/s during atrial systole and sustained in diastole (t = 0.6 s) were only present in real LAAO positionings, which could indicate a higher risk of DRT, compared to the devices covering the PR. No significant differences were found between device types. However, the assumption of Newtonian models showed higher velocities compared to the non-Newtonian strategy (Fig. 3 black block).

High risk of coagulability is associated with the increase of number of wall-adhered platelets, commonly located in the pulmonary ridge region with deep-LAA positioning (Fig. 3 first, second and forth row). Moreover, susceptible areas with platelet accumulation were also seen in the plug type device even in the covered location due to the space left by the device surface curvature with the LAA endothelial wall. In contrast, the disk of the pacifier device allowed a better dispersion of the adhered platelets. Under the Newtonian regime, the highest deposition occurred in the plug device, with almost 20% of the platelets injected attached (Fig. 3 first row).

The ECAP maps in Fig. 3 (last column) display low values achieved for the pacifier device on both configurations (i.e., lowest risk of thrombus formation) due to the higher blood flow velocity values. Meanwhile, indistinctively of the location, the plug device obtained regions with ECAP values higher >1.5, especially the uncovered location where the regions with the peak values matched the low velocities seen in the streamline calculation.

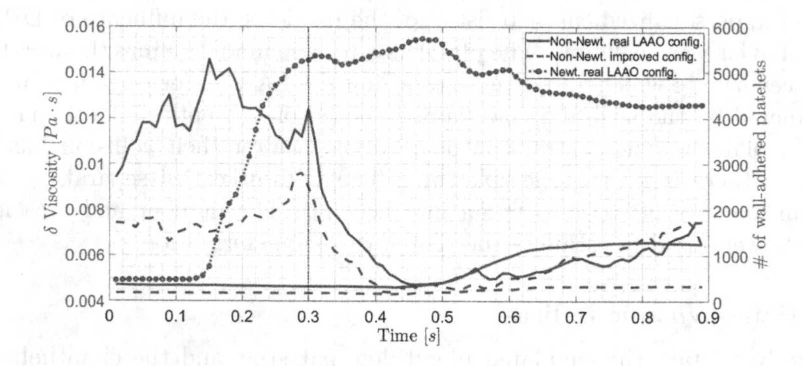

Fig. 2. Impact of viscosity changes on platelet accumulation near the device surface for real left atrial appendage (LAAO) occlusion and improved configuration covering the pulmonary ridge throughout the cardiac cycle with the different rheological strategies tested in Patient 1. Blue color represents the dynamic behavior of the δ viscosity (non-Newtonian viscosity - Newtonian viscosity) and the red color, the accumulation of platelets adhered to the wall. The gray line delimits the systolic (t = 0–0.3 s) and diastolic phases (t = 0.31–0.65 s). (Color figure online)

Finally, quantitative and temporal variations of blood viscosity in the device area can be seen in Fig. 2, showing higher viscosities in uncovered PR configurations within the LAA due to lower velocities than configurations proximal to

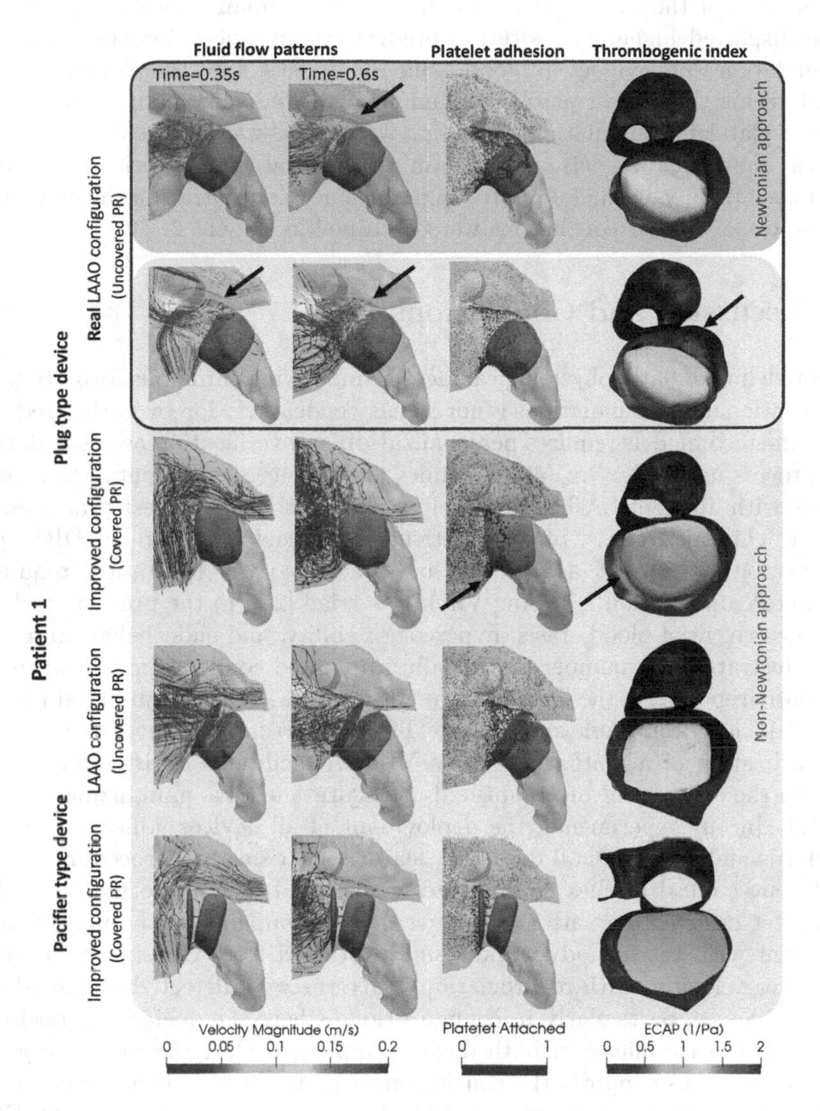

Fig. 3. Patient 1: The first two columns show the simulated blood flow patterns in the left atria (LA) in early- (t = 0.35 s) and late- (t = 0.6 s) diastole evaluated with various configurations of occluder devices. The third column corresponds to the platelet adhesion model at the end of the cardiac cycle (t = 0.89 s) and, the last column represents the endothelial cell activation potential (ECAP) values of the endothelial wall of the appendage near the device surface by cutting at the level of the LAA ostium. A frontal view of the device's surface is shown in white. The black rectangle highlights the comparison between Newtonian and non-Newtonian models in the real LAAO configuration. The black arrows indicate re-circulation at low velocities, platelet accumulation, and high ECAP values on the device surface.

the location of the LAA ostium. In the non-Newtonian approach, the systolic phase displayed higher viscosities. Moreover, significant differences in platelet accumulation were evident in blood behavior models, with the Newtonian models exhibiting the highest platelet accumulation. Non-Newtonian models showed a rise in platelet accumulation coinciding with the E-wave in the diastolic phase, whereas Newtonian models exhibited an exponential increase prior to the opening of the mitral valve and a relatively maintained concentration throughout the cardiac cycle. Similar conclusions were obtained in Patient 2.

4 Discussion and Conclusions

The modeling of pathophysiological factors underlying thrombus formation with computational fluid simulations is not an easy endeavor. Moreover, the credibility of the in-silico models requires personalization and verification as a standardization process inside the V&V40 guidelines [22]. Hence, it is essential to compare models with different LA-based modeling choices to achieve best practices.

The clinical records of the patients analyzed reported a DRT post-intervention. A correct assessment and pathology interpretation requires a parameter combination following Virchow's triad [24]. In the present study, the primary drivers of blood stasis, hypercoagulability, and endothelial injury have been illustrated in a number of virtually implanted configurations that are not frequently reported in the literature. In addition, the virtual web-based platform VIDAA [1] has been employed to create different virtual occlusion configurations. The utilization of a platform such as VIDAA facilitates decision-making and leads to the attainment of an optimal configuration, thus minimizing instances of DRT. In our experiments, the deployment of all devices adhered to clinical standards and experimental data that lead to a satisfactory procedure.

Recent clinical studies [2] indicate a slightly better performance in DRT rates after intervention with the pacifier device compared to the plug, which is consistent with the hemodynamic results obtained from the simulation analysis. Lower velocities with re-circulations patterns were detected at the edges of the plug device, particularly in configurations where the device was positioned deeply. The performance of both devices improved when the pulmonary ridge was covered. This supports the conclusion of Freixa et al. [5] that such an approach may reduce the incidence of DRT. However, the presence of high ECAP values and some complex flow in the plug device even in the covered configuration suggest that device shape has a potential role in blood stasis. Moreover, a proximal location may not be always feasible, as it may lead to improper device compression or insufficient endothelization, rising the risk to suffer a mitral valve leaf's intersection (with the pacifier disk) or device embolization.

Previous studies in modeling fluid flow in atrial fibrillation (AF) have largely relied on the assumption of Newtonian rheology. However, Gonzalo et al. [8] recently discovered that this assumption led to an overestimation of the hypercoagulability state in AF patients. The current study concurs with these findings, revealing that variations in blood rheology significantly influence the platelet adhesion model. The results indicate that there is a correlation between the amount of platelets adhered and increases in flow velocity, with this relationship being especially pronounced in the Newtonian approach (Fig. 2).

In the present study, the benefits of using real post-LAAO configurations in conjunction with non-Newtonian models for DRT stratification in fluid simulations were acknowledged. However, certain assumptions were made regarding the device placement (only the metallic structure was segmented in the plug device), CAD models, and the lack of patient-specific boundary conditions. To improve the realism of the models, further exploration is needed, such as incorporating shear stress in the platelet adhesion or using agent-based models.

Acknowledgement. This project has received funding from the European Union's Horizon 2020 research and innovation programme under grant agreement, No 101016496 (SimCardioTest).

References

1. Aguado, A.M., et al.: In silico optimization of left atrial appendage occluder implantation using interactive and modeling tools. Front. Physiol. **10**, 237 (2019)
2. Alkhouli, M., Ellis, C.R., Daniels, M., Coylewright, M., Nielsen-Kudsk, J.E., Holmes, D.R.: Left atrial appendage occlusion: current advances and remaining challenges. JACC Adv. 100136 (2022)
3. Aminian, A., et al.: Incidence, characterization, and clinical impact of device-related thrombus following left atrial appendage occlusion in the prospective global amplatzer amulet observational study. JACC Cardiovasc. Interv. **12**(11), 1003–1014 (2019)
4. Cresti, A., et al.: Prevalence of extra-appendage thrombosis in non-valvular atrial fibrillation and atrial flutter in patients undergoing cardioversion: a large transoesophageal echo study. EuroIntervention **15**(3), e225–e230 (2019)
5. Flores-Umanzor, E., et al.: Device related thrombosis after left atrial appendage occlusion: does thrombus location always predicts its origin? J. Interv. Card. Electrophysiol. **60**, 347–348 (2021)
6. García-Isla, G., et al.: Sensitivity analysis of geometrical parameters to study haemodynamics and thrombus formation in the left atrial appendage. Int. J. Numer. Methods Biomed. Eng. **34**(8), e3100 (2018)
7. García-Villalba, M., et al.: Demonstration of patient-specific simulations to assess left atrial appendage thrombogenesis risk. Front. Physiol. **12**, 596596 (2021)
8. Gonzalo, A., et al.: Non-newtonian blood rheology impacts left atrial stasis in patient-specific simulations. Int. J. Numer. Methods Biomed. Eng. e3597 (2022)
9. Holmes Jr, D.R., Alkhouli, M., Reddy, V.: Left atrial appendage occlusion for the unmet clinical needs of stroke prevention in nonvalvular atrial fibrillation. In: Mayo Clinic Proceedings, vol. 94, pp. 864–874. Elsevier (2019)
10. Imaging, P.M.: 3mensio Medical Imaging B.V. https://www.3mensio.com/
11. Khalili, E., Daversin-Catty, C., Olivares, A.L., Mill, J., Camara, O., Valen-Sendstad, K.: On the importance of fundamental computational fluid dynamics towards a robust and reliable model of left atrial flows: is there more than meets the eye? (2023). https://arxiv.org/abs/2302.01716
12. Li, A., Ahmadi, G.: Dispersion and deposition of spherical particles from point sources in a turbulent channel flow. Aerosol Sci. Technol. **16**(4), 209–226 (1992)

13. Lim, M.Y., Abou-Ismail, M.Y.: Left atrial appendage occlusion for management of atrial fibrillation in persons with hemophilia. Thromb. Res. **206**, 9–13 (2021)
14. Masci, A., et al.: A proof of concept for computational fluid dynamic analysis of the left atrium in atrial fibrillation on a patient-specific basis. J. Biomech. Eng. **142**(1) (2020)
15. Mill, J., et al.: Sensitivity analysis of in silico fluid simulations to predict thrombus formation after left atrial appendage occlusion. Mathematics **9**(18), 2304 (2021)
16. Mill, J., et al.: Impact of flow dynamics on device-related thrombosis after left atrial appendage occlusion. Can. J. Cardiol. **36**(6) (2020)
17. NV.F.: FEops HeartGuide. https://feops.com/
18. O'Rourke, P.J., Amsden, A.: A spray/wall interaction submodel for the kiva-3 wall film model. SAE Trans. 281–298 (2000)
19. Planas, E., et al.: In-silico analysis of device-related thrombosis for different left atrial appendage occluder settings. In: Puyol Antón, E., et al. (eds.) STACOM 2021. LNCS, vol. 13131, pp. 160–168. Springer, Cham (2022). https://doi.org/10.1007/978-3-030-93722-5_18
20. Pons, M.I., et al.: Joint analysis of morphological parameters and in silico haemodynamics of the left atrial appendage for thrombogenic risk assessment. J. Interv. Cardiol. **2022** (2022)
21. Veronesi, F., et al.: Quantification of mitral apparatus dynamics in functional and ischemic mitral regurgitation using real-time 3-dimensional echocardiography. J. Am. Soc. Echocardiogr. **21**(4), 347–354 (2008)
22. Viceconti, M., Pappalardo, F., Rodriguez, B., Horner, M., Bischoff, J., Tshinanu, F.M.: In silico trials: Verification, validation and uncertainty quantification of predictive models used in the regulatory evaluation of biomedical products. Methods **185**, 120–127 (2021)
23. Wang, Y., Qiao, Y., Mao, Y., Jiang, C., Fan, J., Luo, K.: Numerical prediction of thrombosis risk in left atrium under atrial fibrillation. Math. Biosci. Eng. **17**(3), 2348–2360 (2020)
24. Watson, T., Shantsila, E., Lip, G.Y.: Mechanisms of thrombogenesis in atrial fibrillation: Virchow's triad revisited. Lancet **373**(9658), 155–166 (2009)
25. Weddell, J.C., Kwack, J., Imoukhuede, P., Masud, A.: Hemodynamic analysis in an idealized artery tree: differences in wall shear stress between newtonian and non-newtonian blood models. PLoS ONE **10**(4), e0124575 (2015)
26. Zaccaria, A., et al.: Left atrial appendage occlusion device: development and validation of a finite element model. Med. Eng. Phys. **82**, 104–118 (2020)

Shape-Guided In-Silico Characterization of 3D Fetal Arch Hemodynamics in Suspected Coarctation of the Aorta

Uxio Hermida[1(✉)], Milou P.M. van Poppel[1], Malak Sabry[1], Hamed Keramati[1], David F. A. Lloyd[1,2], Johannes K. Steinweg[1], Trisha V. Vigneswaran[2,3], John M. Simpson[2,3], Reza Razavi[1,2], Kuberan Pushparajah[1,2], Pablo Lamata[1], and Adelaide De Vecchi[1]

[1] School of Biomedical Engineering and Imaging Sciences, King's College London, St Thomas' Hospital, London, UK
uxio.hermida_nunez@kcl.ac.uk
[2] Department of Congenital Heart Disease, Evelina London Children's Hospital, London, UK
[3] Harris Birthright Centre, Fetal Medicine Research Institute, King's College Hospital, London, UK

Abstract. Coarctation of the aorta (CoA) is a prevalent congenital heart defect. Its prenatal diagnosis is challenging, with high false positive rates. The exact cause of CoA is yet not fully understood. Recent research has provided novel insights into the anatomical determinants of CoA based on the in-utero arch anatomy. However, it is also recognized that the pathophysiology of CoA is also intrinsically linked to abnormal flow dynamics. To investigate the interplay between arch anatomy and flow, Computational Fluid Dynamics (CFD) analysis was performed in two fetal cases - a true and a false positive CoA. These anatomies were selected from a population of 108 fetuses with suspected CoA based on a statistical shape analysis score and clinical outcomes. A simplified 0D model of the fetal circulation informed by 2D PC-MRI was used to find patient-specific boundary conditions for an open-loop 3D-0D CFD model. Results from the 3D CFD models were validated against clinical imaging data for each case and provided initial evidence of hemodynamic differences between false positive and true positive CoA cases. These findings demonstrate the potential of using the SSM-guided CFD analysis on a larger cohort of representative cases to better understand the disease mechanisms in CoA and improve its diagnosis before birth.

Keywords: Computational fluid dynamics · Fetal shape analysis · Wall shear stress · Digital twin · Fetal magnetic resonance imaging

1 Introduction

Neonatal coarctation of the aorta (CoA) is a common condition that affects 7–8% of all live births with congenital heart disease. It is characterized by discrete

O. Bernard et al. (Eds.): FIMH 2023, LNCS 13958, pp. 495–504, 2023.
https://doi.org/10.1007/978-3-031-35302-4_51

narrowing of the aortic isthmus following the closure of the arterial duct (AD) after birth. Despite extensive research, prenatal diagnosis of CoA remains difficult, with high false-positive (FP) rates of up to 80% [5,22]. Additionally, the underlying mechanisms leading to neonatal CoA remain poorly understood. Two main hypotheses have been proposed, the hemodynamic hypothesis [9,17], and the Skodaic theory of ductal tissue migration [10,23]. The hemodynamic hypothesis suggests that CoA is caused by abnormal hemodynamic changes during fetal development. The ductal tissue migration theory suggests that CoA is the result of abnormal extension of ductal cells into the aortic isthmus, leading to its constriction after birth. Both hypotheses suggest the mechanistic relevance of blood flow dynamics leading to abnormal wall shear stress (WSS) in CoA.

Computational fluid dynamics (CFD) methods have been widely applied to the quantification and understanding of mechanistic factors driving cardiovascular disease postnatally [14]. However, limited data before birth has hindered the use of these methods to study hemodynamics during human fetal development [3,4,18]. To date, no study has investigated 3D fetal arch hemodynamics combining patient-specific anatomies and boundary conditions.

Recent studies have revealed the anatomical signature in CoA and further suggested that the 3D fetal arch shape is relevant to the pathophysiology of CoA [8,13]. A better understanding of the fetal arch hemodynamics is essential to comprehend the interplay between anatomical remodeling and abnormal blood flow changes in CoA. This study aims to present a methodology for personalized fetal 3D CFD simulations of characteristic phenotypes in CoA based on a statistical shape model (SSM) of the fetal arch anatomy.

2 Methods

To study characteristic hemodynamics in CoA, a SSM-guided workflow was used (see Fig. 1): 1. From a population of cases with acquired 3D fetal cardiovascular magnetic resonance (CMR) images, a SSM is built to capture the anatomical signature in CoA and guide the selection of representative phenotypes for each group. 2. Extraction of 3D anatomy and preparation for CFD. 3: Personalization of boundary conditions with a 0D lumped model of the fetal circulation and 2D Phase-Contrast Magnetic Resonance Imaging (PC-MRI) data. 4: 3D CFD simulations and extraction of biomarkers of interest (e.g., WSS).

2.1 Clinical Data

A cohort of 108 cases with suspected CoA was included, comprising of 65 false positive cases (FP) at a gestational age of 32.2 ± 1.7 weeks and 43 confirmed CoA cases at a gestational age of 32.1 ± 1.7 weeks. T2-weighted 'black-blood' single-shot fast spin echo sequences were acquired for all participants. The acquired 2D data were processed using a motion-corrected slice-volume registration method [12], resulting in high-resolution 3D volumes (see Fig. 1a.). Semi-automatic segmentation was then performed by a clinician with expertise in fetal CMR. Data

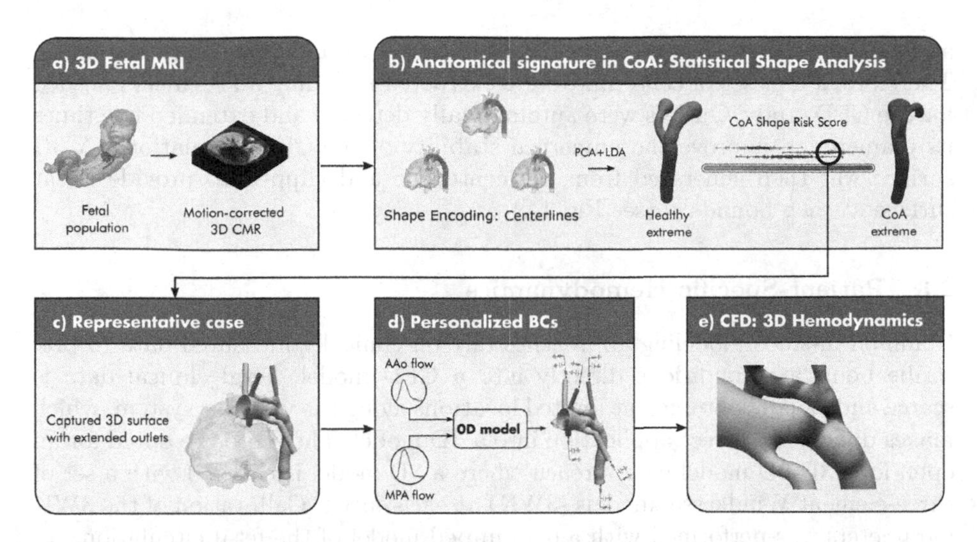

Fig. 1. Study workflow: from a population of fetal cases with suspected coarctation of the aorta (CoA), a statistical shape model (SSM) is built to capture the anatomical signature between false positive (FP) and true positive cases. The SSM is then used to guide the selection of representative cases for each group. Finally, a combination of patient-specific anatomy and boundary conditions is used to perform computational fluid dynamics (CFD) simulations and characterize their 3D hemodynamics. CMR: Cardiovascular magnetic resonance; PCA: Principal component analysis; LDA: Linear discriminant analysis; AAo: Ascending aorta; MPA: Main pulmonary artery.

also included 2D PC-MRI acquired on a 1.5T Ingenia MRI scanner (Philips, Netherlands) retrospectively gated using metric optimized gating [2,11,19]. Data were acquired in six fetal vessels (ascending aorta, descending aorta, superior vena cava, main pulmonary artery, arterial duct and umbilical vein) and processed with CVI42 (Circle Cardiovascular Imaging, Calgary, Version 5.6.4).

2.2 Statistical Shape Model: Representative Phenotypes

In order to construct the SSM, the methodology in [8] was used. Briefly, a set of centerlines were extracted to capture the shape of the three main segments of interest in CoA: the ascending aorta (AAo), descending aorta (DAo), and arterial duct (AD). Then, a combination of PCA and LDA was used to find the anatomical axis that best discriminates between FP and CoA cases. Each anatomy can be then characterized by a single coefficient, referred to as CoA shape risk score (see Fig. 1b.).

The CoA shape risk score allows quantitative characterization of the severity along the disease spectrum. Such information was used to select representative cases and study their 3D hemodynamics. A case with a CoA shape risk score near the average of the CoA population was selected as a representative CoA case. Due to data availability, the selected FP case had a score closer to the healthy population extreme. Centerlines used to build the SSM did not include the right and left pulmonary arteries, as well as the arteries to the upper body and brain. Therefore, a new set of centerlines were extracted including all segments needed for the CFD study. Outlets were automatically detected and extended five times its diameter to improve the numerical stability of the CFD simulations. A 3D surface was then generated from the centerline and clipped to provide a flat surface at each boundary (see Fig. 1c.).

2.3 Patient-Specific Hemodynamics

Common in-silico modeling approaches rely on clinically measured data to prescribe boundary conditions directly into a CFD model. Fetal clinical data is scarce and usually obtained at limited locations along the vascular system, which makes difficult its direct application into a 3D model. Therefore, we opted for an open-loop 3D-0D modeling approach where a 3D model is coupled with a set of three-element Windkessel models (3WK) at each outlet. Calibration of the 3WK parameters was performed with a 0D lumped model of the fetal circulation.

0D Lumped Model. To build an equivalent electrical model of the fetal circulation, we used a simplification of the model in [6] (see Fig. 2a.). The CFD domain was modeled with 8 arterial segments and 6 vascular beds. To capture the local resistance and inertia of blood, each arterial segment is modeled as a resistance (R) in series with an inductance (L). As the 3D model to be simulated has rigid walls, no capacitance was included. For each segment, the patient-specific centerlines were used to calculate the 0D parameters as $R = 8\mu l/\pi r^4$ and $L = \rho l/\pi r^2$, where l and r are the length and mean radius of each arterial segment, μ is the blood viscosity, and ρ is the blood density. Peripheral vascular beds were modeled with a 3WK (i.e., one proximal resistance R_p in series with a capacitance C and a distal resistance R_d). R_p was defined to equal the characteristic impedance of the feeding artery at high frequencies f as

$$R_p = real\{Z_c\} = \sqrt{Z_T Z_L} \tag{1}$$

where $Z_T = 1/j2\pi fC$ and $Z_L = R + j2\pi fL$. R, L and C are the electrical components of the feeding artery. As no capacitance was used for each arterial segment, C was estimated as $C = 3\pi r^3 l/2Eh$, where h is the wall thickness of the artery, assumed to be 10% of the radius, and E the Young's Modulus from [6]. The model was built with Simulink, MATLAB (2022a, The MathWorks Inc., Natick, MA).

Windkessel Parameter Calibration. The goal of the model calibration is to reduce the discrepancy between the model-based output and clinical data from 2D PC-MRI. To start, R_d and C values were initialized from previous research [6]. Then, an optimization process was implemented to find the R_d and C that minimized a cost function based on the clinical data. Specifically, the cost function considered the relative error between the model and clinical

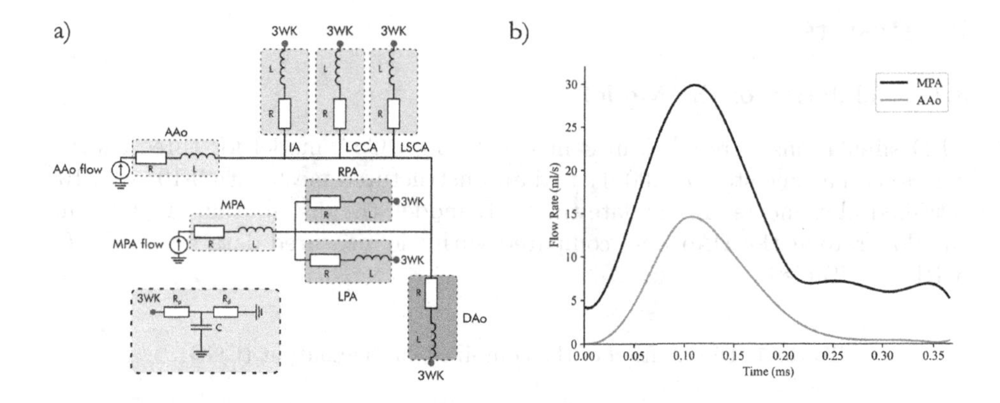

Fig. 2. a) Simplified 0D lumped model of the fetal circulation where each arterial segment is modeled with a resistance (R) in series with an inductance (L). Each peripheral vascular bed is modeled with a three-element Windkessel model (3WK). b) Example flow rate waveforms from a false positive case with 2D phase-contrast magnetic resonance acquisitions at the ascending aorta (AAo) and main pulmonary artery (MPA). AAo and MPA flows are used as input to the 0D model.

data for combined cardiac output (CCO) distribution, flow rate at the DAo, time-to-peak velocity and systolic and diastolic blood pressures derived from [20] as: $P_{sys} = 1.06 \cdot \text{GA} + 15.91$ and $P_{sys} = 0.64 \cdot \text{GA} + 2.47$, where GA is the gestational age of the fetus in weeks.

CFD Simulations: 3D-0D Coupling. The optimal 3WK parameters from the 0D model calibration were used for the 3D simulations. The flow rate waveforms at the AAo and MPA from 2D PC-MRI (see Fig. 2b) were converted to velocity waveforms using the cross-sectional area from the 3D model and imposed at the model inlets. The blood was modeled as an incompressible Newtonian fluid with a density of 1060 kg kg/m^3 and viscosity of 3.6 mPa·s [21]. The flow was assumed to be laminar and the vessel walls to be rigid. All simulations were performed using the finite volume solver in Ansys Fluent 2021 R2 (Ansys Inc. USA) and run until periodic conditions were achieved. The 3WK models were applied at each outlet of the 3D anatomy through C programming language sub-routines. At each time iteration, the instantaneous pressure was calculated as

$$P_{n+1} = \frac{(R_p + R_d + Rp\beta)\, Q_{n+1} - R_p\beta Q_n + \beta P_n}{1 + \beta} \tag{2}$$

where $\beta = R_d C/\Delta t$ [1]. The calculated pressure was then applied as a uniform Dirichlet boundary condition at the relevant output.

3 Results

3.1 Validation of the Model

CFD simulations were performed in a FP and a CoA model for three cardiac cycles with a time-step of 0.0001 s and an unstructured mesh with $9{\cdot}10^5$ to $1{\cdot}10^6$ tetrahedral elements. To validate the CFD model, the distribution of CCO and the flow rate at the DAo were compared with the measured data from 2D PC-MRI - see Table 1.

Table 1. Distribution of the combined cardiac output (CCO).

	FP		CoA	
	$\%CCO_{CFD}$	$\%CCO_{MRI}$	$\%CCO_{CFD}$	$\%CCO_{MRI}$
AAo	25	25	28	28
MPA	75	75	72	72
Upper body/Brain	20	18	29	30
Lungs	49	47	31	27
Descending aorta	31	35	40	43

Figure 3 shows the comparison between simulated and measured flow waveforms at the DAo. In the FP case, a 9% difference in the mean flow-rate and 7% difference in the peak flow rate was obtained. In the CoA case, the difference was 12% and 9% respectively. In both cases, the CFD simulation underestimated the flow rate during peak-systole and diastole.

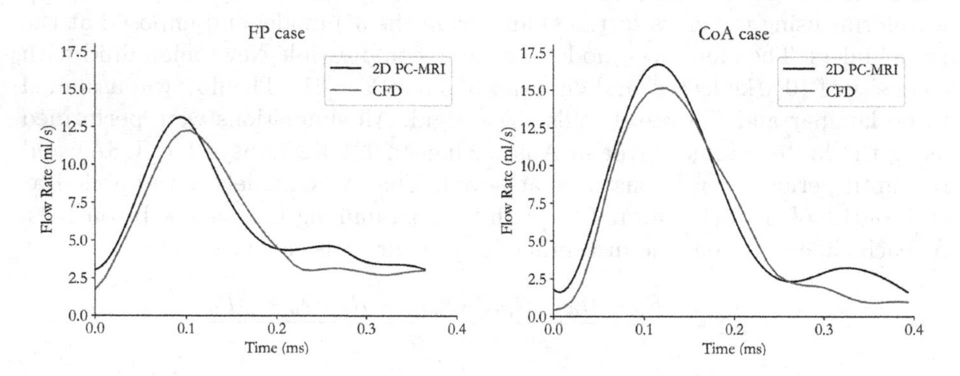

Fig. 3. Comparison of flow rate waveforms at the descending aorta from 3D computational fluid dynamics (CFD) simulations and two-dimensional (2D) Phase-Contrast Magnetic Resonance Images (PC-MRI) for the representative FP and CoA cases.

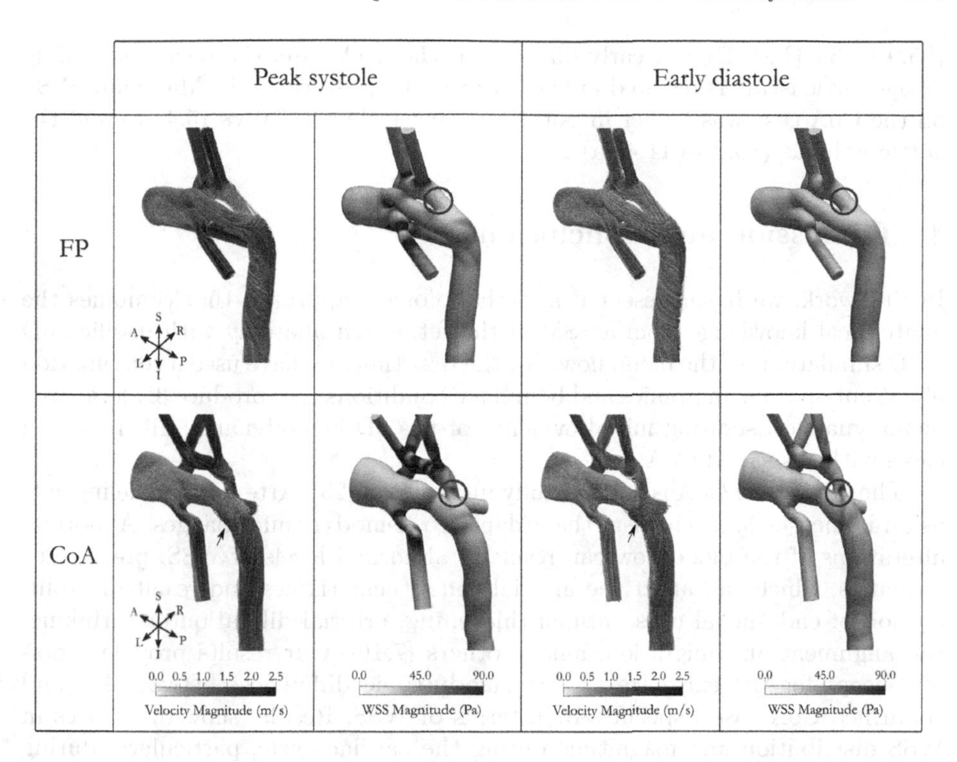

Fig. 4. Comparison of the velocity magnitude and wall shear stress (WSS) at peak systole and early diastole between a representative false positive (FP) case and a confirmed coarctation of the aorta (CoA) case. Black arrows point to the low-velocity recirculating flow at the bifurcation in the CoA case. Blue circles point to differences in WSS at the aortic isthmus. (Color figure online)

3.2 Patient-Specific Hemodynamic Simulations

Figure 4 compares the WSS and velocity magnitude at peak systole and diastole, showing significant differences in hemodynamics between cases. In the FP case, the flow from the arterial duct enters the descending aorta in a helical pattern where it meets the low-velocity flow coming from the aortic isthmus. Conversely, the flow from the arterial duct in the CoA case enters the descending aorta without forming any helical pattern. At the bifurcation, in the CoA case, we observed localized recirculating flow at low velocities (see black arrows in Fig. 4). In the true positive case, during the deceleration part of the systolic phase, there is a physiological retrograde flow from the arterial duct that gets directed into the left subclavian artery through the aortic isthmus.

In terms of WSS, during peak systole, the CoA case showed a ring-shaped pattern of increased WSS in the arterial duct, as well as a localized peak WSS on the posterior side of the aortic isthmus. On the arterial duct, during peak systole, maximum WSS was lower in the CoA case than in the FP (58.9 vs 79.0 Pa). Contrarily, on the aortic isthmus, maximum WSS was higher in the CoA case

(85.9 vs 35.6 Pa). During early diastole, in the CoA case, the retrograde filling of the aortic isthmus resulted in high WSS on its posterior side. Maximum WSS on the CoA case was higher in both the arterial duct (56.0 vs 15.4 Pa) and the aortic isthmus (85.0 vs 11.4 Pa).

4 Discussion and Conclusions

In this work, we have presented a methodological approach that combines the anatomical knowledge from a SSM of the fetal arch anatomy with in-silico 3D CFD simulations of the blood flow. For the first time, we have used a combination of patient-specific anatomies and boundary conditions to reproduce 3D fetal arch hemodynamics, showing initial evidence of the 3D hemodynamic differences in cases with suspected CoA.

The etiology of CoA is still not fully understood [23]. Arterial remodeling acts as an homeostatic mechanism that adapts to hemodynamic changes. Abnormal alterations of the blood flow can result in abnormal levels of WSS, pressure or velocities, which can affect the arterial constituent tissues and result in proliferation of endothelial cells, intimal thickening, arterial dilatation or shrinking, cell alignment and migration, among others [7,16]. Our results provide proof-of-concept for the existence of 3D hemodynamic differences between FP and confirmed CoA cases, specifically in terms of WSS. Results show differences in WSS distribution and magnitude during the cardiac cycle, particularly during early diastole, with a substantial WSS peak in the true positive case that is completely absent in the FP. The FP case showed a more orderly and balanced flow distribution in both systole and diastole. The observed differences suggest the potential significance of WSS in the the development of CoA.

We successfully applied the SSM-guided workflow to study hemodynamics in two representative cases with suspected CoA. The main challenge of this type of CFD analysis is the paucity of clinical data to constrain and validate the numerical models. However, we were able to calibrate our models thanks to the unique combination of high-resolution 3D fetal CMR and 2D PC-MRI data. Their combination allowed us to extract the patient-specific parameters used to tailor the 0D models to each patient's hemodynamics. To further constrain the CFD models, the calibration was also based on the systolic and diastolic pressure using the relationship from [20], and on the time-to-peak velocity. These additional parameters ensured that the model outputs were within physiological ranges. However, given that in clinical practice pressures can not be measured in-vivo, we could not validate the systolic and diastolic pressures from the model. Moreover, only the flow rate at the DAo was available for validation. The use of Doppler velocity waveforms or novel 4D flow MRI methods [15] at the different inlets and outlets could further constrain the model calibration and improve the personalization of the boundary conditions for each case.

Future work will focus on the application of the presented methodology to a larger cohort of representative cases, including control cases. Such analysis may provide further insights into the interplay between shape and blood flow in CoA, potentially unifying the two main hypotheses of the pathophysiology in CoA: hemodynamic theory and Skodaic theory of ductal tissue migration.

References

1. Alimohammadi, M., Agu, O., Balabani, S., Díaz-Zuccarini, V.: Development of a patient-specific simulation tool to analyse aortic dissections: assessment of mixed patient-specific flow and pressure boundary conditions. Med. Eng. Phys. **36**(3), 275–284 (2014). https://doi.org/10.1016/j.medengphy.2013.11.003
2. Bidhult, S., et al.: Independent validation of metric optimized gating for fetal cardiovascular phase-contrast flow imaging. Magn. Reson. Med. **81**(1), 495–503 (2019). https://doi.org/10.1002/mrm.27392
3. Chen, Z., Liu, Z., Du, M., Wang, Z.: Artificial intelligence in obstetric ultrasound: an update and future applications. Front. Med. **8**, 1–9 (2021). https://doi.org/10.3389/fmed.2021.733468
4. Chen, Z., Zhou, Y., Wang, J., Liu, X., Ge, S., He, Y.: Modeling of coarctation of aorta in human fetuses using 3D/4D fetal echocardiography and computational fluid dynamics. Echocardiography **34**(12), 1858–1866 (2017). https://doi.org/10.1111/echo.13644
5. Familiari, A., et al.: Risk factors for coarctation of the aorta on prenatal ultrasound: a systematic review and meta-analysis. Circulation **135**(8), 772–785 (2017). https://doi.org/10.1161/CIRCULATIONAHA.116.024068
6. Garcia-Canadilla, P., et al.: A computational model of the fetal circulation to quantify blood redistribution in intrauterine growth restriction. PLoS Comput. Biol. **10**(6), 9–11 (2014). https://doi.org/10.1371/journal.pcbi.1003667
7. Hahn, C., Schwartz, M.A.: Mechanotransduction in vascular physiology and atherogenesis. Nat. Rev. Mol. Cell Biol. **10**(1), 53–62 (2009). https://doi.org/10.1038/nrm2596
8. Hermida, U., et al.: Learning the hidden signature of fetal arch anatomy: a three-dimensional shape analysis in suspected coarctation of the aorta. J. Cardiovasc. Transl. Res. (2022). https://doi.org/10.1007/s12265-022-10335-9
9. Hutchins, G.M.: Coarctation of the aorta explained as a branch-point of the ductus arteriosus. Am. J. Pathol. **63**(2), 203–214 (1971)
10. Iwaki, R., et al.: Evaluation of ductal tissue in coarctation of the aorta using X-ray phase-contrast tomography. Pediatr. Cardiol. **42**(3), 654–661 (2021). https://doi.org/10.1007/s00246-020-02526-5
11. Jansz, M.S., et al.: Metric optimized gating for fetal cardiac MRI. Magn. Reson. Med. **64**(5), 1304–1314 (2010). https://doi.org/10.1002/mrm.22542
12. Lloyd, D.F., et al.: Three-dimensional visualisation of the fetal heart using prenatal MRI with motion-corrected slice-volume registration: a prospective, single-centre cohort study. Lancet **393**(10181), 1619–1627 (2019). https://doi.org/10.1016/S0140-6736(18)32490-5
13. Lloyd, D.F., Rutherford, M.A., Simpson, J.M., Razavi, R.: The neurodevelopmental implications of hypoplastic left heart syndrome in the fetus. Cardiol. Young **27**(2), 217–223 (2017). https://doi.org/10.1017/S1047951116001645
14. Morris, P.D., et al.: Computational fluid dynamics modelling in cardiovascular medicine. Heart **102**(1), 18–28 (2016). https://doi.org/10.1136/heartjnl-2015-308044

15. Roberts, T.A., et al.: Fetal whole heart blood flow imaging using 4D cine MRI. Nat. Commun. **11**(1), 1–13 (2020). https://doi.org/10.1038/s41467-020-18790-1
16. Roux, E., Bougaran, P., Dufourcq, P., Couffinhal, T.: Fluid shear stress sensing by the endothelial layer. Front. Physiol. **11**(July), 1–17 (2020). https://doi.org/10.3389/fphys.2020.00861
17. Rudolph, A.M., Heymann, M.A., Spitznas, U.: Hemodynamic considerations in the development of narrowing of the aorta. Am. J. Cardiol. **30**(5), 514–525 (1972). https://doi.org/10.1016/0002-9149(72)90042-2
18. Salman, H.E., Yalcin, H.C.: Computational modeling of blood flow hemodynamics for biomechanical investigation of cardiac development and disease. J. Cardiovasc. Dev. Dis. **8**(2), 1–27 (2021). https://doi.org/10.3390/JCDD8020014
19. Schulz, A., et al.: Structured analysis of the impact of fetal motion on phase-contrast MRI flow measurements with metric optimized gating. Sci. Rep. **12**(1), 1–11 (2022). https://doi.org/10.1038/s41598-022-09327-1
20. Struijk, P.C., et al.: Blood pressure estimation in the human fetal descending aorta. Ultrasound Obstet. Gynecol. **32**(5), 673–681 (2008). https://doi.org/10.1002/uog.6137
21. Van Den Wijngaard, J.P., Westerhof, B.E., Faber, D.J., Ramsay, M.M., Westerhof, N., Van Gemert, M.J.: Abnormal arterial flows by a distributed model of the fetal circulation. Am. J. Physiol. Regul. Integr. Comp. Physiol. **291**(5), 1222–1233 (2006). https://doi.org/10.1152/ajpregu.00212.2006
22. Vigneswaran, T.V., Zidere, V., Chivers, S., Charakida, M., Akolekar, R., Simpson, J.M.: Impact of prospective measurement of outflow tracts in prediction of coarctation of the aorta. Ultrasound Obstet. Gynecol. **56**(6), 850–856 (2020). https://doi.org/10.1002/uog.21957
23. Yokoyama, U., Ichikawa, Y., Minamisawa, S., Ishikawa, Y.: Pathology and molecular mechanisms of coarctation of the aorta and its association with the ductus arteriosus. J. Physiol. Sci. **67**(2), 259–270 (2016). https://doi.org/10.1007/s12576-016-0512-x

Showcasing Capabilities of a Hybrid Mock Circulation Loop for Investigation of Aortic Coarctation

Emanuele Perra, Oliver Kreis, and Seraina A. Dual(✉)

KTH Royal Institute of Technology, Stockholm, Sweden
seraina@kth.se

Abstract. Congenital heart diseases are the most frequently diagnosed birth defect of the cardiovascular system (CVS), occurring in 1% of live births globally. Mock circulatory loops (MCLs) replicate the physiological boundary conditions of the CVS, which allow for testing of cardiac assist devices (CADs), but also provide valuable *in vitro* data for optimizing imaging protocols as well as validating computational fluid dynamics simulations. However, innate limitations of traditional MCLs include the difficulty in tuning physical boundary conditions to match the dynamic patient's physiology. To address these limitations, hybrid mock circulatory loops (HMCLs) incorporate elements of both *in vitro* and *in silico* modelling allowing for rapid changes in boundary conditions to be mimicked in closed-loop. In this study, a real-time HMCL testing platform was built, and its use exemplified in the study of aortic coarctation (AoC), a common congenital cardiovascular disorder caused by a narrowing of the descending aorta. Compliant 3D-printed stenosed tubes of varying severity were integrated into the HMCL to represent the AoC model. First their mechanical impedance was quantified using a chirp pressure signal. Second, the effect of the severity of coarctation on the simulated CVS variables (pressure difference, cardiac output) was assessed in dynamic interaction with the closed-loop CVS. This study lays the foundation for future studies into dynamic cardiovascular conditions, imaging improvements, and validation of fluid dynamics modelling.

Keywords: Cardiovascular diseases · Hybrid mock loop · Aortic coarctation

1 Introduction

Cardiovascular diseases (CVDs) are a leading cause of death globally, accounting for over 17 million deaths per year [13]. CVDs include a variety of conditions whose etiology is mostly associated with genetic, lifestyle, and environmental factors [2]. The complexity of the cardiovascular system (CVS) and the variability of physiological behavior among patients can pose challenges in selecting the most appropriate treatment approach. In order to improve currently existing

O. Bernard et al. (Eds.): FIMH 2023, LNCS 13958, pp. 505–514, 2023.
https://doi.org/10.1007/978-3-031-35302-4_52

treatment options, researchers have developed mock circulatory loops (MCLs), which provide a mechanical platform for replicating a specific physiological or pathological condition of the CVS [5]. MCLs further aid in optimizing imaging protocols and in validating computational fluid dynamic models.

MCLs typically consists of hydraulic components such as pumps, pipes, air-filled or fluid-filled reservoirs that are used to mimic the functioning of the heart, the resistance and compliance of blood vessels, and the blood's inertia, closely replicating the physiological behavior of the CVS [10,12,19,21]. Despite the usefulness of MCLs in the study of CVS physiology, one of the major limitations is the complexity in tuning the physical parameters (*e.g.* clamp forces, air reservoir pressure, fluid volumes, *etc.*) to match the dynamically changing physiological properties of the CVS for a specific patient [11,18].

Hybrid MCLs have lately drawn attention in the field of mechanical circulatory support, as they offer a flexible solution for simulating the physiological response to treatments and interventions by incorporating elements of both *in vitro* and *in silico* modelling [7,14–16,20]. This combination allows for the integration of the CVS through numerical simulation, and facilitates the testing of various medical devices, *e.g.* cardiac assist devices (CADs), within a laboratory setting.

Motivated by the need of advancing the general knowledge of CVS, assisting and accelerating the development process of CADs, we have built a real-time HMCL testing platform, and exemplified its potential through the study of the aortic coarctation (AoC), a congenital malformation that usually manifests as a narrowing of the descending aorta. The elevated transmural pressures generated in the aorta [8,17] can pose additional load on the heart, thus limiting the exercise performance of these patients.

In this study, compliant 3D-printed stenosed tubes with varying degrees of narrowing were used to represent an AoC model, and integrated into the HMCL. The rheological properties of the tubes were first characterized and their impact on simulated CVS variables was assessed. This work lays the foundation for future research into CVDs and the development of various CADs [1].

2 Methods

2.1 Hardware - Physical Part

The presented HMCL is based on the two HMCLs from ETH Zurich [15,16]. A representation of HMCL system is depicted in Fig. 1. The physical part of the HMCL used for this study consists of two fluid reservoirs (Fluid reservoir 1 (FR1) and Fluid reservoir 2 (FR2)) (Fig. 1a), in which the fluid pressure is controlled by modulating the pressure of the air cavity present in the hermetically closed reservoir. Air valves (PVQ33-5G-23-01F and PVQ33-5G-40-01F, SMC Pneumatics, Tokyo, Japan) allow exchange of pressurised air or vacuum in the reservoir, while the pressure in the liquid is constantly monitored by a pressure sensor (PN2099,

IFM Electronic Geräte GmbH & Co. KG, Essen, Germany). Both fluid reservoirs are connected *via* PVC tubing (outer diameter = 21.5 mm, wall thickness = 2 mm), in which a 3D-printed (printer model: Formlabs Form 2, Formlabs, Inc., Somerville, MA, USA; printing material: Elastic 50A, MakerBot, Inc., Brooklyn, NY, USA) AoC model is inserted in series (Fig. 1b). The fluid level in the reservoirs is continuously compensated for *via* a backflow pump (Jabsco Pump 23680-4103, Xylem Inc., Washington, USA), which is programmed to generate a unidirectional flow from FR1 to FR2 in order to minimize the fluid level difference in both reservoirs. The actuation of the backflow pump is regulated by feedforward controller based on real-time measurements (sample frequency = 1000 Hz) of the fluid flow (Sonoflow CO.55/190, Sonotec GmbH, Halle, Germany) from FR1 to FR2 (Fig. 1b), and a proportional-integral-derivative (PID) controller which is based on the height of fluid level, as measured by infrared distance sensors (GP2Y0A51SK0F, Sharp K.K., Sakai, Japan) located on the lid of each reservoir. The fluid used for the study consists of a mixture of water and glycerol (volume ratio of 5:3, water/glycerol), equivalent to the viscosity of blood with a hematocrit level of approximately 40%.

2.2 Software - Numerical Model of the CVS

The lumped-parameter numerical model of the CVS used in this study was adopted from a model developed by Colacino et al. [6]. An additional 4-element Windkessel model representing the aortic root segment [4] was inserted in series between the left ventricle and the 5-element Windkessel of the systemic arterial compartment to simulate the occurrence of AoC between the aortic root segment and the systemic arterial compartment (Fig. 1c). Although it is well-known that AoC typically develops as a tight stenosis within the descending aorta [17], this simplification allowed us to easily integrate the physical AoC model into the HMCL setup.

The CVS numerical model and the algorithms acting on the control of the sensors actuators were first implemented in MATLAB Simulink (The MathWorks, Inc., Natick, MA, USA) and ran as a real-time system through the MicroLabBox (MicroLabBox, dSPACE GmbH, Paderborn, Germany) at a sampling frequency of 1000 Hz. The inlet and outlet pressure (p_1 and p_2, respectively) are simulated in the CVS numerical model and are applied to FR1 and FR2, respectively. The measured flow rate q_1 represents the blood flow rate in the AoC as a response to applied pressures and is fed back to CVS numerical model, hence closing the circulation loop (Fig. 1c).

2.3 Experiments and Data Analysis

Compliant 3D-Printed Coarctation Models. Four types of stenosed compliant tubes were 3D-printed to represent different levels of AoC. The selected coarctation indexes (CIs) were defined as the ratio between the widest and narrowest diameters of the stenosed tube and were selected based on previous studies on the effect of AoC on cardiovascular hemodynamics [9]. The selected CIs were:

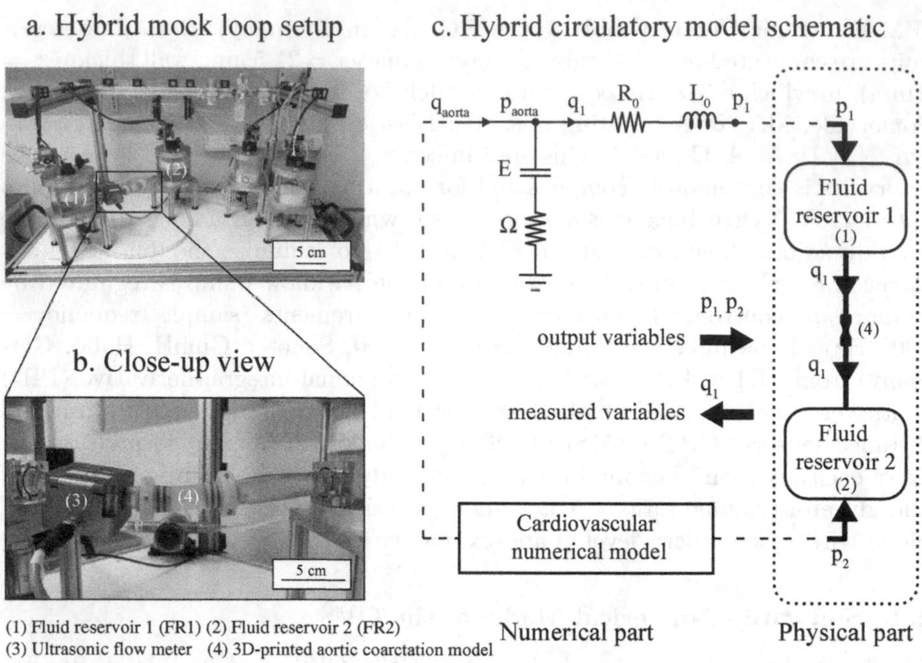

a. Hybrid mock loop setup

b. Close-up view

c. Hybrid circulatory model schematic

(1) Fluid reservoir 1 (FR1) (2) Fluid reservoir 2 (FR2)
(3) Ultrasonic flow meter (4) 3D-printed aortic coarctation model

Numerical part

Physical part

Fig. 1. (**a**) Representation of the HMCL setup and (**b**) close-up view on the compliant 3D-printed coarctation model. (**c**) Schematics of the interfaces between the numerical model of the CVS and the physical part of the HMCL.

a) CI = 1 (control model), b) CI = 0.7 (mild coarctation), c) CI = 0.5 (moderate coarctation), and d) CI = 0.3 (severe coarctation). The geometry of the AoC models used in this study is shown in Fig. 2a.

Characterization of Compliant 3D-Printed AoC Models. The properties of the compliant 3D-printed AoC models were first characterized in terms of mechanical impedance. To achieve this, each AoC model was subjected to a chirp signal with a linear frequency sweep given by Eq. 1. The pressure chirp signal p_1 was applied to FR1, while the pressure p_2 in FR2 was kept at 0 mmHg. This created a pressure difference between FR1 and FR2, and the resulting fluid flow rate q_1 was measured and recorded. The chirp signal was given by:

$$\Delta P = p_1(t) - p_2(t) = P_0 + A \sin\left(2\pi\left(t\frac{f_1 - f_0}{T} + f_0\right)t\right) \text{mmHg}, \quad (1)$$

where $p_1(t)$ is the pressure applied in FR1 and $p_2(t) = 0$ is the pressure applied in FR2, P_0= 10, 37.5 and 56.5 mmHg denotes the offset of the chirp signal, A= 10, 37.5 and 56.5 mmHg the amplitude, f_1=3 Hz the target frequency, f_0= 0.67 Hz the initial frequency and T = 180 s defines the sweeping time. The parameters

P_0 and A were selected based on common pressure drops measurements found in literature [3], while f_1 and f_0 were chosen in order to cover the physiological range of heart rates (HRs) (from 40 to 180 bpm). Finally, the impedance $Z(\omega)$ of each AoC model, at a given pressure drop level, was calculated as $\Delta P(\omega)/Q_1(\omega)$, $\Delta P(\omega)$ and $Q_1(\omega)$ being the fast-Fourier transform of the measured pressure drop and flow rate over time, respectively. The number of measurement repeats for each and sample group was $n = 5$.

Integration of AoC Models with CVS Model. The AoC models were tested in hybrid configuration with the CVS numerical model (Fig. 1c) at different HRs (40, 60, 90, 120 and 180 bpm), the total simulation time being 60 s. The CVS model was kept the same, while the AoC model was exchanged resulting in altered hemodynamics. The mean pressure drop across the coarctation as well as resulting cardiac output was recorded for each AoC model in order to assess the effect of the various levels of coarctation on cardiac function. Finally, the accuracy of the system was evaluated by calculating the root-mean-square (RMS) error between the simulated pressure and the measured pressure in FR1. The time delay between the simulated and actuated pressure was estimated by identifying the location of peak of the cross-correlation function, which corresponds to the time lag between the two signals. The number of experimental repeats per sample group was $n = 3$.

3 Results

3.1 Characterization of Compliant 3D-Printed AoC Models

The results of the mechanical impedance characterization of the AoC models are displayed in Fig. 2b, which shows the impedance as a function of HR for each of the AoC models, at different levels of pressure drop.

It was observed that the control and mild coarctation models have relatively low impedance compared to higher severities. The control model (CI = 1) exhibited an average magnitude of impedance across the explored HR range of 0.02, 0.06 and 0.09 mmHgs/ml, when the applied average pressure drops were 10, 37.5 and 56.5 mmHg, respectively. Similar values were observed for the mild coarctation model (CI = 0.7): 0.03, 0.06 and 0.09 mmHgs/ml. In contrast, the moderate and severe coarctation models (CI = 0.5 and 0.3) displayed elevated impedance, with an increase in impedance for increasing pressure drop, denoting a nonlinear behaviour. The moderate coarctation model (CI = 0.3) exhibited an impedance peak of 0.54 and 0.38 mmHg s/ml at 60 bpm, when the applied average pressure drops were 37.5 and 56.5 mmHg, respectively. While, at ΔP = 10 mmHg, the magnitude of impedance increased from 0.14 to 0.18 mmHg s/ml in the investigated HR range. The severe coarctation model presented a similar behaviour, with an impedance peak of 0.57 and 0.40 mmHg s/ml at 65 bpm (ΔP = 37.5 and 56.5 mmHg), and increasing impedance from 0.12 to 0.18 mmHg s/ml within the explored range of HRs (ΔP = 10 mmHg).

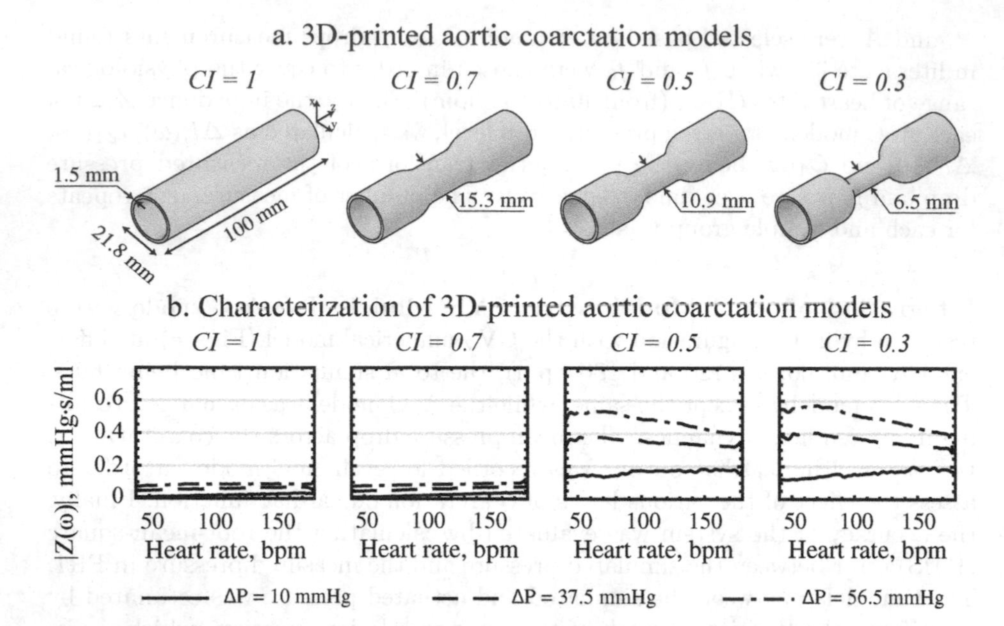

Fig. 2. (**a**) Drawing representing the geometry of the coarctation models at different CI levels and (**b**) their respective impedance curve characterization.

3.2 Integration of AoC Models with CVS Model

Figure 3a depicts the time-average pressure drops across the coarctation models at different HRs in interaction with the CVS model. The pressure drop increased systematically with increasing HR when the severe coarctation model (CI = 0.3) was employed (from 42 to 62 mmHg). Conversely, other models exhibited an average pressure drop peak at HR = 90 bpm (25, 26 and 34 mmHg for CI = 1, 0.7 and 0.5, respectively), being the lowest at HR = 40 bpm (17, 9 and 25 mmHg for CI = 1, 0.7 and 0.5, respectively). The CO was assessed to study the response of the numerical CVS model on coarctation and was found to increase, independently of the coarctation model, with increasing HR in the range 40–120 bpm, being 4.3, 4.4, 4.2 and 3.9l/min at HR = 40 bpm, and 5.1, 5.0, 4.8 and 4.6 l/min at HR = 120 bpm, when CI = 1, 0.7, 0.5 and 0.3, respectively (Fig. 3b). Furthermore, it can be noted how, regardless of the HR, the CO decreases with the severity of the AoC indicating that the heart cannot keep up cardiac output given the increased vascular impedance. However, at HR = 180 bpm, a slight decrease in the CO could be observed for all coarctation models, except for the severe coarctation model.

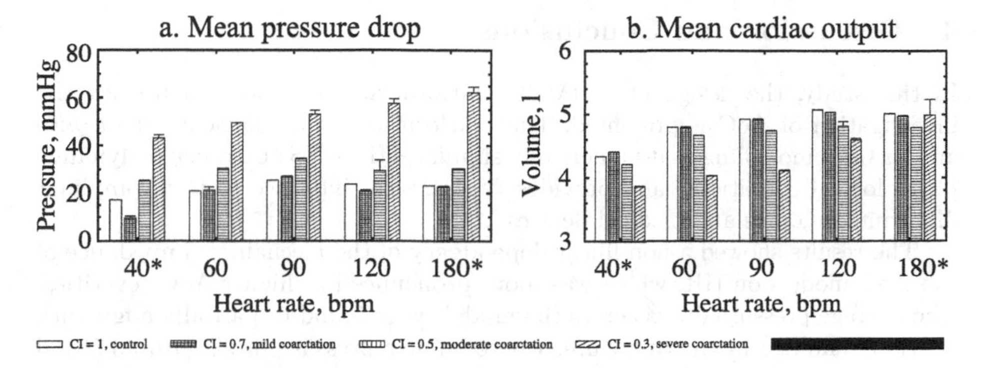

Fig. 3. (a) Time-averaged pressure drop measured across each AoC model and (b) estimated COs as function of HR.

3.3 Evaluation of Pressure Control Performance

An exemplary comparison between the simulated pressure wave to be applied in FR1 and the actuated pressure wave is presented in Fig. 4a. The RMS error between the two signals was found to be less than 20 mmHg for the entire duration of the simulation at HRs of 40 and 60 bpm (Fig. 4b). However, at elevated HRs the RMS error becomes greater, reaching the value of approximately 60 mmHg at HR = 180 bpm.

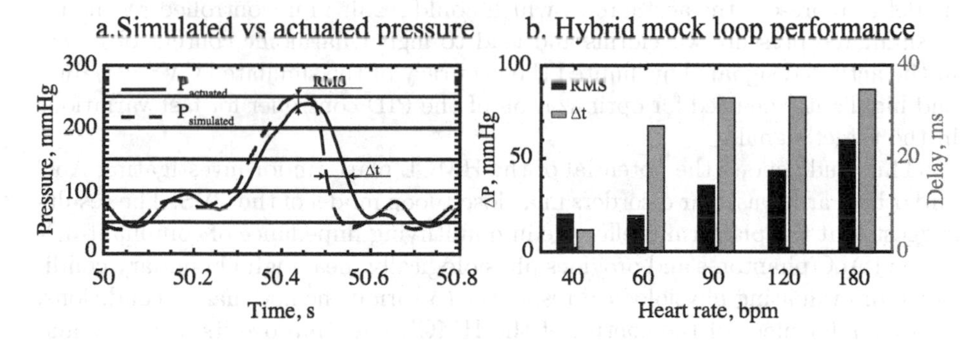

Fig. 4. (a) Exemplary representation of simulated and actuated pressure waveform in FR1. (b) The RMS errors and delay between the two waveforms are calculated at different HRs.

The time delay between the two signals was determined to be less than 4 ms at HR = 40 bpm. On the other hand, the time delay increased up to 34ms when higher HRs were employed in the simulation.

4 Discussion and Conclusions

In this study, the design of a HMCL platform and its potential use for the investigation of AoC is presented. The platform combines elements of *in vitro* and *in silico* modelling as it numerically simulates the CVS at a range of dynamic physiological conditions, and provides the physical interface to test compliant 3D-printed models and medical devices.

The results showed a non-linear dependency of the mechanical impedance of the AoC models on HR, which was more pronounced at higher AoC severities. The average pressure drops across the models were found to partially align with previous findings in the literature, where similar pressure drops (10, 33.5 and 51.7 mmHg) were reported for different CI (1.1, 0.5 and 0.3) levels at a fixed HR of 125 bpm [3].

All COs generated closed-loop numerical model of the CVS were within the physiological range. The resistance to fluid flow increased with lower CI levels, leading to higher pressure drops across the coarctation, and lower COs which can be generated by the heart. However, the pressure difference was compensated for at high HRs due to the increased pumping frequency of the heart, resulting in normalization of CO. The results exemplify, how the HMCL can easily replicate physiological responses to changes in hemodynamic conditions as present in patients with AoC.

Limitations in the current HMCL setup include high RMS error values and delays between the target and actuated pressure waveforms. This discrepancy might be due to the addition of the 4-element Windkessel model in the CVS model to represent the aortic root, which could result in uncontrolled resonance in simulated pressure waveforms and lead to higher harmonic content observed in the actuated signal. This limited the accuracy of the simulated CVS variables and highlights the need for optimization of the PID controller for fast variations in the target signal.

This study shows the potential of the HMCL platform for investigating AoC and other cardiovascular disorders in a closed-loop model of the CVS. The results suggest that the platform is effective in quantifying impedance of compliant 3D-printed AoC phantoms and provides physiologically meaningful boundary conditions for evaluating physiological responses to various hemodynamic conditions. Further refinement of the control of the HMCL may improve its accuracy and allow versatile testing of a wider range of medical devices and cardiovascular conditions. On the other hand, the physiological reliability of the system can be ensured by substituting the *in silico* model with a CVS model of choice, allowing for greater customization and adaptability to specific research or device development requirements. In this way, a HMCL is a very valuable tool in bridging the computational and experimental models thereby allowing for agile choice of the most appropriate tool at any given point in the research and development process.

The herein presented HMCL platform can potentially aid future research into validating patient-specific hemodynamic computational models, as well as clinical imaging modalities and the development of more effective cardiovascular assist devices.

Acknowledgements. We thank and acknowledge VINNOVA (2022-00849), Digital Futures, and the Digital platform of KTH for their financial support; Dr. Marianne Schmid Daners and Dr. Thomas Gwosch at ETH Zurich for sharing their know-how in replicating the hybrid mock circulation loop and provision of the colacino-model implementation; Sara Mettler for the electrical support; Peter Arfert for the mechanical design; Laura Andersson, and Roxanne Rais for their experimental support.

References

1. Functional Imaging and Modeling of the Heart. https://link.springer.com/book/10.1007/978-3-030-78710-3
2. Benjamin, E.J.: Heart Disease and Stroke Statistics-2019 Update: A Report From the American Heart Association. Circulation **139**(10), e56–e528 (2019)
3. Biglino, G., et al.: In vitro study of the Norwood palliation: a patient-specific Mock circulatory system. ASAIO J. **58**(1), 25 (2012). https://doi.org/10.1097/MAT.0b013e3182396847
4. Broomé, M., Maksuti, E., Bjällmark, A., Frenckner, B., Janerot-Sjöberg, B.: Closed-loop real-time simulation model of hemodynamics and oxygen transport in the cardiovascular system. Biomed. Eng. Online **12**(1), 69 (2013). https://doi.org/10.1186/1475-925X-12-69
5. Cappon, F., Wu, T., Papaioannou, T., Du, X., Hsu, P.L., Khir, A.W.: Mock circulatory loops used for testing cardiac assist devices: a review of computational and experimental models. Int. J. Artif. Organs **44**(11), 793–806 (2021). https://doi.org/10.1177/03913988211045405
6. Colacino, F.M., Moscato, F., Piedimonte, F., Arabia, M., Danieli, G.A.: Left ventricle load impedance control by apical VAD can help heart recovery and patient perfusion: a numerical study. ASAIO J. **53**(3), 263–277 (2007). https://doi.org/10.1097/MAT.0b013e31805b7e39
7. Fresiello, L., et al.: A cardiovascular simulator tailored for training and clinical uses. J. Biomed. Inform. **57**, 100–112 (2015). https://doi.org/10.1016/j.jbi.2015.07.004. https://www.sciencedirect.com/science/article/pii/S1532046415001446
8. Itu, L.: Non-invasive hemodynamic assessment of aortic coarctation: validation with in vivo measurements. Ann. Biomed. Eng. **41**(4), 669–681 (2013). https://doi.org/10.1007/s10439-012-0715-0
9. Lemler, M.S., Zellers, T.M., Harris, K.A., Ramaciotti, C.: Coarctation index: identification of recurrent coarctation in infants with hypoplastic left heart syndrome after the Norwood procedure. Am. J. Cardiol. **86**(6), 697–699 (2000). https://doi.org/10.1016/S0002-9149(00)01058-4
10. Li, X., Li, Z., Chen, D.: A mock circulation loop for in vitro haemodynamic evaluation of aorta. J. Phys. Conf. Ser. **1600**(1), 012066 (2020). https://doi.org/10.1088/1742-6596/1600/1/012066
11. Liu, Y., Allaire, P., Wu, Y., Wood, H., Olsen, D.: Construction of an artificial heart pump performance test system. Cardiovasc. Eng. **6**(4), 151–158 (2006). https://doi.org/10.1007/s10558-006-9019-z
12. Malone, A., et al.: In vitro benchtop mock circulatory loop for heart failure with preserved ejection fraction emulation. Front. Cardiovasc. Med. **9** (2022)
13. Mozaffarian, D., et al.: Heart Disease and Stroke Statistics-2016 Update (2016)
14. Nestler, F., Bradley, A.P., Wilson, S.J., Timms, D.L., Frazier, O.H., Cohn, W.E.: A hybrid mock circulation loop for a total artificial heart. Artif. Organs **38**(9), 775–782 (2014). https://doi.org/10.1111/aor.12380

15. Ochsner, G., et al.: A novel interface for hybrid mock circulations to evaluate ventricular assist devices. IEEE Trans. Biomed. Eng. **60**(2), 507–516 (2013). https://doi.org/10.1109/TBME.2012.2230000
16. Petrou, A., Granegger, M., Meboldt, M., Schmid Daners, M.: A versatile hybrid mock circulation for hydraulic investigations of active and passive cardiovascular implants. ASAIO J. **65**(5), 495–502 (2019). https://doi.org/10.1097/MAT.0000000000000851
17. Priya, S., Thomas, R., Nagpal, P., Sharma, A., Steigner, M.: Congenital anomalies of the aortic arch. Cardiovasc. Diagn. Ther. **8**(Suppl 1), S26–S2S44 (2018). https://doi.org/10.21037/cdt.2017.10.15
18. Taylor, C.E., Miller, G.E.: Implementation of an automated peripheral resistance device in a mock circulatory loop with characterization of performance values using Simulink Simscape and parameter estimation. J. Med. Devices **6**(4) (2012). https://doi.org/10.1115/1.4007458
19. Timms, D., Hayne, M., McNeil, K., Galbraith, A.: A complete mock circulation loop for the evaluation of left, right, and biventricular assist devices. Artif. Organs **29**(7), 564–572 (2005). https://doi.org/10.1111/j.1525-1594.2005.29094.x
20. Zielinski, K., Darowski, M., Kozarski, M., Ferrari, G.: The need for hybrid modeling in analysis of cardiovascular and respiratory support. Int. J. Artif. Organs **39**(6), 265–271 (2016). https://doi.org/10.5301/ijao.5000513
21. Zimmermann, J., et al.: On the impact of vessel wall stiffness on quantitative flow dynamics in a synthetic model of the thoracic aorta. Sci. Rep. **11**(1), 6703 (2021). https://doi.org/10.1038/s41598-021-86174-6. https://www.nature.com/articles/s41598-021-86174-6

Hemodynamics in Patients with Aortic Coarctation: A Comparison of *in vivo* 4D-Flow MRI and FSI Simulation

Priya J. Nair[1(✉)], Martin R. Pfaller[2], Seraina A. Dual[3], Michael Loecher[4,5], Doff B. McElhinney[6], Daniel B. Ennis[4,5], and Alison L. Marsden[2]

[1] Department of Bioengineering, Stanford University, Stanford, USA
priyanair@stanford.edu
[2] Department of Pediatrics, Stanford University, Stanford, USA
[3] Department of Biomedical Engineering and Health Systems, KTH Royal Institute of Technology, Stockholm, Sweden
[4] Department of Radiology, Stanford University, Stanford, USA
[5] Division of Radiology, VA Palo Alto Health Care System, Palo Alto, USA
[6] Department of Cardiothoracic Surgery, Stanford University, Stanford, USA

Abstract. The analysis of quantitative hemodynamics provides information for the diagnosis and treatment planning in patients with aortic coarctation (CoA). Patient-specific computational fluid dynamics (CFD) simulations reveal detailed hemodynamic information, but their agreement with the clinical standard 4D-Flow magnetic resonance imaging (MRI) needs to be characterized. This work directly compares *in vivo* CFD fluid-structure interaction (FSI) simulations against 4D-Flow MRI in patients with CoA (N = 5). 4D-Flow MRI-derived flow waveforms and cuff blood pressure measurements were used to tune the boundary conditions for the FSI simulations. Flow rates from 4D-Flow MRI and FSI were compared at cross-sections in the ascending aorta (AAo), CoA and descending aorta (DAo). Qualitative comparisons showed an overall agreement of flow patterns in the aorta between the two methods. The R^2 values for the flow waveforms in the AAo, CoA, and DAo were 0.97, 0.84 and 0.81 respectively, representing a strong correlation between 4D-Flow MRI measurements and FSI results. This work characterizes the use of patient-specific FSI simulations in quantifying and analyzing CoA hemodynamics to inform CoA treatment planning.

Keywords: Aortic coarctation · Hemodynamics · Computational fluid dynamics · Fluid-structure interaction · 4D-Flow MRI

1 Introduction

Coarctation of the aorta (CoA) is a congenital heart defect characterized by a segmental constriction of the aorta. It accounts for 6–8% of all congenital heart defects [8], with an estimated incidence of 3 per 10,000 live births [9]. The

O. Bernard et al. (Eds.): FIMH 2023, LNCS 13958, pp. 515–523, 2023.
https://doi.org/10.1007/978-3-031-35302-4_53

narrowing of the aorta causes a drop in blood pressure (BP) across the CoA. If left untreated, it can result in hypertension, stroke, and aortic rupture [2]. Even after successful repair (defined as correction of the BP drop), patients with CoA continue to experience long-term ventricular and arterial complications, and have significantly lower survival rates than the normal population [4]. Previous studies have related the morbidity in CoA to adverse local hemodynamics in the aorta and its branches [5]. Studying and understanding the hemodynamics associated with CoA and its progression is therefore of high clinical importance.

CFD has proved to be a useful tool for the assessment of local hemodynamics in patients with cardiovascular disease. Non-invasive anatomic and physiological data acquired in the clinic can be used to generate patient-specific hemodynamics simulations to inform diagnosis, treatment and outcomes. While CFD approaches show great potential, the method must be characterized against the clinical standard: *in vivo* 4D-Flow MRI. In this study, we compare qualitative and quantitative hemodynamics from patient-specific CFD FSI simulations to those obtained from 4D-Flow MRI in five patients with CoA.

2 Methods

In this section, we summarize the methods used to create patient-specific CoA models and perform FSI simulations, as well as the methods employed to process the corresponding 4D-Flow MRI datasets.

2.1 CFD FSI Simulation

Patient Data Acquisition. 4D-Flow MRI datasets (IRB approved) were acquired for five patients with CoA (Table 1) who thereafter underwent a catheter-based stenting procedure. Cuff BP was measured non-invasively on the same day. Imaging exams were not conducted on the day of the catheterization but were acquired, on average, 80 days (range = 4–222 days) before catheterization. Average difference in heart rate between the day of imaging and the day of catheterization was 3 bpm, indicating similar physiological states on both days.

Anatomic Model Generation. 3D geometries of patient-specific aortas were generated by importing the MRI magnitude images into SimVascular and manually segmenting the vessels of interest. Centerlines were automatically extracted from the 3D geometry to create the 0D model comprising a network of lumped parameter elements (resistors, capacitors and inductors) [6]. The 3D volume was discretized using a mesh of tetrahedral elements. A three-layer boundary mesh was incorporated to resolve the velocity gradients at the wall. The mesh was further refined at the CoA and the post-stenotic dilation region to capture the jet formed by the CoA. The final mesh had approximately two million tetrahedral elements.

Table 1. Patient characteristics.

Patient	Sex	Age (y)	BSA (m^2)	Cuff BP (mmHg) systolic/diastolic	CO (L/min)	HR (bpm)
P-1	Male	15	1.94	145/86	8.9	68
P-2	Female	54	1.56	154/54	3.8	67
P-3	Male	21	1.58	105/61	3.3	94
P-4	Male	22	1.89	123/47	8.4	63
P-5	Male	44	2.09	152/93	6.4	54

BSA: Body Surface Area, CO: Cardiac Output, HR: Heart Rate

Boundary Conditions. Boundary conditions were tuned using non-invasive cuff BP measurements, and flow measurements derived from 4D-Flow MRI. Eddy current-corrected 4D-Flow MRI datasets were used to measure 2D time-resolved flow at the inlets and outlets using Arterys (Arterys, San Francisco, USA). 0D simulations were run first and used to tune the boundary conditions. The patient-specific temporally varying flow profile was prescribed to the 0D model inlet. Flow splits to each of the aortic branches, along with cuff BP data, were used to tune outflow boundary conditions. Three-element Windkessel boundary conditions (proximal resistance R_p, capacitance C, distal resistance R_d) were imposed at the outlets. Total resistance (R_p + R_d) for each of the branches was calculated using the cuff BP and the flow split determined from MRI. C as well as the R_p/R_d ratio were adjusted to fine-tune the boundary conditions until the calculated pressures matched the patient's systolic and diastolic cuff BP within 5 mmHg. The tuned boundary conditions were then applied to the 3D model outlets (Fig. 1). The patient-specific flow waveform derived from 4D Flow was prescribed to the 3D model inlet, assuming a parabolic flow profile.

Simulations. Simulations were performed with svSolver, SimVascular's finite element solver for fluid-structure interaction between an incompressible, Newtonian fluid and a linear elastic membrane for the vascular wall. The Young's modulus of the aortic wall was defined to be 3×10^6 dyn/cm^2, based on previously reported values of stiffness in a human aorta with CoA [3]. The wall thickness was assumed to be 10% of the vessel diameter. After optimizing the boundary conditions, the 0D simulation was run first for 10 cardiac cycles. The results from the last cardiac cycle of the 0D simulation were projected onto the 3D mesh and used to initialize the 3D simulation, allowing the 3D simulation to converge to a steady solution faster than it would with zero or steady state initialization [7].

The 3D simulations were performed with a deformable wall using the coupled momentum method (CMM). The model was prestressed by applying a diastolic load. For the 3D simulations, Poisson's ratio = 0.5 and density = 1 g/cm^3 were used for the vessel wall. The CMM method assumes a thin-walled structure with the equations of wall deformation coupled with the equations of fluid motion in

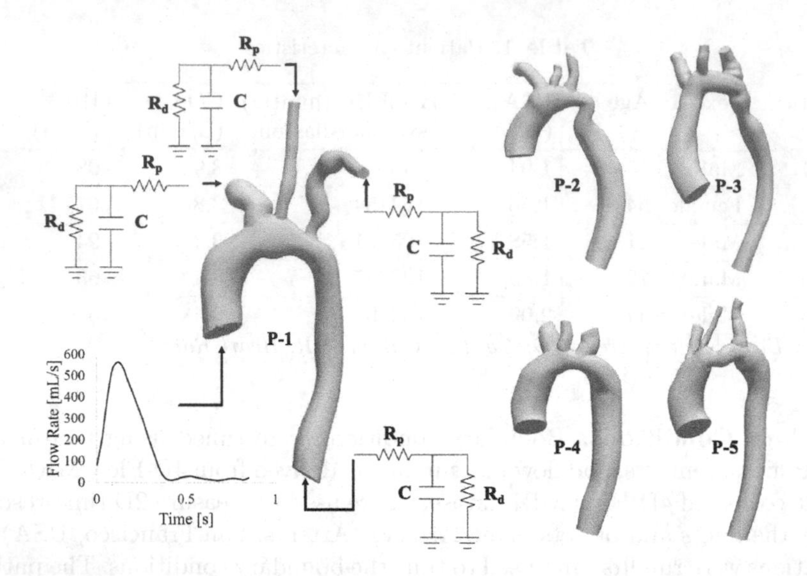

Fig. 1. 3D patient-specific anatomic models generated by segmenting MRI images, with depiction of boundary conditions applied. R_p: proximal resistance, C: capacitance, R_d: distal resistance

such a way that the degrees-of-freedom of the wall and the fluid boundary are enforced to be the same [1]. The phenomenon of locking of the finite elements is therefore not observed. The fluid was prescribed to have density $= 1.06\,\mathrm{g/cm^3}$ and viscosity $= 0.04$ poise to match the properties of blood. The 3D simulations were run for 10 cardiac cycles to ensure convergence; only the last cardiac cycle was analyzed.

Analysis. Time-resolved velocity was extracted from the last simulation cycle. The geometry was sliced at each centerline point and time-resolved cross-sectional average velocities were computed. FSI results had a temporal resolution of 1 ms and were visualized using ParaView.

2.2 4D-Flow MRI

Patient-specific 4D-Flow datasets were accessible through Arterys (Table 2). Eddy current correction was applied using a machine learning-based correction tool available in Arterys and then manually checked. Aorta model geometry from a single timepoint was used to mask 4D-Flow velocity fields. The dicom images were clipped to include only regions that had non-zero velocity and image magnitude. The resulting velocity fields were averaged over the cross-section using the same method described for averaging FSI results and were visualized using ParaView.

Table 2. 4D-Flow MRI imaging parameters.

Patient	VENC (cm/s)	Temporal resolution (ms)	Pixel size (mm^2)
P-1	350	28.4	1.2×1.2
P-2	550	42.5	1.4×1.8
P-3	250	30.3	0.9×1.3
P-4	300	42.3	1.3×1.8
P-5	350	52.8	1.4×1.9

VENC: Velocity encoding

2.3 Statistical Analysis

Errors were computed as the difference between FSI and 4D-Flow-derived measurements, then averaged across patients. R^2 values of flow waveforms from FSI and 4D-Flow MRI at slices in the AAo, CoA and DAo were computed.

3 Results

Qualitative comparisons of flow patterns reveal a good match between the *in vivo* 4D-Flow MRI-derived velocity fields and those obtained from the FSI simulation at peak systole (Fig. 2). At the same time, the velocity magnitude of the 4D-Flow data sometimes appears substantially lower than that of FSI.

Quantitative comparisons of flow rate at the AAo, CoA and DAo slice-levels show a general agreement of flow waveforms between 4D-Flow MRI and FSI simulations (Fig. 3). In P-1 and P-3, a time delay was observed between the occurrence of peak flow in 4D-Flow versus FSI in the CoA and DAo. A lack of conservation of mass in the 4D-Flow MRI flow measurements was also observed, with the mean flow in the DAo exceeding the CoA flow in 3 out of 5 patients. FSI tends to overestimate the mean flow rate, peak flow rate, and peak mean velocity in all slices. The error was minimum at the AAo, where the 4D-Flow MRI-derived flow waveform was prescribed as a boundary condition, and maximum at the CoA (Table 3). Errors in mean flow rate from FSI in the AAo, CoA and DAo were 13.5%, 85.4% and 29% of the mean flow rate measured using 4D-Flow MRI. Similarly, errors in peak flow rate were 10.7%, 68.6% and 37.8%. Errors in peak mean velocity were 18.8%, 88.6% and 53.4%. R^2 for the flow waveforms in the AAo, CoA, and DAo were 0.97, 0.84 and 0.81 respectively. This represents a strong correlation between 4D-Flow measurements and FSI results, with decreasing goodness of fit as distance from the inlet increases.

In P-1 specifically, large errors were observed between 4D-Flow MRI and FSI results, which may be explained by the observation that the vessel significantly expanded/contracted over the cardiac cycle. This large wall deformation caused discrepancies in the measured hemodynamic quantities due to the limits of our current vessel wall registration process. Given that P-1 had large errors even at the inflow (AAo) and therefore could not be expected to have accurate results downstream of it, results from P-1 although reported in the study, are excluded from the summary statistics.

Table 3. Flow characteristics.

Patient	Location	Mean flow rate (L/min)		Peak flow rate (L/min)		Peak mean velocity (cm/s)	
		4D	FSI	4D	FSI	4D	FSI
P-1*	AAo	4.5	9.1	19.7	34.2	98.4	185.0
	CoA	3.3	5.4	5.7	19.3	63.3	240.0
	DAo	4.0	5.4	7.3	18.3	68.3	193.4
P-2	AAo	3.6	4.0	14.6	16.4	50.0	58.3
	CoA	2.2	2.4	9.3	9.4	101.7	131.7
	DAo	2.1	2.4	7.2	7.0	41.7	45.0
P-3	AAo	2.9	2.5	14.9	17.6	56.7	71.7
	CoA	1.1	2.2	3.8	9.7	56.7	140.0
	DAo	1.6	2.2	4.6	8.6	46.7	98.4
P-4	AAo	7.1	8.8	39.7	43.3	98.4	116.7
	CoA	5.2	7.2	29.6	33.6	98.4	121.7
	DAo	3.5	4.8	14.1	18.9	60.0	88.4
P-5	AAo	6.6	6.5	24.3	25.1	36.7	41.7
	CoA	1.7	5.0	8.7	17.8	80.0	203.4
	DAo	3.3	4.2	9.4	12.5	38.3	56.7
Error	AAo	0.65 ± 0.78		2.22 ± 1.20		11.67 ± 6.09	
	CoA	1.65 ± 1.32		4.78 ± 3.76		65.01 ± 47.27	
	DAo	0.78 ± 0.43		2.92 ± 2.20		25.42 ± 20.30	

* *Excluded from summary statistics*

4 Discussion

This study aimed to characterize the agreement of FSI simulations with the clinical standard 4D-Flow MRI in five patients with CoA. Qualitative comparisons between 4D-Flow and FSI showed that patient-specific simulations are capable of capturing the overall flow patterns seen in patients with CoA.

Quantitative FSI results showed discrepancies in mean flow rate, peak flow rate, and peak mean velocity when compared to 4D-Flow. Some of the errors can be explained by the uncertainties associated with the 4D-Flow MRI technique. The sub-optimal spatial and temporal resolution of images acquired using 4D-Flow can contribute to discrepancies we see in the flow fields. Additionally, while several corrections are applied to the acquired velocity field data, some errors

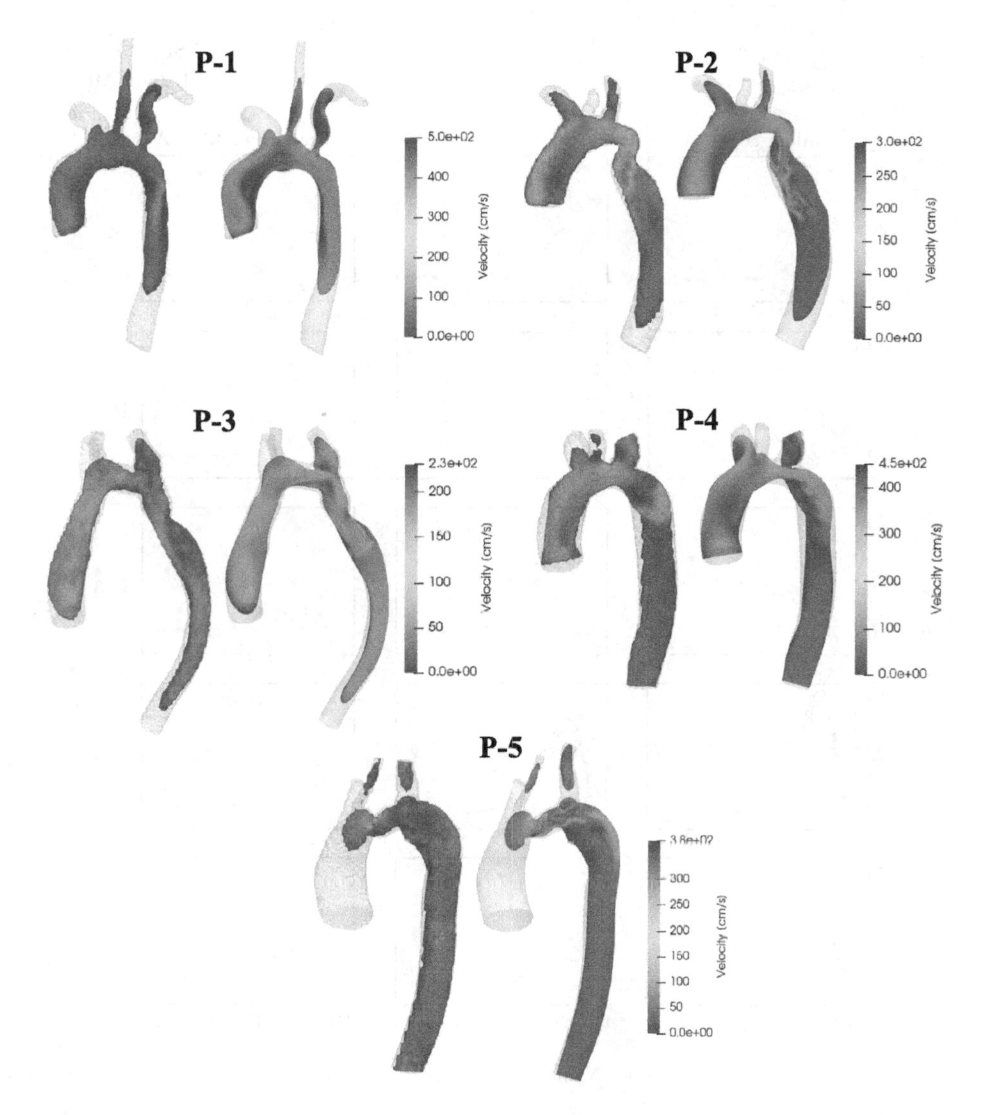

Fig. 2. Velocity magnitude at peak systole on a sagittal plane, a comparison between *in vivo* 4D-Flow MRI (left) and FSI simulation (right).

from phase offsets may still be present. This can be confirmed by our observation of lack of conservation of mass in the flow measured using 4D-Flow MRI in this study. Some of the errors observed in this study can also be attributed to the registration process used to mask the 4D-Flow MRI velocity data. We used the aortic geometry from a single time point to mask velocity information acquired throughout the cardiac cycle. For patients whose vessels experience large deformations over the cardiac cycle, our current registration process would introduce discrepancies in flow measurements.

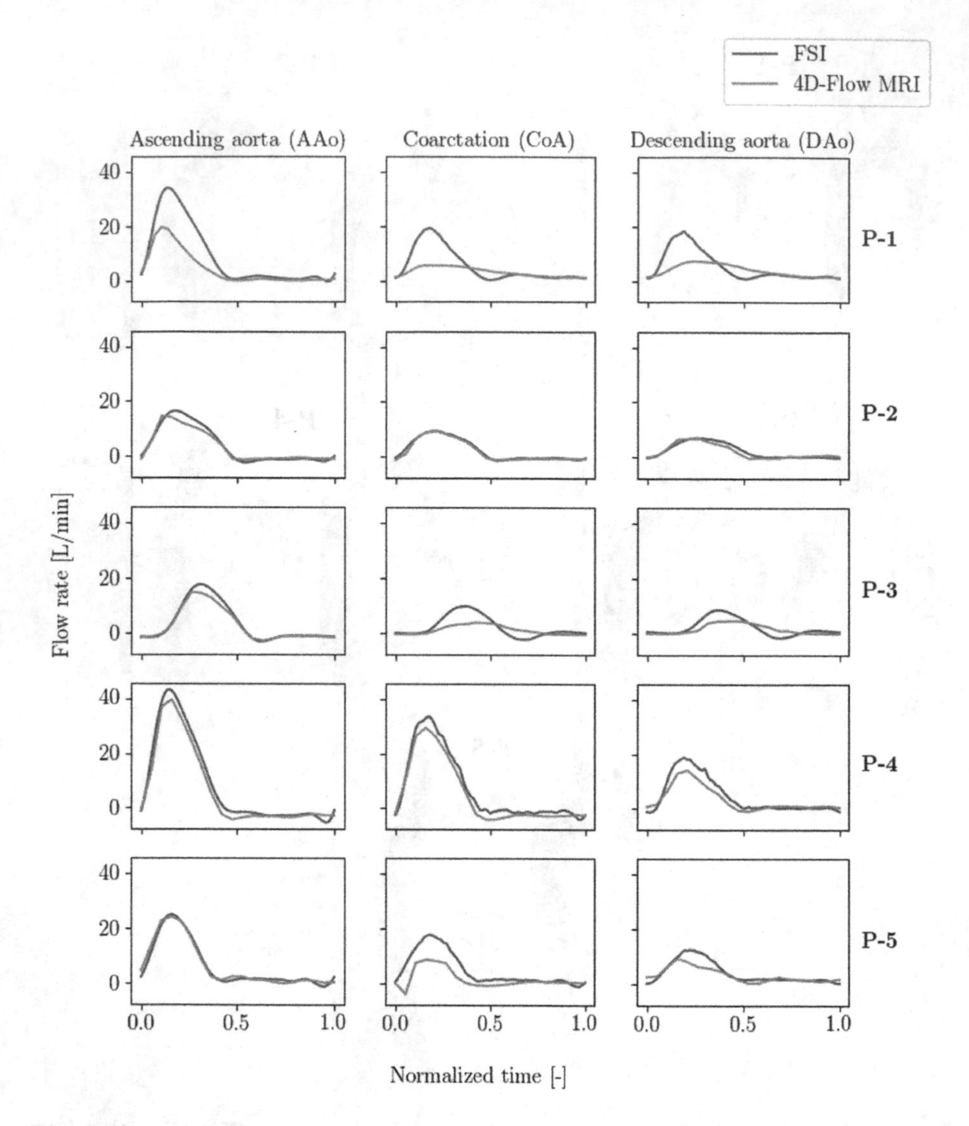

Fig. 3. Comparison of *in vivo* 4D-Flow MRI and simulated volumetric flow rate over time at a slice each in the AAo, CoA, and DAo.

On the FSI side, we used the same vessel wall stiffness (from literature) for all five patients. The mismatch of this defined stiffness with the true stiffness of the patient's aorta may explain the mismatch of the flow waveforms seen at the CoA and DAo. This likely also contributed to the time delay in the flow rate peak between 4D-Flow and FSI in P-1 and P-3. Lastly, the use of parabolic velocity profiles at the inlet could also have resulted in discrepancies downstream. Use of a more comprehensive registration process and methods to determine patient-specific stiffness estimates for the vessel walls can be explored in the future.

Overall, this study characterizes the hemodynamic similarities and differences between 4D-Flow MRI and FSI simulations. It validates FSI simulations as a valuable tool in assessing hemodynamics to inform treatment planning in patients with CoA. These simulations can be utilized to extract important metrics such as wall shear stress that cannot directly be measured in the clinic. Additionally, these simulations can be extended to other hemodynamic states, such as exercise, for a more comprehensive evaluation of a patient's disease burden.

Acknowledgements. The authors thank the Stanford Research Computing Center for computational resources (Sherlock HPC Cluster). This work is funded by a National Science Foundation Graduate Research Fellowship (DGE-1656518) to PJN and by the Stanford Maternal and Child Health Research Institute (award K99HL161313) to MRP.

References

1. Figueroa, C.A., Vignon-Clementel, I.E., Jansen, K.E., Hughes, T.J.R., Taylor, C.A.: A coupled momentum method for modeling blood flow in three-dimensional deformable arteries. Comput. Methods Appl. Mech. Eng. **195**(41–43), 5685–5706 (2006)
2. Jenkins, N.P., Ward, C.: Coarctation of the aorta: natural history and outcome after surgical treatment. Qjm **92**(7), 365–371 (1999)
3. Ladisa, J.F., et al.: Computational simulations for aortic coarctation: Representative results from a sampling of patients. J. Biomech. Eng. 133(9) (2011). https://doi.org/10.1115/1.4004996
4. Lee, M.G.Y., et al.: Long-term mortality and cardiovascular burden for adult survivors of coarctation of the aorta. Heart **105**(15), 1190 LP - 1196 (2019). https://doi.org/10.1136/heartjnl-2018-314257,http://heart.bmj.com/content/105/15/1190.abstract
5. Orourke, M.F., Cartmill, T.B.: Effects of aortic coarctation on pulsatile haemodynamics in proximal aorta. In: Australasian Annals of Medicine. vol. 18, p. 174. Royal Australasian Coll Phys 145 Macquarie Street, Sydney 2000, Australia (1969)
6. Pfaller, M.R., et al.: Automated generation of 0D and 1D reduced-order models of patient-specific blood flow. Int. J. Numer. Meth. Biomed. Eng. **38**(10), e3639 (10 2022). https://doi.org/10.1002/cnm.3639
7. Pfaller, M.R., Pham, J., Wilson, N.M., Parker, D.W., Marsden, A.L.: On the periodicity of cardiovascular fluid dynamics simulations. Ann. Biomed. Eng. **49**(12), 3574–3592 (2021). https://doi.org/10.1007/s10439-021-02796-x
8. Singh, S., et al.: Hypoplasia, pseudocoarctation and coarctation of the aorta-a systematic review. Heart Lung Circ. **24**(2), 110–118 (2015)
9. Torok, R.D., Campbell, M.J., Fleming, G.A., Hill, K.D.: Coarctation of the aorta: management from infancy to adulthood. World J. Cardiol. **7**(11), 765–775 (2015). https://doi.org/10.4330/wjc.v7.i11.765

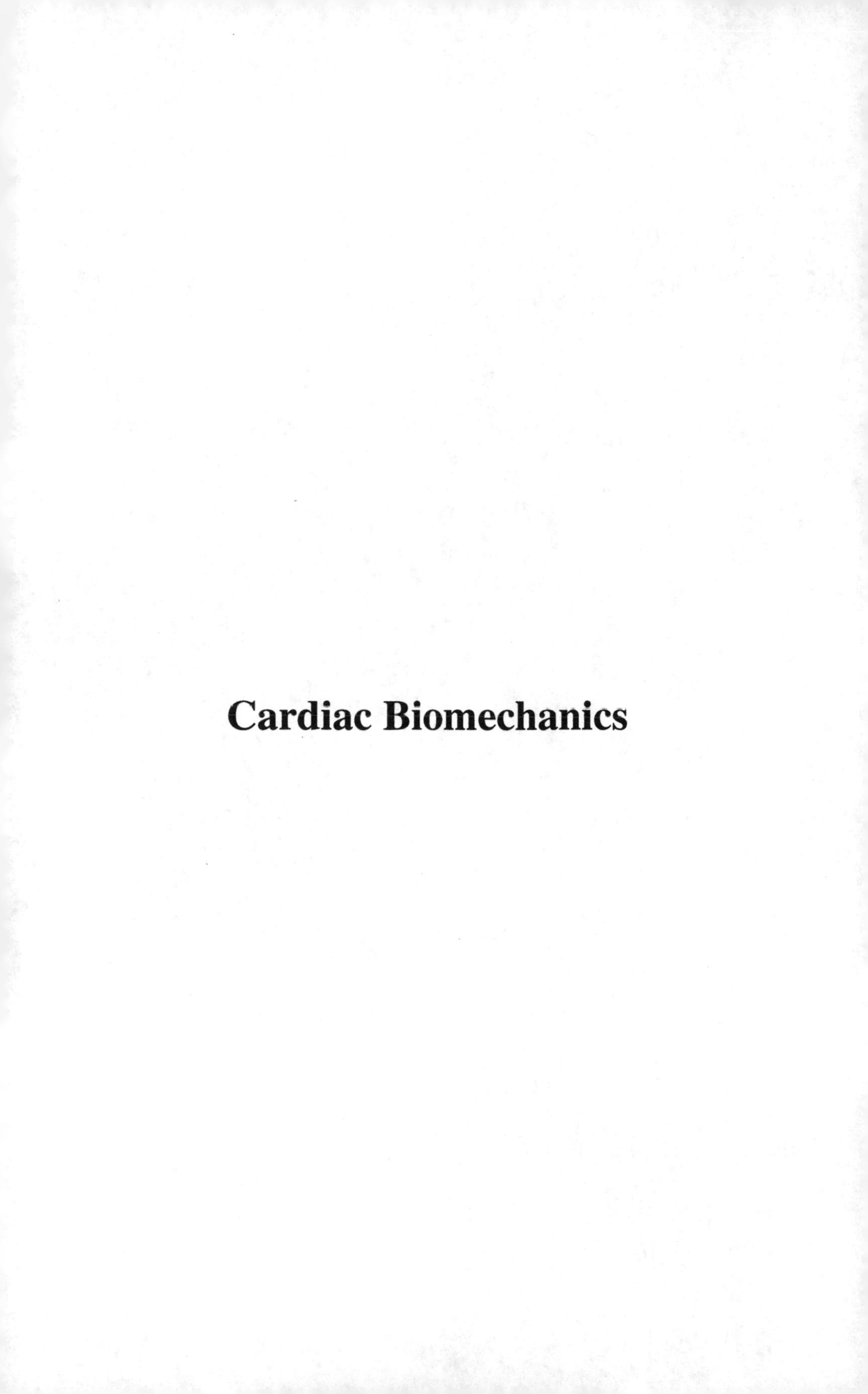

Cardiac Biomechanics

Evaluating Passive Myocardial Stiffness Using *in vivo* cine, cDTI, and Tagged MRI

Fikunwa O. Kolawole[1,2,3(✉)], Vicky Y. Wang[1,2,5], Bianca Freytag[6], Michael Loecher[1,2], Tyler E. Cork[1,2,7], Martyn P. Nash[6], Ellen Kuhl[3], and Daniel B. Ennis[1,2,4]

[1] Radiology, Stanford University, Stanford, CA 94305, USA
fikunwa@stanford.edu
[2] Radiology, Veterans Administration Health Care System, Palo Alto, CA 94304, USA
[3] Mechanical Engineering, Stanford University, Stanford, CA 94305, USA
[4] Cardiovascular Institute, Stanford University, Stanford, CA, USA
[5] Auckland Bioengineering Institute, University of Auckland, Auckland, New Zealand
[6] Univ. Grenoble Alpes, CNRS, TIMC UMR 5525, 38000 Grenoble, France
[7] Bioengineering, Stanford University, Stanford, CA 94305, USA

Abstract. Increased passive myocardial stiffness is implicated in the pathophysiology of many cardiac diseases, and its *in vivo* estimation can improve management of heart disease. MRI-driven computational constitutive modeling has been used extensively to evaluate passive myocardial stiffness. This approach requires subject-specific data that is best acquired with different MRI sequences: conventional cine (e.g. bSSFP), tagged MRI (or DENSE), and cardiac diffusion tensor imaging. However, due to the lack of comprehensive datasets and the challenge of incorporating multi-phase and single-phase disparate MRI data, no studies have combined *in vivo* cine bSSFP, tagged MRI, and cardiac diffusion tensor imaging to estimate passive myocardial stiffness. The objective of this work was to develop a personalized *in silico* left ventricular model to evaluate passive myocardial stiffness by integrating subject-specific geometric data derived from cine bSSFP, regional kinematics extracted from tagged MRI, and myocardial microstructure measured using *in vivo* cardiac diffusion tensor imaging. To demonstrate the feasibility of using a complete subject-specific imaging dataset for passive myocardial stiffness estimation, we calibrated a bulk stiffness parameter of a transversely isotropic exponential constitutive relation to match the local kinematic field extracted from tagged MRI. This work establishes a pipeline for developing subject-specific biomechanical ventricular models to probe passive myocardial mechanical behavior, using comprehensive cardiac imaging data from multiple *in vivo* MRI sequences.

Keywords: Cardiac mechanics · Cardiac MRI · Passive myocardial stiffness

O. Bernard et al. (Eds.): FIMH 2023, LNCS 13958, pp. 527–536, 2023.
https://doi.org/10.1007/978-3-031-35302-4_54

1 Introduction

Altered passive myocardial stiffness is evident in several cardiovascular diseases, including heart failure with preserved ejection fraction [26], hypertrophic cardiomyopathy [3], aortic stenosis [25], and myocardial infarction [9]. As passive myocardial stiffness is an important biomarker of global heart function, a robust method for its reliable identification can improve heart disease management.

Recent research on biomechanical modeling of passive ventricle mechanics has focused on uncertainty quantification [11], validation [10], impact of boundary conditions [17] and residual strains [6], improving the optimization strategy [15], and improving the identifiability of passive myocardial material parameters [19]. Owing to research progress, medical image-based computational constitutive modeling has been applied to clinical data, demonstrating its utility as a clinical prognostic tool [23].

The subject-specific data needed to estimate passive myocardial stiffness using inverse finite element (FE) modeling are best acquired using multiple imaging techniques. For example, the gross geometry of the heart is best determined from cine bSSFP images, regional diastolic kinematics can be measured using cine DENSE imaging or tagging, and cardiac microstructural organization can now be measured non-invasively using *in vivo* cardiac diffusion tensor imaging (cDTI). However, no studies, to our knowledge, have combined *in vivo* cine bSSFP, tagged MRI, and cDTI for estimation of subject-specific passive diastolic mechanical behavior partly due to the lack of comprehensive, high-fidelity datasets. Moreover, there is no established methodology for integrating multiphase *in vivo* geometric, functional, and kinematic data, and single-phase *in vivo* cDTI into a single subject-specific *in silico* model for mechanics modeling.

As demonstrated by a recent study by Stimm *et al.* [22], advancements in *in vivo* cDTI acquisition and post-processing have made incorporating subject-specific *in vivo* microstructure into FE models feasible. While the authors estimated passive myocardial stiffness by matching experimentally measured volumes, progress in MRI-driven kinematics modeling [18] means we can optimize for material parameters using richer objective functions that incorporate the local myocardial deformation kinematics. In addition, absence of a kinematic field meant the authors could not account for the potentially different microstructural configurations between the cDTI acquisition phase (i.e. end-systole) and the phase at which the mechanics model was generated (i.e. diastasis).

Herein, we acquired high fidelity *in vivo* MRI data, including cine bSSFP, tagged MRI, and cDTI, in a healthy adult subject (male, 24 years, BMI 26). We present a framework for integrating multiple sets of subject-specific MRI data to develop an *in silico* FE mechanics model of the subject's left ventricle (LV). Then we demonstrate the feasibility of using local kinematics to calibrate the *in silico* model to estimate passive myocardial stiffness.

2 Methods

2.1 Image Acquisition

The entire MRI dataset, consisting of cine bSSFP, cine diffusion tensor imaging (cDTI) and tagged MRI, was acquired in a 90-minute scanning session using a 3T Siemens MRI scanner at the Veterans Affairs Palo Alto Health Care System (Skyra, Siemens).

In vivo cine bSSFP short axis images were acquired with retrospective gating (TE/TR = 1.49/35.0 ms; FA = 35°; resolution = $1 \times 1 \times 6\,\text{mm}^3$; 6 mm slice thickness; 17 slices; FoV = 250 × 320; 25 cardiac phases). Short axis grid tagged images were acquired using a retrospectively gated sequence (TE/TR = 3.01/31.50 ms; FA = 12°; 7 mm tag spacing; same resolution, slice thickness, slices, FoV, and cardiac phases as cine bSSFP). In long axis, two line tagged images were acquired with tag lines oriented longitudinally and radially (TE/TR = 2.8/46.08 ms; FA = 10°; resolution = $1 \times 1 \times 10\,\text{mm}^3$, 10 mm slice thickness, 3 slices).

In vivo cDTI were acquired after the tagged MRI using a free-breathing, slice-following, spin-echo Echo Planar Imaging (EPI) navigator-based sequence with symmetric first and second-order motion-compensated diffusion gradients [14] (TE/TR = 86/3 R-R intervals; resolution = $2 \times 2 \times 8\,\text{mm}^3$; 8 mm slice thickness; 13 slices; b-values = [0, 350] s/mm^2; number of diffusion directions = 10; Averages = 6; Partial Fourier = 6/8; GRAPPA = 2). Blip-up (right-left) and blip-down (left-right) datasets were acquired at end-systole.

2.2 Geometric Analysis (cine bSSFP)

The cine bSSFP imaging data were processed using Cardiac Image Modeller (CIM, version 8.2, University of Auckland, New Zealand), which provided semi-automated segmentation of the LV surfaces over the entire imaging cycle in the form of FE models. Specifically, landmark points at the LV basal centroid, right ventricular basal centroid and LV apex were manually identified and subsequently used to construct a prior FE model in an orthogonal cardiac coordinate system. The origin of the cardiac coordinate system was located at one-third of the longitudinal distance from the LV base to apex landmark points. The x, y, and z axes were directed towards the LV apex, right ventricular centroid, and posterior LV wall respectively. LV endocardial and epicardial surfaces were interactively generated for all cardiac phases in Cardiac Image Modeler by fitting the prior FE model to guide-points placed by the analyst. The reference ("load free") FE mechanics model geometry was chosen as the diastasis phase corresponding to the start of atrial systole [20], and the timing was determined by the mitral valve motion from the cine bSSFP long-axis images (Fig. 1: middle).

2.3 Kinematics Analysis (tagged MRI)

The grid-tagged images were rigidly registered to the cine bSSFP images. Then, a set of myocardial points were defined in the diastasis phase of the grid-tagged images by masking out the LV myocardium using the diastasis LV surface model mentioned above. This was achievable because the cine bSSFP and tagged images were temporally aligned. Subsequently, we used a validated neural-net-based algorithm [12] to track the local myocardial displacements across the cardiac cycle relative to the diastasis phase (Fig. 1: right).

2.4 Microstructural Analysis (cDTI)

The raw *in vivo* cDTI data were first processed to correct geometric distortion induced by off-resonance phase accrual while traversing k-space over long echo times. Specifically, following 2D rigid registration of the blip-up and blip-down EPI datasets, the complete dataset was distortion corrected using TOPUP and the FSL package [21]. Diffusion tensors at each image voxel were then reconstructed from the distortion corrected diffusion weighted images using least squares linear regression.

To embed the *in vivo* cDTI diffusion tensors measured at end-systole into the LV geometric model at diastasis, a two-step approach was proposed. Step-1 was designed to translate and rotate the raw diffusion tensor data from end-systole to diastasis using the kinematic field derived from tagged MRI analysis. Step-2 was designed to interpret the reconfigured diffusion tensor data within the LV geometric FE model at diastasis using a field-based fitting approach [5]. Prior to Step-1, the cDTI data at end-systole were masked by, first kinematically deforming the cine bSSFP LV geometric model from diastasis to end-systole using the tagged MRI-obtained kinematic field, and then intersecting the deformed model with the cDTI images. Subsequently, at each voxel within the myocardium, a deformation gradient tensor between the end-systole and diastasis frame was interpolated from the kinematics field and decomposed into its rotational and stretch components. The diffusion tensors at end-systole were then subsequently reoriented to their diastasis configurations using the corresponding rotation tensors. The material point positions of the voxels were also translated based on the kinematics field.

To verify this approach of using the kinematics field to reorient myofiber direction data, we simulated the inflation of an LV FE model with a rule-based fiber field description. We then used the FE-simulated kinematics field to reorient the fiber vectors to their deformed states and compared them with the end-inflation directions predicted by the FE solver. The comparison revealed good agreement with an average angle difference of $0.8 \pm 0.8°$ and fiber location difference of 0.007 ± 0.006 mm (mean±std). While in this work the microstructure was reconfigured from the end-systolic to diastasis state, during which the LV undergoes isovolumic relaxation followed by rapid filling, the LV FE model used to validate the microstructure reorientation approach was under passive inflation. However, validation of the method for reconfiguring LV microstructure using deformation kinematics, does not require exact replication of the underlying deformation.

After transforming the *in vivo* DTI data into the appropriate configuration for integration with the geometric model, a microstructural orientation field, defined by the fiber, imbrication, and sheet (Euler) angles, was initialized at each node (a total of 34 nodes) of the FE model and interpolated over the FE model using tricubic Hermite basis functions. The fiber angles, defined as the elevation angle between a fiber vector that was projected to the surface of the FE model and the local circumferential direction, were initially set to be $+60^\circ$ for the endocardial nodes, and -70° for the epicardial nodes, and the imbrication angles, defined as the angle between the fiber vector and its projection to the surface of the FE model, were set to 0°. A local representation of this microstructural orientation field, expressed as the fiber vector $\mathbf{f}_{(v)}$, at each voxel v was computed. The nodal angular parameters describing the fiber vector field were optimized to maximize the alignment of $\mathbf{f}_{(v)}$ with the direction of maximal diffusion. The objective function was expressed as the dot product of $\mathbf{f}_{(v)}$ with the projection of the diffusion tensor ($\mathbf{D}_{(v)}$) in the direction of $\mathbf{f}_{(v)}$. Only fiber and imbrication angles were fitted. Optimization of the fiber orientation field was performed using the software package CMGUI [1] following procedures outlined by Freytag *et al.* [5]. The 3D description of the fiber orientation at diastasis was subsequently incorporated into the *in silico* LV model (Fig. 1).

2.5 Constitutive Relation

The myocardium was modeled using a transversely isotropic exponential constitutive relation [8] described by the following strain energy density function:

$$\begin{aligned} \psi &= \frac{1}{2}c(e^Q - 1) \\ Q &= b_1 E_{ff}^2 + b_2(E_{ss}^2 + E_{nn}^2 + 2E_{sn}^2) + b_3(2E_{fs}^2 + 2E_{fn}^2) \end{aligned} \tag{1}$$

where c, b_1, b_2, and b_3 are the constitutive parameters and E_{ij} denotes the components of the Green-Lagrange strain tensor ($\boldsymbol{E}$). Subscripts f, s, and n correspond to the fiber, sheet, and sheet-normal directions respectively.

Here, we fit c, the parameter that scales the overall passive myocardial stiffness, and kept the anisotropy parameters constant at physiologically realistic values of $b_1 = 8.61$, $b_2 = 3.67$, $b_3 = 25.77$ [24].

2.6 Mechanics Analysis

In this work, the FE mechanics analysis was limited to the passive filling phase between diastasis and end-diastole. The LV diastasis geometric model was used as the reference mechanics model, (Sect. 2.2) and was meshed with standard quadratic tetrahedral elements using GMSH [7] (1.9 mm^3 average element volume, 129,763 nodes, 81,295 elements, 389,289 degrees of freedom). The FE analysis was done using FEBio [13]. In the absence of LV cavity pressure recordings, the *in silico* LV was passively inflated by applying an endocardial pressure of 1 kPa, which is consistent with the expected LV end-diastolic pressure in healthy humans [4,24]. Uncertainty in the simulated material parameter due to a 0.2 kPa uncertainty in LV end-diastolic pressure was also investigated.

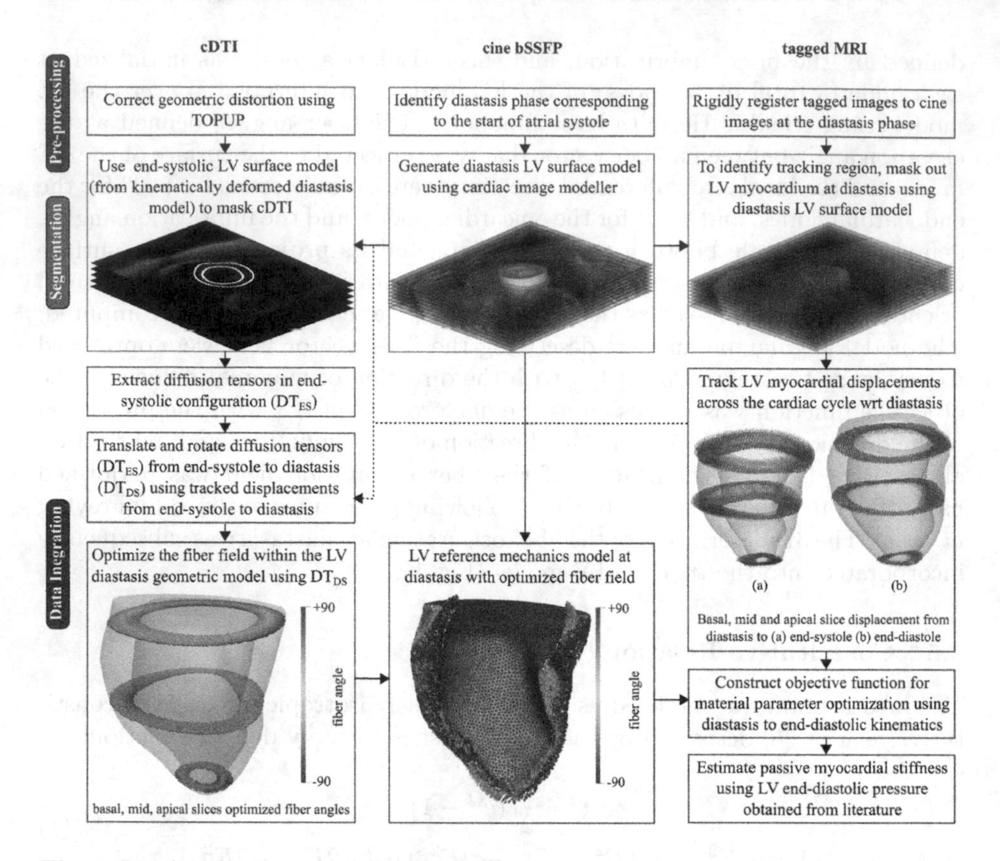

Fig. 1. Pipeline for integrating cine bSSFP, cDTI and tagged MRI data into a mechanics model for estimating passive myocardial stiffness.

Dirichlet boundary conditions were applied at the base of the *in silico* model by constraining the simulated basal surface displacement from diastasis to end-diastole to match that which was obtained from the tagged MRI-obtained kinematics. Then, the experimentally determined LV local displacements from diastasis to end-diastole were used to inversely estimate the optimal transversely isotropic exponential constitutive relation [8] (see Sect. 2.5). An initial estimate of the material parameters was obtained from Wang *et al.* [24]. Then, we iteratively updated the c constitutive parameter to minimize the differences between the experimentally measured nodal displacements (tagged MRI) and the simulated pressure-induced nodal displacements.

3 Results

3.1 Microstructure Fitting

The fitted LV myocardial fiber fields at diastasis and end-systole are shown in Fig. 2. Each fiber vector is color-coded by the fiber angle (i.e. the elevation angle with respect to the short axis plane). Here, we see a variation of fiber

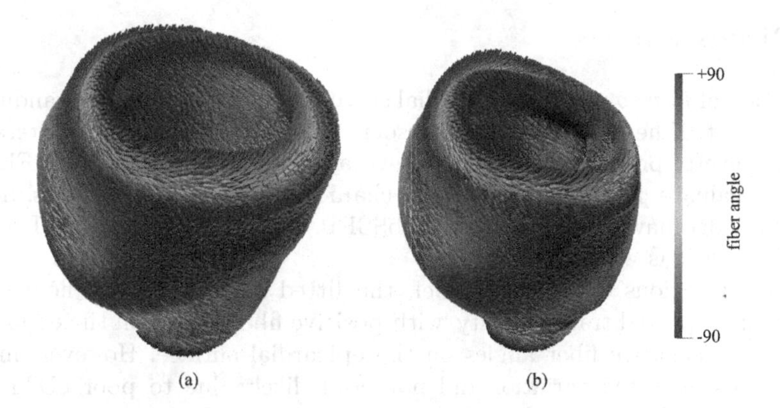

Fig. 2. *In vivo* cardiac DTI fiber vectors after field-fitting showing the LV myocardium mesh elements colored by the fiber angle at (a) diastasis (b) end-systole.

orientation with mostly positive fiber angles on the endocardial surface, and fibers with mostly negative fiber angles on the epicardial surface, in accordance with the expected transmural variation in fiber orientation [23].

3.2 *In Silico* Stiffness Calibration

The optimal bulk stiffness parameter (c) of the transversely isotropic constitutive relation (see Sect. 2.5) was 5.8 ± 1.2 kPa based on a LV filling pressure range of 1 ± 0.2 kPa. The mechanical behavior of the fitted constitutive relation in biaxial extension is shown (Fig. 3a) and compared with results from the Wang *et al.* study on passive stiffness of human myocardium [24]. In Fig. 3b, we show the computed deformed LV and the spatial variation in principal stretch after deformation (mean ± 95%CI = 1.03 ± 0.03 n.d.).

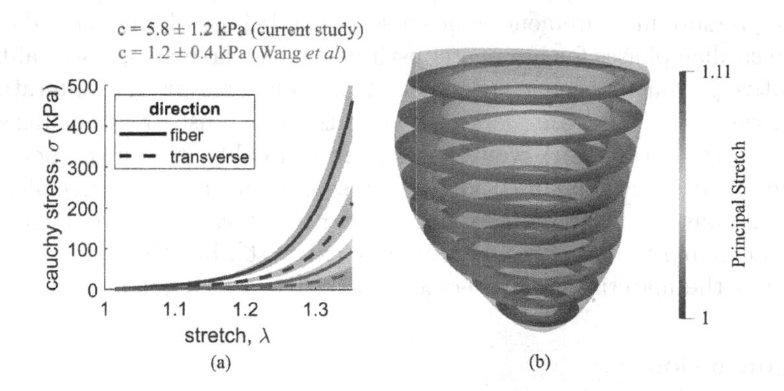

Fig. 3. (a) Myocardial mechanical behavior in simulated equibiaxial extension based on *in vivo* passive myocardial stiffness estimations from this study (blue) and from Wang *et al.* (red) [24]. Uncertainty arises from the range of LV end-diastolic pressure used for the material parameter optimization (1 ± 0.2 kPa). (b) LV computed deformation at maximal inflation. Short-axis slices depict the spatially varying principal stretch.

4 Discussion

Evaluation of *in vivo* passive myocardial stiffness can improve understanding and management of heart diseases with associated myocardial stiffness alterations. Though several previous studies have leveraged medical imaging and FE modeling to evaluate passive myocardial mechanical behavior and stiffness, none to our knowledge have combined *in vivo* bSSFP, tagged MRI, and cDTI to build the FE models as we have done here.

In most regions of the LV model, the fitted microstructural field (Fig. 2) follows the expected transmurality, with positive fiber angles on the endocardial surface and negative fiber angles on the epicardial surface. However, in some regions this expected variation did not hold, likely due to poor cDTI image quality (and subsequently bad diffusion tensors) in these regions. Given the sensitivity of mechanical behavior to microstructure [23], the evaluated material parameter was affected by these regions with anomalous fiber orientations.

Our calibrated passive myocardial stiffness is about 5 times stiffer than the Wang *et al.* study's results [24] (Fig. 3), which could be partly due to fundamental differences between our study and theirs. For example, while Wang *et al.* used catheter pressure measurements, here we used an LV end-diastolic pressure for healthy humans from literature. Additionally, in their study, passive myocardial stiffness was estimated by matching experimentally measured surfaces, whereas, we matched end-diastolic local kinematics. Moreover, Wang *et al.* did not prescribe the microstructure using *in vivo* cDTI.

4.1 Limitations

We only fitted the bulk passive stiffness parameter (c) of the transversely isotropic constitutive relation while keeping the anisotropy parameters constant. Of note, fitting multiple parameters introduces additional considerations related to the identifiability and uniqueness of the parameter set. Secondly, we lacked LV diastolic pressure measurement, which also precluded use of kinematic data from multiple cardiac phases for parameter estimation. Nevertheless, since calibrating constitutive parameters using a generic end-diastolic pressure from literature was feasible, we expect the framework to generalize to the use of personalized diastolic pressures. Future work would explore model calibration using noninvasive estimates of diastolic pressures, such as those derived from Doppler ultrasound imaging [2]. Lastly, we did not consider myocardial *in vivo* pre-stresses [6]. Methods for unloading the pre-stressed geometry exist [16], but they rely on assumptions about the material parameters and have not been extensively validated.

4.2 Conclusion

We have established a method to integrate multi-phase *in vivo* cardiac cine bSSFP and tagged MRI with single-phase *in vivo* cDTI for the development of a subject-specific LV mechanics model. We then demonstrated the feasibility of calibrating the leading material parameter of a transversely isotropic exponential constitutive relation using local deformation kinematics. This framework can also be readily extended to model subject-specific contractile mechanics. Future work will focus on extensive validation and model sensitivity analysis to understand uncertainty in the results. We will also evaluate the passive myocardial stiffness estimation approach in additional human subjects.

Acknowledgements. This work was supported by NSF 2205103 and NIH R01 HL131823 to DBE.

References

1. Bradley, C., et al.: OpenCMISS: a multi-physics & multi-scale computational infrastructure for the VPH/physiome project. Prog. Biophys. Mol. Biol. **107**(1), 32–47 (2011)
2. Channer, K.S., Culling, W., Wilde, P., Jones, J.V.: Estimation of left ventricular end-diastolic pressure by pulsed Doppler ultrasound. Lancet **1**(8488), 1005–1007 (1986)
3. Conrad, C.H., Brooks, W.W., Hayes, J.A., Sen, S., Robinson, K.G., Bing, O.H.: Myocardial fibrosis and stiffness with hypertrophy and heart failure in the spontaneously hypertensive rat. Circulation **91**(1), 161–170 (1995)
4. Feher, J.: Quantitative Human Physiology. Elsevier, Hoboken (2017)
5. Freytag, B., et al.: Field-based parameterisation of cardiac muscle structure from diffusion tensors. In: van Assen, H., Bovendeerd, P., Delhaas, T. (eds.) FIMH 2015. LNCS, vol. 9126, pp. 146–154. Springer, Cham (2015)
6. Genet, M., et al.: Heterogeneous growth-induced prestrain in the heart. J. Biomech. **48**(10), 2080–2089 (2015)
7. Geuzaine, C., Remacle, J.F.: Gmsh: a 3-D finite element mesh generator with built-in pre- and post-processing facilities. Int. J. Numer. Method Biomed. Eng. **79**(11), 1309–1331 (2009)
8. Guccione, J.M., McCulloch, A.D., Waldman, L.K.: Passive material properties of intact ventricular myocardium determined from a cylindrical model. J. Biomech. Eng. **113**(1), 42–55 (1991)
9. Gupta, K.B., Ratcliffe, M.B., Fallert, M.A., Edmunds, L.H., Bogen, D.K.: Changes in passive mechanical stiffness of myocardial tissue with aneurysm formation. Circulation **89**(5), 2315–2326 (1994)
10. Kolawole, F.O., Peirlinck, M., Cork, T.E., Levenston, M., Kuhl, E., Ennis, D.B.: Validating MRI-derived myocardial stiffness estimates using in vitro synthetic heart models. Ann. Biomed. Eng. 1–4 (2023)
11. Lazarus, A., Dalton, D., Husmeier, D., Gao, H.: Sensitivity analysis and inverse uncertainty quantification for the left ventricular passive mechanics. Biomech. Model. Mechanobiol. **21**, 1–30 (2022)
12. Loecher, M., Perotti, L.E., Ennis, D.B.: Using synthetic data generation to train a cardiac motion tag tracking neural network. Med. Image Anal. **74**, 102223 (2021)
13. Maas, S.A., Ellis, B.J., Ateshian, G.A., Weiss, J.A.: FEBio: finite elements for biomechanics. J. Biomech. Eng. **134**(1), 011005 (2012)
14. Moulin, K., et al.: In vivo free-breathing DTI and IVIM of the whole human heart using a real-time slice-followed SE-EPI navigator-based sequence: a reproducibility study in healthy volunteers. Magn. Reson. Med. **76**(1), 70–82 (2016)
15. Nair, A.U., Taggart, D.G., Vetter, F.J.: Optimizing cardiac material parameters with a genetic algorithm. J. Biomech. **40**(7), 1646–1650 (2007)

16. Peirlinck, M., De Beule, M., Segers, P., Rebelo, N.: A modular inverse elastostatics approach to resolve the pressure-induced stress state for in vivo imaging based cardiovascular modeling. J. Mech. Behav. Biomed. Mater. **85**, 124–133 (2018)
17. Peirlinck, M., et al.: Kinematic boundary conditions substantially impact in silico ventricular function. Int. J. Numer. Method Biomed. Eng. **35**(1), e3151 (2019)
18. Perotti, L.E., Magrath, P., Verzhbinsky, I.A., Aliotta, E., Moulin, K., Ennis, D.B.: Microstructurally anchored cardiac kinematics by combining in vivo DENSE MRI and cDTI. In: Pop, M., Wright, G.A. (eds.) FIMH 2017. LNCS, vol. 10263, pp. 381–391. Springer, Cham (2017)
19. Perotti, L.E., Ponnaluri, A.V.S., Krishnamoorthi, S., Balzani, D., Ennis, D.B., Klug, W.S.: Method for the unique identification of hyperelastic material properties using full-field measures. Application to the passive myocardium material response. Int. J. Numer. Method Biomed. Eng. **33**(11), e2866 (2017)
20. Shmuylovich, L., Chung, C.S., Kovács, S.J.: Point: left ventricular volume during diastasis is the physiological in vivo equilibrium volume and is related to diastolic suction. J. Appl. Physiol. **109**(2), 606–608 (2010)
21. Smith, S.M., et al.: Advances in functional and structural MR image analysis and implementation as FSL. Neuroimage **23**(Suppl 1), S208-219 (2004)
22. Stimm, J., et al.: Personalization of biomechanical simulations of the left ventricle by in-vivo cardiac DTI data: impact of fiber interpolation methods. Front. Physiol. **13**, 2485 (2022)
23. Wang, V.Y., Nielsen, P.M.F., Nash, M.P.: Image-based predictive modeling of heart mechanics. Annu. Rev. Biomed. Eng. **17**, 351–383 (2015)
24. Wang, Z.J., Wang, V.Y., Bradley, C.P., Nash, M.P., Young, A.A., Cao, J.J.: Left ventricular diastolic myocardial stiffness and end-diastolic myofibre stress in human heart failure using personalised biomechanical analysis. J. Cardiovasc. Transl. Res. **11**(4), 346–356 (2018)
25. Weidemann, F., et al.: Impact of myocardial fibrosis in patients with symptomatic severe aortic stenosis. Circulation **120**(7), 577–584 (2009)
26. Zile, M.R., Baicu, C.F., Gaasch, W.H.: Diastolic heart failure-abnormalities in active relaxation and passive stiffness of the left ventricle. N. Engl. J. Med. **350**(19), 1953–1959 (2004)

High-Speed High-Fidelity Cardiac Simulations Using a Neural Network Finite Element Approach

Shruti Motiwale[1,2], Wenbo Zhang[1], and Michael S. Sacks[1,2,3](✉)

[1] James T. Willerson Center for Cardiovascular Modeling and Simulation, Oden Institute, University of Texas, Austin, TX, USA
msacks@oden.utexas.edu

[2] Department of Mechanical Engineering, University of Texas, Austin, TX, USA

[3] Department of Biomedical Engineering, University of Texas, Austin, TX, USA

Abstract. A comprehensive image-based computational modelling pipeline is required for high-fidelity patient-specific cardiac simulations. However, traditional simulation methods are a limitation in these approaches due to their prohibitively slow speeds. We developed a physics-based training scheme using differentiable finite elements to compute the residual force vector of the governing PDE, which is then minimized to find the optimal network parameters. We used neural networks for their representation power, and finite elements for defining the problem domain, specifying the boundary conditions, and performing numerical integrations. We incorporated spatially varying fiber structures into a prolate spheroidal model of the left ventricle. A Fung-type material model including active contraction was used. We developed two versions of our model, one was trained on a reduced basis of the solution space, and one was trained on the full solution space. The models were trained against two pressure-volume loops and validated on a third loop (Fig. 1). We validated our implementation against conventional FEM simulation using FEniCS. While the reduced order model was trained faster than the full-order model, we achieved mean and standard deviation of the nodal error between the NNFE solution and the FE solution with 10^{-3} cm, with both models, where the characteristic length was 1 cm (Fig. 2a). The NNFE model predicted each solution within 0.6 ms whereas the FE models took up to 500 ms for each state. The NNFE method can be simultaneously trained over the entire range of physiological boundary conditions. The trained NNFE can predict stress–strain responses for any physiological boundary condition without retraining.

Keywords: Neural Network Finite Element · High-fidelity · Cardiac simulation

1 Introduction

The complex multiphysics and multiscale nature of the human heart presents great challenges for understanding and predicting cardiac biomechanical function. Computational models of the heart based on the finite element method have been developed for biventricular model and left ventricle-mitral valve model [1–3]. High-fidelity computational

O. Bernard et al. (Eds.): FIMH 2023, LNCS 13958, pp. 537–544, 2023.
https://doi.org/10.1007/978-3-031-35302-4_55

models of the heart must involve structurally informed material models, local myofiber structure and imaging-based geometry. Comprehensive patient-specific computational modeling pipeline based on these essential modeling aspects has been developed for a single ovine heart model and successfully applied to study normal and diseased states of the heart [2]. However, traditional finite element methods remain prohibitively slow for patient-specific clinical applications, such as evaluating multiple treatment outcomes for patient-specific treatment planning within clinically relevant timeframes. A digital twin approach has been advocated for a heart model with parameters continuously updated as patient-specific data is acquired [cite]. A computationally efficient approach is required for this application, that does not make approximations in prediction accuracy in favor of efficiency.

To meet the requirements of speed as well as accuracy, we have developed and utilized a novel neural network finite element (NNFE) approach for soft tissue simulations that can produce simulation results within clinically relevant timeframes [4, 5]. The NNFE approach is a physics-based approach for rapid simulations that uses the neural network (NN) to represent the nodal displacements, and finite elements to map the displacement output from the NN on the problem domain, as well as to enforce boundary conditions and perform numerical integrations. In other words, this approach does not rely on data generated from physical experiments or simulations for training, rather, the NNFE model is trained to learn the governing PDE. In this work, we present a feasibility study using an extension of the NNFE approach towards complete organ level cardiac simulations to predict the P-V loop responses of the left ventricle, accounting for active contraction and transmural fiber distributions. We present two versions of the approach, one where a reduced order model is considered, and another version with a full-order approach.

2 Methods

The left ventricle was represented as a prolate spheroid as a first step scenario and was discretized with unstructured tetrahedral quadratic elements with Lagrange basis functions. The basal plane was constrained to in-plane motion only, one point on the basal plane was completely fixed, and an adjacent point was constrained radially, ventricular pressure was applied to the endocardium and the epicardium was traction free. Transmurally varying fiber structures were constructed by varying the fiber directions from –60° with respect to the circumferential direction on the epicardium to 60° on the endocardium (Fig. 1). The normal direction was pointing outwards and the sheet direction orthogonal to the other two directions.

As a step forward from our existing cardiac modeling pipeline [1–3], the passive mechanical properties of the myocardium were modeled with a nearly incompressible transversely isotropic Fung-based hyperelastic material model

$$\Psi = \Psi_{vol}(J) + \Psi_{dev}(\tilde{\mathbf{E}}) \tag{1}$$

$$\Psi_{vol} = \frac{K}{2}\left(\frac{J^2 - 1}{2} - \ln J\right)$$

$$\Psi_{dev} = \frac{c}{2}\exp(\alpha Q - 1)$$

Fig. 1. Fiber orientations in the myofiber and sheet directions respectively.

$$Q = A_1\widetilde{E_{11}^2} + A_2\left(\widetilde{E_{22}^2} + \widetilde{E_{33}^2} + \widetilde{E_{23}^2}\right) + A_3\left(\widetilde{E_{12}^2} + \widetilde{E_{13}^2}\right)$$

where J is the volumetric Jacobian determinant, $\tilde{\boldsymbol{E}} = J^{-\frac{2}{3}}\boldsymbol{E}$ is the deviatoric part of the Green-Lagrange strain tensor $\mathbf{E}$ and K, α, c, A_1, A_2, and A_3 are material parameters. The active contraction was modeled as an additional stress based on the Hunter-McCulloch-Ter Keurs model with minor modifications

$$\boldsymbol{S_{act}} = S_{Ca^{2+}}[1 + \beta(\lambda - 1)]/\lambda^2 \boldsymbol{f}_0 \otimes \boldsymbol{f}_0 \tag{2}$$

$$\lambda = \|\boldsymbol{F}\boldsymbol{f}_0\|_2 \tag{3}$$

where $S_{Ca^{2+}}$ $S_{Ca^{2+}}$ is in the unit of stress resulting from the active contraction usually fitted with the experimental data, β is a material parameter and λ is the fiber stretch. The material constants were taken from an ovine heart model [2], and are summarized in Table 1.

Table 1. Model parameters of Ψ and $\mathbf{S_{act}}$ for active myocardium.

K(Pa)	c(Pa)	α	A_1	A_2	A_3	β
1×10^5	1522.083	2.125	12	8	26	1.4

The NN was trained by minimizing the average norm of residual force vectors of a set of sampled input parameters obtained from the weak form of the governing PDE (Fig. 2). Briefly, the weak form of the governing PDE is

$$\delta W(\boldsymbol{u}, \delta\boldsymbol{u}) = \int_{\Omega_0} \left(\frac{\partial\Psi}{\partial\boldsymbol{E}} + \boldsymbol{S_{act}}\right) : \delta\boldsymbol{E}\, d\Omega - \int_{\Gamma_0} \boldsymbol{T} \cdot \delta\boldsymbol{u}\, d\Gamma = 0 \tag{4}$$

where, $\boldsymbol{u} = \mathbf{UN}$ is the displacement field over the domain Ω, $\mathbf{U}$ are the nodal displacements and $\mathbf{N}$ are the basis functions, $\delta\boldsymbol{u} = \boldsymbol{VN}$ is the virtual displacement field and $\mathbf{V}$

are the virtual nodal displacements, Ψ is the strain energy density function describing the passive behavior of myocardium, $\mathbf{E}$ is the Green-Lagrange strain tensor, $\mathbf{S}_{act}$ is the 2nd Piola-Kirchhoff active contraction stress and $\mathbf{T}$ is the traction vector. The residual force vector of the weak form of the governing PDE satisfies

$$R = \left.\frac{\partial \delta W}{\partial V}\right|_{V=0} = 0 \tag{5}$$

The NNFE model was setup to take pressure p and active contraction $T_{Ca^{2+}}$ $T_{Ca^{2+}}$ as the input and predict the displacement field. Pressure-active stress loops were obtained from the corresponding pressure-volume loops using a secant method. The NN was initialized to output zero displacements to avoid mesh distortion and the neurons were activated with the tangent hyperbolic activation function. The NN was trained over two pressure-volume (P-V) loops and predicted the P-V relationship for a third loop. An identical simulation setup was developed in FEniCS [6–8] for validating the results of the NNFE model.

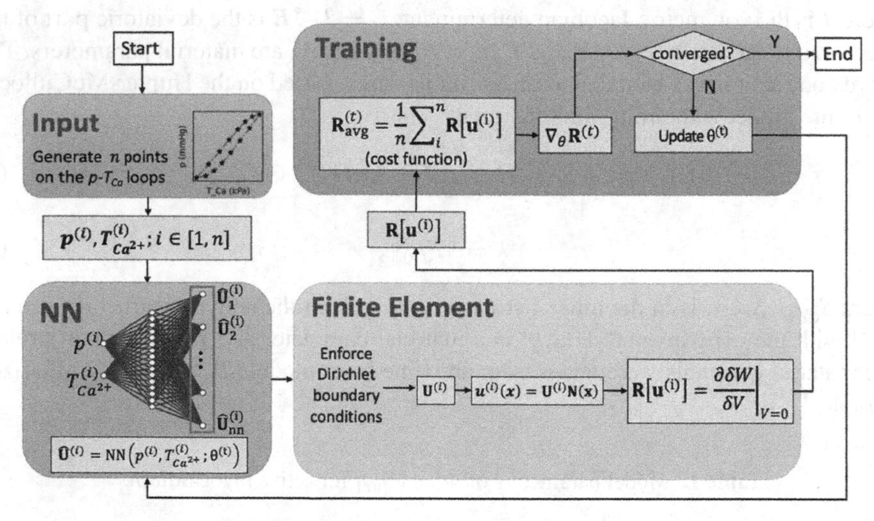

Fig. 2. The NNFE training pipeline

As a first step, a reduced-order NNFE model was developed. In this approach, proper orthogonal decomposition was done on the FE results of the pressure-volume loops used for training to extract the first 20 dominant modes. The output layer of the neural network predicted the weights for these dominant modes to produce the solution displacement field. In addition, the NN has one hidden layer with 20 neurons. The inexact Newton-conjugate-gradient method with Armijo line search was used to train the reduced-order model and the reduced bases were kept fixed during training. The full-order version of the model directly predicted the nodal displacements **U** in the output layer (Fig. 2) and had one hidden layer with 16 neurons. The Dirichlet boundary conditions were enforced strongly on the output of the neural network. Due to the complexity of training over the full space for this model, the model was trained using the first-order gradient based optimization algorithm Adam [9] along with sequential learning rate scheduling in this order: exponential decay, linear decay and constant respectively.

The development of the ventricular model and its physics-based training was facilitated by differentiable finite elements based on open-source software libraries to back-propagate gradient information to neural networks. We leveraged an open-source FE software library DOLFINx[1], to import a general mesh format, mesh coordinates, the map of the degree of freedom and marked boundaries. We developed a code generation procedure for the FE basis functions based on the FInite element Automatic Tabulator (FIAT) [10], which eliminates the need to hard code the basis functions and reuse a broad range of FE families in FIAT. The resulting computational graph for the NNFE was compiled in Jax [11] and the computation was staged out on GPUs.

3 Results

The reduced-order NNFE model as well as the full-order NNFE model accurately predicted the displacement and the corresponding PV loop (Fig. 3). We visualized two states on the validation pressure-volume loop and superimposed the FE solutions and the NN predictions (Fig. 4). The trained NNFE model could accurately produce the twisting induced due to the active contraction in the left ventricle. The histogram of the absolute nodal errors between the NNFE solution and the FE solution demonstrates that the absolute nodal errors clustered around 10^{-3} cm, with the mean nodal error of 2.32×10^{-3} cm, with a standard deviation of 4.53×10^{-3} cm (Fig. 5). The trained NNFE model took 0.6 ms for producing the results for each state, whereas FEniCS took 500 ms. The reduced order model took 3.5 h to train, whereas the full-order model required 168 h to train up to the same level of accuracy as the reduced-order model.

[1] https://github.com/FEniCS/dolfinx.

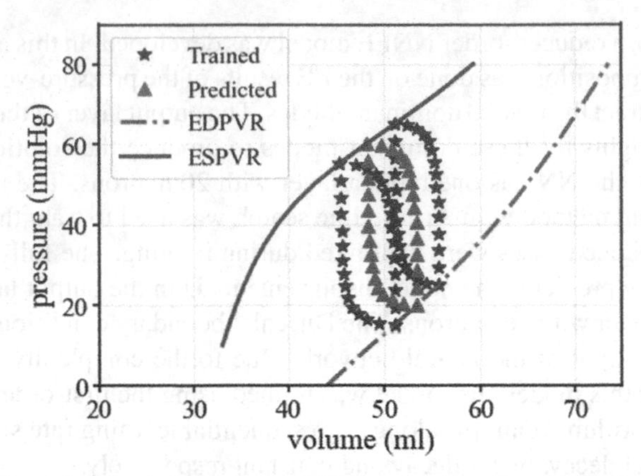

Fig. 3. The NNFE model was trained on two PV loops and predicted the third loop. The trained NNFE model took 0.6 ms for producing the results for each state, whereas FEniCS took 500 ms.

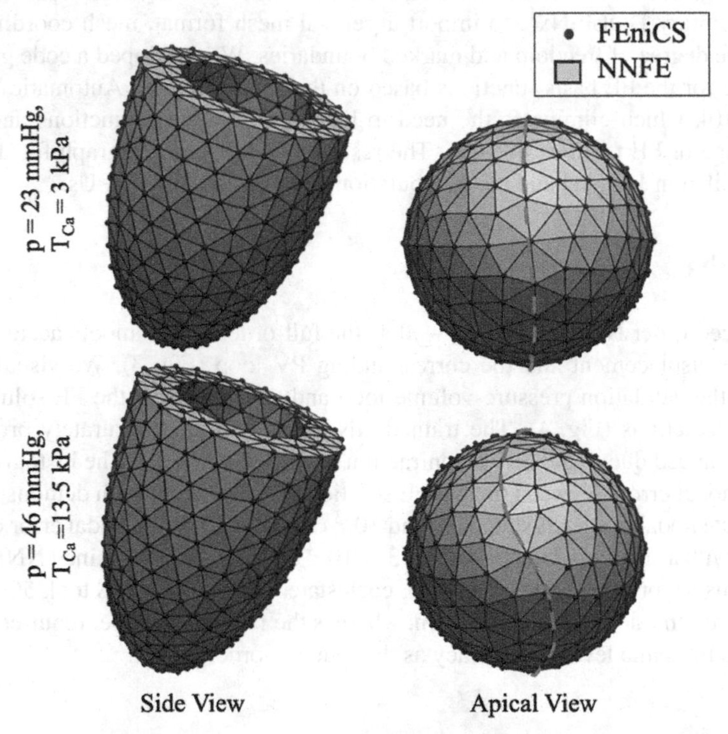

Fig. 4. Two points on the P-V loop as produced by the full-order NNFE method (wireframe) overlaid with FEM solution (dots). The NNFE method can accurately capture the twisting induced by active contraction, as highlighted with the dotted centerline in the apical view. The mean nodal error between the NNFE solution and the FE solution was 2.32×10^{-3} cm, with a standard deviation of 4.53×10^{-3} cm.

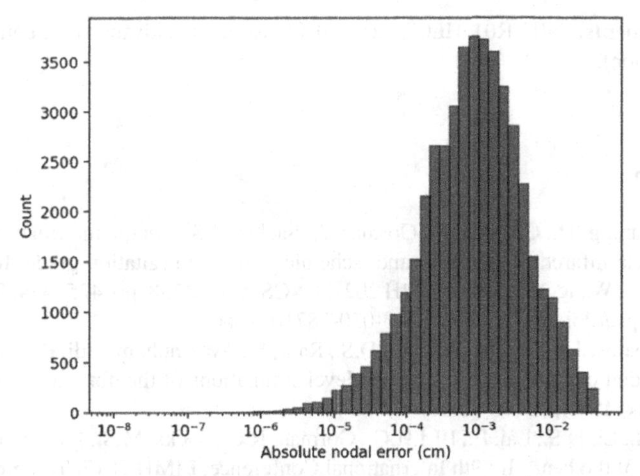

Fig. 5. The distribution of absolute nodal errors of the predictions of the neural network with 1 hidden layer and 16 neurons. The characteristic scale of the geometry is 1 cm.

4 Discussion

We presented a novel NNFE approach to produce the active contractile behavior of a left ventricle under a given PV loop. We presented both a reduced order approach as well as a full-order approach to solve this problem. By training the NN over a reduced subspace the number of training parameters can be reduced, and the training process can be accelerated. However, as the number of input variables to the NN increase, the number of FEM simulations required for basis generations will increase exponentially, and the reduced order approach is not expected to remain feasible for very high number of input variables. Moreover, the reduced bases may impose biases to the training space. It is also important to consider that the long training times of the full-order NNFE method are not a significant limitation for the clinical application, as the NN only needs to be trained once over a range of boundary conditions. Once trained, the NN can produce results for any boundary condition within the training range without a need for retraining. Consequentially, the NNFE approach is well-suited for many-query problems, such as patient-specific surgical planning, where one needs to solve very similar problems repeatedly with only small changes to the inputs. For such applications, the training can be done in advance, and when presented with the patient-specific data, rapid simulation results can be produced with the trained model.

Our results demonstrate the first application of the NNFE approach for a biomedical application at an organ level. We will use this model to study the effect of infarcts in different locations of the ventricle on cardiac behavior. We are also working on extending this method to study the effects of inotropy on cardiac behavior by varying the slope of the end systolic P-V relationship. The present study demonstrates that the NNFE method can model the complex active behavior of the myocardium and serves as an incremental step toward more realistic geometries and clinical applications. While still in its early stages, this approach paves the pathway for high-speed patient-specific clinical simulations.

Acknowledgements. NIH R01 HL073021 and Platform for Advanced Scientific Computing (Swiss Federation).

References

1. Liu, H., Narang, H., Gorman, R., Gorman, J., Sacks, M.S.: On the interrelationship between left ventricle infarction geometry and ischemic mitral regurgitation grade. In: Ennis, D.B., Perotti, L.E., Wang, V.Y. (eds.) FIMH 2021. LNCS, vol. 12738, pp. 425–434. Springer, Cham (2021). https://doi.org/10.1007/978-3-030-78710-3_41
2. Liu, H., Soares, J.S., Walmsley, J., Li, D.S., Raut, S., Avazmohammadi, R., et al.: The impact of myocardial compressibility on organ-level simulations of the normal and infarcted heart. Sci. Rep-uk. **11**(1), 13466 (2021)
3. Soares, J.S., Li, D.S., Lai, E., III J.H.G., Gorman, R.C., Sacks, M.S.: Functional imaging and modelling of the heart. In: 9th International Conference, FIMH 2017, Toronto, ON, Canada, 11–13 June 2017, Proceedings. Lect. Notes. Comput. Sci. 10263, 493–501 (2017)
4. Sacks, M.S., Motiwale, S., Goodbrake, C., Zhang, W.: Neural network approaches for soft biological tissue and organ simulations. J. Biomech. Eng. **144**(12), 121010 (2022)
5. Zhang, W., Li, D.S., Bui-Thanh, T., Sacks, M.S.: Simulation of the 3D hyperelastic behavior of ventricular myocardium using a finite-element based neural-network approach. Comput. Meth. Appl. M. **394**, 114871 (2022)
6. Scroggs, M.W., Dokken, J.S., Richardson, C.N., Wells, G.N.: Construction of arbitrary order finite element degree-of-freedom maps on polygonal and polyhedral cell meshes. ACM T Math Softw. **48**(2), 1–23 (2022)
7. Scroggs, M.W., Baratta, I.A., Richardson, C.N., Wells, G.N.: Basix: a runtime finite element basis evaluation library. J. Open Source Softw. **7**(73), 3982 (2022)
8. Alnæs, M.S., Logg, A., Ølgaard, K.B., Rognes, M.E., Wells, G.N.: Unified form language: a domain-specific language for weak formulations of partial differential equations. ACM Trans. Math Softw. Toms. **40**(2), 9 (2014)
9. Kingma, D.P., Adam, B.J.: A method for stochastic optimization. Arxiv. (2014)
10. Kirby, R.C.: Algorithm 839: FIAT, a new paradigm for computing finite element basis functions. ACM Trans. Math Softw. Toms. **30**(4), 502–516 (2004)
11. Roberts, A., et al.: Scaling Up Models and Data with $\texttt{t5x}$ and $\texttt{seqio}$. arxiv. (2022)

Effect of Varying Pericardial Boundary Conditions on Whole Heart Function: A Computational Study

Justina Ghebryal[1,2](✉), Cristobal Rodero[1], Rosie K. Barrows[1], Marina Strocchi[1], Caroline H. Roney[3], Christoph M. Augustin[4,5], Gernot Plank[4,5], and Steven A. Niederer[1]

[1] Cardiac Electromechanics Research Group, Biomedical Engineering Department, King's College London, London, UK
[2] Eindhoven University of Technology, Eindhoven, The Netherlands
u.ghebryal@tue.nl
[3] Queen Mary University of London, London, UK
[4] Institute of Biophysics, Medical University of Graz, Graz, Austria
[5] BioTechMed-Graz, Graz, Austria

Abstract. Pericardiectomy is recommended therapy for pericarditis, an inflammation of the pericardial layers that surround the heart and play a central role in maintaining cardiac performance. In some cases, the pericardium can be repaired or patched. However, the impact of changes in the pericardium on cardiac function is not clear. The objective of this study is to analyze the effect of the pericardium on whole-heart function by varying normal Robin boundary conditions (BCs) applied on the ventricular epicardium. A piece-wise linear penalty function was defined using two parameters that were varied to describe the regional scaling of normal spring stiffness from apex to base. Gaussian process emulators were used to perform a global sensitivity analysis on how the varying pericardial BCs affect cardiac biomechanics in four-chamber heart models. Our results have shown that pressure- and volume-derived biomarkers change by less than 25% due to variations in the pericardium, with more variation in the right ventricle compared to the left ventricle. On the other hand, measurements for systolic motion exhibited a range of variability greater than 100% of the baseline mean. We predict that the pericardium has limited impact on measures of global function but impacts measures of local cardiac biomechanics.

Keywords: Cardiac electromechanics · Pericardium · Gaussian process emulators · Global sensitivity analysis

1 Introduction

Constrictive pericarditis (CP) is often the result of chronic inflammation of the pericardium, a thin connective tissue membrane that surrounds the heart [18].

O. Bernard et al. (Eds.): FIMH 2023, LNCS 13958, pp. 545–554, 2023.
https://doi.org/10.1007/978-3-031-35302-4_56

Experimental studies have shown that CP can lead to differences in the biomechanical properties of the heart due to an increased pericardial stiffness [1,2], and may lead to heart failure. The removal of the pericardium (pericardiectomy) is the recommended treatment for improving cardiac hemodynamics in CP patients, but it is associated with high prevalence of morbidity and mortality [18]. Currently, alternative treatment options have been focusing on targeting inflammation by using membrane patches following pericardiectomy or on repairing the pericardium [17]. However, the impact of changes in the pericardium on cardiac function remains poorly described. The outcomes of possible treatments for CP would greatly benefit from using computational models to reveal insights into the effect of the pericardium on cardiac function.

Recently, the scientific community has increasingly moved from simulating ventricular to whole-heart models, including the interaction between the ventricles, the atria, and pericardium [29]. These studies have shown that including the effect of the pericardium in computational heart models is important for reproducing physiologically realistic cardiac motion [12]. Several approaches were used to represent the effect of the pericardium by using different types of boundary conditions (BCs) [12]. Strocchi et al. [29] proposed normal Robin BCs on the epicardium, considering spatially varying spring stiffness from apex to base using a penalty formulation. Including the pericardium is important for (i) understanding the physiological function of the heart, (ii) our interpretation of cardiac biomechanics, and (iii) understanding how changes in the pericardium under diseased conditions can impact the heart.

In this study, we use a four-chamber heart model where we represent the pericardium using Robin BCs [29]; where the spring stiffness is scaled with a piece-wise linear penalty function. Gaussian Process Emulators (GPEs) were trained to perform a Global Sensitivity Analysis (GSA) to quantify the biomechanical effects of varying the pericardial BCs.

2 Methods

2.1 Four-Chamber Geometry

This study was performed using a four-chamber heart geometry reconstructed from ECG-gated CT-images of an atrial fibrillation (AF) patient. The preprocessing steps for generating the mesh were performed in an earlier study [5]. Briefly, segmentation of the end-diastolic CT-image was used to generate a linear tetrahedral mesh with an average edge length of 1 mm. The four-chamber heart model had 622,967 nodes and 3,188,446 linear tetrahedral elements. Next, mesh surfaces and volumes were labelled to impose specific electrical or mechanical tissue properties and boundary conditions. To assign ventricular and atrial myofiber orientations to the mesh, universal ventricular/atrial coordinates (UVCs, UACs) [6,24] and the Laplace-Dirichlet rule-based method [7] were used. Finally, as a starting point for the electro-mechanics (EM) simulations, the reference configuration (stress-free state) of the heart geometry was computed using the backward displacement method, described in [9].

2.2 Varying the Pericardial Boundary Conditions

The pericardium was modelled using normal springs with varying stiffness in the apico-basal direction, as proposed in [29]. The spring stiffness was scaled from apex to base using a piecewise penalty function $p(x)$ on the apico-basal UVC coordinate, x:

$$p(x) = \begin{cases} qx + 1, & x \leq \text{penalty threshold} \\ 0 & x > \text{penalty threshold} \end{cases} \tag{1}$$

where q is the slope and x represents the longitudinal UVC in apico-basal direction ($x = 0$ for apex, $x = 1$ for base). To vary the pericardial BCs, we vary two parameters. First, the penalty threshold (PTH) in the range [0.40–0.95], which will also change the slope q (see Fig. 1**A**.1). Secondly, the normal spring pericardial scale stiffness (PSS) which was chosen to vary in the range [10^{-4}–10^{-1}] kPa/μm, as both smaller and larger boundary values lead to numerical instabilities. In the model, the value for PSS is multiplied with the penalty function $p(x)$ for scaling the spring stiffness from apex to base. Both parameters will be varied within the specified range, consistent with literature values [29]: PTH = 0.82 and PSS = 10^{-3} kPa/μm. The model input parameters for PTH and PSS were chosen by sampling the space using Sobol' semi-random sequences [28] using the SALib Python library [14]. A complete overview of the described pipeline is given in Fig. 1**A**.

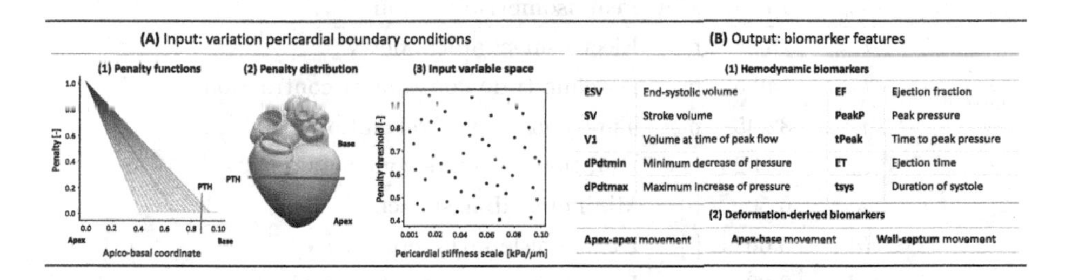

Fig. 1. **A**: Input for varying pericardial boundary conditions. *(1)* penalty functions with a different penalty threshold (PTH); *(2)* mapping on the four-chamber mesh, red line indicates the PTH; *(3)* input parameter space for electro-mechanics simulations. **B**: Output biomarker features. *(1)* hemodynamic biomarkers and *(2)* deformation biomarkers. (Color figure online)

2.3 Modelling Electro-Mechanics

The electrical activation of the heart was simulated using a reaction-Eikonal model [19]. The activation times computed with this model were then used to trigger active stress generation in the cardiac tissue. Ventricular myocardium was modelled as a transversely isotropic conduction medium and conduction velocities were set to the same values as in [5]. Additionally, a fast endocardial layer was defined in the mesh with higher conduction velocities to represent the fast activation of the Purkinje system [5]. The ventricular activation was initiated at the apex.

In the mechanics model, we modelled the ventricles and atria as a transversely isotropic, hyperelastic and nearly incompressible material using the Guccione constitutive law [13]. The passive material parameters were set based on literature for the ventricles (a = 3.0, b_f = 25.0, b_{fs} = 11.0, b_t = 9.0) and atria (a = 2.2, b_f = 15.0, b_{fs} = 11.0, b_t = 9.0) [5]. All other tissues (aortic wall, pulmonary artery, valve planes and vein rings) were modelled as neo-Hookean and material parameters were set according to [5]. In our simulations, we only set the ventricles to contract, while all other structures were set as passive and were included to provide physiological boundary conditions [23].

To simulate the active stress in the ventricles, we used the phenomenological activation-based Tanh Stress model [20]. We calibrated the Tanh model parameters by finding a parameter set leading to a physiological ejection fraction (EF $\approx$ 50%) and left ventricle pressure transient. The Tanh model parameter values were determined using GPEs and using a single wave of Bayesian history matching as done in [22], see Table 1 for the final parameter set used in the simulations.

Table 1. Input parameters for the active Tanh model.

T_{peak}	132.49	kPa	Peak isometric tension
t_{emd}	23.65	ms	Electro-mechanical delay
t_{c0}	149.52	ms	Baseline time constant of contraction
τ_r	32.45	ms	Time constant of relaxation
t_{dur}	634.25	ms	Duration of active contraction
λ_0	0.50	[-]	Minimum fiber stretch
ld	5.66	[-]	Degree of length-dependence
ld_{up}	0.00	ms	Length-dependence of upstroke time
Vm	-60	mV	Membrane potential threshold

All EM-simulations simulations were run with the Cardiac Arrhythmia Research Package (CARP) [4,30], built upon extensions of the openCARP electrophysiology framework [21]. The electrical activation and contraction of the ventricles were coupled with a closed-loop model based on the CircAdapt framework [3]. We simulated three heart beats to reach steady state, with a cycle length of 1000 ms. All simulations were run on TOM2, a high performance computing facility at King's College London.

2.4 Global Sensitivity Analysis

To measure the effect of varying the pericardial BCs on heart function, we computed 10 hemodynamic and 3 deformation output biomarkers for both the left and right ventricle (LV and RV), specified in Fig. 1**B**. These quantities were computed for the last beat. The hemodynamic biomarkers were pressure, volume and flow based. For the deformation biomarkers, we computed the distance of motion between (i) the apex during end-diastole (ED) and end-systole (ES); and the difference in distance between ED and ES for (ii) apex to base and (iii) wall to septum. These three biomarkers were then considered as a measure of systolic motion. To study the relationship between the input (pericardial BCs) and output (biomarkers), we performed a GPE-based GSA using the same pipeline as in Rodero et al. [23]. First, we analyzed the range of variability for all biomarkers for the 50 EM simulations that were performed. Next, we excluded the biomarkers in our analysis for which the range of variability was below 25% of the baseline mean. To make our analysis more robust, we used GPEs in which the GPE was a combination of a linear mean function, a radial basis function kernel and Gaussian noise. We split the data into 80% as training set and 20% as test set. We estimated Sobol' sensitivity first-order indices [28] through the Saltelli method [26]. More details on performing the GPE training can be found in [15].

3 Results

3.1 Identifying Simulation Biomarkers

To evaluate the effect of varying the pericardial BCs on the biomechanical output of the heart, we measured the range of the computed biomarkers normalized by the mean value for each biomarker as done in [23]. This shows the maximum percentage change of each biomarker for the different pericardial BCs. Figure 2 shows the different values for both ventricles. To allow us to focus on simulation outputs that were affected by the pericardial BCs we discarded the biomarkers with a variation below 25%, and focused on the ones above this threshold. This included dPdtmin for the RV and the deformation parameters for both ventricles: apex-apex, apex-base, and wall-septum motion. These results show that variation in all biomechanical parameters and wall-septum deformation is higher for the RV, while for the apex-apex and apex-base deformation the LV shows more variation.

3.2 Global Sensitivity Analysis

We used GPEs to perform a GSA using the pericardial BCs on each biomarker separately. We presented the outcome of the GSA as doughnut charts in Fig. 3. Each piece of the doughnut chart corresponds to the percentage of variance explained globally by the pericardial BCs, considering only first order effects. The median score that was achieved for the emulators was $R^2 = 0.89$. Overall,

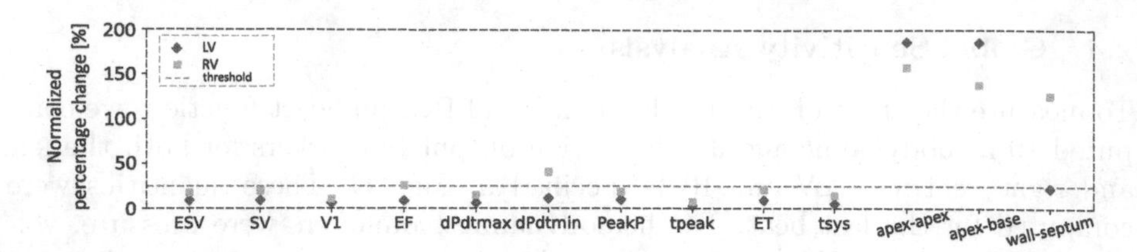

Fig. 2. Range covered by each one of the simulated biomarkers for both ventricles, normalised with the average value of each one of them.

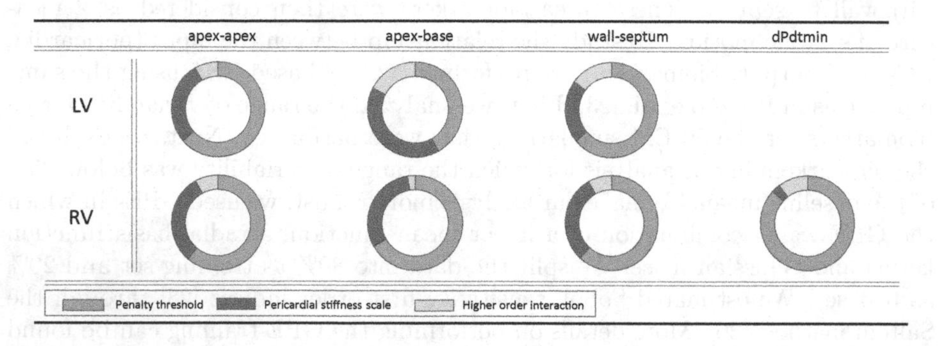

Fig. 3. The impact of the pericardial BCs (penalty threshold and pericardial scale stiffness) on the deformation in three directions for both the LV and RV, and on dPdtmin for the RV.

these results show that PTH explains the most variance for the three deformation biomarkers and for the variability of dPdtmin in the RV.

To better understand the heart motion depending on the different pericardial BCs, Fig. 4 shows a slice of the simulated motion of the four-chamber geometry at the end-diastolic (ED) configuration (light grey) and end-systolic (ES) configuration (dark black). This shows that a high and low PTH lead to more longitudinal and spherical displacement, respectively. For differences in stiffness scale, we did not find a clear effect.

4 Discussion

In this study, we analyzed the effect of varying the pericardial BCs within a predetermined range on the biomechanical output for both the LV and RV. The main observations of our simulations for varying the pericardial BCs are: (i) the hemodynamic biomarkers for both the LV and RV change by less than 25% , (ii) systolic motion for both ventricles exhibit a high variation, (iii) the portion of the heart free of the pericardium has more influence than the stiffness of the pericardium on all the biomarkers, and (iv) the LV and RV exhibit different responses.

Our findings on the small variation in biomechanical outputs for the LV and RV are consistent with those reported in human and animal studies on

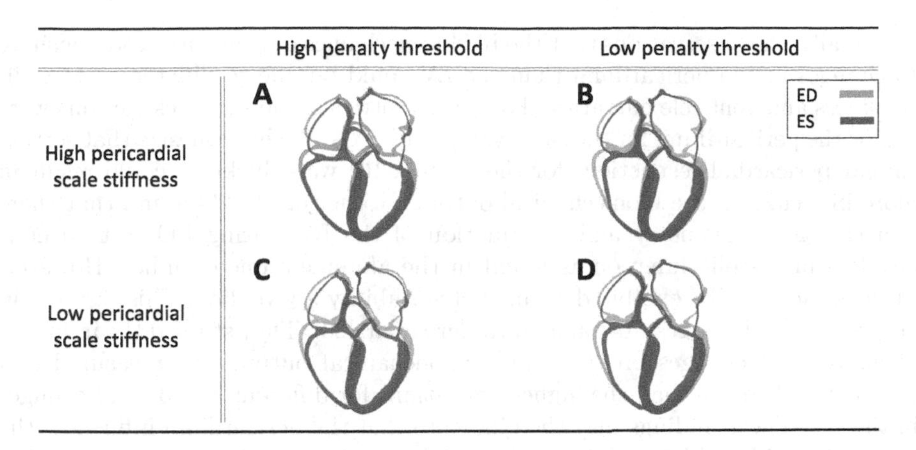

Fig. 4. A slice of the simulated motion for the four-chamber geometry for end-diastole (grey) and end-systole (black) for extremes cases for pericardial threshold (PTH) and pericardial scale stiffness (PSS). **A**: high PTH, high PSS; **B**: high PTH, low PSS; **C**: low PTH, high PSS; **D**: low PTH, low PSS.

cardiac function, before and after pericardial opening (PO) during heart surgery. Mangano [16] has shown that for both the LV and RV, pericardiotomy in the human heart does not substantially influence the systolic function. Daughters et al. [11] only focused on the LV in the human heart and found that the EF remains unchanged upon PO. Bitcon and Tousignant [8] assessed the effect of PO on RV systolic function and found no significant changes upon PO. In our simulations, we can compare PO to setting a low PTH or small PSS. Thus, consistent with our simulations, these studies have shown no significant changes in biomechanical outputs for both ventricles during PO.

The effect of PO on LV global mechanics and local strain has been investigated in the study by Chang et al. [10] in dog hearts. As found in previous studies, they showed that the SV and EF remained the same after PO. However, they found that LV radial strain, which corresponds to the myocardial deformation towards the centre of the LV cavity, was significantly increased after PO. Our results showed a clear correlation between the PTH and LV and RV motion. Comparable to the observations in Chang et al. [10], setting a low PTH in our simulations showed a more spherical deformation of the ventricles. On the other hand, for a high PTH we see that the heart deforms more in longitudinal, consistent with physiological heart motion [27]. Additionally, a computational study by Strocchi et al. showed that the pericardium prevents the heart from spherising during iso-volumic contraction and prevents the apex from moving upwards during ventricular ejection [29]. This means that setting a high PTH leads to a larger portion of the myocardium restricted from deforming spherically (wall-septum motion). On the other hand, a low PTH means a larger portion of the myocardium is free to deform spherically, in agreement with our results.

Finally, the observations that the RV biomechanics outputs are more sensitive to changes in the pericardium than the LV could be due to differences in wall-thickness and ventricle curvature. For both ventricles, the same restrictions were set for the pericardium, for the same values of UVCs. We hypothesize that setting similar pericardial restrictions for the thinner RV wall-thickness could result in more influence on the biomechanical output. Rösner et al. [25] found that there is a change in geometry and deformation of the RV during PO, yet without any loss in systolic function as found in the aforementioned studies. However, dPdtmin for the RV exhibited a range of variability above 25%. This result is in accordance with the results found in Rodero et al. [23]. They studied the influence of anatomical changes on the same biomechanical outputs we measured and found that dPdtmin had the highest variations for different anatomical changes in the RV. These findings may then indicate that the pericardium influences the shape of the RV, which may lead to the higher variations dPdtmin upon changes in pericardial BCs.

Limitations. For this study, we used a geometry generated from an AF patient. In our study, it would be more desirable to use a geometry for a healthy patient and include atrial contraction as well for modelling a more accurate heart deformation. Furthermore, we could move from a linear penalty function to a polynomial function including extra parameters to have a more realistic pericardial stiffness distribution on the epicardium [29]. For the GSA we discarded the biomarkers that showed a variability below 25%. This might exclude clinically relevant features such as the EF, however, we expect that with low variability the variance-based GSA might lead to poor and unreliable outcomes. Thus, this threshold value could be adjusted in future research, although, we anticipate that changing this value will not influence our conclusions. Additionally, no clear reason was found for the higher variations in dPdtmin for the RV compared to the LV. Finally, it is important to note that our results hold for a relatively healthy heart whereas in our comparison to previous clinical and experimental data, animal hearts or diseased hearts that needs surgery were considered. To study how we can model pericarditis and its possible treatments, we would need clinical/experimental data that shows how the pericardial stiffness is distributed so that we can include this in our model.

5 Conclusion

Overall, our work represents a first step in studying the effect of varying the pericardial BCs on the biomechanical output and cardiac deformation in four-chamber heart models. Our results have shown that variations in the pericardium do not have a large effect on pressure and volume derived biomarkers for both the LV and RV. On the other hand, we found that varying the penalty threshold does have a large effect on the systolic motion for both ventricles.

References

1. Afanasyeva, M., Georgakopoulos, D., Fairweather, D., Caturegli, P., Kass, D., Rose, N.: Novel model of constrictive pericarditis associated with autoimmune heart disease in interferon-γ-knockout mice. Circulation **110**(18), 2910–2917 (2004)
2. Alter, P., Figiel, J., Rupp, R., Bachmann, G., Maisch, B., Rominger, M.: MR, CT, and PET imaging in pericardial disease. Heart Fail. Rev. **18**(3), 289–306 (2013)
3. Augustin, C., et al.: A computationally efficient physiologically comprehensive 3D–0D closed-loop model of the heart and circulation. Comput. Methods Appl. Mech. Eng. **386**, 114092 (2021)
4. Augustin, C., et al.: Anatomically accurate high resolution modeling of human whole heart electromechanics: a strongly scalable algebraic multigrid solver method for nonlinear deformation. J. Comput. Phys. **305**, 622–646 (2016)
5. Barrows, R., et al.: The effect of heart rate and atrial contraction on left ventricular function. Comput. Cardiol. **498**, 1–4 (2022)
6. Bayer, J., et al.: Universal ventricular coordinates: a generic framework for describing position within the heart and transferring data. Med. Image Anal. **45**, 83–93 (2018)
7. Bayer, J., Blake, R., Plank, G., Trayanova, N.: A novel rule-based algorithm for assigning myocardial fiber orientation to computational heart models. Ann. Biomed. Eng. **40**(10), 2243–2254 (2012)
8. Bitcon, C., Tousignant, C.: The effect of pericardial incision on right ventricular systolic function: a prospective observational study. Can. J. Anesth./J. Can. d'anesthésie **64**(12), 1194–1201 (2017)
9. Bols, J., Degroote, J., Trachet, B., Verhegghe, B., Segers, P., Vierendeels, J.: A computational method to assess the in vivo stresses and unloaded configuration of patient-specific blood vessels. J. Comput. Appl. Math. **246**, 10–17 (2013)
10. Chang, S., Kim, H., Kim, Y., Cho, G., Oh, S., Sohn, D.: Role of pericardium in the maintenance of left ventricular twist. Heart **96**(10), 785–790 (2010)
11. Daughters, G., Frist, W., Alderman, E., Derby, G., Ingels, N., Jr., Miller, D.: Effects of the pericardium on left ventricular diastolic filling and systolic performance early after cardiac operations. J. Thorac. Cardiovasc. Surg. **104**(4), 1084–1091 (1992)
12. Gerach, T., et al.: Electro-mechanical whole-heart digital twins: a fully coupled multi-physics approach. Mathematics **9**(11), 1247 (2021)
13. Guccione, J., McCulloch, A., Waldman, L.: Passive material properties of intact ventricular myocardium determined from a cylindrical model (1991)
14. Herman, J., Usher, W.: SALib: an open-source python library for sensitivity analysis. J. Open Source Softw. **2**(9), 97 (2017)
15. Longobardi, S., et al.: Predicting left ventricular contractile function via gaussian process emulation in aortic-banded rats. Phil. Trans. R. Soc. A **378**(2173), 20190334 (2020)
16. Mangano, D.: The effect of the pericardium on ventricular systolic function in man. Circulation **61**(2), 352–357 (1980)
17. Marsh, K., et al.: Anti-inflammatory properties of amniotic membrane patch following pericardiectomy for constrictive pericarditis. J. Cardiothorac. Surg. **12**(1), 1–4 (2017)
18. Melo, D., et al.: Impact of pericardiectomy on exercise capacity and sleep of patients with chronic constrictive pericarditis. PLoS ONE **14**(10), e0223838 (2019)
19. Neic, A., et al.: Efficient computation of electrograms and ECGs in human whole heart simulations using a reaction-eikonal model. J. Comput. Phys. **346**, 191–211 (2017)

20. Niederer, S., et al.: Length-dependent tension in the failing heart and the efficacy of cardiac resynchronization therapy. Cardiovasc. Res. **89**(2), 336–343 (2011)
21. Plank, G., et al.: The openCARP simulation environment for cardiac electrophysiology. Comput. Methods Programs Biomed. **208**, 106223 (2021)
22. Rodero, C., et al.: Calibration of cohorts of virtual patient heart models using Bayesian history matching. Ann. Biomed. Eng. **51**, 1–12 (2022)
23. Rodero, C., et al.: Linking statistical shape models and simulated function in the healthy adult human heart. PLoS Comput. Biol. **17**(4), e1008851 (2021)
24. Roney, C., et al.: Universal atrial coordinates applied to visualisation, registration and construction of patient specific meshes. Med. Image Anal. **55**, 65–75 (2019)
25. Rösner, A., et al.: Changes in right ventricular shape and deformation following coronary artery bypass surgery-insights from echocardiography with strain rate and magnetic resonance imaging. Echocardiography **32**(12), 1809–1820 (2015)
26. Saltelli, A., Annoni, P., Azzini, I., Campolongo, F., Ratto, M., Tarantola, S.: Variance based sensitivity analysis of model output. design and estimator for the total sensitivity index. Comput. Phys. Commun. **181**(2), 259–270 (2010)
27. Santiago, A., et al.: Fully coupled fluid-electro-mechanical model of the human heart for supercomputers. Int. J. Numer. Methods Biomed. Eng. **34**(12), e3140 (2018)
28. Sobol, I.: Global sensitivity indices for nonlinear mathematical models and their monte Carlo estimates. Math. Comput. Simul. **55**(1–3), 271–280 (2001)
29. Simulating ventricular systolic motion in a four-chamber heart model with spatially varying robin boundary conditions to model the effect of the pericardium. J. Biomech. **101**, 109645 (2020)
30. Vigmond, E., Hughes, M., Plank, G., Leon, L.: Computational tools for modeling electrical activity in cardiac tissue. J. Electrocardiol. **36**, 69–74 (2003)

An Extended Generalized Hill Model for Cardiac Tissue: Comparison with Different Approaches Based on Experimental Data

Dennis Ogiermann[1(✉)], Daniel Balzani[1], and Luigi E. Perotti[2]

[1] Chair of Continuum Mechanics, Ruhr University Bochum, Bochum, Germany
dennis.ogiermann@ruhr-uni-bochum.de

[2] Department of Mechanical and Aerospace Engineering, University of Central Florida, Orlando, FL, USA

Abstract. In this work we discuss advantages and drawbacks of active mechanics frameworks often used to represent the material response of cardiac tissue. A formal analysis of the models is followed by the application of these frameworks to represent the experimentally measured active and passive response of cardiac tissue. The active strain model is analyzed first and we show that, for commonly used material energies, this framework is not adequate to represent both active and passive tissue responses simultaneously. The active stress model and the generalized Hill model are discussed next. We incorporate the basic idea from Stålhand et al. (2008) to improve the generalized Hill model. Namely we propose to scale the active energy by a function of cross-bridge formation making the model physiologically more accurate. This modification also allows to relate the new Hill type model to the active stress framework. Finally, we show that this extended Hill model best represents the stress-strain response measured in equibiaxial stretch experiments.

Keywords: Active Tissue Model · Extended Hill Model · Invariant Analysis · Cardiac Mechanics · Cardiac Material Modeling

1 Introduction

The active and passive mechanical response of cardiac tissue strongly affects cardiac function and dysfunction. Mathematical frameworks capable of representing the mechanics of cardiac tissue are necessary to simulate normal and pathological conditions, and are therefore a critical component in computational cardiac models aimed at improving patients' diagnosis, prognosis, and therapy planning. In this context, the analysis and validation of the constructed material models is the necessary first step for their deployment in basic and clinical studies.

Three common frameworks exist to model the hyperelastic material response of cardiac tissue: active strain, active stress, and the generalized Hill model. Several authors have studied these different approaches from an analytical standpoint. For example, Ambrosi and Pezzuto [1] investigated conditions for the

O. Bernard et al. (Eds.): FIMH 2023, LNCS 13958, pp. 555–564, 2023.
https://doi.org/10.1007/978-3-031-35302-4_57

existence of solutions to active stress and active strain models. More recently, Giantesio et al. [7] showed that the response of proper active stress and active strain differ significantly, even for simple shear. They also noted that it is conceptually possible to construct active stress tensors such that active stress and active strain coincide in absence of passive response. However, there is no detailed analysis of these frameworks applied to tissue level experimental data, showing that they can predict the active and passive cardiac tissue response simultaneously. In this work, we investigate active mechanics models of cardiac tissue in light of the experimental studies by Lin and Yin [10], which measure both the active and passive response of cardiac tissue. We will highlight possible shortcomings and propose an improved formulation of the generalized Hill model for cardiac mechanics. The proposed approach is inspired by studies in the vascular modeling community and will scale the active energy density using a function of the amount of cross-bridges formed during cardiac contraction and relaxation.

2 Modeling Cardiac Tissue Material Response

The seminal experimental study of Lin and Yin [10] provides simultaneous measures of the active and passive properties of cardiac tissue. In their study, Lin and Yin [10] proposed two different energies to fit the passive and active stress-strain responses separately, namely a Fung-type energy for the passive response and a polynomial energy for the active response. Fung-type models present limitations, e.g., unphysiological coupling of the stretch and shear responses and no straightforward physical interpretation of the parameters. Furthermore, fitting separate energies for the active and passive responses presents another downside as there are no frameworks to merge them into a unified material model to simulate the full cardiac cycle. Nevertheless, the experiments of Lin and Yin [10] allow to probe existing frameworks in cardiac mechanics and investigate which modeling approach can explain the data more accurately. We will start by reviewing the three most common frameworks for the continuum description of active cardiac mechanics. Since the studies by Lin and Yin include only biaxial measurements, we will only investigate transversely isotropic models, although it is known that cardiac tissue is orthotropic due to its mesostructure (see, e.g., [6]).

2.1 Active Strain Models

In active strain models, the deformation gradient tensor is multiplicatively split into an active ($\mathbf{F}^{\mathrm{a}}$) and an elastic ($\mathbf{F}^{\mathrm{e}}$) component, such that $\mathbf{F} = \mathbf{F}^{\mathrm{e}}\mathbf{F}^{\mathrm{a}}$, where $\mathbf{F}^{\mathrm{a}}$ is a function of the contraction model. The elastic part is obtained by $\mathbf{F}^{\mathrm{e}} = \mathbf{F}\mathbf{F}^{-\mathrm{a}}$, whereas the active part $\mathbf{F}^{\mathrm{a}}$ can be written in general form as:

$$\mathbf{F}^{\mathrm{a}} := \lambda_f^{\mathrm{a}} \boldsymbol{f}_0 \otimes \boldsymbol{f}_0 + \lambda_s^{\mathrm{a}} \boldsymbol{s}_0 \otimes \boldsymbol{s}_0 + \lambda_n^{\mathrm{a}} \boldsymbol{n}_0 \otimes \boldsymbol{n}_0 \,, \tag{1}$$

where $\lambda_f^{\mathrm{a}}, \lambda_s^{\mathrm{a}}, \lambda_n^{\mathrm{a}}$ are the active stretches and $\boldsymbol{f}_0, \boldsymbol{s}_0, \boldsymbol{n}_0$ are the corresponding microstructure directions in fiber, sheetlet, and normal to the sheetlet orientations in the reference configuration. For formulations based on $\mathbf{F}^{\mathrm{a}}$, it is standard to define the active stretches based on the contraction and relaxation of cardiomyocytes. The two most common approaches prescribe λ_f^a according to an evolution rule and either constrain $\lambda_s^a = \lambda_n^a = 1$ — i.e., allow for $\det(\mathbf{F}^a) \neq 1$ — or enforce $\det(\mathbf{F}^a) = 1$ by selecting $\lambda_s^a = \lambda_n^a = (\lambda_f^a)^{-1/2}$.

According to this model, the material energy density in the reference configuration is given by $W(\mathbf{F}, \mathbf{F}^a) = W(\mathbf{F}\mathbf{F}^{-a})$. Rossi et al. (2014) [16] proposed a different formulation for the active deformation gradient in the context of the active strain framework based on the argument that cardiac tissue undergoes structural rearrangement at the mesoscale during contraction. Based on this construction, λ_n^a is proportional to λ_f^a and λ_s^a is determined to satisfy the active tissue incompressibility constraint, i.e., $\det(\mathbf{F}^a) = 1$.

2.2 Active Stress Models

Active stress models gained traction over the last few years (see, e.g., [13]) because they allow to directly account for modeling assumptions into the formulation of the stress tensor, circumventing the problem of finding a suitable energy density function corresponding to the sought stress tensor. In active stress models, the total stress is additively decomposed into active and passive components, e.g., the first Piola-Kirchhoff stress tensor can be written as:

$$\mathbf{P}(\mathbf{F}, T^a) := \partial_{\mathbf{F}} W_{\text{passive}}(\mathbf{F}) + \mathbf{P}^a(\mathbf{F}, T^a), \tag{2}$$

where T^a is the tension developed by the contracting cardiomyocytes, $\mathbf{P}^a$ is the active first Piola-Kirchhoff stress tensor, and W_{passive} is the energy density corresponding to the tissue passive response. While this approach allows the advantageous direct decomposition of $\mathbf{P}$, it shifts the burden to the analysis of the material model, e.g., proving the existence of solutions and material stability.

Sack et al. [18] proposed a construction for the active stress similar to the active deformation gradient construction of Rossi et al. [16], where part of the fiber contraction is transferred to the sheetlet direction.

2.3 Generalized Hill Model

The generalized Hill model [8] combines ideas from the active strain and active stress modeling approaches. First, the deformation gradient is multiplicatively decomposed as in the active strain approach. Second, the energy density is additively decomposed into an active and a passive component, leading to an additive decomposition of the stress tensor as in the active stress model. The resulting energy density is:

$$W(\mathbf{F}, \mathbf{F}^a) = W_{\text{passive}}(\mathbf{F}) + W_{\text{active}}(\mathbf{F}\mathbf{F}^{-a}). \tag{3}$$

Several extensions to include viscosity exist [3,14] and, although not directly analyzed in this work, our results will also apply to these models.

2.4 Relation to Smooth Muscle Tissue Models

Due to the orthotropic nature of cardiac tissue and the complex boundary conditions in the heart, a standard generalized Hill model may lead to unphysiological stresses. Indeed, due to residual deformations, the active component may contribute to the total stress even when no cross-bridges are formed.

In general, the active component in the Hill model should only contribute to the total energy and stresses in the tissue when chemical bridges between actin and myosyin filaments are formed, leading to sarcomere shortening and consequent active forces and active stiffness contributions. This mechanism is acknowledged in the vascular and cell modeling communities. For example, in the first active large deformation model for vascular tissue proposed by Stålhand et al. [19], the active energy W_{active} was scaled by a function of the sarcomere model state $\boldsymbol{\alpha}$. Based on this previous work and on the physical meaning of the active component, we propose the following modified energy for modeling cardiac tissue based on a Hill model:

$$W(\mathbf{F}, \mathbf{F}^{\mathrm{a}}) = W_{\text{passive}}(\mathbf{F}) + \mathcal{N}(\boldsymbol{\alpha}) W_{\text{active}}(\mathbf{F}\mathbf{F}^{-\mathrm{a}}) , \tag{4}$$

where $\mathcal{N} : \mathcal{A} \to [0,1]$ is a function describing the amount of formed cross-bridges as in the model of Stålhand et al. [19], and $\mathcal{A}$ is the solution space of the sarcomere model. We note that most active stress models already include a similar mechanism to account for the amount of cross-bridges' formation.

3 Analysis of the Models

We start by summarizing the definition of the invariants of the right Cauchy Green deformation tensor $\mathbf{C} := \mathbf{F}^{\mathrm{T}}\mathbf{F}$ used in the following. The first invariant $I_1(\mathbf{C}) := \mathrm{tr}(\mathbf{C})$ represents the sum of the squares of the principal stretches, i.e., it can be interpreted as a measure of the overall tissue stretch in all directions. The fourth invariant $I_{4,d}(\mathbf{C}) := \boldsymbol{d}_0 \cdot \mathbf{C}\boldsymbol{d}_0 = \lambda_d^2$ measures the square of the stretch along the direction $\boldsymbol{d}_0$ defined in the reference configuration.

3.1 Invariants in the Active Strain Framework

The invariants of the elastic right Cauchy-Green deformation tensor $\mathbf{C}^{\mathrm{e}} = \mathbf{F}^{\mathrm{e}\,\mathrm{T}}\mathbf{F}^{\mathrm{e}}$, corresponding to Eq. 1, can be rewritten as follows:

$$I_1^{\mathrm{e}} = \sum_{d\in\{f,s,n\}} I_{4,d}^{\mathrm{e}} = \sum_{d\in\{f,s,n\}} \frac{1}{\lambda_d^{\mathrm{a}\,2}} I_{4,d} , \tag{5}$$

see Ruiz-Baier et al. [17]. This is a consequence of the orthonormality of the vectors $\boldsymbol{f}_0, \boldsymbol{s}_0, \boldsymbol{n}_0$ defining the cardiac microstructure, which results in the multiplicative split of the stretches $\lambda_d = \lambda_d^{\mathrm{e}}\lambda_d^{\mathrm{a}}$. Considering a strain energy density depending only on $I_{4,f}^{\mathrm{e}}$ and using the transformation of Eq. 5 implies:

$$\mathbf{P}^{\mathrm{a}}(I_{4,f}^{\mathrm{e}}) = \frac{\partial W(I_{4,f}^{\mathrm{e}})}{\partial I_{4,f}^{\mathrm{e}}} \frac{\partial I_{4,f}^{\mathrm{e}}}{\partial I_{4,f}} \frac{\partial I_{4,f}}{\partial \mathbf{F}} = \frac{\partial W(I_{4,f}^{\mathrm{e}})}{\partial I_{4,f}^{\mathrm{e}}} \frac{1}{\lambda_f^{\mathrm{a}\,2}} 2\mathbf{F}\boldsymbol{f}_0 \otimes \boldsymbol{f}_0 . \tag{6}$$

These algebraic manipulations reveal several key implications. First, the common assumption of modeling the active response as a function of only $I^{\mathrm{e}}_{4,f}$ is insufficient to capture the response of the active deformation gradient in the sheetlet and normal directions. Second, the choice of modeling the active deformation gradient directly as incompressible and transversally isotropic leads to an increase in $I_{4,s}$ and $I_{4,n}$ for $I_{4,f} < 1$, corresponding to an expansion in the sheetlet and normal directions, which contradicts experimental findings, e.g., [15]. Interestingly, this drawback could be corrected by constructing $\mathbf{F}^{\mathrm{a}}$ with the mesoscale argument from Rossi et al. [16] when λ^{a}_s and λ^{a}_n are independently related to λ^{a}_f and preserving $\det(\mathbf{F}^{\mathrm{a}}) = 1$. Additionally, according to Eq. 5, an energy function depending on I^{e}_1 and I^{e}_4 only can be written as:

$$W(I^{\mathrm{e}}_1, I^{\mathrm{e}}_4) = W\left(\frac{1}{{\lambda^{\mathrm{a}}_f}^2} I_{4,f}, \frac{1}{{\lambda^{\mathrm{a}}_s}^2} I_{4,s}, \frac{1}{{\lambda^{\mathrm{a}}_n}^2} I_{4,n}\right). \tag{7}$$

This implies that, in this form, the active response is only a 'scaled version' — with respect to the stretches along the microstructure directions — of the passive material response obtained with $\lambda^{\mathrm{a}}_d = 1$. As indicated in the paper by Lin and Yin [10], an energy of such form will likely fail to fit the active and passive response of equibiaxial experiments with a single set of parameters. Indeed the experimentally measured active and passive responses suggest that these are not simply scaled versions of each other with respect to the stretches along $\boldsymbol{f}_0, \boldsymbol{s}_0$, and $\boldsymbol{n}_0$. A representative example of this issue is presented in Fig. 1. We conclude that the active strain model is problematic as a candidate to describe the unified active and passive response of cardiac tissue.

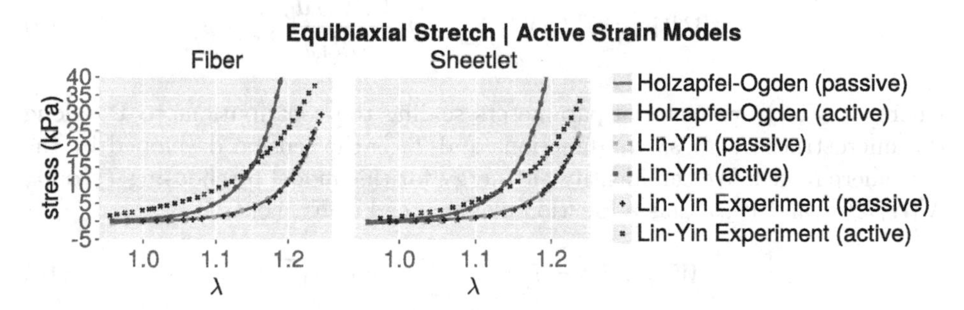

Fig. 1. Stress (P_{11} and P_{22}) strain response: active strain model in fiber and sheetlet directions versus experimental data acquired in an equibiaxial stretch experiment (data shown in Fig. 4 of [10]). The active response is computed using $\lambda_{\mathrm{a}} = 0.9$. Although the tissue passive response is well represented, the active strain framework fails to capture the active response using common material laws for the myocardium. The parameters for the models are given in Table 1.

Table 1. Energies and parameters for the active strain framework described in Sect. 3.1 and used to generate the stress-strain responses shown in Fig. 1.

Holzapfel-Ogden	Lin-Yin
$\frac{a}{2b}\exp(b(I_1-3))+\frac{a_f}{2b_f}\exp(b_f(I_{4,f}-1))$	$C_1\exp(C_2(I_1-3)+C_3(I_1-3)(I_4-1)+C_4(I_4-1))$
$a = 1.580\,\text{kPa}$	$C_1 = 1.436\,\text{kPa}$
$b = 5.801$	$C_2 = 4.214$
$a_f = 0.285\,\text{kPa}$	$C_3 = -0.806$
$b_f = 4.126$	$C_4 = 0.479$

3.2 Relating the Active Stress and the Extended Hill Framework

Based on the modification of the generalized Hill model in Eq. 4, the first Piola-Kirchhoff stress tensor corresponding to the active energy component is:

$$\mathbf{P}^{\mathrm{a}}(\mathbf{F}^{\mathrm{e}}) = \mathcal{N}(\boldsymbol{\alpha}) \sum_{d\in\{f,s,n\}} \frac{2}{\lambda_d^{\mathrm{a}\,2}} \partial_{I^{\mathrm{e}}_{4,d}} W_{\text{active}} \mathbf{F}\boldsymbol{d}_0 \otimes \boldsymbol{d}_0\,, \tag{8}$$

where, for simplicity, we assumed that $W_{\text{active}} = W_{\text{active}}(I^{\mathrm{e}}_{4,f}, I^{\mathrm{e}}_{4,s}, I^{\mathrm{e}}_{4,n})$, as commonly observed in active stress models.

Here we aim to show that, in certain cases, the proposed framework can recover active stress models, while maintaining an energy formulation which allows a simpler formal analysis. For the active stress formulation, we consider the recent model from Piersanti et al. (2022) [13] in which

$$\mathbf{P}^{\mathrm{a}}(\mathbf{F}) = T_{\mathrm{a}}(\boldsymbol{\alpha}) \sum_{d\in\{f,s,n\}} n_d \frac{\mathbf{F}\boldsymbol{d}_0 \otimes \boldsymbol{d}_0}{\sqrt{I_{4,d}}}\,, \tag{9}$$

where $n_{\mathrm{f}}, n_{\mathrm{s}}, n_{\mathrm{n}}$ are material parameters scaling the contributions to $\mathbf{P}^{\mathrm{a}}$ along the microstructure $\boldsymbol{f}_0, \boldsymbol{s}_0, \boldsymbol{n}_0$ directions and T_{a} is the tension developed by the sarcomere model. We can identify an energy for this model by choosing $T_{\mathrm{a}}(\boldsymbol{\alpha}) = \mathcal{N}(\boldsymbol{\alpha}) T_{\mathrm{a}}^{\max}$ and W_{active} as a function of $\mathbf{F}$ instead of $\mathbf{F}^{\mathrm{e}}$, i.e.,

$$W_{\text{active}}(\mathbf{F}) = T_{\mathrm{a}}^{\max} \sum_{d\in\{f,s,n\}} n_d \sqrt{I_{4,d}}\,. \tag{10}$$

This can be interpreted as a model simplification, where it is assumed that the effect of the active stretch on the stress can be neglected, i.e., $\lambda_d^{\mathrm{a}} = 1$. Equivalently, as $\lambda_d^{\mathrm{a}} = 1$, we can assume that Eq. 10 is a function of $\mathbf{F}^{\mathrm{e}}$. This allows to incorporate Eq. 10 in the proposed extended Hill model. Then, the active stress as a function of $\mathbf{F}^{\mathrm{e}}$ has the form:

$$\mathbf{P}^{\mathrm{a}}(\mathbf{F}^{\mathrm{e}}) = T_{\mathrm{a}}(\boldsymbol{\alpha}) \sum_{d\in\{f,s,n\}} \frac{n_d}{\lambda_d^{\mathrm{a}}} \frac{\mathbf{F}\boldsymbol{d}_0 \otimes \boldsymbol{d}_0}{\sqrt{I_{4,d}}}\,. \tag{11}$$

4 Simulation Studies at the Material Point

Having investigated the relation between the active stress and the extended Hill model, we proceed with applying them to the experimental data of Lin and Yin [10]. In their study, the passive response of midwall myocardium is first studied by perfusing the specimens with an excitation contraction decoupler. Afterwards, the excitation contraction decoupler is replaced by barium chloride to trigger maximal steady state contraction and the tissue response is measured. These experiments probe the tissue response at the material level and can be interpreted as a study of the homogenized response at a single material point.

The parameters' estimation for the material response is formulated as a non-linear regression problem solved with the Levenberg-Marquardt algorithm [11]. The regression problem is formulated as the minimization of the differences between computed and measured stresses with respect to the material parameters (denoted by $\boldsymbol{\beta}$):

$$\min_{\boldsymbol{\beta}} \sum_{i=1}^{N_{\text{measures}}} ||\mathbf{P}(\mathbf{F}_i;\boldsymbol{\beta}) - \mathbf{P}_{i,\text{measured}}||_2 \,. \tag{12}$$

$\mathbf{P}_{i,\text{measured}}$ has been measured in [10] while $\mathbf{P}(\mathbf{F}_i;\boldsymbol{\beta})$ is computed from the material energy and imposing the incompressibility constraint as:

$$\mathbf{P}(\mathbf{F};\boldsymbol{\beta}) = \partial_{\mathbf{F}} W(\mathbf{F},\mathbf{F}^{\text{a}};\boldsymbol{\beta}) - p\text{cof}(\mathbf{F})\,, \tag{13}$$

where p is determined so that $\mathbf{P}_{nn} = 0$. We compute the deformation gradient tensor during equibiaxial stretch as:

$$\mathbf{F} = \lambda \boldsymbol{f}_0 \otimes \boldsymbol{f}_0 + \lambda \boldsymbol{s}_0 \otimes \boldsymbol{s}_0 + \frac{1}{\lambda^2}\boldsymbol{n}_0 \otimes \boldsymbol{n}_0\,, \tag{14}$$

where λ is the stretch and the normal component is determined via the incompressibility constraint on $\mathbf{F}$ (i.e., $\det(\mathbf{F}) = 1$). Additionally, during the parameter optimization of the active stress model (Eqs. 9 and 10), we define $s_d = n_d T^{\max}$ to ensure parameter identifiability. We preprocessed the experimental data by shifting the curves on the x-axis, assuming that the first measurement during the passive response in fiber direction corresponds to the reference configuration. The in-house software for the parameters' optimization is implemented in Julia [2] utilizing Tensors.jl [4]. All plots are generated with Makie.jl [5]. All optimizations have been conducted on a standard workstation.

While fitting the passive response of the tissue, in accordance with the experimental setup, we have set $\mathcal{N} = 0$ (Eq. 4), i.e., we set the amount of cross-bridges to 0. As seen in Fig. 1, both the originally proposed Fung-type model [10] and the transversely isotropic Holzapfel-Ogden model [9] represent the passive material response well. For the remainder of the paper we use the latter. To model the active response, we make the simplifying assumption $\mathcal{N} = 1$. However, we note that in the actual experiments, as the tissue is stretched, the filaments would start to slide leading to $\mathcal{N} < 1$. To represent the active material response, we

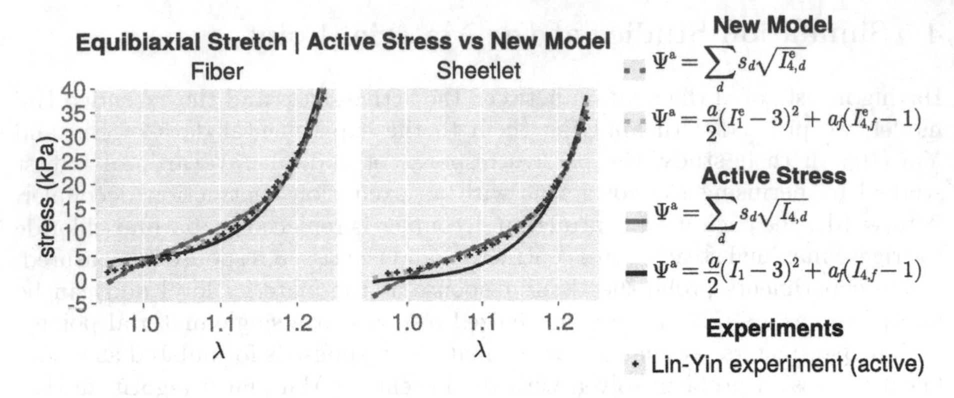

Fig. 2. The active tissue response measured during an equibiaxial stretch experiment (experimental data from Fig. 4 in [10]) is modeled using the active stress and the extended Hill framework. The amount of cross-bridges is maximized, i.e. $\mathcal{N} = 1$, and the active energy formulation is varied as described in the legend. The extended Hill-type framework (Eq. 4) represents well the experimental data in both fiber and sheetlet directions when a polynomial active energy in $(I_1^e - 3)$ and $(I_{4,f}^e - 1)$ is used. The active stress model does not capture the experimental data equally well, especially in the sheetlet direction. The stress-strain response obtained using the active stress model of Piersanti et al. (2022) [13] and our model using Eq. 10 match exactly as expected (see Sect. 3.2).

investigate two representative W_{active}: 1) a polynomial function in $(I_1^e - 3)$ and $(I_{4,f}^e - 1)$; and 2) the energy discussed in Sect. 3.2 to obtain the active stress model of [13]. It can be observed (see Fig. 2) that the extended Hill framework fits the active stress-strain response measured experimentally well when a polynomial function in $(I_1^e - 3)$ and $(I_{4,f}^e - 1)$ is used. More complex energy functions could also be easily incorporated in the future if needed to capture more complex experimental data. On the other hand, the active stress models (and the extended Hill model using the energy function derived to mimic the active stress model in Eq. 10) do not represent the active response equally well, especially in the sheetlet direction. All fitted parameters for the active energies are reported in Table 2 while the parameters for the passive Holzapfel-Ogden model [9] are the same as the ones reported in Table 1.

Table 2. Active energies and parameters employed in the model fittings described in Sect. 4 and shown in Fig. 2.

New Model		Active Stress	
$\frac{a}{2}(I_1^e - 3)^2 + a_f(I_{4,f}^e - 1)$	$\sum_d s_d \sqrt{I_{4,d}^e}$	$\frac{a}{2}(I_1 - 3)^2 + a_f(I_{4,f} - 1)$	$\sum_d s_d \sqrt{I_{4,d}}$
$a = 2.673\,\text{kPa}$	$s_f = 2.504\,\text{kPa}$	$a = 15.020\,\text{kPa}$	$s_f = 3.692\,\text{kPa}$
$a_f = 1.639\,\text{kPa}$	$s_s = 2.146\,\text{kPa}$	$a_f = 4.562\,\text{kPa}$	$s_s = 2.851\,\text{kPa}$
	$s_n = 5.530\,\text{kPa}$		$s_n = 2.823\,\text{kPa}$

5 Concluding Remarks

We have reviewed the most common active mechanics frameworks for cardiac tissue and outlined similarities, advantages, and shortcomings of these approaches. Our analysis suggests that active strain models are likely unable to describe simultaneously the active and passive response of cardiac tissue since the experimental active and passive response are not a scaled version of each other (with respect to the stretches in $\boldsymbol{f}_0, \boldsymbol{s}_0, \boldsymbol{n}_0$) as implied by these models paired with common material energy densities (see Eq. 7 and Fig. 1). Both, the active stress model and the proposed extended Hill model are capable to simultaneously represent the passive and active experimental tissue response, although the newly proposed model offers a better fit, especially in the sheetlet direction (Fig. 2). Furthermore, the extended Hill model can be paired with any material energy density, enabling a high flexibility to fit new experimental data. The remaining minor mismatch in active response is likely due to setting $\mathcal{N} = 1$ (see Eq. 4) and thus fixing the cross-bridge formation to its maximum value throughout the experiment. This is likely not the case as the applied equibiaxial stretch increases. In the future, we will use the energy formulation of the extended Hill model to study its material stability and convexity properties as well as to refine and validate the model against additional experimental data, e.g., in [12].

As a final remark, we note that this work presents only the first step towards improving continuum models to represent the active and passive response of the myocardium. Additional insights will be gained by integrating the proposed and existing models into simulations of ventricular cardiac contraction. These simulations will, for example, highlight the kinematics features of cardiac contraction (e.g., wall thickening, longitudinal shortening, apex to base torsion) that are correctly simulated and which ones are not yet captured consistently and in satisfactory detail using current models. At this stage the presented comparison is not sufficient to determine whether the active stress or the newly proposed model is better to describe the active contraction of cardiac tissue. Determining which framework is best suited to represent the myocardium mechanical response will require additional experimental data for both the active and the passive responses, further computational studies at the tissue and ventricular scales, and more in-depth analyses of the possible contraction frameworks. Nevertheless, the analysis presented in this work contributes toward this goal by highlighting some of the open problems and by suggesting possible solutions to achieve a more accurate active and passive material model for the myocardium.

References

1. Ambrosi, D., Pezzuto, S.: Active stress vs active strain in mechanobiology: constitutive issues. J. Elast. **107**(2), 199–212 (2012)
2. Bezanson, J., Edelman, A., Karpinski, S., Shah, V.B.: Julia: a fresh approach to numerical computing. SIAM Rev. **59**(1), 65–98 (2017)
3. Cansız, B., Dal, H., Kaliske, M.: Computational cardiology: a modified Hill model to describe the electro-visco-elasticity of the myocardium. CMAME **315**, 434–466 (2017)

4. Carlsson, K., Ekre, F.: Tensors.jl — Tensor Computations in Julia. JORS **7**(1), 7 (2019)
5. Danisch, S., Krumbiegel, J.: Makie.jl: flexible high-performance data visualization for Julia. JOSS **6**(65), 3349 (2021)
6. Dokos, S., Smaill, B.H., Young, A.A., LeGrice, I.J.: Shear properties of passive ventricular myocardium. Am. J. Physiol. Heart Circ. Physiol. **283**(6), H2650–H2659 (2002)
7. Giantesio, G., Musesti, A., Riccobelli, D.: A comparison between active strain and active stress in transversely isotropic hyperelastic materials. J. Elast. **137**(1), 63–82 (2019)
8. Göktepe, S., Menzel, A., Kuhl, E.: The generalized Hill model: a kinematic approach towards active muscle contraction. JMPS **72**, 20–39 (2014)
9. Holzapfel, G.A., Ogden, R.W.: Constitutive modelling of passive myocardium: a structurally based framework for material characterization. Philos. Trans. Royal Soc. A **367**(1902), 3445–3475 (2009)
10. Lin, D.H.S., Yin, F.C.P.: A multiaxial constitutive law for mammalian left ventricular myocardium in steady-state barium contracture or tetanus. J. Biomech. Eng. **120**(4), 504–517 (1998)
11. Marquardt, D.W.: An algorithm for least-squares estimation of nonlinear parameters. J. Soc. Ind. Appl. Math. **11**(2), 431–441 (1963)
12. Perotti, L.E., et al.: Estimating cardiomyofiber strain in vivo by solving a computational model. Med. Image Anal. **68**, 101932 (2021)
13. Piersanti, R., et al.: 3D–0D closed-loop model for the simulation of cardiac biventricular electromechanics. CMAME **391**, 114607 (2022)
14. Ponnaluri, A., Perotti, L., Ennis, D., Klug, W.: A viscoactive constitutive modeling framework with variational updates for the myocardium. CMAME **314**, 85–101 (2017)
15. Rademakers, F.E., et al.: Relation of Regional cross-fiber shortening to wall thickening in the intact heart. Three-dimensional strain analysis by NMR tagging. Circulation **89**(3), 1174–1182 (1994)
16. Rossi, S., Lassila, T., Ruiz-Baier, R., Sequeira, A., Quarteroni, A.: Thermodynamically consistent orthotropic activation model capturing ventricular systolic wall thickening in cardiac electromechanics. Eur. J. Mech. A. Solids **48**, 129–142 (2014)
17. Ruiz-Baier, R., et al.: Mathematical modelling of active contraction in isolated cardiomyocytes. Math. Med. Biol. **31**(3), 259–283 (2014)
18. Sack, K.L., et al.: Construction and validation of subject-specific biventricular finite-element models of healthy and failing swine hearts from high-resolution DT-MRI. Front. Physiol. **9**, 539 (2018)
19. Stålhand, J., Klarbring, A., Holzapfel, G.A.: Smooth muscle contraction: mechanochemical formulation for homogeneous finite strains. Prog. Biophys. Mol. Biol. **96**(1–3), 465–481 (2008)

pyheart-lib: A Python Library for LS-DYNA Multi-physics Heart Simulations

Martijn Hoeijmakers, Karim El Houari(✉), Wenfeng Ye, Pierre L'Eplattenier, Attila Nagy, Dave Benson, and Michel Rochette

Ansys Inc., Canonsburg, USA
elhouarikarim@hotmail.com

Abstract. Phyisics-based computer simulations of the heart have huge potential in the medical device industry and clinical practice, for instance to accelerate and improve device designs, assist clinical decision making, or guide treatment planning. For heart simulations the importance of modeling choices with respect to electrophysiology, the structural behavior of the cardiac tissue, the dynamics of blood flow, and their respective coupling strongly depend on the application of interest. LS-DYNA is a finite element solution that offers the necessary multi-physics capabilities and features for heart modeling. Nevertheless, setting up these models and obtaining physiological results is still highly manual and can be cumbersome. Therefore, in this paper we propose a python-based high-level interface to LS-DYNA, that will be free-to-use and dedicated to heart modeling. We introduce the relevant heart modeling features that are available and introduce the modular python library to set up and drive these simulations. Consequently, two example models are presented: a full heart model of electrophysiology and a bi-ventricular model of cardiac mechanics.

Keywords: Cardiac modeling · Finite element analysis · Python

1 Introduction

It is widely recognized that numerical methods-based computer simulations of the heart have tremendous potential, both for the medical device industry as for clinical practice [19].

Over the years several tools dedicated to heart modeling were proposed [13]. However, heart modeling is not yet common practice in the medical device industry. Specific expertise, knowledge, and experience is required to set up workflows that link various tools. E.g. segmentation tools to obtain a suitable geometry, a meshing tool to create a simulation-ready mesh, and a finite element analysis tool for solving the governing equations. Moreover, heart modeling features such as material laws, electrophysiology models, and coupling schemes may not be available in the tool of preference, which further inhibits adoption. Recently, various relevant heart modeling features were added to LS-DYNA, a commercial

O. Bernard et al. (Eds.): FIMH 2023, LNCS 13958, pp. 565–574, 2023.
https://doi.org/10.1007/978-3-031-35302-4_58

general-purpose finite element code known for its (highly) non-linear finite element analysis capabilities. Added features include: mono- and bidomain solvers for electrophysiology modeling, advanced fluid-structure interaction capabilities as demonstrated by [4,15], as well as Purkinje network and fiber generation methods. Nevertheless, creating LS-DYNA heart simulation models still requires an experienced user and manual work. Therefore, we propose a python library – *pyheart-lib* – dedicated to heart modeling, that should help new and existing LS-DYNA users to efficiently set up and post-process these heart models.

2 Methods

The following section is divided into two main parts. The first part describes the available heart modeling features in LS-DYNA, and the second part gives a brief introduction to the python library that was developed to exploit these features (see e.g. Fig. 1). Finally, we give two potential user applications: a full heart electrophysiology model, and a bi-ventricular model of cardiac mechanics.

2.1 LS-DYNA Heart Modeling Features

A heartbeat is the result of three physics that are inherently coupled: electrophysiology (EP), the mechanical response of the tissue, and blood flow [21]. EP drives mechanical deformation by excitation of the myocardial fibers leading to contraction, this mechanical deformation acts to drive blood flow in and out of the cavities. Deformation is however also a function of the pressure exerted by the fluid inside the cavity. Since the influence of the mechanical deformation to the electrophysiological states is less significant, we essentially have a combined one-way (EP-mechanics) and two-way (mechanics-fluids) coupling problem.

Electrophysiology. The bidomain and the monodomain models are widely used to simulate cardiac electrical propagation, and accepted for their physiological relevance [18]. Briefly, the bidomain model assumes that cardiac tissue can be partitioned into two separate conducting media: the intracellular space, located inside the cardiac cells, and the extracellular space that connects cells between them. The monodomain model is a simplification of the bidomain model obtained by assuming an equal anisotropy ratio between the intracellular conductivity tensor and the intracellular conductivity tensor. This assumption considerably reduces the size of the system to be solved while keeping a comparable accuracy for the targeted application [20]. LS-DYNA offers the possibility of using either the bidomain, monodomain, or a mix of both models. In this paper we focus on the monodomain formulation which reads:

$$\beta C_m \frac{\partial V_m}{\partial dt} + \beta I_{ion}(V_m, u) - \nabla \cdot (\sigma \nabla V_m) = \beta I_{stim} \tag{1}$$

where V_m is the transmembrane potential that measures the potential difference between the intracellular and extra-cellular space at time t, β is the cell surface to volume ratio, C_m is the cell membrane capacitance, $I_{ion}(V_m, u)$ is the resulting ionic currents flowing from the extracellular to the intracellular space through cell membranes, σ is an electrical conductivity tensor that expresses the anisotropic inhomogeneous character of the heart tissue, and I_{stim} is the stimulation current applied on a small region of heart tissue in order to initiate the electrical wavefront. Finally, u is a set of variables that satisfy a set of ordinary differential equations that describe how ionic currents evolve with time through a cardiac cell membrane. Such a system can be expressed by the following equation:

$$\frac{\partial u}{\partial t} = f(u, V_m) \tag{2}$$

The choice of the expression of f determines which type of cell model is used to describe ionic kinetics. Users can choose among for instances: Fitzhugh Nagumo, ten Tusscher, ToR-Ord [14] or define their own by providing the expression of f. These equations are completed by Neumann boundary conditions at the domain surface in case of an electrically isolated model, or by flux continuity boundary conditions in the presence of bath. The EP system of equations can be solved using direct or iterative solvers, with several options of operator splitting techniques, for more details see [16].

Mechanics. Cardiac tissue can be characterized by a hyperelastic, nearly incompressible, orthotropic material with a nonlinear stress-strain relationship [9]. The orthotropic property is due to the presence of the myocardial fibers. The deformation is governed by boundary conditions (pericardial constraints), external loads (blood pressure), and active stresses generated along the fiber direction upon contraction. The total stress can be decomposed into passive and active parts as:

$$\boldsymbol{\tau} = \boldsymbol{\tau}_P + \boldsymbol{\tau}_A \tag{3}$$

Passive Stress. $\boldsymbol{\tau}_P$ designates the passive stress which is derived from the strain energy function Ψ. In this work we used the nearly in-compressible Holzapfel-Ogden model with one fiber family [9]:

$$\Psi = \Psi_I(\bar{I}_1) + \Psi_F(\bar{I}_f) + \Psi_V(J) \tag{4}$$

$$= \frac{a}{2b}(exp[b(\bar{I}_1 - 3)] - 1) + \frac{a_f}{2b_f}(exp[b_f(\bar{I}_f - 3)] - 1) + \frac{K}{2}(J-1)^2. \tag{5}$$

Active Stress. Active stress describes the excitation-contraction from the EP model. Multiple active models are available within LS-DYNA, including those presented by [8,10,17]. For the mechanical model presented here, we used the classic model of active stress proposed by Guccione *et al.* [7]. This "Hill"-type active model describes the relationship between the evolution of cytosolic calcium ion concentration and tension developed in the tissue, calcium concentration can be either defined explicitly or can come from a coupled EP simulation.

Circulation Model. The opening/closing state of the heart valves strongly influences the pressures inside the cavity, leading to the characteristic pressure-volume loops with two iso-volumetric phases (iso-volumetric contraction and iso-volumetric relaxation). Fully resolving blood flow and behavior of the heart valves is however not a strict requirement for all applications where the overall cardiac function is of interest. Hence, a uniform cavity pressure is typically assumed [1,3,25]. Similarly, a lumped-parameter approach is available in LS-DYNA, where blood flow in/out of the cavity – represented by an in-compressible volume – can be added as an additional constraint to update the pressure of the cavity.

Fiber Orientation. The spatial arrangement of the ventricular myocytes is crucial for the physiological conduction of the electrical wave and mechanical response. To obtain this typical fiber structure, LS-DYNA uses the rule-based method proposed in [2]. Briefly, the user specifies Laplace-Dirichlet problems for which the solutions gradients define the apex base direction and the transmural direction, for each element of the mesh. These two local vectors are rotated by an angle that varies depending on the transmural depth. The obtained local coordinate system serves as the basis for defining electrical conduction properties and active stress components.

Purkinje Network. The Purkinje network allows for a more appropriate electrical activation sequence. LS-DYNA offers the possibility to construct a set of beams that play the role of Purkinje fibers in EP simulations. LS-DYNA implements a fractal tree algorithm similar to [5] that takes a set of faces and a starting point as inputs in order to construct a set of beam elements that lie on the given surface.

2.2 A Pythonic Interface to LS-DYNA for Heart Modeling

Python is popular in the scientific community due to the large number of third party libraries that are available. Similarly, we developed a python library that integrates some of these third party libraries, such as the Visualization Toolkit [23], and exploits the implemented heart-modeling features in LS-DYNA. Consequently, this library provides pythonic access to LS-DYNA's heart-modeling capabilities, while at the same time reducing complexity by using an abstracted representation of a heart model. In this preliminary work, we introduce two modules: a `preprocessor` module and a `simulator` module.

`preprocessor` The main purpose of the `preprocessor` module is to generate a heart model that contains all geometric heart features relevant for the simulation model, and currently supports models of the isolated left ventricle, bi-ventricular models, and full-heart models (Fig. 1). Moreover, this module provides methods to filter the parts of interest, extract relevant geometric features, and (re)mesh surfaces and volumes if necessary. Examples of relevant heart features include: the endo/epi-cardial, and septal surfaces, apical points, and cavities. For example, this `preprocessor` can currently download and process models from the virtual cohorts of full-heart models by Strocchi *et al.* [24] and Rodero *et al.* [22], but can be extended to other sources.

`simulator` Models defined by the `preprocessor` module can be consumed by the `simulator` module. The `simulator` module contains different simulators, e.g. for EP simulations and electro-mechanics simulations (Fig. 1). Consequently, these simulators contain essential methods for these type of simulations, and include computation of 1) the fiber orientation, 2) the Purkinje network, and 3) the stress-free configuration. Finally, any main simulation can then easily be launched by `simulator.simulate()`. Methods for convenient visualization of resulting fibers or constructed Purkinje networks are also available. A typical code-snippet to generate a full heart Electrophysiology model from [24] is included below.

```
import ansys.heart.preprocessor.models as models
from ansys.heart.simulator.simulator import EPSimulator
# instantiate model information
info = models.ModelInfo(
    database="Strocchi2020",
    path_to_case="01.case",
    work_directory="fullheart",
    path_to_model="fullheart.pickle",
    mesh_size=2.0
)
# instantiate full heart model object
model = models.FullHeart(info)
# extract simulation mesh and save model
model.extract_simulation_mesh()
model.dump_model()
# use extracted simulation mesh to instantiate simulator
simulator = EPSimulator(
    model=model,
    lsdynapath="lsdyna.exe",
    dynatype="smp",
    num_cpus=6,
    simulation_directory="simulation-EP"),
)
# load default settings
simulator.settings.load_defaults()
# compute fiber orientation and Purkinje network
simulator.compute_fibers()
simulator.compute_purkinje()
# visualize fibers and Purkinje
simulator.model.plot_fibers(n_seed_points=2000)
simulator.model.plot_purkinje()
# start main Electrophysiology simulation
simulator.simulate()
```

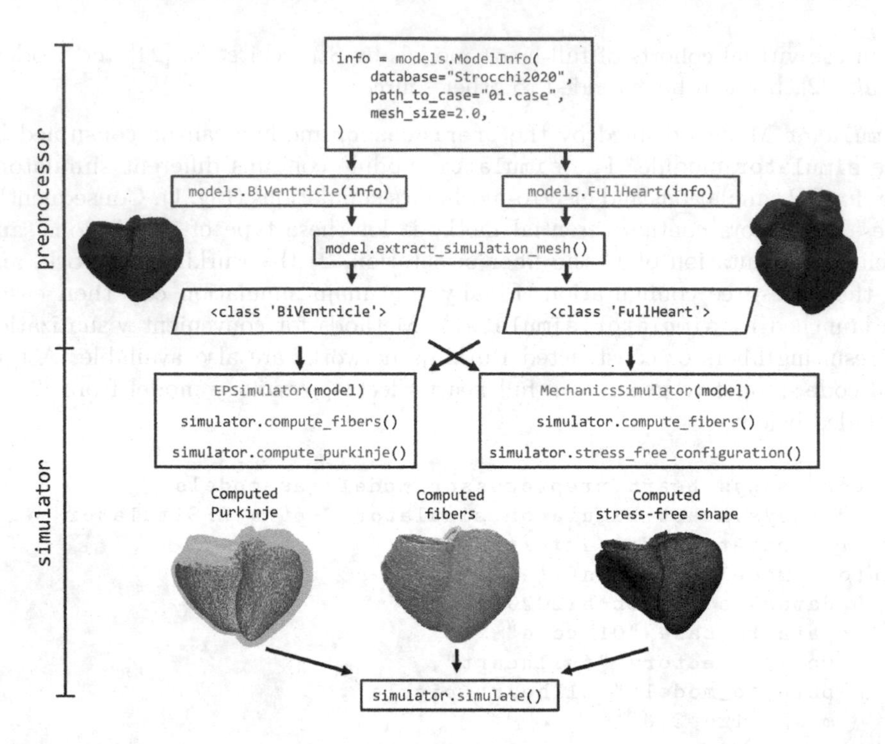

Fig. 1. Flow chart demonstrating the `preprocessor` and `simulator` modules. The preprocessor is used to instantiate a model of a certain geometric type. Consequently, this instantiated model can be used to instantiate a simulator class that defines the physics of interest.

3 Example Models

Section 2.2 describes the `preprocessor` and `simulator` modules. In the following section two example models are presented that were generated with these modules: a full-heart EP model, and a bi-ventricle mechanical model to simulate cardiac mechanics.

3.1 Full Heart Electrophysiology

Methods. The full-heart geometry was obtained from [24], and was re-meshed with an average edge length of 2 mm. The model uses the monodomain framework (Eq. 1) with a 1 ms time step and the ten Tusscher cell model with a finer time step of 0.2 ms, and exploited the implicit first order operator splitting scheme (see appendix D in [16]). The conductivity values in atria and ventricles were 0.5 mS/mm in the fiber direction and 0.1 mS/mm in the normal and sheet directions. The conductive beams were assigned with a conductivity value of 10 mS/mm and the atrioventricular node area with a value of 7 mS/mm. A stimulation current of 50 μA/mm^3 was applied at the sinoatrial node for a duration of 20 ms.

Results. Figure 2b demonstrates that the obtained activation sequence follows a typical healthy wavefront. The initiation of the wavefront occurs at the sinoatrial node, propagates through the right and left atria, is delayed at the atrioventricular node, spreads through the branches, Purkinje network and finally within the ventricular myocardium. The total activation times at the atria and ventricles are 235 ms and 222 ms respectively.

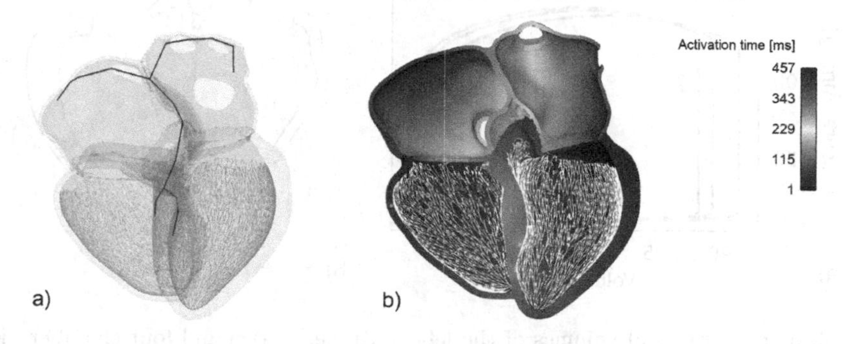

Fig. 2. a) Full heart geometry with the conduction system highlighted in blue. b) activation time visualized in a four-chamber view. (Color figure online)

3.2 Bi-ventricular Mechanical Model

Methods. The bi-ventricular geometry is extracted from a healthy full heart of [22], and represents a healthy case segmented at end-diastole. Fibers were generated on the end-diastolic geometry using a rule based method [2]. The stress free configuration was computed using a left-ventricular and right-ventricular end-diastolic pressure of 15 mmHg and 4 mmHg respectively. The presence of the pericardium was modeled using springs defined in the normal direction of the epicardial surface, and was similar to the implementation by [25]. A constant pre-load (left: p_{venous} = 15 mmHg, right: p_{venous} = 4 mmHg) and Windkessel afterload was added to both cavities, see also [25]. Passive material parameters were based on literature, and chosen as: $a = 2.36$ kPa, $b = 1.75$, $a_f = 0.49$ kPa, and $b_f = 9.01$. Active stress was generated by using a model of active stress proposed in [7], with a maximum active stress of $T_{max} = 125$ kPa. All fibers are triggered simultaneously.

Results. Peak systolic pressure in the left and right ventricles were 146 mmHg and 19 mmHg respectively (Fig. 3a). Multiple cycles were required to obtain

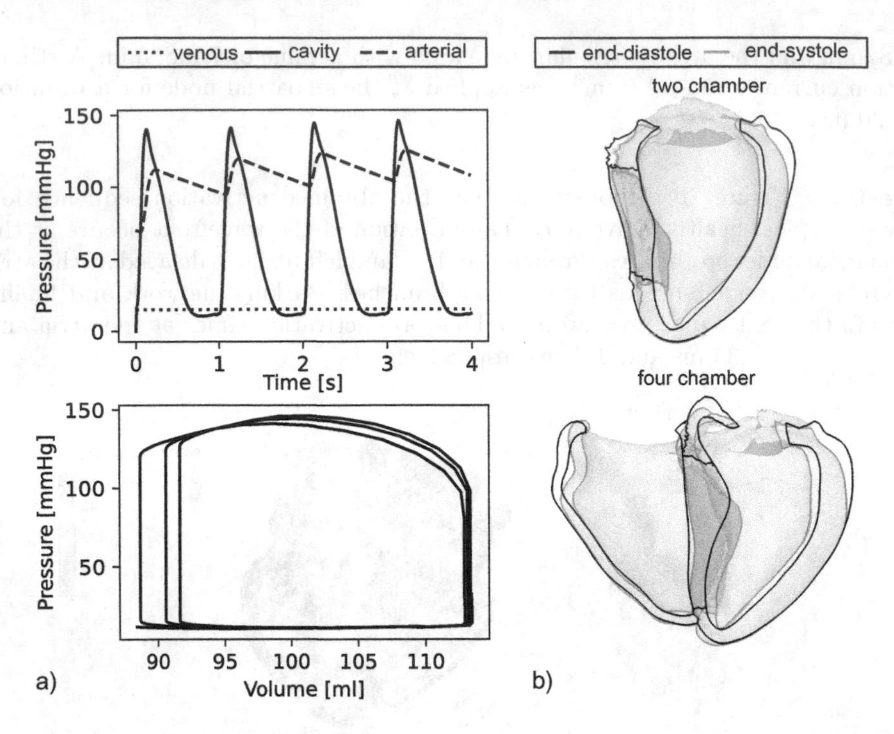

Fig. 3. a) pressures and volumes of the left ventricle. b) two and four chamber views at end-diastole and end-systole. Note: atria not present and thus not visible in the cross-section.

a periodic solution, and in the fourth cycle, a left-ventricular ejection fraction of 19% was obtained. Moreover, Fig. 3b demonstrates that the pericardial constraint limits the motion of the apex and gradually allows more motion toward the base.

4 Discussion

In this manuscript we present a python library to create LS-DYNA heart models. This python library was developed in a modular fashion, that enables a user to create heart models of various levels of complexity through a python interface. Two example models were created: a full-heart electrophysiology model, and a bi-ventricular model of cardiac mechanics. Both models are able to represent typical electrophysiology and mechanical behavior with parameters and model constants from literature.

Even though these models are patient-specific with respect to geometry, calibration with clinical data is necessary. For instance, total activation times observed in Durrer *et al.* [6] are lower than those obtained with the presented model, conductivity would need to be optimized to reach proper activation times. Hence, methods are being developed to optimize conduction velocity given patients' ECG signals, or based on desired conduction velocities.

For the mechanical model, passive material parameters may be personalized by ensuring that the end-diastolic pressure-volume relationship corresponds to empirical relationships found in [11]. To address this need, a `calibration` module, in line with the framework proposed by [12], would be required for further personalizing these models, and is currently being developed.

5 Conclusion

In this manuscript we presented a python library to generate LS-DYNA based heart simulations. We adopted a modular approach, where the geometry of interest and physics of interest are exposed in an abstract way. Consequently, with just a few lines of python code a user can generate example models of different levels of complexity. Hence, we believe this library will facilitate the democratization of cardiac simulation for realistic geometries.

References

1. Baillargeon, B., Rebelo, N., Fox, D.D., Taylor, R.L., Kuhl, E.: The living heart project: a robust and integrative simulator for human heart function. Eur. J. Mech. - A/Solids **48**, 38–47 (2014). https://doi.org/10.1016/j.euromechsol.2014.04.001
2. Bayer, J.D., Blake, R.C., Plank, G., Trayanova, N.A.: A novel rule-based algorithm for assigning myocardial fiber orientation to computational heart models. Ann. Biomed. Eng. **40**(10), 2243–2254 (2012). https://doi.org/10.1007/s10439-012-0593-5
3. Bovendeerd, P., Arts, T., Huyghe, J., van Campen, D., Reneman, R.: Dependence of local left ventricular wall mechanics on myocardial fiber orientation: a model study. J. Biomech. **25**(10), 1129–1140 (1992). https://doi.org/10.1016/0021-9290(92)90069-d
4. Cestariolo, L., Luraghi, G., L'Eplattenier, P., Matas, J.F.R.: A finite element model of the embryonic zebrafish heart electrophysiology. Comput. Methods Programs Biomed. **229**, 107281 (2023). https://doi.org/10.1016/j.cmpb.2022.107281
5. Costabal, F.S., Hurtado, D.E., Kuhl, E.: Generating purkinje networks in the human heart. J. Biomech. **49**(12), 2455–2465 (2016). https://doi.org/10.1016/j.jbiomech.2015.12.025
6. Durrer, D., Dam, R.T.V., Freud, G.E., Janse, M.J., Meijler, F.L., Arzbaecher, R.C.: Total excitation of the isolated human heart. Circulation **41**(6), 899–912 (1970). https://doi.org/10.1161/01.cir.41.6.899
7. Guccione, J.M., Waldman, L.K., McCulloch, A.D.: Mechanics of active contraction in cardiac muscle: part II–cylindrical models of the systolic left ventricle. J. Biomech. Eng. **115**(1), 82–90 (1993). https://doi.org/10.1115/1.2895474
8. Göktepe, S., Acharya, S.N.S., Wong, J., Kuhl, E.: Computational modeling of passive myocardium. Int. J. Numer. Methods Biomed. Eng. **27**(1), 1–12 (2010). https://doi.org/10.1002/cnm.1402
9. Holzapfel, G.A., Ogden, R.W.: Constitutive modelling of passive myocardium: a structurally based framework for material characterization. Philos. Trans. R. Soc. A Math. Phys. Eng. Sci. **367**(1902), 3445–3475 (2009). https://doi.org/10.1098/rsta.2009.0091

10. Hunter, P., McCulloch, A., ter Keurs, H.: Modelling the mechanical properties of cardiac muscle. Prog. Biophys. Mol. Biol. **69**(2–3), 289–331 (1998). https://doi.org/10.1016/s0079-6107(98)00013-3
11. Klotz, S., et al.: Single-beat estimation of end-diastolic pressure-volume relationship: a novel method with potential for noninvasive application. Am. J. Physiol.-Heart Circulatory Physiol. **291**(1), H403–H412 (2006). https://doi.org/10.1152/ajpheart.01240.2005
12. Krishnamurthy, A., et al.: Patient-specific models of cardiac biomechanics. J. Comput. Phys. **244**, 4–21 (2013). https://doi.org/10.1016/j.jcp.2012.09.015
13. Land, S., et al.: Verification of cardiac mechanics software: benchmark problems and solutions for testing active and passive material behaviour. Proc. R. Soc. A: Math., Phys. Eng. Sci. **471**(2184), 20150641 (2015). https://doi.org/10.1098/rspa.2015.0641
14. Livermore Software Technology Corporation: LS-DYNA Keyword User's Manual. Volume I. II, II (2023)
15. Luraghi, G., et al.: On the modeling of patient-specific transcatheter aortic valve replacement: a fluid–structure interaction approach. Cardiovasc. Eng. Technol. **10**(3), 437–455 (2019). https://doi.org/10.1007/s13239-019-00427-0
16. L'Eplattenier, P., Çaldichoury, I., Pin, F.D., Paz, R., Nagy, A., Benson, D.: Cardiac electrophysiology using LS-DYNA. In: 16th International LS-DYNA Users Conference (2020)
17. Martins, J.A.C., Pato, M.P.M., Pires, E.B.: A finite element model of skeletal muscles. Virtual Phys. Prototyp. **1**(3), 159–170 (2006). https://doi.org/10.1080/17452750601040626
18. Pathmanathan, P., et al.: A numerical guide to the solution of the bidomain equations of cardiac electrophysiology. Prog. Biophys. Mol. Biol. **102**(2–3), 136–155 (2010). https://doi.org/10.1016/j.pbiomolbio.2010.05.006
19. Peirlinck, M., et al.: Precision medicine in human heart modeling. Biomech. Model. Mechanobiol. **20**(3), 803–831 (2021). https://doi.org/10.1007/s10237-021-01421-z
20. Potse, M., Dube, B., Richer, J., Vinet, A., Gulrajani, R.: A comparison of monodomain and bidomain reaction-diffusion models for action potential propagation in the human heart. IEEE Trans. Biomed. Eng. **53**(12), 2425–2435 (2006). https://doi.org/10.1109/tbme.2006.880875
21. Quarteroni, A., Lassila, T., Rossi, S., Ruiz-Baier, R.: Integrated heart–coupling multiscale and multiphysics models for the simulation of the cardiac function. Comput. Methods Appl. Mech. Eng. **314**, 345–407 (2017). https://doi.org/10.1016/j.cma.2016.05.031
22. Rodero, C., et al.: Virtual cohort of adult healthy four-chamber heart meshes from CT images (2021). https://doi.org/10.5281/ZENODO.4590294
23. Schroeder, W., Martin, K., Lorensen, B.: The Visualization Toolkit, 4th edn. Kitware, Clifton Park (2006)
24. Strocchi, M., et al.: A publicly available virtual cohort of four-chamber heart meshes for cardiac electro-mechanics simulations (2020). https://doi.org/10.5281/ZENODO.3890034
25. Strocchi, M., et al.: Simulating ventricular systolic motion in a four-chamber heart model with spatially varying robin boundary conditions to model the effect of the pericardium. J. Biomech. **101**, 109645 (2020). https://doi.org/10.1016/j.jbiomech.2020.109645

Evaluation of Mechanical Unloading of a Patient-Specific Left Ventricle: A Numerical Comparison Study

Britt P. van Kerkhof(✉), Koen L. P. M. Janssens, Luca Barbarotta, and Peter H. M. Bovendeerd

Department of Biomedical Engineering, Eindhoven University of Technology, 5600MB Eindhoven, The Netherlands
b.p.v.kerkhof@tue.nl

Abstract. In this study, we use a finite element model of left ventricular (LV) mechanics to evaluate the mechanical unloading of an end-diastolic (ED) geometry by using two different unloading algorithms: a direct method, and an iterative method. Furthermore, we evaluated the effects of using isotropic or anisotropic material properties. One representative ED geometry was derived from an atlas of LV geometries and used for mechanical unloading. We used a volume criterion instead of the more commonly used pressure criterion. The direct and iterative method gave identical results in unloaded geometries. Isotropic versus anisotropic material properties gave only minor differences in geometry. The main effect was found in unloading pressure. Overall, we conclude that both unloading algorithms can be used in further research. However, from a physiological and computational point of view, the direct method is preferable to the iterative method.

Keywords: Finite element modeling · Patient-specific cardiac modeling · Left ventricle · Unloaded configuration

1 Introduction

The geometry of the heart is an important input in finite element models of cardiac mechanics. In these models, the unloaded configuration is usually used as a reference from which the deformation, due to a mechanical load, can be computed. For patient-specific modeling of cardiac mechanics, medical imaging methods can never provide the unloaded configuration of the patient's heart, since physiologically the heart is never in an entirely unloaded state due to a continuously present pressure load. Usually, the end-diastolic (ED) geometry is recorded. Methods to convert this geometry into an unloaded one can be either direct, where the inverse motion from the deformed configuration to the unloaded configuration is solved [1,7,9], or iterative, where the unloaded configuration is determined through multiple forward calculations [2,8,10,14,15]. To our knowledge, these methods have not been compared before. Therefore, the aim of this

O. Bernard et al. (Eds.): FIMH 2023, LNCS 13958, pp. 575–584, 2023.
https://doi.org/10.1007/978-3-031-35302-4_59

study is to compare them and evaluate the differences in the resulting unloaded geometries. Both methods depend on the in vivo measured geometry and the corresponding blood pressure at the moment of imaging. However, this pressure is often uncertain or unknown. Furthermore, the results are then also dependent on the properties of the material, which are also not well known. Therefore, in the current study, both methods are implemented using a volume criterion, and the effects of the material properties are evaluated by using isotropic and anisotropic material properties. We use an atlas-based average patient-specific left ventricular (LV) geometry.

2 Methods

2.1 Model of Cardiac Mechanics

The finite element model of LV mechanics presented in [3] was used as a basis for computing tissue deformations during mechanical unloading. Material properties were kept identical. However, the simplified ellipsoidal shape in the latter study was replaced by an average LV ED geometry (Sect. 2.2). Since we focus on the passive pressure-volume relation during filling, we only include passive material properties. Passive material behavior is assumed non-linearly elastic, transversely isotropic, and nearly incompressible. It is described through a strain energy density function W, composed of a shape part W_s and a volumetric part W_v:

$$W_s = a_0 \left(e^{Q_s} - 1\right) \tag{1}$$

$$\begin{aligned} Q_s = a_1 \left(E_{ff}^2 + E_{ss}^2 + E_{nn}^2\right) + \frac{1}{2} a_2 \left(E_{ns}^2 + E_{sn}^2\right) \\ + a_3 E_{ff}^2 + \frac{1}{2}\left(a_2 + a_4\right)\left(E_{fs}^2 + E_{sf}^2 + E_{fn}^2 + E_{nf}^2\right) \end{aligned} \tag{2}$$

$$W_v = a_5 [\det \left(\boldsymbol{F}^T \cdot \boldsymbol{F}\right) - 1]^2 \tag{3}$$

where W_s is formulated in terms of components of the Green-Lagrange strain tensor $\boldsymbol{E}$ with respect to the material-bound fiber basis $\{\vec{e}_f, \vec{e}_s, \vec{e}_n\}$, Q_s is a function of the Green-Lagrange strain invariants for a transversely isotropic material [4], $\boldsymbol{F}$ is the deformation gradient tensor, and a_0 till a_5 are material parameters.

Cardiac deformation is computed by solving for equilibrium between forces related to passive stress $\boldsymbol{\sigma}$ and cavity pressure:

$$\vec{\nabla} \cdot \boldsymbol{\sigma} = \vec{0} \tag{4}$$

Dirichlet boundary conditions were used to fix the LV base in space, i.e. the basal plane is prevented from moving in axial direction, and rotation of the subendocardial basal ring is suppressed. Neumann boundary conditions were applied at the LV endocardial and epicardial surfaces. The endocardial surface was subjected to a uniform cavity pressure, while the epicardial surface was traction free.

2.2 Geometry

We used the average ED left ventricular shape, obtained from geometric data from the Cardiac Atlas Project (CAP, http://www.cardiacatlas.org), based on a population of about 2000 asymptomatic subjects [12]. The associated ED surface mesh of the atlas was interpolated onto an approximate volume-matched ellipsoidal template mesh, after aligning the two basal planes. Then, the interpolated ED surface mesh was used to define nodal displacements relative to the template mesh, which were imposed as Dirichlet boundary conditions on the endocardial and epicardial surfaces. Subsequently, mechanical equilibrium (Eq. 4) was solved, in order to move the internal nodes of the mesh.

In the template mesh, a rule-based method was used to define the fiber orientation in terms of a helix angle and a transverse angle [3]. The spatial deformation gradient of the geometric mapping from the template mesh to the ED mesh was used to reorient the fiber field defined over the template mesh. The generated ED geometry was further used in the mechanical unloading simulations. The approximate volume-matched template mesh and the derived ED mesh with their fiber orientations can be seen in Fig. 1.

2.3 Computation of the Unloaded Configuration

Direct Method. The direct method to compute the unloaded geometry, as used in [1], is based on the concept of Inverse Design [7]. Usually, in solving (4) the unloaded reference configuration Ω_0 is known, and the displacement field $\boldsymbol{u}$ to a loaded configuration Ω is sought. In our case, the loaded configuration Ω_{ED} is known, and the displacement -$\boldsymbol{u}$ to the reference configuration Ω_0 is sought. So, we reformulated the Cauchy stress in terms of the deformation gradient, associated with the inverse displacement -$\boldsymbol{u}$.

In order to solve (4) an ED pressure load is needed. Since this pressure was unknown, the pressure load was incremented until an estimated unloaded cavity volume was reached. This volume $V_{c,0}$ can be approximated by the cavity volume at end-systole $V_{c,ES}$. Therefore, we assumed a physiological ejection fraction (EF) of 60% and estimated $V_{c,0}$ as follows:

$$V_{c,0} \approx V_{c,ES} = \left(1 - \frac{\mathrm{EF}}{100}\right) \cdot V_{c,ED} \tag{5}$$

where $V_{c,ED}$ is the known end-diastolic cavity volume.

The first estimate of the unloaded geometry is chosen to be the ED geometry. For each pressure increment, the cavity volume is calculated with the inverse displacement $-\boldsymbol{u}$. When the calculated volume reaches $V_{c,0}$, the total inverse displacement field is used to recover the unloaded geometry. Finally, the fiber orientation in the unloaded geometry was obtained by reorienting the fiber orientation with the deformation gradient.

Iterative Method. The iterative method to recover the unloaded geometry is similar to that described in [2]. In its original form, the first estimate of the unloaded geometry is chosen to be the ED geometry, which is then inflated to the measured ED pressure. The key step of this algorithm is the backward application of the nodal displacements, resulting from this forward inflation, to obtain a new estimate of the unloaded geometry. The final estimate is assumed to be the one that, when loaded to the ED pressure, results in the ED geometry within set error limits.

We adapt this method to use a volume criterion instead of a pressure criterion since the pressure was unknown. We use the volume criterion as in (5) to estimate the unloaded cavity volume. To determine the filling pressure, we used filling pressures of 8 mmHg, 10 mmHg, and 12 mmHg as initial guesses. We inflated the LV to the desired pressure by gradually increasing the pressure load using small pressure increments and computed the corresponding unloaded geometries by subtracting the nodal displacements from the original LV ED geometry. Deviations of the volumes of these geometries from the unloaded cavity volume satisfying (5) were used to find the pressure resulting in the desired volume. As in [10], the convergence criterion was set to 0.1 mm. Each iteration, fiber orientation was reoriented by using the deformation gradient.

2.4 Numerical Implementation and Simulations Performed

The model was implemented in the FEniCS open-source computing platform and solved using a 2.6 GHz 16-core Intel processor. The mesh contained 23232 tetrahedral elements and 111078 degrees of freedom, yielding an average element volume of 5.5 mm^3, corresponding to an element length of about 3.0 mm.

In order to transform the template geometry into the ED geometry material parameters in (1–3) were set to $a_0 = 0.4$ kPa, $a_1 = 3.0$, $a_2 = 6.0$, $a_3 = 0.0$, $a_4 = 0.0$, and $a_5 = 0$ kPa. Setting a_5 to zero allows volume change of the myocardial wall. In addition, isotropic material properties were used ($a_3 = 0.0$) while solving for mechanical equilibrium (4).

During mechanical unloading, material parameters in (1–3) were set to $a_0 = 0.4$ kPa, $a_1 = 3.0$, $a_2 = 6.0$, $a_4 = 0.0$, and $a_5 = 55$ kPa. When assuming isotropic material properties, a_3 was set to 0.0, referred to as simulation ISO. Additionally, a_3 was set to 3.0 for anisotropic material properties, which we will refer to as ANISO.

To apply the direct method, the weak form of (4) was formulated with respect to the known configuration Ω_{ED}. For the iterative method, the weak form was written with respect to the iteratively updated unloaded configuration.

A total of 8 simulations were performed, two simulations to evaluate the direct method and 6 to evaluate the iterative method. In the direct method, we gradually decreased the pressure increments, enabling an accurate approximation of the unloaded cavity volume (deviation of less than 0.05 ml). To quantify the differences between the unloaded geometries resulting from both methods, we compute the deviation in the nodal position of corresponding nodes in the computed geometries.

3 Results

3.1 End-Diastolic Geometry

The approximate volume-matched template mesh and the derived ED mesh with their fiber orientations are shown in Fig. 1. Mechanical equilibrium (Eq. 4) was solved in two steps to transform the template geometry to the ED configuration. For numerical stability, each step involved half of the total displacement from template to ED mesh. LV cavity volume changed from 111.3 ml in the template geometry to 110.9 ml in the ED configuration, and wall volume changed from 124.7 ml to 127.8 ml.

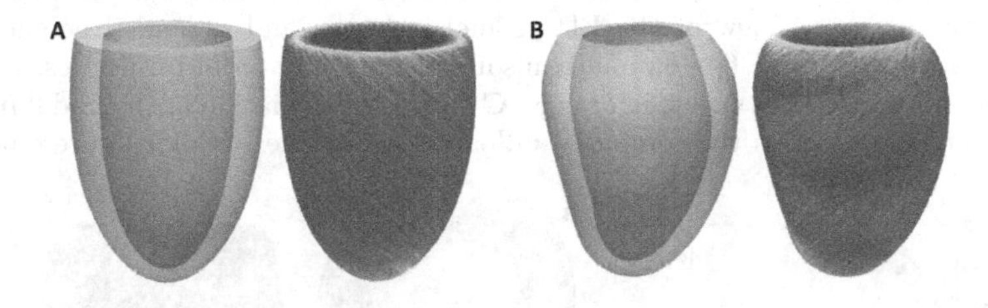

Fig. 1. View of meshes and fibers for the template geometry (**A**), and the end-diastolic configuration (**B**). The fiber vector is indicated with streamlines, whereas the color indicates the base-to-apex component, expressed as $\sin(\alpha_h)$, with α_h the helix angle. The color is ranging from −1 (blue) on the epicardial surface, to +1 (red) on the endocardial surface. (Color figure online)

3.2 Computation of the Unloaded Configuration

Direct Method. Figure 2 shows the passive pressure-volume relation simulated with the direct method. The required unloaded cavity volume satisfying (5) was reached at a cavity pressure of 8.7 mmHg when using isotropic material properties (simulation ISO) and a cavity pressure of 12.2 mmHg when using anisotropic material properties (simulation ANISO). In both cases, the diameter and length (base-to-apex) decrease, with an increase in wall thickness (Fig. 2**B**). No apparent difference in geometry can be noticed. In simulation ISO the decrease in length is slightly larger compared to simulation ANISO. In general, an average difference in geometry of 0.74 ± 0.32 mm was found. The highest deviations can be seen in the apical region, with a maximum of 1.5 mm (Fig. 2**C**).

Iterative Method. Figure 3**A** shows the computed unloaded cavity volumes for the three initial pressure estimates. The unloaded cavity volume satisfying (3) was reached at a cavity pressure of 8.6 mmHg and 11.4 mmHg in simulations ISO and ANISO, respectively. Similar to the direct method, LV length and diameter decrease, with an increase in wall thickness (Fig. 3**B**). Again, no clear difference can be seen between simulations ISO and ANISO, but an average difference in geometry of 0.70 ± 0.32 mm was found, with a maximum of 1.3 mm (Fig. 3**C**).

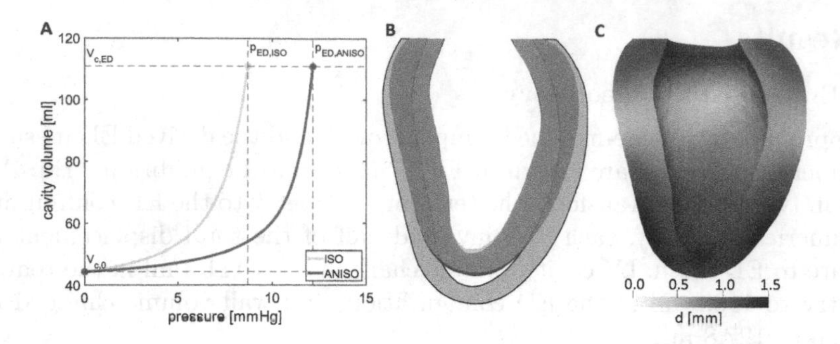

Fig. 2. **A** Simulated passive pressure-volume relation when using the direct method. **B** shows the long-axis view of the ED geometry (grey) and unloaded geometry for simulations with isotropic (yellow) and anisotropic (red) material properties, with the resemblance in geometries shown in orange. **C** shows the deviation in the nodal position of corresponding nodes in the computed unloaded geometries. (Color figure online)

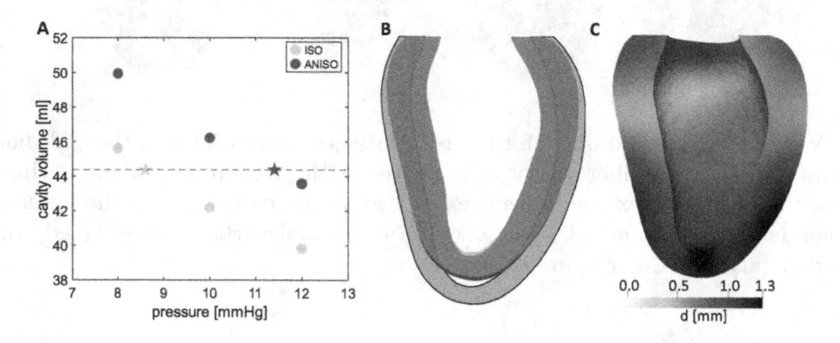

Fig. 3. **A** Pressure estimates and computed unloaded cavity volumes, filled circular dots represent the initial pressure estimates, and the stars represent the final pressure estimates resulting in the unloaded cavity volume satisfying (5). **B** shows the long-axis view of the ED geometry (grey) and unloaded geometry for simulations with isotropic (yellow) and anisotropic (red) material properties, with the resemblance shown in orange. **C** shows the deviation in the nodal position of corresponding nodes in the computed unloaded geometries. (Color figure online)

Comparison. To better evaluate the geometrical differences between the unloaded geometries resulting from both methods, we computed the deviation between the geometries, expressed in deviations in nodal positions (Fig. 4). In simulations ISO (Fig. 4**A**), the highest deviation can be seen in the apical region of both cross sections, with a maximum deviation of 0.30 mm, which decreases towards the basal region. An average deviation of 0.06 ± 0.04 mm was found. In simulations ANISO (4**B**), this gradient from apex to base is less apparent. However, the highest deviations can still be noticed near the apical region, with a maximum discrepancy of 0.65 mm. In general, an average deviation of 0.28 ± 0.12 mm was found.

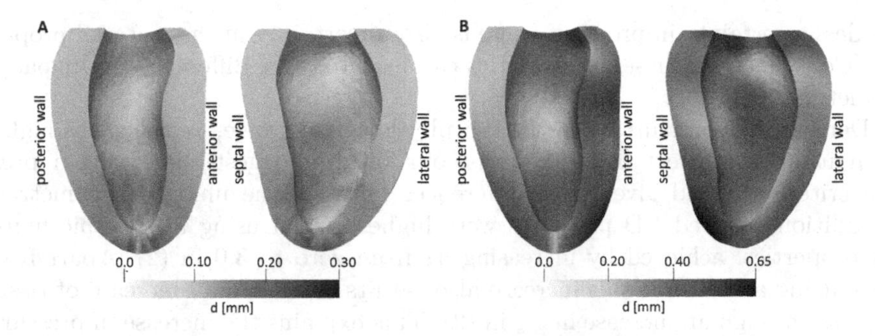

Fig. 4. Deviation in the nodal position of corresponding nodes in the unloaded geometries obtained with the direct method relative to the iterative method. **A**: simulations with isotropic material properties; **B**: simulations with anisotropic material properties.

4 Discussion

In this study, we assessed the differences in computed unloaded geometries when using either a direct or iterative method. In addition, the effects of isotropic and anisotropic material properties during mechanical unloading were evaluated.

A total of four unloaded geometries were obtained. All geometries showed a decrease in LV length and diameter, with subsequent increases in wall thickness to preserve wall volume. We found an average deviation in nodal position of 0.06 ± 0.04 mm and 0.28 ± 0.12 mm, for simulations ISO and ANISO respectively. Hence, the differences in the derived unloaded geometries are considered small as compared to the base-to-apex length of 90 mm in the end-diastolic state.

For both algorithms, a volume criterion was used assuming that the unloaded cavity volume is comparable with the end-systolic cavity volume. In addition, we assumed a physiological EF of 60% (Eq. 5). These assumptions led to an average ratio of unloaded cavity volume to wall volume $V_{c,0}/V_w$ of 0.34, consistent with experimental findings [5,11,13]. However, in a pathological heart, the EF may be different, and the assumption that $V_{c,0}$ approximates $V_{c,ES}$ may be questionable. On the other hand, the EF can still be measured, facilitating the applicability of the volume criterion.

Since we used a volume criterion, derived unloaded geometries cannot be very different from each other because they satisfy the same volume. We evaluated the effect of tightening the convergence criterion in the iterative method by tightening it 10-fold, which led to a difference in volume of 0.19 ml. Thus, tightening the convergence criterion might decrease the difference in geometries resulting from both methods. On the other hand, tightening the convergence criterion increases computational costs.

Usually, measured ED pressure is used in mechanical unloading. However, this necessitates the measurement of the ventricular pressure which requires the placement of a catheter in the ventricle, with all its risks and expense. Moreover, ED pressure is low and therefore fairly sensitive to noise and catheter placement.

Besides uncertainty in pressure, there is also uncertainty in the material properties. Consequently, these uncertainties can lead to large differences in unloaded geometries.

Derived ED pressures were in the physiological range, but slightly higher when using the direct method. Thus, our results suggest that using a pressure criterion would give larger differences between the unloaded geometries. In addition, derived ED pressures were highest when using anisotropic material properties, achieved by increasing a_3 from zero to 3.0 in (2). Apart from introducing anisotropy, this increase also results in an overall increase of tissue stiffness through an increase in Q_s in (2). This explains the increase in pressure, needed to unload the ED geometry. The switch to anisotropic material properties does not have a major effect on the unloaded geometry. This shows that the computed unloaded geometries do not depend strongly on the rule-based fiber field that we used, which is useful since determination of the patient-specific fiber field is challenging.

When comparing both methods, the direct method is preferred over the iterative method, as we can then directly simulate the LV passive pressure-volume relation. However, the implementation required access to the finite element code, as reparameterization of the weak form is required, which can be considered a drawback. Despite the elegance of this approach, to our knowledge, it is not available in any widely used commercial simulation software, emphasizing the usefulness of an open-source finite element computing platform.

On the contrary, the iterative method has been widely used because of its algorithmic simplicity and straightforward implementation, as only an update of the mesh coordinates needs to be performed. However, the iterative method requires multiple iterations to reach convergence, involving higher computational costs compared to the direct method. In our study, 12 iterations were needed to reach convergence. Each iteration, depending on the filling pressure, we used 30 to 50 pressure increments. In contrast, the direct method requires only one iteration. In the direct method, we gradually decreased the pressure increments from 0.2 mmHg to 0.06 mmHg, leading to 60 and 100 pressure increments for simulation ISO and ANISO, respectively, allowing the volume criterion to be achieved with an accuracy of 0.05 ml. Overall, the direct method was about 10 times faster than the iterative method.

In addition, it has been demonstrated that with its original formulation, the iterative method may not converge for problems with large deformation [10]. Moreover, such large deformations can lead to buckling of the more compliant right ventricle (RV) [6]. Computational costs and convergence problems could be decreased by the augmented method proposed by Rausch et al. [15]. In this augmented method, the subtracted displacement is multiplied by a scalar α. However, optimal values for α appeared to be problem specific, which complicates the implementation of the augmented method.

Finally, in our study, we only used an average LV geometry. Therefore, further research is needed to see whether our findings also hold for biventricular geometries.

5 Conclusion

We have demonstrated that both the direct and iterative unloading algorithm can be used to compute the unloaded LV geometry from a measured ED geometry. We used a volume criterion instead of the more commonly used pressure criterion. Our results suggest that using a pressure criterion would give larger differences in the unloaded geometries. Derived ED pressures were in the physiological range, but slightly higher when using the direct method. ED pressures were highest when using anisotropic material properties. However, introducing anisotropy did not have major effects on the unloaded geometry. The direct and iterative method gave comparable results in unloaded geometries, with differences on the order of 0.1 mm, which is considered small as compared to the base-to-apex length of 90 mm in the end-diastolic state. However, using the iterative method is computationally more expensive. Moreover, with the direct method we can directly simulate the LV pressure-volume relation. Therefore, from a physiological and computational point of view, the direct method is preferable to the iterative method.

Acknowledgements. This work is funded by European Union's Horizon 2020 research and innovation program under grant agreement 874827 (BRAVƎ).

References

1. Barbarotta, L., Bovendeerd, P.: A computational approach on sensitivity of left ventricular wall strains to geometry. In: Coudière, Y., Ozenne, V., Vigmond, E., Zemzemi, N. (eds.) FIMH 2019. LNCS, vol. 11504, pp. 240–248. Springer, Cham (2019). https://doi.org/10.1007/978-3-030-21949-9_26
2. Bols, J., Degroote, J., Trachet, B., Verhegghe, B., Segers, P., Vierendeels, J.: A computational method to assess the in vivo stresses and unloaded configuration of patient-specific blood vessels. J. Comput. Appl. Math. **246**, 10–17 (2013). https://doi.org/10.1016/j.cam.2012.10.034
3. Bovendeerd, P.H.M., Kroon, W., Delhaas, T.: Determinants of left ventricular shear strain. Am. J. Physiol. Heart Circ. Physiol. **297**(3), H1058–H1068 (2009). https://doi.org/10.1152/ajpheart.01334.2008
4. Bovendeerd, P., Arts, T., Huyghe, J., van Campen, D., Reneman, R.: Dependence of local left ventricular wall mechanics on myocardial fiber orientation: a model study. J. Biomech. **25**(10), 1129–1140 (1992). https://doi.org/10.1016/0021-9290(92)90069-d
5. Delhaas, T., Arts, T., Bovendeerd, P.H., Prinzen, F.W., Reneman, R.S.: Subepicardial fiber strain and stress as related to left ventricular pressure and volume. Am. J. Physiol. Heart Circ. Physiol. **264**(5), H1548–H1559 (1993). https://doi.org/10.1152/ajpheart.1993.264.5.h1548
6. Finsberg, H., et al.: Efficient estimation of personalized biventricular mechanical function employing gradient-based optimization. Int. J. Numer. Meth. Biomed. Eng. **34**(7), e2982 (2018). https://doi.org/10.1002/cnm.2982

7. Govindjee, S., Mihalic, P.A.: Computational methods for inverse deformations in quasi-incompressible finite elasticity. Int. J. Numer. Meth. Eng. **43**(5), 821–838 (1998). https://doi.org/10.1002/(sici)1097-0207(19981115)43:5¡821::aid-nme453¿3.0.co;2-c
8. Krishnamurthy, A., Villongco, C.T., Chuang, J., Frank, L.R., Nigam, V., Belezzuoli, E., Stark, P., Krummen, D.E., Narayan, S., Omens, J.H., et al.: Patient-specific models of cardiac biomechanics. J. Comput. Phys. **244**, 4–21 (2013). https://doi.org/10.1016/j.jcp.2012.09.015
9. Lu, J., Zhou, X., Raghavan, M.L.: Inverse elastostatic stress analysis in pre-deformed biological structures: demonstration using abdominal aortic aneurysms. J. Biomech. **40**(3), 693–696 (2007). https://doi.org/10.1016/j.jbiomech.2006.01.015
10. Marx, L., Niestrawska, J.A., Gsell, M.A., Caforio, F., Plank, G., Augustin, C.M.: Robust and efficient fixed-point algorithm for the inverse elastostatic problem to identify myocardial passive material parameters and the unloaded reference configuration. J. Comput. Phys. **463**, 111266 (2022). https://doi.org/10.1016/j.jcp.2022.111266
11. McCulloch, A.D., Smaill, B.H., Hunter, P.J.: Regional left ventricular epicardial deformation in the passive dog heart. Circ. Res. **64**(4), 721–733 (1989). https://doi.org/10.1161/01.res.64.4.721
12. Medrano-Gracia, P., et al.: Left ventricular shape variation in asymptomatic populations: the multi-ethnic study of atherosclerosis. J. Cardiovasc. Magn. Reson. **16**(1), 1–10 (2014). https://doi.org/10.1186/s12968-014-0056-2
13. Nikolić, S.: Passive properties of canine left ventricle: diastolic stiffness and restoring forces. Circ. Res. **62**(6), 1210–1222 (1988). https://doi.org/10.1161/01.res.62.6.1210
14. Nikou, A.: Effects of using the unloaded configuration in predicting the in vivo diastolic properties of the heart. Comput. Meth. Biomech. Biomed. Eng. **19**(16), 1714–1720 (2016). https://doi.org/10.1080/10255842.2016.1183122
15. Rausch, M.K., Genet, M., Humphrey, J.D.: An augmented iterative method for identifying a stress-free reference configuration in image-based biomechanical modeling. J. Biomech. **58**, 227–231 (2017). https://doi.org/10.1016/j.jbiomech.2017.04.021

A Modelling Study of Pulmonary Regurgitation in a Personalized Human Heart

Debao Guan[1(✉)], Yingjie Wang[1], Xiaoyu Luo[1], Mark Danton[2], and Hao Gao[1]

[1] School of Mathematics and Statistics, University of Glasgow, Glasgow, UK
Debao.Guan@glasgow.ac.uk

[2] Golden Jubilee National Hospital Glasgow, Glasgow, UK

Abstract. Pulmonary regurgitation is a frequent outcome of repaired tetralogy of Fallot, and its chronicity could have a detrimental effect on right ventricle function. To evaluate the effects of PR on cardiac functions, a subject-specific bi-ventricular model is developed in this study coupled with a closed-loop parameter model to depict systemic and pulmonary circulations. After calibrating the normal bi-ventricle model with in vivo measurement, pulmonary regurgitation is introduced by a one-directional channel connecting the right ventricle and pulmonary artery. We have demonstrated that this developed bi-ventricular model can reproduce the main characteristics of cardiac functions under pulmonary regurgitation, such as increased end-diastolic pressure and volume, diminishment of isovolumetric relaxation, enhanced active tension in the right ventricle, and the bulged septum towards the left ventricle. Once validated, it is believed this cardiac model will be very useful to provide insights into cardiac biomechanical function under pulmonary regurgitation.

Keywords: Pulmonary regurgitation · Bi-ventricular model · Closed-loop circulation model · Cardiac function

1 Introduction

Incompetence of pulmonary valve (PV) leads to backward or retrograde flow across the pulmonary valve in diastole. Although trivial and mild pulmonary regurgitation is commonly found in healthy individuals and well tolerated by right ventricle (RV), chronic severe pulmonary regurgitation can have detrimental effects on cardiac function [4]. Pathologic pulmonary regurgitation commonly occurs after surgical valvotomy or repaired tetralogy of Fallot (TOF) due to the absence of PV. Persisting pathological pulmonary regurgitation may lead to RV volume overload and dilation in order to preserve RV cardiac output. In repaired TOF patients, pulmonary valve replacement has yielded mixed results in cardiac function [16]. Due to the complex RV structure and clinical surgery, effective

O. Bernard et al. (Eds.): FIMH 2023, LNCS 13958, pp. 585–593, 2023.
https://doi.org/10.1007/978-3-031-35302-4_60

surgical plans are needed to improve the outcomes of pulmonary valve replacement, in particular in TOF patients. Therefore it is essential to quantify cardiac function under pulmonary regurgitation before pulmonary valve replacement to predict the benefits of its replacement.

Although recent development in computational modelling has facilitated the adoption of patient-specific ventricle models to improve patient management in cardiac diseases [2,11], biomechanical modelling of pulmonary valve regurgitation and its effects on RV pump function is least appreciated and most poorly studied. There are only a few studies on pulmonary valve regurgitation-related cardiac models. For example, by using patient-specific in vivo imaging derived RV and left ventricle (LV), Tang's group [14,15] have explored various surgical options for pulmonary valve replacement in TOF patients, by studying the impact of patch size and scar tissue trimming on RV function. In a recent study, Gusseva et al. [10] modelled pulmonary valve replacement in repaired TOF patients using a lumped parameter heart model in order to assess the influences of pulmonary regurgitation on ventricle function. They concluded that a personalized biomechanical cardiac model can potentially optimize the clinical management of pulmonary valve replacement. Moreover, the role of LV in a heart with pulmonary regurgitation (PR) remains less understood, with few studies devoted to RV-LV interactions. Excessive RV volume or pressure overload can have a deleterious effect on LV function, which in turn may have an adverse impact on RV function [12]. Therefore, a real bi-ventricular model is necessary to quantify the impact of PR. In this study, a patient-specific bi-ventricular heart model was employed to evaluate the effects of PR on cardiac functions. Two cases were studied, the normal baseline case and the PR case. Qualitative comparisons with experimental studies and clinical observations were analyzed, including cardiac pump function and PR volume. To our best knowledge, this is the first study combing a three-dimensional (3D) finite-element model and a lumped-parameterized closed-loop systemic model to study how PR affects RV function and in turn LV function, which could further serve as a proof-of-concept study to assess and predict RV failure.

2 Method

2.1 Pulmonary Regurgitation Model

A three-dimensional bi-ventricle human heart model was constructed from an in vivo cardiac magnetic resonance (CMR) imaging study of a healthy volunteer (33 years, female) with no prior history of cardiovascular disease. The study was approved by the National Research Ethics Services, and written consent was obtained before the CMR study. CMR image protocol involved steady-state precession cine images at both the short-axis and long-axis views for ventricular structure and functional assessment. Short-axis and long-axis cine images at early-diastole were chosen to reconstruct the 3D bi-ventricle model by assuming that the ventricular pressure is at its lowest at early-diastole when the mitral valve just opens [5]. The ventricular wall boundaries were manually segmented at each image plane using in-house Matlab code including both the RV and LV boundaries. The short-axis wall boundaries were further aligned to the boundaries in long-axis views. Manually extracted wall boundaries were imported into SolidWorks (Dassault Systemes, MA USA) for geometry reconstruction that was finally meshed using ICEM (ANSYS, Inc. PA USA), as shown in Fig. 1(a). The layered myofibre architecture (Fig. 1(b)) was generated using a rule-based method [9] with the fibre rotation angles varying linearly from −60° at epicardium to 60° at endocardium in both LV and RV.

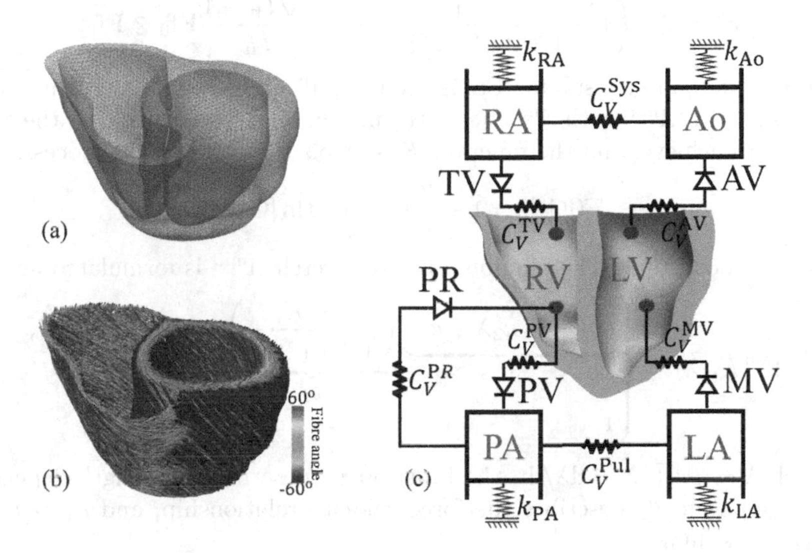

Fig. 1. The human bi-ventricle geometry (a), the rule-based myofibre structure (b) where the myofibre angle varies from −60° at epicardium to 60° at endocardium, and (c) schematic illustration of the bi-ventricular model coupled with a circulatory system. MV: mitral valve; AV: aortic valve; RA: right atrium; TV: tricuspid valve; PV: pulmonary valve; LA: left atrium; Ao: aorta; Sys: systemic circulation; Pul: pulmonary circulation; PA: pulmonary artery and PRV: pulmonary regurgitation valve. A grounded spring with a stiffness (k) is tuned to provide the appropriate pressure-volume response (i,e., compliance) for that cavity. C_V is the viscous resistance coefficient to describe resistance between cavities. Note pulmonary regurgitation is controlled by the resistance C_V^{PRV}.

The total Cauchy stress in myocardium is additively split into a passive part ($\boldsymbol{\sigma}^{\mathrm{p}}$) and an active part ($\boldsymbol{\sigma}^{\mathrm{a}}$). By assuming myocardium is hyperelastic and incompressible, the passive property of the myocardium is described by an invariant-based strain energy function [7]

$$\begin{aligned} W = & \frac{a_{\mathrm{g}}}{2b_{\mathrm{g}}}\{\exp[b_{\mathrm{g}}(I_1 - 3)] - 1\} + \frac{a_{\mathrm{f}}}{2b_{\mathrm{f}}}\{\exp[b_{\mathrm{f}}(\max(I_{4\mathrm{f}}, 1) - 1)^2] - 1\} \\ & + \frac{a_{\mathrm{n}}}{2b_{\mathrm{n}}}\{\exp[b_{\mathrm{n}}(\max(I_{4\mathrm{n}}, 1) - 1)^2] - 1\} + \frac{a_{\mathrm{fs}}}{2b_{\mathrm{fs}}}\{\exp(b_{\mathrm{fs}} I_{8\mathrm{fs}}^2) - 1\}, \end{aligned} \tag{1}$$

in which $a_{(\mathrm{g,\,f,\,n,\,fs})}$, $b_{(\mathrm{g,\,f,\,n,\,fs})}$ are material parameters, $I_1 = \mathrm{trace}(\mathbf{F}^T\mathbf{F})$ with $\mathbf{F}$ being the deformation gradient tensor, $I_{4\mathrm{f}} = \mathbf{F}\mathbf{f}_0 \cdot \mathbf{F}\mathbf{f}_0$, $I_{4\mathrm{n}} = \mathbf{F}\mathbf{n}_0 \cdot \mathbf{F}\mathbf{n}_0$, $I_{8\mathrm{fs}} = \mathbf{F}\mathbf{f}_0 \cdot \mathbf{F}\mathbf{s}_0$, and $\mathbf{f}_0$, $\mathbf{s}_0$, $\mathbf{n}_0$ are the unit myofibre, sheet, and sheet-normal directions in the reference configuration. Finally, $\boldsymbol{\sigma}^{\mathrm{p}} = \mathbf{F}\partial W/\partial \mathbf{F} - p\mathbf{I}$ in which p is the Lagrange multiplier to enforce incompressibility and $\mathbf{I}$ is the identity matrix.

A new hybrid active contraction model [8] is used here to model the active stress as

$$\boldsymbol{\sigma}^a = K_a^p \left(1 - \frac{1}{1 + \mathcal{F}([\mathrm{Ca}])(\xi - 1)}\right) \frac{\eta\sqrt{I_{4\mathrm{f}}} - 1}{I_{4\mathrm{f}}} \mathbf{F}\mathbf{f}_0 \otimes \mathbf{F}\mathbf{f}_0, \tag{2}$$

where K_a^p and η are constants, [Ca] is the normalized intracellular calcium transient, and $\xi = 1/\Lambda^{\min}$ with $\Lambda^{\min}$ being the minimum stretch ratio that the active element can achieve, and the function $\mathcal{F}$ controls the contraction process,

$$\mathcal{F}([\mathrm{Ca}]) = 1 + \frac{2}{\pi} \arctan(\beta \ln [\mathrm{Ca}]), \tag{3}$$

where β is a constant. The minimum active stretch $\Lambda^{\min}$ is formulated as

$$\Lambda^{\min}(\lambda, \dot{\lambda}) = \begin{cases} \underbrace{(\kappa_1\lambda^2 + \kappa_2\lambda + \kappa_3)}_{\vartheta_1} \underbrace{\left(\frac{1 + \kappa_4}{1 + \kappa_4\, e^{\kappa_5 \dot{\lambda}}}\right)}_{\vartheta_2} & \text{for} \quad \vartheta_1\vartheta_2 \le 1, \\ 1, & \text{otherwise}, \end{cases} \tag{4}$$

in which $\lambda = \sqrt{I_{4\mathrm{f}}}$, $\dot{\lambda} = \mathrm{d}\lambda/\mathrm{d}t$, the function ϑ_1 describes the length-dependent relationship, and ϑ_2 describes the force-velocity relationship, and $\kappa_1 \sim \kappa_5$ are positive constants.

This bi-ventricle model is implemented in ABAQUS (Dassault Systemes, Johnston RI, USA), which is further connected to a simplified closed-loop lumped parameter model representing the systemic and pulmonary circulations, as shown in Fig. 1(c), to provide physiologically accurate pressure boundary conditions for LV and RV. Specifically, pulmonary regurgitation is realized by a one-directional valve (denoted PRV) parallel to the one-directional pulmonary valve (denoted PV) as shown in Fig. 1(c). Thus blood from the PA can flow back to RV through PRV when the PA pressure is higher than the RV pressure. The functional PV allows RV to eject blood to PA when the RV pressure is higher than PA pressure. In this study, we first calibrated the bi-ventricular model without pulmonary regurgitation by setting the PRV resistance C_V^{PRV} with a very large value, i.e. $1000\,\mathrm{MPa} \cdot \mathrm{s/m^3}$, which is the baseline case. The passive parameters in Eq. (1) were firstly inversely estimated to match in vivo measured end-diastolic volumes of the LV and RV, respectively, similar to our previous study [6]. Then, the active contractility K_a^p was adjusted to match the measured ejection fractions and end-systolic pressures of this healthy volunteer. Other parameters of the active contraction model and the close-loop models were adapted from [8,9] to ensure pressures at different compartments were in their physiological ranges. Finally we set the PRV resistance of $C_V^{\mathrm{PRV}} = 15\,\mathrm{MPa{\cdot}s/m^3}$ to allow for pulmonary regurgitation, which is the PR case.

3 Result

Figure 2(a) shows the pressure-volume loops of LV and RV. The stroke work of RV in the PR case is larger than the baseline case but not in LV. Furthermore, the RV isovolumetric relaxation disappears in the PR case due to the pulmonary regurgitation, i.e. increased RV cavity volume when the RV is relaxing even with decreased RV pressure. The RV filling volume is much larger in the PR case (167 mL) compared to the normal case (129 mL), and the corresponding end-diastolic pressure in the RV increases to 8 mmHg, which is 4 mmHg in the normal case. This may be partially explained by the regurgitation-induced volume overload. Interestingly, the LV pump function is also affected by the excessive filling of RV, the LV end-diastolic volume is reduced from 118 mL to 98 mL in the PR case and further with reduced end-diastolic/systolic pressures. The minimum diastolic pressure in PA reduces from 17 mmHg to 8 mmHg in the PR case as shown in Fig. 2(b). The excessive filling of RV due to pulmonary regurgitation leads to higher systolic pressures in RV and PA. It also seems that excessive RV filling lowers the after-load of LV, which leads to a smaller end-systolic volume in the PR case as shown in Fig. 2(a).

Figure 2(c) shows the accumulated pulmonary regurgitation volume within one cardiac cycle with respect to the pressure difference between the RV and PA. The regurgitation starts when the PA pressure is higher than the RV pressure at late systole, the accumulated regurgitation volume increases due to fast myocardial relaxation. After reaching the maximum pressure difference (around 20 mmHg), the RV starts to be filled from the right atrium with increased RV pressure, because of the positive pressure difference between the PA and RV, the regurgitation volume keeps increasing until the RV pressure is higher than the PA pressure, which occurs at the beginning of isovolumetric contraction.

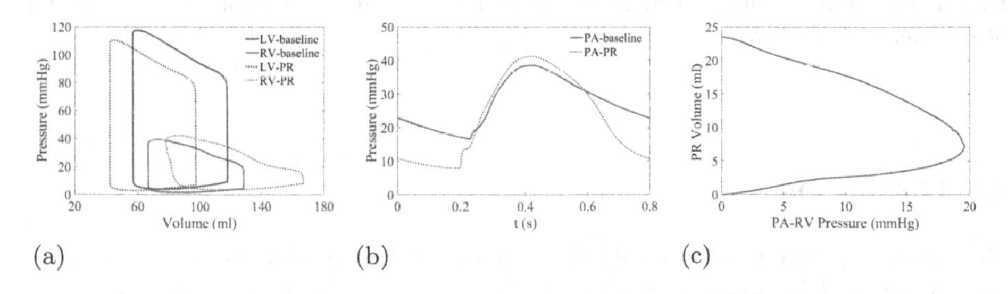

Fig. 2. Comparison of simulated cardiac functions during one cardiac cycle between the normal baseline case and the PR case. (a) Pressure-volume loops, (b) pulmonary artery pressure, (c) pulmonary regurgitation volume with respect to the pressure difference between PA and RV.

Deformations of the bi-ventricle heart at both end of diastole and end of systole are shown in Fig. 3, including distribution of stress component along the fibre direction ($\sigma_{\rm ff}$). At end of diastole, the septum in the PR case moves toward

LV, resulting in a bigger RV cavity but a smaller LV cavity when compared to the normal case. Much higher σ_{ff} can be found in the RV of the PR case at end of diastole, indicating excessive myofibre stretch due to volume overload, which hinders the filling of LV, with much lower myofibre stress compared to the RV. At end of systole, again much higher myofibre stress can be found at the RV in the PR case as shown in the right panel of Fig. 3. Because high stretch would lead to high active tension according to the Frank-Starling law of the heart, suggesting improved contractility in the RV as an adaptive compensation under PR. Similarly, the smaller LV cavity volume at end-diastole would lead to reduced active tension and small stroke volume for the PR case. The stress responses in LV and RV free walls are very different at end-diastole and end-systole, but not for the septum, which may suggest that the stress pattern in septum may be insignificant in characterizing the severity of PR.

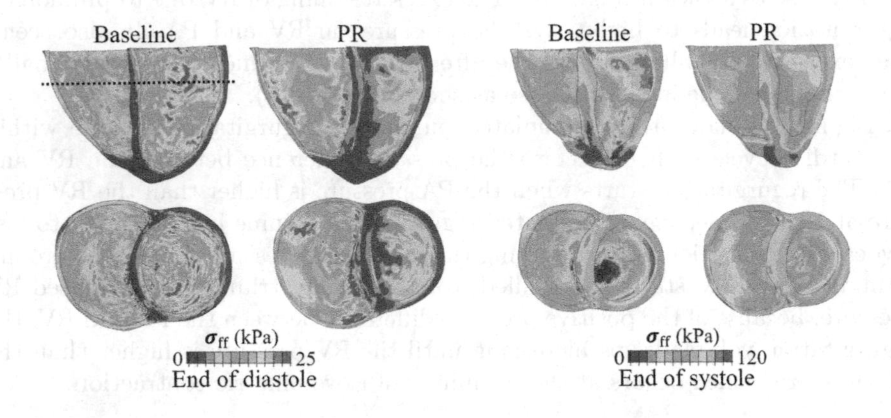

Fig. 3. Distributions of stress along the fibre direction (σ_{ff}) at end of diastole and end of systole, respectively.

4 Discussion

This study aims to develop a pulmonary regurgitation model based on a subject-specific human bi-ventricle model. A closed-loop lumped parameter model is employed here to provide hemodynamic boundary conditions for this 3D bi-ventricle model, in which the blood flow and pressure in each cavity serve as boundary conditions for the connected upstream and downstream cavities. This closed-loop model has been widely used in the literature [13]. The leakage from PA to RV is realized by a leaking valve which operates when the PA pressure is higher than the RV pressure. Severity of pulmonary regurgitation can be further adjusted by changing the resistance of the leaking valve (C_V^{PRV}) as shown in Fig. 1(c). Our simplified leaking valve model is similar to [10], in which pulmonary regurgitation of Tetralogy of Fallot patients was studied using a reduced order heart model.

In the PR case, the total regurgitated volume from the pulmonary artery is 24 mL in one cardiac cycle, which is in the range reported by Wald et al. [17], i.e. the median pulmonary regurgitation volume is 39 mL within a range of 0–129 mL from 64 patients. Studies have shown that pulmonary regurgitation inevitably overloads RV in diastole with increased end-diastolic volume and end-diastolic pressure, further leading to RV dilation in order to accommodate pulmonary regurgitation and maintain normal pump function, and eventually RV failure if without intervention [4].

Enhanced active tension has been reported in RV under pulmonary regurgitation [3], which could be partially explained by the high diastolic stretch due to volume overload in diastole. Our model further shows that pulmonary regurgitation in RV pushes the septum towards LV, especially in the long-axis view, and a "D-shaped" profile in the show-axis view (Fig. 3). The bulge of septum is consistent with the reported findings [12], and it could be an indicator of the severity of pulmonary regurgitation.

Under chronic pulmonary regurgitation, growth and remodelling mostly occur in RV because of volume overload in RV, leading to RV dilation and ultimately pump function failure [1]. The severity of PR may be the main determinant of RV functional and structural adaptation. By quantifying biomechanical behaviours of RV under PR, it is possible to understand how PR triggers myocardial growth and remodelling.

Finally, we would like to mention limitations in this study. Firstly, the rule-based fibre structure may limit the sensitivity of cardiac function response to pulmonary regurgitation, while obtaining subject-specific myofibre structure from in vivo CMR data is still very challenging. Secondly, a simplified circulation model is adopted which cannot describe the realistic blood flow like the one-dimensional model with rich arteries and veins. The PR case is not calibrated from real patient data but derived from a healthy bi-ventricular model. We are working towards patient-specific PR modelling using clinical measurements of the Tetralogy of Fallot patients.

5 Conclusion

A pulmonary regurgitation model based on a patient-specific bi-ventricular model has been developed in this study, and closed-loop systemic and pulmonary circulation models are further included to provide physiologically accurate pressure boundaries for the bi-ventricle model. Pulmonary regurgitation is realized by allowing the blood in PA to flow back into RV when PA pressure is higher than RA pressure. We have demonstrated that this developed bi-ventricular model can reproduce the main characteristics of cardiac functions under pulmonary regurgitation, such as increased end-diastolic pressure and volume, diminishment of isovolumetric relaxation, enhanced active tension in RV, and the bulged septum towards LV. Furthermore, this model can quantify biomechanical responses of both RV and LV under pulmonary regurgitation, the computed stresses could be very useful to understand pathological growth and remodelling in the myocardium. Once validated, this cardiac model will have the potential to provide insights into RV biomechanics under pulmonary regurgitation.

Acknowledgements. We are grateful for the funding provided by the British Heart Foundation (ref: PG/22/10930), and the UK Engineering and Physical Sciences Research Council (EP/S030875, EP/S020950/1, EP/S014284/1, EP/R511705/1).

References

1. Bouzas, B., Kilner, P.J., Gatzoulis, M.A.: Pulmonary regurgitation: not a benign lesion. Eur. Heart J. **26**(5), 433–439 (2005)
2. Corral-Acero, J., et al.: The 'digital Twin'to enable the vision of precision cardiology. Eur. Heart J. **41**(48), 4556–4564 (2020)
3. El-Harasis, M.A., et al.: Progressive right ventricular enlargement due to pulmonary regurgitation: clinical characteristics of a "low-risk" group. Am. Heart J. **201**, 136–140 (2018)
4. Fathallah, M., Krasuski, R.A.: Pulmonic valve disease: review of pathology and current treatment options. Curr. Cardiol. Rep. **19**, 1–13 (2017)
5. Gao, H., Aderhold, A., Mangion, K., Luo, X., Husmeier, D., Berry, C.: Changes and classification in myocardial contractile function in the left ventricle following acute myocardial infarction. J. R. Soc. Interface **14**(132), 20170203 (2017)
6. Gao, H., Wang, H., Berry, C., Luo, X., Griffith, B.E.: Quasi-static image-based immersed boundary-finite element model of left ventricle under diastolic loading. Int. J. Numer. Methods Biomed. Eng. **30**(11), 1199–1222 (2014)
7. Guan, D., Ahmad, F., Theobald, P., Soe, S., Luo, X., Gao, H.: On the AIC-based model reduction for the general Holzapfel-ogden myocardial constitutive law. Biomech. Model. Mechanobiol. **18**, 1213–1232 (2019)
8. Guan, D., Gao, H., Cai, L., Luo, X.: A new active contraction model for the myocardium using a modified hill model. Comput. Biol. Med. **145**, 105417 (2022)
9. Guan, D., Yao, J., Luo, X., Gao, H.: Effect of myofibre architecture on ventricular pump function by using a neonatal porcine heart model: from DT-MRI to rule-based methods. R. Soc. Open Sci. **7**(4), 191655 (2020)
10. Gusseva, M., et al.: Biomechanical modeling to inform pulmonary valve replacement in tetralogy of fallot patients after complete repair. Can. J. Cardiol. **37**(11), 1798–1807 (2021)
11. Mangion, K., Gao, H., Husmeier, D., Luo, X., Berry, C.: Advances in computational modelling for personalised medicine after myocardial infarction. Heart, **104**(7), 550–557 (2018)
12. Naeije, R., Badagliacca, R.: The overloaded right heart and ventricular interdependence. Cardiovasc. Res. **113**(12), 1474–1485 (2017)
13. Sack, K.L., et al.: Construction and validation of subject-specific biventricular finite-element models of healthy and failing swine hearts from high-resolution DT-MRI. Front. Physiol. **9**, 539 (2018)

14. Tang, D., Del Nido, P.J., Yang, C., Zuo, H., Huang, X., Rathod, R.H., Gooty, V., Tang, A., Wu, Z., Billiar, K.L., et al.: Patient-specific mri-based right ventricle models using different zero-load diastole and systole geometries for better cardiac stress and strain calculations and pulmonary valve replacement surgical outcome predictions. PLoS ONE **11**(9), e0162986 (2016)
15. Tang, D., et al.: A multiphysics modeling approach to develop right ventricle pulmonary valve replacement surgical procedures with a contracting band to improve ventricle ejection fraction. Comput. Struct. **122**, 78–87 (2013)
16. Vliegen, H.W.: Magnetic resonance imaging to assess the hemodynamic effects of pulmonary valve replacement in adults late after repair of tetralogy of fallot. Circulation **106**(13), 1703–1707 (2002)
17. Wald, R.M., et al.: Refining the assessment of pulmonary regurgitation in adults after tetralogy of fallot repair: should we be measuring regurgitant fraction or regurgitant volume? Eur. Heart J. **30**(3), 356–361 (2009)

Pump and Tissue Function in the Infarcted Heart Supported by a Regenerative Assist Device: A Computational Study

Koen L. P. M. Janssens(✉), M. van der Knaap, and Peter H. M. Bovendeerd

Eindhoven University of Technology, Eindhoven, The Netherlands
{k.l.p.m.janssens,p.h.m.bovendeerd}@tue.nl

Abstract. Adverse ventricular remodelling following acute myocardial infarction (MI) may induce ventricular dilation, fibrosis and loss of global contractile function, possibly resulting in heart failure. Cardiac patches, composed of living cardiac tissue, may be able to restore cardiac function and reduce adverse ventricular remodelling post-MI. The design of these devices is complex and computational modelling may provide means to create an efficient design from a mechanical point of view. In this study, we used a finite element approach to analyze the effect of a cardiac patch, mounted over a chronic infarct region, on left ventricular pump function and myocardial tissue function.

The infarct, 15% in size, was found to reduce stroke work by 30% compared to the healthy heart. This disproportional loss was attributed to unfavourable mechanical interactions in healthy tissue, adjacent to the infarction. The effect of the patch was investigated for a circumferential, an oblique, and a longitudinal orientation. In its most favourable oblique configuration, the cardiac patch was able to recover 6% of stroke work lost. This specific configuration was found to best restore the transmission of active force, that was lost due to the myocardial infarction.

Keywords: Computational modelling · Myocardial infarction · Cardiac patch

1 Introduction

Acute myocardial infarction (MI) is one of the leading causes of death worldwide with an estimated global prevalence approaching 3 million deaths annually [1]. In MI, a prolonged lack of oxygen and nutrient supply to part of the myocardium causes the irreversible death of cardiomyocytes. In patients that survive the acute phase of MI, deceased cardiomyocytes are gradually replaced by fibrotic scar tissue over the course of several weeks. This tightly cross-linked, collageneous tissue has significant tensile strength and helps prevent rupture of the myocardial wall [2–4]. However, pathological fibrosis and excessive deposition of collagen fibers

O. Bernard et al. (Eds.): FIMH 2023, LNCS 13958, pp. 594–603, 2023.
https://doi.org/10.1007/978-3-031-35302-4_61

within the scar tissue are important contributors to left ventricular dysfunction and play an important role in LV remodelling [5]. Abnormal mechanical loading of the myocardium is thought to drive the remodelling process and promote infarct expansion and wall thinning. Long-term adverse ventricular remodelling post-MI is associated with LV dilation, diastolic dysfunction, ventricular tachycardia and eventually heart failure (HF) [6,7].

Left ventricular assist devices (LVAD), in the form of mechanical propeller pumps, are increasingly used as destination therapy in patients suffering from end-stage HF that are not eligible for transplantation. However, these patients are often re-hospitalized within the first year post-implantation with high mortality and complication rates [8]. Following these observations, regenerative approaches have received an increasing amount of attention. Implantation of tissue-engineered myocardial patches over the infarct area may provide passive mechanical support and limit the abnormal loading conditions that drive adverse LV remodelling. It may also provide an active contribution to restoring cardiac function. Several experimental studies already demonstrated that implantation of a cardiac patch in both small and large animal models of acute MI can reduce infarct expansion, thinning of the infarct wall and increase cardiac function in remote tissue and the border zone over a period of several weeks [9–12]. The comparison of these studies is complicated by the use of different animal models, patch sizes, cell sources or infarct characteristics. For example, improvements in cardiac function were not observed in rat and guinea pig models of chronic MI [13,14]. Furthermore, changes in cardiac function are often based on hemodynamic indices which lack information on local tissue mechanics in the form of stress and strain patterns and are important cues for LV remodelling.

Computational models can assist in the interpretation of experimental studies and provide means to isolate contributions of certain design elements to overall patch functionality. For example, Yu et al. used patient-specific computational models of an infarcted heart to assess whether direct remuscularization would increase arrythomogeneity after transplantation of pluripotent stemcell-derived cardiomyocyte patches [15].

In this study, we present a computational framework that allows for the analysis of cardiac patch mechanics and its interaction with the host myocardium by extending the finite element model of cardiac mechanics in [16]. In this model, MI was represented by a region of fibrotic tissue with complete loss of contractile function while LV geometry was kept constant. We will quantify the active mechanical support that the cardiac patch provides in terms of global pump function and local myofiber mechanics. Subsequently, we assess whether the orientation of the patch with respect to the LV is an important contributor to overall cardiac function that should be considered in the surgical procedure.

2 Methods

The finite element model of cardiac mechanics in MI from [16] was extended to include a contractile cardiac patch fixed over the infarct site. A short description of the existing model is provided here. The left ventricular geometry in its unstressed state is approximated using a thick-walled truncated ellipsoid with wall and cavity volumes of 136 ml and 44 ml respectively (Fig. 1a). Rigid body motion was suppressed using a nullspace boundary condition on all nodes located in the basal plane of the geometry. The myocardium is modelled as a non-linearly elastic, transversely isotropic, incompressible material with an active stress component that acts parallel to the fiber direction. The Cauchy stress tensor $\boldsymbol{\sigma}$ is given by:

$$\boldsymbol{\sigma} = f_{pas}\boldsymbol{\sigma}_{pas} + f_{act}\sigma_{act}\vec{e}_f\vec{e}_f \tag{1}$$

Here $\vec{e}_f$ represents the unit vector in fiber direction and $\boldsymbol{\sigma}_{pas}$ and σ_{act} represent the passive and active components of the total stress tensor respectively. Factors f_{pas} and f_{act} were introduced to vary the material properties throughout the geometry spatially. Both were set to a value of 1 to model healthy tissue in simulation REF. A rule-based method was used to define the fiber orientation in terms of a helix angle α_h and transverse angle α_t [17]. The region of infarcted tissue was modelled to be representative of that resulting from an occlusion of the circumflex artery, as a circular area with a radius of 2.5 cm and its midpoint located roughly 1 cm below the LV equator. We assumed a case of chronic MI with a 10-fold increase in passive tissue stiffness and no active stress generation. Therefore, within the infarct region, factors f_{pas} and f_{act} were set to 10 and 0 respectively. We will refer to this simulation as CMI. Overall, the infarct region comprised 15% of the total LV wall volume. The pressure within the LV during filling and ejection phase is governed by a 0D closed-loop lumped parameter model of the systemic circulatory system [17].

2.1 Modelling the Cardiac Patch

The cardiac patch was modeled as a rectangular strip of contractile myocardial tissue measuring 6 by 4 cm with a thickness of 2 mm. It was centred on top of the infarct region. In the patch, fibers were aligned along the long side. Three different patch orientations were modelled, with the long side oriented at an angle of 0, 45 and 90° with the LV circumferential direction (Fig. 1b). These models are referred to as P0, P45 and P90, respectively. In P45, the orientation of fibers within the cardiac patch roughly coincides with epicardial fibers in the healthy myocardium. We assumed material properties within the cardiac patch to be equal to those of healthy myocardium and set factors f_{pas} and f_{act} to a value of 1.

2.2 Simulations and Postprocessing

The model was implemented in the FEniCS open-source computing platform and solved using a 2.6 GHz 16-core Intel processor [18]. A timestep of 2 ms was used. The mesh contained 40954 quadratic tetrahedral elements with 185184 degrees-of-freedom and 61728 nodes, yielding an average element size of 4.01 mm^3. For every simulation of model REF, CMI, P0, P45 and P90, 5 cardiac cycles were computed with a heart rate of 75 bpm such that a hemodynamic limit cycle was

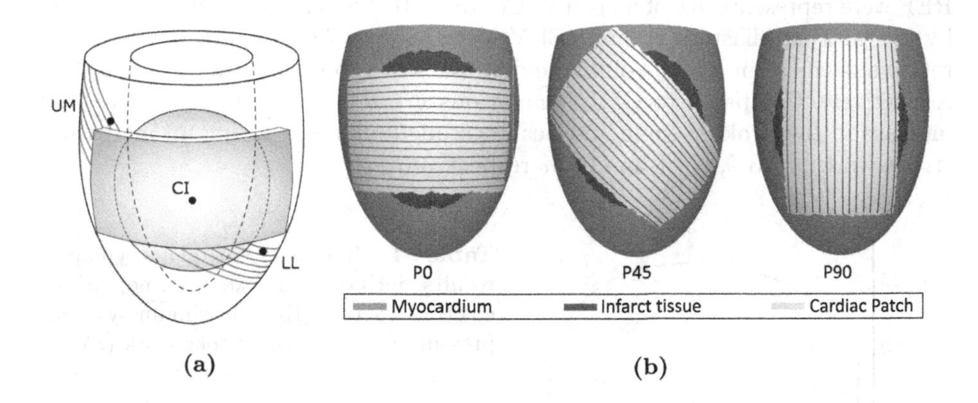

Fig. 1. (**a**) LV geometry with fiber orientation in healthy myocardium, infarct region and cardiac patch. Sites CI, UM and LL are positioned in the infarct, upper medial and lower lateral regions of the LV respectively. (**b**) The different orientations of the cardiac patch are depicted in P0, P45 and P90.

achieved with a change in stroke volume of less than 1% between cycles. Cardiac function was assessed using the pressure-volume signals of the last cardiac cycle and stroke work W was computed according to:

$$W = \int p\mathrm{d}V \tag{2}$$

To analyze local mechanical function on a tissue scale, Cauchy fiber stress σ_f and fiber strain ε_f were computed over the duration of the cardiac cycle. Subsequently, local tissue function was assessed by computing work density using:

$$w = \int \sigma_f \mathrm{d}\varepsilon_f \quad \text{where} \quad \varepsilon_f = \ln\left(\frac{l_s}{l_{s,0}}\right) \tag{3}$$

where l_s and $l_{s,0}$ represent the actual and reference sarcomere length respectively.

We report tissue function for 3 different sites. Site CI was placed in a central, subepicardial position in the infarct region at 8% of the total epi- to endocardial distance. Sites UM and LL were placed in the healthy tissue at the same subepicardial position, 4 cm in the upper medial and lower lateral direction from the infarct center respectively (Fig. 1a). Fiber stress and strain within the cardiac patch were computed as the average of 25 centrally spaced sites, referred to as P.

3 Results

3.1 Hemodynamics

Figure 2 shows the pressure-volume loops for all simulations and a quantitative hemodynamic summary is given in Table 1. The hemodynamics of simulation

REF were representative of a healthy LV. In CMI, systolic LV pressure decreased by 12% and end-diastolic volume (EDV) decreased by 7% compared to REF. The relative decrease in stroke volume and stroke work were equal to 18% and 29% respectively. Compared to CMI, simulations P0, P45 and P90, showed a small increase in LV stroke volume and peak systolic pressure, causing an increase in stroke work of 2.3%, 3.3% and 0.7% respectively.

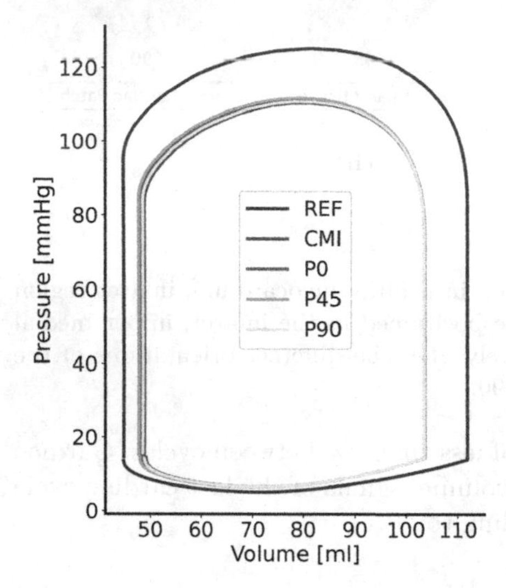

Fig. 2. Left ventricular pressure-volume plot for simulation REF, CMI, P0, P45 and P90.

Table 1. Summary of hemodynamic results including stroke volume (SV), ejection fraction (EF), maximum systolic pressure ($p_{lv,max}$) and stroke work (SW).

sim	SV [ml]	EF [-]	$p_{lv,max}$ [mmHg]	SW [J]
REF	67.0	59.9	125.2	0.984
CMI	54.8	52.8	110.4	0.695
P0	55.5	53.6	111.4	0.711
P45	55.8	53.9	111.8	0.718
P90	55.0	53.1	110.8	0.700

3.2 Local Mechanics

In Fig. 3, fiber stress is plotted vs. logarithmic fiber strain in sites UM, CI, LL and P. In the healthy heart (REF), the stress-strain loops at sites CI, UM and LL were almost identical. Maximum fiber strain was equal to 0.13 at end-diastole (denoted by ○) and peak fiber stress during ejection equalled roughly 60 kPa. Work density, represented by the surface area enclosed by the stress-strain loop, equalled 6.7 kPa.

In the chronically infarcted heart (CMI), work density decreased to 0 kPa in site CI. Sites UM and LL both showed stress-strain loops that are skewed to the left. Here, end-diastolic fiber strain increased to 0.16 and fibers shortened during isovolumic contraction. Peak fiber stresses decreased to a value of 38 kPa and 43 kPa for sites UM and LL respectively and fibers shortened further towards the end of systole (denoted by ◇). Work density was reduced to 5.0 kPa and 5.7 kPa in UM and LL respectively.

Simulations P0 and P90 showed stress-strain loops in sites UM and LL that are similar to those in simulation CMI with the exception of a minor increase in peak fiber stress. In P45, both sites UM and LL showed less shortening of fibers during isovolumic contraction and increased peak fiber stress, leading to work densities of 5.9 kPa and 7.6 kPa respectively. Both values were higher when compared to simulations P0 and P90. In site CI, the strain excursion over the cardiac cycle decreased in all patch simulations. Stress-strain loops within the cardiac patch (P) generally showed low fiber stress, and reduced strain excursion and, in the case of simulation P90, a shift leftward. Overall, work density was equal to 1.7 kPa, 2.4 kPa and 0.8 kPa in P0, P45 and P90 respectively.

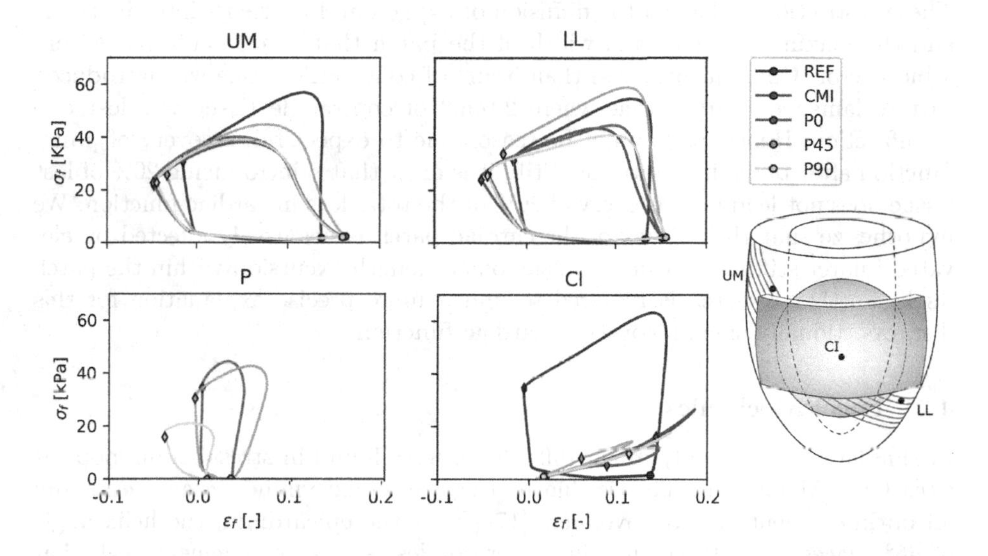

Fig. 3. Cauchy fiber stress vs. fiber strain plots for all simulations in the infarct site (CI), medial border zone (UM), lateral border zone (LL) and cardiac patch (P).

4 Discussion

In this study, we attached a contractile cardiac patch over a region of chronically remodeled infarct tissue and analyzed the results in terms of global cardiac function and local fiber mechanics.

4.1 Hemodynamics

A chronically remodeled infarct comprising 15% of the total LV wall volume was found to cause a 30% loss of pump function compared to the healthy case. This disproportional loss in cardiac function may be attributed to disadvantageous mechanical interactions between the infarct tissue and the healthy tissue neighbouring the infarct region resulting in the smaller stress-strain loops in Fig. 3 [16]. Including a cardiac patch was found to slightly increase stroke volume and maximum systolic pressure but also slightly impair ventricular filling. The amount of functional improvement was dependent on the orientation of the cardiac patch. Simulation P90 did not show any significant changes in stroke work while P45 presented with an increase that was 40% larger compared to P0. This suggests that patch orientation is an important determinant of cardiac function.

In absolute values however, these increases in stroke work are only a fraction of the initial loss of function found in CMI. This can be partially related to the size of the cardiac patch, which is limited by constraints in tissue engineering. These restrictions relate to the diffusion of oxygen and nutrients into the tissue and the maximum length and width of the patch that can be cultured within a bioreactor. Consequently, less than $5\,\text{cm}^3$ of contractile tissue was introduced to the damaged heart muscle where $20\,\text{cm}^3$ of contractile tissue was lost due to infarction. Hence, it may not be reasonable to expect full recovery of pump function after patch implantation. Still, it is clear that reintroducing 20% of lost tissue does not lead to a recovery of 20% of the total loss in cardiac function. We hypothesize that the efficacy of the cardiac patch is negatively affected by elevated infarct stiffness which limits sarcomere length excursion within the patch itself. Further research is required to find a more precise explanation for this disproportionally small recovery of cardiac function.

4.2 Local Mechanics

In simulations REF, only minor differences were found in stress-strain loops at sites CI, UM and LL. This is due to the use of axisymmetric geometry and an optimal orientation of myofibers [17]. Near the epicardium, the helix angle of 45° places these three sites in a more or less series arrangement such that they are mechanically linked to each other. In the healthy case, during myofiber activation, the similar increase in stiffness at these sites allows for the transmission of high active stress. In CMI, the lack of contractility in the infarct region leaves compliant tissue that solely depends on its passive material properties. The stress needed to achieve mechanical equilibrium and counteract active stress from the healthy tissue can only be achieved by stretching the infarct. Consequently, fibers at sites UM and LL can shorten during isovolumic contraction as they pull on the infarcted tissue. Finally, this leads to less stress generation due to the Frank-Starling principle.

When comparing the simulations of different patch orientations, the largest increase in cardiac function was found in P45. As the amount of contractile tissue within the patch remains equal between simulations, these differences can only be attributed to mechanical interactions with the LV. Differences in local mechanics show that with an orientation of 45°, fibers within the cardiac patch as well as the healthy myocardium surrounding the infarct region are able to generate more work. The helix angle of roughly 45° near the epicardium places the cardiac patch and epicardial fibers in a series arrangement with each other. Consequently, the activated fibers in the patch contribute to the transmission of force that is generated by fibers in sites UM and LL, on top of the force transmission through the infarct. This mechanism counteracts the early shortening of fibers observed in CMI which eventually contributes to higher overall cardiac function. Reduction of early fiber shortening appears to be more effective near site LL compared to site UM (Fig. 3). Work density in LL even exceeds that of the healthy heart. This effect is related to the increased end-diastolic sarcomere length that allows for higher stress generation through the Frank-Starling mechanism. Further research is required to explain why this effect is less pronounced at site UM.

4.3 Study Limitations and Clinical Implications

In this study, we chose to model myocardial infarction through a simplified geometry of the left ventricle, local absence of active stress development and increased passive tissue stiffness. We did not include thinning of the ventricular wall, which is reported to reduce to about 60% of its initial thickness [19]. Further research is required to assess whether local tissue mechanics are significantly affected by wall thickness. We expect that the effects of a decreased wall thickness would be comparable to lowering infarct stiffness through its material properties.

We also did not take into account the right ventricle (RV) and pulmonary circulation which has implications for pressure and flow dynamics in the 0D circulatory model. Adding the RV may increase LV filling pressure after MI and lead to increased end-diastolic volume and pressure. Adding the RV may also change local stress and strain patterns near the RV attachment sites. In our current model, we positioned the infarcted region on the LV free wall, relatively far away from the RV. Therefore we believe the spatial inhomogeneities observed near the cardiac patch would still hold when adding the RV.

Furthermore, we only considered the acute effect of patch placement on cardiac and tissue function and neglected the long-term effects of LV growth and remodeling following patch implantation. Abnormal levels of stress and strain near the infarct region are thought to drive the remodeling process. Our study shows that the cardiac patch can alter these stresses and strains. Dedicated research should point out whether these changes can positively affect global cardiac function in the long term.

In the most favourable situation P45, only 6% of the 0.29 J loss of cardiac function post-MI is recovered after implantation of the cardiac patch. For clinical applications, the risks associated with surgery would outweigh the potential increase in cardiac function and render this treatment non-beneficial. Nevertheless, our analysis suggests that the orientation of the patch is an important aspect given the relative increase in cardiac function of 40% between P45 and P0. Future developments in tissue engineering may allow for the fabrication of a larger cardiac patch that may further improve cardiac function. At organ level, it seems logical that the amount of tissue within the patch would approach the amount of contractile tissue lost due to the infarction. However, our results suggest that this amount of newly introduced tissue should even exceed the amount of lost tissue in order to fully restore cardiac function.

5 Conclusion

We have shown that a cardiac patch, composed of living cardiac tissue, can provide a contribution to cardiac pump and tissue function. The patch is able to provide this support through the addition of contractile tissue but also through the reduction in unfavourable mechanical interactions between healthy and infarct native tissue. In this respect, the orientation of the patch was important as these interactions were counteracted best when epicardial fibers and fibers in the patch were aligned.

Acknowledgements. This research was supported by the European Research Council (ERC) under the European Union's Horizon 2020 research innovation programme under grant agreement no. 874827 (BRAVƎ).

References

1. Mechanic, O.J., Gavin, M., Grossman, S.A.: Acute myocardial infarction. In: StatPearls. StatPearls Publishing (2022)
2. Gupta, K.B., Ratcliffe, M.B., Fallert, M.A., Edmunds, L.H., Jr., Bogen, D.K.: Changes in passive mechanical stiffness of myocardial tissue with aneurysm formation. Circulation **89**(5), 2315–2326 (1994). https://doi.org/10.1161/01.CIR.89.5.2315
3. McGarvey, J.R., Mojsejenko, D., Dorsey, S.M., Nikou, A., Burdick, J.A., Gorman, J.H., et al.: Temporal changes in infarct material properties: an in vivo assessment using magnetic resonance imaging and finite element simulations. Ann. Thorac. Surg. **100**(2), 582–589 (2015). https://doi.org/10.1016/j.athoracsur.2015.03.015
4. Fomovsky, G.M., Macadangdang, J.R., Ailawadi, G., Holmes, J.W.: Model-based design of mechanical therapies for myocardial infarction. J. Cardiovasc. Transl. Res. **4**(1), 82–91 (2010). https://doi.org/10.1007/s12265-010-9241-3
5. Burchfield, J.S., Xie, M., Hill, J.A.: Pathological ventricular remodeling: mechanisms: part 1 of 2. Circulation **128**(4), 388–400 (2013). https://doi.org/10.1161/CIRCULATIONAHA.113.001878
6. Frantz, S., Hundertmark, M.J., Schulz-Menger, J., Bengel, F.M., Bauersachs, J.: Left ventricular remodelling postmyocardial infarction: pathophysiology, imaging, and novel therapies. Eur. Heart J. **43**(27), 2549–2561 (2022). https://doi.org/10.1093/eurheartj/ehac223
7. French, B.A., Kramer, C.M.: Mechanisms of post-infarct left ventricular remodeling. Drug Discov. Today Dis. Mech. **4**(3), 185–196 (2007). https://doi.org/10.1016/j.ddmec.2007.12.006
8. Dunlay, S.M., Strand, J.J., Wordingham, S.E., Stulak, J.M., Luckhardt, A.J., Swetz, K.M.: Dying with a left ventricular assist device as destination therapy. Circ. Heart Fail. **9**(10), e003096 (2016). https://doi.org/10.1161/CIRCHEARTFAILURE.116.003096
9. Gao, L., et al.: Large cardiac muscle patches engineered from human induced-pluripotent stem cell-derived cardiac cells improve recovery from myocardial infarction in swine. Circulation **137**(16), 1712–1730 (2018). https://doi.org/10.1161/CIRCULATIONAHA.117.030785

10. Wendel, J.S., Ye, L., Zhang, P., Tranquillo, R.T., Zhang, J.: Functional consequences of a tissue-engineered myocardial patch for cardiac repair in a rat infarct model. Tissue Eng. Part A **20**(7–8), 1325–1335 (2014). https://doi.org/10.1089/ten.tea.2013.0312
11. Kawamura, M., et al.: Feasibility, safety, and therapeutic efficacy of human induced pluripotent stem cell-derived cardiomyocyte sheets in a porcine ischemic cardiomyopathy model. Circulation **126**(11 Suppl 1), S29–S37. https://doi.org/10.1161/CIRCULATIONAHA.111.084343
12. Montgomery, M., Ahadian, S., Davenport Huyer, L. et al.: Flexible shape-memory scaffold for minimally invasive delivery of functional tissues. Nat. Mater. **16**, 1038–1046 (2017). https://doi.org/10.1038/nmat4956
13. Riegler, J., et al.: Human engineered heart muscles engraft and survive long term in a rodent myocardial infarction model. Circ. Res. **117**(8), 720–30 (2015). https://doi.org/10.1161/CIRCRESAHA.115.306985
14. von Bibra, C., et al.: Human engineered heart tissue transplantation in a guinea pig chronic injury model. J. Mol. Cell. Cardiol. **166**, 1–10 (2022). https://doi.org/10.1016/j.yjmcc.2022.01.007
15. Yu, J.K.: Assessment of arrhythmia mechanism and burden of the infarcted ventricles following remuscularization with pluripotent stem cell-derived cardiomyocyte patches using patient-derived models. Cardiovasc. Res. **118**(5), 1247–1261 (2022). https://doi.org/10.1093/cvr/cvab140
16. Janssens, K.L.P.M., Kraamer, M., Barbarotta, L., Bovendeerd, P.H.M.: Post-infarct evolution of ventricular and myocardial function. [manuscript submitted for publication]
17. Bovendeerd, P.H.M., Kroon, W., Delhaas, T.: Determinants of left ventricular shear strain. Am. J. Physiol. Heart Circ. Physiol. **297**(3), H1058–H1068 (2009). https://doi.org/10.1152/ajpheart.01334.2008
18. Logg, A., Mardal, K.-A., Wells, G.N.: Automated Solution of Differential Equations by the Finite Element Method. The FEniCS Book, vol. 84. Springer, Heidelberg (2012). https://doi.org/10.1007/978-3-642-23099-8
19. Richardson, W.J., Clarke, S.A., Quinn, T.A., Holmes, J.W.: Physiological implications of myocardial scar structure. Compr. Physiol. **5**(4), 1877–909 (2015). https://doi.org/10.1002/cphy.c140067

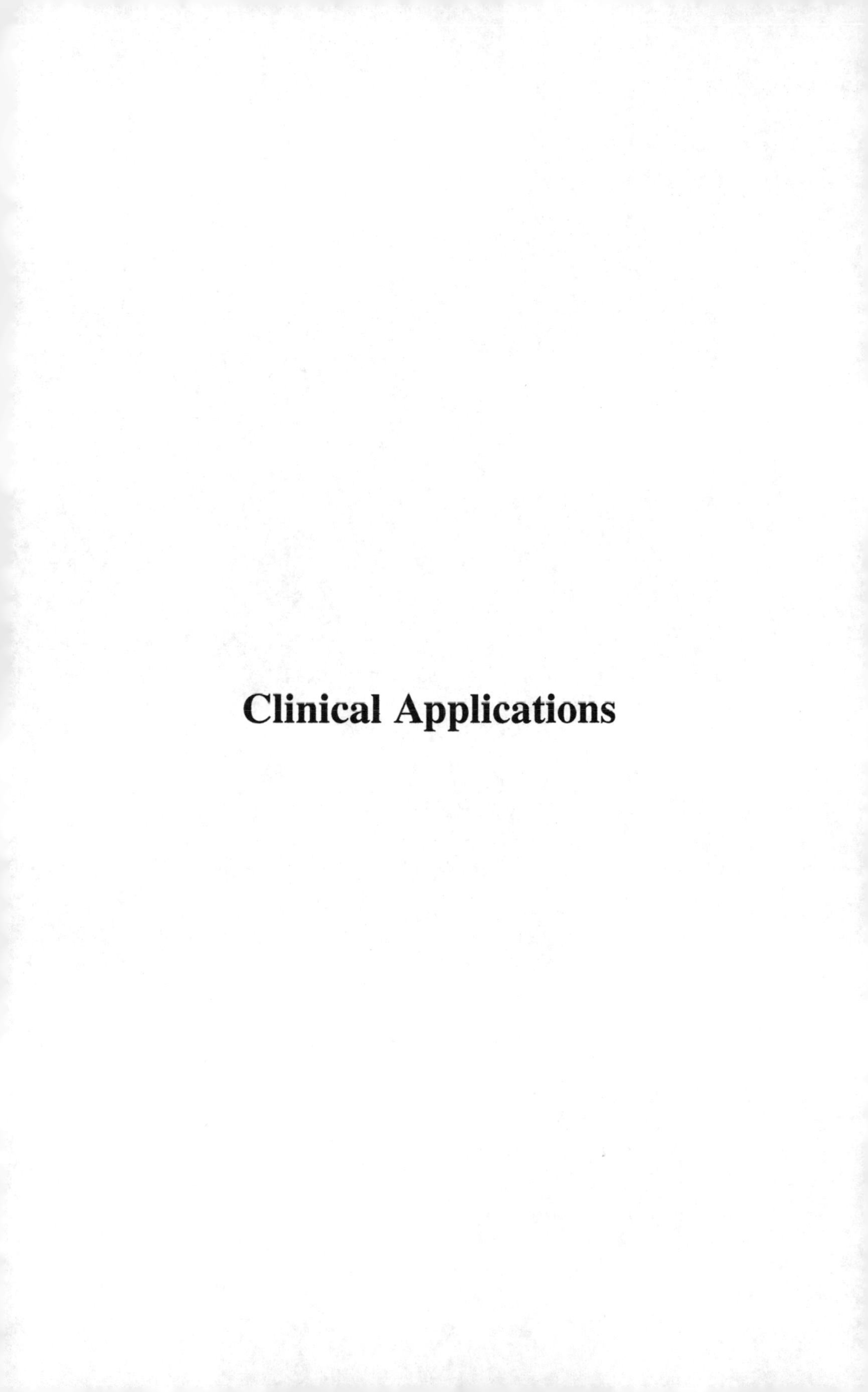

Clinical Applications

Which Anatomical Directions to Quantify Local Right Ventricular Strain in 3D Echocardiography?

Maxime Di Folco[1,2(✉)], Thomas Dargent[1,3], Gabriel Bernardino[1,4], Patrick Clarysse[1], and Nicolas Duchateau[1,5]

[1] Univ Lyon, Université Claude Bernard Lyon 1, INSA-Lyon, CNRS, Inserm, CREATIS UMR 5220, U1294, 69621 Villeurbanne, France
maxime.difolco@helmholtz-muenchen.de
[2] Institute of Machine Learning in Biomedical Imaging, Helmholtz Munich, Munich, Germany
[3] Laboratoire de Météorologie Dynamique (UMR8539), École Polytechnique, IPSL, CNRS, Palaiseau, France
[4] BCNmedtech, DTIC, Universitat Pompeu Fabra, Barcelona, Spain
[5] Institut Universitaire de France (IUF), Paris, France

Abstract. Technological advances in image quality and post-processing have led to the better clinical adoption of 3D echocardiography to quantify cardiac function. However, the right ventricle (RV) raises specific challenges due to its specific half-moon shape, which led to a lack of consensus regarding the estimation of RV motion and deformation locally. In this paper, we detail three ways to estimate local anatomically-relevant directions at each point of the RV surface, in 3D, and the resulting Green-Lagrange strain projected along these directions. Using a database of RV surface meshes extracted from 3D echocardiographic sequences from 100 control subjects, we quantified differences between these strategies in terms of local anatomical directions and local strain, both at the individual and population levels. For the latter, we used a specific dimensionality reduction technique to align the latent spaces encoding the strain patterns obtained from different computations of the anatomical directions. Differences were subtle but visible at specific regions of the RV and partially interpretable, although their impact on the population latent representation was low, which sets a preliminary quantitative basis to discuss these computation standards.

Keywords: Cardiac imaging · right ventricle · 3D echocardiography · myocardial strain · standardization

1 Introduction

Assessing the cardiac function is complex given the variety of factors affecting the heart geometry and its dynamics. Three-dimensional quantification of its shape and deformation is particularly relevant for the right ventricle (RV),

O. Bernard et al. (Eds.): FIMH 2023, LNCS 13958, pp. 607–615, 2023.
https://doi.org/10.1007/978-3-031-35302-4_62

given its asymmetric shape not properly captured with standard 2D imaging planes [1]. Myocardial strain can be decomposed along three orthogonal directions (radial, circumferential, and longitudinal) related to the local arrangement of fibers within the myocardium, and undergoes specific changes along these directions depending on disease. However, this local assessment remains complex and sensitive to differences in computations. As a result, for echocardiography, clinical studies focus on global or regional assessment in 2D and mostly along the longitudinal direction, which may be critical to detect subtle abnormalities related to disease.

Despite recent promising post-processing of 3D echocardiographic sequences of the RV [2], there is currently no standard for the definition and computation of anatomical directions for the RV [3] and therefore local directional strain, contrary to the left ventricle (LV) [4]. A group of authors proposed to decompose cardiac motion along global axes aligned to its main dimensions [5,6]. Other works used local anatomical coordinates to decompose deformation, defining the longitudinal and circumferential directions from the long-axis and radial directions [7]. Coordinates-independent analysis was also proposed by considering principal strain, obtained by eigendecomposition of the strain tensor [8].

In this paper, we detail three relevant strategies to estimate local anatomical directions over the RV in 3D echocardiography, two of them not being implemented in the literature. Furthermore, we quantify their differences and impact on local strain computations against the RV geometry, both at the individual and population levels, with the underlying aim to foster discussion around such standards.

2 Methods

2.1 Data and Pre-processing

We processed RV surface meshes from 100 control subjects, obtained from semi-automatic endocardial segmentation by an expert clinician and tracking of 3D echocardiographic sequences using commercial software (4D RV Function 2.0, TomTec Imaging Systems GmbH, Germany), and exported for post-processing as VTK files. These meshes consisted of 822 points and 1587 triangular cells, after cropping out the tricuspid and pulmonary valves. Point-to-point mesh correspondences were provided by the commercial software, allowing comparisons between subjects at each location of the RV surface. We also realigned them across the population using generalized Procrustes analysis with a rigid transform.

In the following, the local directions and strain are defined at each cell of the RV surface mesh. The radial direction is defined as the normal to the RV surface at each point. We therefore focus explanations on the other two directions: longitudinal and circumferential.

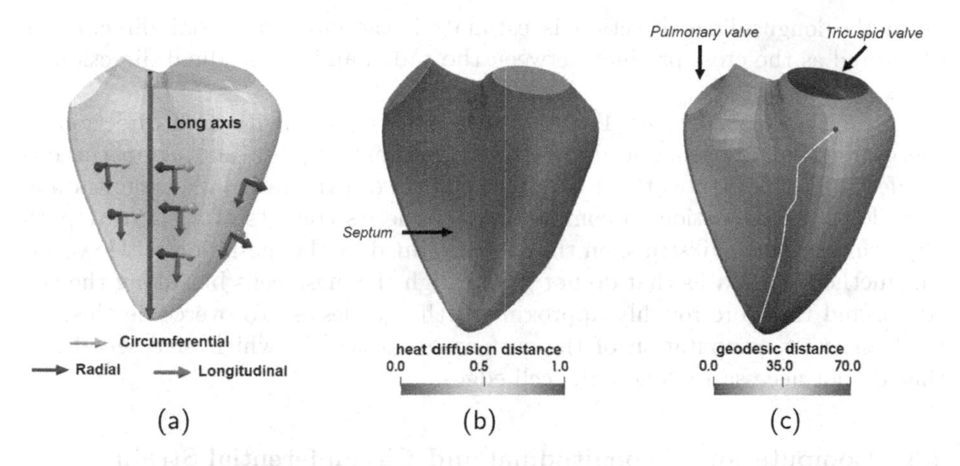

Fig. 1. The three ways to compute circumferential and longitudinal directions we evaluated in this paper. (a) Long-axis computations, (b) Heat diffusion computations (the turquoise disks represent the cold point, while the apex stands as the hot point), (c) Geodesic distance computations (the white line corresponds to the geodesic joining the two purple dots). (Color figure online)

2.2 Estimation of Circumferential and Longitudinal Directions

Long-Axis Method: The first implemented method (Fig. 1a) uses the long-axis, which joins the apex and the basal point equidistant from the valves centers. The circumferential direction is obtained locally from the cross product between the radial direction and the long-axis, then the longitudinal direction is obtained from the cross product between the radial and circumferential directions, as reported in [7].

The RV is actually bi-axial, considering the axes joining the apex and the center of each valve; the single axis definition is therefore not fully anatomically-relevant. In other words, the longitudinal direction estimated with this approach is a rough approximation.

Heat Diffusion Method: This method estimates the longitudinal direction as the gradient of the map u, which is defined by solving the following partial differential equation:

$$\nabla \cdot (\nabla u) = 0. \quad (1)$$

This corresponds to the stationary solution of a diffusion process from hot to cold points, which in our case were set to the apex ($u = 1$) and the valves ($u = 0$), respectively (Fig. 1b). The map u can be estimated iteratively, by (at each iteration) setting the value at each point as the weighted average of the values at neighboring points in the graph defined by the RV mesh, updating all points, and then restoring the original values 1 and 0 to the apex and the valves. An alternative can be to find a direct solution using the Laplace Beltrami operator.

Once the longitudinal direction is estimated, the circumferential direction is computed as the cross product between the radial and longitudinal directions.

Geodesic Distance Method: This method defines the longitudinal direction as the gradient of the geodesic distance to the apex (Fig. 1c), and again the circumferential direction as the cross product between the radial and longitudinal directions. A first option to compute the geodesics consists of a shortest path algorithm such as Dijkstra's, on the graph defined by the mesh points. However, this method uses paths that do not go through the mesh cells but along the cell edges, and therefore roughly approximate the geodesics. To overcome this, we used an exact computation of the surface geodesics [9], which gives geodesics that do not necessarily follow the cell edges.

2.3 Computation of Longitudinal and Circumferential Strain

Once the local directions are computed, the local strain tensor is estimated as the Green Lagrangian strain:

$$\mathbf{E} = \frac{1}{2}(\mathbf{J}^T \cdot \mathbf{J} - \mathbf{I}), \tag{2}$$

where $\mathbf{J} = \nabla v + \mathbf{I}$, with ∇v the displacement gradient at a given point, and $\mathbf{I}$ is the identity matrix.

Then, longitudinal and circumferential directional strains are obtained by projecting the strain tensor along these two directions, as:

$$\mathbf{E_h} = \mathbf{h}^T \cdot \mathbf{E} \cdot \mathbf{h}, \tag{3}$$

where $\mathbf{h}$ is the unit vector defining the considered direction. As only endocardial meshes were available due to the thin RV myocardial wall, the radial strain was not computed.

This slightly differs from the computations in [7], which estimated the relative change of length of $5mm$ segments along each anatomical direction separately. Although equivalent in theory for uni-axial deformation, the Green Lagrangian strain is better defined in practice (computations based on the gradient, and not on a direction whose definition may be arguable).

In our database, all the strain patterns were available at each point of the RV surface and at each instant of the cycle. Nonetheless, we focused the evaluation on end-systolic strain patterns, of higher magnitude. In all figures, results are displayed on end-diastolic meshes, which better render anatomical differences between subjects before deformation.

2.4 Quantifying Differences in Strain Patterns

For each individual, we first quantified the point-to-point differences between the anatomical directions estimated with each of the three methods described above,

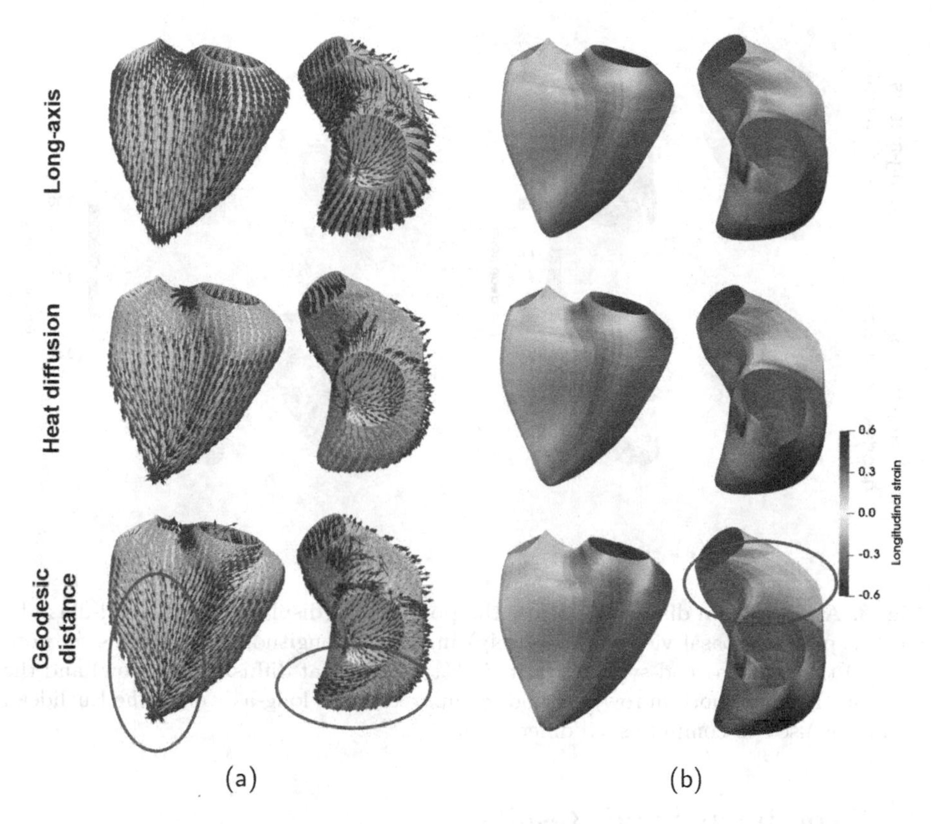

Fig. 2. Illustration on a representative case of the local differences (displayed on the end-diastolic mesh/septal and basal views, respectively) in (a) the longitudinal directions, and (b) the resulting end-systolic strain, in %, for the three different computations: long-axis, heat diffusion and geodesic distance.

and between the strain values obtained from these directions. The Euclidean norm was used to compute such differences.

We also performed comparisons at the population level, by examining the latent spaces encoding the strain patterns, obtained by a specific dimensionality reduction technique that performs latent space alignment (multiple manifold learning (MML) algorithm, as used in [10]). We fed this algorithm with longitudinal strain obtained by the long-axis method and either the heat diffusion or the geodesic distance method, for the whole population. Dimensionality reduction in MML is based on a generalization of the Laplacian eigenmaps framework [11] to several descriptors. It estimates a latent space for each descriptor, and simultaneously brings together in the latent space samples for which the descriptors are close (regarding a given sample and its neighbors). This is relevant to examine for which samples in the population the strain computations differ and their impact on a low-dimensional representation of the population.

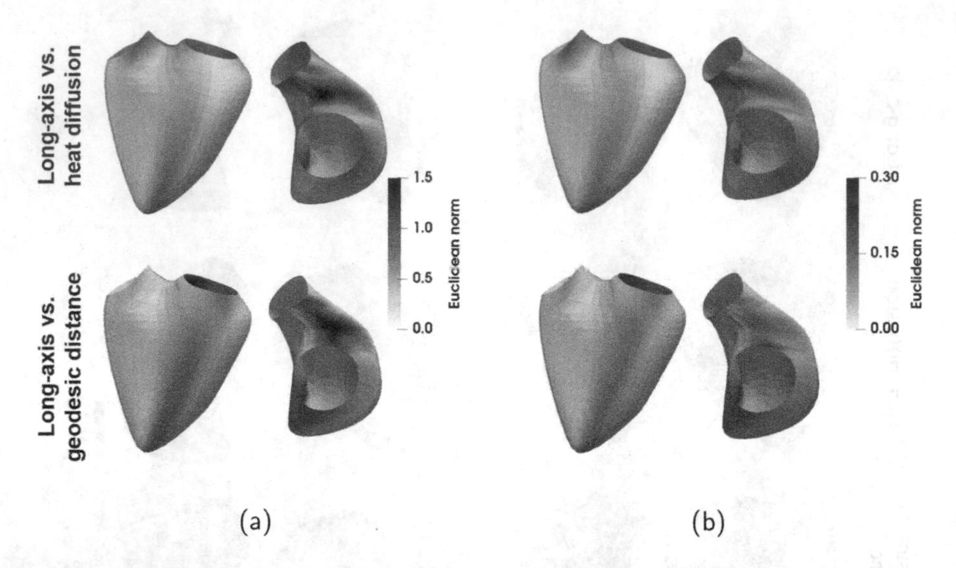

Fig. 3. Average local differences across the population (displayed on the end-diastolic mesh/septal and basal views, respectively) in (a) the longitudinal directions, in *mm*, and (b) the resulting end-systolic strain, in %, for the heat diffusion (top row) and the geodesic distance (bottom row) methods compared to the long-axis one. The Euclidean norm was used to compute such differences.

3 Experiments and Results

3.1 Local Differences

We first examined local differences regarding the anatomical directions and strain. Figure 2 illustrates this on a representative individual. Only the longitudinal direction and longitudinal strain are displayed to remain concise.

The three methods observed independently provide directions that may sound plausible. However, differences in the longitudinal direction are visible between the long-axis method and the two other methods on the RV septum, in particular near the apex where the former tends to be straight, while the other two tend to point at the apical point (see red circle in the first column). The geodesic distance method is rather sensitive to local specificities of the RV shape, namely that the shortest path taken to reach the apex is not exactly what one might (wrongly) imagine, in particular near the valves (see red circle in the second column). These differences mostly impact strain computations near the valves, where differences in the directions are combined to smaller and more elongated triangular cells, and where segmentation and tracking are also more challenging.

Figure 3 complements these observations by showing the average differences in the longitudinal direction and longitudinal strain across the whole population

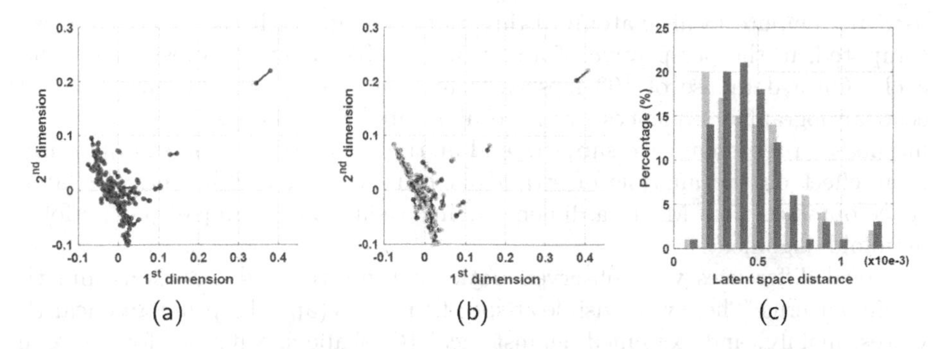

Fig. 4. Latent space estimated by manifold alignment (MML algorithm) for (a) the long-axis (in purple) and heat diffusion (in blue) methods or for (b) the long-axis and geodesic distance (in yellow) methods. The black lines join pairs of samples from the same subject. (c) Distribution of distance between samples from the same subject in blue and yellow respectively for the latent spaces in (a) and (b). (Color figure online)

(long-axis method compared to the other two), displayed over the average mesh across the population. Again, differences are mostly located near the valves. They are much more limited over the RV walls regarding the longitudinal direction, while slight differences are noticed regarding longitudinal strain near the apex at the borders of the septal wall (long-axis vs. the two other methods), and the valve at the borders of septal wall (long-axis vs. geodesic distance method). This corresponds to the observations highlighted with the red circles in Fig. 2.

3.2 Population Differences

Figure 4 summarizes the distribution of samples in the latent space estimated by MML, fed with the longitudinal strain for the long-axis and heat diffusion methods (Fig. 4a), and the long-axis and geodesic distance methods (Fig. 4b). Colored dots stand for the distribution of individual samples in the estimated latent space (one color for each method), and black lines join the two samples corresponding to the same subject. The first two dimensions of the latent spaces are displayed. The MML algorithm brings together samples whose input descriptors are close, but does not necessarily bring together samples whose input descriptors differ. Here, we observe that the two latent spaces are rather close, even for individuals a bit out of the distribution, meaning that the different computations of strain have low effect on the analysis of this population. This is confirmed by the histogram of the distances between samples in the two latent spaces (computed across all dimensions).

4 Discussion and Conclusion

We have presented three methods to estimate local anatomical directions over the RV, for which no consensus exists in the literature. These directions have been used to compute local anatomical directions, along which the local strain was computed, in the perspective of analyzing the RV function across the cardiac cycle. On a database of 100 meshes from control subjects obtained from 3D echocardiographic sequences, we have illustrated the differences between these methods on representative subjects and at the population level, and quantified their effect on the analysis of strain patterns using a low-dimensional latent space obtained from MML, a dimensionality reduction technique that performs manifold alignment.

Local differences were observed both on the estimated directions and the strain patterns. These were visible at specific regions (apical septum and near the valves, mainly) and examined against local RV shape specificities for representative individuals. These trends were confirmed by observing these local differences at the population level, but their impact on the latent representation learnt was low and these differences have a minor effect on the main relevant patterns present in the population. This may come from the fact that they only spread across a limited portion of the full RV area, while MML compares subjects across the whole RV (here, using Euclidean distances between the data at all mesh points, considered as a column vector). More differences would be observed by focusing MML on the regions identified in Fig. 3, or with a different metric. Besides, the apex and valve regions are often subject to higher noise and lower image quality, and therefore less reliable for clinical interpretations.

Reaching a consensus to define these anatomical directions, on the specific 3D RV geometry, is challenging. The long-axis method is rather simple to compute but its output may be arguable in particular near the apical septum. The geodesic distance method may seem the most intuitive to match the abstract definition of the longitudinal direction (shortest path to reach the apex), but its output does not seem relevant at specific regions of the RV. The heat diffusion method seems to provide relevant outputs across the whole RV. Computation times were in comparable orders of magnitude for the long-axis ($0.64 \pm 0.01s$ per subject) and geodesic distance ($1.31 \pm 0.03s$) methods, and slightly longer for the heat diffusion method ($4.2 \pm 0.05s$).

Reaching a clear consensus to define such directions should be addressed by our scientific community. Meanwhile, it could be interesting to develop representation learning methods (for the analysis of populations) that could explicitly incorporate several ways to compute a given descriptor (e.g. strain obtained by different methods) and therefore enrich the analysis with such "uncertainties". We will address this in future work.

Acknowledgements. The authors acknowledge the support from the French ANR (LABEX PRIMES of Univ. Lyon [ANR-11-LABX-0063], the JCJC project "MIC-MAC" [ANR-19-CE45-0005]), and the European Union - NextGenerationEU, Ministry of Universities and Recovery, Transformation and Resilience Plan, through a call from Pompeu Fabra University (Barcelona). They are also grateful to P. Moceri (CHU Nice, France) for providing the imaging data related to the studied population, and to M. Sermesant (INRIA Epione, Sophia Antipolis, France) for initial discussions on this topic.

References

1. Sanz, J., Sánchez-Quintana, D., Bossone, E., Bogaard, H.J., Naeije, R.: Anatomy, function, and dysfunction of the right ventricle: JACC state-of-the-art review. J. Am. Coll. Cardiol. **73**, 1463–82 (2019)
2. Addetia, K., Muraru, D., Badano, L.P., Lang, R.M.: New directions in right ventricular assessment using 3-dimensional echocardiography. JAMA Cardiol. **4**, 936–44 (2019)
3. Duchateau, N., Moceri, P., Sermesant, M.: Direction-dependent decomposition of 3D right ventricular motion: beware of approximations. J. Am. Soc. Echocardiogr. **34**, 201–3 (2021)
4. D'hooge, J., et al.: Regional strain and strain rate measurements by cardiac ultrasound: principles, implementation and limitations. Eur. J. Echocardiogr. **1**, 154–70 (2000)
5. Lakatos, B.K., Nabeshima, Y., Tokodi, M., Nagata, Y., Tösér, Z., Otani, K., et al.: Importance of nonlongitudinal motion components in right ventricular function: three-dimensional echocardiographic study in healthy volunteers. J. Am. Soc. Echocardiogr. **33**, 995-1005.e1 (2020)
6. Tokodi, M., Staub, L., Budai, A., Lakatos, B.K., Csákvári, F.I., Suhai, M., et al.: Partitioning the right ventricle into 15 segments and decomposing its motion using 3D echocardiography-based models: the updated ReVISION method. Front. Cardiovasc. Med. **8**, 622118 (2021)
7. Moceri, P., Duchateau, N., Gillon, S., Jaunay, L., Baudouy, D., Squara, F., et al.: Three-dimensional right ventricular shape and strain in congenital heart disease patients with right ventricular chronic volume loading. Eur. Heart J. Cardiovasc. Imaging **22**, 1174–81 (2021)
8. Satriano, A., Pournazari, P., Hirani, N., Helmersen, D., Thakrar, M., Weatherald, J., et al.: Characterization of right ventricular deformation in pulmonary arterial hypertension using three-dimensional principal strain analysis. J. Am. Soc. Echocardiogr. **32**, 385–93 (2019)
9. Mitchell, J., Mount, D., Papadimitriou, C.: The discrete geodesic problem. SIAM J. Comput. **16**, 647–68 (1987)
10. Di Folco, M., Moceri, P., Clarysse, P., Duchateau, N.: Characterizing interactions between cardiac shape and deformation by non-linear manifold learning. Med. Image Anal. **75**, 102278 (2022)
11. Belkin, M., Niyogi, P.: Laplacian eigenmaps for dimensionality reduction and data representation. Neural Comput. **15**, 1373–96 (2003)

Biomechanical Model to Aid Surgical Planning in Complex Congenital Heart Diseases

Maria Gusseva[1], Nikhil Thatte[2], Daniel A. Castellanos[2], Peter E. Hammer[3], Sunil J. Ghelani[2], Ryan Callahan[4], Tarique Hussain[1], and Radomír Chabiniok[1](✉)

[1] Division of Pediatric Cardiology, Department of Pediatrics, UT Southwestern Medical Center, Dallas, TX, USA
radomir.chabiniok@utsouthwestern.edu
[2] Department of Cardiology, Boston Children's Hospital, Boston, MA, USA
[3] Department of Cardiac Surgery, Boston Children's Hospital, Boston, MA, USA
[4] Division of Cardiology, Department of Pediatrics, Children's Hospital of Philadelphia, Philadelphia, PA, USA

Abstract. In the patients with congenitally corrected transposition of great arteries (ccTGA) it is important to evaluate the function of the left ventricle (LV) prior to performing a complex surgery establishing normal anatomical connection of LV, RV with systemic, pulmonary circulation, respectively. Current clinical techniques have proven inadequate to promptly assess the LV functionality. We propose using a biomechanical model to (1) estimate the mechanical properties of the LV at baseline, and (2) predict the functional adaptation of the LV to the repair through *in silico* modeling of surgery. The catheterization and cardiac magnetic resonance imaging data of two patients with ccTGA were used to create patient-specific models of LV – $\mathcal{M}^{\mathrm{LV}}_{\mathrm{baseline}}$. For an *in silico* repair, the model $\mathcal{M}^{\mathrm{LV}}_{\mathrm{baseline}}$ was used while imposing the increased afterload conditions as predicted by the model of systemic circulation. Our results showed that LV contractility at the baseline vs. predicted repaired state was 93 kPa vs. 143 kPa and 136 kPa vs. 145 kPa for Patients 1 and 2, respectively. Therefore, the LV of Patient 1 would require a 54% augmentation in LV contractility if the surgery was to be successful. In contrary, the model suggests that the LV of Patient 2 is already at a contractile state adequate to sustain its function after the surgery. This work demonstrates that biomechanical modeling is a promising tool to test various hemodynamic conditions *in silico*. Such predictions have a potential to provide additional insights into procedural success in this population.

Keywords: In silico medicine · surgery planning · biomechanical heart modeling · myocardial contractility · congenitally corrected transposition of great arteries (ccTGA)

O. Bernard et al. (Eds.): FIMH 2023, LNCS 13958, pp. 616–625, 2023.
https://doi.org/10.1007/978-3-031-35302-4_63

1 Introduction

Congenitally corrected transposition of great arteries (ccTGA) is a complex congenital heart disease that is characterized by abnormal atrio-ventricular and ventriculo-arterial connections. Namely, the morphological left ventricle (LV) is connected to the right atrium and is ejecting into the pulmonary circulation, while the right ventricle (RV) is connected to the left atrium and is ejecting into the systemic circulation. These patients can undergo an atrial and arterial switch operation – i.e. double switch operation (DSO) – to restore the morphological LV and RV to their systemic and subpulmonary functions, respectively [11]. However, there are non-negligible risks during the complex surgical procedure and the long-term outcomes are difficult to predict by current techniques.

The prolonged period of the LV working as a subpulmonary ventricle leads to its progressive deconditioning. This puts the LV at risk of failure to sustain the function of the systemic ventricle once the LV is connected with the aorta and systemic circulation. Surgical pulmonary artery banding (PAB) is often an intermediate step in order to retrain the LV prior to the final DSO. The standard clinical assessment of LV retraining and preparedness for DSO after some time of PAB (i.e. 6–18 months) involves cardiac catheterization and cardiac magnetic resonance imaging (MRI) or 3D echocardiography. These techniques provide somewhat arbitrary 'cut-off' values of LV pressure, mass, and mass-to-volume ratio [11]. However, evaluating the mechanical properties of the LV by separate pressure (catheterization) and volumetric (cardiac MRI) analyses provides an inadequate understanding of the ventricle's performance. In addition to that the properties of the LV must be compared to the existing systemic RV and the aortic circulation. Therefore, there is a need to reconstruct LV and RV pressure-volume (PV) loops in ccTGA using non-simultaneously acquired catheterization and cardiac MRI data. Finally, the predictive capability of clinical indices is still limited as the long-term outcomes of DSO on LV health are not clearly understood.

We hypothesize that a biomechanical model could augment clinical evaluation of LV preparedness for the DSO. Firstly, the model can assist in clinical data processing by allowing construction of the PV loops from non-simultaneously acquired cardiac MRI and catheterization data [5]. Secondly, the model could provide an estimate of patient-specific LV and RV contractility for a given functional state of the ventricles. Model-derived contractility has been shown to reflect ventricular inotropic state and thus is a promising patient-specific measure of myocardial energetics in various cardiac pathologies [6,13] including ccTGA [2]. Finally, the biophysical formulation of the model renders its suitable to perform an *in silico* surgical operation [7] to e.g. predict the change in LV contractility required to sustain increased afterload conditions such as in the post-DSO state.

The aim of the present study is to construct LV and RV models from cardiac MRI and catheterization data, and perform *in silico* DSO in order to predict the contractile adaptation required of the LV to the increased afterload of the systemic circulation. In addition to that we aim to construct a model of *in silico*

PAB tightening in order to explore the effect of increasing the LV-to-pulmonary artery (PA) pressure difference on LV contractility at baseline.

2 Methods

2.1 Data

The datasets of two ccTGA patients were included in the study. Patients 1 and 2 underwent an interventional follow-up exam 16 and 5 months after PAB was performed, respectively. The follow-up exam consisted of sequential acquisition of left and right heart catheterization pressures – LV, RV, PA and aorta (AO) – and cardiac MRI data. A retrospective ECG-gated cine bSSFP sequence (SENSE = 2, spatial resolution $2 \times 2 \times 10$ mm, temporal resolution $\sim$ 30 msec) was used. Ventricular volumes from cardiac MRI were extracted by using the CVI42 software (Circle Cardiovascular Imaging Inc., Calgary, Canada) and the motion tracking algorithm of [4]. Table 1 shows patient demographics and characteristics of the cardiovascular system. These post-PAB datasets will be further referred to as baseline data.

Table 1. Patient demographics and characteristics derived from catheterization pressure and cardiac MRI data. AO: aorta, BSA: body surface area, EDP: end-diastolic pressure, PSP: peak systolic pressure, FW: free wall, LV: left ventricle, PA: pulmonary artery, RV: right ventricle.

Pt #	age year	BSA m^2	LV FW mass g	RV FW mass g	Septum mass g	LV PSP mmHg	LV EDP mmHg	PA PSP mmHg	RV PSP mmHg	RV EDP mmHg	AO PSP mmHg
1	6.0	0.87	20.3	21.2	7.7	76.0	9.0	26.0	94.0	9.0	94.0
2	1.5	0.50	8.7	9.7	3.5	78.0	10.0	22.0	75.0	9.0	77.0

2.2 Biomechanical Model Calibration to Patient-Specific Clinical Data

A biomechanical model of a single ventricle [1] and a lumped parameter model of a circulation were employed. A ventricular cavity was represented by a sphere with an inner radius R and a wall thickness d (Fig. 1) [1]. The constitutive mechanical laws described an active (contractile) and passive (viscoelastic) behavior of the myocardium [3]. The active law described a sarcomere-level contraction of the myocardium – i.e. inspired by Huxley's sliding filament theory [8]. Myosin heads were modeled as series of springs with active stress ($\dot{\tau}$) and active stiffness ($\dot{k}$) produced for a certain sarcomere extension (e_{fib}):

$$\begin{cases} \dot{k}_c = -(|u| + \alpha|\dot{e}_{\text{fib}}|)k_c + n_0(e_{\text{fib}})k_0|u|_+ \\ \dot{\tau}_c = -(|u| + \alpha|\dot{e}_{\text{fib}}|)\tau_c + \dot{e}_{\text{fib}}k_c + n_0(e_{\text{fib}})\sigma_0|u|_+, \end{cases} \quad (1)$$

where $u(t)$ is an electrical activation function representing a calcium transport into the myocytes ($u > 0$ is during activation and $u \leq 0$ is at repolarization).

The parameter α is a bridge destruction rate upon rapid change in sarcomers' length [3]. The function $n_0(e_{\text{fib}}) \in [0,1]$ models the Frank-Starling law [3]. The parameter σ_0 is the level of active contractility developed by the sarcomeres during ventricular systole.

Atrioventricular valve (AV) and ventricular outflow tract (VOT) mechanics were represented by a system of diodes with forward and backward resistances ($R^{\text{AV}}_{\text{for}}, R^{\text{AV}}_{\text{back}}$ and $R^{\text{VOT}}_{\text{for}}, R^{\text{VOT}}_{\text{back}}$, respectively). The outflow from the ventricle (Q) was coupled to the pressures in the atrium, ventricle and the artery ($P_{\text{at}}, P_{\text{V}}, P_{\text{ar}}$, respectively) as described in [14]. The circulation system was represented by a two-stage Windkessel system.

The model was coupled to the patient data by using a sequential calibration procedure [6,10]. Note that the model is not a closed-loop system, hence we prescribed pressures in the atria (P_{at}) and venous system (P_{vs}), Fig. 1, and created separate models of LV and RV with pulmonary, systemic circulation, respectively. Briefly, the circulation system was decoupled from the heart to calibrate proximal and distal resistances, and distal capacitance (R_{p} and R_{d}, and C_{d}, respectively) by imposing the measured arterial flow. The venous volume pressure P_{vs} was assumed to be 5 mmHg for all simulations. Ventricular sphere thickness was prescribed from cardiac MRI. LV and RV mass in the model were considered as LV and RV free wall mass, respectively, increased by the mass of a half of the septum. The stress-free reference volume was derived as in [9], similarly to previous pre-clinical studies [6,7,10]. Passive constitutive law was adjusted to match cardiac MRI end-diastolic volume (EDV) by imposing measured end-diastolic ventricular pressure (EDP). Note that atrial pressure as on Fig. 1 – i.e. filling pressure for the ventricle – is a prescribed EDP from the data. AV valve and VOT mechanics were calibrated to match the AV valve regurgitation fraction (i.e. by tweaking $R^{\text{AV}}_{\text{back}}$) and VOT pressure difference (i.e. by tweaking $R^{\text{VOT}}_{\text{for}}$), respectively. Electrical activation was prescribed as a stepwise function for the active constitutive law (Eq. 1), where the timings of QRS and ST durations were directly imposed from electrocardiogram measurements. Myocardial contractility was calibrated to match the ventricular peak systolic pressure (PSP) and volume.

For each patient we built baseline models of the LV ejecting blood into the modeled pulmonary (PA) circulation and the RV contracting into the aorta (AO) and systemic circulation. These models will be in the sequel denoted as $\mathcal{M}^{\text{LV}}_{\text{baseline}}$, $\mathcal{M}^{\text{PA}}_{\text{baseline}}$, $\mathcal{M}^{\text{RV}}_{\text{baseline}}$, $\mathcal{M}^{\text{AO}}_{\text{baseline}}$, respectively. Patient-specific mechanical properties of the LV and RV were given by myocardial contractility and stiffness, $R^{\text{AV}}_{\text{back}}$ and $R^{\text{VOT}}_{\text{for}}$, and $R_{\text{p}}, R_{\text{d}}, C_{\text{d}}$ of the circulation. The computation time of a single cardiac cycle for the model implemented in Matlab is around 30 s.

2.3 Left and Right Ventricular Pressure-Volume Loops

The LV and RV model-derived pressure and volume waveforms were used as a template to synchronize corresponding clinical pressure-volume signals in time. Time-synchronization was achieved by minimizing the distance between model-

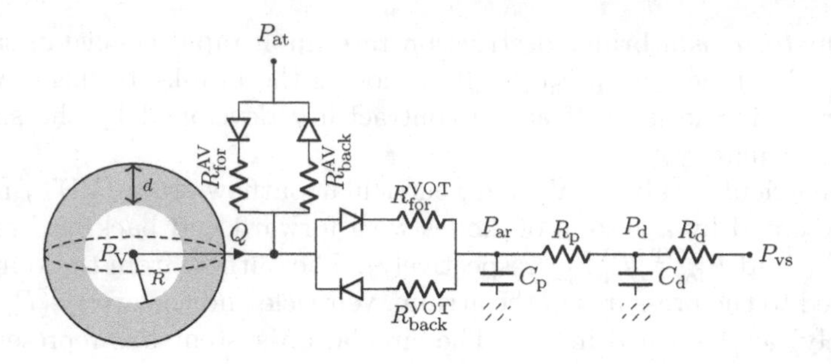

Fig. 1. Biomechanical model of ventricular cavity coupled with the atrioventricular and arterial valves via system of diodes and with circulation via two-stage Windkessel.

and data-derived signals as in [5]. Baseline RV and LV PV loops were constructed from time-synchronized clinical data.

2.4 *In silico* Double Switch Operation

The models $\mathcal{M}_{\text{baseline}}^{\text{LV}}$ and $\mathcal{M}_{\text{baseline}}^{\text{AO}}$ were used to build an *in silico* switch model of LV ($\mathcal{M}_{\text{switch}}^{\text{LV}}$) and predict the change in LV contractility required to maintain AO pressure following the switch operation. Specifically, the model $\mathcal{M}_{\text{baseline}}^{\text{LV}}$ was coupled to $\mathcal{M}_{\text{baseline}}^{\text{AO}}$. Ventricular geometry and passive mechanical law were preserved as in $\mathcal{M}_{\text{baseline}}^{\text{LV}}$. $R_{\text{for}}^{\text{VOT}}$ was set to its minimum value to represent no LV-to-AO pressure difference following the switch operation. There are no published data regarding the change in LV EDP following the DSO. Thus, we assumed the variation of LV EDP between a physiological range of 6 mmHg and 15 mmHg. We manually recalibrated the LV contractility for the corresponding value of imposed LV EDP such that (1) AO PSP was preserved as in the baseline data and (2) LV PSP was equal to AO PSP.

The $\mathcal{M}_{\text{switch}}^{\text{LV}}$-predicted contractility was taken as an average of LV contractilities calibrated for LV EDP of 6 mmHg, 9 mmHg, 12 mmHg and 15 mmHg.

2.5 *In silico* Pulmonary Artery Band Tightening

The models $\mathcal{M}_{\text{baseline}}^{\text{LV}}$ and $\mathcal{M}_{\text{baseline}}^{\text{PA}}$ were used to model an *in silico* PAB tightening of LV ($\mathcal{M}_{\text{banding}}^{\text{LV}}$) and predict a change in LV contractility for a range of LV-to-PA pressure differences. Modifying the method developed in [7] (where applied for predicting the effect of intervention on right ventricular outflow tract for patients with repaired Tetralogy of Fallot), the ventricular geometry and passive mechanical law were preserved as in $\mathcal{M}_{\text{baseline}}^{\text{LV}}$. There are no published data regarding the change in LV EDP following the PAB tightening. Thus, we assumed the variation of LV EDP between a physiological range of 6 mmHg and 15 mmHg. To model an increase in LV-to-PA pressure difference the resistance $R_{\text{for}}^{\text{VOT}}$ was sequentially increased. While preserving PA PSP as in baseline data

LV contractility was manually recalibrated for corresponding increase in resistance and the range of imposed LV EDP. The procedure was performed until the level of PAB tightening (i.e. LV-to-PA pressure difference) provided the level of LV contractility required to perform an *in silico* switch as predicted in Sect. 2.4.

The contractility predicted by the model $\mathcal{M}_{\text{banding}}^{\text{LV}}$ was taken as an average of LV contractilities calibrated for LV EDP of 6 mmHg, 9 mmHg, 12 mmHg and 15 mmHg for every imposed value of $R_{\text{for}}^{\text{VOT}}$.

3 Results

Figures 2 and 3 show an example of LV and RV model calibration to the baseline data. Figures 4 and 5 show the PV loops predicted by the model $\mathcal{M}_{\text{switch}}^{\text{LV}}$ vs. LV and RV baseline PV loops.

The LV contractility at baseline vs. average predicted by the model $\mathcal{M}_{\text{switch}}^{\text{LV}}$ was 93 kPa vs. 143 kPa and 136 kPa vs. 145 kPa for Patients 1 and 2, respectively (Table 2). Figure 4 shows an increase in baseline LV contractility predicted by the model $\mathcal{M}_{\text{banding}}^{\text{LV}}$ for the associated increase in LV PSP for Patient 1. In Patient 2 the baseline LV contractility was close to the value predicted by the model $\mathcal{M}_{\text{switch}}^{\text{LV}}$. Therefore, no further *in silico* PAB tightening was performed (Fig. 5).

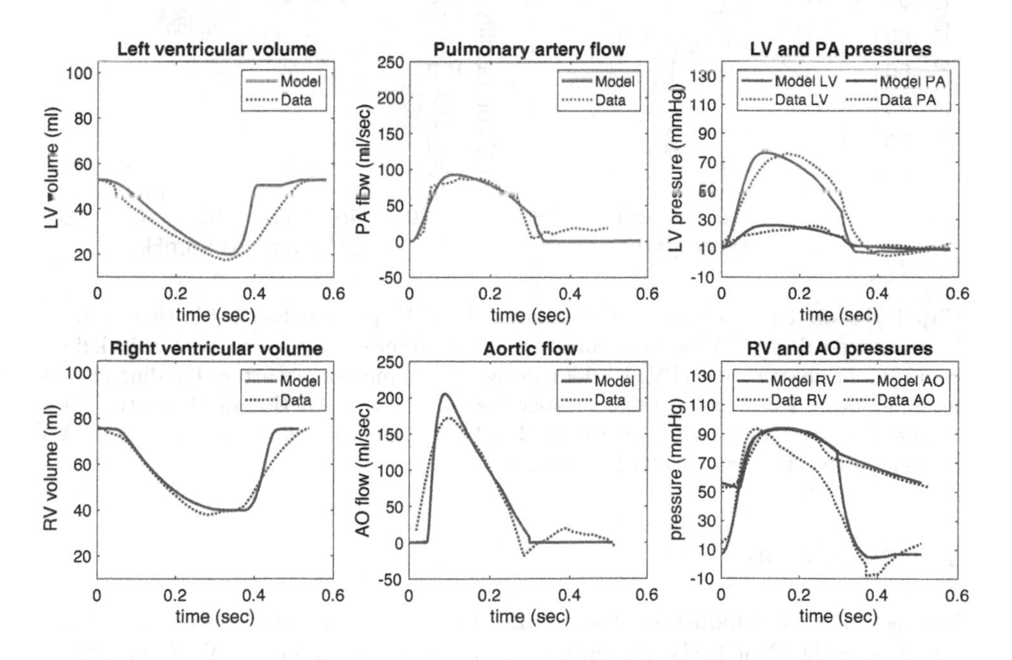

Fig. 2. Example of model calibration for left and right ventricles of Patient 1. Note that clinical data were time-synchronized with respect to the corresponding model waveforms. AO: aorta, LV: left ventricle, PA: pulmonary artery, RV: right ventricle.

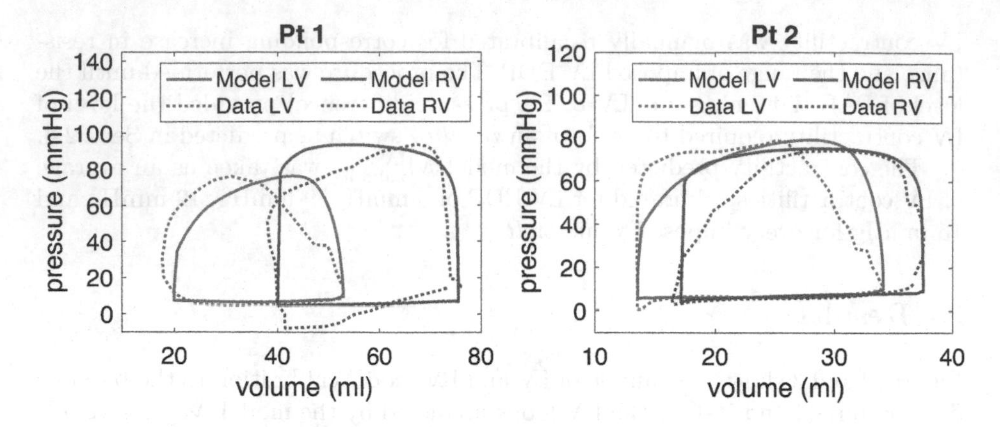

Fig. 3. Data- and model-derived pressure-volume loops for Patient 1 and Patient 2. Note that clinical data were time-synchronized with respect to the corresponding model-derived waveforms. LV: left ventricle, RV: right ventricle.

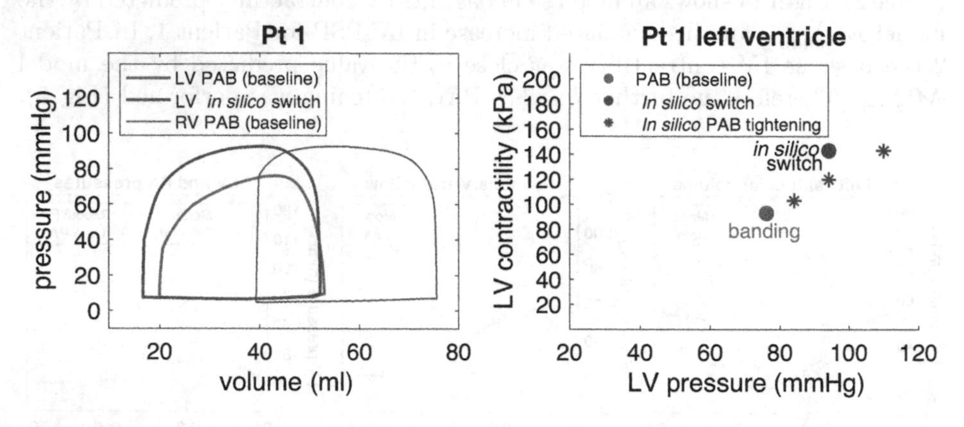

Fig. 4. Prediction of a change of left ventricular (LV) performance for Patient 1. Left: LV pressure-volume (PV) loop obtained by *in silico* double switch operation (black line) vs. LV and right ventricle (RV) PV loop derived from pulmonary artery banding (PAB) baseline model (blue and red lines, respectively). Right: prediction of LV contractility by the *in silico* double switch operation (black dot) and various levels of *in silico* PAB tightening (black stars). (Color figure online)

4 Discussion

In this study we demonstrate the potential of a biomechanical model to perform an *in silico* DSO or PAB tightening for patients with ccTGA. We demonstrate the ability of the models to predict the required contractile adaptation of the LV in response to (1) changed afterload conditions – LV exposure to systemic afterload after DSO, or (2) increasing the level of LV-to-PA pressure difference by PAB tightening.

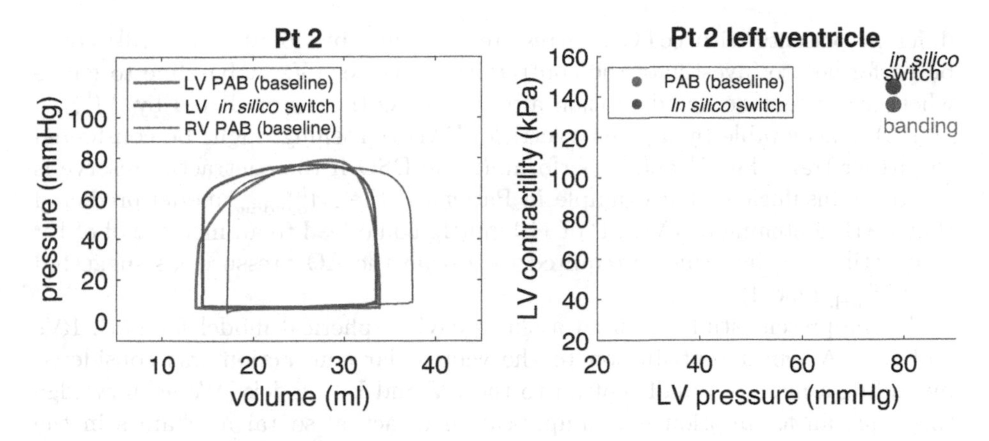

Fig. 5. Prediction of a change of left ventricular (LV) performance for Patient 2. Left: LV pressure-volume (PV) loop obtained by *in silico* double switch operation (black line) vs. LV and right ventricle (RV) PV loop derived from pulmonary artery banding (PAB) baseline model (blue and red lines, respectively). Right: prediction of LV contractility by the *in silico* switch operation (black dot). (Color figure online)

Table 2. Prediction of left ventricular (LV) contractility (kPa) and stroke work (J) by *in silico* switch model for the various levels of LV end-diastolic pressures (EDP). RV: right ventricle.

	Baseline		*In silico* double switch operation			
	RV	LV	LV	LV	LV	LV
LV EDP mmHg			6	9	12	15
Pt 1						
Contractility (kPa)	150	93	152	145	140	136
Stroke work (J)	0.41	0.25	0.33	0.36	0.37	0.37
Pt 2						
Contractility (kPa)	155	136	152	147	142	138
Stroke work (J)	0.17	0.16	0.16	0.17	0.17	0.17

The LV contractility predicted by the model $\mathcal{M}^{LV}_{switch}$ was higher by 54 % and by 7 % than the LV contractility at baseline for Patients 1 and 2, respectively. This suggests that the LV of Patient 1 will be forced to generate substantially higher levels of myocardial contractility once connected to the aorta and systemic circulation compared to its baseline contractility. In contrary, the LV of Patient 2 at baseline is almost functioning at the contractility level required to pump the blood into the aorta once the double switch is performed. Therefore, we hypothesize that (1) Patient 1 is likely to be less ready for DSO than Patient 2, and (2) Patient 1 might benefit from additional interventions. This demonstrates the vast variation in the performance of the morphologic LV in different patients with ccTGA. Stress testing (e.g. by dobutamine [13]) could be performed to investigate the contractile reserve of LVs at baseline to gauge whether the contractile adaptation after the DSO (i.e. as predicted by $\mathcal{M}^{\text{LV}}_{\text{switch}}$ model) is achievable by a given ventricle. PAB tightening might be considered to further train the LV before performing the DSO if the contractile reserve is felt to be insufficient. For example in Patient 1, the $\mathcal{M}^{\text{LV}}_{\text{banding}}$ model predicted that PAB tightening at LV PSP of 110 mmHg could lead to an increase of of LV contractility to the value as required to sustain the AO pressure (as suggested by $\mathcal{M}^{\text{LV}}_{\text{switch}}$ model).

In the present study we used a single cavity spherical model for both RVs and LVs. A septal contribution to the ventricular contraction was considered by adding a mass of a half septum to the RV and LV models. We acknowledge that such an assumption is a simplification of actual septal mechanics in the intact heart. In the future, 3D modeling setup with the biventricular geometry could be considered to account for septal mechanics for overall biventricular output. It will be important to compare the results of the present study with 3D biventicular simulations. However, 3D simulations are computationally costly, which limits their direct implementation in clinical settings. The simplification of the geometry allows to significantly reduce the computation time while preserving multiscale representation of myocardial contraction and essential valvular mechanics. Thus, such a model has a potential to be included in a clinical workflow.

The current model predictions are valid for a given geometry of the heart – i.e. to evaluate the readiness of the LV for DSO at a given time point. However, in the case no intervention is performed during next month, tissue- and organ-level remodeling is likely to take place. New clinical data should be collected and coupled to the model to re-assess the altered mechanical state of the heart. In the future, a model of long-term myocardial remodeling (e.g. a kinematic growth theory [12]) could be combined with the current instantaneous model to provide an augmented mechanical description of myocardial remodeling.

5 Conclusion

Personalized biomechanical modeling is a promising tool of the *in silico* medicine to predict the functional state of the ventricle(s) in various hemodynamic conditions. The predictions of such models have the potential to augment our understanding of pathophysiology of patients and contribute to the optimal clinical management.

Acknowledgments. The authors would like to acknowledge Dr. Philippe Moireau and Dr. Dominique Chapelle, Inria research team MΞDISIM (France), for the development of the cardiac simulation software CardiacLab used in this work. Research reported in this publication was supported by Children's Health^{SM} but the content is solely the responsibility of the authors and does not necessarily represent the official views of Children's Health^{SM}.

References

1. Caruel, M., Chabiniok, R., Moireau, P., Lecarpentier, Y., Chapelle, D.: Dimensional reductions of a cardiac model for effective validation and calibration. Biomech. Model. Mechanobiol. **13**(4), 897–914 (2014)
2. Chabiniok, R., Moireau, P., Kiesewetter, C., Hussain, T., Razavi, R., Chapelle, D.: Assessment of atrioventricular valve regurgitation using biomechanical cardiac modeling. In: Pop, M., Wright, G.A. (eds.) FIMH 2017. LNCS, vol. 10263, pp. 401–411. Springer, Cham (2017). https://doi.org/10.1007/978-3-319-59448-4_38
3. Chapelle, D., Le Tallec, P., Moireau, P., Sorine, M.: Energy-preserving muscle tissue model: formulation and compatible discretizations. Int. J. Multiscale Comput. Eng. **10**(2), 189–211 (2012)
4. Genet, M., Stoeck, C.T., Von Deuster, C., Lee, L.C., Kozerke, S.: Equilibrated warping: finite element image registration with finite strain equilibrium gap regularization. Med. Image Anal. **50**, 1–22 (2018)
5. Gusseva, M., et al.: Time-synchronization of interventional cardiovascular magnetic resonance data using a biomechanical model for pressure-volume loop analysis. J. Magn. Reson. Imaging **57**(1), 320–323 (2023)
6. Gusseva, M., et al.: Biomechanical modeling to inform pulmonary valve replacement in tetralogy of Fallot patients after complete repair. Can. J. Cardiol. **37**(11), 1798–1807 (2021)
7. Gusseva, M., Hussain, T., Hancock Friesen, C., Greil, G., Chapelle, D., Chabiniok, R.: Prediction of ventricular mechanics after pulmonary valve replacement in tetralogy of Fallot by biomechanical modeling: a step towards precision healthcare. Ann. Biomed. Eng. **49**(12), 3339–3348 (2021)
8. Huxley, A.: Muscle structure and theories of contraction. Prog. Biophys. Biophys. Chem **7**, 255–318 (1957). https://ci.nii.ac.jp/naid/10005175870/en/
9. Klotz, S.: Single-beat estimation of end-diastolic pressure-volume relationship: a novel method with potential for noninvasive application. Am. J. Physiol. Heart Circulatory Physiol. **291**(1), H403–H412 (2006)
10. Le Gall, A., et al.: Monitoring of cardiovascular physiology augmented by a patient-specific biomechanical model during general anesthesia. A proof of concept study. PLoS ONE **15**(5), e0232830 (2020)
11. Marathe, S.P., et al.: Contemporary outcomes of the double switch operation for congenitally corrected transposition of the great arteries. J. Thorac. Cardiovasc. Surg. **164**(6), 1980–1990 (2022)
12. Rodriguez, E.K., Hoger, A., McCulloch, A.D.: Stress-dependent finite growth in soft elastic tissues. J. Biomech. **27**(4), 455–467 (1994)
13. Ruijsink, B., et al.: Dobutamine stress testing in patients with Fontan circulation augmented by biomechanical modeling. PLoS ONE **15**(2), e0229015 (2020)
14. Sainte-Marie, J., Chapelle, D., Cimrman, R., Sorine, M.: Modeling and estimation of the cardiac electromechanical activity. Comput. Struct. **84**(28), 1743–1759 (2006)

Automated Estimation of Left Ventricular Diastolic Chamber Stiffness: Application to Patients with Heart Failure and Aortic Regurgitation

Abdallah I. Hasaballa[1](✉), Debbie Zhao[1], Vicky Y. Wang[1], Thiranja P. Babarenda Gamage[1], Stephen A. Creamer[1], Gina M. Quill[1], Peter N. Ruygrok[2,3], Satpal S. Arri[2], Robert N. Doughty[2,3], Malcolm E. Legget[3], Alistair A. Young[4,5], and Martyn P. Nash[1,6]

[1] Auckland Bioengineering Institute, University of Auckland, Auckland, New Zealand
abdallah.hasaballa@auckland.ac.nz
[2] Green Lane Cardiovascular Service, Auckland City Hospital, Auckland, New Zealand
[3] Department of Medicine, University of Auckland, Auckland, New Zealand
[4] Department of Biomedical Engineering, King's College, London, London, UK
[5] Department of Anatomy and Medical Imaging, University of Auckland, Auckland, New Zealand
[6] Department of Engineering Science, University of Auckland, Auckland, New Zealand

Abstract. Diastolic dysfunction of the heart is present in most forms of cardiac failure. Left ventricular (LV) diastolic chamber stiffness has been proposed as a metric for obtaining insights into the progression of this disease and help to inform treatment decisions. However, the challenges in robustly estimating chamber stiffness have limited the evaluation of its prognostic value. This study aimed to develop an automated workflow that enables routine estimation of chamber stiffness from haemodynamic measurements and real-time 3D echocardiographic data to enable such investigations. The workflow was demonstrated on a cohort of 20 patients with heart failure (HF), 7 patients with aortic regurgitation (AR) without HF, and 12 control subjects. A mixed-effects linear regression model was used to examine the differences in diastolic chamber stiffness among the patient groups taking into account the beat-to-beat variations in chamber stiffness estimates. The variances of the standard deviation in chamber stiffness estimates for each patient groups were also evaluated to investigate the influence of beat-to-beat variations in LV pressure on diastolic chamber stiffness estimates. Overall, chamber stiffness was found to be significantly higher int the heart failure with preserved ejection fraction (HFpEF) group (2.4 ± 0.9 mmHg/mL, $p = 0.02$) and the heart failure with reduced ejection fraction (HFrEF) group (2.1 ± 1.7 mmHg/mL, p = 0.017) compared to the control group (1.1 ± 0.5 mmHg/mL). The lowest estimates were observed in the AR without heart failure group (1 ± 0.4 mmHg/mL, $p = 0.84$). HFrEF patient group exhibited the largest variance of the standard deviation in chamber stiffness estimates, followed by the HFpEF group, suggesting the beat-to-beat variations in LV pressures had a substantial effect in these groups. Future work will seek to apply this novel automated methodology to support estimation

O. Bernard et al. (Eds.): FIMH 2023, LNCS 13958, pp. 626–635, 2023.
https://doi.org/10.1007/978-3-031-35302-4_64

of chamber stiffness in a robust and reproducible manner in larger clinical studies to further elucidate its benefits for patient diagnosis and management.

Keywords: Diastolic chamber stiffness · Heart failure · Aortic regurgitation · Cardiac catheterisation · 3D echocardiography

1 Introduction

Quantitative characterisation of diastolic function is important in clinical research and medicine [1]. Left ventricular (LV) diastolic chamber stiffness has been proposed as a metric for obtaining insights into the progression of this disease and help to inform treatment decisions. However, despite decades of research, the role of LV diastolic chamber stiffness and its prognostic value in cardiac patients remains unclear and controversial [2]. Possible reasons include differing methods of measuring LV chamber stiffness with varying accuracy and reproducibility, and the multi-factorial nature of diastolic dysfunction. Several factors, including ventricular architecture and geometry, catheter calibration, beat-to-beat variability in LV pressures and volume estimates, and blood viscosity, may affect chamber stiffness estimates. Understanding the implications of these factors remains an area of active investigation [3]. This study aimed to: 1) develop an automated workflow for efficient estimation of patient-specific LV diastolic chamber stiffness from LV catheterisation recordings and real-time 3D echocardiography; 2) demonstrate the application of this workflow on a cohort of patients with HF and aortic regurgitation (AR) to determine whether there are differences in chamber stiffness between these patient groups; and 3) apply the workflow to assess the sensitivity of chamber stiffness estimates to beat-to-beat pressure variability.

2 Methods

2.1 Patient Population

We studied patients undergoing interventional haemodynamic investigations and real-time 3D echocardiography (less than one hour between assessments) at Auckland City Hospital. All participants provided written informed consent, and the study protocols were approved by the Health and Disability Ethics Committee of New Zealand (17/NTB/46). Participants were grouped into control, HFpEF, HFrEF, and AR without HF. Diagnostic criteria for the HF groups were based on the presence of symptoms and/or signs of HF. HF patients with a left ventricular ejection fraction ≥50% were defined as HFpEF, and those with an ejection fraction <50% were defined as HFrEF. The AR group was selected based on patients with AR (diagnosed with Doppler echocardiography), without symptoms associated with HF. The control group included patients referred for cardiac catheterisation due to chest pain or shortness of breath, but were found to have no coronary artery disease stenosis <25%, and for whom transthoracic echocardiography indicated normal cardiac structure and function.

2.2 3D Echocardiogram Acquisition and Processing

Transthoracic real-time 3D echocardiograms (echo) were acquired using a Siemens SC2000 ultrasound system with a 4Z1c matrix array transducer (Siemens Medical Solutions, Issaquah, WA, USA) from the apical window in a left lateral decubitus position. LV geometry throughout each cardiac cycle was segmented using a fully automated B-Spline Explicit Active Surfaces algorithm [4]. The resultant LV geometries were refined using our novel bias correction method based on dynamic time warping and partial least squares regression [5], to improve agreement with cardiac magnetic resonance imaging as the reference modality for cardiac chamber volume quantification. The imaging frame at ED was defined based on the R peak of the electrocardiogram (ECG), while end-systolic (ES) imaging time point was defined when the LV volume was minimal. To account for any differences in heart rate during the short period between catheterisation and imaging, we used the ECG PR/RR interval ratio (recorded during catheterisation) to identify the diastasis (DS) timepoint in the 3D volumetric imaging data.

2.3 LV Catheterisation and Haemodynamic Analysis

A fluid-filled pigtail catheter (Boston Scientific, Marlborough, MA) was inserted into the radial or femoral artery and guided using x-ray fluoroscopy to access the LV through the aorta. A contrast agent was injected through the catheter for ventriculography and to measure regional myocardial blood flow as part of the clinical procedure. Four continuous snapshots of LV pressure recordings during free-breathing (~9 heartbeats per snapshot) were recorded and simultaneous ECGs were sampled at 240 Hz using the Mac-Lab Haemodynamic Recording System (GE Healthcare, Chicago, IL); see Fig. 1a.

For diastolic chamber stiffness analysis, two cardiac time points (DS and ED) were identified on the LV pressure waveforms. The first step in the analysis was to identify individual cardiac cycles by detecting R peaks in the ECG signal. Then the ECG trace was divided into time segments (representing separate cardiac cycles), with the beginning of each segment defined to be 450 ms prior to the R peak, and the end of each segment defined to be 500 ms following the R peak. The choice of the time segment was made to ensure the inclusion of both P and R waves of each heartbeat. The R peaks were detected as local maxima in the QRS complexes, which were identified using a discrete wavelet transform-based noise removal and feature extraction method [6], also used to identify peaks of the P-waves for further analysis below.

An automated method adapted from techniques by Mynard et al. [7] was used to identify ED using the maximum curvature identified in the LV pressure traces. It should be noted that the R peak is widely used to identify the ED imaging frame in 3D echocardiographic data as it is a well-known method for determining the timing of cardiac events. However, its use may not always result in an ideal identification of ED pressure in pressure waveforms, particularly in patients with HF, due to the variability of the delay between electrical and mechanical events in the heart [8]. Hence, the peak-curvature method offers a more dependable means of detecting ED in LV pressure traces [7]. Next, the DS time point on the LV pressure trace was determined by subtracting the ECG PR interval from the ED time point (see Fig. 1b). For patients in atrial fibrillation, where the P-wave was absent, the DS time point was defined using a fixed PR interval of

120 ms, based on the average PR interval in a healthy population [8]. It should be noted that the ED state on volume and pressure traces are identified differently; in the volume traces, the ED frame is based on the R peak of the ECG (as described in Sect. 2.2), while in the pressure traces, ED is identified using the maximum curvature method.

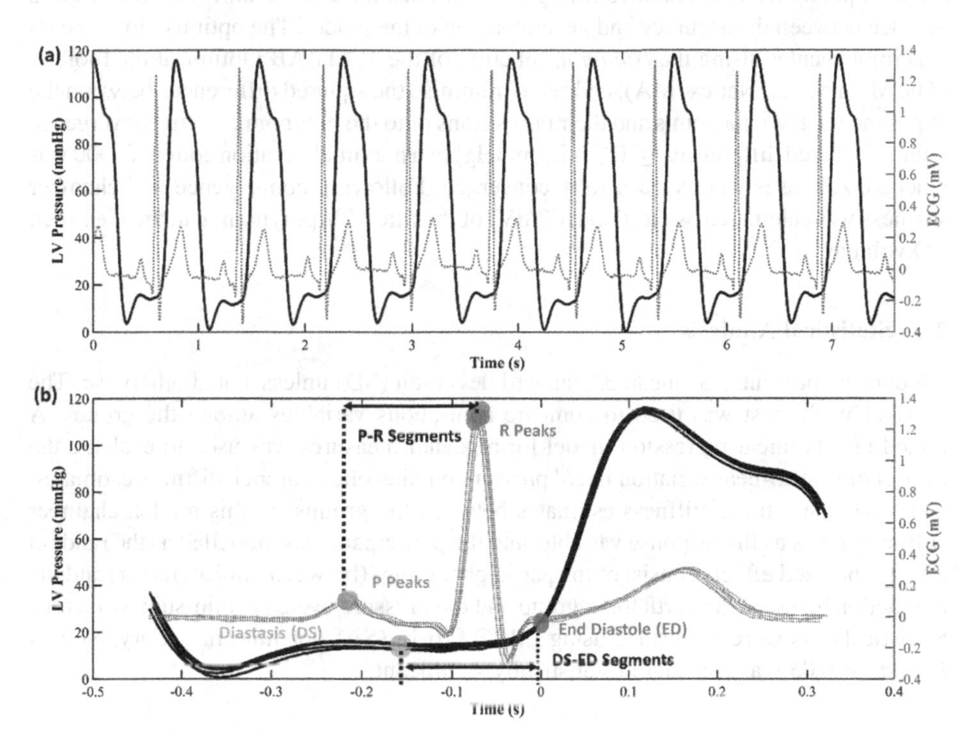

Fig. 1. (**a**) Representative continuous left ventricular (LV) pressure traces (black) with concurrent electrocardiogram (ECG; red) recordings. (**b**) Individual cycle LV pressure traces were aligned at their end-diastolic (ED) time points. For each LV pressure trace, the interval between diastasis (DS; pink circles) and ED (blue circles) was defined to be equal to the PR interval derived from the ECG.

2.4 Diastolic Pressure-Volume Curve Generation

LV pressure measurements and imaging data were not recorded simultaneously. To account for any difference in heart rate, we temporally registered the LV volume curves (derived from 3D echo) with the LV pressure traces for individual cycles. LV pressure was sampled at a much greater rate than that of the 3D echo (240 samples per second or ~189 points per cycle, versus 30 image frames per cycle, respectively). Temporal registration was achieved by linearly scaling the DS-ED volume traces to match the duration of the pressure traces between DS and ED. Around 35 to 45 LV pressure samples were typically acquired during late diastolic filling (DS to ED), so for consistency, we interpolated the pressure data to 40 regular time points between DS and ED.

2.5 Chamber Stiffness Quantification

To estimate chamber stiffness, a 7th-order polynomial was fitted to the diastolic PV data (Fig. 2c). The selection of a 7th-order polynomial was based on the characteristics of the data points from the passive filling PV relationship, and the aim was to achieve a balance between the accuracy and generalisation of the model. The optimisation process was implemented using the *lsqnonlin* function of the MATLAB Optimisation Toolbox (The MathWorks, Natick, MA), and set to minimise the squared differences between the experimental PV data points and their projections onto the polynomial until convergence (sum of squared differences $\leq 10^{-5}$ mL · mmHg) or a maximal iteration count of 1000 was reached. All cases were successfully converged. Following convergence, LV chamber stiffness was calculated as the slope (dP/dV) of the fitted PV polynomial at the identified ED volume.

2.6 Statistical Analysis

All data are presented as mean ± standard deviation (SD) unless stated otherwise. The Kruskal-Wallis test was used to compare continuous variables among the groups. A mixed-effects linear regression model for repeated measures was used to evaluate the impact of beat-to-beat variation in LV pressure on diastolic chamber stiffness estimates, and to compare these stiffness estimates between the groups. In this model, chamber stiffness serves as the response variable, and the participant was modelled as the random effect. The fixed effects consist of the participant group (between-subject factor) and the interaction between the participant group and the pressure cycle (within-subject factor). Statistical tests were performed using JMP® Pro 16 (SAS Institute Inc., Cary, NC). A P-value < 0.05 was considered statistically significant.

3 Results and Discussion

3.1 Patient Characteristics

A summary of patient characteristics is presented in Table 1. Overall, the median age was 61 years (interquartile range 54–68 years), and 62% of the subjects were male. High blood pressure was present in 16 patients (41%), and cardiomyopathy was present in 12 patients (31%). AR was graded trivial (n = 1), mild (n = 4), moderate (n = 1) and severe (n = 1).

3.2 A Case Study

Figure 2 shows a typical set of LV pressure traces (Fig. 2a) and the LV volume trace (Fig. 2b) for the same patient, along with the fitted PV polynomial curves (Fig. 2c) for late diastolic filling, and estimates of the diastolic chamber stiffness from each fitted PV polynomial curve (Fig. 2d). The proposed approach for identifying DS timepoints in the LV pressure traces (as described in Sect. 2.3) successfully positioned them on the plateaus of the LV pressure traces for all patients. The 7th-order polynomial curves provided an excellent representation of the experimental PV data in all cases (as seen

Table 1. Patient characteristics.

Characteristics	Control (n = 12)	HFpEF (n = 9)	HFrEF (n = 11)	AR without HF (n = 7)
Age in years (number of males)	60±11(6)	65±10 (5)	64±11 (8)	58±16 (5)
Obesity (BMI ≥ 30, kg/m^2)	2	5	7	2
Hypertension	4	3	5	4
Valvular disease	–	6	7	7
Cardiomyopathy	–	2	10	0
Atrial fibrillation	–	5	5	0
NT-proBNP (pg/mL)	8.5 (6–12)	42$^{*,\$}$ (11–133)	120$^{\#}$ (72–487)	48^{+} (21–102)
LVEF (%)	63±2	57±18$^{\$}$	47±2$^{\#}$	58±2^{+}
LVEDVI (mL/m^2)	63±15	68±12$^{\$}$	102±19$^{\#}$	86±15$^{+,¥}$
LV mass index (g/m^2)	60±14	60±13$^{\$}$	103±20$^{\#}$	91±15^{+}
GLS (%)	−27±3	−25±6$^{\$}$	−16±2$^{\#}$	−22±2^{+}
LVED pressure (mmHg)	14±5	15±4	19±8	15±2
Heart rate (beats/min)	72±13	68±10	78±14	70±9

* $p < 0.05$ control vs HFpEF, $^{\#}$ $p < 0.05$ control vs HFrEF, $^+$ $p < 0.05$ control vs AR with no HF, $^{\$}$ $p < 0.05$ HFpEF vs HFrEF, $^{¥}$ $p < 0.05$ HFpEF vs AR with no HF. Mean ± SD presented except for NT-proBNP, for which the median is presented, followed by the interquartile range in parentheses. BMI: body-mass index; LVEF: left ventricular ejection fraction; LVEDVI: left ventricular end-diastolic volume index; GLS: global longitudinal strain; LVED: left ventricular end-diastolic.

in Fig. 2c). In comparison, several functions have been proposed in the literature to represent the diastolic PV relationship, such as exponential, power law, and sigmoidal functions [9, 10]. While linear regression is a common approach due to its simplicity and robustness to data noise, it can often result in poor representations of the raw (generally non-linear) pressure-volume data, particularly at ED. An example of beat-to-beat variation of diastolic chamber stiffness estimates is shown as a violin plot in Fig. 2d. Across all patients, the median and interquartile range of per-patient SDs of diastolic chamber stiffness estimates were 0.26 mmHg/mL and (0.16–0.36) mmHg/mL, respectively, showing small inter-cycle variation.

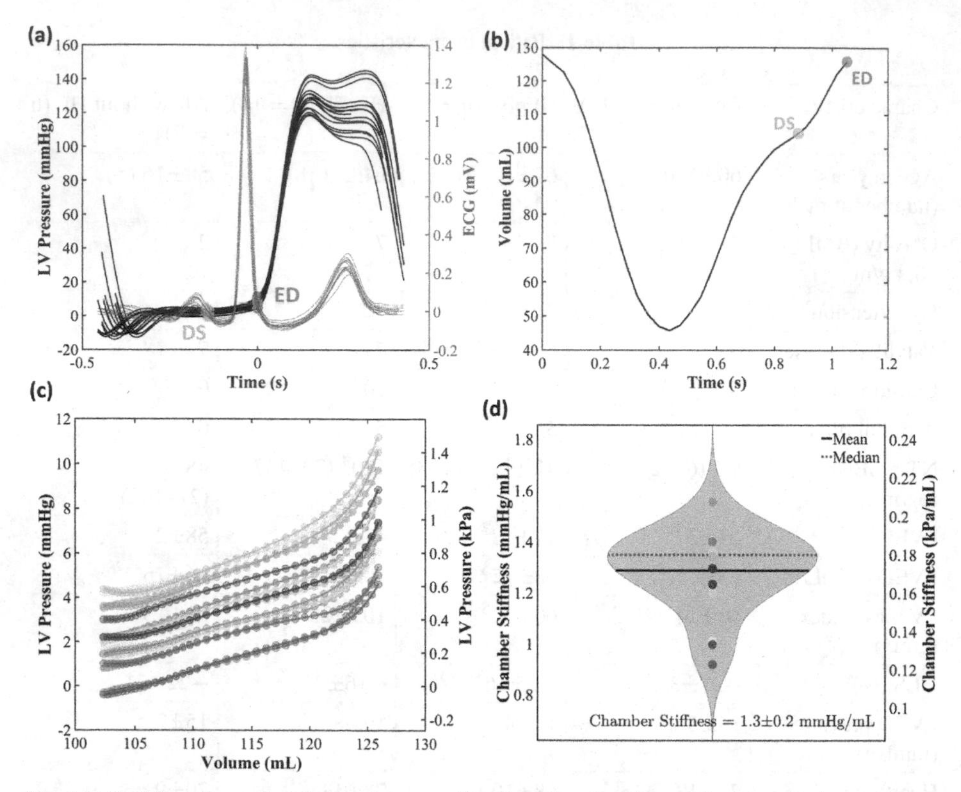

Fig. 2. Workflow for estimating patient-specific diastolic chamber stiffness applied to data from one patient. (**a**) Left ventricular (LV) pressure traces (black lines) and ECG traces (red lines). (**b**) LV volume trace derived from real-time 3D echocardiographic volumetric imaging data. (**c**) Diastolic pressure-volume raw data (symbols) and fitted curves (lines). (**d**) Violin plot of diastolic chamber stiffness estimates derived from each fitted pressure-volume curve.

3.3 Diastolic Chamber Stiffness Estimates

Figure 3 shows diastolic chamber stiffness estimates across all cycles for each patient. Results indicate small variability within most patients, except for two HFrEF cases. The AR without HF group had the least beat-to-beat variation, followed by the control group, and then the HFpEF group. Table 2 presents the mean and variance of SD in diastolic chamber stiffness values, and the *P*-values from the mixed-effects linear regression models. The mean values were determined by averaging chamber stiffness estimates within each group. The variance was calculated by finding the difference between individual stiffness estimates and the patient average, then determining the 25th and 75th percentiles. Group comparisons used *P*-values from mixed-effects models, adjusted for clustering within groups. Overall, estimates of diastolic chamber stiffness were nearly double in the HF patient groups compared to the control and AR without HF groups. Statistically significant differences were found among all groups, except for the two pair comparisons of (HFpEF vs HFrEF) and (control vs AR without HF). There was no relationship between severity of AR and chamber stiffness (mild or trivial vs moderate or severe, $p = \text{NS}$).

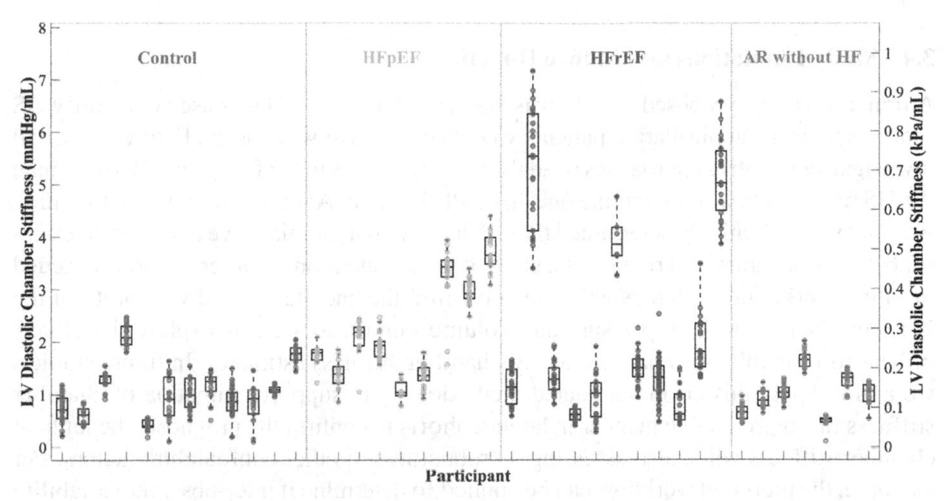

Fig. 3. Multi-cycle estimates of left ventricular (LV) diastolic chamber stiffness for each patient.

Table 2. Mean diastolic chamber stiffness estimates across all patients. The variance of the SD was calculated by subtracting the patient mean stiffness from individual estimates, then computing the 25th and 75th percentiles of these differences from the mean. *P*-values were obtained from mixed effects linear models, adjusted for clustering with groups.

		Control (n = 12, points = 240)	HFpEF (n = 9, points = 171)	HFrEF (n = 11, points = 238)	AR without HF (n = 7, points = 142)
Mean chamber stiffness ± SD (mmHg/mL)		1.1 ± 0.5	2.4 ± 0.9	2.1 ± 1.7	1 ± 0.4
Variance of SD (mmHg/mL)		−0.2 to 0.1	−0.2 to 0.2	−0.3 to 0.3	−0.1 to 0.1
P-values	Control	–	0.017	0.020	0.84
	HFpEF	–	–	0.83	0.023
	HFrEF	–	–	–	0.026

Some studies have found higher diastolic stiffness in AR compared to healthy individuals, but this can vary widely [11, 12]. However, some studies have reported decreased chamber stiffness but normal muscle stiffness in patients with AR when systolic performance is preserved [13, 14]. The variability in diastolic stiffness estimates among AR patients is likely due to multiple factors, such as the cause of AR, degree of regurgitation, duration of AR, and presence of comorbidities. Our study found no statistically significant differences between the AR without HF and control groups. It is important to note that the mean EF in the AR without HF group in this study was 58%, indicating normal systolic performance.

3.4 Study Limitations and Future Directions

A limitation of the proposed workflow is that a fixed PR interval was used to identify DS time points in atrial fibrillation patients where the P-wave was absent. Future work will investigate alternative approaches to analyse relationships in the ECG signals to estimate the DS time points without relying on a fixed PR interval. Another limitation of the study was that we qualitatively determined that a 7th-order polynomial gave a consistent trade-off between accuracy and robustness. Quantitative evaluation in larger cohorts is needed in future work. Future studies will also delve into the uncertainty and variability of the ED identification in both pressure and volume curves, as well as explore the effects of beat-to-beat volume variability on LV chamber stiffness estimates. In future studies, we aim to apply this novel automated methodology to support estimation of chamber stiffness in a reproducible manner in larger cohorts to confirm the prognostic benefits of chamber stiffness while also assessing its robustness to other confounding factors. For example, the proposed workflow can be applied to determine if inter-observer variability in LV geometric model creation significantly influences estimates of chamber stiffness. It would also be beneficial to investigate whether combining diastolic chamber stiffness with other cardiac biomarkers improves the prognostic value in different patient groups.

4 Conclusion

We developed an automated workflow that enables routine estimation of patient-specific diastolic chamber stiffness using LV catheterisation and real-time 3D echocardiography. The application of this workflow to a cohort of patients with HF and AR showed that the beat-to-beat variation in LV pressure did not have a significant effect on the estimated chamber stiffness when comparing different patient cohorts. However, within each subject, beat-to-beat pressure variability may still have a notable effect on LV chamber stiffness estimates. A mixed-effects linear regression model revealed that diastolic chamber stiffness was higher in the HFrEF and HFpEF groups compared to the AR without HF group and control subjects. This novel automated methodology will enable larger clinical studies to be conducted in a robust and reproducible manner to confirm the reliability of diastolic chamber stiffness for routine evaluation of patients.

Acknowledgements. We gratefully acknowledge the financial support from the Health Research Council of New Zealand (17/608). We also acknowledge the important roles of our research nurses Mariska Oakester Bals, Jane Hannah, Anna Taylor, and Gracie Hoskin for their invaluable assistance in participant recruitment and data collection.

References

1. Hultkvist, H., et al.: Evaluation of left ventricular diastolic function in patients operated for aortic stenosis. PLoS ONE **17**, e0263824 (2022)

2. Kamimura, D., Suzuki, T., Fox, E.R., Skelton, T.N., Winniford, M.D., Hall, M.E.: Increased left ventricular diastolic stiffness is associated with heart failure symptoms in aortic stenosis patients with preserved ejection fraction. J. Cardiac Fail. **23**, 581–588 (2017)
3. Hieda, M., et al.: One-year committed exercise training reverses abnormal left ventricular myocardial stiffness in patients with stage B heart failure with preserved ejection fraction. Circulation **144**, 934–946 (2021)
4. Pedrosa, J., et al.: Fast and fully automatic left ventricular segmentation and tracking in echocardiography using shape-based b-spline explicit active surfaces. IEEE Trans. Med. Imag. **36**, 2287–2296 (2017)
5. Zhao, D., et al.: Correcting bias in cardiac geometries derived from multimodal images using spatiotemporal mapping. Sci. Rep. **13**, 8118 (2023)
6. Lin, H.-Y., Liang, S.-Y., Ho, Y.-L., Lin, Y.-H., Ma, H.-P.: Discrete-wavelet-transform-based noise removal and feature extraction for ECG signals. Innov. Res. BioMed. Eng. **35**, 351–361 (2014)
7. Mynard, J.P., Penny, D.J., Smolich, J.J.: Accurate automatic detection of end-diastole from left ventricular pressure using peak curvature. IEEE Trans. Biomed. Eng. **55**, 2651–2657 (2008)
8. Aro, A.L., et al.: Prognostic significance of prolonged PR interval in the general population. Eur. Heart J. **35**, 123–129 (2014)
9. Hoit, B.D.: Left ventricular diastolic function. Crit. Care Med. **35**, S340–S347 (2007)
10. Freytag, B., et al.: In vivo pressure-volume loops and chamber stiffness estimation using real-time 3D echocardiography and left ventricular catheterization–application to post-heart transplant patients. In: International Conference on Functional Imaging and Modeling of the Heart, pp. 396–405. Springer
11. Anand, V., et al.: Predictive value of left ventricular diastolic chamber stiffness in patients with severe aortic stenosis undergoing aortic valve replacement. Eur. Heart J. Cardiovasc. Imag. **21**, 1160–1168 (2020)
12. Reil, J.-C., et al.: Reduced left ventricular contractility, increased diastolic operant stiffness and high energetic expenditure in patients with severe aortic regurgitation without indication for surgery. Interact. Cardiovasc. Thorac. Surg. **32**, 29–38 (2021)
13. Gaasch, W.H., Levine, H.J.: Chronic Aortic Regurgitation. Springer Science & Business Media (2012)
14. Hess, O., Ritter, M., Schneider, J., Grimm, J., Turina, M., Krayenbuehl, H.: Diastolic stiffness and myocardial structure in aortic valve disease before and after valve replacement. Circulation **69**, 855–865 (1984)

A Computational Pipeline for Patient-Specific Prediction of the Post-operative Mitral Valve Functional State

Hao Liu[1], Natalie T. Simonian[1], Alison M. Pouch[2], Joseph H. Gorman, III[3], Robert C. Gorman[3], and Michael S. Sacks[1(✉)]

[1] James T. Willerson Center for Cardiovascular Modeling and Simulation, The Oden Institute for Computational Engineering and Sciences and the Department of Biomedical Engineering, The University of Texas at Austin, Austin, TX 78712, USA
msacks@oden.utexas.edu

[2] Departments of Radiology and Bioengineering, University of Pennsylvania, Philadelphia, PA 19104, USA

[3] Gorman Cardiovascular Research Group, Department of Surgery, Smilow Center for Translational Research, Perelman School of Medicine, University of Pennsylvania, Philadelphia, PA 19104, USA

Abstract. Mitral valve (MV) repair is safer than replacement for mitral regurgitation (MR) treatment, but long-term outcomes remain suboptimal and poorly understood. Moreover, preoperative optimization is complicated due to the heterogeneity of MR presentations and potential repair configurations. We thus developed a patient-specific MV computational pipeline to quantitatively predict the post-repair MV functional state using standard-of-care preoperative imaging data alone. First, we built a finite-element model of the full patient-specific MV apparatus by quantifying the MV chordae tendinae (MVCT) distributions from 5 CT-imaged excised human hearts and incorporating this data with patient-specific MV leaflet geometries and and MVCT origin displacements from preoperative 3D echocardiography. We then calibrated the leaflet and MVCT pre-strains by simulating preoperative MV closure in order to tune the functionally equivalent, patient-specific mechanical behavior. With this fully calibrated MV model, we simulated undersized ring annuloplasty (URA) by modifying the annular displacement to match the applied ring size. In all patient cases, the postoperative geometries were predicted to within 1 mm of the target, and the MV leaflet strain fields demonstrated very good global and local correspondence to results from a previous heavily validated pipeline. Additionally, our model predicted increased postoperative posterior leaflet tethering in a recurrent patient, which is the likely driver of long-term MV repair failure. This pipeline allows us to predict postoperative outcomes using strictly preoperative clinical data, which lays the foundation for quantitative surgical planning, personalized patient selection, and ultimately, more durable MV repairs.

O. Bernard et al. (Eds.): FIMH 2023, LNCS 13958, pp. 636–647, 2023.
https://doi.org/10.1007/978-3-031-35302-4_65

Keywords: Mitral valve disease · valve repair · cardiac simulation · surgical planning

1 Introduction

Mitral regurgitation (MR) is a highly prevalent cardiac disease affecting approximately 2% of the United States population. MR of any type is a major prognostic factor of mortality, with even mild MR increasing the risk of mortality by 17% in patients with ischemia but no MR [21]. Furthermore, MR remains severely undertreated despite its deadly outcomes, even in patients with low comorbidity, suggesting a critical need for treatment planning optimization [12].

Currently, MV repair is the preferred option for treating mitral regurgitation (MR) as it preserves the native MV structure and function and has better survival rates compared to replacement [1,17,19]. Undersized ring annuloplasty (URA) is a surgical repair technique that aims to improve leaflet approximation by reducing the annular orifice area and restoring sufficient coaptation [28]. However, up to 30% of patients may still experience recurrent MR six months after surgery [17], and long-term outcomes of MV repair remain unpredictable, suboptimal, and not well understood. Therefore, there is an urgent need for a more comprehensive and quantitative understanding of the functional consequences of MV repair in order to develop more effective and durable treatments for MR.

The earliest finite element (FE) models of the mitral valve (MV) were developed in the 1990s by Kunzelman et al., who simulated closure of excised porcine valves and incorporated various tissue properties such as fiber orientation, anisotropic mechanical properties, and variable tissue thickness. Annular and papillary muscle displacements were included as Dirichlet boundary conditions, and the simulations were validated against measurements of leaflet coaptation and stress distributions [15]. Subsequent studies by Salgo et al. investigated the influence of annular shape on MV kinematics, demonstrating that a saddle-shaped annulus reduced peak stress better compared to a flat configuration [26].

With the advancement of imaging techniques such as computed tomography (CT) and magnetic resonance imaging (MRI), in-vivo structural models of the MV were developed using boundary conditions extracted directly from patient imaging data [31]. More recently, more complex MV models were created using CT imaging, where the MV chordae tendinae (MVCT) insertions, branching patterns, and origins on the papillary muscles were segmented and incorporated into finite element models [14,22,32]. Real-time 3D echocardiography (rt-3DE) has become the clinical gold standard for MV imaging due to its high temporal resolution and contrast. In 2006, a pilot study demonstrated the feasibility of reconstructing patient-specific FE meshes of MV leaflet geometry from clinical echocardiographic datasets [30]. More recently, patient-specific rt-3DE imaging has been used in computational modeling of the MV to assess MV geometric parameters under various MR repair conditions [3,10,16,18].

Advances in rt-3DE imaging have enabled the acquisition of high-resolution MV images over the entire cardiac cycle using standard-of-care imaging techniques. In this work, we have developed a patient-specific MV simulation pipeline *based strictly on preoperative, standard-of-care imaging* to predict post-repair MV functional behavior. The ability to preoperatively *predict* MV repair outcomes on a patient-specific basis will enable the quantitative assessment of repair strategies, tailored treatment plans, and ultimately improved patient outcomes.

2 Methods

2.1 Patient-Specific Full MV Apparatus

Population-Averaged MVCT Origins and Insertions. As the MVCT are of low mass and narrow profile, it is not presently possible to reliably visualize the MVCT structure on rt-3DE imaging. Therefore, we first analyzed 5 CT-imaged ex vivo human hearts from the Visible Heart Lab at the University of Minnesota to establish a population-averaged basis for the MVCT (Fig. 1A). First, we generated representative anatomical MVCT origins by quantifying their positions in the ex vivo human hearts (Fig. 1B). Next, we defined an MVCT insertion density map to guide the distribution of the MVCT insertions on given patient-specific MV leaflet geometries (Fig. 1E).

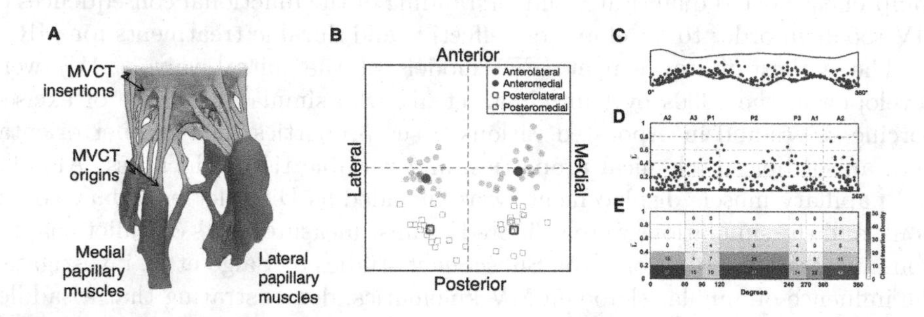

Fig. 1. The MVCT origins and MVCT insertion density map from the ex vivo human hearts. (A) The subannular structure of a representative CT-imaged ex vivo cardiac reconstruction. MVCT are colored in cyan, and papillary muscles are highlighted in red. Black arrows indicate a subset of MVCT insertions and origins. (B) The MVCT origins from all 5 hearts are shown in grey and are grouped based on the location of their respective insertions (anterolateral, anteromedial, posterolateral, posteromedial). The averages of these groups define the four representative origins. (C) All insertions from the 5 ex vivo human hearts plotted on the averaged 2D MV geometry. Note the higher density of insertions near the free edge, and the relatively sparse clear zone closer to the annulus. (D) The same data plotted on a normalized domain, where $\bar{L}$ represents the distance of a given insertion point from the free edge normalized by the distance of the free edge to the annulus. (E) The normalized MVCT insertion map, which can then be used to generate an anatomical distribution of MVCT insertions on a given patient-specific ED leaflet geometry. (Color figure online)

Patient-Specific MV Leaflets, Identifiable MVCT Origins, and Boundary Conditions. Full-volume intraoperative rt-3DE scans were acquired as standard of care during URA MV repair in three patients. The pipeline for segmenting and meshing the end-diastolic (ED) and end-systolic (ES) MV geometries from clinical rt-3DE images has been extensively detailed in [23]. Ultimately, this procedure results in meshes with uniform, unstructured nodal distributions, nodes spaced approximately 1 mm apart, and approximately 2000 elements per MV mesh, which are defined as 3-node triangular shell elements (S3) in Abaqus/Explicit (Dassault Systèmes).

A critical limitation of rt-3DE is that the full coaptation zone cannot be directly visualized, and the two leaflets cannot be distinguished in this region, which precludes the accurate computation of leaflet strain and geometry mismatch [29]. A previously described and extensively validated shape-morphing technique was used to morph the ED leaflet mesh to match the ES leaflet shape, which recovers the full coaptation zone and ensures material point correspondence between the ED and ES state meshes [23,29]. This fully reconstructed ES geometry was used as the target geometry in the calibration simulations (preoperative) and for validation (postoperative).

Though it is impossible to identify all MVCT origins from rt-3DE imaging, we segmented all possible patient-specific MVCT origins to determine their ED to ES displacements. These displacements and the ED to ES annular displacements computed from the MV leaflet geometries were defined as Dirichlet boundary conditions. Next, we used the MVCT insertion density plot defined from the *ex vivo* human cardiac data to map MVCT insertion points on the patient-specific ED leaflet geometry and connected each insertion to one of the four representative MVCT origins from Sect. 2.1 based on their location on the MV leaflet. After defining its connectivity, each individual MVCT was discretized using 1.5 mm long two-node truss elements (T2D3) in Abaqus (Dassault Systèmes). The integrated patient-specific ED MV leaflet geometry, MVCT structure, MVCT origin displacements, and annular displacement constitutes the final full patient-specific MV apparatus (Fig. 2).

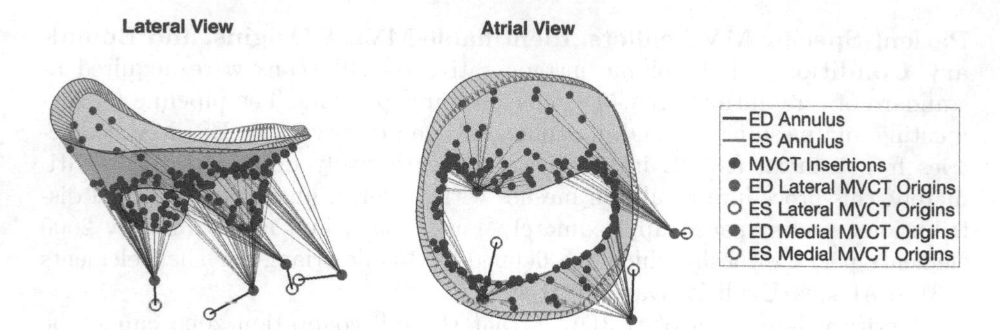

Fig. 2. The final patient-specific full MV apparatus, shown in the lateral and atrial views. The ED MV leaflet geometry is segmented from rt-3DE imaging; the MVCT insertions are distributed according to the insertion density map (shown in grey); and the four representative MVCT origins and their ED to ES displacements (guided by the rt-3DE traced MVCT origin displacements) are shown in blue (lateral) and red (medial). The black boundary shows the preoperative ED-state annulus, the red boundary denotes the preoperative ES-state annulus, and the arrows indicate the annular displacement boundary condition. (Color figure online)

2.2 Preoperative Calibration of MV Leaflet and MVCT Prestrain

Initial MV Leaflet and MVCT Material Models. Patient-specific material properties are impossible to obtain using standard-of-care rt-3DE imaging alone, so, rather than matching the MV leaflet and MVCT tissue properties exactly, we aimed to instead tune their prestrains by iteratively simulating MV closure and minimizing the error relative to the target, rt-3DE segmented ES geometry, which would allow us to arrive at functionally equivalent, calibrated material properties for each patient. The mechanical response of the MV leaflet tissue was modeled using a well-established isotropic incompressible hyperelastic structural constitutive model [8,23]. The prestrains in the circumferential and radial material directions were utilized to update the deformation matrix after each iteration of the closure simulation. The entire constitutive model as well as the prestrain update was implemented in Abaqus/Explicit using a VUMAT subroutine. The mechanical response of the MVCT was modeled using an Ogden model with parameters fitted from data on aged human MVCT, and the MVCT prestrain was handled in Abaqus as a thermal expansion [34].

Calibration Simulations. The full calibration process consists of three steps:

1. A transmural pressure of 100 mmHg was applied on the ventricular surface of the MV leaflets in the patient-specific full MV apparatus as a loading condition, and the initial material properties were defined as described above. MV closure was simulated under these conditions and the final ES geometry was used in the next step to calibrate the MVCT prestrains.
2. In the second step, we adjusted the MVCT prestrain to minimize the difference between the simulated and target geometries by correcting for the

length difference between the simulated and target MVCT. This process was repeated until the change in mean l^2 norm between the simulated and target ES geometries was less than 5%.

3. Due to the nonuniform distribution of MVCT insertions in the human MV, it is also necessary to adjust the MV leaflet prestrain to maximize the match between the simulated and target geometries. The material properties of the MV leaflets have been shown to be highly heterogeneous, with the anterior belly region found as anisotropic [9,25], and the coaptation region as nearly isotropic, likely due to the high density of MVCT insertions [5,20]. Therefore, we separated the MV leaflets into three zones and used fminsearch in MATLAB to find the values of the six prestrains that minimized the l^2 norm. This procedure was repeated until convergence. The anisotropy in the belly zones was validated in that the prestrain in the circumferential direction was much larger than the radial direction. Both prestrains in the coaptation region were around 1, which corresponds with prior findings that this zone is nearly isotropic. The final calibration results for all three patients are presented in (Fig. 3A).

2.3 Postoperative Prediction

Finally, using the fully calibrated material properties from the preoperative calibration step, we simulated URA repair by modifying the annular boundary condition such that the ES state annular dimensions corresponded with the ring type and size each patient received. We assumed that the transmural pressure was unchanged between the preoperative and postoperative states; that the MVCT origin displacements were the same immediately after repair; and that the MV leaflet and MVCT prestrains do not change from the preoperative state. The simulation result was validated by comparing both the geometry and MV leaflet strain fields with the target *postoperative* ES geometry, which was obtained in a similar manner to the target preoperative ES geometry via rt-3DE segmentation and then shape-morphing to recover the full oaptation zone.

3 Results

3.1 Postoperative Prediction Results

After incorporating the updated MV leaflet and MVCT prestrains and updating the annular boundary condition to match the annuloplasty ring dimensions, we observed that our predictive simulation was able to match the target post-ES geometry to within 1 mm in all three patients (Fig. 3B). We also analyzed the axial S11 stress in the MVCT and noted substantial differences in the chordal stress distributions pre- and postoperatively between a patient who had recurrent MR at six months and a patient who did not (Fig. 4A). Namely, though both patients had significantly elevated anterior MVCT stress prior to repair, in the nonrecurrent patient, this tension was alleviated after the ring was implanted,

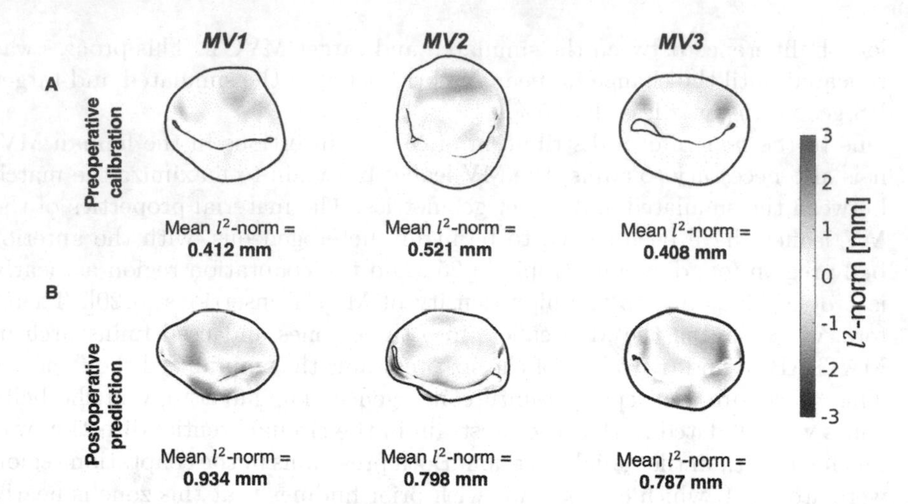

Fig. 3. The final results for the preoperative calibration and predictive postoperative simulations in all 3 patient MVs (MV1: recurrent; MV2: recurrent; MV3: non-recurrent). (A) The preoperative calibration results after tuning MVCT and MV leaflet prestrains. (B) The l^2 norm between the predicted and target geometries, demonstrating submillimeter agreement.

while in the recurrent patient, the posterior MVCT stress increased nearly three-fold (Fig. 4B). This result confirms findings from other groups [3] and indicates exaggerated posterior leaflet tethering due to excessive displacement of the posterior annulus towards the LV septum by the ring. The predictive postoperative simulation is also able to capture a substantial reduction in von Mises leaflet stress after URA repair in all three patients, a finding which has been echoed in previous literature [33]. In conjunction with the results of the MVCT stress, these findings suggest that normalization of MV leaflet stress alone may not be an adequate target for repair optimization, and that an integrated approach to the functional behavior of the full MV apparatus is crucial to predicting outcomes.

3.2 Validation

The sub-millimeter l^2 norms between the predicted and shape-morphed ES geometries confirm the technique's capability to reproduce the immediate post-operative geometry with high fidelity. We also found minimal difference in mean strains in both states circumferentially and radially when compared to the results of our existing shape-morphing pipeline [23], as well as close correspondence of local tissue deformation.

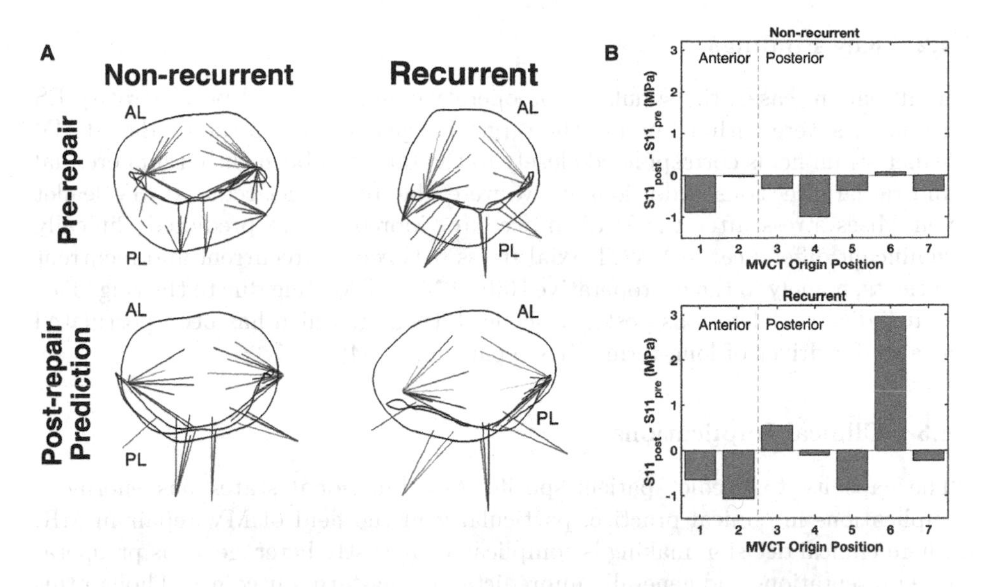

Fig. 4. MVCT axial S11 stress in 1 nonrecurrent and 1 recurrent patient MV before and after repair (A) Ventricular views with MVCT axial stress above 4 MPa highlighted in red and MVCT with stress less than 0.5 MPa not plotted. Both MVs demonstrate high MVCT stress in the anterior leaflet pre-repair. After repair, this stress is offloaded in the non-recurrent MV, but in the recurrent MV, this stress is transferred to the posterior leaflet, likely due to exacerbated posterior leaflet tethering imposed by the ring. (B) Change in MVCT stress, demonstrating an overall decrease in the non-recurrent patient, but a significant *increase* on the posterior side of the recurrent MV. (Color figure online)

4 Discussion

4.1 General Overview

The main objective of this study was to demonstrate that the patient-specific post-repair MV state can be predicted from strictly preoperative clinical data, predicated on a rigorous understanding of MV mechanics following decades of prior work [6,7,11,13,23]. This work is motivated by an urgent clinical need for quantitative treatment planning strategies, particularly in the face of rapidly proliferating MV repair approaches and devices, the highly combinatorial nature of MV repair, and suboptimal outcomes reported in major randomized controlled trials. Previous work has demonstrated that even the uniform stress applied by an annuloplasty ring contributes to biosynthetic changes in the MV leaflets, and that altered homeostasis and mechanical loading of the MV can induce plastic deformation of the leaflets, which are likely drivers of repair failure [2,11,24]. Consequently, the ability to *predict the functional state* of the complete MV apparatus enables a multidimensional and direct assessment of the proposed repair, as well as the possibility for preoperative optimization of the treatment strategy for each patient.

4.2 Key Findings

In all patient cases, the simulated preoperative and predicted postoperative ES geometries were within 1 mm of the target ES geometry, and the predicted MV leaflet strain fields corresponded closely to the target in both the circumferential and radial directions. Additionally, we were able to predict reduced MV leaflet von Mises stress after the URA procedure. Moreover, we predicted clinically significant differences in MVCT axial stress between nonrecurrent and recurrent patients, namely in the postoperative shift of MVCT loading due to the ring. This shift indicates deleterious posterior leaflet tethering, which has been postulated as a major driver of long-term URA repair failure [4].

4.3 Clinical Implications

The capacity to predict patient-specific MV functional states has enormous implications in clinical practice, particularly in the field of MV repair in MR, where clinical decision making is complicated by vastly heterogeneous preoperative presentations and generally unpredictable long-term outcomes. Though this challenge has spurred prolific inquiry into prognostic factors of repair success, these metrics largely reduce the complexities of MV function and behavior to rt-3DE-based geometric measurements and as such are at once overly restrictive and inadequate [27]. A functional approach, like the one presented in this study, which relies strictly on the same preoperative data, may be able to overcome these critical limitations and provide a more nuanced and insightful understanding into MV behavior in disease and after repair. For example, our observations regarding exacerbated posterior chordal stress and posterior leaflet tethering correlate strongly with previous findings explicating a possible mechanism underlying long-term post-repair MR recurrence [3]. Therefore, minimizing posterior chordal stress could potentially be used as a target in patient selection or to optimize the repair configurations (i.e. ring size, shape, etc.) and indeed already motivates adjunctive procedures such as papillary muscle relocation whose aim is to relieve the excess stress imposed on the MVCT by URA. Furthermore, though this pipeline was initially tested on surgical URA repair, the pipeline itself is entirely repair-agnostic, and we are currently applying it to patients who have undergone repair using more contemporary approaches, such as transcatheter edge-to-edge repair. Overall, this pipeline addresses an urgent clinical need for quantitative treatment planning strategies and thus lays the foundation for tailored patient selection and ultimately, a more durable repair.

Acknowledgements. This material is based upon work supported by the National Institutes of Health grants HL129077, HL119297 to MSS and RCG, and an American Heart Association pre-doctoral fellowship to NTS.

References

1. Acker, M.A., et al.: Mitral-valve repair versus replacement for severe ischemic mitral regurgitation. N. Engl. J. Med. **370**(1), 23–32 (2014)
2. Ayoub, S., et al.: Regulation of valve interstitial cell homeostasis by mechanical deformation: implications for heart valve disease and surgical repair. J. R. Soc. Interface **14**(135), 20170580 (2017)
3. Bouma, W., et al.: Preoperative three-dimensional valve analysis predicts recurrent ischemic mitral regurgitation after mitral annuloplasty. Ann. Thorac. Surg. 101(2), 567–575; discussion 575 (2016). https://doi.org/10.1016/j.athoracsur.2015.09.076, www.ncbi.nlm.nih.gov/pubmed/26688087
4. Bouma, W., et al.: Preoperative three-dimensional valve analysis predicts recurrent ischemic mitral regurgitation after mitral annuloplasty. Ann. Thorac. Surg. **101**(2), 567–575 (2016)
5. Drach, A., Khalighi, A.H., Sacks, M.S.: A comprehensive pipeline for multi-resolution modeling of the mitral valve: validation, computational efficiency, and predictive capability. Int. J. Numer. Methods Biomed. Eng. **34**(2), e2921 (2018)
6. Drach, A., Khalighi, A.H., Sacks, M.S.: A comprehensive pipeline for multi-resolution modeling of the mitral valve: validation, computational efficiency, and predictive capability. Int. J. Numer. Methods Biomed. Eng. **34**(2), e2921 (2018). https://doi.org/10.1002/cnm.2921
7. Eckert, C.E., Zubiate, B., Vergnat, M., Gorman, 3rd, J.H., Gorman, R.C., Sacks, M.S.: In vivo dynamic deformation of the mitral valve annulus. Ann Biomed Eng **37**(9), 1757–71 (2009). https://doi.org/10.1007/s10439-009-9749-3, http://www.ncbi.nlm.nih.gov/entrez/query.fcgi?cmd=Retrieve&db=PubMed&dopt=Citation&list_uids=19585241
8. Fan, R., Sacks, M.S.: Simulation of planar soft tissues using a structural constitutive model: finite element implementation and validation. J. Biomech. **47**(9), 2043–2054 (2014)
9. Grashow, J.S., Sacks, M.S., Liao, J., Yoganathan, A.P.: Planar biaxial creep and stress relaxation of the mitral valve anterior leaflet. Ann. Biomed. Eng. **34**(10), 1509–1518 (2006)
10. Hart, E.A., et al.: Transoesophageal echocardiography-based computational simulation of the mitral valve for mitraclip placement. Improving Treat. Plann. Card. Interv. 53 (2019)
11. Howsmon, D.P., et al.: Mitral valve leaflet response to ischaemic mitral regurgitation: from gene expression to tissue remodelling. J. R. Soc. Interface **17**(166), 20200098 (2020)
12. Iung, B., et al.: Contemporary presentation and management of valvular heart disease: the eurobservational research programme valvular heart disease ii survey. Circulation **140**(14), 1156–1169 (2019)
13. Khalighi, A.H., Rego, B.V., Drach, A., Gorman, R.C., Gorman, J.H., Sacks, M.S.: Development of a functionally equivalent model of the mitral valve chordae tendineae through topology optimization. Ann. Biomed. Eng. **47**(1), 60–74 (2019)
14. Kong, F., Caballero, A., McKay, R., Sun, W.: Finite element analysis of mitraclip procedure on a patient-specific model with functional mitral regurgitation. J. Biomech. **104**, 109730 (2020)
15. Kunzelman, K., Cochran, R., Chuong, C., Ring, W., Verrier, E.D., Eberhart, R.: Finite element analysis of the mitral valve. J. Heart Valve Dis. **2**(3), 326–340 (1993)

16. Mansi, T., et al.: An integrated framework for finite-element modeling of mitral valve biomechanics from medical images: application to MitralClip intervention planning. Med. Image Anal. **16**(7), 1330–46 (2012)
17. McGee, E.C., et al.: Recurrent mitral regurgitation after annuloplasty for functional ischemic mitral regurgitation. J. Thorac. Cardiovasc. Surg. **128**(6), 916–24 (Dec 2004). https://doi.org/10.1016/j.jtcvs.2004.07.037, http://www.jtcvsonline.org/article/S0022-5223(04)01143-2/pdf
18. Meijerink, F., et al.: Intraoperative post-annuloplasty three-dimensional valve analysis does not predict recurrent ischemic mitral regurgitation. J. Cardiothorac. Surg. **15**(1), 1–8 (2020)
19. Members, W.C., et al.: 2020 ACC/AHA guideline for the management of patients with valvular heart disease: a report of the American college of cardiology/American heart association joint committee on clinical practice guidelines. J. Am. Coll. Cardiol. **77**(4), e25–e197 (2021)
20. Padala, M., Sacks, M.S., Liou, S.W., Balachandran, K., He, Z., Yoganathan, A.P.: Mechanics of the mitral valve strut chordae insertion region (2010)
21. Perrault, L.P., et al.: Optimal surgical management of severe ischemic mitral regurgitation: to repair or to replace? J. Thorac. Cardiovasc. Surg. **143**(6), 1396–1403 (2012)
22. Pham, T.: Finite element analysis of patient-specific mitral valve with mitral regurgitation. Cardiovasc. Eng. Technol. **8**(1), 3–16 (2017)
23. Rego, B.V., et al.: A noninvasive method for the determination of in vivo mitral valve leaflet strains. Int. J. Numer. Methods Biomed. Eng. **34**(12), e3142 (2018)
24. Rego, B.V., et al.: Remodeling of the mitral valve: an integrated approach for predicting long-term outcomes in disease and repair. Ph.D. thesis (2019)
25. Sacks, M.S., Yoganathan, A.P.: Heart valve function: a biomechanical perspective. Philos. Trans. R. Soc. B: Biol. Sci. **362**(1484), 1369–1391 (2007)
26. Salgo, I.S., et al.: Effect of annular shape on leaflet curvature in reducing mitral leaflet stress. Circulation **106**(6), 711–717 (2002)
27. Selection of the optimal candidate to Mitraclip for secondary mitral regurgitation: beyond mitral valve morphology. Front. Cardiovasc. Med. **8**, 585415 (2021)
28. Schubert, S.A., Mehaffey, J.H., Charles, E.J., Kron, I.L.: Mitral valve repair: the French correction versus the American correction. Surg. Clin. **97**(4), 867–888 (2017)
29. Simonian, N.T., Liu, H., Pouch, A.M., Gorman, J.H., III., Gorman, R.C., Sacks, M.S.: Quantitative in vivo assessment of human mitral valve coaptation area after undersized ring annuloplasty repair for ischemic mitral regurgitation. JTCVS Tech. **16**, 49–59 (2022)
30. Verhey, J.F., Nathan, N.S., Rienhoff, O., Kikinis, R., Rakebrandt, F., D'Ambra, M.N.: Finite-element-method (FEM) model generation of time-resolved 3D echocardiographic geometry data for mitral-valve volumetry. Biomed. Eng. Online **5**(1), 1–9 (2006)
31. Votta, E., Caiani, E., Veronesi, F., Soncini, M., Montevecchi, F.M., Redaelli, A.: Mitral valve finite-element modelling from ultrasound data: a pilot study for a new approach to understand mitral function and clinical scenarios. Philos. Trans. A Math. Phys. Eng. Sci. **366**(1879), 3411–34 (2008). https://doi.org/10.1098/rsta.2008.0095, http://www.ncbi.nlm.nih.gov/pubmed/18603525
32. Wang, Q., Sun, W.: Finite element modeling of mitral valve dynamic deformation using patient-specific multi-slices computed tomography scans. Ann. Biomed. Eng. **41**(1), 142–153 (2013)

33. Wong, V.M., Wenk, J.F., Zhang, Z., Cheng, G., Acevedo-Bolton, G., Burger, M., Saloner, D.A., Wallace, A.W., Guccione, J.M., Ratcliffe, M.B., et al.: The effect of mitral annuloplasty shape in ischemic mitral regurgitation: a finite element simulation. Ann. Thorac. Surg. **93**(3), 776–782 (2012)
34. Zuo, K., Pham, T., Li, K., Martin, C., He, Z., Sun, W.: Characterization of biomechanical properties of aged human and ovine mitral valve chordae tendineae. J. Mech. Behav. Biomed. Mater. **62**, 607–618 (2016)

Automatic Aortic Valve Pathology Detection from 3-Chamber Cine MRI with Spatio-Temporal Attention Maps

Y. On[1](✉), K. Vimalesvaran[1], C. Galazis[1], S. Zaman[2], J. Howard[1], N. Linton[1], N. Peters[1], G. Cole[2], A.A. Bharath[1], and M. Varela[1]

[1] Imperial College London, Exhibition Road, London SW7 2AZ, UK
yu.on16@imperial.ac.uk
[2] Imperial College Healthcare NHS Trust, Du Cane Road, London W12 0HS, UK

Abstract. The assessment of aortic valve pathology using magnetic resonance imaging (MRI) typically relies on blood velocity estimates acquired using phase contrast (PC) MRI. However, abnormalities in blood flow through the aortic valve often manifest by the dephasing of blood signal in gated balanced steady-state free precession (bSSFP) scans (Cine MRI). We propose a 3D classification neural network (NN) to automatically identify aortic valve pathology (aortic regurgitation, aortic stenosis, mixed valve disease) from Cine MR images. We train and test our approach on a retrospective clinical dataset from three UK hospitals, using single-slice 3-chamber cine MRI from N = 576 patients. Our classification model accurately predicts the presence of aortic valve pathology (AVD) with an accuracy of 0.85 ± 0.03 and can also correctly discriminate the type of AVD pathology (accuracy: 0.75 ± 0.03). Gradient-weighted class activation mapping (Grad-CAM) confirms that the blood pool voxels close to the aortic root contribute the most to the classification. Our approach can be used to improve the diagnosis of AVD and optimise clinical CMR protocols for accurate and efficient AVD detection.

Keywords: Cardiac MRI · Aortic Valve · Convolutional Neural Networks · Cine MRI · Classification · Aortic Regurgitation · Aortic Stenosis · Blood Flow

1 Introduction

Aortic Valve Disease. Aortic valve disease (AVD) is one of the most common cardiovascular pathologies [1]. It affects 4.5 million people in Europe, with a prevalence of 0.18% [2,3]. The most common manifestations of AVD [4,5] are:

- aortic stenosis (AS) - a narrowing of the aortic valve, leading to a reduced aortic valve cross-sectional area and increased blood velocity during systole;
- aortic regurgitation (AR) - a leaking of the aortic valve, causing the blood to flow backwards into the left ventricle during diastole;

O. Bernard et al. (Eds.): FIMH 2023, LNCS 13958, pp. 648–657, 2023.
https://doi.org/10.1007/978-3-031-35302-4_66

– mixed valve disease (MVD) - simultaneous presentation of AR and AS.

The assessment of AVD using MRI typically relies on blood velocity estimates acquired using 2D cardiac gated phase contrast (PC) MRI [6], which can provide maps of blood velocity perpendicular to the aortic valve.

Cine MRI. Cine MRI is arguably the most commonly used cardiac magnetic resonance (CMR) sequence, as it provides valuable information about cardiac function, anatomy and blood flow. Cine MRI usually employs a 2D bSSFP (balanced steady state free precession) readout, with thick slices (8–10 mm thick), each acquired in one breathhold and is typically gated to image the cardiac cycle at 20 or more cardiac phases [7]. Single slices are commonly acquired in 4-chamber, 2-chamber and 3-chamber (also known as left ventricular outflow) orientations, as well as stacks of short axis slices for full ventricular coverage [8].

Blood is typically bright in bSSFP images given its comparatively high T_2/T_1 ratio and bSSFP's relative insensitiveness to flow due to the inherent first-order velocity compensation of most of its gradients. Turbulent flow or high blood speeds can however lead to the loss of phase coherence between consecutive radio-frequency pulses, creating dark regions in the blood pool [9]. These flow void regions can be exploited to help diagnose blood flow pathology [10], but to date this assessment has relied on subjective visual assessment or, more recently, machine learning tools reliant on expert manual annotations [5].

Aim. We propose a deep learning approach to automatically detect aortic valve pathology (AS, AR, and MVD) from 3-chamber Cine CMR, using multi-binary labels without the need for any expert annotation of imaging features (such as aortic valve insertion points).

2 Methods

Imaging Data. We train and evaluate our method on a retrospective clinical CMR dataset, obtained from different scanner manufacturers across three hospitals in the UK. We use $N = 576$ single-slice bSSFP 3-chamber Cine MR images. The resolution of the images is 1.17–1.56 × 1.17–1.56 × 8 mm^3 and the number of frames is 22 ± 10 (range: 12–32). Each image is labelled into four different classes using information from clinical records. In total, 387 patients (67%) have no aortic valve abnormalities ("no pathology"). Of the remaining 189, 83 (14%) have AR, 56 (10%) AS, and 50 (9%) MVD. We train a 3D convolutional neural network (CNN) for two separate classification tasks: 2-class (labels: "no pathology" or "AVD") and 4-class classification (labels: "no pathology", "AR", "AS" or "MVD").

Our pipeline consists of three main steps, as shown in Fig. 1: (i) adaptive heart extraction in all 3 – *chamber* Cine MRI frames, (ii) image classification into either 2 or 4 classes, and (iii) identification the voxels that contributed the most to the classification using Grad-CAM (Gradient-weighted Class Activation Mapping) heatmaps [11].

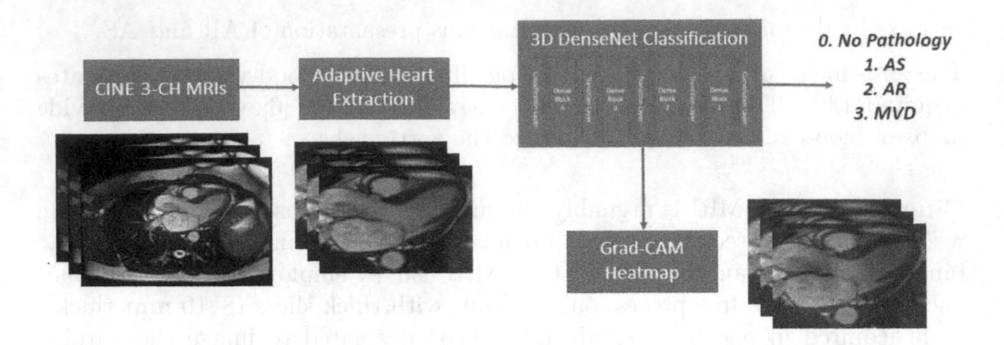

Fig. 1. Overview of our automatic method for aortic valve pathology detection from cine MRI. Each of the steps is described in the text in detail.

Adaptive Heart Extraction. We propose a heuristic algorithm (see Fig. 2) to automatically identify the cardiac structure in all Cine MRI frames, based on the assumption that the heart is the largest image structure that moves across the cardiac cycle. First, the algorithm calculates the absolute difference of the early systolic and late systolic cardiac phases. A Canny edge detector ($\sigma = 2.0$) [12] is then applied to identify the edges separating the different structures in the difference image, which undergo morphological dilation with a 2-pixel radius diamond structuring element. The largest connected component in the dilated edge image is identified as the heart and a tight bounding box is drawn around it. To standardise the images for preparation to the classification model, two post-processing steps are applied. The extracted region is re-sampled to a spatial resolution of 1×1 mm, the images are padded with zeros or cropped further for a consistent image size of $224 \times 224 \times D$ (D: original number of frames).

Classification Neural Network. A 3D DenseNet model [13] based on the prototype available in the MONAI library [14] is used to classify the 2D+time Cine MR images. As shown in Fig. 3, the model has 4 identical dense blocks separated by transitional layers. The dense blocks extract image features with a given size by concatenating the outputs of five successive convolutional operations. They are separated by transitional layers which combine and downsample the features outputted by each dense block before passing them to the next one. After the final dense block, an adaptive average pooling layer is implemented with a n-class softmax layer to perform the classification (see Fig. 3).

Prior to the classification, we perform z-score normalisation [15] and 'on-the-fly' data augmentation. We use data augmentation techniques such as rotation, contrast and bias field adjustment [16] to minimise overfitting [14]. Each augmentation is individually applied with a probability of 0.2. In order to cope with the variability in the number of frames, the input is area interpolated [17] to a depth of 30 to synchronise all the input dimensions to $224 \times 224 \times 30$.

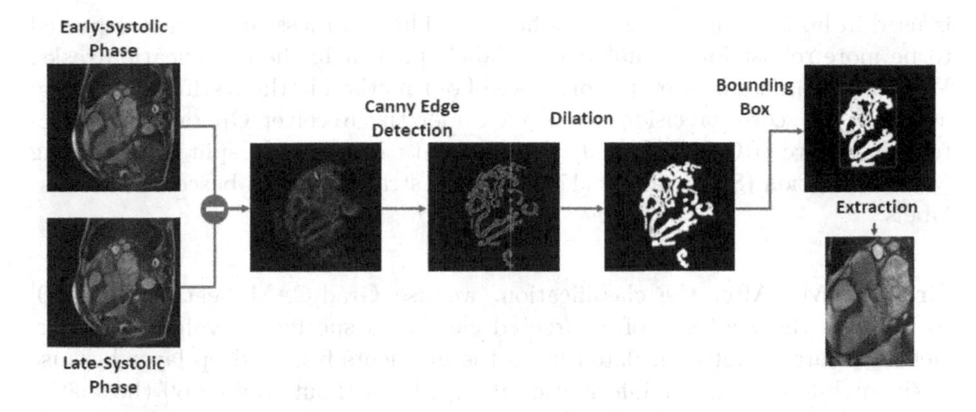

Fig. 2. The adaptive heart detection algorithm. The early-systolic phase is subtracted from the late-systolic phase, before undergoing Canny edge detection and morphological dilation. The largest connected component in this image is extracted and used to create a bounding box around the heart.

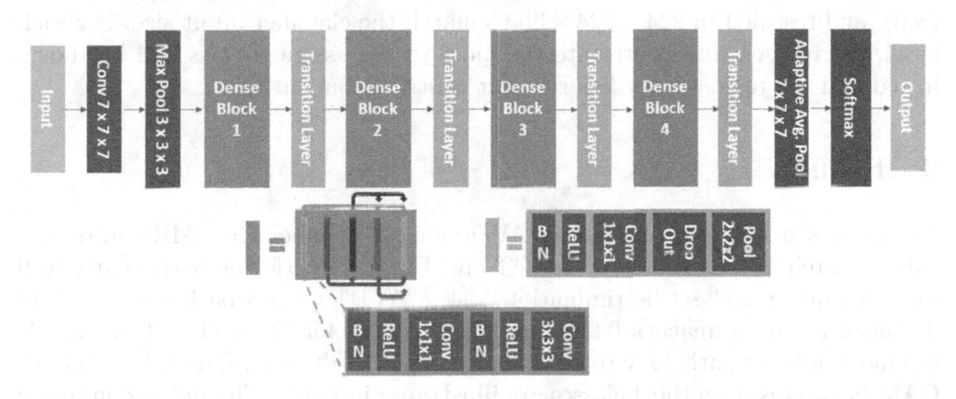

Fig. 3. A diagram of the 3D DenseNet architecture used. It consists of 4 identical dense blocks separated by transitional layers. Each dense block includes 5 consecutive operations, each including $1 \times 1 \times 1$ and $3 \times 3 \times 3$ convolutional layers ("Conv"). Within the dense block, the feature maps from each convolutional layer are concatenated to the output of the next layers. The transition layers downsample the feature maps produced by the dense blocks and include batch normalisation ("BN"), a $1 \times 1 \times 1$ convolutional layer, a $2 \times 2 \times 2$ average pooling layer and is placed between two consecutive dense blocks to facilitate the downsampling operation. Rectified linear units ("ReLU") are used throughout as activation layers. The feature vector is then flattened and classified into one of the classes with a softmax function.

Experimental Setting and Performance Criteria. We train the CNN on 1 GeForce RTX6000 graphical processing units (NVIDIA, Santa Clara, California) for approximately 11 h. We optimise our CNN using the Adam (learning rate: 0.0001) and train it for 500 epochs with a batch size of 2. Focal loss [18] function is used in both 2- and 4-class classification. The focal loss function is expected to be more robust for the imbalanced labels present in the classification tasks. We assess the classification performance of our method in the testing data using: accuracy, F_1 score, precision and Area Under the Receiver Operating Characteristics Curve (ROC AUC) [19]. The 576 data is randomly split into training (322), validation (81), and test (173) sets in a stratify fashion based on the class labels.

Grad-CAM. After the classification, we use Grad-CAM heatmaps [11,20] to visualise the gradients of a targeted class at a specific convolutional layer. Heatmaps are usually calculated using the gradients from a deep layer because of the high-level semantic information it encodes without trading off the spatial information. Also, because of the class discriminant nature of Grad-CAM, only the gradients from the targeted class are used while the gradients from other classes are set to zero. We use the last convolutional layer of the final transitional layer (see Fig. 3) to generate the heatmap. An open-source library [21] is utilised. The returned heatmap is normalised to a range between 0 (blue) and 1 (red), and rescaled to $224 \times 224 \times 30$ to match the classifier input size. For each label, the regions that contribute the most to the assigned class will be shown in red and the regions with less relevant contributions, in blue.

3 Results

Our proposed method can identify AVD using 3-chamber Cine MRI automatically and reliably, as shown in Table 1 and Fig. 4. It performs particularly well for the simpler 2-class discrimination task ("AVD" vs "no pathology") where its mean accuracy reaches 0.85, compared to 0.75 for the 4-class task. AR is the most difficult pathology to identify, as shown in Fig. 4d). Examples of Grad-CAM heatmaps from the 4 classes are illustrated in Fig. 5. The red region in the heatmaps shows the region contributed the most to the chosen label, which are mostly in the region of the aortic root and along the flow jet, as expected. In Fig. 5b and d, which corresponds to AS and MVD, the red region mostly highlights the high velocity blood flow and aortic valve; whereas in Fig. 5c, which correspond to AR, the red regions focus at the aortic root and the area that blood flows backward into the left ventricle. These regions are similar to the regions that clinicians use to visually inspect to assist diagnosis. This is further illustrated in Fig. 6. In the AS cases, GradCAM focuses on the aortic root region, especially during systole. In AR, on the other hand, the GradCAM tracks the backwards flow of blood into the ventricle in diastole.

Table 1. Evaluation metrics for the 2-class and 4-class classification. The values shown are mean ± standard deviation.

Metrics	2-Class	4-Class
Accuracy	0.85 ± 0.03	0.75 ± 0.03
F_1 Score	0.83 ± 0.03	0.57 ± 0.03
Precision	0.83 ± 0.03	0.57 ± 0.03

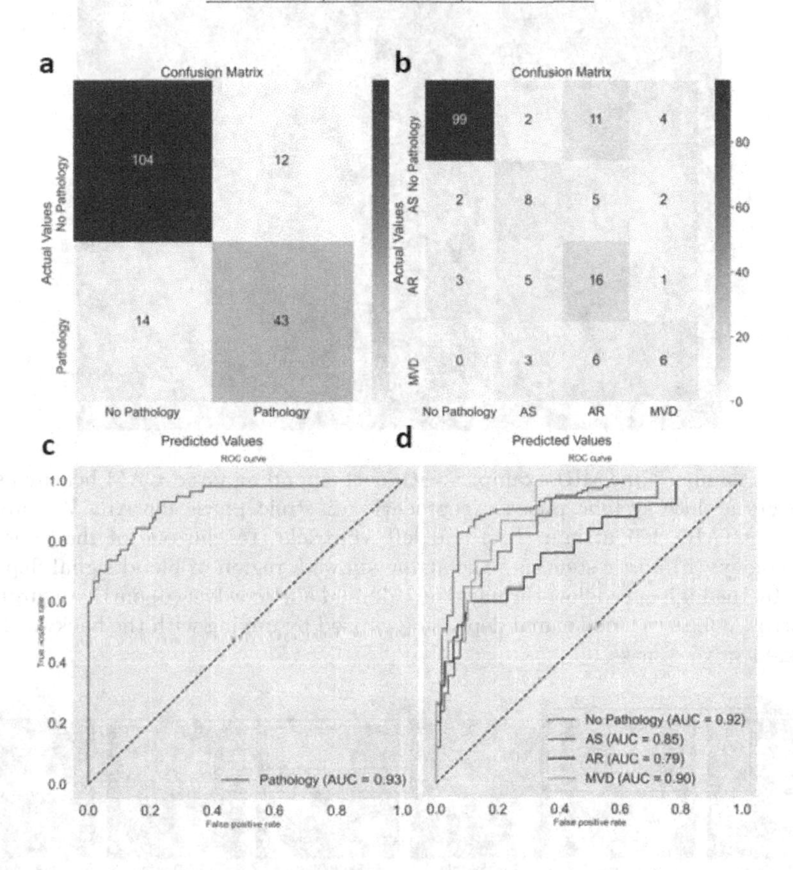

Fig. 4. The confusion matrix and ROC AUC of the classification results. (a, c) Confusion matrix and ROC AUC of the 2-Class classification (b, d) Confusion matrix and ROC AUC of the 4-Class classification. The x-axis of the confusion matrix shows the predicted labels versus the y-axis as the ground truth labels. The ROC in the multiclass classification is one versus all: the given class is treated as positive class with the rest treated as negative class.

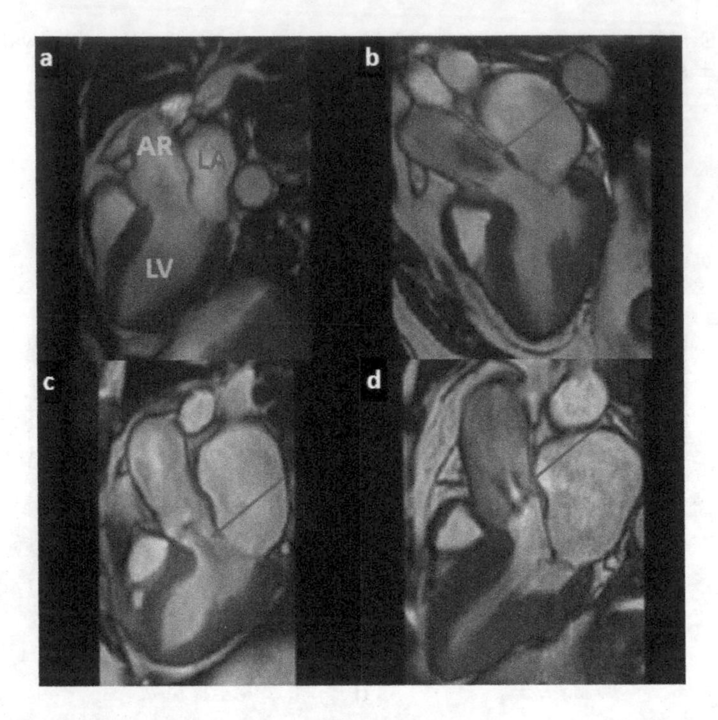

Fig. 5. 3-chamber Cine MRI examples with corresponding Grad-CAM heatmaps from 4 subjects at late systolic phase except early diastolic phase for AR. We label the aortic root (AR), left atrium (LA) and left ventricle (LV) in one of the panels. (a) no pathology; (b) aortic stenosis - the arrow shows a region of blood signal dephasing caused by the high-speed flow through the calcified aortic valve; (c) aortic regurgitation - the arrow points to blood signal dephasing caused by mixing with the backwards flow; (d) mixed valve disease.

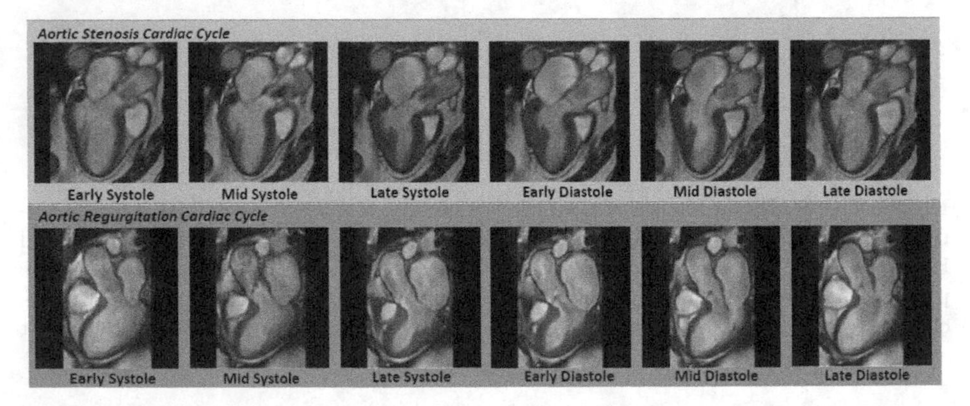

Fig. 6. Examples of 3-chamber Cine MR images across the cardiac cycle, overlaid with corresponding Grad-CAM heatmaps from 2 subjects with AS (top) and AR (bottom). (Top - AS) The heatmaps show a strong contribution of the aortic root region throughout, especially during systole. (Bottom - AR) During the diastolic phases, the classification network focuses on the backwards flow from the aorta to the LV.

4 Conclusions

We present a robust pipeline that can identify aortic valve pathology automatically from 3-chamber Cine MRI, without the need for manual annotations of the aortic valve features and blood flow and delineation of the heart. The classification pipeline (with visual inputs provided by the Grad-CAM heatmaps) can be used a triage tool to identify patients with suspected AV pathology who may benefit further CMRI investigations such as PC-MRI. The proposed analysis tools are part of an effort to personalise clinical CMR protocols, making them less time consuming and costly.

Previous work has used machine learning methods to identify AVD from 3-chamber Cine MRI, but relied on expert manual annotations as inputs to the classification method [5]. In particular, Vimalesvaran *et al* propose a multi-level network to identify aortic valve landmarks, followed by a random forest model for AVD pathology binary classification. Their study reached an accuracy and F_1 score of 0.93 ± 0.03 and 0.91 ± 0.04 respectively, comparable to 0.85 ± 0.03 and 0.83 ± 0.03 we achieve respectively. Our approach, however, relies on the direct application of a CNN to the images, avoiding the need for the time-consuming manually-annotated image features as inputs.

Underlying the good performance of our method is the proposed adaptive heart extraction algorithm. This is able to identify the heart in Cine MRI in a computationally low-cost, automatic setup without any need for segmentation labels. Our offline experiments suggested the proposed algorithm is robust even for patients with low cardiac function. This step may not be necessary when training the classifier on much larger datasets using a network with a higher capacity. Other options consist in using a supervised heart identification method or unsupervised approaches using Grad CAM [11].

To mitigate the 'black-box' nature of applying NN in the clinical environment, we apply Grad-CAM after the classification process. For all the considered labels, as illustrated in Figs. 5 and 6, the model consistently relies on voxels from the blood pool in the aortic root and, for AR, also in the LV. This is reassuring as AVD patients often present with morphological changes in the heart [22], which the CNN could have relied upon for its classification. Moreover, GradCAM (see Fig. 6) suggests that the classifier may also perform well when receiving a smaller number of cardiac phases as inputs, which could reduce the computational burden of the network.

Once trained, our approach can be used in real time at the scanner side, to identify patients with likely AVD who would benefit from additional dedicated imaging, such as aortic root PC-MRI scans. It can therefore contribute to the personalisation and optimisation of CMR protocols, leading to improved more efficient patient diagnosis. Future work will include the testing of the proposed tools on larger datasets from other sites, a comparison of our approach with different state-of-the-art CNNs, and classification of valvular pathology using Cine MRIs acquired in different views.

Acknowledgements. This work was supported by the British Heart Foundation Centre of Research Excellence at Imperial College London (RE/18/4/34215). K. Vimalesvaran and S. Zaman were supported by UKRI Centre for Doctoral Training in AI for Healthcare grant number EP/S023283/1. The project was partly supported by a Rosetrees Interdisciplinary Award.

References

1. Guglielmo, M., et al.: The role of cardiac magnetic resonance in aortic stenosis and regurgitation. J. Cardiovasc. Dev. Dis. **9**(4), 108 (2022). https://doi.org/10.3390/jcdd9040108
2. Timmis, A., et al.: European society of cardiology: cardiovascular disease statistics 2021. Eur. Heart J. **43**(8), 716–799 (2022). https://doi.org/10.1093/eurheartj/ehab892
3. Iung, B., et al.: A prospective survey of patients with valvular heart disease in Europe: the euro heart survey on Valvular Heart Disease. Eur. Heart J. **24**(13), 1231–1243 (2003). https://doi.org/10.1016/S0195-668X(03)00201-X
4. Vahanian, A., et al.: 2021 ESC/EACTS guidelines for the management of valvular heart disease: developed by the task force for the management of valvular heart disease of the European Society of Cardiology (ESC) and the European Association for Cardio-Thoracic Surgery (EACTS). Eur. Heart J. **43**(7), 561–632 (2021). https://doi.org/10.1093/eurheartj/ehab395
5. Vimalesvaran, K., et al.: Detecting aortic valve pathology from the 3-chamber cine cardiac mri view. In: Medical Image Computing and Computer Assisted Intervention - MICCAI 2022, pp. 571–580. Springer, Cham (2022). https://doi.org/10.1007/978-3-031-16431-6_54
6. Troger, F., et al.: A novel approach to determine aortic valve area with phase-contrast cardiovascular magnetic resonance. J. Cardiovasc. Magn. Reson. Off. J. Soc. Cardiovasc. Magn. Reson. **24**, 7 (2022)
7. Sechtem, U., et al.: Cine mr imaging: potential for the evaluation of cardiovascular function. Am. J. Roentgenol. **148**(2), 239–246 (1987). pMID: 3492096. https://doi.org/10.2214/ajr.148.2.239
8. Kramer, C., et al.: Standardized cardiovascular magnetic resonance imaging (cmr) protocols: 2020 update. J. Cardiovasc. Magn. Reson. **22** (2020)
9. Bieri, O., Scheffler, K.: Flow compensation in balanced ssfp sequences. Magn. Reson. Med. **54**(4), 901–907 (2005). https://onlinelibrary.wiley.com/doi/full/10.1002/mrm.20619
10. Sommer, G., Bremerich, J., Lund, G.: Magnetic resonance imaging in valvular heart disease: clinical application and current role for patient management. J. Magn. Reson. Imaging **35**(6), 1241–1252 (2012). https://onlinelibrary.wiley.com/doi/full/10.1002/jmri.23544
11. Selvaraju, R.R., et al.: Grad-cam: why did you say that? visual explanations from deep networks via gradient-based localization. CoRR, vol. abs/1610.02391 (2016). https://arxiv.org/abs/1610.02391
12. Canny, J.: A computational approach to edge detection. IEEE Trans. Pattern Anal. Mach. Intell. PAMI **8**(6), 679–698 (1986)
13. Huang, G., Liu, Z., Weinberger, K.Q.: Densely connected convolutional networks. CoRR, vol. abs/1608.06993 (2016). https://arxiv.org/abs/1608.06993
14. Cardoso, M.J., et al.: Monai: an open-source framework for deep learning in healthcare (2022). https://arxiv.org/abs/2211.02701

15. Patro, S.G.K., Sahu, K.K.: Normalization: a preprocessing stage. CoRR, vol. abs/1503.06462 (2015). https://arxiv.org/abs/1503.06462
16. Sudre, C.H., Cardoso, M.J., Ourselin, S.: Longitudinal segmentation of age-related white matter hyperintensities. Med. Image Anal. **38**, 50–64 (2017). https://www.sciencedirect.com/science/article/pii/S1361841517300257
17. Caruso, C., Quarta, F.: Interpolation methods comparison. Comput. Math. Appl. **35**(12), 109–126 (1998). https://www.sciencedirect.com/science/article/pii/S0898122198001011
18. Lin, T., et al.: Focal loss for dense object detection. CoRR, vol. abs/1708.02002 (2017). https://arxiv.org/abs/1708.02002
19. McNeil, B.J., Hanley, J. A.: Statistical approaches to the analysis of receiver operating characteristic (roc) curves. Med. Decis. Mak. **4**(2), 137–150 (1984). pMID: 6472062. https://doi.org/10.1177/0272989X8400400203
20. Gotkowski, K., González, C., Bucher, A., Mukhopadhyay, A.: M3d-cam: a pytorch library to generate 3d data attention maps for medical deep learning. CoRR, vol. abs/2007.00453 (2020). https://arxiv.org/abs/2007.00453
21. Gotkowski, K., et al.: M3d-cam: a pytorch library to generate 3d data attention maps for medical deep learning (2020)
22. Thubrikar, M.: The Aortic Valve. Routledge, Abingdon (2018)

Uncertainty to Improve the Automatic Measurement of Left Ventricular Ejection Fraction in 2D Echocardiography Using CNN-Based Segmentation

Antonio Sánchez-Puente[1,2,3](✉), Pablo Pérez-Sánchez[1,2,3], Víctor Vicente-Palacios[4], Alberto García-Galindo[5,6], Pedro Pablo Vara[1], Candelas Pérez del Villar[1,2,3], and Pedro L. Sánchez[1,2,3]

[1] Department of Cardiology, Hospital Universitario de Salamanca, Salamanca, Spain
{asanchezpu,pperezsanc,ppvara,mcperezvi,plsanchez}@saludcastillayleon.es
[2] CIBERCV, Instituto de Salud Carlos III, Madrid, Spain
[3] Institute for Biomedical Research of Salamanca, IBSAL, Salamanca, Spain
[4] Philips Healthcare, Amsterdam, The Netherlands
victor.vicente.palacios@philips.com
[5] Universidad de Navarra, Institute of Data Science and Artificial Intelligence (DATAI), Pamplona, Spain
agarciagali@unav.es
[6] Universidad de Navarra, Tecnun School of Engineering, San Sebastián, Spain

Abstract. The echocardiographic measurement of the left ventricular ejection fraction (LVEF) is the accepted clinical way to assess the cardiac function of a patient. For this measurement, a physician needs to identify the end-systole and end-diastole, segment the left ventricle in those frames, obtain the volume from the masks, and compute the LVEF. Naive implementations of convolutional neural network (CNN) based segmentation algorithms to perform this measurement might encounter problems identifying the end-systole and end-diastole if there is a single poorly segmented frame in the whole echocardiogram, which would ruin the measurement of LVEF and require manual review by a human operator. In this research article, we present how to use different uncertainty metrics to identify poorly segmented frames and quantify how these techniques improve the concordance between algorithm and human operator measurements in a population-based cohort of echocardiographic examinations.

Keywords: Echocardiography · Deep Learning · Uncertainty

1 Introduction

Echocardiography is the most widely used cardiac imaging procedure in the world, because it is accessible, fast and inexpensive. In clinical practice, it is used along electrocardiography for patient triage. This process usually lasts between

O. Bernard et al. (Eds.): FIMH 2023, LNCS 13958, pp. 658–667, 2023.
https://doi.org/10.1007/978-3-031-35302-4_67

10 and 20 min and is performed by expert technicians. Afterwards, cardiologists or radiologists extract measurements from the study [17]. The workflow from the image acquisition to the analysis lasts between 30 and 45 min. Nonetheless, it is possible to train non-experts to perform echocardiograms and alleviate the workload of clinical centres [4], which is also important for centres with a lack of echocardiographic specialists or isolated rural areas.

An easy way to know the patient's overall cardiac condition is through the calculation of the echocardiographic left ventricular ejection fraction (LVEF). Automatic calculation of LVEF would allow these non-expert technicians to have a quick indicator to refer patients to expert physicians. In this context, deep learning techniques are being used for this automation through automatic segmentation of the LV [8,11,13,15].

LVEF calculation is performed from two different echocardiographic views: two chambers or four chambers. End-systole and end-diastole must be identified and segmented. Cardiac volume is obtained from the segmentation masks and the LVEF measurement is provided [17]. However, poorly segmented frames can lead to end-systole and diastole misidentification and/or to a erroneous LVEF measurement. This can happen due to the shortcomings of the neural network, or because of a low quality image acquisition. In order to mitigate the harmful effect of a bad automatic segmentation, the addition of a module that measures the uncertainty of the model will allow to obtain better results.

We found in the literature different approaches to measure uncertainty in automatic echocardiogram segmentations. Different ensemble methods (MC drop-out, Test-time data augmentation and Horizontal and vertical ensemble) have been compared for the calculation of uncertainty in LV segmentation. The resulting uncertainty measures were used to reject low-quality images [3]. Another original approach is to construct segmentation uncertainty estimates by learning a joint latent space of echocardiograms and ground-truth segmentations. Based on this latent space, plausible neighbouring segmentations are sampled [9].

Furthermore, most of the automatic segmentation applications found in the current literature use open datasets such as CAMUS [11] or EchoNet-Dynamic [13], which are well-curated and homogeneous. Meanwhile, databases from a cardiology department might contain more challenging images with an increased number of cardiac cycles, or intermittent quality through the acquisition, besides not having identified end-systole and end-diastole.

This publication pursues two objectives. On the one hand, an automated deep learning system will be developed to segment the left ventricle from two-chamber and four-chamber apical views of an echocardiographic examination, identify the end-systolic and end-diastolic frames, and obtain the LVEF. On the other hand, uncertainty metrics will be implemented and tested to reject poorly segmented frames and improve the identification of end-systolic and end-diastolic frames.

2 Methodology

2.1 Database Summary

Two different sets of images were used in this study: For training the neural network, the echocardiographic images from the cardiology department of the University Hospital of Salamanca were used, consisting of 11,087 2-chamber view images and 17,720 4-chamber view images collected from 30/05/18 to 27/12/19. A subset of 214 of these images were manually reviewed to ensure good image quality and accurate segmentation, and served as a validation set during training. For evaluating the automated system, the echocardiographic images from the SALMANTICOR study [12] study were used, consisting of 1,915 4-chamber view images and 1,547 2-chamber view images. This study is a population-based study carried out in the province of Salamanca, Spain, between 2015 and 2018, that provides a representative sample of the local population.

2.2 Neural Network

The architecture of the neural network is a U-Net [14]. It takes a monochrome 384×384 image as an input, and consists of 5 descending blocks, a bottom block, and 5 ascending blocks. Each block contains two convolutional layers with 3×3 filters with ReLu activation, a batch normalization layer and a max pooling layer at the end (for the descending blocks) or an upsampling layer at the beginning (for the ascending blocks). The first block has 10 filters and the numbers of filters is doubled as we descend, and halved as we ascend. The output is obtained from a final sigmoid activation layer that brings a final 384×384 image with values between 0 and 1, with 0.5 being the threshold for a pixel to be considered LV (Fig. 1).

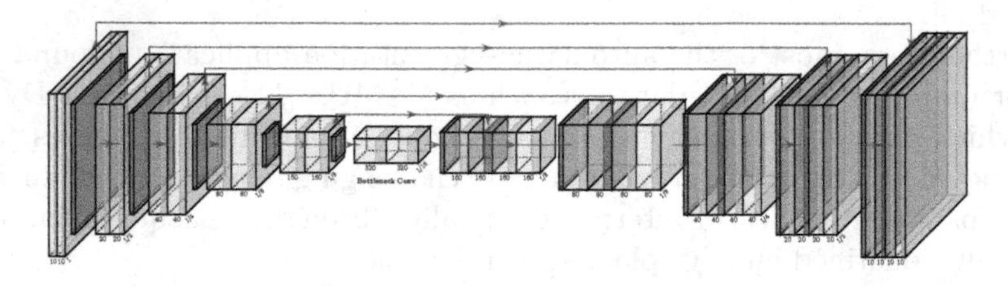

Fig. 1. Diagram of the U-Net architecture used [7]

The images from the training dataset were augmented by rotations up to 20 degrees, shearing up to 5 degrees, zooming between 70% and 140%, horizontal reflection, and gamma transformation with a coefficient between 0.5 and 2. The network was trained using batches of 16 images, consisting of 4 echocardiographic images with 4 augmentations each [6]. The loss function was the generalized dice loss [16] and the optimizer was Adam [10]. The learning rate was set at 10^{-4}

and halved after 3 epochs without improvement [1]. The training was stopped after 10 epochs without improvement [5, Chapter 7], the training went for 35 epochs (the model weights after epoch 25 were selected).

2.3 Segmentation and Uncertainty Modelling

Echocardiographic images are cropped to square format and resized to 384 × 384 pixels, the image is segmented by the neural network, and then, postprocessed to remove possible artifacts. The post-processing consisted of an opening the image (morphology dilation of the binary mask followed by erosion) using a 21-pixel square structuring element, and then closing the image (erosion followed by a dilation). This process removes small holes and islands and smoothes the perimeter of the mask. Finally, the largest connected component is kept.

To understand how confident the neural network is in the segmentation, we would like to map the probability of a pixel to be segmented as LV by the network. However, given an echocardiographic image, the network is deterministic and gives a unique mask. An alternative to create different segmentations for an image is to make changes that should not affect the way it is segmented, for example, change the brightness, contrast, gamma, or add some noise. Further possible processing techniques are spatial transformations, such as rotations, reflections, shear or zooming, that can be reverted back in the segmentation mask. We created a probability map in this way, using 20 different augmentations (of the same type as the ones used during the network training), and assigning a probability for each pixel as the percentage of augmentations in which that pixel was segmented as LV. This probability map is then used to measure the uncertainty of the segmentation through different metrics.

Also, this probability map is used to improve on the original segmentation, considering the mask of all pixels that have been segmented as LV at least in half of the augmentations (equivalently, a probability of 0.5 or higher in our map). Intuitively this is the average of all segmentations. Both original and improved segmentations are compared in the results section.

2.4 Uncertainty Metrics

The probability map is transformed into a score of whether the segmentation of an echocardiographic frame is reliable. This is done with the following metrics:

- *Probability map DSC* [3]: We threshold the probability map for a value of 0.16 (that is to say, we consider LV the pixels that at least 3 augmentations considered to be LV) and 0.84 (we consider LV the pixels that at least 17 augmentations considered to be LV) and obtain two masks that, intuitively, represent the confidence interval on how big and how small the segmentation can be. In this metric we measure the dice similarity coefficient (DSC) between them.
- *Probability map volume differences*: This metric is similar to the previous one, but instead of measuring the DSC, the volume of both masks is calculated, and then the difference in volumes over the mean volume is obtained.

- *Sample variance*: The variance of each pixel is calculated from the probability map, it is summed over all pixels, and normalized by the area of the mask.
- *DSC frame over frame*: For each mask, we compute the DSC with the masks from the previous and the following frame, and use the minimum of the two values as a metric. This metric is different from the previous ones as it does not make use of the probability map.

These metrics assign a reliability score to the segmentation of each frame. Afterwards, we can set a threshold, so that all segmentations of frames with scores below that threshold are rejected and not used for the LVEF measurement.

2.5 End-Systole and End-Diastole Identification and LVEF Measurement

It is common that segmentation networks are evaluated just on end-systolic and end-diastolic frames already identified by a cardiologist, or that a dataset has these instants identified as the first and last frames of the video. However, in clinical practice, the start of the echocardiographic acquisition might not happen at end-systole or end-diastole, and it is also usual to record several cardiac cycles of the patient, which is especially true in the case of patients with irregular heartbeat. Our goal is that our algorithm should be able to measure the LVEF without any other input than a two or four chamber echocardiographic view, and therefore, has to identify the end-systole and end-diastole without additional information from a cardiologist.

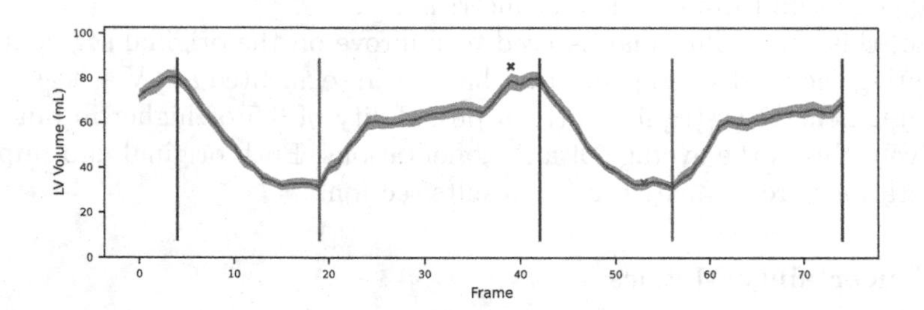

Fig. 2. Volume curve of an echocardiogram as obtained by the automatic system. Vertical blue and red lines represent the heuristically obtained end-diastoles and end-systoles. Blue and red crosses represent the cardiologist identified end-diastole and end-systole frames with the volume from the ground truth mask. The shaded region represents the variability in volume as measured in the probability map. (Color figure online)

The LV segmentation masks of each frame are translated to a volume measurement using Simpson's method, resulting in a volume curve along all frames of the echocardiogram (Fig. 2). To obtain the systolic and diastolic volume from the curve, we have considered two methods:

- A naive method, in which we identify one end-systole and one end-diastole as the frames of greatest and lowest volume.
- A heuristic method, in which we identify each end-systole and end-diastole based on prominence, then average the corresponding volumes.

For the latter method, we have identified the end-systole and end-diastole based on the prominence of the peaks and valleys of the volume curve. The prominence of a peak is defined as the vertical distance between the peak and its lowest contour line [18]. If the prominence of the peak is 1/4th of the maximum difference in volume, it was considered an end-diastole. If the prominence was not achieved because the peak was close to the beginning or the end of the echocardiographic acquisition, but the one-sided prominence is at least 2/3rd of the maximum difference in volume, then we also consider it to be an end-diastole. We use the same procedure for the end-systole, using valleys instead of peaks.

2.6 Experiments

To thoroughly evaluate the deep learning model we have created, and evaluate the performance of the segmentation, systole and diastole identification, and unreliable frame rejection, we have conducted the following 4 experiments:

Experiment 1. *DSC at the cardiologist marked end-systole and end-diastole*: This test evaluates the frame-by-frame segmentation of our network against the segmentation carried out by the specialist, which is only done in end-systole and end-diastole. We compare both the original segmentation by the network and the improved one by averaging the 20 augmentations.

Experiment 2. *LVEF similarity at the cardiologist marked end-systole and end-diastole*: The LVEF measured by the cardiologist is compared to the LVEF by the original and improved segmentations of our network in the same frames.

Experiment 3. *LVEF similarity between the cardiologist and the automated system*: In this experiment, the system does not receive any input on which are the end- systolic and diastolic frames. The system segments the whole echocardiographic video, automatically identifies the end- systole and diastole, and derives the LVEF. Only the improved segmentation is used in this experiment.

Experiment 4. *LVEF similarity between the cardiologist and the automated system rejecting frames using uncertainty*: In this experiment, an uncertainty threshold will be set, and the frames falling behind that threshold will be rejected, using only the remaining ones for the systole and diastole detection, and the LVEF measurement. We will choose each one of the uncertainty metrics, vary the threshold, and measure the similarity of the LVEF to the cardiologist versus the number of discarded frames.

To measure the similarity between the automated system and the cardiologist, the RMSE (root mean square error) is reported, taking the cardiologist measurement as the ground truth we are aiming for.

3 Results

Results from Experiment 1 (Table 1) are consistent with the training, and show that the probability map improved segmentation obtains more accurate results.

Table 1. Experiment 1

	DSC respect to cardiologist mask				
segmentation	All	4-chamber View	2-Chamber View	Diastole Frames	Systole Frames
Automatic Original	0.9207	0.9233	0.9175	0.9369	0.9046
Automatic Improved	0.9251	0.9274	0.9222	0.9407	0.9094

In experiment 2 (Table 2) the neural network LVEF values are slightly lower respect to the cardiologist measurements. The probability map improved segmentation leads to better results than the original one, as in the case before. In experiment 3, the automated selection of systole and diastole leads to significantly greater LVEF values, especially using the naive method of selecting the maximum and minimum volume as diastole and systole. The averaging method produces lower RMSE than experiment 2, despite the overestimation.

Table 2. Experiment 2 and Experiment 3

	segmentation	sys./dias. identification	LVEF	Similarity (RMSE)		
				All	4-chambers	2-Chambers
Experiment 2	Cardiologist	Cardiologist	61.9 ± 7.7	–	–	–
	Automatic Original	Cardiologist	60.1 ± 8.5	7.90	7.75	8.09
	Automatic Improved	Cardiologist	60.7 ± 8.4	7.84	7.89	7.78
Experiment 3	Automatic Improved	Naive	68.9 ± 6.9	9.82	9.87	9.75
	Automatic Improved	Heuristic average	64.9 ± 6.9	7.70	7.58	7.86

The results from experiment 4 can be seen in Fig. 3. No matter the strategy for identifying the systolic and diastolic volume, using uncertainty to reject unreliable frames improves the LVEF measurement and reaching a minimum of 8.81 RMSE for the naive (minimum and maximum volume) strategy, and 7.43 RMSE for the heuristic averaging strategy.

4 Discussion

We have developed a method to segment the LV from two- and four-chamber echocardiographic images and obtain the LVEF. End-systole and end-diastole are automatically identified, and an uncertainty-aware procedure has been implemented to discard poorly segmented frames. The use of any of the uncertainty

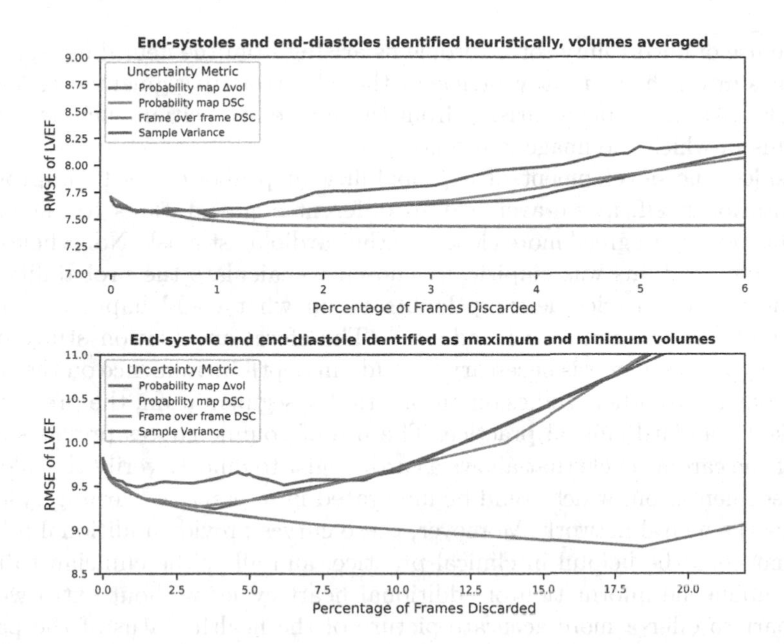

Fig. 3. Evolution of the similarity of the automatic LVEF measurement respect to the cardiologist, as unreliable frames are rejected by putting a threshold on different uncertainty metrics, for both end- systole and diastole identification strategies.

metrics tested improves the measurement of LVEF. As unreliable frames are discarded, the similarity of the results towards the cardiologist measurement increases. However, beyond a certain percentage of frames discarded, the results worsen as a result of losing too much information.

The *sample variance* metric gave the worst results and *DSC frame over frame* gave the best results. Since *DSC frame over frame* is a different approach from the other three methods, a logical step would be to combine both approaches to further improve results.

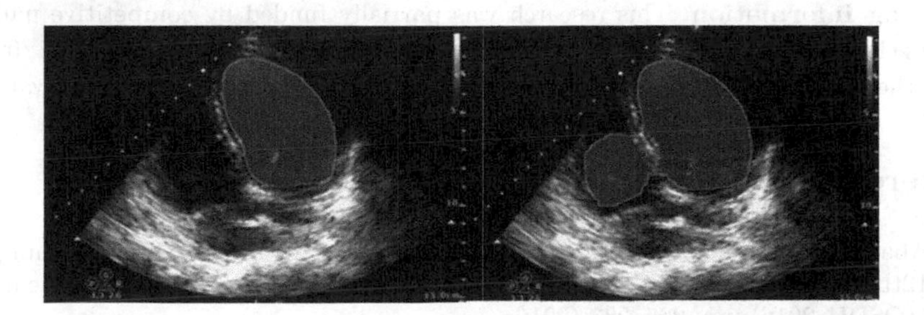

Fig. 4. Two consecutive frames from an echocardiogram, one with the right ventricle incorrectly segmented as LV. Uncertainty addresses this problem rejecting that frame.

The use of uncertainty solves problems arising from the neural network, such as frames in which the right ventricle or the left atria are segmented as left ventricle (Fig. 4), or problems arising from the image acquisition such as echocardiograms in which the image is too noisy or lost.

Besides, the development of a probability map allowed us to improve the segmentation itself, as we averaged 20 different augmentations to offer a new segmentation that agreed more closely to the cardiologist mask. Nonetheless, this set of augmentations was empirically chosen to calculate the probability maps and uncertainty metrics, leaving doubts as on what could happen if another number of augmentations were adopted. Therefore, an ablation study of this specific hyperparameter is necessary to study in depth its influence on the results.

In contrast to other studies on automatic LV segmentation, the present work mimicked standard clinical practice. The use of volume curves provides simple view of the cardiac cycle that allows a cardiologist to quickly verify the adequacy of the segmentation, which could be integrated in interactive learning system to improve the neural network. Moreover, these curves provide additional information that could be helpful in clinical practice, and allow the clinician to record and combine the information of additional heart cycles without extra work on their part to offer a more accurate picture of the health status of the patient. This issue is clinically relevant if we considered the potentially adverse influence of premature ectopic beats or tachyarrhythmias on LVEF calculations.

The system performs well, having a 7.4 RMSE respect to the cardiologist. If instead of considering the cardiologist to be the ground truth, we put both the cardiologist and the system on the same grounds, we would have reported an inter-observer variability of 7.3, which is not far from the 6.6 reported in the literature among human experts [2]. However, the study was not designed to measure this variability and these numbers should be taken with care.

Finally, a system as the one presented in this publication might be useful for less expert users. In addition to serving as a preliminary filter for more in-depth analysis by expert cardiologists of cases in which the calculated LVEF is below standard values, it could also be used to refer those echocardiographic studies in which the calculation of LVEF has a high uncertainty.

Funding Information. This research was partially funded by competitive national grants (PI14/00695, PI17/00145, PI21/00369) and by the CIBERCV (CB16/11/00374) from the Institute of Health Carlos III, Spanish Ministry of Science and Innovation.

References

1. Abadi, M., et al.: Tensorflow: a system for large-scale machine learning. In: 12th {USENIX} Symposium on Operating Systems Design and Implementation ({OSDI} 2016), pp. 265–283 (2016)
2. Baron, T., Berglund, L., Hedin, E.M., Flachskampf, F.A.: Test-retest reliability of new and conventional echocardiographic parameters of left ventricular systolic function. Clin. Res. Cardiol. **108**(4), 355–365 (2019)

3. Dahal, L., Kafle, A., Khanal, B.: Uncertainty estimation in deep 2d echocardiography segmentation. arXiv preprint arXiv:2005.09349 (2020)
4. Fernandez, M.A.G.: Is it possible to train non-cardiologists to perform echocardiography? (2014)
5. Goodfellow, I., Bengio, Y., Courville, A.: Deep Learning. MIT press, Cambridge (2016)
6. Hoffer, E., Ben-Nun, T., Hubara, I., Giladi, N., Hoefler, T., Soudry, D.: Augment your batch: Improving generalization through instance repetition. In: Proceedings of the IEEE/CVF Conference on Computer Vision and Pattern Recognition, pp. 8129–8138 (2020)
7. Iqbal, H.: Harisiqbal88/plotneuralnet v1.0.0 (2018). https://doi.org/10.5281/zenodo.2526396
8. Jafari, M.H., Van Woudenberg, N., Luong, C., Abolmaesumi, P., Tsang, T.: Deep bayesian image segmentation for a more robust ejection fraction estimation. In: 2021 IEEE 18th International Symposium on Biomedical Imaging (ISBI), pp. 1264–1268. IEEE (2021)
9. Judge, T., Bernard, O., Porumb, M., Chartsias, A., Beqiri, A., Jodoin, P.M.: *CRISPL* -reliable uncertainty estimation for medical image segmentation. In: International Conference on Medical Image Computing and Computer-Assisted Intervention, pp. 492–502. Springer, Heidelberg (2022). https://doi.org/10.1007/978-3-031-16452-1_47
10. Kingma, D.P., Ba, J.: Adam: a method for stochastic optimization. arXiv preprint arXiv:1412.6980 (2014)
11. Leclerc, S., et al.: Deep learning for segmentation using an open large-scale dataset in 2d echocardiography. IEEE Trans. Med. Imaging **38**(9), 2198–2210 (2019)
12. Melero-Alegria, J.I., et al.: Salmanticor study. rationale and design of a population-based study to identify structural heart disease abnormalities: a spatial and machine learning analysis. BMJ Open **9**(2), e024605 (2019)
13. Ouyang, D.: Video-based AI for beat-to-beat assessment of cardiac function. Nature **580**(7802), 252–256 (2020)
14. Ronneberger, O., Fischer, P., Brox, T.: U-Net: convolutional networks for biomedical image segmentation. In: Navab, N., Hornegger, J., Wells, W.M., Frangi, A.F. (eds.) MICCAI 2015. LNCS, vol. 9351, pp. 234–241. Springer, Cham (2015). https://doi.org/10.1007/978-3-319-24574-4_28
15. Smistad, E., et al.: Real-time automatic ejection fraction and foreshortening detection using deep learning. IEEE Trans. Ultrasonics Ferroelectr. Freq. Control **67**(12), 2595–2604 (2020)
16. Sudre, C.H., Li, W., Vercauteren, T., Ourselin, S., Jorge Cardoso, M.: Generalised dice overlap as a deep learning loss function for highly unbalanced segmentations. In: Cardoso, M.J., et al. (eds.) DLMIA/ML-CDS -2017. LNCS, vol. 10553, pp. 240–248. Springer, Cham (2017). https://doi.org/10.1007/978-3-319-67558-9_28
17. Toussaint, W., et al.: Design considerations for high impact, automated echocardiogram analysis. arXiv preprint arXiv:2006.06292 (2020)
18. Virtanen, P., et al.: SciPy 1.0 Contributors: SciPy 1.0: fundamental algorithms for scientific computing in python. Nat. Methods **17**, 261–272 (2020). https://doi.org/10.1038/s41592-019-0686-2

Automated Estimation of Motion Patterns of the Left Ventricle Supports Cardiomyopathy Identification

Edmond Astolfi[1], Athira Jacob[1], Indraneel Borgohain[1], Akos Varga-Szemes[2], Puneet Sharma[1], and Tiziano Passerini[1](✉)

[1] Siemens Healthineers, Digital Technology and Innovation, Princeton, NJ, USA
tiziano.passerini@siemens-healthineers.com
[2] Medical University of South Carolina, Charleston, SC, USA

Abstract. We propose a method to extract 3D shape and motion features of the left ventricle from 2D cine scans and investigate their utility in a cardiomyopathy detection task. To this aim, we develop an automatic processing pipeline that builds a 3D model of the left ventricular endocardium and epicardium from 2D cine cardiac MRI exams. We analyze a database of 1,045 clinical MRI studies including a combination of healthy subjects and cardiomyopathy patients with clinically reported disease labels. We use manifold learning techniques to extract shape and motion features from sequences of 25 3D shapes generated for each patient data set and representing a cardiac cycle. Finally, we train a disease classifier using a combination of simple metrics, shape features and motion features. On a testing set of 197 subjects, this classifier achieves the best performance in identifying patients from healthy subjects (AUC 0.922), as well as distinguishing 4 different classes of cardiac diseases (AUC 0.909).

Keywords: cardiac motion · shape analysis · heart disease classification

1 Introduction

Left ventricular (LV) function is a key predictor of cardiac disease. Cardiovascular magnetic resonance imaging (MRI) is regarded as the gold standard modality for the assessment of LV function. The most significant metrics for the evaluation of cardiac function are LV volumes and ejection fraction (EF). However, the shape and motion patterns of the LV can vary significantly between subjects, independently of the variation of simple global metrics. This is especially the case in various forms of cardiomyopathies. Several studies have indicated that regional variations in the LV shape can be distinctively associated to various cardiac conditions, thus potentially enabling a more fine-grained patient stratification or more accurate disease staging [1]. Additionally, contraction patterns of the LV can also be distinctively associated to cardiac conditions [2].

Due to the heart's motion and complex morphology, accurate estimation of LV function requires images with high spatial and temporal resolution. Cine balanced steady-state free-precession (bSSFP) cardiac MRI is routinely used in clinical practice, with a

O. Bernard et al. (Eds.): FIMH 2023, LNCS 13958, pp. 668–677, 2023.
https://doi.org/10.1007/978-3-031-35302-4_68

protocol including the acquisition of time-resolved, cardiac gated, 2D images in multiple imaging planes, including short-axis and long axis projections. We hypothesize that this data allows the accurate estimation of the 3D shape and motion patterns of the LV, thus it enables the identification of various cardiomyopathies with higher accuracy compared to approaches using simpler, global metrics such as ventricular volumes and EF. This work aims at defining and evaluating an automated processing method for the estimation of LV 3D shapes and motion patterns. Furthermore, a classifier is trained to separate normal subjects from heart failure patients using input features derived from the estimated shapes and motion. The accuracy of the classifier is evaluated on a cohort of patients with clinically reported disease condition.

2 Methods

The image processing workflow comprises image segmentation, contour-based 3D shape estimation and motion analysis. Details on each step of the workflow are provided in the subsequent sections.

We consider cine bSSFP cardiac MR data acquired with a standard protocol and stored in DICOM format. Each DICOM object is an electrocardiography-gated sequence of temporally resolved 2D images of the heart acquired in a cross-sectional plane. For each subject, a combination of short axis (SAX) and long axis views (apical 2-chamber A2C, apical 3-chamber A3C and apical 4-chamber A4C) are available. Each sequence is processed by a fully automatic machine learning algorithm to segment the LV endocardium and epicardium in all temporal frames [3].

Using the position and orientation information stored in each DICOM header, the contours are positioned in 3D space. A slice alignment algorithm is applied to correct for inter-slice motion. The mean shape of an atlas of the LV based on manual annotations is aligned to the set of contours by an affine transformation followed by a non-rigid registration step. For each subject data set, a sequence of 3D shapes is independently estimated (one for each temporal frame). Shape and motion analysis is performed on the entire database through manifold learning.

Multiple shape and motion descriptors are derived, and their association with disease labels is studied. The potential application for patient stratification is demonstrated on a cohort of patient data including disease markers. Multiple classification models are defined including different sets of input features. The performance of each classifier is measured and compared.

2.1 Slice Alignment

We assume that SAX acquisitions are performed in a short time window which allows to consider a uniform orientation of all SAX imaging planes with respect to the LV long axis. We therefore assume that misalignment in the SAX stack of images is limited to mutual in-plane translation (no rotation).

Each long axis imaging sequence is acquired after re-positioning the image slice, which allows for a more significant likelihood of out-of-plane misalignment with respect to the SAX stack. We account for this uncertainty by initially positioning each long axis sequence in the 3D spatial position and orientation indicated by the DICOM header. We then assume that the long axis imaging plane is translated in space by an unknown displacement vector. Mis-alignment due to plane rotation is difficult to assess. In this work we make the simplifying assumption that the effect of rotation can be approximated by an in-plane rotation of the long axis contour around its barycenter.

To compute the optimal rotation and translation, for each long axis slice we first test multiple plane translations along the in-plane and out-of-plane directions. Plane positions which reduce the number of intersections with SAX contours are dropped. For each remaining plane position, we compute the optimal in-plane rotation and in-plane translation, minimizing the cumulative distance between the traces of SAX contours in the long axis plane and vice-versa, for all temporal frames. Finally, we select for each temporal frame the optimal out-of-plane translation.

2.2 3D Shape Estimation

The 3D shape of the cardiac chamber is estimated with a multi-step process. In the shape initialization step, the expected shape is generated based on a mean shape model. This model is obtained by Procrustes analysis [4] on a large database of annotated chambers. The mean shape (represented by a triangular mesh) is then aligned with the current contours based on anatomical landmarks (LV apex and two points on the mitral annulus in A4C view). In brief, an affine transformation is estimated based on the position of the landmarks and the position of corresponding points in the shape model. Thin plate splines interpolation is used to apply the transformation to the remaining mesh points. The same mean shape is used as a starting point to represent all shapes, thus all shapes share the same representation as triangular meshes with the same number of points and point-wise correspondence.

Finally, a shape refinement step further deforms the aligned chamber shape to maximize its similarity with the aligned contours provided as an input. The algorithm proceeds with an iterative sequence of three steps. In the first step, the SAX contours are aligned with in-plane translation to minimize the average distance of contour points to mesh points. This addresses SAX slice misalignment due to different LV location in subsequent breath holds. In the second step, we non-rigidly deform the mesh by a point-based elastic registration method using thin plate spline interpolation; this registration method uses the set of contour points and their closest points from the mesh as landmarks. In the third step, the deformed mesh is projected on a shape space for regularization. The shape space is defined as the span of 80 generator shapes obtained from a principal component analysis of a database of manually annotated shapes.

We assess the quality of the obtained 3D shapes by taking as a reference the set of 2D contours produced by a previously validated automatic contouring method [3]. We compute multiple metrics comparing the 2D contours and the cross section of the 3D mesh obtained with the corresponding imaging plane: the dice similarity coefficient, and the mean and Hausdorff distance. Additionally, we compare the estimated LV end diastolic (EDV) and end systolic volumes (ESV) against those derived from the automatically generated 2D contours using the standard summation-of-disks method. A good agreement in these metrics guarantees that the obtained 3D shapes provide an accurate approximation of the LV contour in all imaging planes and allow the estimation of clinically relevant measurements consistent with the previously validated method.

2.3 Motion Mode Analysis

The analysis of motion modes follows the manifold learning technique detailed in [5]. In brief, a sequence of 3D shapes of the LV (both endocardial and epicardial walls) is estimated with the described method for each data set in a large database of 2D cardiac MRI acquisitions. A quality control step is applied to the 3D shapes. An isolation forest algorithm with 0.01 contamination is used to detect and remove outliers. Sequences with more than one consecutive outlier shapes are removed from subsequent analysis. All remaining sequences are resampled by linear interpolation to a standard number of 25 equispaced frames per heart cycle. All the shape sequences are transformed to a normalized and centralized reference system by Procrustes analysis. We then use unsupervised manifold learning to identify the nonlinear degrees of freedom that describe the shapes and their motion. Each shape is mapped to a low dimensional embedding space by isometric mapping, using as distance metric between different shapes the mean Euclidean distance between corresponding points. The dimension of the embedding space is chosen as 2 based on previous analysis [5]. Shapes coming from the same temporal sequence define closed-loop trajectories in the embedding space. For each sequence, the ED shape is first considered and hierarchical k-mean clustering is applied to identify 4 distinct clusters of initial shapes. For each cluster, motion modes are identified as follows. First, all trajectories in the cluster are shifted in the embedding plane so that the initial shape coincides with the cluster center. In this way, sequences with similar ED shapes are aligned removing translations of the initial shape, and the remaining variability represents different motion patterns. Then, k-means clustering is applied to group the trajectories in 6 distinct sub-classes within the same ED shape cluster. Each motion mode is defined as the linear combination of all sequences in the group, weighted by the inverse distance from the group center. Only motion modes composed of at least 10 trajectories are kept.

After the motion modes have been identified, any sequence of shapes can be associated to a motion mode based on minimization of a distance metric. The distance metric chosen is the sum of the distances between corresponding shapes in the sequence and in the motion mode. Distances are computed as Euclidean distances in the embedding space.

2.4 Shape and Motion Features for Patient Stratification

We evaluate the use of shape- and motion-derived metrics as markers of pathological conditions. We focus on different variants of heart failure: dilated cardiomyopathy (DCM), hypertrophic cardiomyopathy (HCM), ischemic heart disease (IHD), and myocarditis (MYO) [6, 7]. Gold standard disease labels are based on clinical reports. We train Random Forests classifiers using 5-fold cross-validation and consider multiple different sets of input features. We first select features that have been reported in clinical literature as predictive of disease progression: EDV, ESV, EF, myocardial mass at ED (EDM) and ES (ESM), average myocardial thickness at ED (EDt) and ES (ESt). We then add shape-based metrics based on the proposed embedding: coordinates of the ED shape and of the ES shape in the embedding space, and Euclidean distance between the ED and ES shapes in the embedding space. Finally, we add features based on the motion mode analysis: a vector of features, each measuring the average distance of the given sequence of shapes from a motion mode. The motion modes considered to compute these features are selected based on the performance of the classifier.

3 Results

3.1 Data Description

The data consisted of cardiac cine MRI of 1,045 patients acquired from Medical University of South Carolina, Charleston, SC, United States. The institutional IRB approved the retrospective study protocol with a waiver for informed consent. A2C, A3C, and A4C views covering the entire LV were obtained using a bSSFP sequence. A stack of slices in SAX orientation covering the entire LV were obtained using the same bSSFP sequence (1.5T and 3T, Magnetom Avanto and Skyra, Siemens Healthcare, Erlangen, Germany), with a variable slice thickness of 7.63 ± 0.15 mm and a slice gap of 8 ± 0.1 mm, with or without compressed sensing, with a TR in the range of 35.71 ± 5.7 ms. The resulting images had an in-plane resolution of 1.91×1.91 mm. 45% of the cohort were female.

3.2 3D Shape Estimation

All the 1,045 cardiac MRI datasets could be processed by the image segmentation method with no failures. After quality control, 983 sequences were retained for subsequent analysis. The results of the quality assessment for the generated 3D shapes are reported in Table 1, showing that the 3D shapes have an excellent agreement with the 2D contours on all imaging planes. Moreover, global indices of cardiac function (EDV, ESV, and EF) are highly reproducible from 3D shapes as shown by minimal bias and tight limits of agreement.

Table 1. Accuracy of 3D meshes on all imaging planes, using automated 2D contours as reference

Structure	Metric	Value
LV endocardium	Dice (mean +/− std)	0.95 ± 0.03
	Mean point-to-point distance (mean +/− std)	0.74 ± 0.37 mm

(continued)

Table 1. (*continued*)

Structure	Metric	Value
	Hausdorff distance (mean +/− std)	3.24 ± 1.61 mm
LV epicardium	Dice (mean +/− std)	0.97 ± 0.02
	Mean point-to-point distance (mean +/− std)	0.75 ± 0.37 mm
	Hausdorff distance (mean +/− std)	3.56 ± 1.88 mm
EDV	Pearson's correlation coefficient	0.925
	Bias (3D mesh based – 2D contours based)	2.538 ml
	Limits of Agreement	[−75.937, 49.453] ml
ESV	Pearson's correlation coefficient	0.954
	Bias (3D mesh based – 2D contours based)	3.409 ml
	Limits of Agreement	[−58.903, 39.638] ml

3.3 Motion Modes

A total of 22 distinct motion modes are extracted from the database. The 4 primary motion modes are defined as the weighted sum of all the sequences in each of the ED shape clusters, the weights being proportional to their distance from the cluster center. The 4 primary motion modes are depicted in Fig. 1. A sequence of 5 3D shapes sampled from the center of each cluster is shown (after sequence normalization and centralization) for each primary motion mode, together with the representation of the cluster center in the embedding plane. In the embedding plane, each sequence is represented by a closed trajectory: each trajectory begins in the bottom extremity (corresponding to the ED shape), proceeds to the top extremity (corresponding to the ES shape) and returns to the bottom. Normalized volume curves of the 4 motion modes (after sequence normalization and centralization) are also shown. The corresponding computed EF shows significant differences between the motion modes: 35% (motion mode red), 44% (motion mode black), 53% (motion mode blue), 59% (motion mode green). Based on EF and visual inspection, the different motion modes are characterized by clearly distinct motion patterns: weak contraction and minimal thickening for motion mode red; strong contraction and thickening for motion mode green; intermediate contraction strength and thickening for motion modes blue and black, with more elongated shape and thicker ED myocardium in model mode blue.

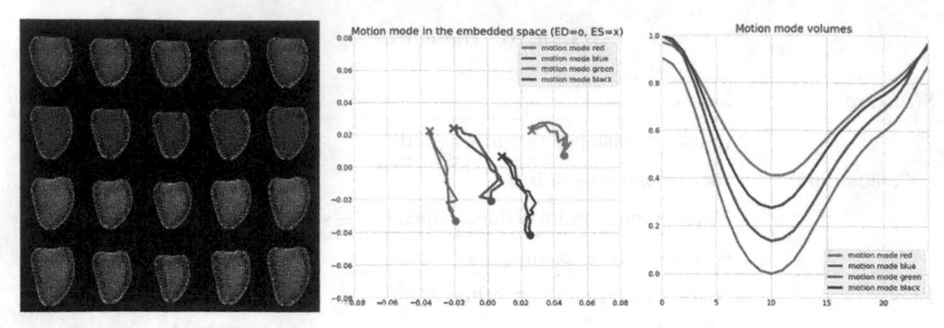

Fig. 1. Left: primary motion modes (after normalization and centralization) in the original shape space. Center: primary motion modes represented as trajectories in the embedding space. Each trajectory starts from the lower right end (marked by a circle representing the ED frame), proceeds to the top left end (marked by a cross representing the ES frame) and returns to the lower right end. The axes represent the nondimensional coordinates in the embedding plane. Right: volume curves of the primary motion modes (after normalization and centralization).

3.4 Disease Classification

The 983 sequences retained after quality control were associated to the following disease labels: 512 normals, 91 DCM, 93 HCM, 260 IHD, 27 MYO. We split them into training and testing sets with an 80%-20% ratio. The distribution of disease labels in the training and testing set matched the distribution in the full data set. We first assessed the performance of the classifiers in distinguishing normal versus abnormal subjects. The AUC results are summarized in Table 2. The classifiers using features from motion analysis used the average distance of each sequence from the 4 primary motion modes: adding the average distances from other motion modes did not improve their performance.

Table 2. AUCs of binary classification with different input features

Classifier type	Input parameters	AUC
Clinical metrics	ESV, EDV, EF, EDM, ESM, EDt, Est	0.904
Shape analysis	Clinical metrics + Embedding	0.909
Motion analysis	Clinical metrics + Distances from primary motion modes	0.889
Combined	All of the above	**0.922**

We then trained the classifiers to distinguish different kinds of heart failure (HCM, DCM, ICM, MYO). The average and per-class AUCs are summarized in **Table 3**. The confusion matrix for the best classifier (using the combination of all input features) is reported in Fig. 2.

Table 3. AUCs of multi-class classification with different input features

Classifier type	Average AUC	Per-class AUC				
		Normal	DCM	HCM	ICM	MYO
Clinical metrics	0.885	0.899	0.965	0.906	0.848	0.600
Shape analysis	0.893	0.901	0.976	0.903	0.869	0.625
Motion analysis	0.898	0.907	0.988	0.912	0.867	0.646
Combined	**0.909**	0.913	0.985	0.892	0.907	0.630

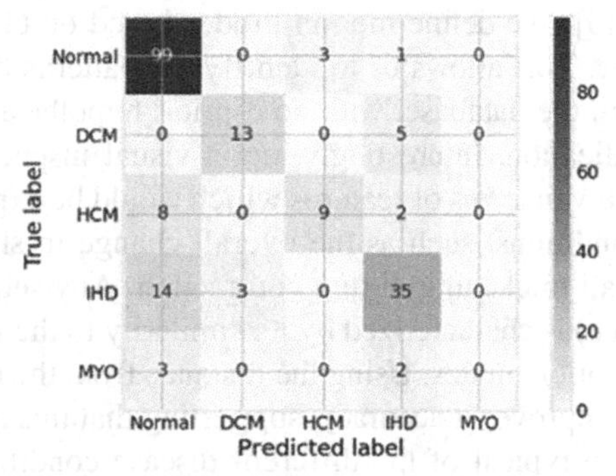

Fig. 2. Confusion matrix for multi-class classification using the best classifier. The numbers and color scale represent the case count for each combination of predicted and true labels.

4 Discussion

Different forms of heart failure are characterized by alteration of the ventricular motion patterns, in addition to alteration of the ventricular shape. However, the alterations could be subtle, or can have complex inter-dependency (e.g., in dilated ischemic cardiomyopathy) and a large body of literature has indicated that relatively simple morphology markers (volumes, thicknesses, basic shape descriptors) may not be specific enough to capture differences between the typical presentation of cardiac diseases. Following this line of reasoning, we followed a data-driven approach based on manifold learning to extract compact representations of complex shape and motion patterns of sequences of 3D shapes extracted from cardiac imaging. We demonstrate improved disease classification performance on addition of the proposed shape and motion features to the clinical measures.

A precondition for the proposed processing method is availability of cardiac-gated multi-plane cardiac MRI data. We have not explored in this study the case in which data may be only partially available (e.g., missing slices in the stack of SAX images, or lack of one or more long axis views). We also assume that temporal resolution is relatively high (25 frames per beat in the considered database).

Estimation of the 3D shape of the LV from sparse 2D sampling is subject to intrinsic uncertainty associated with the position in space of the imaging planes. Our method uses a simple model of slice misalignment which assumes that the orientation of the imaging planes is known exactly, misalignment being explained by in-plane translations of the SAX images and a combination of translation and in-plane rotations for long axis planes. This model proves robust in large scale testing, producing 3D shapes consistent with automatically drawn contours in 2D images, as measured by a combination of global (dice coefficient, volume) and local (point-to-point distance) metrics. However, we may underestimate the effect of tilts of the imaging planes affecting their mutual orientation in space.

We follow a standard approach to embedding sequences of 3D shapes to a low-dimensional space under the hypothesis that the motion of the LV can be described by a small number of free parameters. We use a bidimensional embedding space consistent with previous studies [5]. We define motion modes based on clustering of the shapes and sequences of shape. This allows us to identify the patterns of motion of the shape points that emerge from the database, with no a-priori hypothesis on their significance towards disease classification. Interestingly, upon visual inspection the main motion modes differ qualitatively in terms of features which would be expected to be associated to different disease conditions, such as the overall change in shape volume over one cardiac cycle or the wall thickening during contraction. Any sequence in the database (or any unseen sequence) is characterized by its similarity to the main motion modes or any of the secondary motion modes. Using the distance from the main motion modes as input for the classifier improves it accuracy, suggesting that this approach may uncover motion patterns that are typical of the different disease conditions. This is observed consistently when applying the classifiers to the binary classification problem of distinguishing normal from abnormal cases (as seen in Table 2) or the more complex problem of differentiating multiple disease conditions (as seen in Table 3).

For the latter problem, the analysis of AUCs for each class (Table 3) reveals that addition of shape and motion features benefit classification performance for the IHD class the most (5.9-point improvement), followed by DCM and Normal classes (1.4- and 2-point improvement respectively). Both IHD and DCM are known to cause wall motion abnormalities and thinning of the myocardial wall [7, 8], which could be captured by the extracted features. Despite the comparative advantage of shape and motion features, the confusion matrix for the best model (Classifier 4) indicates the biggest failure mode to be in distinguishing IHD from Normal class. Based on encouraging preliminary results, we believe that the expansion of the training set and the inclusion of additional secondary motion modes into the definition of the input features could further boost classification accuracy. On the other hand, clinical diagnosis of IHD often requires additional late gadolinium enhanced images which are not included in our analysis. Similarly, current clinical guidelines require more advanced MRI protocols (T1, T2 mapping) to detect myocarditis [9], which is currently not well classified by any of the considered classifiers. We believe that addition of features based on these advanced protocols could lead to improved classification performance.

Research reported in this publication was supported by the National Institute of Biomedical Imaging and Bioengineering of the National Institute of Health under award number R01EB027774. The content is solely the responsibility of the authors and does not necessarily represent the official views of the NIH. The concepts and information presented in this paper are based on research results that are not commercially available. Future commercial availability cannot be guaranteed.

References

1. Vadakkumpadan, F., Trayanova, N., Wu, K.C.: Image-based left ventricular shape analysis for sudden cardiac death risk stratification. Heart Rhythm **11**(10), 1693–1700 (2014)
2. Corral Acero, J., et al.: Understanding and improving risk assessment after myocardial infarction using automated left ventricular shape analysis. JACC: Cardiovasc. Imaging **15**(9), 1563–1574 (2022)
3. Toupin, S., et al.: Incremental prognostic value of fully-automatic LVEF by stress CMR using machine learning. Arch. Cardiovasc. Dis. Suppl. **15**(1), 63 (2023)
4. Dryden, I.L., Mardia, K.V.: Statistical Shape Analysis. John Wiley and Sons, Chichester (1998)
5. Yang, L., Georgescu, B., Zheng, Y., Wang, Y., Meer, P., Comaniciu, D.: Prediction based collaborative trackers (PCT): a robust and accurate approach toward 3D medical object tracking. IEEE Trans. Med. Imaging **30**(11), 1921–1932 (2011)
6. McKenna, W.J., Maron, B.J., Thiene, G.: Classification, epidemiology, and global burden of cardiomyopathies. Circ. Res. **121**(7), 722–730 (2017)
7. Sirajuddin, A., et al.: Ischemic heart disease: noninvasive imaging techniques and findings. Radiographics **41**(4), 990–1021 (2021)
8. Wallis, D.E., O'Connell, J.B., Henkin, R.E., Costanzo-Nordin, M.R., Scanlon, P.J.: Segmental wall motion abnormalities in dilated cardiomyopathy: a common finding and good prognostic sign. J Am Coll Cardiol. **4**(4), 674–679 (1984)
9. Al-Akchar, M., Shams, P., Kiel, J.: Acute myocarditis. In: StatPearls [Internet]. StatPearls Publishing, Treasure Island (FL) (2023)

Assessment of the Evolution of Temporal Segmental Strain in a Longitudinal Study of Myocardial Infarction

Bianca Freytag[1(✉)], Nicolas Duchateau[2,3], Lorena Petrusca[2,4], Jacques Ohayon[1,5,6], Pierre Croisille[2,4], and Patrick Clarysse[2]

[1] Univ. Grenoble Alpes, CNRS, TIMC UMR 5525, Grenoble, France
bianca.freytag@univ-grenoble-alpes.fr
[2] Univ. Lyon, INSA-Lyon, Université Claude Bernard Lyon 1, CNRS, Inserm, CREATIS UMR 5220, U1294, Lyon, France
[3] Institut Universitaire de France (IUF), Paris, France
[4] Department of Radiology, Hôpital Nord, Univ. Hospital of Saint-Étienne, Saint-Priest-en-Jarez, France
[5] Center for Cardiovascular Regeneration, Department of Cardiovascular Sciences, Houston Methodist Research Institute, Houston, TX, USA
[6] Savoie Mont-Blanc Univ., Polytech Annecy-Chambéry, Le Bourget du Lac, France

Abstract. In this study, we propose an approach for the assessment of the evolution of heart function in a population of ST-elevation myocardial infarction (STEMI) patients. The patients were imaged using cine MRI with late gadolinium enhancement at 1 month after STEMI and at follow-up 12 months later. We reduced the dimensionality of the temporal strain data in order to represent the segmental strain patterns found in these individuals in a more compact manner. Then, this compact description of strain was used to identify left ventricular segments with abnormal function, as compared to a control group of healthy patients, and to investigate if 1) the compact representation of the patterns is more effective in predicting scarring than traditional measurements; and 2) how the strain patterns evolve over the following months. To that end, we offer a method for tracking strain evolution and, in particular, comparing infarcted regions with remote regions and healthy hearts. On a population of 29 STEMI patients and 18 controls, we found that, despite considering the extent of the lesion, we were not able to identify a clear mechanism of evolution; nonetheless, the technique may be beneficial in subsequent, larger longitudinal studies to quantitatively characterise patient outcomes.

Keywords: Temporal strain · Myocardial Infarction · Dimensionality Reduction · Abnormality Quantification

1 Introduction

Feature tracking based on routinely collected non-contrast cine magnetic resonance imaging (MRI) sequences may quantify myocardial deformation during

O. Bernard et al. (Eds.): FIMH 2023, LNCS 13958, pp. 678–687, 2023.
https://doi.org/10.1007/978-3-031-35302-4_69

cardiac contraction [5,6]. Myocyte necrosis and subsequent scarring following myocardial infarction alter the mechanical characteristics of the myocardium, resulting in altered global and segmental strain [15]. Chronic myocardial scars, in particular those with wall thinning and obvious wall motion abnormalities, cause severe segmental strain impairment, which may be exploited to discriminate scar tissue from remote myocardium [14].

late gadolinium enhancement (LGE) identifies infarcted myocardium in vivo, where myocardial contractility is unlikely to be restored during coronary revascularization [7]. However, the use of the contrast agent gadolinium poses a risk to patients with severe renal impairment [13]. As a result, it is desirable to accompany and eventually replace LGE images with other means, such as strain information derived from feature tracking in cine imaging [16].

The formation of scars in the myocardium after ST-elevation myocardial infarction (STEMI) can occur within a few days as the acute phase of the infarction resolves and the tissue begins to heal. The transition from acute to sub-acute infarction typically occurs within the first 2–3 weeks after the initial injury [12]. However, the healing process can continue for several months, and factors such as the size of the infarct, the patient's age and overall health, and the presence of comorbidities can all impact the rate of healing.

The consequence of scarring on segmental strain over time is rarely investigated. A few studies looked at segmental myocardial strain as a predictor of cardiac function and mortality over longer time periods (such as one year) [1,2,9]. These studies used echocardiography, which is highly dependent on image quality and inconsistent in accuracy and reproducibility, specifically when measuring segmental strains [10]. Furthermore, *Mollema et al.* [9] and *Antoni et al.* [2] used a semi-quantitative wall motion score index and peak strains rather than the full temporal strain curves. Only *Mollema et al.* and *Abate et al.* [1] compared changes in basic strain measurements at follow-up. Another group compared strain in acute versus sub-acute infarcts (4–5 weeks after STEMI) assessed by MRI, which may be too short to see remodelling because of still ongoing pathophysiologic processes [11].

In this study, the changes in segmental, temporal strain patterns from non-contrast cine images were analysed in patients 1 and 12 months after STEMI. The primary focus was on the changes in these patterns between the two visits, but the study also investigated whether using a compact representation of the patterns is more effective than traditional measurements, such as peak strains.

2 Methods

2.1 Experimental Data

This study investigated left ventricular (LV) strain in two cohorts.

The first dataset was a sub-group of the HIBISCUS-STEMI cohort, for which STEMI patients, treated with primary percutaneous coronary intervention in the cardiac intensive care unit of the Hospices Civils de Lyon, France, were prospectively enrolled. Patients underwent cardiac MRI 1 month and 12 months

Table 1. Demographic data (mean ± standard deviation). BSA: Body surface area. LVM: left ventricular mass. EDV: end-diastolic volume. ESV: end-systolic volume. EF: ejection fraction. A p-value ≥ 0.05 was considered not significant (NS).

	Control	STEMI	P-value (t-test)
N (N male)	18 (13)	29 (25)	
Age (years)	52.9 ± 16.0	56.3 ± 8.8	NS
BSA (m^2)	1.91 ± 0.18	1.91 ± 0.19	NS
LVM (g)	95.5 ± 24.9	106.9 ± 26.8	NS
EDV (ml)	142.9 ± 28.3	165.6 ± 42.1	<0.05
ESV (ml)	59.9 ± 15.9	86.6 ± 37.9	<0.01
EF (%)	58.4 ± 6.2	49.5 ± 10.9	<0.01

after their admission for STEMI. 2-chamber, 4-chamber, and ventricular short-axis cine images, as well as LGE images, were acquired. Further details can be found in [3]. 29 patients with infarcts in the left anterior descending (LAD) artery were selected, while patients with infarcts in other territories were excluded.

The second dataset, the control group, consisted of 18 subjects and was part of the MARVEL cohort. These participants were healthy, sedentary volunteers of various ages. They were also scanned following the same protocol as above.

The demographics of both groups are listed in Table 1. The two groups were similar in age, body surface area, and LV mass. The gender proportions were matched as closely as the base cohorts allowed. Both studies were approved by the local ethics committee and are registered at ClinicalTrials.gov (NCT03064503 and NCT03070496). All patients provided written, informed consent.

2.2 Calculation of Strain and LGE Extent

In both studies, LV segmentation was performed using the CVI42 software (Circle Cardiovascular Imaging, Calgary, Canada) on short-axis, 2-chamber, and 4-chamber long-axis images. LV mass, ejection fraction (EF), and cardiac index were calculated from this segmentation. The software's Feature Tracking module was used to estimate circumferential, radial, and longitudinal strains and strain rates over the epicardial and endocardial surfaces throughout the cardiac cycle. Strains were calculated using the short-axis images and the 2- and 4-chamber long-axis views with respect to the end-diastolic configuration. End-diastole was defined to co-occur with the frame of the largest LV volume, and the strain curves were temporally aligned such that they all started at end-diastole with 0% strain. Longitudinal strain curves were found to be of poor quality or lacking coverage over some segments, so they were disregarded in further analysis. The strains and strain rates were exported on a per-segment basis, using 16 of the 17 AHA segments (excluding the apical segment). In addition, peak strains in circumferential and radial directions were also exported, as were wall thickening

Table 2. List of all calculated features, detailing the step in which they were calculated and if they had a high loading in the component loadings of the embedding PCA.

Calculated in	Feature	High loading in 2D embedding
PCA	PCs of circ./radial strains	1st and 2nd PCs
	PCs of circ./radial strain rates	no
CVI42	Ejection fraction	yes
	Cardiac Index	no
	Wall thickening	yes
	End-diastolic wall thickness	no
	Peak circ./radial strain	Radial peak strain only
	Time delay of peak circ./radial strain	no
	Percentage of LGE enhanced segment area	not incl. in embedding

and end-diastolic wall thickness. CVI42 defines the frame with the smallest LV volume as the frame at which end-systole occurs. This timepoint was confirmed by visually checking for maximum LV contraction. A time delay was calculated for each segment's strain curve that signifies how much earlier or later the peak value for this strain curve was reached in comparison to end-systole.

In the STEMI group, infarcts were semi-automatically segmented from the LGE images using CVI42's Tissue Characterization Module with a full-width at half maximum algorithm and guided by previous segmentation results [4]. Infarct size was assessed on a segment basis as a percentage of the segment area and written out for each of the 16 AHA segments.

2.3 Dimensionality Reduction of Temporal Strain Data

To reduce the dimensionality of the temporal strain data, a Principal Component Analysis (PCA) was applied to the circumferential and radial strains and strain rates individually. This approach allowed for the compact description and comparison of the major temporal patterns and has been shown to be useful for the identification of LV segments with abnormal function [14].

First, the PCA was fitted to the cases in the control group to establish the baseline patterns seen in a healthy population. Then, the fitted PCA was used to transform the control and STEMI strain and strain rate curves into their respective latent spaces. The number of components was chosen such that 95% of the variance was captured (5 principal components for circumferential strain, 4 for radial strain, and 11 and 12 for circumferential and radial strain rates, respectively).

The fitting of the PCA and any further steps were implemented in Python using the scikit-learn library.

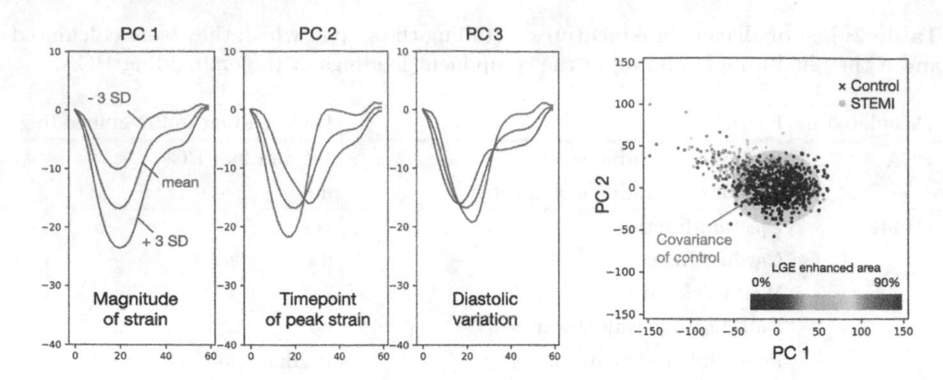

Fig. 1. Result of the PCA for circumferential strain. On the left, the first three principal components, fitted to the control group, are shown. On the right, each segment is represented by its first two principal components and colour-coded with the percentage of enhanced area for this segment (STEMI group). The majority of segments with an LGE enhanced area > 50% lie outside the distribution of control cases as defined by the covariance of the control group.

2.4 Embedding of Strain Patterns to Visualise Evolution

As the number of features (9 and 23 latent dimensions for strain and strain rates, respectively, plus functional measurements as listed in Table 2) was too large to facilitate a simple, intuitive observation of the changes in strain patterns, a second PCA was fitted to the set of features to produce a 2-dimensional embedding. This second PCA will further be called the "embedding PCA" to avoid confusion with the first PCA, which reduced the temporal strain curves to the strain patterns.

In a first step, all features were averaged over the LAD territory (AHA segments 1, 2, 7, 8, 13, 14) and the remaining "remote" segments. Then, the embedding PCA was fitted to both the control and the STEMI groups to create the two-dimensional map shown in Fig. 2. The last step consisted of the calculation of 1) the Euclidean distance between the two samples representing the strain at months 1 and 12 for the LAD territory and remote regions of each individual heart in the 2D embedding plane, and 2) the Mahalanobis distance of each sample with respect to the control group distribution.

3 Results

The results of the first PCA, which reduced the temporal strain curves to strain patterns, were well aligned with previous findings by *Tabassian et al.* [14]. Cardiac strain in infarcted segments was hypothesised to be 1) smaller in amplitude (hypokinetic) or even dyskinetic (positive strain where negative was expected and vice versa) and 2) dyssynchronous (variations in timing of shortening and lengthening of the segments) [15]. These patterns were reflected in the first and

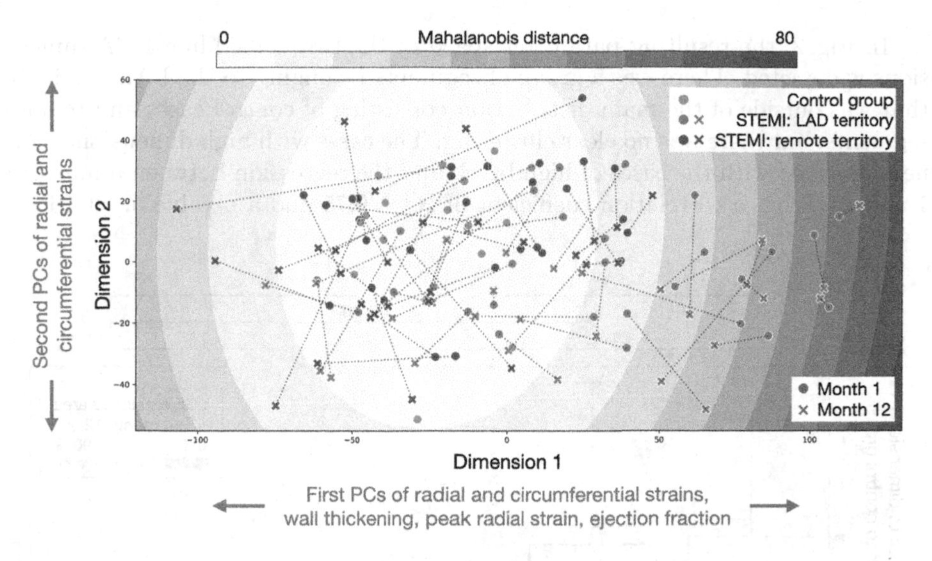

Fig. 2. Two-dimensional embedding of strain patterns in LAD, remote, and control segments. For the STEMI group, the evolution in strain features is represented by dashed lines between months 1 and 12. The Mahalanobis distance with respect to the control group is indicated by the grey contours. The features that vary strongly in the two dimensions are listed along the axes.

second principal components in both radial and circumferential strain, as illustrated in Fig. 1. When applying the condition that 95% of the variance should be explained by the principal components, the number of components needed to be increased to more than 10 for the two strain rates, but when looking at their importance for the component loadings of the embedding PCA, all were deemed to have little variation in the dimensions of the embedding PCA.

Previous studies have attempted to detect the presence of infarcts with a single, strain-related (global, not segmental) value [8], but in the present study, there was no individual feature and no small number of combined features, calculated in the primary PCA, that correlated highly with the extent of LGE.

With two extracted dimensions, the embedding PCA covered 93.5% of the variance in the features. The embedding PCA component loadings showed that dimension 1 (which explained 78.2% of variance) best explained the variance in the first principal components of radial and circumferential strain, wall thickening, peak radial strain, and ejection fraction, while the second principal components of radial and circumferential strain varied most along the direction of dimension 2 (which explained 15.3% of variance). This meant that the two main temporal strain patterns (amplitude of strain and time delay of peak strain) were well encoded in the resulting 2-dimensional embedding. Interestingly, the two classic measurements, circumferential and radial peak strain, showed little variation in dimension 1 and only weak importance in dimension 2. The loading values for all of the other features listed in Table 2 were also low.

In Fig. 2, the resulting parameter space of the two embedding PCA dimensions is depicted. There was a group of segments belonging to the LAD territory that was outside of the main distribution consisting of control cases and remote segments, but there was no clear clustering. The cases with high dimension 1 values coincided with the cases of high LGE, and the regression between dimension 1 and LGE has a correlation coefficient of $R^2 = 0.75$ and a p-value of <0.01.

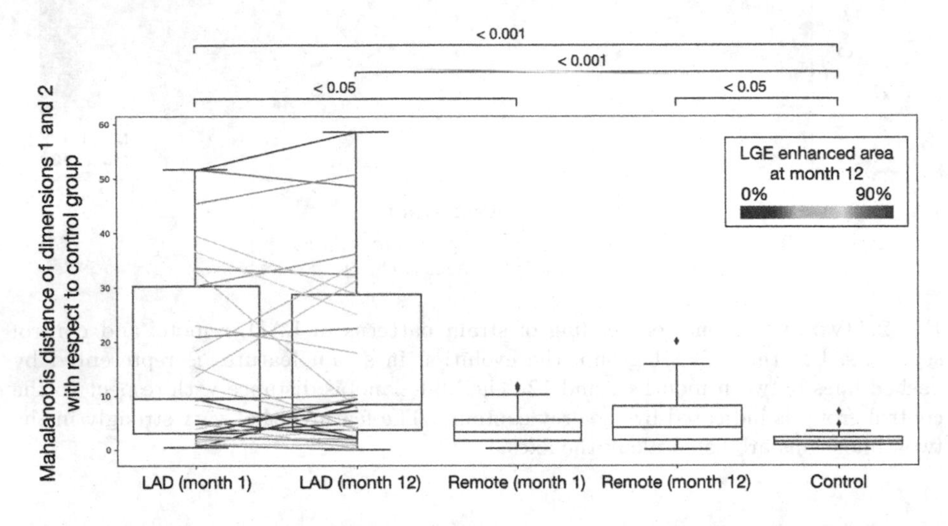

Fig. 3. Comparison of the Mahalanobis distance for the three regions at months 1 and 12. For the LAD (STEMI) territory, the evolution between the two timepoints of each segment in terms of the Mahalanobis distance is presented with a line, colour-coded by the percentage of LGE enhanced area at month 12. Higher LGE scores were associated with a greater Mahalanobis distance, but there was no evidence that the percentage of LGE had an effect on the direction of evolution.

Figure 3 shows the Mahalanobis distance of the LAD and remote segments versus the control group segments. The Mahalanobis distance in the segments within the LAD (STEMI) territory was significantly elevated in comparison to the control group segments. The plot allows one to look at the direction in which segments evolved, such as recovering towards the control group (negative slope) or evolving further away from the normal strain (positive slope). There was no clear direction of evolution, either when looking at cases of high or low percentages of LGE-enhanced area.

When considering the Euclidean distance between timepoints month 1 and month 12, remote segments tended to move more than infarcted segments (Fig. 4), and, overall, segments within hearts that had a high percentage of LGE enhanced area moved less than their counterparts in hearts with a low LGE score.

While dimension 1 of the embedding PCA correlated well with the amount of LGE ($R^2 = 0.77$), the question remains if this complex indicator, which takes into account the strain patterns as well as some global functional measurements, is better than commonly used strain indicators of reduced cardiac contraction, such as peak strains. Indeed, peak circumferential strain ($R^2 = 0.85$) and peak radial strain ($R^2 = 0.8$) correlated well with LGE, too.

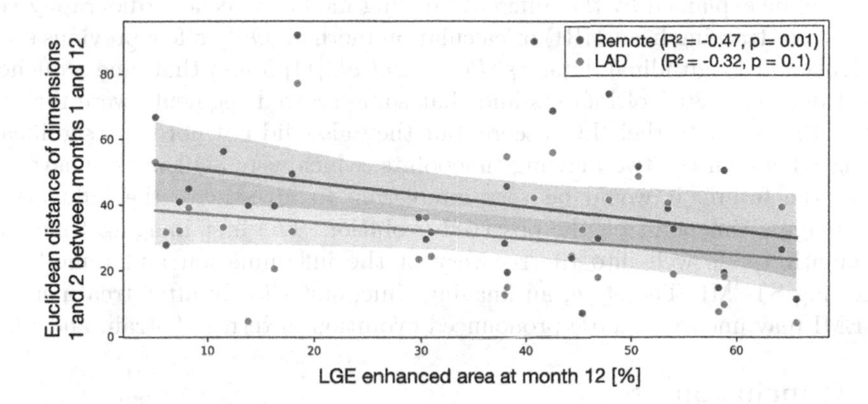

Fig. 4. Euclidean distance between months 1 and 12, representing the change each segment underwent in terms of strain.

4 Discussion

Infarcted segments in the LAD territory were thought to evolve differently than segments in the remote territory at the outset of the study. At month 1, the infarct can be considered stable and the inflammation resolved, so that the extent of the infarct and strains at month 12 remain unchanged. For the remote segments, their evolution was not as clear; one hypothesis was that in hearts with small infarcts, strain may change very little over time as the overall ability of the heart to contract is not much reduced. Vice versa, in hearts with severe infarcts, the remote segments may have to contract harder to generate sufficient cardiac output. In this paper, there was no evidence to support such hypotheses.

There was also no clear indication as to why segments evolved favourably or worsened in terms of strain. A few reasons for this, apart from the fact that such evolution may in fact not be reliably predicted by current measurements of strain, lie in the shortcomings of this study: 1) The size of the two groups was small; 2) only the LAD territory was considered; and 3) longitudinal strain was omitted for quality reasons but is commonly, at least as a global measurement, regarded as a good predictor of the presence of infarction. Further validation of the strain calculations, with the aim to include longitudinal strains, is needed to ensure that the inability to detect a clear evolution of cardiac dysfunction in STEMI patients is not due to inaccuracies with the overall strain calculations or the omission of longitudinal strain information.

In another paper that followed a very similar dimensionality reduction scheme, strain and strain rates were found to have higher predictive power for LGE [14] than classically used end-systolic strains. Interestingly, here, strain rates played an inferior role, and there was no single, individual feature or small number of combined features that correlated better with LGE than peak strains; this may be explained by the different imaging modality (echocardiography versus feature tracking from MRI) or calculation method. Only a few previous studies looked at longitudinal datasets. *Polacin et al.* [11] found that cine sequences could detect 75–80% of infarcts and that some scarred segments were reclassified with respect to their LGE score, but they also did not note any significant changes between the two imaging timepoints, which were 3-12 weeks apart.

In the future, it would be very interesting to investigate the influence of the imaging timepoint on the observed evolution. The first imaging timepoint at month 1 was well into the recovery of the inflammation and remodelling following STEMI. Therefore, an imaging timepoint shortly after treatment for STEMI may uncover a more pronounced evolution in terms of strain and LGE.

5 Conclusion

This study is exploratory in that it aims to investigate the evolution of temporal, segmental strain in STEMI patients over a time period long enough to ensure concluded and stable remodelling. It employed the entire temporal strain curves of each AHA segment rather than a single (semi-quantitative) measurement of wall motion or peak strains, and it described these strains in a concise manner by relating the strain to the most important temporal patterns.

On the considered set of STEMI cases, there was no clear trend in the evolution of strains between the two timepoints. The study shows that, when maintaining the temporal information in strain patterns, no better predictive power of LGE extend was achieved compared to classic measurements like peak strains.

However, the proposed approach to tracking the evolution of strain across time may be used in other longitudinal studies to help clinicians better understand the patient's outcome.

Acknowledgements. The authors would like to thank Michel Ovize, Thomas Bochaton, and Nathan Mewton for sharing in-vivo data from the HIBISCUS cohort. We also thank Circle Cardiovascular Imaging (Calgary, Canada) for making the CVI42 software package available for research purposes. We acknowledge the support of the French Agence Nationale de la Recherche (ANR) under grants ANR-19-CE45-0020 (SIMR project), ANR-11-LABX-0063 (LABEX PRIMES of Univ. Lyon), and ANR-19-CE45-0005 (MIC-MAC project), and the Fédération Francaise de Cardiologie (MI-MIX project, Allocation René Foudon).

References

1. Abate, E., et al.: Value of three-dimensional speckle-tracking longitudinal strain for predicting improvement of left ventricular function after acute myocardial infarction. Am. J. Card. **110**(7), 961–967 (2012)

2. Antoni, M.L., et al.: Prognostic importance of strain and strain rate after acute myocardial infarction. Eur. Heart J. **31**(13), 1640–1647 (2010)
3. Bernelin, H., et al.: Neprilysin levels at the acute phase of ST-elevation myocardial infarction. Clin. Cardiol. **42**(1), 32–38 (2019)
4. Bochaton, T., et al.: Impact of age on systemic inflammatory profile of patients with ST-segment-elevation myocardial infarction and acute ischemic stroke. Stroke **53**(7), 2249–2259 (2022)
5. Cao, J.J., et al.: Left ventricular filling pressure assessment using left atrial transit time by cardiac magnetic resonance imaging. Circ. Cardiovasc. Imaging **4**(2), 130–138 (2011)
6. Claus, P., Omar, A.M.S., Pedrizzetti, G., Sengupta, P.P., Nagel, E.: Tissue tracking technology for assessing cardiac mechanics. JACC Cardiovasc. Imaging **8**(12), 1444–1460 (2015)
7. Kim, R.J., et al.: The use of contrast-enhanced magnetic resonance imaging to identify reversible myocardial dysfunction. N. Engl. J. Med. **343**(20), 1445–1453 (2000)
8. Mangion, K., McComb, C., Auger, D.A., Epstein, F.H., Berry, C.: Magnetic resonance imaging of myocardial strain after acute ST-segment-elevation myocardial infarction: a systematic review. Circ Cardiovasc. Imaging **10**(8), e006498 (2017)
9. Mollema, S.A.: Viability assessment with global left ventricular longitudinal strain predicts recovery of left ventricular function after acute myocardial infarction. Circ. Cardiovasc. Imaging **3**(1), 15–23 (2010)
10. Pedrizzetti, G., Claus, P., Kilner, P.J., Nagel, E.: Principles of cardiovascular magnetic resonance feature tracking and echocardiographic speckle tracking for informed clinical use. J. Cardiovasc. Magn. Reson. **18**(1), 51 (2016)
11. Polacin, M., et al.: Segmental strain for scar detection in acute myocardial infarcts and in follow-up exams using non-contrast CMR cine sequences. BMC Cardiovasc. Disord. **22**(1), 226 (2022)
12. Richardson, W.J., Clarke, S.A., Quinn, T.A., Holmes, J.W.: Physiological implications of myocardial scar structure. Compr. Physiol. **5**(4), 1877–1909 (2015)
13. Rogosnitzky, M., Branch, S.: Gadolinium-based contrast agent toxicity: a review of known and proposed mechanisms. Biometals **29**(3), 365–376 (2016)
14. Tabassian, M., et al.: Machine learning of the spatio-temporal characteristics of echocardiographic deformation curves for infarct classification. Int. J. Cardiovasc. Imaging **33**(8), 1159–1167 (2017)
15. Voigt, J.U., Cvijic, M.: 2- and 3-dimensional myocardial strain in cardiac health and disease. JACC Cardiovasc. Imaging **12**(9), 1849–1863 (2019)
16. Yu, S., et al.: Correlation of myocardial strain and late gadolinium enhancement by cardiac magnetic resonance after a first anterior ST-segment elevation myocardial infarction. Front. Cardiovasc. Med. **8**, 705487 (2021)

Localizing Cardiac Dyssynchrony in M-mode Echocardiography with Attention Maps

Marta Saiz-Vivó[1(✉)], Isaac Capallera[1], Nicolas Duchateau[2,3], Gabriel Bernardino[1], Gemma Piella[1], and Oscar Camara[1]

[1] Physense, BCN Medtech, Department of Information and Communication Technologies, Universitat Pompeu Fabra, Barcelona, Spain
{marta.saiz,gabriel.bernardino,gemma.piella,oscar.camara}@upf.edu, isaac.capallera01@estudiant.upf.edu

[2] Univ Lyon, Université Claude Bernard Lyon 1, INSA-Lyon, CNRS, Inserm, CREATIS UMR 5220, U1294, 69621 Lyon, France
nicolas.duchateau@creatis.insa-lyon.fr

[3] Institut Universitaire de France (IUF), Paris, France

Abstract. Cardiac Resynchronization Therapy (CRT) is a treatment aimed at restoring the electrical synchronization in patients with heart failure and intraventricular conduction delay. However, over 30% of patients do not respond to CRT. Septal Flash (SF), an abnormality characterized by a rapid inward-outward abnormal motion at early systole, has been linked to an improved response to CRT in patients with Left Bundle Branch Block (LBBB). In clinical practice, the detection of SF is usually performed manually through echocardiographic acquisitions, which is subjective and dependent on the operator's experience. To address this issue, a deep classification model for automatic SF detection from 2D anatomical M-mode images has been proposed. Additionally, this work focuses on SF localization from gradient-based attention maps, which provide a visual explanation of the output prediction of the model. Two models based on convolutional neural networks (CNNs) were trained with original and cropped M-modes from 143 patients, and achieved an accuracy of 0.83 and 1.0 respectively, on 29 testing patients. The attention map visualization showed that in SF cases, the models effectively identified the discriminant regions, while in non-SF cases, the maps appeared more dispersed. Further research is necessary to quantitatively evaluate the attention map results.

Keywords: Localization · M-mode echocardiography · heart failure · cardiac resynchronization therapy · attention maps

1 Introduction

Cardiac resynchronization therapy (CRT) is a medical treatment developed to recover electrical synchrony in heart failure patients with intraventricular conduction delay, characterized by disordered ventricular contractions and adverse

O. Bernard et al. (Eds.): FIMH 2023, LNCS 13958, pp. 688–697, 2023.
https://doi.org/10.1007/978-3-031-35302-4_70

clinical prognosis [3]. However, a significant amount of patients do not respond well to the treatment, with CRT non-response rate still over 30% [15]. Among the identified mechanisms amenable to CRT response, septal flash (SF) is indicative of electrically mediated dyssynchrony such as left bundle branch block (LBBB) [17]. It is characterized by a fast inward-outward motion of the septum that occurs during early systole or isovolumetric contraction period and mostly ends before aortic valve opening. Due to the early contraction of the right ventricle (RV), a transseptal pressure difference is created that tethers the septum towards the left ventricle (LV) [3]. Several studies [1,6] have shown SF to be a strong predictor of CRT response in patients with LBBB, thus accurate identification of SF is of interest for CRT patient selection and response prediction.

In clinical practice, SF is commonly assessed through simple 'eye-balling' from 2D transthoracic echocardiography [3]. However, visual assessment methods rely heavily on operator's experience, which could lead to subjective diagnosis. Consequently, several works have focused on more automatic detection of SF. For example, statistical atlases and representation learning were proposed to automatically detect and quantify abnormal ventricular motion, including SF [7,8,11,16]. With recent computational advancements, deep learning methods have emerged as powerful tools for image analysis tasks such as object classification and detection, segmentation and registration, offering new perspectives for the automatic characterization of mechanisms relevant to CRT response [13]. In the field of medical image classification, several authors employ convolutional neural network (CNN)- based methods. However, these models are often seen as 'black boxes' offering little to no explanation on why the output prediction was chosen. In the context of our application, a CRT response prediction model was proposed by Puyol-Antón et al. [12] from cardiac magnetic resonance images with an interpretable variational autoencoder. More generally, many works build upon gradient-based class activation mapping (CAM) approaches, such as Grad-CAM++ [5,18], to offer a visual explanation of the output prediction via localization maps.

Regarding SF detection, Qu et al. [13] recently proposed a CNN-based approach to classify 2D+t B-mode sequences with SF. To capture both spatial and temporal context (required to detect the abnormal septal motion) the authors proposed a linear attention cascaded net (LACnet) with CNN-based encoders and a LSTM- based decoder for temporal feature extraction. However, the design of a complex deep learning architecture with temporal units capable of handling long-term dependencies was required, and handling sequential image data may lead to increased computational complexity compared to simple 2D image processing. Furthermore, the localization of the SF region, employed in the model's output prediction, was not provided.

Motion-mode (M-mode) is an echocardiography technique that provides a one-dimensional view of the tissue along a specific ultrasound beam in different temporal instances, enabling the analysis of tissue motion from 2D images, where the x-axis represents time. Due to its high temporal resolution, it is often used for septal motion analysis and SF detection [6]. A variant of M-mode, known as anatomical M-mode [4], allows the extraction of M-mode images along freely specified lines as a post-processing.

In this work, we obtain localization maps of SF prediction from 2D echocardiographic sequences. We take advantage of virtual M-mode images to use gradient-based class activation maps that consider both spatial and temporal aspects of the sequence. We thoroughly evaluate their ability to localize SF by their provision of discriminant image regions employed in the model's output prediction, which can help to better interpret echocardiographic images, in a longer term objective of assisting less experienced clinicians to find SF and develop an interactive process.

2 Methods

2.1 Dataset

In this study, the dataset was provided by the Hospital Clínic de Barcelona. It consisted of 2D ultrasound sequences (GE Vingmed Ultrasound A.S., Horten, Norway) with the corresponding electrocardiogram (ECG) from 89 CRT patients acquired both at baseline (n = 89) and 12 months follow-up (n = 89) after the CRT implant. The CRT inclusion criteria corresponded to the international guidelines: symptomatic heart failure with QRS duration >120 ms, and NYHA classification III-IV or NYHA II who covered less than 500 m in the 6 min walking test. The transthoracic echocardiography sequences were acquired in an apical four-chamber view, useful for the assessment of abnormal septal motion. After the deletion of corrupted sequences, we analyzed a total of 143 sequences of two subpopulations, with (n = 55) and without SF (n = 88).

The SF/non-SF labels were annotated by one experienced observer and controlled by another experienced observer. Septal segmentations at the end-diastolic frame were provided as a sequence of 62 points with spatial and temporal normalization, also manually marked.

2.2 Virtual M-mode Generation

Ultrasound data was provided in echoline format and a B-mode scan reconstruction was performed with isotropic pixel size of 0.03 mm. The average frame rate was 60 fps. Using the R peaks of the ECG, the frames corresponding to one complete cardiac cycle were extracted (30–150 frames). The reconstructed image sequence was resized to 500×500 pixels, reducing computational complexity for posterior M-mode generation.

The generation of a virtual M-mode image post-hoc from a B-mode sequence implies reading 2D pixel samples along a specified line for each frame. To achieve this whilst capturing the septal motion, a cut plane in the direction perpendicular to the septum was applied to slice the B-scan considering the concatenated temporal frames as third dimension. Figure 1 illustrates the steps involved in the M-mode generation. Two points above and below the septal point of interest (mid-basal region), automatically extracted from the septum segmentation

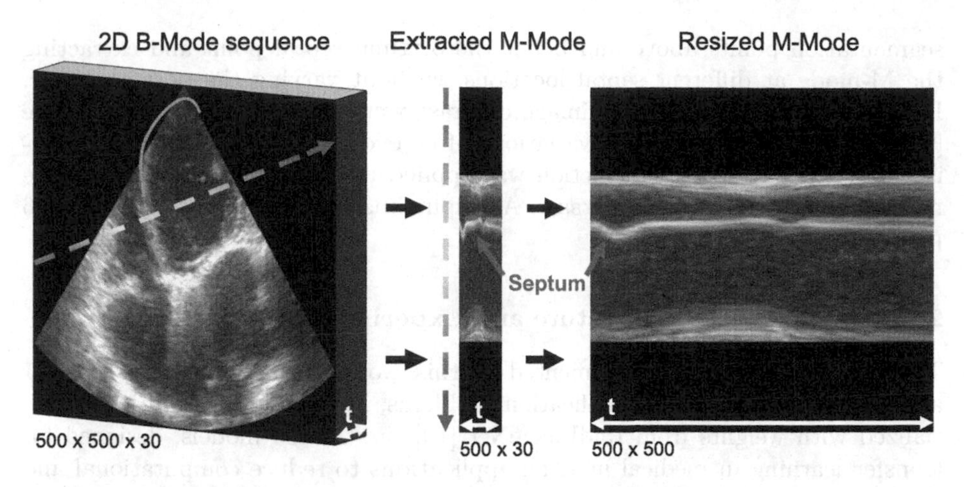

Fig. 1. Steps for virtual M-mode generation. Left: B-mode sequence with septum perpendicular vector (arrow) defined from two points (in red) sampled from septum segmentation (in green). Middle: synthetic M-mode. Right: resized M-mode (Color figure online)

as the 10th and 20th point, were used to define the tangent direction of the mid-myocardium, from which the normal vector, in the direction towards the LV blood pool, was computed (Fig. 1-left). To slice the B-scan sequence as a 2D+t volumetric image, the *vtkImageReslice()* function from the VTK library in Python was employed. Figure 1-middle illustrates the cut image, where the x-axis corresponds to the number of frames and the y-axis to the grayscale pixel intensities along the vector perpendicular to the septum, from the RV region (above) to the LV lateral wall (below). Finally, a temporal resampling was applied to generate the final M-mode image with 500 × 500 pixels (Fig. 1-right). The proposed approach enabled the semi-automatic extraction of anatomical M-modes, in contrast to conventional methods [4].

2.3 Preprocessing and Data Augmentation

To evaluate the impact of extracting the region of interest on SF recognition rate, an additional dataset was created by cropping the synthetic M-modes. Specifically, the M-mode images were cropped along the spatial axis, considering only the upper half region of the image scan that contains the septum. Moreover, given that SF occurs very early in the cardiac cycle, the M-mode was cropped along the temporal axis to retain the initial half of the cardiac cycle. Both the original and cropped M-mode datasets were used to develop SF classification models described in Sect. 2.4.

Data augmentation allows increasing the performance of image classification tasks. In this work, data augmentation was applied both offline and online for each dataset. The offline data augmentation increased the number of M-mode image training samples by a factor of 11 through sampling 5 additional septum segmentation points above and below the original septal point and extracting the M-mode at different septal locations, without varying the normal vector. Random perturbations of the image contrast were applied online in the range [0.5, 2], adding further image variations after every epoch. For image normalization purposes, mean subtraction was applied to every input image from the mean image of the training dataset. All input images were resized to 256×256 pixel size.

2.4 CNN Model Architecture and Experimental Settings

The CNN-based model implemented in this work adopted the DenseNet121 architecture, relevant for classification problems, as a backbone, and it was initialized with weights from RadImageNet [10] pre-trained models, designed for transfer learning in medical imaging applications to reduce computational and data expenses [2]. A fully connected layer with 2 output units and a softmax activation function were added to the DenseNet121 backbone to obtain the final binary classification output. After initializing with the pre-trained weights, the full architecture was retrained. The model was trained for 5 epochs (which was enough for convergence, probably thanks to pre-training) with an initial learning rate of 0.001, batch size of 4 and Adam optimization.

Two experiments were performed, on the cropped and original datasets with similar settings. The stratified training/test dataset split was performed with a ratio of 80:20 based on patient identifier to avoid placing highly correlated samples of the same patient in different groups and reduce overfitting. The same procedure was followed to generate the validation set from the training set. All experiments were implemented on NVIDIA Tesla T4 GPU with Python and Keras library. Once trained, the testing samples were employed to evaluate the diagnostic performance of the model with the area under the curve (AUC) and the accuracy as evaluation metrics.

2.5 Implementation of Grad-CAM++ Attention Method

To localize SF, the discriminative image regions used by the CNN to predict the output were identified through Grad-CAM++ attention maps [5] on the test samples. This algorithm is a visualization tool to better understand CNN model predictions through class-specific activation maps. The activation maps represent regions (with higher intensity) of the input image where the CNN model has "looked" to output the predicted class.

To estimate these regions, Grad-CAM++ computes the gradients of the predicted class with respect to the last convolutional layer. In this work the last convolutional layer of DenseNet121 was selected with feature map pixel dimensions of 8×8. The obtained heatmap was normalized in the range [0,1] and upsampled to input image resolution. The attention algorithm was implemented through the 'tf-keras-vis' visualization toolkit [9].

Table 1. Test accuracy and area under the curve (AUC) of convolutional neural network (CNN) models trained with virtual original M-mode and cropped M-mode images

CNN Model	Accuracy	AUC
w. original M-mode	0.83	0.95
w. cropped M-mode	1.00	1.00

3 Results

The models were trained and evaluated with 91 training patients and 23 validation patients (respectively 5005 and 253 samples due to off and on-line data augmentation), and 29 testing patients.

3.1 Results on the Test Dataset

To evaluate the trained models, the test dataset was preprocessed as done for the training set but without the data augmentation. Table 1 shows the test results for the models trained with the original M-mode and the cropped M-mode. The network trained with cropped images obtained the highest accuracy (100%), although accuracy for the model trained with original M-mode was reasonably high (83%). The AUCs for the cropped and original M-mode models were 1.0 and 0.95, respectively.

3.2 Grad-CAM++ Visualization on the Test Dataset

Figure 2 shows Grad-CAM++ visualization results on 4 test patients with correct output predictions from original M-mode (Fig. 2, 1st and 3rd column) and cropped M-mode (Fig. 2, 2nd and 4th column) CNN models. On patients with SF (top rows) we observe that both CNN models accurately localize the region of SF occurrence, both along the spatial (vertical) and temporal (horizontal) axis, as the regions with high activation occur between mitral valve closure (MVC) and aortic valve opening (AVO). In the non-SF cases (bottom rows) the heatmaps appear with less localized attention, specially in the cropped M-mode model where other areas towards the RV region have been activated.

Figure 3 shows Grad-CAM++ visualization results for patients that were incorrectly predicted by the original M-mode model. As observed in the corresponding attention map, the network failed to detect the discriminant image features related to SF, thus leading to activated regions far from the septal region. In the first patient (Fig. 3, 1st column) the M-mode exhibits low contrast between septum and LV blood pool, which might explain the failure in the localization. The last two columns correspond to incorrect predictions of SF for non-SF cases. In the first one, the attention map appears focused on the septum, likely due to the detection of a perturbation similar to SF. On the second one, regions corresponding to the LV lateral wall appear mistakenly activated.

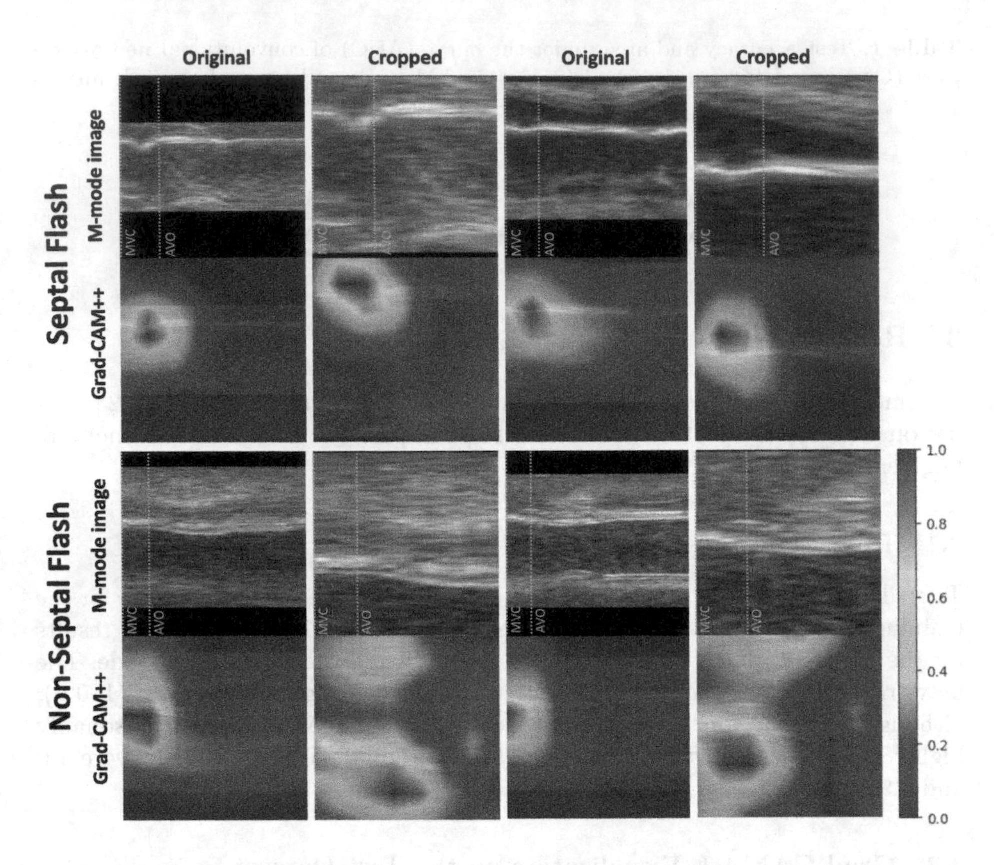

Fig. 2. Grad-CAM++ heatmap results for septal flash (SF) and non-SF correct predictions. 1st and 3rd column: extracted from original M-mode trained model. 2nd and 4th column: extracted from cropped M-mode trained model. MVC; mitral valve closure. AVO; aortic valve opening

4 Discussion

In this work we proposed the development of 2D SF detection models from generated anatomical M-mode images and the localization of SF through gradient-based attention maps.

In the test dataset, the model trained with original virtual M-modes obtained a reasonably high accuracy, with a higher precision for non-SF cases. The network's performance further improved to a perfect classification accuracy of 100% when trained with cropped images around the temporal and spatial region of interest. However, this increased (and perfect) accuracy may be due to the over-simplification of the SF classification problem by reducing the search area. The experiment demonstrated the feasibility of SF detection from the generated M-modes. Nonetheless, it is crucial to acknowledge the potential limitations of the

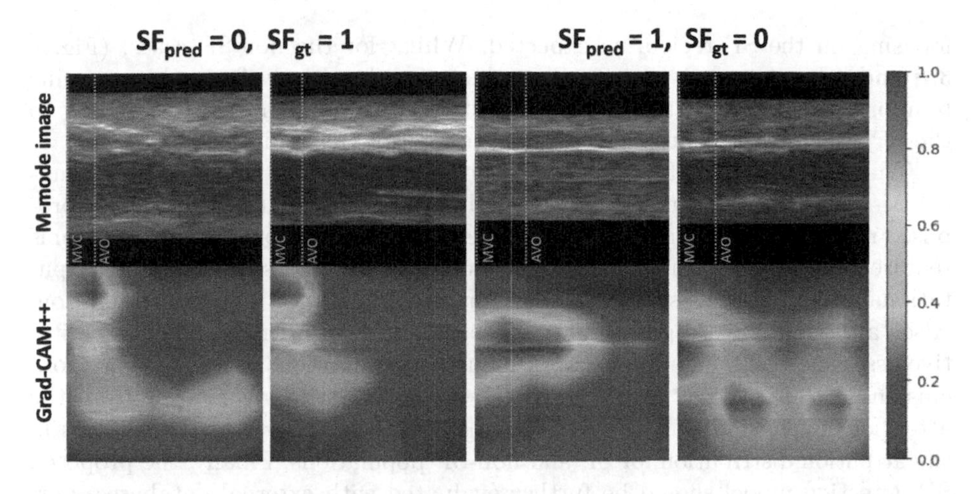

Fig. 3. Grad-CAM++ heatmap results for septal flash (SF) and non-SF incorrect predictions from original M-mode model. pred: predicted class, gt: ground-truth class. MVC; mitral valve closure. AVO; aortic valve opening

experiment and model, including the generalizability of the results to different datasets.

The accuracy of the cropped M-mode classification model surpasses the performance reported in previous studies, including state-of-the-art CNN-based detection from US sequences (LACNet, 91% [13]) or machine learning-based classification from myocardial strain (linear discriminant analysis (LDA), 94% [16]). These findings indicate that the cropped M-mode model has the potential to be an effective approach for SF detection in medical imaging. However, each test dataset is unique and further evaluation with large and external testing databases is necessary to comprehensively understand the effectiveness and limitations of the model for SF classification.

Gradient-based class activation maps were employed to visualize the differentiating image regions for the network's output prediction. As observed in Fig. 2, for SF prediction (Fig. 2, 2nd row) the attention maps show a localized activation in the septum, specifically in the motion abnormality temporal region (between mitral valve closure and aortic valve opening). This suggests that the network has successfully learnt the discriminant image features characterizing SF in images with different septum localization and different septum/LV blood pool contrast, a necessary requirement for generalization purposes. On the other hand, for the case of non-SF patients (Fig. 2 bottom row), the attention maps show higher activations in regions other than the septum and in general, a higher dispersion. This is especially observed on the cropped M-mode activation maps (Fig. 2, 2nd and 4th column). Our hypothesis is that upon not finding the characteristic motion perturbation of SF, the networks focuses on random locations; however, further testing is needed to confirm this. Regarding the wrongly classified SF cases (Fig. 3, 1st and 2nd column), the attention maps have difficulty focusing on the SF region, as expected. Whilst for the non-SF cases, (Fig. 3, 3rd and 4th column), it appears that the network wrongly focused on specific temporal (3rd column) and spatial regions (4th column) not corresponding to the SF region, which could have led to the incorrect classification as SF.

The development of 2D SF detection models from generated anatomical M-modes has shown promising results. Nonetheless, concerning the proposed methodology of anatomical M-mode generation, the processing algorithms required to downsize the original reconstructed B-mode sequence for computational complexity reasons, with the corresponding loss of spatial resolution. Also, although superimposing gradient-based activation maps allows qualitative assessment of where the network focused its attention, to draw deeper conclusions, we propose as future work to extract quantitative metrics from the attention maps, similar to *Schöttl et al.* [14], and to analyse the differences in the attention distribution for SF and non-SF populations. Finally, the proposed SF detection model should be further evaluated with external databases. Further work will be conducted to evaluate SF localization from gradient-based approaches in models trained for CRT response prediction.

Acknowledgements. This project has received funding from the European Union's Horizon 2020 research and innovation programme under grant agreement No 101016496 (SimCardioTest), from the French ANR (LABEX PRIMES of Univ. Lyon [ANR-11-LABX-0063] and the JCJC project "MIC-MAC" [ANR-19-CE45-0005]). G. Piella is supported by ICREA under the ICREA Academia programme. They are also grateful to M. Sitges and A. Doltra (Hospital Clínic de Barcelona, Spain) for providing the imaging data related to the studied population, and to B. Bijnens (ICREA Barcelona, Spain) for initial discussions on this topic.

References

1. Bennett, S., et al.: Septal flash as a predictor of cardiac resynchronization therapy response: a systematic review and meta-analysis. J. Cardiovasc. Echogr. **31**(4), 198 (2021)
2. Cadrin-Chênevert, A.: Moving from imagenet to radimagenet for improved transfer learning and generalizability. Radiol. Artif. Intell. **4**(5), e220126 (2022)
3. Calle, S., Delens, C., Kamoen, V., De Pooter, J., Timmermans, F.: Septal flash: at the heart of cardiac dyssynchrony. Trends Cardiovasc. Med. **30**(2), 115–122 (2020)
4. Carerj, S., et al.: Anatomical m-mode: an old-new technique. Echocardiography **20**(4), 357–361 (2003)
5. Chattopadhay, A., Sarkar, A., Howlader, P., Balasubramanian, V.N.: Grad-CAM++: generalized gradient-based visual explanations for deep convolutional networks. In: 2018 IEEE Winter Conference on Applications of Computer Vision (WACV), pp. 839–847. IEEE (2018)
6. Doltra, A., et al.: Mechanical abnormalities detected with conventional echocardiography are associated with response and midterm survival in CRT. JACC: Cardiovasc. Imaging **7**(10), 969–979 (2014)
7. Duchateau, N., De Craene, M., Piella, G., Frangi, A.F.: Constrained manifold learning for the characterization of pathological deviations from normality. Med. Image Anal. **16**(8), 1532–1549 (2012)

8. Duchateau, N., et al.: A spatiotemporal statistical atlas of motion for the quantification of abnormal myocardial tissue velocities. Med. Image Anal. **15**(3), 316–328 (2011)
9. Kubota, Y.: tf-keras-vis (2022). https://keisen.github.io/tf-keras-vis-docs/
10. Mei, X., et al.: Radimagenet: an open radiologic deep learning research dataset for effective transfer learning. Radiol. Artif. Intell. **4**(5), e210315 (2022)
11. Peressutti, D., et al.: Prospective identification of CRT super responders using a motion atlas and random projection ensemble learning. In: Navab, N., Hornegger, J., Wells, W.M., Frangi, A.F. (eds.) MICCAI 2015. LNCS, vol. 9351, pp. 493–500. Springer, Cham (2015). https://doi.org/10.1007/978-3-319-24574-4_59
12. Puyol-Antón, E., et al.: Interpretable deep models for cardiac resynchronisation therapy response prediction. In: Martel, A.L., et al. (eds.) MICCAI 2020. LNCS, vol. 12261, pp. 284–293. Springer, Cham (2020). https://doi.org/10.1007/978-3-030-59710-8_28
13. Qu, M., Wang, Y., Li, H., Yang, J., Ma, C.: Automatic identification of septal flash phenomenon in patients with complete left bundle branch block. Med. Image Anal. **82**, 102619 (2022)
14. Schöttl, A.: Improving the interpretability of gradcams in deep classification networks. Procedia Comput. Sci. **200**, 620–628 (2022)
15. Sieniewicz, B.J., et al.: Understanding non-response to cardiac resynchronisation therapy: common problems and potential solutions. Heart Fail. Rev. **24**, 41–54 (2019)
16. Sinclair, M., et al.: Myocardial strain computed at multiple spatial scales from tagged magnetic resonance imaging: estimating cardiac biomarkers for CRT patients. Med. Image Anal. **43**, 169–185 (2018)
17. Smiseth, O.A., Russell, K., Skulstad, H.: The role of echocardiography in quantification of left ventricular dyssynchrony: state of the art and future directions. Eur. Heart J. Cardiovasc. Imaging **13**(1), 61–68 (2012)
18. Zhang, Y., Hong, D., McClement, D., Oladosu, O., Pridham, G., Slaney, G.: Grad-cam helps interpret the deep learning models trained to classify multiple sclerosis types using clinical brain magnetic resonance imaging. J. Neurosci. Methods **353**, 109098 (2021)

The Extent of LGE-Defined Fibrosis Predicts Ventricular Arrhythmia Severity: Insights from a Preclinical Model of Chronic Infarction

Terenz Escartin[1,2], Philippa Krahn[1,2], Cindy Yu[3], Matthew Ng[2], Jennifer Barry[2], Sheldon Singh[2], Graham Wright[1,2], and Mihaela Pop[2](✉)

[1] Department of Medical Biophysics, University of Toronto, Toronto, Canada
[2] Sunnybrook Research Institute, Toronto, Canada
mihaela.pop@utoronto.ca
[3] University of Waterloo, Waterloo, Canada

Abstract. Abnormal propagation of cardiac electrical impulses in hearts with fibrosis developed post-infarction often lead to rapid ventricular tachycardia (VT) and sudden cardiac death, a major cause of mortality. Certain values of ejection fraction (e.g. EF < 35%) are used clinically to refer scar-related VT patients for ICD implantation; however, these values are not an indication of VT severity (i.e., how fast is the heart rate in VT). Our aim here is to use a preclinical model of chronic fibrosis to determine whether the extent of fibrosis defined by MRI correlates better than EF with the heart rate during VT (a measure known as *VT cycle length*). Specifically, n = 10 pigs with prior infarct underwent MR imaging (i.e., cine scans to calculate EF, and high resolution late gadolinium enhancement (LGE) to measure the extent of fibrosis, followed by an X-ray guided VT inducibility study to determine VT cycle length. The total infarct size in LGE images was given by the extent of dense scar plus that of *gray zone,* GZ (a mixture of viable muscle and collagen fibers, located at the scar periphery), as defined by two clinically accepted segmentation thresholding methods: 5SD (standard deviation) and FWHM (full-width at half maximum), respectively. Overall, LGE-defined scar/GZ corresponded well to infarct heterogeneities observed in collagen-sensitive histological stains. Our quantitative results showed that the amount of LGE-defined fibrosis (relative to the left ventricular volume) correlated well with VTCL (R ~ 0.78), suggesting that it could be a potential clinical predictor of dangerous VT.

Keywords: myocardial infarct · MR imaging · late gadolinium enhancement LGE · heterogeneous fibrosis · ventricular tachycardia substrate

1 Introduction

Abnormally high heart rhythms (tachyarrhythmias) are often associated with structural heart disease, such as myocardial infarction. In particular, scar-related ventricular tachycardia (VT) is a major cause of sudden cardiac death (SCD) [1]. The VT substrate (a mixture of viable myocytes and collagen bundles) allows the electrical impulse to traverse dense scars through these viable paths and to loop around the dense scars with a

O. Bernard et al. (Eds.): FIMH 2023, LNCS 13958, pp. 698–707, 2023.
https://doi.org/10.1007/978-3-031-35302-4_71

VT cycle length (VTCL) depending directly on the scar size [2]. The geometrical configuration of unexcitable dense scars encasing viable paths (i.e., critical channels with reduced conduction velocity of the electrical impulse), forms a reentry circuit (Fig. 1) that facilitates a VT wave to repeatedly enter from one side of the circuit and exit from the opposite side, overriding the normal sinus node-driven heart rhythm [3].

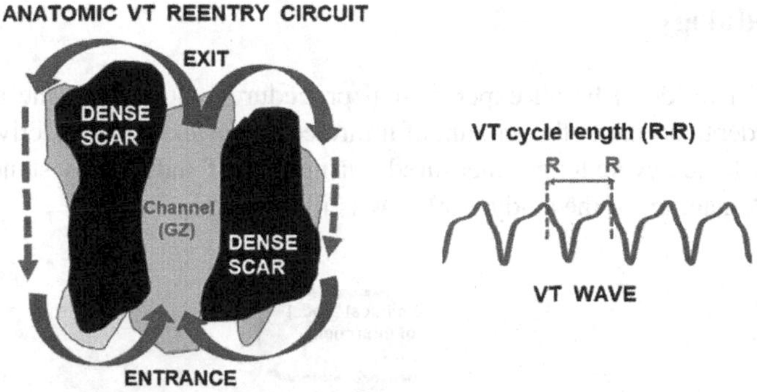

Fig. 1. Reentry circuit in scar-related VT and its associated wave, where the VTCL is determined from consecutive R-R intervals.

In practice, a VT episode can be terminated either by an implantable defibrillator device (ICD) or by an RF ablation procedure that interrupts the circuit by thermally ablating the critical channels. Thus, accurate localization of infarct heterogeneities is critical for VT diagnosis [4]. Moreover, prior to ICD implantation or RF ablation procedures, VT is evaluated in the EP lab during an invasive test inducing VT through rapid pacing.

Several studies have shown that cardiac MR imaging can identify the location and extent of the scar and VT substrate using late gadolinium enhanced (LGE) imaging. Due to its intermediate MR signal intensity (between the signal intensity of healthy myocardium and dense scar), the VT substrate is usually called *gray zone* (GZ). The amounts of scar and GZ are useful clinical predictors of mortality rates [5] or arrhythmia events (as demonstrated by the VT/VF episodes recorded by ICDs) [6]. Unfortunately, most clinical MR imaging exams use 2D LGE methods with spatial resolution limited to ~ 8–10 mm slice thickness, leading to partial volume effects and inadequate quantification of scar and GZ. This also affects the sudden cardiac death (SCD) risk stratification as well as the selection of candidates for ICDs. With respect to scar/GZ quantification from LGE images, several semi-automatic algorithms have been developed, including: standard-deviation (n-SD); full-width-at-half-maximum (FWHM); graph cut/continuous max-flow; and, gaussian-mixture clustering [7]. However, among them, FWHM and n-SD approaches remain the most clinically accepted methods.

In this study, we hypothesized that LGE-derived fibrosis is a better predictor of the VT severity (indicated by VTCL), than the ejection fraction (EF). Specifically, here we aimed to use a high-resolution free breathing 3D LGE method (at 1.4 mm isotropic spatial resolution), in order to identify and adequately quantify the infarct location and size (i.e., amount of dense scar and GZ). To achieve this, we proposed to use a pig model of chronic infarction, and to correlate the amount of LGE-derived fibrotic infarct with the VTCLs recorded during VT inducibility tests performed under x-ray guidance. Additionally, we also correlated the recorded VTCLs with the EF values derived from cine MR imaging scans, and then studied which MRI-based biomarker is a better predictor of fast (dangerous) VTs.

2 Methodology

In this section we describe all experimental procedures, along with the analysis performed in order to correlate the amount of infarcted tissue and EF, respectively, with the severity of VT (i.e., cycle length measured during the VT induced by standard clinical protocols). A diagram of the study workflow is illustrated in Fig. 2.

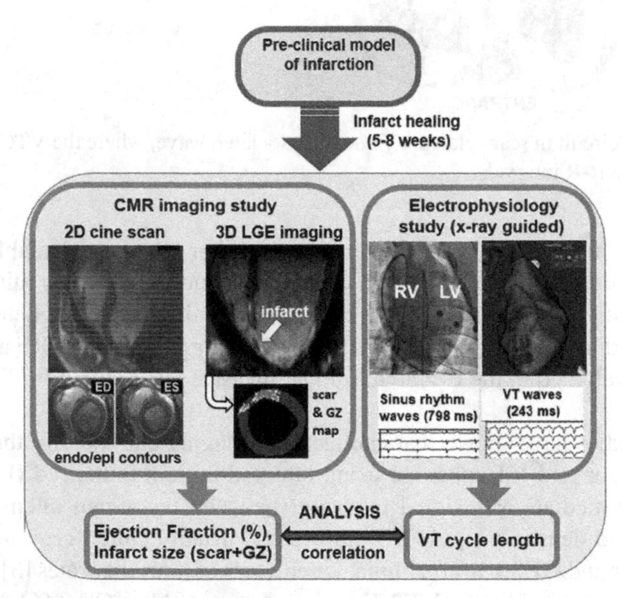

Fig. 2. Schematic diagram of the MR-EP studies and analysis performed in pigs with chronic infarction, to correlate the left ventricular EF and infarct size with VTCL

2.1 Preclinical Model of Infarction

This study included n = 10 swine (each weighing ~30kg prior to the infarct creation). All procedures (i.e., infarct creation, MRI scans, and x-ray guided arrhythmia studies) were conducted with approval from the Animal Care Committee at Sunnybrook Research Institute, Toronto (Canada). During each procedure, the pigs were placed under a cocktail anesthesia (0.05 mg/kg atropine and 33 mg/kg ketamine), maintained by 1–5% isoflurane delivered via a mechanical ventilator. In order to create heterogeneous infarctions comprised of dense scar and GZ (i.e., VT foci), the pigs underwent a left anterior descending artery occlusion via balloon inflation for 90–100 min, followed by reperfusion, as in [8, 9]. The infarcted animals were allowed to heal during the following 5–8 weeks. Note that the infarct heterogeneity was verified by histological examination in select samples, at the completion of MR-EP studies.

2.2 MR Imaging and Associated Image Analysis

All MR scans were performed on a 3T MR750 scanner (GE Healthcare, Waukesha, Wisconsin) using an 8-channel cardiac coil. Amiodarone was injected to avoid spontaneous arrhythmia occurrence and to stabilize the heart rate during the scans.

a) *Cardiac function* was evaluated using a 2D steady-state-free-precession (SSFP) sequence in cine mode. Approximately 18 short-axis slices were prescribed per animal to fully cover the heart ventricles. The following MR imaging parameters were used: 8 views/segment, 20 phases per slice, TR = 4ms, TE = 1.7 ms, flip angle = 45°, FOV = 24 × 21.6 cm, matrix = 224 × 160 and slice thickness = 5 mm.
   The cine SSFP images were analyzed using the *CVI42* software (Circle Cardiovascular Imaging, Calgary, CA). Specifically, we manually draw the endocardial and epicardial contours of the left ventricle at *end systole* (ES) and *end diastole* (ED) phases, which allowed us to compute the ejection fraction, EF, using the formula: EF (%) = (EDV-ESV)/EDV, where EDV and ESV are the blood volumes corresponding to ED and ES phases, respectively.
b) For *infarct imaging,* a bolus of Gd-DTPA (0.2 mmol/kg, Magnevist) was intravenously administered and imaging started ~ 10min after the contrast agent injection. For the LGE scan, we used a free-breathing high resolution 3D inversion recovery (IR) prepared spoiled gradient echo (SPGR) sequence with respiratory navigation. The following MR imaging parameters were used: TE = 1.5 ms, TR = 3.5 ms, flip angle = 15°, BW = 100 kHz, optimized inversion TI ~ 250 ms (to maximize the contrast between the infarcted area and healthy tissue), and voxel size = 1.4 × 1.4 × 1.4mm^3.

In this work we focused only on the left ventricle (LV) analysis, for which we first manually delineated the epicardial and endocardial contours. We then characterized infarct heterogeneities and computed *tissue maps* by defining 3 classes: dense scar; grey zone (GZ); and, healthy myocardium. For this, we performed a thresholding of the signal intensity in the LGE images using the two widely used segmentation approaches: standard deviation (n-SD) method and FWHM method, respectively.

i) For the n-SD method, we used the mean and standard deviation (SD) of the signal intensity (SI). For each LGE slice, an ROI was drawn in the remote/healthy myocardium zone. The mean (mean_remote), peak (peak_remote), and standard deviation (SD_remote) of the SI within the selected remote ROI were calculated. To obtain the aforementioned classes (scar, GZ and healthy tissue) in the resulting *tissue map*, we used cut-off thresholds of the SI as per clinical guidelines [10]:

$$SI\_scar > mean\_remote + 5 * SD\_remote \quad (1)$$

$$mean\_remote + 2 * SD\_remote < SI\_gz < mean\_remote + 5 * SD\_remote \quad (2)$$

where SI_scar is the SI of a pixel classified as dense scar (i.e., infarct core), and SI_gz is the SI of a pixel classified as gray zone. Henceforth, the n-SD method will be referred as the 5SD method.

ii) The second method was based on the full-width half-maximum (FWHM) approach [11], using a 50% threshold and the following definitions:

$$SI\_scar > 0.5 * peak\_scar \tag{3}$$

$$peak\_remote < SI\_gz < 0.5 * peak\_scar \tag{4}$$

where peak_scar is the peak of signal intensity of all fibrotic (infarcted) pixels.

The entire analysis pipeline including all processing steps and both semi-automatic segmentation algorithms is presented in Fig. 3. It employed in-house written scripts and was implemented in Matlab using the *image processing toolbox* and a GUI interface. Notably, approximately 600 LGE images acquired in the 10 pigs were segmented by each method. The LV binary masks were used to calculate the LV volume, per heart.

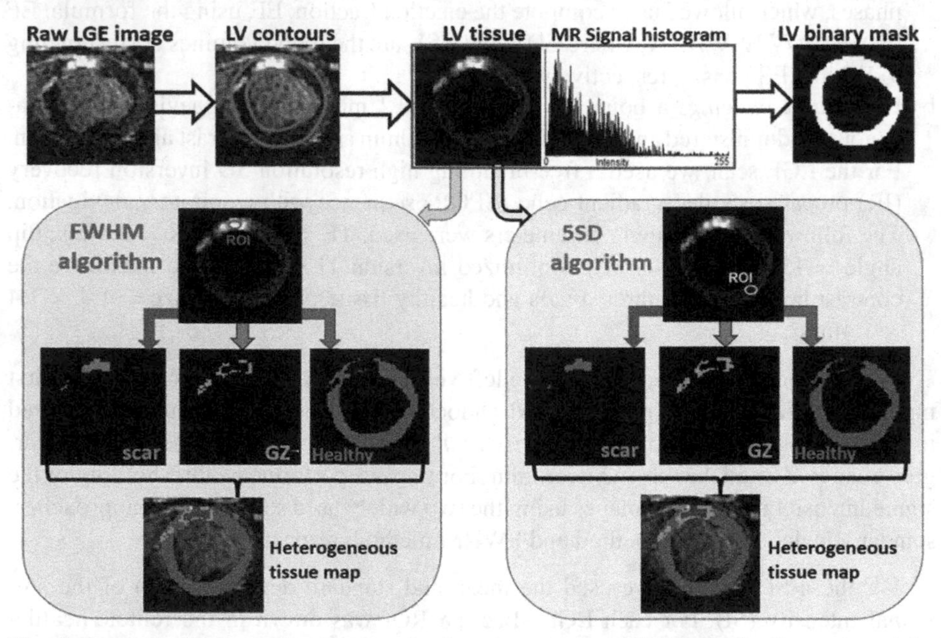

Fig. 3. Image analysis pipeline presented using an exemplary LGE image in short-axis and the associated pixel-wise *tissue maps* obtained by the 5SD and FWHM methods (pink = healthy myocardium, green = GZ, blue = scar).

2.3 Electrophysiology Study and VT Inducibility Protocol

All pigs underwent x-ray guided EP studies the day after the MR scans, to ensure that amiodarone was effectively out of the body and did not impede VT inducibility. All EP studies were carried out under X-ray guidance, using a C-arm Toshiba INFINIX VF-I/SP-S. For the VT inducibility tests, we inserted an SF Thermocool catheter (Biosense Webster Inc., Irvine, CA, USA) into the right ventricle (RV) of each pig, and then performed rapid pacing, mimicking the clinical diagnostic protocols. Specifically, we started with a train of 8 × S1 pacing stimuli at 400–600 ms, followed by S2-S3 extra-stimuli with reduced coupling interval, until VT was induced. The VT waves were recorded and VTCL was calculated from the R-R intervals (peak-to-peak).

2.4 Histology

Select samples (4 mm thickness) from hearts were cut to align with the short-axis view of LGE images and embedded in paraffin. Thin slices (4 μm) were cut from each paraffin block, fixed on large glass slides and stained with collagen-sensitive Masson Trichrome stain. The stained slides were matched to their corresponding short-axis LGE images using anatomical markers (RV/LV insertion point, papillary muscle), and were used for a qualitative comparison with the LGE-derived tissue maps.

2.5 Statistical Analysis

For the quantitative analysis, linear fits were performed in order to derive the correlation between VTCL and % fibrosis (scar + GZ) relative to LV volume, as well as between VTCL and EF (%). In addition, the Bland–Altman analysis was used to evaluate bias and the two-tailed P value was used to statistically compare the amount of fibrosis (scar + GZ) determined by the 5SD method vs. the FWHM method.

3 Results

Figure 4 shows exemplary outputs of the scar/GZ segmentations (note: in the raw LGE image the infarct is indicated by a hyperenhanced area). Both *tissue maps* identified similar infarct patterns and extent (although the 5SD-defined dense scar was slightly smaller), as well as surviving sub-endocardial GZ pixels and healthy pixels. Qualitatively, the infarct heterogeneities corresponded well to those identified in the histology stain (see white rectangles in tissue maps and magnified area in MT slide).

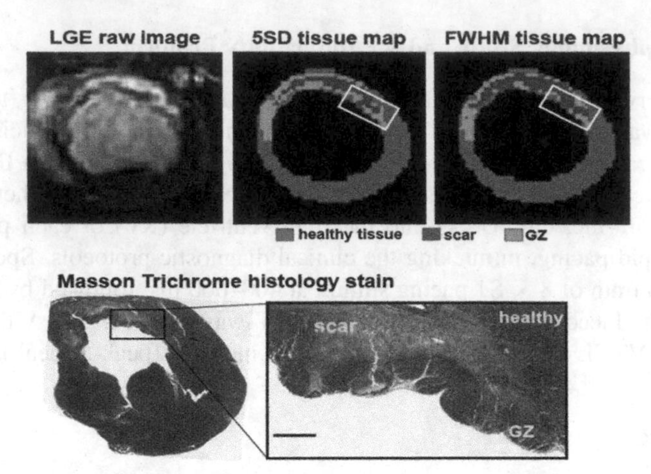

Fig. 4. Qualitative example of the segmentation comparison: *(upper row)* a raw LGE image and resulted pixel-wise tissue maps obtained using 5SD and FWHM methods (pink = healthy myocardium, green = GZ, blue = scar); *(bottom row)* corresponding histology MT slide (collagenous fibrosis stained blue and healthy tissue in dark red). (Color figure online)

Figure 5a presents the results obtained in all animals, relating the VTCL and % fibrosis (scar + GZ) calculated relative to LV volume. We observed that smaller amounts of fibrosis clearly related to faster VT rates (i.e., more dangerous arrhythmia events).

The linear fits yielded good correlations between the VTCL and the % fibrosis defined by either segmentation method ($R > 0.78$). Furthermore, there was no bias between the two segmentation methods, as indicated by the Bland-Altman plot (see Fig. 5b). In addition, the difference in % fibrosis derived by the two segmentation methods was not statistically significant ($P < 0.05$), with a correlation coefficient close to unity.

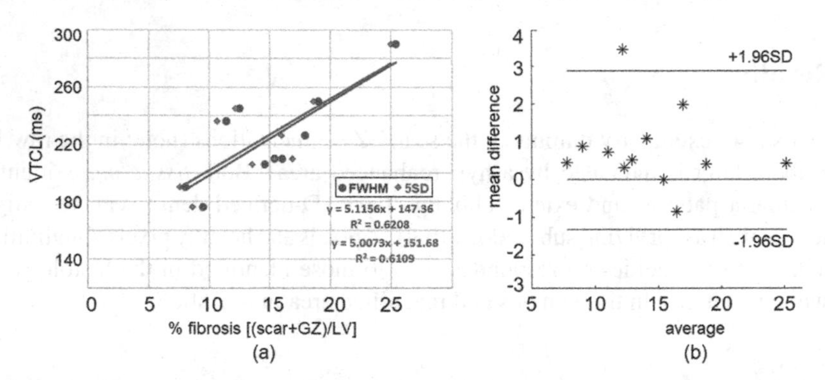

Fig. 5. Results from quantitative analysis: (a) linear fitting of VTCL vs % fibrosis defined by 5SD and FWHM methods; (b) Bland-Altman plot (overall bias ± 95% limits of agreement) showing no bias between the 5SD and FWHM segmentation methods.

Figure 6 shows results from the correlation analysis between the VTCL (ms) and the measured EF values (%) of the left ventricle. Overall, the linear fit yielded a rather poor correlation (R = 0.13) between these two measurements, suggesting that EF is not a good indicator of how fast the VT rate might be in a given post-infarction case.

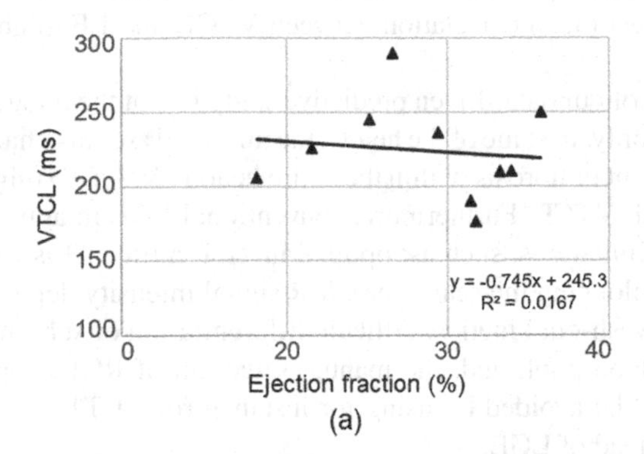

Fig. 6. Linear fitting demonstrating a poor correlation between VTCL (ms) and EF (%).

4 Discussion

Accurate prediction of ventricular arrhythmia in post-infarction patients is a clinical burden. Especially, the prediction of how fast the heart rate might be in the settings of scar-related VT is of paramount importance. With this respect, contrast-enhanced MR imaging represents an exquisite tool for the evaluation of fibrotic infarct heterogeneity and the identification of GZ, where the substrate of potentially lethal VT is harboured.

First, the work presented in this study successfully demonstrated the capability of a high resolution 3D LGE method (with isotropic voxel size) to distinguish dense scar from GZ in a pig model of chronic infarction. This allowed us to accurately assess the extent of fibrosis by two segmentation methods based on SI thresholding (i.e., 5SD and FWHM), both methods yielding similar results. Our preclinical model mimicked the human pathophysiology of scar-related VT, thus highlighting its translational aspect and the potential clinical utility of the 3D LGE method in predicting the severity of VT episodes. As noted in Fig. 4, the ischemic GZ had fibrotic bundles intermingled with viable myocytes (salvaged by reperfusion), while in the dense scar most necrotic areas were replaced by mature collagenous fibrosis. This heterogenous pattern (as revealed by the MR signal intensity enhancement within the infarcted area) likely favoured the VT inducibility. Typically, the VT waves rotate around unexcitable patches (scars).

Second, our results showed that the correlation between VT cycle length and the amount of heterogeneous infarct (scar + GZ), was superior to the correlation between VTCL and EF. To the best of our knowledge this is the first quantitative study suggesting that the amount of LGE-defined fibrosis might be a better predictor (than EF) of the severity of VT. Previous preclinical and clinical studies have suggested only the utility of MRI-defined scar and GZ, respectively, as predictors of mortality (based on SCD evidence) [5], or predictors of possible occurrence of arrhythmia (shown by recorded VT/VF episodes in patients with implantable defibrillator devices) [6]. The negative slope obtained in our correlation analysis between VTCL and EF, could be explained by the fact that longer VTCLs correspond to larger infarcts, which are usually associated with reduced EF values. However, in the clinics, many patients present with preserved EF; thus, we suggest that a correlation between VTCL and EF might not be clinically useful.

One limitation of our data-driven predictive study is that the image analysis was performed for the LV only. In some of the hearts, the infarcted area also had RV involvement; thus, the total amount of fibrosis within the entire heart (LV + RV) might slightly change the correlation with VTCL. Furthermore, conventional LGE imaging is known to have several technical limitations, such as: optimal inversion time TI is found through trial-end-error tests (prolonging the scan time); MR signal intensity depends on the gadolinium wash-in and wash-out kinetics; difficult differentiation of sub-endocardial scar/GZ from the bright blood pool; and, the manual selection of ROI is operator-dependent. These issues could be avoided by using for instance robust T1/T1* imaging mapping methods [8, 9] instead of LGE.

In this work we specifically correlated the total amount of infarct (scar + GZ), since the VT wave around dense (unexcitable) patches propagates through both healthy tissue and slow conduction GZ areas (located at the periphery of dense scar). As per the simplified VT circuit illustrated in Fig. 1, the reentry wave will have different speeds in different tissue segments, each contributing to the final cycle length of one complete rotation. Moreover, the main channels often traverse large scars (partitioning them into smaller dense patches that define the VT pattern), while some circuits present with secondary channels [2] that can alter the VT morphology and CL during the inducibility procedure. Thus, VTCL cannot exactly correlate with the total scar circumference; instead, the length of each circuit component needs to be measured. For these reasons, we also did not expect to obtain a perfect correlation in Fig. 5. Additionally, while the amount of GZ is an independent predictor of the arrhythmia [5, 6], the GZ size alone would not have a strong correlation with VTCL.

Considering these complex aspects, our future work will focus on reconstructing the 3D geometry of the circuits, as such knowledge is critical in planning the RF ablation procedures [4, 12] in order to successfully terminate VT. We also plan to further employ biophysical modelling tools to accurately predict the electrical wave propagation and VT inducibility per individual heart. Such virtual 3D heart models can integrate image-defined scar/GZ areas [13, 14], from which one will be able to precisely compute the CL of dangerously fast (potentially lethal) VTs via *in silico* computer simulations.

5 Conclusion

Our preclinical study suggests that the LGE-defined extent of infarct is a better predictor of VTCL than the ejection fraction, which was poorly correlated with VTCL. Although Gd-DTPA is not a collagen-specific contrast-agent, it can be used with confidence to evaluate the extent of fibrosis (scar + GZ) in remodelled chronic infarcts by both 5SD and FWHM standard methods, as they yielded similar results in segmenting the infarct heterogeneities such as dense scar and GZ (where the VT substrate resides).

Acknowledgements. The authors would like to thank: Mr. Adebayo Adeeko for processing the explanted hearts and for staining the histology slides; Ms. Melissa Larsen for help with the animal preparation during the MRI-VT studies; and, Dr. Xiuling Qi and Mr. Venkat Ramanan for help with MR imaging. This research was funded by the Canadian Institutes of Health Research (CIHR): Project grant PJT-153212 awarded to Dr Mihaela Pop and Project grant PJT-178299 awarded to Dr. Graham Wright.

References

1. John, R.M., et al.: Ventricular arrhythmia and sudden cardiac death. Lancet **380**, 1520–1529 (2012)
2. Kebler, A., Rudy, Y.: Basic mechanisms of cardiac impulse propagation and associated arrhythmias. Physiol. Rev. **84**, 431–488 (2004)
3. Kanagasundram, A.N., Richardson, T.D., Stevenson, W.G.: The heart rate of ventricular tachycardia. Circulation **143**(3), 227–229 (2021)
4. Kuck, K.H., et al.: VTACH study group. Catheter ablation of stable VT before defibrillator implantation in patients with coronary heart disease (VTACH): a multicentre randomised controlled trial. Lancet **375**, 31–40 (2010)
5. Yan, A.T., Shayne, A.J., Brown, K.A., Gupta, S.N., Chan, C.W., et al.: Characterization of the peri-infarct zone by contrast-enhanced cardiac MRI is a powerful predictor of post-infarction mortality. Circulation **114**(1), 32–39 (2006)
6. Roes, S.D., et al.: Infarct tissue heterogeneity assessed with contrast-enhanced MRI predicts spontaneous ventricular arrhythmia in patients with ischemic cardiomyopathy and implantable cardioverter-defibrillator. Circ. Cardiovasc. Imaging **2**(3), 183–190 (2009)
7. Wu, Y., Tang, Z., Li, B., Firmin, D., Yang, G.: Recent advances in fibrosis and scar segmentation from cardiac MRI: a state- of-the-art review and future perspectives. Front. Physiol. **12**, 709230 (2021)
8. Pop, M., et al.: High resolution 3D T1* mapping and quantitative image analysis of the '*gray zone*' in chronic fibrosis. IEEE Trans. Biomed. Eng. **61**(12), 2930–2938 (2014)
9. Zhang, L., Lai, P., Pop, M., Wright, G.: Accelerated multi-contrast volumetric imaging with isotropic resolution for improved peri-infarction characterization using parallel imaging, low-rank and spatially varying edge-preserving sparse modeling. Magn. Reson. Med. **79**(6), 3018–3031 (2018)
10. Schulz-Menger, J., et al.: Standardized image interpretation and post processing in cardiovascular MRnce: soc. for cardivascular magnetic resonance (SCMR) board of trustees task force on standardized post processing. J. Cardiovasc. Magn. Reson. **15**, 35 (2013)
11. Flett, A.S., et al.: Evaluation of techniques for the quantification of myocardial scar of differing etiology using cardiac MR. JACC Cardiovasc. Imaging **4**, 150–156 (2011)
12. Krahn, P.R.P., et al.: MRI-Guided cardiac RF ablation for comparing MRI characteristics of acute lesions and associated electrophysiologic voltage reductions. IEEE Trans. Biomed. Eng. **69**(8), 2657–2666 (2022)
13. Pop, M., et al.: Correspondence between simple 3D MRI-based heart models and in-vivo EP measures in swine with chronic infarction. IEEE Trans. Biomed. Eng. **58**(12), 3483–3486 (2011)
14. Cedilnick, N., et al.: Fast personalized electrophysiological models from CT images for ventricular tachycardia ablation planning. Europace **20**, iii94–iii101 (2018)

Left Ventricular Work and Power are Constant Despite Varying Cardiac Cycle Length—Implications for Patients with Atrial Fibrillation

Debbie Zhao[1(✉)], João F. Fernandes[2], Stephen A. Creamer[1], Abdallah I. Hasaballa[1], Vicky Y. Wang[1], Thiranja P. Babarenda Gamage[1], Malcolm E. Legget[3], Robert N. Doughty[3,4], Peter N. Ruygrok[4], Pablo Lamata[2], Alistair A. Young[2], and Martyn P. Nash[1,5]

[1] Auckland Bioengineering Institute, University of Auckland, Auckland, New Zealand
debbie.zhao@auckland.ac.nz

[2] School of Biomedical Engineering and Imaging Sciences, King's College London, London, UK

[3] Department of Medicine, University of Auckland, Auckland, New Zealand

[4] Green Lane Cardiovascular Service, Auckland City Hospital, Auckland, New Zealand

[5] Department of Engineering Science, University of Auckland, Auckland, New Zealand

Abstract. Atrial fibrillation (AF) is associated with stroke and heart failure, and poses a significant global health burden. Consequently, efforts remain ongoing in better characterising and understanding AF and its underlying mechanisms. This study explores cardiac energetics associated with AF by testing the hypothesis that left ventricular stroke work and systolic power are conserved despite changes in cardiac cycle duration. By combining invasive haemodynamic data and 3D echocardiography, we generated two *in vivo* pressure-volume loops (corresponding to a short and long cardiac cycle within the same subject) in a sample of 20 patients exhibiting sinus arrhythmia. Subsequently, we found no statistically significant differences in work (0.10 ± 0.22 J) or power (0.03 ± 0.56 W), despite significant differences in stroke volume (7 ± 13 ml) and cardiac output (1.08 ± 0.98 L/min) between short and long cycles (differing by 274 ± 145 ms). Given the repeatability in work and power despite substantial R-R variability, left ventricular energetics may provide more reliable metrics for cardiac function in the presence of AF to better guide patient management.

Keywords: Atrial fibrillation · Haemodynamics · Ventricular energetics

J. F. Fernandes—Joint first authorship.

O. Bernard et al. (Eds.): FIMH 2023, LNCS 13958, pp. 708–717, 2023.
https://doi.org/10.1007/978-3-031-35302-4_72

1 Introduction

Atrial fibrillation (AF) is the most common form of cardiac arrhythmia, and a major risk factor for stroke and heart failure. Consequently, AF is associated with substantial cardiac-related morbidity and mortality, posing a significant burden to patients and healthcare systems worldwide. The current estimated global prevalence of AF is 50 million [17], and this is expected to rise alongside the aging population. Research efforts have thus been directed at better understanding the mechanisms underlying AF, with outcomes contributing to informing guidelines for screening, diagnosis, and subsequent clinical management [8].

Assessment of the left ventricle (LV) as the primary pumping chamber, and its ability to drive the systemic circulation, is important in AF, and relevant indices are needed to monitor for dysfunction. Traditionally, LV volumetry has been used to derive metrics such as stroke volume (SV), ejection fraction (EF), and cardiac output (CO), for evaluating systolic function. However, there are several technical limitations related to image acquisition, reproducibility, and interpretation—in addition to physiological factors such as load dependence—that can be particularly problematic in AF. Clinical guidelines for chamber quantification by echocardiography [10] therefore recommend that relevant measurements should be averaged across a minimum of five beats in the presence of AF, to obtain a representative assessment of overall cardiac function. This is generally a time-consuming process, and resulting measurements may still vary depending on the cardiac cycles selected for analysis.

Recently, cardiac indices associated with ventricular energetics have been shown to provide increased diagnostic value in various patient populations, such as those with aortic stenosis [4], heart failure with preserved ejection fraction [1], and in cardiac intensive care units [9], where loading conditions can fluctuate as a result of sustained pharmacological effects. Previous basic physiology experiments have also concluded that LV stroke work is unaffected by changes in heart rate as the result of a steady-state response [11,13]. Given the analogous complexities associated with rhythm irregularity, we postulate that such indices may also be appropriate for the characterisation of LV function in AF.

In the present study, we aimed to test the hypothesis that patient-specific LV systolic energy consumption remains constant despite substantial heart rate variability at rest, and can therefore provide a more reliable estimate of LV function than conventional volumetry. By analysing *in vivo* pressure-volume (PV) loops in patients with R-R irregularity, we derived LV stroke work and power to provide further insight into cardiac energetics.

2 Methods

2.1 Patient Recruitment and Data Acquisition

Patients presenting for coronary angiography at Auckland City Hospital, who exhibited irregular cardiac rhythm during catheterisation, were prospectively recruited for haemodynamic assessment and subsequent imaging with transthoracic three-dimensional echocardiography (3DE). Ethical approval was granted by the Health and Disability Ethics Committee of New Zealand (17/NTB/46), and written informed consent was obtained from each participant.

Continuous LV pressure was measured using a fluid-filled pigtail catheter (Impulse^{TM} Diagnostic Catheter, Boston Scientific, Marlborough, MA, USA) via radial access. Simultaneous pressure and electrocardiogram (ECG) measurements were recorded at a sampling frequency 240 Hz (see Fig. 1), using the MacLab Hemodynamic Recording System (GE Healthcare, Chicago, IL, USA).

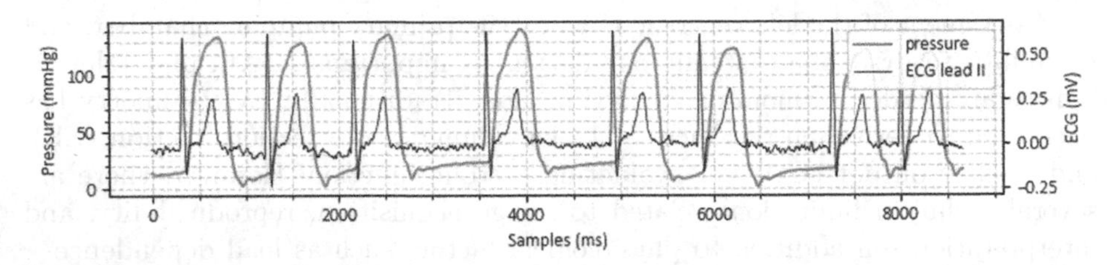

Fig. 1. An example left ventricular pressure waveform and concurrent electrocardiogram (ECG) recorded for a patient with atrial fibrillation and R-R irregularity.

Following catheterisation, real-time LV-focused 3DE images were acquired using a Siemens ACUSON SC2000 Ultrasound System and a 4Z1c matrix array transducer (Siemens Medical Solutions, Issaquah, WA, USA), within an hour after invasive pressure measurement. Imaging was performed in a steep left lateral decubitus position under breath-holds, and acquisition parameters were optimised on a per-patient basis to maximise the temporal resolution while maintaining an adequate spatial resolution for analysis. For each patient, 4–8 single-cycle, retrospective R-wave gated acquisitions were obtained.

Subjects for inclusion in this study were manually selected based on ECG appearances featuring irregular R-R intervals over a period of $\geq$30 s, in accordance with AF guidelines [8]. To ensure sufficient R-R variability, rhythm irregularity was defined as having a minimum difference in R-R interval of 100 ms (and a proportional change of at least 15% in duration) between the shortest and longest observed cycle.

2.2 Data Analysis and Generation of Pressure-Volume Loops

As invasive haemodynamic measurement and imaging could not be performed concurrently, pressure and volume data were independently analysed and subsequently paired according to R-R interval (computed from the respective ECGs of each modality). To maximise R-R variability while ensuring an acceptable match in cardiac cycle duration between pressure and volume, the shortest and longest cycles were prioritised for analysis, and a tolerance of 10% mismatch in relative R-R interval between paired cycles was set. In cases where the matching criteria were not satisfied at the extremes, data corresponding to the next shortest or longest cycle were selected.

To account for the electromechanical delay between the ECG R-peak and measured pressure at end-diastole (ED), the location of ED was separately determined using an automated pressure foot detection algorithm based on a simple shear transform [2]. The continuous haemodynamic recordings were subsequently spliced at these locations to extract individual cardiac cycles of interest.

To extract volume-time curves for each cardiac cycle, the LV was initially segmented from 3DE using a fully automated B-Spline Explicit Active Surfaces algorithm [14]. To improve agreement with cardiac magnetic resonance imaging as the reference modality for chamber volume quantification, a previously devised bias correction method (based on dynamic time warping and partial least squares regression) [19] was subsequently applied to refine LV segmentations. Conventional volumetric indices representing LV systolic function, including SV, EF, CO, as well as global longitudinal strain (GLS), were computed from the LV segmentations corresponding to the maximum and minimum volumes over the cardiac cycle.

Finally, PV loops were generated by normalising both pressure and volume data (using linear interpolation) across the entire cardiac cycle (from ED to ED), and subsequently verified by visual inspection. This process yielded two PV loops per patient, corresponding to short and long cardiac cycles (Fig. 2).

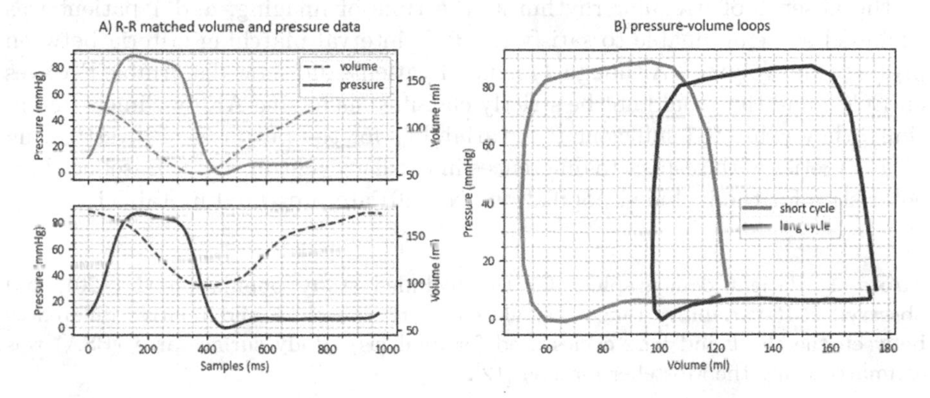

Fig. 2. An example of *in vivo* pressure-volume loops generated from pressure and volume data corresponding to short and long cardiac cycles within the same patient. A) Aligned pressure and volume data after matching of R-R intervals. B) Resultant pressure-volume loops after normalisation of pressure and volume data.

2.3 Calculation of Stroke Work and Power

The stroke work of the LV (corresponding to the component of ventricular energy consumption required to eject blood) was computed using the area enclosed within the PV loop (multiplied by a conversion factor to obtain work in units of joules). Stroke power (representing the average energy consumption per unit time during contraction) was subsequently derived by dividing stroke work by the systolic duration from ED to end-systole (ES) [5], determined by the time value corresponding to the minimum volume.

2.4 Statistical Analyses

Paired-samples *t*-tests were used to identify statistically significant differences in cardiac indices derived from short and long cycles within the same patient (with continuous variables expressed as mean ± standard deviation). To further assess the reliability of functional indices despite R-R variability, an intraclass correlation coefficient (ICC) was calculated using a two-way, mixed-effects model for absolute agreement. After quantitative analysis, Bland-Altman plots were used to visualise the agreement in LV energetics between short and long cycles. All statistical tests were two-tailed and deemed significant for p-values < 0.05, and analyses were performed using IBM SPSS Statistics for Windows (Version 26.0, IBM Corp., Armonk, NY, USA).

3 Results

3.1 Patient Characteristics

A total of 27 patients were initially recruited for invasive pressure measurement, of which 20 patients were included in the study (6 patients were excluded due to the absence of irregular rhythm at the time of imaging, and 1 patient was excluded as being unable to satisfy the R-R interval matching criteria between pressure and volume). Although 4 included patients did show discernible P-waves on ECG (and thus could not be strictly classified as having AF by clinical definition), they exhibited sufficient R-R variability for the purposes of investigating LV energetics with respect to differences in cardiac cycle duration. Baseline characteristics for the included population (N = 20) are presented in Table 1.

Table 1. Patient demographics for the included study population (N = 20), and observed R-R variability range (calculated as the absolute and relative differences between the short and long cycles used for analysis). Body surface area (BSA) was estimated using the Mosteller formula [12].

	Male (N = 13)	Female (N = 7)	Total (N = 20)
Age (years)	62 ± 13	69 ± 6	64 ± 12
Height (cm)	177 ± 7	164 ± 5	172 ± 9
Weight (kg)	99 ± 24	82 ± 22	93 ± 25
BSA (m^2)	2.19 ± 0.28	1.91 ± 0.26	2.09 ± 0.30
R-R variability (ms)	278 ± 139	268 ± 155	274 ± 145
R-R variability (%)	33 ± 17	33 ± 15	33 ± 17

3.2 Effect of Patient-Specific R-R Variability

Differences in cardiac indices derived from conventional volumetry and LV energetics corresponding to short and long cycles within the same patient are presented in Table 2. Statistically significant differences were found in LV end-diastolic volume (EDV) and end-systolic volume (ESV), as well as in SV and CO, but not in EF. Meanwhile, no significant differences in LV stroke work or power were observed between short and long cardiac cycles.

Table 2. Left ventricular (LV) indices (mean ± standard deviation (SD)) derived from short and long cycles in N = 20 patients. For each parameter, the difference is calculated based on the value derived from $\text{cycle}_{long} - \text{cycle}_{short}$. Asterisks (*) indicate statistically significant differences where corresponding p-values < 0.05 (in bold).

LV index	Short	Long	Difference	p-value
EDV (ml)	146 ± 49	163 ± 44	*17 ± 28	**0.020**
ESV (ml)	67 ± 35	76 ± 32	*9 ± 18	**0.037**
SV (ml)	79 ± 18	86 ± 16	*7 ± 13	**0.028**
EF (%)	55 ± 7	54 ± 7	−1 ± 6	0.392
CO (L/min)	5.92 ± 2.11	4.84 ± 1.50	*−1.08 ± 0.98	**<0.001**
GLS (%)	−17.5 ± 3.1	−16.5 ± 3.7	−1.0 ± 3.0	0.176
Stroke work (J)	0.96 ± 0.30	1.06 ± 0.27	0.10 ± 0.22	0.057
Stroke power (W)	2.58 ± 0.89	2.55 ±0.71	−0.03 ± 0.58	0.837

3.3 Agreement in Work and Power

Bland-Altman plots showing the bias and 95% limits of agreement in LV stroke work and power between short and long cycles are presented in Fig. 3. The ICC values for work and power were 0.803 and 0.868, respectively, suggesting good-to-excellent reliability in measurements from the same patient. In comparison, lower ICC values were obtained for EF and GLS, of 0.777 and 0.757, respectively.

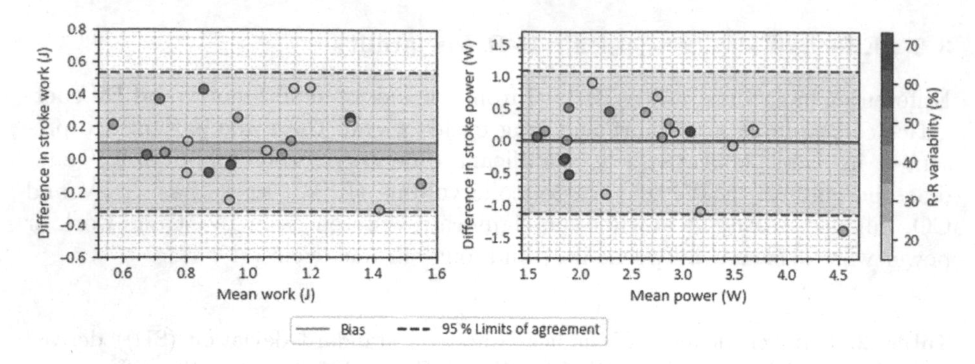

Fig. 3. Bland-Altman plots showing biases and 95% limits of agreement between patient-specific short and long cardiac cycles (coloured with respect to the magnitude of R-R variability) from which left ventricular stroke work and power are derived.

4 Discussion

With the increasing incidence and prevalence of AF, having already been classified as an epidemic, robust methods are needed to accurately determine and monitor LV function and contractility. However, volumetric indices such as SV are sensitive to preload as a result of the Frank-Starling mechanism, and thus possess limitations when applied to patients exhibiting R-R irregularity. In this study, we hypothesised that LV energy consumption per contraction is constant regardless of R-R interval, and subsequently found no statistically significant differences in stroke work and power between short and long cardiac cycles within the same patient. Although significant differences in the means of EF and GLS between short and long cycles were likewise not observed, the magnitudes of the standard deviation of differences were comparable to the variability observed within the population itself (see Table 2). These results suggest that indices derived from LV energetics may be more reliable markers of ventricular function in patients with sinus arrhythmia, compared to those derived from conventional volumetry. Furthermore, as several medications used to control AF are negatively inotropic, such indices may also be useful to monitor the effects of treatment in this population.

The hypothesis that LV work and power is constant across cardiac cycles of varying length may be perceived as counter-intuitive, as it may be assumed that shorter cycles are associated with a lower energy consumption. However, for efficient performance and functioning of the heart, the myocardium must retain its energetic output on the blood pool in order to maintain cardiac and circulatory homeostasis. This occurs by maintaining constant energy expenditure over a range of R-R intervals independent of the cardiac load of each heartbeat. Furthermore, the full cardiovascular system balance is maintained through pressure gradients that originate in the contracting heart, as sudden variations in systolic pressure would potentially impair the system [18]. Therefore, a healthy heart (even in AF) should preserve its energy consumption and pressure output with varying cycle duration.

For non-simultaneous recordings of pressure and volume, the independent measurements must be combined and temporally aligned for the generation of PV loops. As it is known that the various phases of the cardiac cycle do not scale uniformly with changes in R-R interval [7], previous studies account for this by adopting phase-specific scaling approaches based on the identification of temporal fiducial landmarks [6,15]. In this work, by normalising pressure and volume data across the entire cardiac cycle, there is an assumption that the relative proportions of systole and diastole do not change provided that the R-R intervals of independent cycles are sufficiently similar. Based on a qualitative assessment of the resultant PV loops, we found that phase-specific alignment was not required under our set tolerance of 10% mismatch in total cycle duration. This finding is supported by previous work which concluded that the duration of ventricular contraction (as denoted by the Q-T interval) with respect to a given R-R interval, is highly consistent within the same subject [3].

Although the subjects included in this study were primarily AF patients, outcomes from this work may be extended to other pathologies which likewise exhibit R-R irregularity (such as in the case of bigeminy or Wolff-Parkinson-White syndrome). The ability to leverage *in vivo* PV loops for ventricular energetics not only provides more reliable indices in cases where preload may fluctuate, but also the means of a broader assessment of LV function.

4.1 Future Work

While *in vivo* PV loops can provide a comprehensive assessment of cardiac energetics, the invasive nature of catheterisation limits its widespread use. Non-invasive measures such as pressure estimation from Doppler echocardiography, or the generation of PV loops using time-varying elastance models [16], may therefore facilitate clinical uptake. Although we have shown that LV stroke work and power can be reliably assessed despite R-R irregularity, the degree in which these indices relate to the extent of LV dysfunction has not yet been tested. Thus, further investigations (with larger sample sizes) are needed to determine the prognostic value of LV energetics as determinants of cardiac function in AF. Furthermore, the trajectory of work and power may contribute additional information regarding the pathophysiological mechanisms of AF, and represents a natural next step in determining the clinical utility of LV energetics.

5 Conclusion

This study explored the effect of varying cardiac cycle duration on LV stroke work and power in patients exhibiting R-R irregularity. Having found no statistically significant differences despite large changes cardiac cycle length within the same patient, we conclude that the energetic output of the LV is likely conserved as a result of myocardial adaptation in response to arrhythmia. As an adjunct to conventional volumetry, LV energetics may provide additional and more reliable information on function and contractility for the effective monitoring of patients with AF.

Acknowledgements. This research was funded by the Health Research Council of New Zealand (grant 17/608). We gratefully thank our cardiac sonographer, Gina Quill, and research nurses, Mariska Oakes-ter Bals, Jane Hannah, Anna Taylor, and Gracie Hoskin, for their clinical expertise and invaluable assistance with patient recruitment and data collection.

References

1. AbouEzzeddine, O.F., et al.: Myocardial energetics in heart failure with preserved ejection fraction. Circ. Heart Fail. **12**(10), e006240 (2019). https://doi.org/10.1161/CIRCHEARTFAILURE.119.006240
2. Balmer, J., et al.: Pre-ejection period, the reason why the electrocardiogram Q-wave is an unreliable indicator of pulse wave initialization. Physiol. Meas. **39**(9), 95005 (2018). https://doi.org/10.1088/1361-6579/aada72
3. Batchvarov, V.N., et al.: QT-RR relationship in healthy subjects exhibits substantial intersubject variability and high intrasubject stability. Am. J. Physiol. Heart Circ. Physiol. **282**(6), H2356–H2363 (2002). https://doi.org/10.1152/ajpheart.00860.2001
4. Ben-Assa, E., et al.: Ventricular stroke work and vascular impedance refine the characterization of patients with aortic stenosis. Sci. Transl. Med. **11**(509), eaaw0181 (2019). https://doi.org/10.1126/scitranslmed.aaw0181
5. Fernandes, J.F., et al.: Right ventricular energetics and power in pulmonary regurgitation vs. stenosis using four dimensional phase contrast magnetic resonance. Int. J. Cardiol. **263**, 165–170 (2018). https://doi.org/10.1016/j.ijcard.2018.03.136
6. Freytag, B., et al.: *In Vivo* pressure-volume loops and chamber stiffness estimation using real-time 3D echocardiography and left ventricular catheterization – application to post-heart transplant patients. In: Ennis, D.B., Perotti, L.E., Wang, V.Y. (eds.) FIMH 2021. LNCS, vol. 12738, pp. 396–405. Springer, Cham (2021). https://doi.org/10.1007/978-3-030-78710-3_38
7. Fridericia, L.S.: The duration of systole in an electrocardiogram in normal humans and in patients with heart disease*. Ann. Noninvasive Electrocardiol. **8**(4), 343–351 (2003). https://doi.org/10.1046/j.1542-474X.2003.08413.x
8. Hindricks, G., et al.: 2020 ESC guidelines for the diagnosis and management of atrial fibrillation developed in collaboration with the european association for cardio-thoracic surgery (EACTS): the task force for the diagnosis and management of atrial fibrillation of the Europea. Eur. Heart J. **42**(5), 373–498 (2021). https://doi.org/10.1093/eurheartj/ehaa612
9. Jentzer, J.C., Anavekar, N.S., Burstein, B.J., Borlaug, B.A., Oh, J.K.: Noninvasive echocardiographic left ventricular stroke work index predicts mortality in cardiac intensive care unit patients. Circ. Cardiovasc. Imaging **13**(11), e011642 (2020). https://doi.org/10.1161/CIRCIMAGING.120.011642
10. Lang, R.M., et al.: Recommendations for cardiac chamber quantification by echocardiography in adults: an update from the American society of echocardiography and the european association of cardiovascular imaging. Eur. Heart J. Cardiovasc. Imaging **16**(3), 233–271 (2015). https://doi.org/10.1093/ehjci/jev014

11. Mitchell, J.H., Wallace, A.G., Skinner, N.S.: Intrinsic effects of heart rate on left ventricular performance. Am. J. Physiol.-Legacy Content **205**(1), 41–48 (1963). https://doi.org/10.1152/ajplegacy.1963.205.1.41
12. Mosteller, R.D.: Simplified calculation of body-surface area. N. Engl. J. Med. **317**(17), 1098–1098 (1987). https://doi.org/10.1056/NEJM198710223171717
13. Noble, M.I.M., Wyler, J., Milne, E.N.C., Trenchard, D., Guz, A.: Effect of changes in heart rate on left ventricular performance in conscious dogs. Circ. Res. **24**(2), 285–295 (1969). https://doi.org/10.1161/01.RES.24.2.285
14. Pedrosa, J., et al.: Fast and fully automatic left ventricular segmentation and tracking in echocardiography using shape-based b-spline explicit active surfaces. IEEE Trans. Med. Imaging **36**(11), 2287–2296 (2017). https://doi.org/10.1109/TMI.2017.2734959
15. Puyol-Antón, E., et al.: A multimodal spatiotemporal cardiac motion atlas from MR and ultrasound data. Med. Image Anal. **40**, 96–110 (2017). https://doi.org/10.1016/j.media.2017.06.002
16. Seemann, F., et al.: Noninvasive quantification of pressure-volume loops from brachial pressure and cardiovascular magnetic resonance. Circ. Cardiovasc. Imaging **12**(1), e008493 (2019). https://doi.org/10.1161/CIRCIMAGING.118.008493
17. Tsao, C.W., et al.: Heart disease and stroke statistics-2022 update: a report from the American heart association. Circulation **145**(8), e153–e639 (2022). https://doi.org/10.1161/CIR.0000000000001052
18. Wang, R., et al.: Blood pressure fluctuation during hospitalization and clinical outcomes within 3 months after ischemic stroke. Hypertension **79**(10), 2336–2345 (2022). https://doi.org/10.1161/HYPERTENSIONAHA.122.19629
19. Zhao, D., et al.: Correcting bias in cardiac geometries derived from multimodal images using spatiotemporal mapping. Sci. Rep. **13**, 8118 (2023). https://doi.org/10.1038/s41598-023-33968-5

Correction to: Simulated Excitation Patterns in the Atria and Their Corresponding Electrograms

Joshua Steyer, Lourdes Patricia Martínez Diaz, Laura Anna Unger, and Axel Loewe

Correction to: Chapter "Simulated Excitation Patterns in the Atria and Their Corresponding Electrograms" in: O. Bernard et al. (Eds.): *Functional Imaging and Modeling of the Heart*, LNCS 13958, https://doi.org/10.1007/978-3-031-35302-4_21

The original version of this chapter, the acknowledgment text section in Chapter 21 was not updated. However, this has now been rectified, and the updated acknowledgment text has been included.

The updated original version of this chapter can be found at
https://doi.org/10.1007/978-3-031-35302-4_21

O. Bernard et al. (Eds.): FIMH 2023, LNCS 13958, p. C1, 2023.
https://doi.org/10.1007/978-3-031-35302-4_73

Author Index

A
Abell, Emma 3
Alajrami, Eman 283, 394
Alblas, Dieuwertje 356
Albors, Carlos 485
Alemany, Ignasi 54
Aleshaiker, Sama 394
Alluri, Prasanna G. 74
Amaro, Diego 194
Ardekani, Siamak 320
Argus, Finbar 475
Arri, Satpal S. 626
Aslanidi, Oleg 435
Astolfi, Edmond 668
Ataullakhanov, Fazoil 435
Augustin, Christoph M. 545
Avazmohammadi, Reza 34, 74
Axel, Leon 403
Azarmehr, Neda 283, 394

B
Babarenda Gamage, Thiranja P. 708
Baiges, Anna 465
Balmus, Maximilian 435
Balzani, Daniel 555
Banduc, Tomás 87
Banerjee, Abhirup 223
Barbarotta, Luca 575
Barbaroux, Hugo 412
Barrows, Rosie K. 545
Barry, Jennifer 698
Bartolucci, Chiara 175
Bear, Laura 3
Benoist, David 3, 44
Benson, Dave 565
Berger, Marie-Odile 455
Bernard, Olivier 245
Bernardino, Gabriel 338, 465, 607, 688
Bernus, Olivier 3, 44
Bertrand, Ambre 213
Bharath, Anil A. 366, 648
Bijnens, Bart 465
Bloomfield, Frank H. 475
Bohoran, Tuan A. 292
Borgohain, Indraneel 668
Botti, Sofia 175
Bovendeerd, Peter H. M. 97, 147, 575, 594
Brune, Christoph 356, 445
Busuttil, Oliver 3

C
Cabanis, Pierre 3, 44
Callahan, Ryan 616
Caluori, Guido 184
Camara, Oscar 338, 465, 485, 688
Camprecios, Genis 465
Camps, Julia 213, 223
Capallera, Isaac 688
Carvalho, Bruno M. 403
Castellanos, Daniel A. 616
Cedilnik, Nicolas 329
Chabiniok, Radomír 235, 616
Choudhary, Gaurav 74
Clarysse, Patrick 607, 678
Cluitmans, Matthijs J. M. 97, 147
Cochet, Hubert 485
Cole, G. 648
Collin, Annabelle 184
Constantin, Marion 3
Cork, Tyler E. 527
Correas, María 127
Costabal, Francisco Sahli 137
Coudière, Yves 194
Courand, Pierre-Yves 245
Coutinho, Edson A. G. 403
Creamer, Stephen A. 626, 708
Croisille, Pierre 678
Cui, Junning 25

D
Danton, Mark 585
Dargent, Thomas 338, 607
De Vecchi, Adelaide 425, 495
de Vecchi, Adelaide 435

O. Bernard et al. (Eds.): FIMH 2023, LNCS 13958, pp. 719–723, 2023.
https://doi.org/10.1007/978-3-031-35302-4

Dhiver, Philippe 194
Di Folco, Maxime 607
Diaz, Lourdes Patricia Martínez 204
Dileep, Drisya 14
Dinmohammadi, Fateme 283
Dogrusoz, Yesim Serinagaoglu 166
Doorly, Denis J. 54
Dos Santos, Pierre 44
Dos-Santos, Pierre 3
Doughty, Robert N. 626, 708
Dual, Seraina A. 505, 515
Dubes, Virginie 3, 44
Duchateau, Josselin 3, 44
Duchateau, Nicolas 338, 607, 678, 688

E
Ennis, Daniel B. 64, 412, 515, 527
Escartin, Terenz 698

F
Fernandes, João F. 708
Ferreira, Pedro F. 54
Feuerstein, Delphine 194
Francis, Darrel P. 283, 394
Freytag, Bianca 527, 678
Fyrdahl, Alexander 425

G
Galazis, C. 648
Gamage, Thiranja P. Babarenda 626
Gander, Lia 87, 137
Gao, Hao 585
Garcia, Damien 245
García, I. 107
Garcia-Cañadilla, Patricia 465
Garcia-Criado, Angeles 465
García-Galindo, Alberto 658
García-Pagan, Juan Carlos 465
Gassa, Narimane 157
Genet, Martin 235
Gentles, Thomas L. 475
Ghebryal, Justina 147, 545
Ghelani, Sunil J. 616
Giannakidis, Archontis 292
Gorman, III, Joseph H. 636
Gorman, Robert C. 636
Grau, Vicente 213, 223
Guan, Debao 585
Gusseva, Maria 616
Guillem, María S. 127

H
Hadadi, Azadeh 265
Haïssaguerre, Michel 3, 44
Hammer, Peter E. 455, 616
Han, Q. Joyce 64
Hasaballa, Abdallah I. 626, 708
Helbing, Wim A. 347
Hermida, Uxio 495
Hernandez-Gea, Virgina 465
Hlivak, Peter 166
Hoeijmakers, Martijn 565
Hofman, Marieke 356
Houari, Karim El 565
Howard, J. 648
Hu, Pengpeng 310
Hussain, Tarique 235, 616

I
Iriart, Xavier 485

J
Jacob, Athira 668
Jaïs, Pierre 184
Janssens, Koen L. P. M. 97, 575, 594
Jelenc, Matija 301
Jevsikov, Jevgeni 283, 394
Jodoin, Pierre-Marc 245

K
Kalinin, Vitaly 157
Kampaktsis, Polydoros N. 292
Keramati, Hamed 425, 495
Kerkhof, Britt P. van 575
Khaledian, Nariman 455
Knaap, M. van der 594
Kolawole, Fikunwa O. 527
Krahn, Philippa 698
Krause, Rolf 117, 137, 175
Kreis, Oliver 505
Kruithof, Evianne 97, 147
Kuhl, Ellen 527
Kunze, Karl P. 412

L
L'Eplattenier, Pierre 565
Labrousse, Louis 3, 44

Lalande, Alain 265
Lamata, Pablo 425, 495, 708
Lane, Elisabeth 283
Lane, Elisabeth S. 394
Leatherman, Abby 34
Leb, Jay 292
Leclerc, Sarah 265
Legget, Malcolm E. 626, 708
Leguèbe, Michael 194
Li, Lei 223
Li, Zongfeng 25
Liberos, A. 107
Ling, Hang Jung 245
Linte, Cristian A. 375
Linton, N. 648
Lip, Gregory Y. H. 435
Liu, Hao 636
Lloyd, David F. A. 495
Loecher, Michael 412, 515, 527
Loewe, Axel 204
Lozano, M. 107
Luo, Xiaoyu 585
Lupon, Nestor Pallares 44

M
Ma, YingLiang 310
Magat, Julie 3, 44
Marlevi, David 425
Marsden, Alison L. 515
Maso Talou, Gonzalo D. 475
May, Robyn W. 475
McCann, Gerry P. 292
McElhinney, Doff B. 515
McLaughlin, Laura 292
Mehdi, Rana Raza 74
Mehta, Vishal S. 310
Mendiola, Emilio A. 34, 74
Michel, Cindy 3
Mill, Jordi 485
Motiwale, Shruti 537
Moustakidis, Serafeim 292
Muffoletto, Marica 255
Mukherjee, Tanmay 74

N
Nagy, Attila 565
Naidoo, Preshen 283, 394
Nair, Priya J. 515
Nash, Martyn P. 527, 626, 708
Nati Poltri, Simone 184
Nechipurenko, Dmitry 435
Neelakantan, Sunder 34
Neji, Radhouene 412
Ng, Matthew 698
Niederer, Steven 310
Niederer, Steven A. 255, 545
Nielles-Vallespin, Sonia 54, 412
Nordsletten, David 435

O
Ogbomo-Harmitt, Shaheim 435
Ogiermann, Dennis 555
Ohayon, Jacques 74, 678
Olivares, Andy L. 485
On, Y. 648
Ondrusova, Beata 166
Ouadah, Cylia 265
Ozenne, Valéry 3, 44

P
Painchaud, Nathan 245
Pallares-Lupon, Nestor 3
Pannetier, Valentin 194
Pasdois, Philippe 3
Passerini, Tiziano 668
Pavarino, Luca F. 175
Peirlinck, Mathias 347
Pentenga, Puck 347
Pérez del Villar, Candelas 658
Perez-Campuzano, Valeria 465
Pérez-Sánchez, Pablo 658
Pernot, Mathieu 3
Perotti, Luigi E. 64, 274, 555
Perra, Emanuele 505
Perrin, Douglas P. 455
Peters, N. 648
Petrusca, Lorena 678
Peyrat, Jean-Marc 329
Pezzuto, Simone 87, 117, 137
Pfaller, Martin R. 515
Piella, Gemma 688
Plank, Gernot 545
Poignard, Clair 184
Pop, Mihaela 698
Pordoy, Jamie 283
Pouch, Alison M. 636
Pourtau, Line 3
Pushparajah, Kuberan 495

Q

Quesson, Bruno 3
Quill, Gina M. 626
Qureshi, Ahmed 435

R

Ramlugun, Girish 3, 44
Rasoolzadeh, Nika 166
Razavi, Reza 310, 495
Rinaldi, C. Aldo 310
Rochette, Michel 565
Rodero, Cristobal 545
Rodrigo, M. 107
Rodriguez, Blanca 213, 223
Rogier, Julien 3
Romero, P. 107
Romitti, G. S. 107
Roney, Caroline H. 545
Rosilho de Souza, Giacomo 117
Ruygrok, Peter N. 626, 708

S

Sabry, Malak 425, 495
Sacks, Michael S. 537, 636
Sadayappan, Sakthivel 74
Safaei, Soroush 475
Sahli Costabal, Francisco 87
Sainz, Angela Lopez 465
Saiz-Vivó, Marta 688
Sánchez, Jorge 127
Sánchez, Pedro L. 658
Sánchez-Puente, Antonio 658
Santos, Selan R. dos 403
Scott, Andrew D. 54, 412
Sebastian, R. 107
Serej, Nasim Dadashi 283, 394
Sermesant, Maxime 384
Serra, D. 107
Severi, Stefano 175
Sharma, Puneet 668
Shontz, Suzanne M. 375
Shun-shin, Matthew J. 283, 394
Siddiqi, Kaleem 14
Sigfridsson, Andreas 425
Simon, Richard 375
Simonian, Natalie T. 636
Simpson, John M. 495
Singh, Sheldon 698
Sirajuddin, Minhajuddin 14
Škardová, Kateřina 235
Škrlj, Luka 301
Steinweg, Johannes K. 495
Steyer, Joshua 204
Stowell, Catherine C. 394
Strocchi, Marina 545
Stroh, Ashley 347
Suk, Julian 445
Svehlikova, Jana 166
Syed, Tabish A. 14

T

Thatte, Nikhil 616
Tipre, Dnyanesh 320
Turon, Fanny 465

U

Ugander, Martin 425
Unger, Laura Anna 204
Upendra, Roshan Reddy 375
Usman, Muhammad 74

V

Vaillant, Fanny 3
van der Sluis, O. 97
van Genuchten, Wouter 347
van Poppel, Milou P. M. 495
Vanderslice, Peter 34
Vara, Pedro Pablo 658
Varela, Marta 366, 648
Varga-Szemes, Akos 668
Varray, François 25
Verhoosel, C. V. 97
Viana, Felipe A. C. 274
Vicente-Palacios, Víctor 658
Vigmond, Edward 3
Vigneswaran, Trisha V. 495
Villanueva, M. Inmaculada 465
Villard, Pierre-Frédéric 455
Vimalesvaran, K. 648
Vrtovec, Tomaž 301

W

Walton, Richard 3, 44, 194
Wang, Eric 34
Wang, Shunli 25
Wang, Vicky Y. 527, 626, 708
Wang, Yanan 14
Wang, Yingjie 585

Wang, Zhinuo 223
Wang, Zhisheng 25
Wei, Jinchi 320
Weiser, Martin 137
Weiss, Robert G. 320
Whitt, Emily 274
Willems, R. 97
Williams, Michelle C. 255
Williams, Steven E. 255, 435
Wilson, Alexander J. 64
Wolterink, Jelmer M. 356, 445
Wright, Graham 698

X

Xiang, Qian 34
Xu, Hao 255

Y

Yacoub, Magdi H. 425
Yang, Yingyu 384
Ye, Wenfeng 565
Yeung, Kak Khee 356
Younes, Laurent 320
Young, Alistair A. 255, 412, 626, 708
Yu, Cindy 698

Z

Zaman, S. 648
Zemzemi, Nejib 157
Zhang, Ruiyi 320
Zhang, Wenbo 537
Zhao, Debbie 626, 708
Zolgharni, Massoud 283, 394
Zuben, Andre Von 274

Printed in the United States
by Baker & Taylor Publisher Services